THE
FLIGHT
INSTRUCTOR'S
MANUAL

THIRD EDITION

WILLIAM K. KERSHNER

THE FLIGHT INSTRUCTOR'S MANUAL

THIRD EDITION

 IOWA STATE UNIVERSITY PRESS / AMES

TO MY WIFE, BETTY

and to the flight instructors, pilots, and others in aviation who tried to improve my aeronautical knowledge over the years: Pat Howell, Arch Agee, Horace Draughon, Waldo Rassas, Frank Knapp, Wade Hadley, Buford Ledbetter, Baxter Lehman, Clyde Brown, Piedmont Poindexter, L. M. DeRose, Truman W. Finch, E. B. Salsig, and many others to whom I owe much thanks.

WILLIAM KERSHNER has over five decades of experience as pilot in command in more than 100 types and models of airplanes from 40 HP Cubs to jets. As a ground instructor and flight instructor he has taught many students—using the principles that he stresses in his books. He also holds commercial and airplane transport pilot certificates, and he is the author of *The Student Pilot's Flight Manual, The Instrument Flight Manual, The Advanced Pilot's Flight Manual,* and *The Basic Aerobatic Manual* (Iowa State University Press). Kershner was named the 1992 FAA/General Aviation Flight Instructor of the Year and the 1997 Elder Statesman of Aviation and was inducted into the Flight Instructor Hall of Fame in 1998.

ILLUSTRATED BY THE AUTHOR

The cover photo by ELIZABETH MOTLOW is a reenactment of Bill Kershner soloing his son Bill some years ago at the Tullahoma, Tennessee, airport. Bill, the younger, is now flying for a major airline. The airplane is a Cessna Aerobat owned by Ace Aerobatic School; William K. Kershner, Chief (and only) Instructor.

© 1993 William K. Kershner. Previous editions © 1981 and 1974 Iowa State University Press, Ames, Iowa 50014
All rights reserved
♾ Printed on acid-free paper in the United States of America

Authorization to photocopy items for internal or personal use, or the internal or personal use of specific clients, is granted by Iowa State University Press, provided that the base fee of $.10 per copy is paid directly to the Copyright Clearance Center, 222 Rosewood, Danvers MA 01923. For those organizations that have been granted a photocopy license by CCC, a separate system of payments has been arranged. The fee code for users of the Transactional Reporting Service is 0-8138-0634-8/93 $.10.

First edition, 1974 (seven printings); second edition, 1981 (nine printings)

Third edition, 1993

Library of Congress Cataloging-in-Publication Data

Kershner, William K.
    The flight instructor's manual / William K. Kershner.—3rd ed.
        p.   cm.
    Includes bibliographical references and index.
    ISBN 0-8138-0634-8 (alk. paper)
    1. Flight training. I. Title.
TL712.K47   1993
629.132'52—dc20   92-15750

Last digit is the print number: 9 8 7 6

CONTENTS

INTRODUCTION

THIS BOOK is a reference for those in the process of working on the flight instructor's certificate and a guide for the new instructor.

I believe that too often newly certificated instructors feel they are left "on their own" and need a written reference to which they can turn for information about students' and more advanced pilots' problems that they will encounter. I've tried to list common errors in each area or maneuver and have suggested methods of coping with them.

The manual is set up so that an instructor who is starting to teach advanced stalls, for instance, can refer to that chapter for suggestions without reading the rest of the book. While each chapter is a part of the whole book, I've tried to write each so that it can be used by itself as a reference for a particular phase of instruction. As an example, the advanced stall material repeats some of the common errors noted earlier in the chapter on elementary stalls. Or a chapter on other advanced flying may cover a review and lead-in from the more elementary sessions.

The reader will note that I advocate the teaching of integrated ground and flight instruction, rather than a block of ground school that doesn't apply to the flying at that time. Lest it appear that I am a late convert to the system, I have been using this method of instructing with every student I've worked with since 1949, and *The Student Pilot's Flight Manual* was published in 1960 using this approach, I believe, for the first time in print. I feel that this is the best technique to keep the students' interest and to help them learn.

The Flight Instructor's Manual, with minor variations, parallels the material as presented in *The Student Pilot's Flight Manual, The Advanced Pilot's Flight Manual, The Basic Aerobatic Manual,* and *The Instrument Flight Manual.* This might be considered the instructor's text for those four manuals; and it is a "how to instruct" guide rather than a detailed text on the basics. The other manuals should be used for details as they would be presented to the new pilot.

Again, as in my other books, changes were made by the FAA in forms and releases during the process of writing. I've tried, in most cases, to stick to the general idea of teaching people how to become safe pilots. As you work on the flight instructor's certificate, use this manual for the principles, but refer to your Flight Safety District Office for the latest specific requirements.

I've been fortunate in that a number of knowledgable individuals aided in the process. Any errors are mine, however.

I wish to thank the following people:

Bill Whitmore of the Nashville FSDO, whose knowledge of and common sense approach to instrument flying I particularly respect. Many of his suggestions were used here.

The late Lonnie Thurston of the Nashville FSDO gave me many good ideas about ground reference material.

Stanley Mohler, M.D., former head of the Aeromedical Applications Division of the FAA at Washington, who has visited Sewanee several times and with whom I have flown aerobatics in the checking of physiological effects of g forces on the pilot. His suggestions and information sent on that subject were very valuable to me in writing on aerobatics instruction.

Leslie McLaurin, airport manager at Sewanee, who read the complete manuscript and made good comments and suggestions that were incorporated.

Glen McNabb, flight instructor of Jasper, Tennessee, who let me use him as a "test project" while he was working on his certificate several years ago. Some ideas from his notes were used here.

Evelyn Bryan Johnson, flight instructor at Morristown, Tennessee, who reviewed the first half of the book and whose comments were much valued.

Genie Rae (Mrs. Dave) O'Kelley, flight instructor of Knoxville, whose attitude toward instructing should be copied by more of us.

Bob Bomar, veteran flight instructor of Shelbyville, Tennessee, with whom I've discussed flying and instructing many times over the years and whose opinions I respect highly.

Major Gary L. Moore, U.S. Army, for his paper "Integrated Flight Instruction," as part of his postgraduate work at Middle Tennessee State University.

John Anderton, former flight student at Sewanee, now a captain with U.S. Air, whose work with a camera in several spin sessions was above and beyond the call of duty.

The late Louis Cassels and *Nation's Business* for permission to refer to Mr. Cassels's article "Eight Steps to Better Training" (published March 1961, *Nation's Business,* 1615 H Street N.W., Washington, D.C. — check for reprints).

William E. Wells, who came to take my aerobatic program and left me with very useful copies of his own work in setting up flight instruction syllabuses.

I must acknowledge the help I received from *Cessna Aircraft Company.* My work with Ed McKenzie, Joyce Case, and Jim Kemper on the *Cessna 150 Aerobat Manual* remains one of the most enjoyable times of my writing career. Bill Thompson, Engineering Testpilot at Cessna, gave me the benefit of his knowledge and he furnished concrete help in the form of reports. I appreciate very much *Cessna Aircraft Company's* permission to use sections of the *Aerobat Training Manual* in the part on aerobatics.

James S. Bowman, Jr., of the NASA Spin Tunnel at Langley answered my questions on the subject, making me realize how much there is to learn.

Thanks must go to Callie Hood, Barbara Hart, Denise Childers, and Judy Rickman for typing the smooth copy of the manuscript; to my wife, Betty, who typed the rough copy; and to my daughter, Cindy, who helped arrange illustrations. Nan Thomas at the University of the South was helpful with facilities there.

In the second edition, Rowena Malone and Judy Timberg made further improvements in presentation and readability.

Appreciation is expressed to James W. "Pete" Campbell of the FAA for his efforts on behalf of flight instructor refresher clinics through the years and most of all for making FARs interesting. Figure 17-8 was inspired by one of his talks; other ideas on principles of instructing germinated at Pete's clinics.

Thanks to Scot Oliver of Sewanee for his aid in making movies of spins.

Special thanks to George Naff who came to take my aerobatic course and left me most useful ideas of syllabuses for this book.

The term *Pilot's Operating Handbook* is used here as a term to include the older airplanes' Owner's Manuals and/or Airplane Flight Manuals and other manufacturers' operating information.

As this book has to be a general approach to flying, the *Pilot's Operating Handbook* procedures for a particular airplane will naturally take precedence over pro-

cedures indicated — for example, use of carburetor heat, flaps, spin entries and recoveries.

The instrument markings in the illustrations are generally close to those expected for the type of airplane being discussed but in most cases the "numbers" (V_{so}, etc.) have been fictionalized.

A NOTE ON THE THIRD EDITION

The latest FARs and government publications have been included in this edition, but you should check with your FSDO and other sources for changes in FARs! (Fortunately flying and instructing basics will remain basically the same, however.)

In the flight manuals I've often had to use the words "he" and "him" to refer to students, just because it's impossible to mention both men and women every time. However, an increasing number of flight instructors and students are women, and these manuals are meant to be for and about them as well as their male counterparts.

I would appreciate comments from flight instructors who use the book.

William K Kershner

1 FUNDAMENTALS OF FLIGHT INSTRUCTION

ON BEING A FLIGHT INSTRUCTOR

BACKGROUND

THE FLIGHT INSTRUCTOR'S CERTIFICATE IS THE MOST IMPORTANT ONE ISSUED. Unfortunately, it is not always viewed that way, but instead it is often thought of as a "license to build up time" for other flying jobs—or other certificates and ratings. If you're going to be a flight instructor (for whatever reason), be the best while you are doing it.

THE BEST FLIGHT INSTRUCTORS HAVE THE FOLLOWING THINGS IN COMMON:

1. A KNOWLEDGE OF THE SUBJECT. They continually study to update themselves.

2. THE ABILITY TO TEACH. They know how people learn and provide instruction appropriate to the individual and the circumstance. Their instruction is accurate and properly sequenced.

3. A GENUINE INTEREST IN THE LEARNER. They like working with people. The instructor-student relationship is good and they have the confidence of those who are learning. They have consideration of the student's point of view.

4. PROFESSIONALS IN THE AIR. They have skill and are self-disciplined in the airplane. Their relationship with students is that of friendly authority. (If they are authoritative only, that's bad; if they are friendly without authority they are ineffective and the student will pay later.)

5. ADAPTABILITY. If the old "proven" methods don't work with a particular student, professional flight instructors use new techniques, understanding that these new approaches apply only to these individuals.

6. CONSISTENCY. They don't change personality but use the same standards from one flight to another.

7. UPDATED INSTRUCTING TECHNIQUES. Every once in a while they take a look at themselves to see if they've gotten in a rut and change their outlook to rejuvenate their attitudes toward instructing.

Successful instructors' goals of teaching are to:
   a. instill high standards in students;
   b. teach precision habits;
   c. reduce tolerances as instruction proceeds;
   d. detect unsafe habits and correct them (or, more positively, teach safe habits from the beginning).

YOUR INFLUENCE

The flight instructor exerts more influence on flight safety than any other pilot. You may ask, "What about airline captains who fly thousands of passengers every year? Don't they have more influence than people who may instruct, at most, 30 people in that time?" Remember that those airline captains didn't spring fully rated into the left seat; much of their attitude toward flying, and their flying habits, are the result of the first few hours of their flight instruction. If they have been flight instructors at some point in their careers, you may be assured that they passed on some of the ideas they got in *their* early learning period. This passing of information by flight instructors includes bad ideas as well as good ones, unfortunately.

The student pilot tends to imitate his instructor. If you are the kind of person who thinks that Federal Aviation Regulations are for less-experienced pilots than you, think about the effect on the student if you buzz and show off at low altitudes. You might get away with it, but if you had a student get in trouble because he was influenced by your actions, you'd feel low enough to put on a 10-gallon hat and walk under a snake's belly (and that's *low*).

It will bring the idea of your influence directly home when a student of 10 or 15 years ago stops by to say, "I'll always remember that time we were doing stalls, and you taught me that . . ." You don't even remember flying with him, it's been so long ago, but it *has* stuck with him. Another time, you may hear about one of your ex-students who, after inadvertently flying into IFR conditions, saved himself and his passengers by using the emergency instrument flying instruction you hammered so hard at him. Your next reaction, though, after feeling that you've accomplished *something,* will be to wonder why you weren't able to instill enough judgment so that he would avoid getting into such a situation requiring the emergency training.

Speaking of judgment, throughout this book scenarios will be used to show you how to help the student or low time pilot set up his own ideas of aeronautical decision-making (ADM) or judgment calls.

You will be an obvious example, and how you act during flight training will affect how the student pilot makes his or her decisions in later years. You, the instructor, may lecture all you please about aviation safety, but unless you fly the way you talk there will be faint hope of student pilots taking you seriously.

You may later see one of your private or commercial (or instrument, or instructor) trainees getting careless. Will your influence be enough to straighten him out?

It's good to know that people come from miles around to get the word on flying because they know that you are the person who will give it to them straight. *Use your influence* to make these people fly safely.

GENERAL RESPONSIBILITIES

As a flight instructor you will be responsible for:

1. Starting a student, giving him ground and flight instruction so that he may safely solo.

2. Overseeing his solo flights in the local area.

3. Giving him periodic dual instruction to add to his knowledge and to check his progress.

4. Giving him proper ground and flight instruction to assure him of safely flying the prerequisite amount of solo cross-country for the private certificate.

5. Preparing him to pass the private written test and check ride.

That simple layout of requirements has a certain *now* aspect as it stands. It boils down to (a) assuring that he is safe to fly solo locally, (b) making sure that he is safe to fly solo cross-country, and (c) seeing to it that he can pass the private check ride and not embarrass you. What's not listed in the earlier list is implied. You'll make sure that (a) 2 years from now he doesn't neglect to preflight the airplane properly, resulting in problems in flight, (b) three years from now he uses judgment and cancels a flight even at great inconvenience because of weather he feels he can't handle, or (c) 4 years after he's left you, he gets several hours of dual because he's hit a slump and knows he needs instruction. *Teach him to recognize his own limitations.*

The same thing applies when you are working with a person on the commercial, multiengine, or instrument rating or flight instructor's certificate. You'll actually have three goals to accomplish:

1. Teaching him the particular item so that he can move on to the next phase.

2. Giving him knowledge so that he can continue to complete the requirements to get the particular certificate or rating.

3. Teaching him an attitude toward flying that is fixed in his mind long after he's forgotten exactly how to do a specific maneuver for the flight test.

You teach an instrument student *basic* instrument flying so that he will be ready to go on to radio work and take the flight test. After he takes the flight test he will have to cope with the weather and ATC systems, so your real objective is for him to be successful in doing that, not just the short range aim to pass the flight test. The flight test *is* necessary because it covers areas of knowledge needed to cope with the system, but it's poor policy for an instructor to train an individual strictly for a flight test and even worse to train for a *particular check pilot.* ("This check pilot is heavy on ADF work so we will work on that and won't have to spend much time working on the VOR . . ." Both the instructor and check pilot are wrong in this case.)

You'll see that your personal responsibility can be heavy. A long distance call from a town to which your student left on a cross-country an hour ago can be quite a jolt until you find out he's okay, except that he had a flat tire taxiing in there and will be delayed. Or, you'll worry about the student who should have been back 20 min ago until (finally) you get his call on the Unicom, or he comes into view over the horizon. You'll be relieved and just a little peeved at him but will give him a chance to explain before you start to discuss his shortcomings.

You'll feel the pressure, too, when you've worked with a person for several weeks or months, recommended him for a particular certificate, and sent him up for the flight test. It can be a very long interval from the time he taxis out with the check pilot until you find out he passed. You can remember all sorts of information that should have been covered in more detail. Of course, his busting of a flight test is not a grave event—as compared to his having an accident—but you'll sweat the flight test anyway.

Another of your responsibilities to the student is to be truthful if one of your demonstrations goes awry. Most of the times you goof will be obvious to the student (or at least that wasn't the way you explained it on the ground), so tell him that was not the way to do it. Then, when you do it slickly and say that is the way, he'll trust your statements.

Always be ready to add to your instruction. Many times an unplanned event gives a good opening to add some knowledge of flying. ("Notice how smooth the air got when we climbed out of the haze layer? Well, that's because . . . ") Sometimes your throat is sore and you're hoarse from talking, you're beat and just want to get back on the ground; but then you see the airplane ahead of you on final catching wake turbulence and doing violent maneuvers to get out of it, and you have a graphic way to make a point. So you talk about wake turbulence (and avoid it) while the sight of that airplane is still fresh.

Here's an example to sum up the personal and moral responsibility you'll feel as an instructor:

Some years ago a flight instructor was asked by another, "Do you remember John R., who was your student a couple of years ago and moved away?" The instructor recalled him very well. The other instructor said, "Well, he got killed last week."

The instructor, who had worked with John R. from the first flight until he had gotten his private certificate, was shaken and tried to think of things he might have done wrong during the flight training. His thoughts were interrupted by the giver of bad tidings. "Yeah, they say that it was one of the worst freeway pileups in L.A. in years. Must have been 40 cars involved, and he was just riding as a passenger." Needless to say, in addition to feeling bad about John R.'s demise, there was a genuine feeling of relief that he wasn't killed as the pilot of an airplane.

You'll concentrate on *decision-making skills* with all the people you teach. All too often instructors forget that this is the most important skill of all. It's a nebulous quality, changing from situation to situation, but this book will try to give ideas on how to convince new pilots of its importance. The most important times for decision making will be after the pilot has gotten out on his own and has to make a go/no-go decision without your help.

YOUR PERSONAL CONSIDERATIONS

■ **Integrity.** Take a look at an area of most importance—personal integrity.

One of the biggest sources for gripes by flight students is for them to come to the airport at the scheduled time to find that the instructor has departed on a more lucrative charter flight. Nobody bothered to contact the student, who may have driven many miles and changed his own schedule to be there at that time.

You know that if some individuals made the statement that the sun will definitely be coming up tomorrow you'd rush out and stock up on candles. On the other hand, you'd bank your life on what some other people say. *Keep your word.*

■ **Appearance and Other Things.** Let's face it, some of us were born to look like five miles of bad road and that can't be helped, but the airplane cabin is a small world. Nowadays many of your students will be successful professional men and women and they won't put up with a guy in greasy overalls smelling like he's been cleaning hangars for the past five days and nights. Clean slacks and sport shirts or a neat, clean flight suit are fine. A suit and tie may sometimes be carrying it a little too far, but it's a lot better than the other extreme.

Pilots (and instructors) have been known to take a very

small drink of alcoholic beverage on very rare occasions, as anyone who has been to one of their parties will testify. That's fine, but it goes without saying that meeting a student for flying with beer or whiskey on your breath would hardly inspire confidence. Imagine your reaction if you took up scuba diving and on your first lesson the instructor leaned over, breathed 90-proof fumes in your face, looked at you with bloodshot eyes, and said, "Okay, let's dive." Give yourself at least *eight hours of sleep* between *any* drink and flying. After the plane is in the hangar, it's your business what you do—as long as you'll be completely ready for tomorrow's flights. Some instructors won't come to the airport after having a beer even if they aren't going to fly at all, because if somebody smells their breath the final magnified story will likely be that they were "loaded to the gills and making inverted passes at the hangar." (All he did was drive back over to pick up some charts for tomorrow's cross-country after having a beer at the airport lounge at the end of the day's flying.)

■ **Actions.** Even more important than your appearance will be your actions. Some instructors think that the louder they yell and the more they shake the student up, the more apt he will be to learn. Instructors could get away with it in olden days in military flight programs, but nobody learning to fly on his own is going to pay today's prices to hear himself ranted at. A person who uses personal abuse has no business being a flight instructor.

It's probably most difficult to be patient in a group training program; you'll have five or six students and they're at the same stage. The first student of the morning can do a pretty grim job of flying and gets nothing more than a humorous correction and careful extra instruction from you. The situation tends to deteriorate until the student who rides with you in the afternoon makes a minor bobble of the nature you've been facing all day, and you want to snap his head off (sure you have a headache, but that's the first time *he* has made that mistake). So it boils down to this: Patience is one of the finest virtues of a flight instructor and a student shouldn't suffer for your, or other students', problems—only his (it says here).

Another point to remember is that sarcasm is no training aid and *has no place in the syllabus at any time.*

As far as your language is concerned, an occasional "hell" or "damn" may slip out to emphasize a point, but you'll have to know your student and you should generally avoid it. As for those four-letter Anglo-Saxon words, forget it. Sure, sometimes it may seem that they help when you hit your thumb with a hammer or some other such calamity occurs, but nobody is paying $50 an hour to hear you swear. You'd be leery of a doctor who used obscenities as he treated you, right?

There's the strong, silent type of instructor who grunts occasionally and that's about all. Too much talking distracts a student, but often these guys rightly realize that if they ever open their mouths people will find out they aren't good instructors. The students can fly with a pig that gives an occasional grunt or snort a lot more cheaply and learn just about as much. Some of these extreme cases only grunt or make no sound at all until the student goofs (which he is likely to do, since he hasn't been getting any instruction on the proper way to do it) and then the instructor violently grabs the controls, jerking the airplane around shouting something like, "You call that a &%*#!! 720° turn?!?" (No, actually the student thought they were supposed to be doing spins; or was it slow flight?) Shouting and yanking the controls out of the student's hands shows a definite lack of self-control. See Chap. 2.

■ **Prejudices.** All people have prejudices of one kind or another, and some of your students may possibly be the combination of *all* of yours. The problem with prejudice is that

learning can't proceed if the prejudice is getting in the way. If you don't like long hair (on men) and beards, your student having these tonsorial accomplishments will sense it and your instructor-student relationship will be bad, as the education people say. Or maybe you don't like bald-headed men, or think women should be kept from the airport. You'll have to look at people as individuals, and usually after you fly with them awhile you'll get along well together.

Some people are permanently antagonistic to each other from first sight and you may feel after a couple of flights with a student that it won't work at all. Transfer him to another instructor; there's no point in both of you getting an ulcer (and wasting the student's money). The necessity for letting a student fly with somebody else because of "personality clashes" will arise occasionally, so don't worry about it—unless you start to have a steady turnover.

KEEPING UP TO DATE

After becoming an instructor *you should attend a Flight Instructors' Seminar every year if at all possible.* This gives you an inner authority that will show in your instructing. When you come back from a seminar you won't be wondering whether the FAA might have changed some requirement in the last year that you haven't heard about. It will confirm that you are (1) teaching the items the FAA will want on the flight test, and (2), more important, getting the information on flight safety and latest techniques. You'll also learn a great deal informally as you and the other instructors exchange information (much of it in the form of hangar flying). As you attend the seminars over the years you'll see the same group of instructors there—the ones who need the information most are always absent.

Reading aviation magazines is also a good way to keep current on what's happening in flight training, as well as proposed and actual FAR changes. Aviation organizations such as the Aircraft Owners and Pilots Association and National Association of Flight Instructors publish newsletters for their members.

■ **Instructor's Library.** You should have available for your use the following materials:

1. *Basic Flight Information and ATC Procedures; Airport/Facility Directory* for your area.

2. Federal Aviation Regulations Parts 1, 23, 25, 61, 67, 71, 91, 95, 97, 135 (if Air Taxi), 141, 143, and National Transportation Safety Board (NTSB) Part 830.

3. The latest Practical Test Standards for Private, Commercial, Instrument, Multiengine, Flight Instructor (Airplane), and Flight Instructor (Instrument).

4. The latest Knowledge Test information available.

5. Notices of Proposed Rule Making (you need to get on the mailing lists for these).

You should have all available FAA and Weather Service pilot training manuals as well as texts published commercially (see the Bibliography).

THE RIGHT SEAT

This chapter has been on the theoretical side, but one practical aspect of being a flight instructor you'll have to face is that you must become fully proficient in flying the plane from the right (or rear) seat. Airplanes are designed to be flown from the left side (or front) and that's where nearly everything of importance is located. You'll find that the first time or two in the right seat you feel as graceful as an elephant doing ballet. For one thing, the flight instruments are on the left side and the airspeed indicator is sometimes set so far back in the panel that the lower (and critical) speeds can't be read. Fig. 1-1 shows how things will appear on your first flight from the right seat.

Fig. 1-1. Things sure can look bare on your side of the instrument panel at first.

You will soon get used to the relative position of the needle in the various airspeed regimes, but you could help yourself during flight instructor training by putting various colored strips of tape on the panel by the important airspeeds (Fig. 1-2).

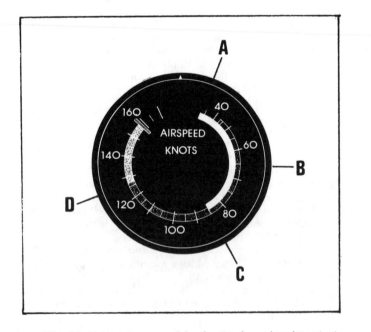

Fig. 1-2. Using tape as a quick reference for various important airspeeds. *A* is the bottom of the white arc; *B* is the recommended approach speed; *C* is the top of the white arc; and *D* is the maximum structural cruising speed. (Usually you would be able to see the numbers on the left side of the airspeed indicator.)

Naturally, such things as switches, starters, and some fuel selectors tend to be left-oriented, but generally this will only be a problem to you on the student's first flight. Under your supervision he will be doing the starting and pretakeoff checks on all the following flights.

FLIGHT INSTRUCTOR REQUIREMENTS

Here are some *specific* knowledge and skill requirements you'll consider in working for the flight instructor's certificate, as listed in the FARs.

Flight Instructors

§ 61.181   Applicability.
This subpart prescribes the requirements for the issuance of flight instructor certificates and ratings, the conditions under which those certificates and rating are necessary, and the limitations on those certificates and ratings.

§61.183   Eligibility requirements.
To be eligible for a flight instructor certificate or rating a person must:
(a) Be at least 18 years of age;
(b) Be able to read, speak, write, and understand the English language. If the applicant is unable to meet one of these requirements due to medical reasons, then the Administrator may place such operating limitations on that applicant's flight instructor certificate as are necessary;
(c) Hold either a commercial pilot certificate or airline transport pilot certificate with:
(1) An aircraft category and class rating that is appropriate to the flight instructor rating sought: and
(2) An instrument rating, if the person holds a commercial pilot certificate that is appropriate to the flight instructor rating sought, if applying for—
(i) A flight instructor certificate with an airplane category and single-engine rating;
(ii) A flight instructor certificate with an airplane category and single-engine class rating;
(iii) A flight instructor certificate with a powered-lift rating; or
(iv) A flight instructor certificate with an instrument rating.
(d) Receive a logbook endorsement from an authorized instructor on the fundamentals of instructing listed in §61.185 of this part appropriate to the required knowledge test;
(e) Pass a knowledge test on the areas listed in §61.185(a) of this part, unless the applicant:
(1) Holds a flight instructor certificate or ground instructor certificate issued under this part;
(2) Holds a current teacher's certificate issued by a State, county, city, or municipality that authorizes the person to teach at an educational level of the 7th grade or higher; or
(3) Is employed as a teacher at an accredited college or university.
(f) Pass a knowledge test on the aeronautical knowledge areas listed in §61.185(a)(2) and (a)(3) of this part that are appropriate to the flight instructor rating sought;
(g) Receive a logbook endorsement from an authorized instructor on the areas of operation listed in §61.187(b) of this part, appropriate to the flight instructor rating sought;
(h) Pass the required practical test that is appropriate to the flight instructor rating sought in an:
(1) Aircraft that is representative of the category and class of aircraft for the aircraft rating sought; or
(2) Approved flight simulator or approved flight training device that is representative of the category and class of aircraft for the rating sought and used in accordance with an approved course at a training center certificated under part 142 of this chapter.
(i) Accomplish the following for a flight instructor certificate with an airplane or a glider rating:
(1) Receive a logbook endorsement from an authorized instructor indicating that the applicant is competent and possesses instructional proficiency in stall awareness, spin entry, spins, and spin recovery procedures after providing the applicant with flight training in those training areas in an airplane or glider, as appropriate, that is certificated for spins; and
(2) Demonstrate instructional proficiency in stall awareness, spin entry, spins, and spin recovery procedures. However, upon presentation of the endorsement specified in paragraph (i)(1) of this section an examiner may accept that endorsement as satisfactory evidence of instructional proficiency in stall awareness, spin entry, spins, and spin recovery procedures for the practical test, provided that the practical test is not a retest as a result of the applicant failing the previous test for deficiencies in the knowledge or skill of stall awareness, spin entry, spins, or spin recovery instructional procedures. If the retest is a result of deficiencies in the ability of an applicant to demonstrate knowledge or skill of stall awareness, spin entry, spins, or spin recovery instructional procedures, the examiner must test the person on

stall awareness, spin entry, spins, and spin recovery instructional procedures in an airplane or glider, as appropriate, that is certificated for spins;

(j) Log at least 15 hours as pilot in command in the category and class of aircraft that is appropriate to the flight instructor rating sought; and

(k) Comply with appropriate sections of this part that apply to the flight instructor rating sought.

§61.185 Aeronautical knowledge.

(a) A person who is applying for a flight instructor certificate must receive and log ground training from an authorized instructor on:

(1) Except as provided in paragraph (b) of this section, the fundamentals of instructing, including:

(i) The learning process;
(ii) Elements of effective teaching;
(iii) Student evaluation and testing;
(iv) Course development;
(v) Lesson planning; and
(vi) Classroom training techniques.

(2) The aeronautical knowledge areas for a recreational, private, and commercial pilot certificate applicable to the aircraft category for which flight instructor privileges are sought; and

(3) The aeronautical knowledge areas for the instrument rating applicable to the category for which instrument flight instructor privileges are sought.

(b) The following applicants do not need to comply with paragraph (a) of this section:

(1) The holder of a flight instructor certificate or ground instructor certificate issued under this part;

(2) The holder of a current teacher's certificate issued by a State, county, city, or municipality that authorizes the person to teach at an educational level of the 7th grade or higher; or

(3) A person employed as a teacher at an accredited college or university.

§61.187 Flight proficiency.

(a) *General.* A person who is applying for a flight instructor certificate must receive and log flight and ground training from an authorized instructor on the areas of operation listed in this section that apply to the flight instructor rating sought. The applicant's logbook must contain an endorsement from an authorized instructor certifying that the person is proficient to pass a practical test on those areas of operation.

(b) *Areas of operation.*

(1) *For an airplane category rating with a single-engine class rating:*

(i) Fundamentals of instructing;
(ii) Technical subject areas;
(iii) Preflight preparation;
(iv) Preflight lesson on a maneuver to be performed in flight;
(v) Preflight procedures;
(vi) Airport and seaplane base operations;
(vii) Takeoffs, landings, and go-arounds;
(viii) Fundamentals of flight;
(ix) Performance maneuvers;
(x) Ground reference maneuvers;
(xi) Slow flights, stalls, and spins;
(xii) Basic instrument maneuvers;
(xiii) Emergency operations; and
(xiv) Postflight procedures.

(2) *For an airplane category rating with a multiengine class rating:*

(i) Fundamentals of instructing
(ii) Technical subject areas;
(iii) Preflight preparation;
(iv) Preflight lesson on a maneuver to be performed in flight;
(v) Preflight procedures;
(vi) Airport and seaplane base operations;
(vii) Takeoffs, landings, and go-arounds;
(viii) Fundamentals of flight;
(ix) Performance maneuvers;
(x) Ground reference maneuvers;
(xi) Slow flight and stalls;
(xii) Basic instrument maneuvers;
(xiii) Emergency operations;
(xiv) Multiengine operations; and
(xv) Postflight procedures.

(3) *For a rotorcraft category rating with a helicopter class rating:*

(i) Fundamentals of instructing;
(ii) Technical subject areas;
(iii) Preflight preparation;
(iv) Preflight lesson on a maneuver to be performed in flight;
(v) Preflight procedures;
(vi) Airport and heliport operations;
(vii) Hovering maneuvers;
(viii) Takeoffs, landings, and go-arounds;
(ix) Fundamentals of flight;
(x) Performance maneuvers;
(xi) Emergency operations;
(xii) Special operations; and
(xiii) Postflight procedures.

(4) *For a rotorcraft category rating with a gyroplane class rating:*

(i) Fundamentals of instructing;
(ii) Technical subject areas;
(iii) Preflight preparation;
(iv) Preflight lesson on a maneuver to be performed in flight;
(v) Preflight procedures;
(vi) Airport operations;
(vii) Takeoffs, landings, and go-arounds;
(viii) Fundamentals of flight;
(ix) Performance maneuvers;
(x) Flight at slow airspeeds;
(xi) Ground reference maneuvers;
(xii) Emergency operations; and
(xiii) Postflight procedures.

(5) *For a powered-lift category rating:*

(i) Fundamentals of instructing;
(ii) Technical subject areas;
(iii) Preflight preparation;
(iv) Preflight lesson on a maneuver to be performed in flight;
(v) Preflight procedures;
(vi) Airport and heliport operations;
(vii) Hovering maneuvers;
(viii) Takeoffs, landings, and go-arounds;
(ix) Fundamentals of flight;
(x) Performance maneuvers;
(xi) Ground reference maneuvers;
(xii) Slow flight and stalls;
(xiii) Basic instrument maneuvers;
(xiv) Emergency operations;
(xv) Special operations; and
(xvi) Postflight procedures.

(6) *For a glider category rating:*

(i) Fundamentals of instructing;
(ii) Technical subject areas;
(iii) Preflight preparation;
(iv) Preflight lesson on a maneuver to be performed in flight;
(v) Preflight procedures;
(vi) Airport and gliderport operations;
(vii) Launches, landings, and go-arounds;
(viii) Fundamentals of flight;
(ix) Performance speeds;
(x) Soaring techniques;
(xi) Performance maneuvers;
(xii) Slow flight, stalls, and spins;
(xiii) Emergency operations; and
(xiv) Postflight procedures.

(7) *For an instrument rating with the appropriate aircraft category and class rating:*

(i) Fundamentals of instructing;
(ii) Technical subject areas;
(iii) Preflight preparation;
(iv) Preflight lesson on a maneuver to be performed in flight;
(v) Air traffic control clearances and procedures;
(vi) Flight by reference to instruments;
(vii) Navigation aids;
(viii) Instrument approach procedures;
(ix) Emergency operations; and
(x) Postflight procedures.

(c) The flight training required by this section may be accomplished:

(1) In an aircraft that is representative of the category and class of aircraft for the rating sought; or

(2) In an approved flight simulator or approved flight training device representative of the category and class of aircraft for the rating sought, and used in accordance with an approved course at a training center certificated under part 142 of this chapter.

§61.189 Flight instructor records.

A flight instructor must sign the logbook of each person to whom that instructor has given flight training or ground training.

(b) A flight instructor must maintain a record in a logbook or a separate document that contains the following:

(1) The name of each person whose logbook or student pilot certificate that instructor has endorsed for solo flight privileges, and the date of the endorsement; and

(2) The name of each person that instructor has endorsed for a knowledge test or practical test, and the record shall also indicate the kind of test, the date, and the results.

(c) Each flight instructor must retain the records required by this section for at least 3 years.

§61.191 Additional flight instructor ratings.

(a) A person who applies for an additional flight instructor rating on a flight instructor certificate must meet the eligibility requirements listed in §61.183 of this part that apply to the flight instructor rating sought.

(b) A person who applies for an additional rating on a flight instructor certificate is not required to pass the knowledge test on the areas listed in §61.185(a) of this part.

§61.193 Flight instructor privileges.

A person who holds a flight instructor certificate is authorized within the limitations of that person's flight instructor certificate and ratings, and that person's pilot certificate and ratings, to give training and endorsements that are required for, and relate to:

(a) A student pilot certificate;
(b) A pilot certificate;
(c) A flight instructor certificate;
(d) A ground instructor certificate;
(e) An aircraft rating;
(f) An instrument rating;
(g) A flight review, operating privilege, or recency of experience requirement of this part;
(h) A practical test; and
(i) A knowledge test.

§61.195 Flight instructor limitations and qualifications.

A person who holds a flight instructor certificate is subject to the following limitations:

(a) *Hours of training.* In any 24-consecutive-hour period, a flight instructor may not conduct more than 8 hours of flight training.

(b) *Aircraft ratings.* A flight instructor may not conduct flight training in any aircraft for which the flight instructor does not hold:

(1) A pilot certificate and flight instructor certificate with the applicable category and class rating; and

(2) If appropriate, a type rating.

(c) *Instrument rating.* A flight instructor who provides instrument flight training for the issuance of an instrument rating or a type rating not limited to VFR must hold an instrument rating on his or her flight instructor certificate and pilot certificate that is appropriate to the category and class of aircraft in which instrument training is being provided.

(d) *Limitations on endorsements.* A flight instructor may not endorse a:

(1) Student pilot's certificate or logbook for solo flight privileges, unless that flight instructor has—

(i) Given that student the flight training required for solo flight privileges required by this part; and

(ii) Determined that the student is prepared to conduct the flight safely under known circumstances, subject to any limitations listed in the student's logbook that the instructor considers necessary for the safety of the flight.

(2) Student pilot's certificate and logbook for a solo cross-country flight, unless that flight instructor has determined the student's flight preparation, planning, equipment, and proposed procedures are adequate for the proposed flight under the existing conditions and within any limitations listed in the logbook that the instructor considers necessary for the safety of the flight;

(3) Student pilot's certificate and logbook for solo flight in a Class B airspace area or at an airport within Class B airspace unless that flight instructor has—

(i) Give that student ground and flight training in that Class B airspace or at that airport; and

(ii) Determined that the student is proficient to operate the aircraft safely.

(4) Logbook of a recreational pilot, unless that flight instructor has—

(i) Given that pilot the ground and flight training required by this part; and

(ii) Determined that the recreational pilot is proficient to operate the aircraft safely.

(5) Logbook of a pilot for a flight review, unless that instructor has conducted a review of that pilot in accordance with the requirements of §61.56(a) of this part; or

(6) Logbook of a pilot for an instrument proficiency check, unless that instructor has tested that pilot in accordance with the requirements of §61.57(d) of this part.

(e) *Training in an aircraft that requires a type rating.* A flight instructor may not give flight training in an aircraft that requires the pilot in command to hold a type rating unless the flight instructor holds a type rating for that aircraft on his or her pilot certificate.

(f) *Training received in a multiengine airplane, a helicopter, or a powered-lift.* A flight instructor may not give training required for the issuance of a certificate or rating in a multiengine airplane, a helicopter, or a powered-lift unless that flight instructor has at least 5 flight hours of pilot-in-command time in the specific make and model of a multiengine airplane, helicopter, or powered-lift, as appropriate.

(g) *Position in aircraft and required pilot stations providing flight training:*

(1) A flight instructor must perform all training from in an aircraft that complies with the requirements of §91.109 of this chapter.

(2) A flight instructor who provides flight training for a pilot certificate or rating issued under this part must provide that flight training in an aircraft that meets the following requirements—

(i) The aircraft must have at least two pilot stations and be of the same category, class, and type, if appropriate, that applies to the pilot certificate or rating sought.

(ii) For single-place aircraft, the presolo flight training must have been provided in an aircraft that has two pilot stations and is of the same category, class, and type, if appropriate.

(h) *Qualifications of the flight instructor for training first-time flight instructor applicants.*

(1) The ground training provided to an initial applicant for a flight instructor certificate must be given an authorized instructor who—

(i) Holds a current ground or flight instructor certificate with the appropriate rating, has held that certificate for at least 24 months, and has given at least 40 hours of ground training; or

(ii) Holds a current ground or flight instructor certificate with the appropriate rating, and has given at least 100 hours of ground training in an FAA-approved course.

(2) Except for an instructor who meets the requirements of paragraph (h)(3)(ii) of this section, a flight instructor who provides training to an initial applicant for a flight instructor certificate must—

(i) Meet the eligibility requirements prescribed in §61.183 of this part;

(ii) Hold the appropriate flight instructor certificate and rating;

(iii) Have held a flight instructor certificate for at least 24 months;

(iv) For training in preparation for an airplane, rotorcraft, or powered-lift rating, have given at least 200 hours of flight training as a flight instructor; and

(v) For training in preparation for a glider rating, have given at least 80 hours of flight training as a flight instructor.

(3) A flight instructor who serves as a flight instructor in an FAA-approved course for the issuance of a flight instructor rating must hold a current flight instructor certificate with the appropriate rating and pass the required initial and recurrent flight instructor proficiency tests, in accordance with the requirements of the part under which the FAA-approved course is conducted, and must—

(i) Meet the requirements of paragraph (h)(2) of this section; or

(ii) Have trained and endorsed at least five applicants for a practical test for a pilot certificate, flight instructor certificate, ground instructor certificate, or an additional rating, and at least 80 percent of those applicants passed that test on their first attempt; and

(A) Given at least 400 hours of flight training as a flight instructor for training in an airplane, a rotorcraft, or for a powered-lift rating; or

(B) Given at least 100 hours of flight training as a flight instructor, for training in a glider rating.

(i) *Prohibition against self-endorsements.* A flight instructor shall not make any self-endorsement for a certificate, rating, flight review, authorization, operating privilege, practical test, or knowledge test that is required by this part.

(j) A flight instructor may not give training in Category II or Category III operations unless the flight instructor has been trained and tested in Category II or Category III operations, pursuant to §61.67 or §61.68 of this part, as applicable.

§61.197 Renewal of flight instructor certificates.

(a) A person who holds a flight instructor certificate that has not expired may renew that certificate for an additional 24 calendar months if the holder:

(1) Passes a practical test for—

(i) Renewal of the flight instructor certificate; or

(ii) An additional flight instructor rating; or

(2) Presents to an authorized FAA Flight Standards Inspector—

(i) A record that shows that within the preceding 24 calendar months the flight instructor has endorsed at least five students for a practical test for a certificate or rating, and at least 80 percent of those students passed that test on the first attempt.

(ii) A record that shows that within the preceding 24 calendar months, the flight instructor has served as a company check pilot, chief flight instructor, company check airman, or flight instructor in a part 121 or part 135 operation, or in a position involving the regular evaluation of pilots, in which that authorized FAA Flight Stan-

dards Inspector is acquainted with the duties and responsibilities of the position, and has satisfactory knowledge of its current pilot training, certification, and standards; or

(iii) A graduate certificate showing the person has successfully completed an approved flight instructor refresher course consisting of ground training or flight training, or both, within the 90 days preceding the expiration month of his or her flight instructor certificate.

(b) If a person accomplishes the renewal requirements of paragraph (a)(1) or (a)(2) of this section within the 90 days preceding the expiration month of his or her flight instructor certificate:

(1) That person is considered to have accomplished the renewal requirement of this section in the month due; and

(2) The current flight instructor certificate will be renewed for an additional 24 calendar months from its expiration date.

(c) The practical test required by paragraph (a)(1) of this section may be accomplished in an approved flight simulator or approved flight training device if the test is accomplished pursuant to an approved course conducted by a training center certificated under part 142 of this chapter.

§61.199 Expired flight instructor certificates and ratings.

(a) *Flight instructor certificates.* The holder of an expired flight instructor certificate may exchange that certificate for a new certificate by passing a practical test prescribed in §61.183(h) of this part.

(b) *Flight instructor ratings.*

(1) A flight instructor rating or a limited flight instructor rating on a pilot certificate is no longer valid and may not be exchanged for a similar rating or a flight instructor certificate.

(2) The holder of a flight instructor rating or a limited flight instructor rating on a pilot certificate may be issued a flight instructor certificate with the current ratings, but only if the person passes the required knowledge and practical test prescribed in this subpart for the issuance of the current flight instructor certificate and rating.

SUMMARY

This chapter has covered briefly what it means to be a flight instructor in general. Additional points will be made throughout this book on your responsibilities, influence, and other items as applicable to particular situations.

To be a good flight instructor you will have to know your subject, understand how people learn, and be able to put yourself in the student's shoes—to treat the student as you would like to be treated.

Remember these points:

1. *Air discipline* or professionalism means that as an instructor you'll teach and strictly observe systematic rules and procedures adopted for flying. You won't compromise for an easy out in a particular situation. (For instance, cutting out part of the preflight check with a student because you're behind schedule.)

2. *Know your limitations.* Later, don't let the fact that you are experienced get you pushed into flying under marginal mechanical or weather conditions.

3. *Watch for overconfidence.* You'll probably hit this point after 200–300 hr instructing in the same airplane. You have the airplane and student situation wired, you figure, and nothing could happen that you couldn't handle.

4. *Stay current.* Keep up your studying and flying.

5. *Study emergency procedures.* Review from time to time what you would do in the event of an electrical fire or other actual emergencies. You owe it to the student to be able to get you both out of a problem.

6. *Don't surprise the student.* In a normal flight lesson he should not encounter something unexpected from the instructor. You should tell him beforehand of the possibilities of errors during a maneuver or flight. If he has heard about what he's encountering, his anxiety level will decrease. Your student, like a good experienced pilot, should have surprises kept to a minimum. Let him know what to expect. Don't do something and then tell him what it was; *tell him about it* and *then do it.*

7. *Don't try to memorize many different airplanes' "numbers,"* but learn as much as possible about the principles of aerodynamics and engines so that this knowledge may be applied to a particular airplane as necessary. Be sure that you know the numbers of the airplane you are using for instruction, but don't worry if somebody asks you the landing speed of the Fokker D-7 (at max certificated weight, of course) and you don't have an immediate answer. *Know where to look for information.*

8. *Impress on your students from the beginning* that the first requirement is control of the airplane at all times. Too many accidents, fatal and otherwise, have happened because a pilot is distracted by things such as the smell of smoke (imagined or real), wasps or bees in the cabin, or a door coming open in flight. A wasp sting is less of a catastrophe than flying into the hangar because of distraction, and the airplane won't "fall" because the door is open. Planes have been landed gear-up because the pilot smelled a smoldering cigarette in the ashtray and got in a prodigious hurry to get back to the ground. *Teach them to fly the airplane—and to use the checklist.*

9. *If you don't teach the student something new every flight* then you've wasted both his time and yours.

10. One thing of interest you will discover if you don't already know about it is that *no matter how well you've trimmed the airplane,* when you turn it over to one of your advanced students or another instructor, *he will retrim it*—and so will you in the reverse situation. Perhaps this is an innate tendency to "claim territory," to establish control of the airplane by retrimming.

11. *When you have to analyze* what the airplane is doing and explain it in understandable terms to a beginner you'll start really learning to fly.

Finally, opinion will vary widely among instructors as to what is the ultimate truth in flight instruction. A brand new flight instructor who was concerned about how he would be as a teacher approached a veteran of many years of rear- and right-seat experience. Could the experienced instructor sum up the most important factors of flight instructing, based on his great experience, in one sentence?

The crow's-feet at the corners of the older man's eyes deepened and his tanned face took on a very thoughtful look.

There was a long pause, and as he started to speak the new instructor took out a pencil and notebook to catch the ultimate truth. The old instructor cleared his throat and spoke.

"Don't let 'em eat chili and hot dogs just before you go up to do spins."

HOW PEOPLE LEARN – TO FLY

BACKGROUND

Since a flight instructor's job is teaching, you should know as much as possible about how people learn and what affects the learning process. Unfortunately, many texts on education tend to use complicated language and base a lot of the techniques on classroom situations with children. Your teaching situation will obviously be different; you will be teaching adults, mostly on a one-to-one basis, and generally not in a classroom (although you will use classrooms for ground and preflight instruction). It's been said that an airplane is the worst place possible to try to learn anything; the student is taken up in what seems to be a flimsy contraption with vibration and noise and often has unintelligible directions shouted over that noise. You may tell a person he's "doing a great job," but if you have to shout it at the top of your lungs, it's going to contribute to tensing him up. (This could be a good argument for using an intercom, if possible.) This chapter will take a general look at how people learn, with some examples of how this knowledge can be applied to flight training.

BASICS OF LEARNING

The first requirement, as obvious as it may seem, is communication. If the student doesn't get the information (by listening, reading, watching, etc.) he isn't going to gain new knowledge. So, there'd better be some exchange of information.

EDUCATION HAS THREE MAIN PURPOSES:
(1) DISCOVERY OF NEW KNOWLEDGE,
(2) DISSEMINATION OF KNOWLEDGE, AND
(3) TRANSLATION OF FACTS AND KNOWLEDGE INTO ACTION.

The Einsteins of this world are usually given job (1). You may discover something new in flying, but this won't be your primary purpose.

As far as (2) is concerned, you will disseminate knowledge (spread the word) but eventually this dissemination will turn into (3) when the student translates the facts and knowledge you gave him into action and gets lost on his first solo cross-country. Dissemination of "knowledge" doesn't even have to be accurate if it doesn't get translated into action. You can tell anybody you want to that it's okay to fly a small airplane into a tornado and, as long as nobody tries it, you're doing fine with your spreading of "facts and knowledge."

That's where flight instructing is different from some college courses. You may rest assured that the information you put out will be acted upon. *Your main purpose in education will be to help the student translate facts and knowledge into action* (Fig. 2-1).

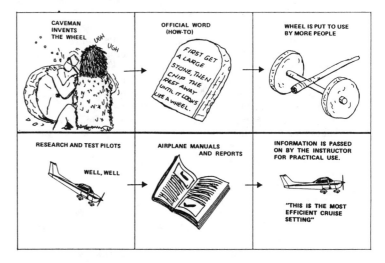

Fig. 2-1. Translating facts and knowledge into action. Two different applications.

■ **The Senses and Learning.** Educators believe that the various senses contribute to learning by the following approximate amounts: sight—75%, hearing—13%, touch or feel—6%, smell—3%, and taste—3%.

There's quite a gap between sight and hearing, but notice that together they make up 88 percent of the learning receptors. If a student is having problems in getting a proper landing attitude it would be good to take him out to the end of the runway and have him watch landings—*as you describe what's happening*. He will get more out of watching actual landings for 15 min, listening to you as it happens, than he would by just listening to an hour of explanation.

The top three senses in terms of percentage of learning intake (*sight, sound,* and *touch*) are the primary ones used in flying; every one of these three is used. In instrument flying, sight is an even more predominant part of the whole than in the VFR phase of instruction. In flying by instruments, a change in *sound* (because of an increase or decrease in airspeed for instance) would have likely been shown by the instrument indications before it actually built up enough to be noticeable.

Smell is another method of learning. Your nose will tell you that an aroused skunk is under your house long before any of the other senses even suspect he's there. So you've gotten a *fact* through smell. Your next move would be to get someone with the *knowledge* to get him out without further demonstrations of his unique reactions to stress. Smell can be very important in flying, too; it will tell you of the presence of an electrical fire before you *see* smoke or *hear* the radios fade out. Smell also might alert you that the fuel being put in your airplane is jet fuel and not the 100LL octane you require.

(*Taste* is not as valuable in flying as the other senses, but it could *confirm* what you had *heard* from other pilots about the food at the airport cafe, and now *see* and *smell,* as the Blue Plate Special is slid in front of you. You might also have the "coppery taste of fear" in a tight situation, but it's likely that the other senses would have already let you know you were scared.)

■ **Perception and Insight.** *Perception* is defined as an awareness of the environment through physical sensation, or penetration, or discernment. In other words, are you getting the full message?

Insight is understanding. Do you understand what's getting to you through your perception? You can have perception without insight; you are aware that something is happening but don't know *why.* So, *insight depends on the accuracy of perception, and the person's ability to understand.*

The flight instructor is there to help students develop both *perception and insight* in things pertaining to flight. At first glance, it might seem that the instructor's job only concerns *insight,* or understanding what is going on during the flight, but this is not the whole job. You will direct the attention of the student's perception to an item, then lead his insight to it. Your directing of attention to what the airplane (or wind) is doing is extremely important in the early stages of training because the student pilot doesn't know what to look for. Most of your instruction of the presolo student will consist of developing his perception (and, with it, his insight and understanding) of what are elementary things to you. He will not see the aircraft traffic that is perfectly obvious to you. As you taxi out for that first flight, he won't see the large orange tetrahedron practically in front of the hangar until it's called to his attention and explained. After that the student knows its use and looks for it.

When you first start instructing you will be surprised at how much the new students miss and how often you'll have to direct their attention to sight, sound, and feel of the airplane and its environment. Take the simple process of hearing and understanding what is being said on the radio. With your direction the student will learn to pick out his number and add information pertaining to him (including other traffic that might affect his flight), while letting the rest go by. It's been said that the two most common questions from the student pilot starting to use aviation communication are: "Was he talking to us?" and "What did he say?" (These are used by experienced pilots, also.) You can expect this to happen on the first flight(s) of every student when he has to use the radio.

So you will (1) train his perceptions as far as the flight environment is concerned and (2) ensure that he has understanding (insight) of the facts and knowledge he gets.

ATTENTION. There's the old story of the man who went to buy a mule. The owner said that this was without a doubt the smartest mule ever encountered by mankind, and he could do all sorts of tricks on voice command, which made him very valuable. "Fine," said the prospective buyer. "Order him to do something." Whereupon the owner picked up a 4-ft length of 2-by-4 and hit the mule solidly between the eyes. While the animal was staggering around, trying to focus properly again, the buyer-to-be said, "I thought you said the mule would perform on voice command; why the 2-by-4?" "It's true," said the owner, "but *first* you have to get his attention (Fig. 2-2)." There are, of course, more subtle means of getting the student's attention (although there may be occasions when you will more than half-seriously consider that method). *If you direct his perception in an interesting way, he'll likely give you his attention.*

Fig. 2-2. Using a 2-by-4 as a training aid is fine (?) for mules, but it has its drawbacks in training humans.

If you asked some flight instructors if they would hit a student over the head with a rubber chicken to get his attention, they would answer quickly that they certainly would not because, "It would shake the student up too much." Yet, they may do the equivalent verbally by shouting and using sarcasm; or physically, by abruptly snatching the controls away (or both shouting *and* snatching the controls away) as was indicated in Chap. 1. This shocks the recipient of this delicate attention, and the training situation goes from bad to worse, which probably leads to more shouting and more deterioration, in a vicious cycle. Some students have been known to "get the instructor's attention" when they got back on the ground after this type of "training."

The biggest aid in keeping the student's attention is to tell him (her) during ground or flight instruction *why* you are going to spend that time at the chalkboard covering certain maneuvers or instruments. If he understands how it will directly affect him, you'll have his attention. Bring the student—or in the case of a formal ground school, students—into the discussion. One big advantage you'll have is that the people you are working with when teaching flying are nearly all "volunteers" and are spending money to learn about the subject. Flight instructors who go to public high schools as guest speakers are usually appalled at the lack of attention by many of the listeners, who may be dozing or reading while the instructor is talking. The students are there because the law says they have to be there to a certain age, and an extremely interesting subject (to you, anyway) is something that has to be sat through until the end of the period.

■ The Laws of Learning

Readiness—Is the student ready for this segment of the course?

Exercise—Repeat and practice what has been introduced.

Effect—A favorable effect is naturally best. Are you, as an instructor, making the learning process enjoyable?

Primacy—It's best to teach it right the first time. You'll find that the first 5 or 6 hr of a student's instruction are the most important in setting good habits of good preflight checks, using a checklist, looking around in flight, etc. More about this later.

Intensity—An intense and/or colorful approach to the subject helps the learning process. Don't drone at the student(s) when you are lecturing or giving a pre- or postflight briefing.

Recency—The most recently learned is best remembered. Items learned many years ago and not since used may be vague indeed.

Okay, moving on . . .

■ How Adults Learn. *Your job as a flight instructor will mostly be the teaching of adults on a one-to-one basis,* although you will also be called upon to teach from two to forty in a classroom.

Many new instructors don't know what route they should take in teaching people to fly. Perhaps they had a rather perfunctory course in the required maneuvers when getting ready for the flight instructor's flight test, and maybe their written preparation was a quickie course on how people learn and theory of instruction (which consisted mostly of sample tests with the answers as expected to be found on the FAA test). Since the new instructor has not had actual experience in teaching, he may hark back to his grammar and junior high school days and imitate the techniques of his teachers there. (Should he be the harsh disciplinarian like Mr. Strong, or be a buddy to his students like Mrs. Love? Or should he alternate techniques so the students can't predict him?)

An instructor may feel the need to be in control all the time and so act like a Nazi SS officer to assure there is no letdown of discipline. But trying to teach adults with attitudes and techniques that are often used for children just won't work. It seems that common sense would tell the instructor, after some bad times in teaching, that adults must be treated differently.

Following are some ideas about adults and how they learn:

1. Adults Must Want to Learn. Adults must personally want to learn the material. Outside pressure (grades, parents, etc.) do not affect them; they have to be motivated from within and don't like compulsory courses. In the case of the flight student who is paying for flight training, his or her motivation to learn is self-evident. However, you may run into some flight scholarship recipients who are taking flying instruction because "it's there." This is sometimes the case with people who are taking ROTC in college. They meet the mental and physical requirements, and the flying (flight instruction program) is available, but they are not motivated to fly. These people are eliminated, usually at the early stages, but nearly always before completing the postgraduate air force or navy flight training *unless somebody motivates them to learn.* They may not consciously drag their heels but the result is the same—if they don't want to learn, they won't cut it.

2. Adults Will Learn Only What They Feel They Need to Learn. For some reason, many flight manuals and how-to-fly textbooks feel compelled to begin with a chapter or two on the history of flight. You've seen it; there's the story of Icarus, and copies of Leonardo da Vinci's flying machine sketches, plus a picture of Montgolfier's balloon, followed by a coverage of the Wright brothers' flight. This is fine for a text for schoolchildren (who learn because that material is told to learn), but most adults would consider that material as "stuffing." The person who wants to learn to fly is interested in just that, whether it is a short-term goal such as soloing or getting a private license, or a long-term goal of being a professional pilot. If he knew that he would have to answer questions on the history of aviation for a certificate, or in order to get a job, he might look at it in a little better light, even though secretly he would see little or no application as far as actual flying is concerned. A student who intends to teach aeronautics in public school will no doubt see the benefit in a knowledge of aviation history but will see no application to getting a pilot certificate or rating.

The adult expects immediate benefits. He doesn't want to hear, "You will be glad you learned this—someday." In short, adults don't like being given a lot of baloney.

3. Adults Learn by Doing. You may lecture for many hours covering the exact step-by-step procedures of flying every required maneuver in preparation for passing the private practical test, and you'll find that without putting this into use in the airplane you are wasting your time. (It's estimated that learners retain only 20 percent of what they hear and don't do.) This is the problem with some ground schools that teach irrelevant or ill-timed material. For instance, it's not unusual for schools to teach ground school in subject blocks. This two weeks, meteorology will be covered in vast and laborious detail, including the reading of sequence reports, weather theory, and other "dry" subjects. But none of the people in the classroom have even soloed yet and hence are far more interested in takeoffs and landings and traffic patterns, so they can't see any immediate application. They are not *doing* anything associated with the weather.

This also harks back to the fact that *adults will learn only what they feel they need to learn.* They're being taught about weather but are doing nothing where it affects them. They will be very much interested in weather when they get to the cross-country stage and expect to actually encounter it. *Then* meteorology should be taught.

In the block type of training program, after students finish the weather study they have a block session of FARs (they probably are still completely under the control of the instructor and couldn't break an FAR if they wanted to). This subject block of learning is too often followed by a mass exodus from aviation to areas of more interest such as sailing or skiing. *The ground instruction should parallel the flight training.* This means that in certain phases of flight training at a session of ground instruction you may cover the parts of theory of flight, meteorology, navigation, and federal aviation regulations that would apply to particular situations the student is encountering at that time. *The student will learn by doing* and will also remember the material better.

The often-used practice of separating ground school from flight training shows a lack of knowledge of how adults learn. Too often the student attitude is, "Whew, I've gotten all the ground school and passed the written; now I can forget all that stuff and concentrate on the flight test." An onerous chore is over and he starts to forget (and it's been found that adults who have learned by listening to lectures, seeing slides, reading books, or by other such passive types of learning will forget 50 percent of it within 1 yr and 80 percent within 2 yr). The retention rate is much higher if they *participate and use* immediately what is being talked about.

4. Adult Learning Is Concentrated on Problems and These Problems Must Be Realistic. The adult will want to know *why* a particular problem or situation is presented, so you might as well explain from the beginning where you are going and how the knowledge can be applied.

Set up your problems realistically. Teach navigation by using a flight from the local airport to another one reasonably nearby — a problem that will be followed by actual use of the material. It's true that the principle of navigation can be taught very well using a theoretical flight between two airports in Africa, but the student will not learn as readily as in a here-and-now situation. And besides, he will see little value in learning to navigate in Africa, no matter how much principle is involved. This leads to the point that adults learn by getting specific problems and working out the answers, from which the principles may be learned. The specific cross-country flight in the "local" area can be used to bring out the point that the same general navigation principles are used for any cross-country flight.

5. Experience Affects Adult Learning. The adult's learning must be related to his or her past learning experiences. As a flight instructor, you should find out the student's background as soon as possible so that you will know how to teach that person. This will be covered more in detail shortly, but you can see that if you launch into a full-fledged aerodynamic lecture with math and graphs for a person who only graduated from high school or who was a music major in college, you've lost him. If the flight student is a graduate aerospace engineer, you can better keep his interest by using a more technical approach (but be sure *you* know what you are talking about). So, to repeat, *before you can teach anyone about flying, you'll have to find out his or her background so you can know which route to take to do an effective job.* Your sessions will be give and take with discussions; that way you'll learn about your students' background. More about this shortly (Fig. 2-3).

Fig. 2-3. Know the students' backgrounds and levels of flight experience.

6. Adults Learn Best in an Informal Environment. Treat them like adults; they don't need discipline such as "no questions" or "no talking." If *you* went to a class and were treated like a fifth-grader, you would be so busy thinking about what a Mickey Mouse setup it was that you wouldn't be taking in anything that was being said. You'd also be antagonistic toward the instructor and this would hamper your learning process. You should let the students express their ideas as the class or discussion goes on; however, if a question is raised about a subject you plan to cover shortly anyway, tell them that this will be discussed soon. Tell the questioner that it's a point that definitely needs to be covered. This will flatter him and insure his attention, and you haven't broken your train of instruction.

Adults like humor in their instruction; it fosters the feeling of an informal atmosphere and helps learning. Use it, but don't overdo it; you are an instructor — not a comedian.

7. A Variety of Methods Should Be Used in Teaching Adults. This is common sense, but is too often overlooked. Use training aids (slides, models, movies) whenever possible to make your point. People learn better when several of their senses are being used at the same time. To stand up and drone at them without any visual aids means that *hearing* is the only one of their senses being used (and sleep will probably mercifully turn that one off). Memory also depends on the use of more than one of the senses. Educators say that learners retain:

10% of what they *read*
20% of what they *hear*
30% of what they *see*
50% of what they *see and hear*
70% of what they *say*
90% of what they *say as they do a thing*

You'll find that some students will talk to themselves as they fly a maneuver after you've demonstrated it, and while this may be disconcerting when you first encounter it, let them do it (it's a valid learning procedure) even if it bugs you.

If you are merely interested in imparting facts and bits of knowledge, the lecture-type approach is useful, but if you want to change the student's way of thinking, a discussion session is best. If you want to change his position on a subject you may find that as he discusses it he'll adopt your views.

You will naturally pick the method and place of instruction most suited to the situation. While an airplane is the final place of training (after all, you are trying to teach him to fly one), it is the worst place to teach a course on FARs. Probably worse places could be found, such as inside an operating steel mill or the median strip of Los Angeles freeway, but this is debatable. Sure, he needs to know the FARs so he can use them when he's flying, but it's best that he be introduced to them on the ground. On the other hand, to give a 2-hr detailed lecture on a flight maneuver on the ground (if you could talk 2 hr on *any* flight maneuver) is certainly a waste of talent, too. These examples are exaggerated, but you could be off on your time or place to present material.

Not only should you vary your approach to instructing for the students' sakes, but for your own as well. After teaching a particular subject many times, you'll find that you tend to drone on and let your mind wander. The students' minds will wander, too. Instead of just sticking to the *what, why,* and *how* approach in, say, your preflight discussion, why not approach it when possible from a *who, where,* and *when* aspect as well? (For instance, *where* do you plan to do turns around a point? "Well, there's this nudist colony five miles north of the airport and . . .")

8. Adults Want Guidance, Not Grades. The people you will be teaching have been out of school a long time and will most likely feel that they've lost some of their ability to study. If you push the idea of grades too hard, they'll back off because they are afraid that they will be shown up. This doesn't mean that they aren't interested in knowing how they are do-ing, but they want to be told privately. Many adults set such high goals for themselves, they get discouraged. *You* will have to keep their spirits up and you can do this by letting them know *in advance* what problems everybody has with a particular maneuver or area of learning. As will be implied throughout this book, your psychology of instructing should be that of "pessimistic optimism." In introducing a new maneuver, for instance, you set the scene by introducing and explaining it, with *specific* errors that you have found other people make when they first attempt it. Your prediction of errors based on what other people have done, *before the student has done the maneuver,* puts him more at ease since he realizes you aren't criticizing him personally and that other people have made mistakes. If no mention is made of probable errors beforehand he magnifies any mistakes he makes, and the learning process suffers because he's worrying about his comparable performance rather than listening to what you are saying. If he avoids many of the errors you've mentioned, he's that much further ahead—his morale is better (and of course you will certainly mention it if *he* didn't make some of the errors usually encountered). If he makes *all* of the mistakes you mentioned, he's still doing as well as everybody else.

You may run into an individual who discovers new knowledge and makes mistakes that nobody has ever heard of. It's considered bad form to break up laughing and say, "Good grief, I didn't know that *anybody* could mess up that way!" Encourage him as best you can, and add his exotic mistakes to your list of common errors to give to the *next* student.

■ **The Student's Background.** You will tailor your instruction (preflight, inflight, and postflight) to a particular student. An explanation that would hold the interest of the college graduate would likely be over the head of the guy who only finished the sixth grade. Talking to the well-educated person in sixth grade terms could be boring, but this is less apt to happen than first thought, since flying is a new area for the student. *If in doubt, keep it on the simple side.*

Taking two *extremes* of people who have never handled the controls of an airplane—an ex-paratrooper turned engineer-businessman, and a timid soul who doesn't particularly care for flying but is taking lessons to please her husband (or his wife) who is learning to fly (and wants him or her to do the same)—make a comparison of your training approaches.

THE EX-PARATROOPER/ENGINEER-BUSINESSMAN. As an ex-paratrooper, this student pilot has been dealing with airplanes in the military (although he will likely have had more takeoffs than landings in them) and in addition, he's probably a regular rider of the airlines. His background of action, technical knowledge, and business experience make the following three things of importance: (1) you won't have to do a lot of assuring him about "how safe flying is, so please relax"; (2) your explanations can be fairly technical and full without a lot of background information to bring him up to date; and (3) he's a businessman and will want to get on with the flying. On the first flight you will definitely let him taxi the airplane in a clear area, make the pretakeoff check (with briefing from you and reference to the checklist) and follow you through the takeoff (or use the rudders himself).

A total of about 45 minutes at the chalkboard and at the airplane before the first flight would not be out of order with this man. With some students, covering too much before the first flight could confuse them; but they all should understand the functioning of the controls. Also, if this sample student has an idea of "torque" in the climb and how to make a turn, you can even be giving instructions in this area as you climb. You can talk him through the traffic pattern, turns, and departure and then take over when you wish to show him the practice area. In short, with such a student you should begin letting him do as much of the work as possible. He will enjoy the challenge.

Professional people as a rule can be a joy to instruct, particularly in the earlier stages. Because of their aggressiveness and intelligence, they are used to making their own decisions and sticking their necks out. They will go into learning to fly with a will—and will do their homework. They can ask some mighty sharp questions. You'll find them challenging to work with, and you may have to do some boning up in certain areas of aviation in which you've been a little weak. Problems can arise. The student may catch you in a mistake, or you may act uncertain about a particular point, and if you're not careful, you'll lose control of the situation. When that happens you'd better get it back fast or have the student use another instructor.

When you lose control this way you are under pressures that could cause you to make a wrong, or perhaps fatal, decision:

For example, *he's pushing you to let him go on that long cross-country.* You personally think he needs some more dual in pilotage; he was a little weak in maintaining his course and keeping up with checkpoints.

He maintains that this is his only day off from the office (or hospital) for the next 5 days. You think, in addition to his slight problem of pilotage, he'll be rushed in his planning in order to complete the trip before dark. Besides, even though the sky is clear, visibilities are only 5 to 6 mi, a handicap since he has a *slight* tendency to wander and lose checkpoints. Still, as he says, it would sure speed up training if this long solo cross-country could be gotten out of the way. So you make the wrong decision and let him go. He gets lost and lands in a plowed field and a big repair bill results. Who answers for this? You do, of course, and his ire will probably be turned on you for letting him go.

He's pressuring you to recommend him before you feel he is ready to pass the flight test. He may feel that he can pass it but you feel that, say, more stall practice is needed. Again comes the factor of his being a "busy person" and this is the only time he can take the flight test, etc. Or it's possible that a high grade on the written could be "accidentally" flaunted at you. Maybe you feel your grammar isn't so hot and he may be pressuring you with his higher education.

You'll find that the vast majority of professional people will be conscientious and good students. *These points are only mentioned to cover the very few problem types you might encounter.* Incidentally, some problems might be brought on by yourself. Bluffing, lack of instructing ability, or ill-prepared material will be quickly spotted by these intelligent and able people.

Don't be afraid to say "I don't know, but I'll find out" to these (or any) students. Business and professional people refer to books continually, so why shouldn't you? You are a professional also, even if you probably don't charge enough money for your time.

THE TIMID STUDENT. Some students, both men and women, seem naturally timid. The engineer–businessman won't have to be convinced of your competence; he can get a pretty good idea from your attitudes and actions even though at this stage he knows nothing of how to go about actually flying the airplane. The timid students may have to have their confidence in flying (and you) built up. They may have higher IQs than the more aggressive students and may turn out to be better and safer pilots, but because of a lack of technical background, or anxiety, you will approach their initial training in a different manner.

Their first flight will likely be no more than an orientation to the idea of flying and getting used to the controlling of an airplane (at altitude), particularly if they have had little or no experience in a small airplane. You'll find in some cases people who have flown in airliners still have some suspicion of "little planes." The impression given to some people by movies and television is that a plane will "fall" unless the pilot uses all of the superhuman skill available to him, and of course if the engine quits, the whole outfit—airplane, people, baggage, and all other things attached to the plane—will drop with the glide ratio of a free-falling safe. You'll make more effort to demonstrate the inherent stability of the airplane during the flight and will work on a general building up of confidence. You will do a lot of the work on this flight, explaining the steps and not turning over the controls to the student until you're up in smooth air with the airplane trimmed and the power set.

THE AVERAGE STUDENT. Every maneuver and description in this book will be aimed at the average student with any problem types mentioned as they come up. If the average student can be categorized, it could be said that he or she is of average intelligence, has a fair amount of mechanical ability, can drive a car, but has comparably little experience with aviation, even as a passenger. Most students are more worried about fouling up than getting hurt but expect to have a certain amount of prestige among their nonflying associates for having the guts to do such a thing as take flying lessons. Because of this, you can help by mentioning that a student is doing well (if that is true) to wife, husband, and/or friends. This will give the student added enthusiasm for flying.

Psychologists have suggested that ego-building or pride can originate from two main sources:

1. *Pride because of the group.* The student is proud to be known as a "flier" or "pilot." There's still some glamour about flying, and people are proud to be a member of what is still known as a select group. You can help this feeling by bringing students into hangar flying sessions and generally making them feel at home around the airport. Even people who are big wheels in their own right enjoy being members of this new "inside."

You know full well that the young male student will casually mention within the hearing of some of the females of his acquaintance that he is "flying." This pride because of the group is a very important part of the learning process. As he gains experience and has years of flying behind him, it will be less of a factor than the second kind of pride.

2. *Pride within the group.* Professional pilots like to be the best in their job and appreciate the respect of their flying contemporaries. To hear in a roundabout way that you are one of the "best instrument (or aerobatic, etc.) pilots in the business" is worth more than a brass band and 10,000 non-pilots shouting, "What a good pilot you are!" This pride-within-the-group approach to instruction is effective for the advanced learner who happens to be taking a flight course (multiengine, instrument, etc.), if he is not participating in a group program and there is no direct competition involved. In direct competition comparing progress may antagonize or discourage the person who isn't doing so well. This can happen in group primary flight-training programs. To make the statement, "Jones is a topnotch student, but Smith can't find the ground with both hands," so that both students can hear it, is to make Jones cocky and lazy and Smith discouraged and lazy. Don't ever compare students directly with each other.

This doesn't mean that you can't let it be "overheard" when you tell a student that "your turns around a point were just right that last flight." You are not comparing him directly with another student in that case.

MILITARY STUDENTS. You may not get to instruct in the military, but it might be well to take a look at military flight students and compare them with their civilian counterparts.

At NAS Pensacola during one period it was found that the age of the average flight student was 22 years, half of the students were married, and 40 percent had had previous military experience.

As you can imagine, these pilots were at the peak of physical and mental aptitudes, but even they experienced some problems with the flight program.

There were problems of *overconfidence* (they would tend to cut the instructor off while he was talking). *Underconfident* trainees needed boosting, and some trainees were *forgetful*, just like that 50-year-old you're instructing.

Apathy was found, too, usually caused by personal problems, dislike of the military, or accidents in the training command. And *inconsistency*, hot one day and cold the next, hits these picked people as well as the average man or woman who comes out to your airport for flight training. The point is that people tend to have the same basic problems in learning to fly.

The military students were broken down into five basic types (the following are navy terms):

1. *Plodder.* He's on his way, but not setting any records.
2. *Oddball.* He's marching to a different drummer.
3. *Fireball.* Hopes to be a triple ace within a month of getting his wings and is above average in the flight program.
4. *Dilbert.* Uses poor headwork and is behind the airplane all the time. He's marginal to unsatisfactory.

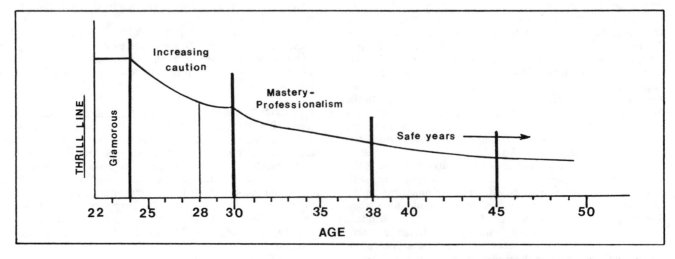

Fig. 2-4. Career phases of naval aviators.

5. *Average.* Pretty close to the civilian average. He'll have some problems at various stages of training but will complete it.

A study of life phases of naval aviators could very well follow the route of civilian pilots of equal ages. Fig. 2-4 shows the "thrill line" versus age of pilots as seen by the military. The ages may be moved slightly in the various stages, but this chart follows closely the maturation of a (male) pilot.

EFFECTIVE INSTRUCTION

You will develop the student's *perceptions* and *insights* and establish proper *habit patterns* in his flying. Your instruction should be such that he will not fly without a thorough preflight inspection or won't make a turn without clearing the area first. You should establish such strong habit patterns that he will feel extremely uncomfortable if he starts to neglect some item you taught him was necessary for safe flight. So—he doesn't fly until he checks it.

The student takes past knowledge and understanding (you can substitute *perception* and *insight* if you like), as well as *past habit patterns, and he transfers this to new knowledge.* Sometimes you'll find that his past knowledge and habit patterns are transferred to a new situation and they don't work. For instance a student who has done all of his flying on the front side of the power curve cannot transfer that knowledge to flying on the back side of the power curve, and he has to get new knowledge from you. On the front side he may have unconsciously been thinking (even though you may have mentioned energy effect) that when he moves the control wheel back, the airplane climbs—which it does when cruising at the higher airspeeds. If he had no instruction on low speed/high angle of attack flight, and if he were in a tight situation, he could transfer old knowledge and bring the nose up to climb, which the airplane would not do. On the contrary, the sink rate would increase. As the adage goes, "Pull back on the wheel a little to climb, pull it back a little more to descend."

You will run into a *transfer of past habits* when the new student starts taxiing for the first time. He sees (or you mention) the need for a turn, and he has a control wheel in his hand. You've told him that an airplane is steered on the ground with the rudder pedals, but he has driven a car too much to ignore that wheel he's holding, so you have to remind him again about the rudder pedals. As covered in Chap. 4 on the first flight, you will probably have him keep his hand off the control wheel while taxiing until you can teach him a new habit pattern applying to airplanes. He will carry over "old" knowledge and actions to a new situation. Your job will be to direct him safely to new knowledge.

■ **Introduction of Material.** Educators suggest that the introduction of material should be on a *whole-part-whole* basis. You'll find yourself following this procedure in giving ground or flight instruction.

As an example, when introducing the *rectangular course* at the chalkboard to a presolo student, you don't just start by telling a step-by-step method of doing it, but introduce it by explaining *what* it is, and *why* you are doing it (this maneuver will be covered in more detail in a later chapter), but is given here as an example of the whole-part-whole principle.

GROUND INSTRUCTION—WHOLE

What—Explain that the rectangular course is a maneuver in which the ideas of correcting for wind drift are used to stay in a given geographic area (around a field, or fields, forming a square or rectangle). Using the chalkboard, draw a rectangular course, as shown in Fig. 2–5.

Why—This is normally a part of the *whole* idea. You would explain that the rectangular course is a good training ma-

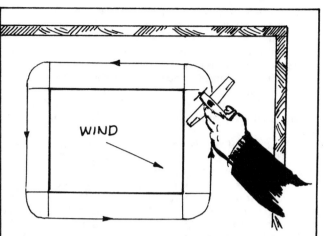

Fig. 2-5. The rectangular course as a whole. You can "fly" a model around the course drawn on the chalkboard to further illustrate a point.

neuver for wind drift correction in preparation for the *traffic pattern,* as well as for cross-country flying. (You are giving him the word that this is not just a one-shot practice maneuver but you are preparing him for greater things.) You could say that the straight-flight portions fall in with the same idea as tracking on a prechosen course on a 100-mile-leg cross-country. The turns serve as an intoduction to more complex maneuvers that require flying a circular path while correcting for wind drift.

GROUND INSTRUCTION—PART. You would then proceed to explain the maneuver on a step-by-step basis, including the entry, covering the unique aspects of each leg and turn as affected by the wind. You will mention common errors that may arise in various parts of the maneuver. The general tendency is to undercorrect for drift, which can affect the legs (and turns) differently since on one leg it may mean that the airplane drifts into the field, and the same error may cause it to move out too far on another leg.

You should cover the problem of a too steep or too shallow bank at *each* turn.

GROUND INSTRUCTION—WHOLE. After you have covered the maneuver step by step at the chalkboard (and you can use a model to "fly" around it to help make it clear, as shown in Fig. 2-5), sum it all up by repeating the maneuver as a whole, covering all parts, but not stopping to concentrate on a particular part. Basically, this second *whole* coverage is a summation and review of what you've covered—it ties it all together.

INFLIGHT INSTRUCTION—WHOLE. In flight, the initial *whole* part of your instruction will be a demonstration of one complete circuit of the field or fields pointing out the pattern that he is supposed to fly, noting specific landmarks such as those at corners. Correct for wind, mentioning the wind direction and how you are correcting for it, but don't grind him down with details. This demonstration is a transfer from the chalkboard discussion to an actual situation—as a *whole.*

INFLIGHT INSTRUCTION—PART. Your move here will be to instruct and correct his errors as each part of the maneuver is done. You will be instructing on *specific items* as they are encountered. He may be having a problem with drift correction or altitude on a particular leg. So, you will concentrate on instructing the *parts* of the maneuver.

INFLIGHT INSTRUCTION—WHOLE. The last part of your inflight introduction to the rectangular course is again to take an overall look at it. Your student may be having *general* problems with altitude control, he may not *see* the whole maneuver and perhaps is concentrating so hard on correcting for wind drift or holding altitude that he continually flies past the point of turn. You should again draw his attention to the distant corners or points or turn; *he may not see the forest for the trees.* (But try to avoid clichés like this in your instructing.)

Parts don't have the same properties when they are isolated. If you just take part of the straightaway of the rectangular course, it will be the same as flying down a straight road. In the turn, it will be as if the only problem is to correct for wind drift in a turn. Stress that *both* abilities are requirements for a full maneuver. It's like the old, old story of the blind men and the elephant: The one holding the trunk described the animal as a snake, the one at the leg said the elephant was like a tree trunk, the one holding the tail said it was a rope, and the man touching its side said it had to be a wall. The final decision was that it was an elephant—when the committee got together about it.

■ **Ground Instruction Techniques.** This whole-part-whole technique is best for formal ground school as well. Too many ground instructors rely on slides and other material (which sometimes need better organization) rather than thinking things through and setting up the lesson logically. Training aids are valuable in instructing, but that's what they are—*aids.* They are no substitute for a good instructor.

As a specific example, you are going to teach a ground school course in the theory and operation of flight instruments for some new instrument students. Your equipment is a chalkboard, and better yet, you have a set of old instruments that are cut away to expose the insides. (If you are going into the instrument-training business, you *should* have such a set of instruments—a local machine shop can cut them open for a reasonable price under your guidance.) You also have a slide series on the instruments, but how or if you will use it depends on how the series presents the material. You are going to discuss the following instruments:

1. *Airspeed indicator*
2. *Attitude indicator**
3. *Turn and slip or turn coordinator**
4. *Altimeter*
5. *Heading indicator**
6. *Vertical speed indicator*

Should you teach them in the order given above? Should you lump them into groups? Note that three of the instruments start with an A and the others don't. Would this be a good way to split up your coverage? No, you should look for the common factors that tie certain instruments together; considering their first letters wouldn't be the way to do it.

The mode of operation of some of the instruments depends on air pressure or change of air pressure, and the others (indicated by asterisks) depend on the operation of a gyro. You decide to talk about the different systems and their instruments, separating your main subject periods into the pitot-static system instruments and the gyro-driven instruments. Which group should you discuss first? It's suggested that the *pitot-static system instruments be discussed first* because new instrument students have been using at least two of them (airspeed and altimeter) since they started flying and are using them as VFR pilots every time they go up. They haven't had so much experience with the gyro instruments. You'll find that their current use of the pitot-static and static system instruments will whet their interest in how these instruments operate and will make them want to know about the other instruments they'll be using in IFR work. Of course, they've been flying

with the gyro instruments in sight all along and used them in the emergency instrument training, but this will be the first time they will really "see" them.

Your notes might look like this:

The Pitot-Static System
1. *The pitot-static system*
2. *The pitot-static instruments*
   a. *Airspeed indicator*
   b. *Altimeter*
   c. *Vertical speed indicator*
3. *The system and the instruments*

Why this order? It's the *whole-part-whole* concept again. You will look at the pitot-static system and how it works and then look at the instruments within it, one by one. After you've done this you pull everything back together and see how problems in the system would affect the instrument operations.

What are you doing with cutaway instruments? If you have any, keep them out of sight until you've given everybody a good understanding of how the system and instruments work and *then* let them look at the instruments to see the components you've been discussing in simple terms. If you let the students see the instruments first, they will see what appears to be a conglomerate of springs, levers, and diaphragms, and this could result in inward groans and the turning off of the thought process of the less technically minded members of the class. *Always operate on the premise that even the most complicated mechanical systems usually have simple principles, and the springs and levers are not needed at this stage to get this simple theory to people.*

Going back to the earlier idea that *adults will learn only what they need to learn,* if you try to get people to memorize all the names of the parts of the instruments, they will back off. They don't plan to be instrument technicians, they want to get information they will use *as pilots.* Stick to this idea, and keep reminding them that's the way you are teaching by occasionally prefacing your remarks with, *"As a pilot,* you'll find that . . .," or "You'll discover *as you fly instruments* . . . "* This is not to say that you'll preface useless information with such remarks (and you shouldn't give out useless information anyway), but it does keep them satisfied that the knowledge you are imparting will be used *in a practical way.* It keeps them aware that you are working toward a particular goal—which will be used in flight.

We now know that adults want to learn specifics that can be used to show the principles. In showing them the specifics of the simple pitot-static system that they will be using, you also are giving them the principles needed to see how all pitot-static systems work as well as teaching them the theory of the interchanges of pressure, density, and velocity in flight. If you had given an hour lecture on the physical principles of fluid dynamics and statics, you would have lost them no matter how valuable the material could be for future use.

A diagram of the pitot-static system should be drawn on the board as shown in Fig. 2-6.

You could set up an outline for instruction as follows:

I. *The Pitot-Static System*
   A. *Static sources.* Discuss the system, pointing out the probable locations of the static sources on the airplane, noting that the locations are selected because it is desired to have as true a value of the outside static pressure (the actual atmospheric pressure) as possible. The fact that the airplane is moving through the air can complicate this. Later in the discussion you can show them that a 1 percent error in static pressure may be many times as critical for airspeed accuracy as the same

error in the dynamic pressure. Discuss position error effects.

B. *Pitot tube.* Point out that of the three instruments in the pitot-static system, the airspeed is the only one using the pitot tube. Note the effects of yaw and pitch on the pitot-static system.

II. *The Pitot-Static Instruments*

A. *Airspeed indicator.* Introduce this first because it better illustrates the need for a pitot-static system. Fig. 2-7 shows that you are adding the workings of the airspeed indicator so they can see how it works generally as well as seeing it as a part of the system. Start out by showing the system at rest at sea level, noting that the static pressures in the case (surrounding the diaphragm) and within the diaphragm are equal at 2116 psf, the standard pressure. No airspeed indication is given since a balance exists. Emphasize that this static pressure exists in both the pitot tube and static tube. Assume that the airplane is now moving at 100 K at sea level. The dynamic pressure is added to the static pressure in the pitot tube and since the

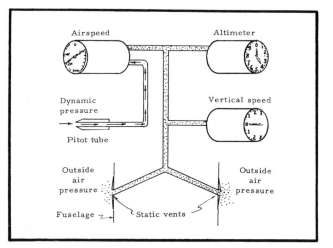

Fig. 2-6. A simple pitot-static system. You will shortly show that the pitot tube has both static *and* dynamic pressures in flight.

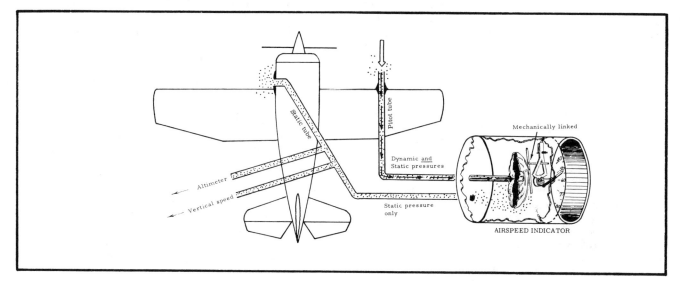

Fig. 2-7. A schematic of the airspeed indicator as part of the pitot-static system. (See *The Student Pilot's Flight Manual.*)

static pressure surrounding the diaphragm remains the same, the two static pressures are canceled, the diaphragm expands, and through mechanical linkage the airspeed is indicated at, say, 100 K. The airspeed indicator is a pressure instrument calibrated for sea level operation. It measures the dynamic pressure, which is a combination of the density and velocity of the air molecules. This should lead to a discussion of why a correction is needed to find the true airspeed. You should also cover these airspeeds: IAS, CAS, EAS, and TAS.

B. *Altimeter*
1. Draw in a simple diagram of the workings of an altimeter in your chalkboard drawing.
2. Note that the diaphragm is sealed and is mechanically linked to the hands. The diaphragm expands as the altitude increases because the pressure in the case (which is connected to the static system and thus gets the outside air pressure) decreases.
3. Discuss the following:
   a. Temperature—(nonstandard) effects of the altimeter (corrected altitude)
   b. Pressure variations flying from high- to low-pressure areas and vice versa

   c. Indicated altitude
   d. True altitude
   e. Absolute altitude
   f. Density altitude
   g. Pressure altitude

C. *Vertical speed indicator*
1. Draw in a simple view of the interior (Fig. 2-8).

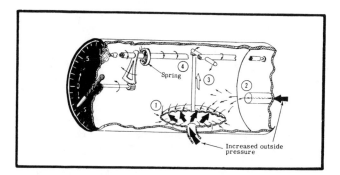

Fig. 2-8. A schematic of the vertical speed indicator. You need not go into as much detail on the board as shown here. (See *The Student Pilot's Flight Manual.*)

2. Note that this instrument is a valuable aid but is too often ignored. (It is not required by FARs for IFR flying.)
3. The vertical speed indicator operates because of change of pressure. It indicates a vertical rate of climb or descent because the pressures in the case and diaphragm are not the same if the airplane is descending or climbing and the outside pressure is changing. Discuss the restriction in the instrument and how it works during climb and descent.
4. Discuss the indication lag of 6–9 sec and point out that the vertical speed indicator is a good trend instrument. Note that there is an instantaneous vertical speed indicator available that does not have a lag.

You've covered the pitot-static system as a whole, looked at the instruments, and now will take a look at the whole system again.

III. *Pitot-Static System Problems*
  One thing that must be discussed in an instrument course is what happens to the system and its instruments if ice or other material closes the pitot tube or static ports.
  A. *Alternate static source.* This usually has cockpit pressure. Errors to watch for if the alternate source is used:
   1. Airspeed—will read slightly high because of lowered pressure in the instrument case resulting from the air moving by the airplane and lowering the cabin pressure. (Static pressure from the pitot tube is greater than the pressure in the case.)
   2. Altimeter—will read slightly higher than correct altitude because of the lower-than-ambient static pressure in the cabin. You should also mention that while the conditions in 1 and 2 usually prevail, a look at some current *Pilot's Operating Handbook*s shows that there are exceptions to the "rule" just given and that the *POH* is the final criterion in this area.
   3. Vertical speed indicator—will indicate a rate of climb immediately as the alternate source is turned on (the pressure drops) but will settle back to normal. It will operate normally after this, because it measures *change* of pressure.

You should also discuss what would happen if the static system is suddenly and completely sealed at, say, an altitude of 10,000 ft. Mention that the system cannot be perfectly sealed, but for the sake of the example it will be assumed so. You may check your students' understanding of the system by asking them how each instrument would be affected. This is a form of learning called *skill to apply;* that is, can they apply what they've been told about the system to this new situation. If they fully understand the principles of operations they can see how new factors can affect the instruments.

LEVELS OF LEARNING. You have probably encountered all *four levels of learning* in your teaching of the pitot-static system and instruments:
  1. *Rote.* This is the lowest type of learning and consists merely of repeating back the information given. Some of your students may still be at this stage when you are through with this particular discussion. They can repeat back almost verbatim but they lack . . .
  2. *Understanding.* This is the next step. The people you've been talking to understand what you've been saying, it makes sense to them but this does not necessarily mean that they have the . . .

3. *Skill to apply that understanding.* In the example of the completely plugged static system you'll find out who can apply the previous knowledge. When you hear a student say, after you've discussed the pitot-static system, "So, that's why my airspeed indicator reads high in a right slip and low in a left slip. It's because of the location of my *one* static port," you know that he has reached the fourth level of learning, the . . .

  4. *Ability to correlate with other instruction.* He may have been taught (or found out himself) that the airspeed indicator does read differently in left or right slips but was not given a reason. Even though you haven't mentioned it yet he sees the relationship once the principles of the system are explained.
  If cutaway instruments are available use them now to briefly repeat the principles.
  Have a period of questions and answers.
  At the next session, or after a break, you should go into the gyro system and instruments and might set up your notes as follows:

  I. *The Gyro System—Theory and Background*
   A. *Venturi.* Disadvantages for instrument flying.
   B. *Engine-driven vacuum and pressure (pneumatic) pump systems.* Advantages over venturi. Suction and pressure indications.
   C. *Electric.* Advantages at high altitudes over vacuum-driven. Most airplanes have some electric instruments and some vacuum-driven as mutual backups.

You are starting with the more primitive systems and progressing to the electrical instruments used by high-speed, high-altitude airplanes.

  II. *The Gyro Instruments*
   A. *Attitude indicator.* Sketch on blackboard (show gyro). The only instrument that gives both pitch and bank information, therefore it is centrally placed on the attitude instrument flying panel. Operates on theory of rigidity in space. Vacuum, pressure (pneumatic), and electric types. Theory of precession and how precession affects the instrument under certain conditions. Pitch and bank limits. Nonspillable types. RPM required for various types.
   B. *Heading indicator.* This instrument operates on the principle of rigidity in space, as does the attitude indicator. Effects of precession. Limits of pitch and bank. RPMs required. Sketch the instrument on the blackboard or show an overhead projection as you talk.
   C. *Turn and slip.* This instrument, unlike the other two, operates on the principle of precession. Sketch it. The other vacuum-driven gyros use about 4.5- to 5.0-in. Hg suction; vacuum-driven turn and slips use 1.8- to 2.1-in. Hg. Usually the T/S is electric to act as a backup to the other two vacuum-driven gyro instruments. The 2- and 4-min turn instruments. Discuss the *turn coordinator* and its operations.
  III. *The Gyro Instruments and System as a Whole.*
    The entire system should be reviewed with special emphasis on troubleshooting from a pilot's standpoint. Discuss what the pilot should do in the event of a high or low suction or pressure gauge reading; the probable causes. Preventive maintenance. Repeat the principles briefly, using pictures and cutaway instruments, if available. Have a question and answer period on the system.

GROUND SCHOOL PRESENTATIONS. Getting back to the idea of attention, you've sat through lectures that were dull, even though the subject was not supposed to be. Some speakers could make their audience go to sleep in the middle of a lecture on sex. A knowledge of the subject is essential, but your presentation is very important also.

You might take a look at the F-O-U-R items to get some ideas as to how you can make the most of your lecture time.

1. F—Familiarity. Do you know your subject and audience? Probably one of the toughest talks to give is to a flying club with a range of listeners from people who may have 1 or 2 hr flying time to flight instructors and airline pilots. You'll have to make the subject understandable to the low-time pilots and keep the interest (and accuracy) in it for the professionals. Too often, people try to make the presentation to the professionals and lose the less-experienced people.

2. O—Organization. What is the best way to explain the subject? One approach would be to start off with the *objectives* of your talk. ("I'm going to cover the flight instruments so that you will have a better understanding of how they work.") A *conclusion* would be that knowing how they work would help the listeners use them better. You're saying that the adults will make practical use of the information you are about to impart.

You'll *introduce* the subject and go into the *body* of the lecture. At the end you should have a brief summary of the subject and, as noted earlier, allow a question and answer period.

3. U—Urgency. Is there a need to know the material or are you and the audience putting in required time? Adults want useful information, not hot air. By imparting useless information, you may hurt yourself later when you really have something to say.

4. R—Routine. How is your routine presentation? Do you stand up and speak in a low monotone at your listeners? You should be enthusiastic about your subject and use plenty of eye contact. *Don't* pick one person to speak to; bring everybody into it by talking to and looking at different people in the audience. A lecturer who gazes up at a far corner of the ceiling and drones along may end his lecture to find that the audience has quietly gone. (He wasn't looking at them and so didn't know they were leaving.) Raise your voice from time to time when making a point. Vary your tone. Physical discomfort or worse, too much physical comfort, is bad for the listeners. A lecture given in a too-warm room immediately following a meal is going to have listeners drifting off. Of course, a too-cold room or hard chairs and other discomforts can distract from the subject, too. More about this shortly.

A pacer who strides from one side to the other as he talks can draw attention from what he's saying, and of course there is the chalk pitcher-upper who has his audience at the edge of their seats wondering when he'll inevitably drop it.

Think about all the dull lectures you've ever attended and avoid the distracting mannerisms those speakers had (including the "uh's" and "you know's" we all tend to use when talking).

When you organize your material, you'll generally break it down to the *Introduction* (tell your listeners where you're going), the *Development* (build up your "argument"), and the *Conclusion* (review what you've covered—pull it together).

USE OF ANALOGIES. If you can tie in something well known with the new maneuver or material, it makes learning a lot easier for the student. For instance, when introducing a student (or private pilot) to the controllable or constant-speed-propeller idea, the analogy of the various gears of an automobile can be used. The rpm control is like a gearshift except there are many more gears (rpm) available instead of just a

few fixed ones as in cars. Low pitch corresponds to low gear (or D1 and D2 in an automatic transmission in a car), and the throttle does the same job as the foot pedal. The fixed-pitch-propeller airplane is like a car with only one forward gear and is most effective at some particular speed or power situation.

Weight and balance computations may be tough for students without a background in physics (and remember, you may be instructing people with backgrounds varying from a sixth-grade education to doctorates in aeronautical engineering). You'll have to start off with a simple lever and fulcrum system and work into the application to airplane weight and balance and simple stability and control. Your steps might be:

1. Discuss arms, moments, inch-pounds or pound-inches and how they are derived.

2. Draw a simple system (with even numbers) and note that the moments must be equal for the system to balance. The idea of a large child and a small one on a seesaw is a good illustration.

3. From this, you can move on into finding the empty-weight center of gravity and running a weight and balance with fuel and occupants.

4. Or, get into the idea of the vertical forces in straight and level flight and why the center of gravity must be limited (hence the need for weight and balance computations).

5. Explain that for the airplane with a tail-down force existing in flight, closing the throttle decreases the tail-down force and temporarily upsets the balance of the moments until the airplane noses down to pick up enough free stream velocity to make up for the loss in slipstream (and other factors; how far you go will depend on the stage of the trainee's progress). This last idea is also good for explaining why the airplane is inherently stable and, contrary to some preconceived notions your students may have, it does not suddenly flip into a spin when the engine is throttled back.

Other analogies may be used in your instruction, such as comparing the automobile fuel and oil systems, if warranted. However, some of your students who've never looked beneath the hood of a car might find such comparisons confusing. If you do use the comparison of autos and airplanes, be sure to stress that unlike the car, the airplane's ignition system is not affected if the battery goes dead; the battery is only needed for starting and to furnish power for radios and electrical equipment.

You'll likely work out some of your own comparisons as your instructing experience increases. If not carefully chosen and used, however, an analogy does more harm than good.

WRITTEN TESTS. As a flight instructor you'll be working with written tests, both those made up by somebody else and your own, and you should look at some points concerning them.

The multiple choice question has been found to be the best all-around for written tests and at this stage of your flying you've had several of these.

There are true-false tests, too, but these don't check the test taker's knowledge of the subject as well. Flipping a coin could be a good way to get answers for these since there's a 50-50 chance of getting a correct answer.

The essay-type test is not a good test to be used for flying knowledge because it depends on a person's ability to express himself on paper rather than showing whether he knows the subject or not.

Completion tests allow some leeway, too, as to whether the person answering "meant" the correct answer. In other words, the true-false and multiple choice have only one answer and are objective, whereas the other two types may be subjective in varying degrees.

A good written test, whether yours or someone else's, should have these factors:

1. Reliability. Is it consistent? Can you rely on this test to do what it's designed to do for a number of people going through the course?

2. Usability. Can it be used readily without complications or extra problems?

3. Validity. The test should measure what it's supposed to measure. That is, it should measure practical aeronautical knowledge that can be used by the person taking the test. Getting back to how adults learn, you wouldn't want to throw in questions on aviation history to a group of people in an instrument ground school. Does the test result show that the person taking it is or is not getting and retaining the knowledge needed to achieve the goals of the course or program?

4. Comprehensiveness. The written test should cover all areas of the information discussed and studied.

5. Discriminating. The test should detect differences in student knowledge. Mr. Flyte may need more help on FARs while Mr. Walker may have the knowledge he needs on this subject.

In short, the written test should consistently, accurately, and thoroughly check the students' knowledge of the area covered so that you will know where and how to review the subject needed.

■ **Rates of Learning.** You'll find that your students will have different rates of learning. This, of course, doesn't come as a world-shattering surprise to anyone out of kindergarten, but it should be looked at to see *how* and *why* this is the case, and *what* you might do to help matters. Rates of learning vary because of:

1. Intelligence. This is obvious, but you can use this fact (which the average person has suspected for years) to keep the student's attention. If he is highly intelligent you won't have to hit him with what is known as "meaningful repetition." You can move on. If he's the type who has trouble reading the comics without moving his lips, you can expect that his rate of progress will be something less than spectacular. The intelligence of the people you will be teaching will have a direct effect on . . .

2. The Instructor's Technique. You can directly control a student's rate of learning by your technique of instructing. If you are obviously interested in teaching and know the subject, this can be a technique in itself. You'll see in the next chapter that every flight lesson should include three parts: (a) preflight briefing, (b) infight instruction, and (c) postflight review. *If you don't use all three parts, your technique of instructing is not good.* Also, do you use an easy, logical approach, or do you believe in the 2-by-4 method of getting attention? The rate of learning will be controlled by how "shook-up" you get a student, which brings up . . .

3. Fear and Anxiety. The belief that the use of fear and anxiety helps people learn is one of the biggest fallacies existing in flight instructing. A person who is afraid or anxious isn't going to be paying attention to anything except his own problems. If you contribute to his fear by shouting or threatening him, you are hurting his learning rate, not helping it. *At all times, an instructor should be working to alleviate fear and anxiety.* This doesn't mean that you won't teach him to respect weather (or even fear some types of weather) or to respect the airplane, but he doesn't have to be threatened or discouraged. ("At the rate you're going, you'll bust that private flight test flat as a fritter!") This, of course, makes the student immedi-

ately relax and stop worrying about the flight test. You may be assured that his rate of learning will go to zero after that bit of psychology by the instructor. Unfortunately, some instructors figure that they were taught that way and they turned out okay, didn't they? (Did they?) *Instructors tend to teach as they have been taught.*

4. Physical State. The rate of learning depends on this as well as the mental states mentioned above. If the student is feeling under the weather he won't be learning very well. If the student over there in the left seat is wondering after that stall series if he is going to have to ask for a "comfort bag," he is not going to be listening with great interest while you are talking up and demonstrating the 720° power turn. Aside from the obvious perils of continuing your instruction when the student is quietly getting a yellow complexion, the learning process has come to a definite halt. There will be more about this problem later in the book. *At all times, during ground or flight instruction try to be aware of the onset of student fatigue. Tired people don't learn well.*

5. Motivation. If a person is motivated he will learn much more quickly. *Your technique of instructing* can help motivate the student. A part of the motivation is in the *what, why,* and *how* approach. If he isn't sure *what* you're going to do that lesson or *why* or *how,* he's not going to go into the project with enthusiasm. He wants to know *where* you are going and *what* are the long-term goals. One flight instructor, known for his oratory, had his students' faces wet with tears before he was halfway through his preflight briefing and at the finish would stand aside as they dashed to the plane, shouting, "Let's get out there and do a perfect rectangular course in memory of the Wright brothers." (*That's motivation.*)

Praise helps motivate people to learn. Constructive criticism helps learning but will have to be presented properly to be sure it's taken as intended.

The student *will* be motivated if he sees where you are going with your instruction. Developing motivation for learning will be one of your biggest (and sometimes toughest) jobs.

LEARNING RATE VARIATIONS – PLATEAUS. Fig. 2-9 shows what can happen as the student goes through a flight course. Assuming that he knew nothing about flying before he started the course, the learning rate the first hour should be the highest he will have in his flying career.

In Fig. 2-9 a leveling off of rate of learning is shown. The "plateau" indicates that the student isn't learning. The three

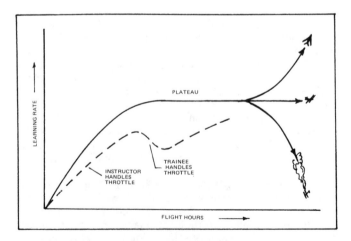

Fig. 2-9. A typical learning rate curve. The dashed line shows the pattern in teaching an aerobatic maneuver (loop).

airplanes indicate which directions he may take from that point.

Certain plateaus are normal in flight training and should be expected. You should tell the student *when* to expect them. For instance, in teaching aerobatics, certain airplanes with high elevator forces may require that the student do the first few loops using both hands on the wheel or stick with the instructor handling the throttle, so that he can get an idea of the maneuver without too many distractions. Later, when the student uses the throttle for the first time, his performance will suffer. After practice, he will be doing acceptable loops while handling the throttle. He might even suffer a definite "reversal" of the learning rate curve at the first use of throttle, as shown in Fig. 2-9.

A more usual case of temporary leveling-off is when something extra is added to the current area of knowledge in the *presolo stage*. For instance, the beginning student has been practicing straight climbs and level turns; when you have him combine the two, you may expect that he will experience a temporary decline in his performance. He should pick up shortly, however, and continue to improve his climbing turns. The same thing will happen when glides and turns are combined, but the situation is usually exaggerated in that case because the lack of power and less-effective rudder are more apt to cause coordination problems.

When the student has a longer plateau it will be up to you to move him out of it. A long term slump often happens during the presolo phase when the student pilot is shooting takeoffs and landings, because the instructor may, for several straight sessions, have had him shoot a full hour of takeoffs and landings without leaving the traffic pattern. This is where the "six (or seven, or eight) hour slump" occurs. If the student is making no progress in finding the ground, or has other problems on landing, take him out of the pattern for some high work. Perhaps a review of stalls and slow flight is in order; at least some sort of a break is needed. Some instructors have their students depart briefly and reenter the traffic pattern several times during a period of takeoffs and landings. This not only gives a break but also helps smooth out his entries and departures. Others in a calm wind situation may have the student shoot landings in the opposite direction (traffic permitting). Shooting too many touch-and-go landings without a break can cause a slump. Shoot full-stop-and-taxi-back landings if you think this is starting to happen. If you decide that stopping the landing practice for that period is the best move, let him know that it has happened to other people. The worst thing you could do is to blister him for his faults and then not schedule him for a flight for several days "to let him rest." If you think he was having problems on the last flight, wait until you fly with him after he's had a week or so to stew over his predicament. His motivation will have experienced a definite injury. He might even cancel that later appointment and ease out of flying because he feels "what's the use." He hasn't been told that the problem he's having is a rather common one and that it can be worked out. He doesn't know that *learners differ in degree, but not in kind.* He may be having difficulties at a particular stage in training, but all people have some problems (some just like his, or even worse) at some time during a flight course. The fact that he is having troubles is not unique; people tend to magnify their own problems and assume that others are coasting on through.

The main thing is *not* to keep hammering at a maneuver, hoping that somehow by pure repetition he'll make progress. Take a break and review the maneuver; ask him questions to check his understanding of what he is trying to do.

Because of variations in learning rate between persons, a syllabus cannot be followed rigidly. Jim Brown may be ready for stalls in a particular flight as the syllabus says, but Bob Green will need at least another hour of the Four Fundamen-

tals and low work before getting seriously into stalls. The syllabus is a guide—that's all.

Motivation is the dominant force governing the learning process.

■ Obstacles to Learning

FEAR, ANXIETY, AND TENSION. As mentioned earlier, fear and anxiety are sometimes *mistakenly* considered aids to learning. Some instructors' methods of introducing stalls can affect the learning of other maneuvers because the students are so afraid of getting into a stall while being instructed in such things as 720° power turns, slow flight, or low work that they aren't listening or really trying to learn.

It won't be unusual for you to see such physical evidences of anxiety as face sweating (even when it's cool); excessive hand wiping on trousers, indicating palm sweating; or trembling. Usually the latter is not noticeable until after the flight is over, and it can usually be attributed to tenseness. It's natural for student pilots to be tense, but if they freeze up they aren't learning anything. Watch for signs of abnormal stress, because you need to project the student's performance here to later flights when you aren't along. His reactions throughout the presolo period will be a major factor in deciding when, or even *if,* he will solo. His performance on the dual cross-country will decide when, or if, he will make the solo cross-country.

You will build up a good idea about how he will perform on the first solo by checking his mental reactions (as well as his physical performance) in the several hours you'll be flying with him in the presolo phase. Of course, you won't really know how he will react to being alone in an airplane until he is shooting those solo landings. It's possible that being alone could trigger off a panic that had absolutely no evidence of existing while you were with him, but this would be an extremely rare occurrence.

Your best indication of his mental attitude is how he acts or reacts to a change or when something out of the ordinary occurs. You've been shooting landings on a particular runway for half an hour and change to another runway, for instance. You might expect him not to do quite so well on the first pattern for the new runway, but if he's getting close to the solo stage the next pattern and landing should fall right into place. If the presence of another airplane makes him go around (or you tell him to go around without an explanation) he should not have an "abnormal reaction." He shouldn't overreact and move wildly, starting to add power then going to trim and back to the power, but he should, without delay, add power and make his move. On the other hand, he shouldn't sit there doing nothing until you have to take over.

If the student pilot appears to be *too* calm under stress this could be a warning, too. There's the saying, "If you are calm when all others around you have lost their heads, maybe you just don't understand the situation." One flight instructor, flying with a 3-hr student, had a partial power failure at a critical part of the climbout after takeoff. The student was flying it at the time, and the instructor took over and headed back to the airport. He had to get back in a rather roundabout way to avoid quarries, woods, and other obstacles and just managed to touch down (still in a shallow-banked turn) on a cross-runway. During the roll-out as the engine was breathing its last, the student asked, "Why are we coming back so soon?" After they had turned off the runway and stopped, he was informed of the problem. The student had not developed his perception well enough to recognize the trouble and be afraid, and the instructor was too busy at the time to give him a running account.

So teach him to recognize problems and respect the airplane (and the elements), but don't use fear as a method of instruction.

ERRATIC SCHEDULING. If your arrangement is such that the student is working very hard for several days and then is not scheduled to fly for a week or more, you've set up a couple of obstacles to learning. First, by scheduling too heavily you don't give the student time to absorb the instruction; it's coming at him too fast. Second, by following this overscheduling with a long layoff you'll give him a good opportunity to forget what little he did absorb.

You've read of people starting to learn to fly from scratch at, say, 0600 in the morning and soloing by sunset. This is a publicity gimmick, and there is a question whether it would be possible for the student to safely solo the next day without a great deal more instruction. If early solo is the only goal, it's possible to solo some people in 3 hr if conditions are just right. The student would be given just enough information to solo and probably would not even leave the pattern after the first 30 min. The airport would have no other traffic and the wind would be steady, not too strong, and right down the runway. The student would be learning by *rote* (and during the solo flight the instructor would likely be quietly praying that no outside factors would be introduced, such as a wind shift or transient aircraft arriving).

For presolo students and for those up through the first solo cross-country, it would be best to schedule for no more than once a day, and no less than twice a week. A couple of exceptions to the not-more-than-once-a-day scheduling might be on the day when you find he's about ready for the first solo and could use another flight to complete the job, or if he had a *short* dual cross-country in the morning and you'd like to get him out on his first solo cross-country to the same places in the afternoon.

Winds and weather respect no schedule and in the winter you may find that several days or a couple of weeks go by when you're not able to fly with some of your student pilots. After the weather breaks, the common tendency is to *over-schedule* a student to take advantage of the sun. Of course you have to make a living, but try to fly with as many different people per day as possible rather than scheduling one for several times in that period.

■ **Memory Aids.** No matter how much knowledge is crammed into a person's head, if he can't remember it for use later, it's useless. If you mistakenly schedule too many flights for a student in a given period, all may not be lost if your instructing techniques include good memory stimulators. If he remembers most of the information you give him in that hectic period, you can pin it down and work on assuring that he understands it when things get a little quieter. You should always instruct in a manner that not only gives him the information but also will help him to remember it.

Below are several ideas you should keep in mind:

1. Praise. It not only motivates but also helps your students to remember. *People remember pleasant things or pleasant circumstances longer than unpleasant ones.* When the student does a good stall recovery, your "That's the right way exactly" will help him to remember the best way to recover from a stall. However, if you praise him when he knows he didn't do so well, he will start questioning your ability to instruct.

One of the most important rules you should remember as a flight instructor is: PRAISE IN PUBLIC, REPRIMAND IN PRIVATE.

2. Association. People remember by associating new facts with other facts already known. You may subconsciously do this when you want to remember people's names. (Sometimes it can be embarrassing, for example, if you meet a man to whom you were introduced earlier, and your memory asso-

ciation nudges you about "building materials." His name is that of something used to build houses. You say, "It's good to see you again, Mr. Wood," and he snaps, "My name is *Stone.*")

3. Repetition. As an instructor, you will cover each maneuver *at least* three times each flight: (1) in the preflight discussion, (2) during the flight, and (3) in the postflight review. You may go over the information a *couple of times* in *each* of these parts of the flight. In repeating the material you'll use learning techniques and other memory aids as applicable. Repetition must be meaningful repetition. To have a student pilot write 500 times, "I will not bounce on landing," is of no value whatsoever—and as an adult, he wouldn't write it anyway.

If you are teaching in a good atmosphere (friendly and interested in the student), he will remember better than if you mechanically use all the educational techiques but apparently couldn't care less whether he learns or not.

INSTRUCTING FLIGHT INSTRUCTORS

When you are working on your instructor's certificate or are getting someone else ready for that certificate, the usual route is for the instructor-to-be to act as the instructor and the old-timer to be the student. This doesn't always work, because the instructor-trainee knows his "student" is already more than familiar with the material and so is reluctant to go into the details on maneuvers or other areas that would be necessary in the real situation.

If you are using a three- or four-place airplane, one idea is to get a friend who has average abilities but has never flown and give him the whole first lesson (or more) under the supervision of the old-time instructor (you, later), who will sit in the back and add or correct as necessary. The nonpilot will be glad to get a free lesson(s), and the instructor-to-be (who should be well checked out in the right seat) will be operating under actual conditions since he will have to present material and explanations so that the person will understand. The questions asked will be realistic, too, because the beginning pilot will be asking about what really puzzles him. The instructor-supervisor can review the flight afterward with the instructor-trainee.

A postsolo student pilot who is ready to be introduced to, say, advanced ground reference maneuvers might be pleased to save money and still have the services of two instructors with him. A private pilot who would like to learn chandelles and/or lazy eights might be interested in sharing expenses or getting the introduction free. The people approached should be given the full story about what you are doing and if they feel at all reluctant about a "double dual," drop the idea.

If you are working with a large number of student pilots the instructor-trainee can get realistic instructing experience at any point of preprivate (and further) training. Be sure that the student or private pilot does not suffer either from a financial or training standpoint; to leave a student confused or with bad information is something you must avoid, even if he gets the lesson free. On the other hand, the person who gets a free first lesson might decide to continue flying at the usual rates.

INSTRUCTING FOREIGN STUDENTS

You may get a chance to instruct foreign students (this is written for the U.S. flight instructor), and will find it interesting and more challenging than teaching people who have a lifetime of experience with American English.

The first thing that you should know when working with

foreign students is that just because their English isn't as good as yours doesn't mean they are slow. Their IQ and educational background may be higher than yours. One way to look at it is to stop and think how good *you* are at using *their* language.

Secondly (and we all tend to make this mistake), talking extra loudly does not help comprehension by the object of your "instruction."

Thirdly, you may catch yourself lapsing into "Pidgin" English. It's usually the result of the frustration of not getting a particular point across. You've explained turns around the point a couple of times but he or she still doesn't seem to quite get that idea of *why* the steepest bank is required when the airplane is headed directly downwind. So you make it extra simple by very slowly (and sometimes very loudly) repeating the information word by word.

Instructing foreign students is sometimes pretty rough for the new flight instructor, because he's not sure that his *American* students are getting the information, and he is still unsure of himself.

As you do with your U.S. students, find out the backgrounds of your foreign students. What cities in their particular countries are they from? It wouldn't hurt for you to do some reading on their country and go over maps so that you can talk with them about it. We Americans are usually very bad in geography of foreign nations, and you can pleasantly surprise them by breaking out an encyclopedia a few days early and getting a good look at the geography and cultural background of their country and discussing it with them.

■ **English and English.** It's possible that your foreign students have learned British English in their own country rather than American English, and this could cause some problems too. If you are a person who thinks British movies should have subtitles, you'll know what this means.

Our language is full of slang that is changing constantly, and this can be puzzling to the foreigner who is having problems because the English he learned in school was a more formal type—just as the foreign languages taught in our schools stick to the formal approach rather than worry about *their* slang.

Some flight schools teaching foreign students in the United States charge a 25-cent fine if the students are caught using their own language on the flight line. English is the language of aviation around the world, and you should encourage them to get as proficient as possible in the use of the U.S. (or British) aeronautical terminology.

■ **Special Nonaviation Technical Terms.** The Naval Air Training Command trains a comparatively large number of foreign students who have to learn not only everyday English but aviation and nautical terms as well.

An example might be: "Use the scuttlebutt just aft of the athwartships passageway, starboard side, main deck." Translated to more landlubberly terms, this would be: "Use the drinking fountain just behind the hall that runs across the ship, on the right side of the main floor." (Navy and exnavy personnel who read that translation are requested not to laugh.)

■ **Bureaucracy.** Then there is bureaucratic jargon. This language use exceeds the condition that most Americans can correlate with in actuality and presents manifest problems for nondomestic individuals, since it is usually the epitome of obfuscation and is normally executed so that the perpetrator of this philological endeavor cannot be subject to punitive actions in the event that some negative response occurs. (See how easy it is?) The foreign student is no doubt familiar with the bureaucratic language of his own country but now has to learn the English version of a nonlanguage.

■ **Technical Terms.** The Naval Air Training Command booklet *Flight Instruction of Foreign Students* mentions that we have words that can mean several things. For instance, "gear" can mean a wheel with cogs on it, undercarriage of an aircraft (the landing gear), or articles of clothing (flight gear).

You may have to expand your terms with these words, for example always saying the full "landing gear" when you mean the structure of the airplane that goes from it to the ground and has wheels on the bottom part of it.

■ **Grammar.** You remember your own struggles with English grammar, diagramming sentences and checking what participles and conjunctions are all about. Your grammar may not be the best, but keep it plain, anyway. The foreign student has studied formal English ("It is I" rather than "It is me"), and there may a transition for him, so patience in both of you is indicated.

■ **Contractions and Spelling.** "I'll" sounds like "aisle." We'll (wheel), he's (ease), we'd (weed), we've (weave) can also be confusing.

Spelling: Pronounce the following English words ending in OUGH:
Though—oh
Through—ooh
Trough—off
Rough—uff
Bough—ow
How about pear, pare, and pair?

■ **Slang and Exaggeration.** There's an old pilot saying: "If you can't be good, be colorful!" In the earlier part of this chapter, it was noted that your instructions should be given in simple terms, even if you are working with a U.S. student who is a sharp professional in his own field. And you should certainly keep it simple—not colorful—for the foreign student.

It's been said that the British are masters of the understatement and Americans use exaggeration. To say to a foreigner, "Touch that mixture again and I'll throw you out of the airplane!" could cause consternation and dismay (or get you a punch in the mouth). Exaggeration can ease over into sarcasm sometimes. Saying to a foreign student after he's dropped the airplane in 3 or 4 ft with a resounding crash, "Great landing. *That's* the way to do it all right!" could cause some "minor" misunderstanding. Sarcasm isn't good for morale for your U.S. students and could be mistaken for real instruction by foreign students. As indicated earlier, sarcasm has no place in flight instruction.

Slang such as "ease off a little gun" for reducing power is certainly colorful but could be interpreted as "He's (ease) not on (off) a small (little) rifle (gun)." "He's not on a small rifle" is what he might think you said, which is an interesting observation to make when he's discovered that he's too high on final. "Reduce power" would be a lot better.

You're introducing stalls, and at the end of the talk you say, "Okay, summed up: you'll ease the nose up. Then it will break and, in this model airplane, the right wing falls off. Be ready for it." You may look up to see that a student is missing. You're requiring too much of a sacrifice, and besides, you haven't mentioned anything about parachutes to survive the catastrophe you've just described.

KISS ("Keep it simple, stupid") should be the criterion for flight instruction. "All right, now you can set up a turn to the south" could be replaced with "Turn south now."

■ **Do They Understand?** Some foreign students may be ashamed to admit that they don't understand and will nod in agreement or say "yes" even when they don't understand anything you're saying. This can be a dangerous situation, particularly if you send such a student out on a solo flight to do maneuvers he doesn't fully comprehend.

You will need to take more time briefing and debriefing foreign students. Ask questions that require them to show whether they understand what you've been saying. As with U.S. students, a good method of instructing during the preflight briefing is have the student explain back to you what will be done during the flight. This normally will bring out any points of misunderstanding.

The chalkboard and training aids assume even more importance in briefing and debriefing, since a picture is less apt to be misunderstood than a verbal description.

In covering emergency procedures, you should be sure that the student fully understands, since you'll not likely provide an actual electrical fire or systems failure to demonstrate what you mean. You can simulate situations in flight after a good discussion on the ground, and during the simulation make sure that he understands the steps required.

If you can work it, talk to your foreign students in situations other than flying. They can get accustomed to the sound of your voice and speech habits. It will help them understand better when you are talking in ground school or in the airplane.

Expect that the foreign student will take slightly longer to solo or get the various certificates because he may have to think about what you're saying. This is particularly true during landing practice, where split-second response is required. If you are prepared from the beginning to take a little more time on each step, you won't find late in training that you

have problems and need a large block of extra time.

Foreign students may have problems with multiple choice written tests because of trouble reading the questions, particularly when two or more of the choices are nearly the same. You might prepare extra testing experience for them before sending them up for the FAA exam. (U.S. students can also have problems reading the questions.)

Okay, maybe waving the flag is uncalled for here, but the effect you make as an individual on foreign students may do more for U.S. relations with their countries than six State Department papers. It's hoped that they will look back on their experience here with good feelings.

SUMMARY

This was a general approach to how people learn to fly. In later chapters various maneuvers will be covered, with specific application of these general facts about learning. This book is written so that you, as an instructor, can review a chapter or section concerning a particular maneuver or lesson that you haven't taught in a long time. An attempt was made to include all information needed to teach it, even at the cost of repeating material given here (call it *meaningful* repetition). You will note that some of the learning techniques mentioned in this chapter have been used throughout this book. For example, the phrase "As a flight instructor, you will find that . . ." appears often. You'll use the same technique when you start instructing.

ANATOMY OF A FLIGHT LESSON

THREE PARTS

All flight lessons must have three parts: (1) preflight briefing, (2) inflight instruction, and (3) postflight review. *If the flight lesson does not have all of these parts, it is incomplete.* This doesn't mean that you will spend a minimum of 15 min before and after every flight, *but* you will have a period before for review or introduction of maneuvers, and you'll have a period after for review or questions.

The flight lesson *could* consist of 45 min of preflight briefing, 45 min of flight, and 5 min of review, questions, and discussion of the next lesson. This division of time would be most likely at a point of training where the student enters a new area of instruction, such as the first flight, the first flight of an instrument syllabus, or the first dual or solo cross-country flight. You'd give a thorough preflight briefing so that the student understands what's going to happen, fly the flight, and keep the postflight time down because of possible fatigue on his or her part. (You answer any questions and suggest reading material.) Basically, *you* would set the length of the preflight and the flight itself (unless the student is getting queasy or tired), but generally the number of questions the student asks or the need for review controls the length of the postflight period.

Because of tight scheduling and the fact that they are

paid for flight time only, some instructors' preflight briefing consists of, "Let's go," and the postflight discussion goes into details such as, "You need more work on turns. I'll schedule you for 0900 Tuesday."

Too few instructors charge for ground *and* flight instruction time. If the student gets an hour of ground time and an hour of flight time, he should be charged for 2 hr of your time at, say, $15 an hour. One student, who was used to the old procedure of only paying for flight time, asked his new instructor, "Do you mean I'm paying you $15 an hour for this preflight instruction?" The flight instructor answered that he would be glad to give him the same information in the airplane in flight at $50 an hour. The student quickly saw the logic of paying for ground instruction. In many cases, *an hour of ground instruction and an hour of flight instruction* for a total of $65 is more valuable than *2 hr of unbriefed flight instruction* at a cost of $100 and the instructor still gets $30 for his time. Either flight instructors are conscientious and are giving sufficient ground instruction but *aren't* being paid for it, hence aren't getting the proper income for their skills, or if instructors are not getting paid for ground instruction, this area is neglected due to financial pressure—which means as tight flight scheduling as possible.

THE SIX UNITS—PEDPER

There are six units to be covered in every flight lesson: (1) preparation, (2) explanation, (3) demonstration, (4) practice, (5) evaluation, and (6) review.

■ **1. Preparation.** Every flight lesson will need some preparation on your part; many times it will mean only a quick mental review of the maneuvers to be covered and a check to make sure that you have the needed chalkboard or other equipment—including the airplane. Airplanes make flying easier.

Greeting a student with "What do *you* want to do today?" is no way to impress him with your authority as an instructor. His flying lessons are very important to him and if he's met with that, or an "Uh, what did we do last time," he may be turned off by your apparent lack of interest in him. He will know you are interested and are a sharp instructor if you can mention items that he did well (or had some problem with) on the last flight. Try to mention something that happened on earlier flights. ("The wind isn't so gusty as it was last Tuesday, so our crosswind landing practice ought to go better.")

It's best if you refer to students a few years or more older than you are as "Mr. Smith" or "Mrs. Jones" and keep it on a more formal basis, but first names on both sides are fine for contemporaries as long as you both know which is the pilot in command. You might ask some general questions to relax the student if this is your first meeting. Incidentally, your greeting to the student will set the stage for the flight. If he or she is slightly late, accept an apology in good humor and drop the subject. You can ruin the efforts of the entire flight by getting into an argument at the beginning. If it's a flagrant and repeated offense, mention it at the end of the postflight briefing as an apparent afterthought.

Most flight schools keep the student pilots' logbooks in the office. Not only is this convenient for the student, who doesn't have to worry about hauling the book back and forth, but it's also available for the instructor to review to prepare for the upcoming flight. *It's better for you to look at it before he gets there, if possible, but it should be checked, even in his presence.* The same thing goes if you are using flight sheets. A *pencil* notation ("X-wind") may give you a clue to mention that as an opener for the next flight.

You'll want to be sure that the student is prepared. Does he have the background for this flight lesson? You should have prepared him for *this* flight in the postflight review of the previous one, and given him reading to do and told him what you plan to cover next. He should also have been told what equipment (protracters, computers, etc.) to bring.

Get the student involved in planning the flight. Ask him to explain maneuvers (giving airspeeds and control functions) or to describe what he will be doing in the airplane during various maneuvers, including emergency procedures covered in earlier flights.

LESSON PLANS. It's sometimes suggested that a lesson plan be made, and you'll have to show on the instructor's flight test that you are able to make one, and use it (see Fig. 6-1 for a sample lesson plan). Many instructors, however, feel that a formal lesson plan is too stiff and artificial and so they don't use one after they get a little instructing experience. It is a good aid at first and may be used for all flight lessons if you desire. One problem is that all rigid lesson plans tend to turn off the average pilot. They smack too much of the classroom. Your later lesson plans can be notes on a scrap of paper—anything that helps you set up the lesson can be a lesson plan.

At the end of this chapter, the six units will be brought back together to show a simple form of lesson plan.

■ **2. Explanation.** This is one of the most important units of flight instruction. The student should have a clear idea of what he is going to do, and here is where he should find out. In normal instruction, the explanation is always *given on the ground before the flight.* To try to introduce a maneuver in the airplane is a waste of time and money. However, you may find that something unplanned happens during the flight and a brief explanation at that point would be more valuable for learning than waiting to get back on the ground to talk about it.

Your preflight explanation should include *what, why,* and *how* (and sometimes *when* and *where* are applicable).

What is the maneuver or flight lesson to be covered? Laying it out for the student in specifics will help him to remember the maneuvers later when he is reviewing it on his own. It will also help him remember the *why* and *how* parts.

Why is it discussed and practiced? (Because it's required on the flight test and *that's* why it's done is not the proper reply.) Tell the student not only about *why* a maneuver is done now, but how it fits into the flight syllabus (and later flying), and the principles involved.

If possible, tie in the maneuver being explained with previous maneuvers or discussions. This may seem hard at first, but as you do more instructing you'll be able to bring in points discussed in earlier lessons.

How is the final part of the explanation and you would use the *whole-part-whole* method of presentation. Show him, using a model, the *wingover* as a whole, then the parts step by step, then a review of the whole. In the *how* portion of your discussion be sure to mention common errors to be expected; he must know what to avoid as well as what to do. As was noted in Chap. 2, people are put more at ease if they realize that others have made certain errors on a particular maneuver.

Try not to skip any important items when discussing a maneuver or lesson. Sometimes you may find that in discussing a maneuver such as the chandelle, you've done a great job (you think) giving him the whole-part-whole story, perhaps with an anecdote thrown in, and are quite satisfied. The student asks, "What do we do with power in the chandelle?" (You forgot to mention it.) Tell him when he will add climb power and then *repeat* the explanation of the maneuver as a whole, with power use inserted at the proper place. After you've covered the maneuver and fixed it in his mind *without* power use, he will likely forget the power if you toss it in as an afterthought *without repeating the maneuver as a whole.* You can check this in flight; the late, semicasual notice of throttle use may result in your having to remind him several more times.

When will a certain maneuver or principle be used? Obviously it will be used *now* (since you are explaining it for use in the upcoming flight), but at what other times will it come up in his flying career? This point will likely be covered in the *why* part of your presentation but there may be occasions when you want to hit this idea separately.

Keep your explanations as simple as possible (Fig. 3-1). The *explanation* part of your flight lesson may contain a *review* of an earlier flight as well as covering new material. For instance, you would probably review the rectangular course when introducing and explaining S-turns across a road, or you might review the simpler stalls when introducing advanced stalls.

Again, when explaining the maneuver or lesson, be sure to cover the common errors.

Fig. 3-1. The student is wondering if he'll be able to do all that calculating during a stall.

■ 3. Demonstration. Your philosophy of instruction should always be to keep your demonstrations to a minimum. The student learns by doing, and more than one demonstration of a maneuver before he gets a chance to do it is normally a waste of time. You may have to demonstrate it again if he is having problems, but keep the number of initial demonstrations as low as possible (Fig. 3-2).

If a student is going to have an airsickness problem, it will often occur after a series of demonstrations. *The more the student flies the airplane, the better for the learning process, as well as for his physical well-being.*

With experience, you'll know how much demonstration is normally required for various maneuvers. You will also see that for a particular student to go on to a simple next step you should only demonstrate a *part* of a maneuver, such as only completing one 360° of a 720° power turn or, in teaching aerobatics, showing only one-half of a Cuban eight before letting the student practice it.

Don't be too hasty in demonstrating again if he is having minor problems. Give him a couple of tries, while you explain it, before taking over. In your initial demonstration explain the maneuver as you perform it, keeping your explanation as close as possible to the one you made during the preflight briefing.

For primary-type flying the "follow me through" method of instruction is *not valid in most cases.* The student may not be able to feel the real control pressures and is not sure whether the airplane's response was due to your or his efforts. In introducing certain aerobatic maneuvers, however, the following-through method has some validity, since he may need to see as well as feel the control movement at first.

■ 4. Practice. Depending on the circumstances, you may talk the student through his initial try at a maneuver, or you may prefer to wait until he has completed it and discuss any errors he made.

If he is flying a rectangular course, you will have time to make comments as he is actually doing it. A stall recovery might be a little fast for you to keep up with easily, but if you are instructing snap rolls you couldn't get a word in before one of those maneuvers was complete.

Don't fail to make corrections the first time an error is made, whether you are able to do it during the maneuver or have to wait until right afterwards. His errors will "set" like concrete if he commits them several times without correction. You may bring his attention, in flight, to a particular item to be discussed when you are back on the ground.

If he has a good idea of the maneuver but is still making small errors, describe them and allow him to iron them out himself, but don't overdo the practice. If he is proceeding well and is getting the idea, move on to something else. You'll find that by working too hard at minor problems you can set him back on the maneuver as a whole. Also, *you* might start feeling uncomfortable (Fig. 3-3).

■ 5. Evaluation. Sometimes the *evaluation* and *review* parts of the lesson are treated separately but in most cases may be combined. Discussing *evaluation,* you have two main items to consider:

1. Does the student have the idea of the maneuver and the principle involved?

2. Is his progress on the maneuver itself satisfactory or does he need more practice in the mechanics?

A student may have the principle well in mind, but his actual flying of the airplane needs work. Maybe he can set up

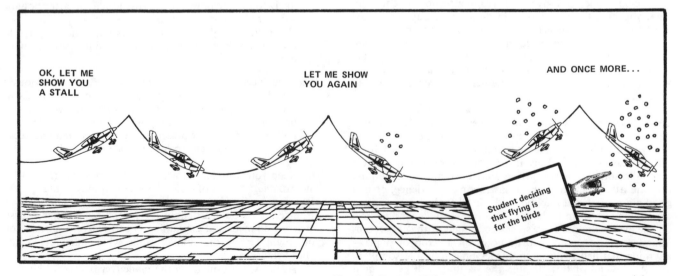

Fig. 3-2. Demonstrations can be overdone. Keep them to a minimum.

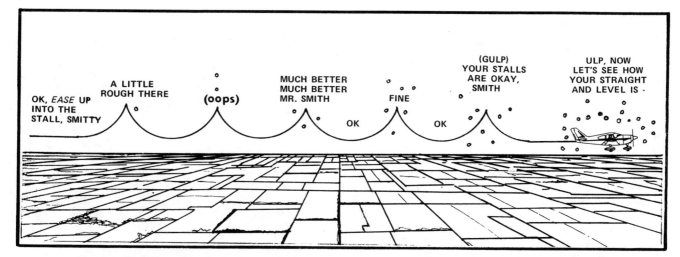

Fig. 3-3. The advantage of being an instructor is that when things start getting rough you can always change maneuvers.

a perfect wind drift correction but is unable to maintain a constant altitude in a rectangular course. If that's the case, at the end of the period your evaluation would be that he needs more work on altitude control when he is flying the airplane and coping with other requirements such as wind drift correction. You would decide that he's not ready for traffic pattern work and, in your evaluation, would find out *why* he is having that problem. It could be poor use of trim, tension (he's just a little ground shy this first time and is tending to climb), or maybe he's been carrying too much (or too little) power and this is causing altitude problems. Of course, some of these problems will be evident to you as they are happening; you will correct them at the time. But, if he continues to repeat these errors your *evaluation* may be that he needs more work on the Four Fundamentals and the division of attention required by the rectangular course is too much for him at this point. This means, of course, that your *evaluation* on the last flight—that he was ready for this maneuver—was wrong.

You'll do a short informal evaluation after each maneuver is completed *in flight*. The student's rectangular course is good, you can now introduce the S-turns across a road you discussed in the preflight briefing.

■ **6. Review.** You'll *review* each maneuver briefly after it has been completed. ("On that right chandelle, you needed slightly more right rudder to keep the turn coordinated.") When you are evaluating or reviewing in flight, don't get so absorbed that you forget to look around for other traffic. A *review* should be done at the end of every flight even if you take only a few minutes. Here you'll also establish the mood for the next flight. Cover both good and bad areas of the flight and, if possible, mention specific instances. You should also cover briefly your plans for the next lesson, schedule it, and assign reading for it.

CRITIQUE. Here are some suggestions for a postflight critique you might consider. (It is not a step in the grading process.) The critique should be:

1. Objective. It should have no personal opinions or biases.

2. Flexible. It must fit the situation; each flight is different, even though the same maneuvers were covered in previous flights.

3. Acceptable. It must be fair and given with authority, conviction, sincerity, and competence.

4. Comprehensive. It must cover strengths as well as weaknesses.

5. Constructive. The student must profit from it.

6. Organized. It's invalid if not.

7. Thoughtful. The value of the individual must never be minimized.

8. Specific. The critique should cover specific points. Don't make it so general that the student doesn't know his weaknesses.

The critique may be written by the instructor; told personally to the student by the instructor; or the student may be asked to write a self-analysis.

SUPERVISED SOLO FLIGHTS

One complaint students have is that they don't know what to do on a local solo flight, and so they do a desultory maneuver or two or wander around sightseeing.

Some flight schools pay instructors for supervised solo flights of their students—a good idea. The instructors are expected to brief the students and oversee the flights. For *local* solo you may set up the flight during the postflight briefing of a dual flight, while his weak maneuvers are still fresh in your mind. Have him make a list, or have a syllabus lesson, to stick on the instrument panel or carry on a knee board as a reference during the solo practice. If possible, you should talk to him just before and after the flight, but your schedule may not allow it. Encourage him to make some form of shorthand notes during the flight so that he can ask questions as necessary when you get together again, but not to keep his head down in the cockpit writing a book. You might, for instance, fly a dual with him on Monday, make a list of maneuvers for his solo flight that he will fly Wednesday, and review that Wednesday solo flight with him on Friday. Or, you may fly a dual, brief him for a solo flight to be flown right after he's had a rest break, and be there to talk to him as soon as he gets back; it depends on how busy you both are.

For solo cross-country flights you should *always brief the student just before he leaves.* You should go over his flight planning with him and actually supervise his weather check for the first couple of solo cross-country flights. To sign him off today for a flight that won't take place for maybe a week is really sticking your neck out (and his, too). The weather might be marginal the day of the solo, and if nobody stops to check what kind of flight he's going on, problems occur.

Supervised solo is a solo flight for which the student has been briefed, and the instructor is aware of the exact time, place, and conditions of the flight. No student pilot should fly solo without instructor supervision.

SUMMARY

Preparation will obviously be that part of the lesson done on the ground *before the flight.*

Explanation will be done on the ground *before the flight, and in flight* as necessary (usually during the *demonstration*). By far the greatest part of time spent in *explanation* should be done on the ground.

Demonstration and *practice* are the only two lesson units normally restricted to actual flight, but as the use of simulators increases, these parts will be done on the ground, also.

Evaluation may be done in flight briefly (you'll decide from the student's performance whether to go on to other maneuvers while you are up there), but the biggest time spent on *evaluation* will be on the ground after the flight as you take a look at his general progress as well as that for this specific flight. Check his progress (and your own) by asking questions.

Review may be done in flight, for a specific maneuver, and should always be done for the flight itself after you get back on the ground. To help you remember, review the flight in the order it occurred (preflight, pretakeoff, etc.).

As stated earlier, sometimes *evaluation* and *review* (or review and evaluation) are combined as one part of a lesson. You can decide as you instruct more what way works best for you.

A quick mental checklist to see if you have considered all parts of the flight lesson is PEDPER, the first letters of the parts. This is not meant to replace a more formal lesson plan if you like that route. That, plus remembering the *What, Why,* and *How* parts of the explanation (the first "E" in PEDPER), will give you a good framework for the lesson.

The lessons following in this book will use the PEDPER idea but occasionally will deliberately stray from the formal approach to make for a change of pace and more interesting reading.

PRESOLO INSTRUCTION

PREFLIGHT INSTRUCTION

Try to get the student to a quiet, informal spot for your preflight instruction, and put him at ease. Some flight schools have separate rooms or cubicles for instructors to use for pre- and postflight briefing. This is far superior to trying to have a discussion in a busy airport office. (Hey Joe! Where's that #%&!! gas ticket for 41 Tango?" "I dunno, you had the #%&!! thing last!")

As was mentioned in Chap. 2, your route of instruction depends on the student's background. You might give him a (very) brief history of *your* background. One of the first things we like to know about a new acquaintance is his job — what does he do for a living?

Some may be uncertain about starting to fly and may say, "I'm too old to be doing this, but I've always wanted to fly," or "I'm not very mechanically minded." It's usually because they are afraid of failure, and your ground discussions, particularly for the first couple of flights, should be as private as possible, to help them relax.

They want assurances and should get them. Briefly laying out the steps they will cover in completing the training course will help put them at ease. So, some of the new students you meet will express some form of uncertainty.

Your preflight instruction for the first flight should consist of three basic parts: (1) classroom discussion, (2) cockpit familiarization, and (3) preflight check and discussion.

Circumstances may be such that a better learning situation would be established by getting the student into the air without unreasonable delay. For instance you may have a timid student who will require a session, or sessions, of familiarization with flight before getting on with the actual training. There's no point in going into detail this first flight, but a discussion of the airport and local area and what you plan to do during the flight will put him at ease. After you've been in the air for a while and he is "beginning to put all his weight down," you may introduce effects of controls and the Four Fundamentals in an easy manner. Or you may spend the whole first flight just letting him get used to the idea.

You may shorten the classroom discussion and/or cockpit familiarization to fit the situation, *but never rush or cut out part of the preflight check of the airplane.*

■ **Classroom Discussion.** A chalkboard and model should be used here and you generally shouldn't spend more than 30 min. Below are some suggestions in the form of notes for items to be covered. You may decide that covering all the items mentioned here could tire the student; adjust your instruction to the student and situation, and save some of the information for the second session.

1. *Discuss what the student can expect this flight.* Lay it out in general terms such as describing the airport and local

practice area and what you will be doing—give a brief step-by-step coverage.

2. *Discuss the controls and how they operate* (including the throttle, trim, and flaps) and *effects of controls* and how airspeed is a factor in control effectiveness. Point out that the elevators don't "elevate" the airplane.

3. *Discuss the Four Fundamentals* (turns, climbs, descents, and straight and level) in general terms, noting that nearly all normal flying consists of these or combinations of these.

4. *Discuss the turn.* (Use the model and chalkboard.)
   a. Discuss use of ailerons and rudder. Demonstrate a turn entry and roll-out. Discuss the slip and skid.
   b. Explain why back pressure is necessary in the turn. (The airplane would turn indefinitely except for this one detail.)
   c. Review the turn entry and roll-out with *elevator* use included.
   d. Tell him that *three* things are done, for example, when rolling into a left turn: (1) apply left aileron pressure, (2) left rudder (simultaneously), and (3) back pressure as the bank increases. To roll out: (1) apply right aileron, (2) right rudder (simultaneously), and (3) relax back pressure as the bank decreases. This idea of *three* items to be considered will be covered for each of the Four Fundamentals and will help him to remember.
   e. Discuss load factors in a turn.
   f. Mention the common errors to be expected.

Introduce the turn before the other three Fundamentals, both on the ground and in the air in a normal lesson, because you will want to see that the student retains as much enthusiasm as possible during and after each flight, and this is particularly important after the first flight. Assuming a maximum time in flight of 1 hr for this first flight (and for most students a maximum of 45 min is even better), you may not get to introduce all of the Four Fundamentals, so the turn should be introduced first. Why? The turn shows constantly changing references and the student is "doing something." He is "actually flying the airplane."

5. *Introduce the climb next* because it is a dynamic and an "optimistic" maneuver; he is gaining altitude and "doing something." (A descent or glide might be considered a "pessimistic" maneuver—he's coming back to earth and the flight is ending.)
   a. Discuss "torque." Cover the factors that tend to make the airplane yaw to the left on takeoff, in a climb, or in level slow flight. Use the chalkboard to illustrate the idea of an offset fin or offset thrust line.
   b. Tell him the *three* steps used in going into a climb: (1) ease the nose up to the climb attitude, (2) increase to climb power, and (3) correct for "torque."
   c. Tell him the *three* steps used in leveling off from the climb: (1) ease the nose over to the level flight attitude and as the speed builds up to cruise, (2) ease off the right rudder, and (3) adjust power.
   d. Discuss using elevator and rudder trim in the climb and level-off.
   e. Cover the common errors encountered in climbs.
   f. Discuss the climbing turn—a combination of two Fundamentals covered thus far.

6. *The glide or descent* (clean)
   a. Mention the fact that in a *dive* the airplane will tend to yaw right, but in a power-off glide the yaw effect is considered to be zero.
   b. Explain loss of slipstream effect. The rudder will be less effective in the glide than in straight and level flight and the climb.

   c. Discuss carburetor heat briefly.
   d. Note the *three* steps required to set up a power-off glide or descent: (1) carburetor heat (or check for the need for carburetor heat), (2) smoothly close the throttle, and (3) hold the nose in the level flight attitude until the airspeed approaches the recommended value, then move it down to the proper attitude.
   e. To recover from the glide, lead the altitude by about 50 ft and (1) add cruise power (a lead of 50 rpm will be discussed later), (2) ease the nose to the level flight attitude, and (3) carburetor heat OFF, if used.
   f. Discuss using trim in glide and level-off.
   g. Talk about common errors.
   h. Discuss combining two of the Fundamentals (gliding turn).

7. *Straight and level flight*
   a. Note that the secret of straight and level flight is proper power setting and trim; in smooth air the pilot should not have to do anything once it's established.
   b. *Three* things he will have to consider in straight and level flight are: (1) bank attitude, (2) pitch attitude (trim), and (3) power setting.
   c. Explain that more time will be spent in straight and level flying than any other condition, and some experienced pilots still have trouble with it.
   d. Tell about common errors.

Leave time for the student's questions and you may have him or her go through the Four Fundamentals, using the model. However, you may decide with some students that the first preflight discussion is a little early to put them on the spot by having them "perform." If in doubt, ask for questions only.

■ **Cockpit Familiarization.** After you've completed the chalkboard work and had a short break, move out to the airplane for a look at the controls and instruments. In the summer, the airplane cockpit can be an oven, so you may want to use another trainer in the hangar or move the one you'll be using to a shady spot. In the winter an airplane parked out on the ramp also can be a very uncomfortable place, so this will be a factor in how long you spend on the cockpit familiarization. *If the student is physically uncomfortable, he won't be learning.*

Assuming that the temperature is reasonable, you should have him get into the airplane (in the left seat in a side-by-side trainer, naturally) and have him adjust the seat so that he can comfortably move all controls their full travel. Make sure that he can see out well; short people may have to use a cushion under and behind them. Once the best arrangement is made, tell him that this is the way he should be seated each flight. If extra cushions are necessary he should *always* use them, because his view of the airplane's attitudes will not be the same without them.

Using the best order of introduction for your airplane's cockpit arrangement, discuss the instruments and controls to be used during the flight. Tell him you will discuss the other instruments (attitude indicator, heading indicator, etc.) later.

Have him move the controls and change the elevator tab setting as he watches the surfaces. Remind him that he will be flying control *pressures,* not movements. Tell him that for some nosewheel airplanes the rudder pedals are "firm" and he shouldn't try to move them when the airplane is sitting still. Explain the use of throttle, trim, carburetor heat, flaps, fuel selectors, brakes, and other controls.

You should note that the airspeed is marked in knots (if it is) and very briefly compare knots and nautical miles to mph and statute miles and indicate that he'll get more about this later. If you say "climb at 80," he'll use that number and not try to convert from knots to mph.

If you feel that it won't be stretching the time out, making him tired later during the flight, you might explain and have him simulate turns (rolling in, the turn itself, and rolling out), climbs, glides, and straight and level while calling his attention to trim and power requirements.

■ **The Preflight Check.** Carefully do the check, as always, and have the student follow with you. Some instructors, when flying by themselves or making a charter flight with nonpilot passengers, do nothing more than check the fuel and oil, and sometimes not even that, and hop in. (What the heck, you came down in this airplane not 2 hr ago.) In 2 hr several tractors could have rammed the elevators, or a mechanic could have removed the rudder. It *has* happened. There will be sharp-eyed students taking in all this "inspection" from their vantage points in the chairs in front of the office.

You might tell the student during the preflight check that not only is his neck involved, but he could be spared a great deal of embarrassment in instances of a less serious nature.

The first (and subsequent) preflight checks should be used to give instruction, while not wasting any time.

As you move around the plane, tack down points brought up during the cockpit familiarization (the various control movements, trim tabs and how they work, pitot and static systems, etc.).

Some flight instructors feel you shouldn't give the student details such as this on the first flight, saying that it's too much to remember. (This is probably true if, after you got down, you asked him questions on every phase covered.) *But* the second preflight check will make more sense. You might explain to the student that he may not consciously remember more than a part of this first flight after he gets down, but the items will be familiar when he meets them again. During the preflight tell the student *what* you are doing, *why* you are doing it, and show him *how* to do it.

Use a good clockwise (or counterclockwise) check, not missing any items (Fig. 4-1).

Some instructors have their students check the fuel and oil items *first* (*1–6* in Fig. 4-1), then do the structural check, starting at the flaps on the pilot's side. For instance, one popular two-place trainer has five fuel items: Check the *two* tanks for volume, drain fuel from the *two* tanks and then drain the (lower) main strainer for water (the fifth item) and check the oil. You will set up the best program, one that will not only cover your specific trainer but can be used on other airplane types as well.

You may want to calibrate and mark a blunt wooden stick in gallons for use in checking fuel in your airplane, since it's very difficult to look into a tank and know how much fuel is in it. Certainly the fuel gauges shouldn't be depended on for checking the quantity before a flight.

■ **Starting.** Unless the student has had plenty of experience around airplanes, it would be best for you to start it the first time. Go through the *prestart checklist.* Call off the items, making sure your "Clear!" is loud and clear.

Have the student check the oil pressure after starting, and follow through with the *poststart checklist.*

■ **Taxiing.** Most students should taxi the airplane on the first flight. This doesn't mean that you're going to let him weave his way amongst those $50,000 (and more) airplanes the first time around (no instructor has strong enough nerves for *that*), but once you're out on an uncongested taxiway it's best to start him right in (Fig. 4-2).

As the airplane starts to move forward you should demonstrate checking the brake effectiveness before the airplane picks up enough speed to cause problems if they aren't working.

Sometimes you'll get a King Kong type who has the strength of ten, and he's using those ten people to hold full right rudder. Unfortunately, right rudder is the *wrong* rudder as far as that parked Jetstar is concerned. Of course, he sees that he is boresighting a pile of dollars and has the control wheel turned fully to the left to stop the turn. As you know now, and he'll learn shortly, turning the control wheel at this point, at least for the majority of airplanes, has as much effect on turning as pulling the carburetor heat. You, meanwhile, are pushing on the left rudder pedal with only the strength of one (or maybe one and a half because of increased adrenalin flow) and telling him gently and quietly TO GET HIS FOOT OFF THE RIGHT RUDDER!! (Remember: No four-letter Anglo-Saxon words.) This brings up a point that you should have mentioned during the preflight discussion; when you say "I got it!" he's to let you have the airplane back — *NOW.*

The new instructor learns during the first taxiing session with a student that:

1. Some students get confused as to how the throttle works and will wind up thundering down the taxiway and moving the throttle the wrong way in trying to slow down.

2. People who've been great sled riders in their youth want to push the right rudder to go left, and vice versa.

3. *All* students tend to turn the control wheel to steer the airplane, and it would be best for the first couple of sessions to have your student keep his hand completely off the wheel while taxiing. Otherwise, he'll be *transferring* automobile knowledge to the airplane. Later, after he's pretty well got it in his mind that the rudder pedals steer the airplane *on the ground,* you can show him the niceties of holding the ailerons and elevators in various wind conditions; on this first flight, however, it would normally only add to the confusion.

One problem that may crop up later in the student's taxiing is carrying too much power and holding the speed down with the brakes. Sometimes the tendency is for the instructor to be thinking about the flight ahead and allowing so-called "minor" taxiing problems to slip by. The power-and-brakes method can burn out brakes, particularly at large airports where long-distance taxiing is necessary — and the brakes could fail at an inopportune moment.

Another thing you should watch for is too-fast taxiing by the student. This sneaks up also (you want to get into the air

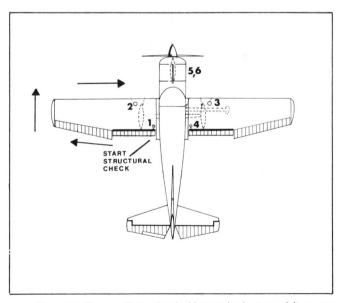

Fig. 4-1. The preflight check. Your attitude toward it as an instructor will be very important.

START STRUCTURAL CHECK

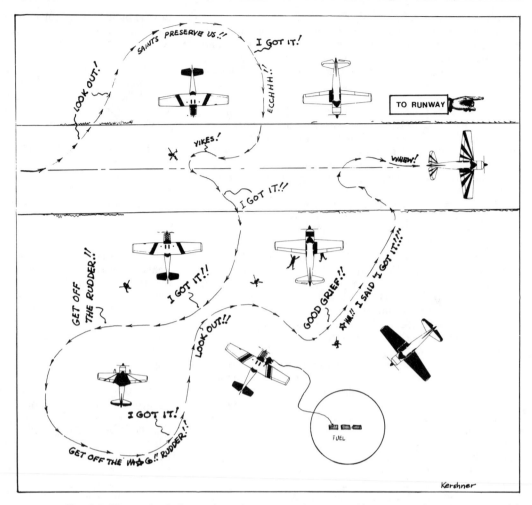

Fig. 4-2. The student's first taxi session can turn into an exciting event unless you are careful.

as soon as possible) and even though the speed is not too fast for *you* to cope with, remember that he'll be going solo later and it just might be too fast for him. This is a particularly bad situation when he's taxiing downwind; he gets too fast and comes up to the place to turn or stop before he realizes it. His reaction is to jam on the brakes, and he'll have all the problems that result from such a practice. You have to be instructing every minute in an airplane, even with a good student. One instructor half-jokingly stated that he kept as alert "as if the student were planning at every second to do away with both of us"—not a bad idea.

Keep the taxi speed down to a fast walk in uncongested areas, and much slower when later you let him taxi in congested areas. Older instructors can remember when "fully functioning controls" meant no brakes on their side—in fact, it meant no brakes on *anybody's side.*

PRETAKEOFF CHECK

You'd better do this first one with the student watching you. Expect to take longer than normal because you are explaining the process. *Use a checklist.* You can let him know that he'll do the check next time, with your help.

You should strongly suggest that following the checklist item by item is the best way to avoid problems like doors being left unlatched, leaving the ignition on one mag, or trying to take off with the trim in the full nose-up setting.

If you're operating from a controlled field, don't have the

student contact the tower for takeoff *on this first flight.* On the second and following flights it would probably be a good idea—with your coaching—to have him do all communications. Here is probably one of the parts of flight training you'd definitely rather do for yourself. You'll find after cringing through some of the worst voice procedures you've ever heard, you'll want to "help" him by making *all* of the reports. *Don't.* In critical situations you may have to make the radio calls since there's no point in fouling up the rest of traffic badly. You can tell him you'll explain what was happening after the flight.

Radio work is pretty easy for you now, but as mentioned in Chap. 2, one of the things you'll notice is that the student only picks up parts of any tower transmission at first. This happens particularly when the ATIS, tower, or ground control gives several items such as wind, runway, and altimeter, and most students will turn to you and ask, "What runway did he say?" or ask about other items that were perfectly comprehensible to you because you knew what was coming. A brief chalkboard talk sometime in the early flying stages, outlining the functions of local control and ground and approach controls can help a great deal in clearing things up so he'll know what to expect. The problem with the new student is that he tries to listen to everything and so hears nothing. He'll learn later to let the instructions to other planes fade by—unless they might also affect him—and normally only come alert when his number is called.

At airports that have only a Unicom, the student should get in the habit of getting airport advisories before taxi and when approaching the field for landing. At small airports that have no Unicom, and even for those that have one, the instructor should, in later flights, act as *ground control,* etc. (not really broadcasting), to get the student used to the idea of using the radio before going into a controlled airport.

TAKEOFF AND DEPARTURE

This first takeoff should be yours (with a description of what you're doing). In some cases you might let him use the rudders to keep it straight. You can explain the idea of the best climb speed and traffic pattern as you climb out and depart. Tell the student that he can climb the airplane "for a couple of minutes." If he is having problems after that time you can take over again without his feeling that he's failed (you said "for a couple of minutes"). On the other hand, if he is doing well, you can say so and let him continue.

Your first tendency will be to be impatient with the student's clumsiness with the controls. The airplane wallows and slips and skids and you get impatient to take it back. Let him work it out while you explain.

Let the student experiment during the climb to the practice altitude; he'll learn something about the feel of the controls. Discuss the left-turning tendency, and if possible, give him some idea of the climbing turn. Explain that shallow climbing turns are necessary to keep the nose from obscuring vision along the general route of climb. Get up to smooth air if at all practicable.

■ **Leveling Off.** Here is a good chance to further clarify the idea of elevator (and rudder) trim. Make clear the idea of trim taking elevator (or aileron, or rudder) *pressure* off the controls.

The first level-off is a good time to demonstrate how stable the airplane really is. Be at least 3000 ft above the surface during your early high work.

EFFECTS OF CONTROLS

■ **Elevators.** The student may have seen some of the not-so-good flying movies that are popular on the Late Show. The hero-pilot hollers to the copilot, "Help me get out of this stall before it's too late." Both of them using both hands somehow wiggle the elevators, or something, until after a tough 5 min everything is under control. Another familiar scene is when the plane is barely staggering along, trying to clear a ridge (or houses or trees), the pilot again turns to his faithful copilot for aid, and both of them use both hands and full strength to pull the nose up some more to clear the obstacle. *You* know

that when the airplane is on edge, you *don't* pull the nose up but ease it down to get more climb—which probably means that you'll never be a director of aviation movies.

Let the student use the elevators at both cruise and lower speeds. He can see later that the elevators don't always "elevate" the airplane, but now he can take your word for it. You can prove your point later in the course during slow flight.

This is a good time to let him see how the trim tab operates, although he will not be using it until he has an understanding of the changing elevator pressure with changing airspeeds.

■ **Rudder.** He should get a chance to use the rudder at cruise and at lower speeds, including the power-off conditions. This can make clear the idea of the effects of the slipstream (or lack of it) on coordination in gliding turns.

The average student will prefer to enter turns using rudder alone rather than using the ailerons alone. This is probably because the feeling in the turn entry using only the rudder is close to that in an automobile during a turn. The yawing force as the turn is started tends to move him to the outside (or upper) side of the turn, which is a "natural" feeling. He is again transferring experiences in the automobile to the airplane.

You can show him shortly that the turn, once established, is the same whether originated by either rudder or aileron. Demonstrate roll-outs with rudder alone.

■ **Ailerons.** There may be a great deal of scrambling up the side of the cockpit by the student when the ailerons are used alone to start a turn. The initial slip, resulting from the adverse yaw, is uncomfortable and he feels he is going to "fall" down to the inside of the turn. Excessive and/or abrupt banking using ailerons alone can bring out a subconscious fear of heights and could cause the student to tense up. You'd best keep your demonstrating, and the student's practicing, to milder-type roll-ins and roll-outs. You can demonstrate adverse yaw at cruise and explain *why* the rudder is used with ailerons. You know that at slow speeds adverse yaw can be pretty bad so it's better not to get into this at first. He will get the idea later during slow fight. Have him use the rudder *and* ailerons at slower-than-cruise speeds here, though.

■ **Coordination Exercise.** The coordination exercise is a good way for the student to get the feel of the flight controls and how they may be used to achieve the required effect (not necessarily to keep the ball in the middle). Give a short demonstration to two or three cycles. A too-long session of this maneuver can mean an onset of queasiness, which can set the student back psychologically. Besides, *you* may have a problem (Fig. 4-3).

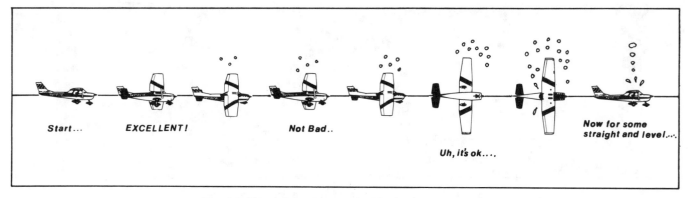

Start... EXCELLENT! Not Bad.. Uh, it's ok.... Now for some straight and level....

Fig. 4-3. The coordination exercise. Don't use such steep banks unless the student is obnoxious. Too much coordination exercise at a time is not good for either of you. (*The Basic Aerobatic Manual*)

Have the student line up with a good reference on the horizon and roll the airplane from bank to bank, keeping the nose pinned on the reference. For instance, have the student start (with the nose lined up with the reference) by using left aileron and left rudder, as in starting a turn. Before the nose has a chance to move he should use right rudder so that the point is held. Then he applies right aileron and *more* right rudder to roll the airplane to the right, stopping any turn tendency with opposite (left) rudder, and so on.

Elevators are part of the exercise and the altitude should stay within 100 ft of the starting value.

You can also use the coordination exercise at lower airspeeds (you set the power and trim the airplane) so that the student can see the effects of adverse yaw, particularly in slow flight. Again, don't have long sessions of this exercise and be prepared for some wild swingings of the nose when he first does the maneuver (and maybe later, too).

And, for Pete's sake, *don't* call this maneuver a "Dutch roll." Dutch roll, a stability and control term, is a condition of a coupling of lateral-directional oscillations with the nose yawing as the airplane rolls from bank to bank; the object here is to keep the nose on the point.

FOUR FUNDAMENTALS

■ **The Turn.** Some instructors advocate starting the practice of the Four Fundamentals with straight and level, and of course when you trim the airplane the first time this first flight (and the student trims the airplane on this and following flights) you are practicing the first principle of good straight and level flight. *But* the whole idea of straight and level is getting the airplane properly trimmed and leaving it alone as much as possible. To the average student, at first there's nothing to straight and level flight, so start off with turns as suggested in the preflight instruction section. The power and trim should be for straight and level and he should return to this between turns, but don't go into details yet.

After he has the turn pretty well in mind, make it a point to keep your hands and feet well away from the controls so that he knows he's doing it all.

At first have him do turns without worrying about rolling out on a specific heading or making a precise turn of 90°. He'll be busy enough just trying to coordinate ailerons, rudder, and elevators without worrying about particular headings.

Your procedure could be like the following:

1. Demonstrate entering a turn, talking to the student about your use of aileron and rudder (and the elevator) as the turn is set up. Make the turn, using 20–30° of bank.

2. Let the turn proceed and point out that once the proper bank is established the ailerons and rudder are "neutralized" and the only pressure held is back pressure. (Use one finger crooked around the wheel to keep the nose up.)

3. Roll out with coordinated controls and make a special point to mention that the back pressure being held in the turn must be relaxed as the airplane is rolled out of the turn. Repeat the entrance into the turn but don't hold any back pressure; this shows the student that the nose will drop and altitude is lost. Shallow out, reapply back pressure, and establish a normal turn, explaining the procedure again. As you roll out don't relax the back pressure, and show the student how the nose rises in this case.

4. Demonstrate the turn as it should be done, from roll-in to roll-out, explaining again.

Let the student make a turn of medium bank (20–30°). A too-steep turn would more than likely make him uncomfortable this early in training and the airplane could get away from him (much back pressure needed – but not held), which would just about guarantee he'd feel uncomfortable.

In the side-by-side trainer, he won't be sitting on the center line, so you'll have to point out how the nose attitude appears different in left and right turns. *Your* view of the nose attitude will be opposite to his, so make sure you tell him the proper pitch correction technique (Fig. 4-4).

Make the point at this time that his reference over the nose is directly in front of *him,* not over the center of the nose (Fig. 4-5).

A very shallow bank doesn't give the idea of good control, as the student sneaks or drifts into a turn. In this situation you're not really sure whether a wing happened to go down and the turn started of its own accord, or the student did it. Neither is he.

The first time a student tries turns, you will feel that they are the most horrendous maneuverings ever imposed on a self-respecting airplane. Well, patiently suffer through them. Remember *your* first few turns? (Your opinion that student pilots' first turns are grim is not likely to change, even after 20 years of instructing).

Let him practice a number of turns. Tell him the good things he's done during each turn, while you correct the errors.

These first turns should be done by outside references. Fig. 4-6 shows the method of making 90° turns by reference to the wing tip and nose.

You may expect that he will have trouble rolling out on a specific point. He may take several tries before becoming reasonably accurate in this regard.

You may let him do a few roll-outs with reference to the heading indicator, but emphasize the practice of *looking out*

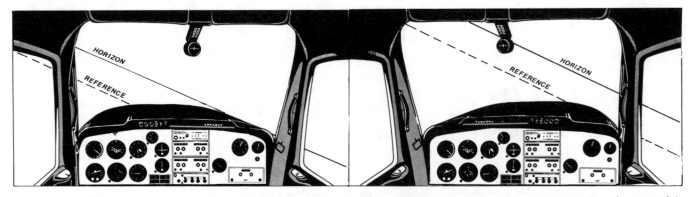

Fig. 4-4. A left turn as seen from left and right seats. Remind the student that the reference point *directly in front of him* will be the same distance below the horizon in left and right turns for a given bank, even if the overall view of the nose shows an apparent difference.

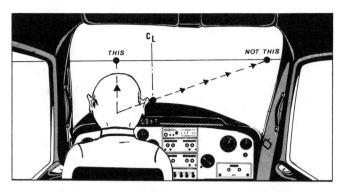

Fig. 4-5. The student will tend to use the center of the nose as a reference in a side-by-side airplane. Watch for this.

and clearing the area before starting the turn as well as during the turn. It's a good idea at first to have him confirm it and say out loud that it's "clear left" or "clear right" before starting the turn. One drawback to the integrated visual instruments program is the tendency of the student to keep his gaze in the cockpit rather than checking for other airplanes.

The tendency is for him to descend in a left turn and gain altitude in a right turn because of his position in the side-by-side trainer. More about this later.

Once he has the idea of the turn, a good practice maneuver is to do a series of 90° turns at 15° bank and 180° turns at 30° bank. If he is doing very well, you may then let him do a series of 45° banked turns (360° of turn) to show the need for increased back pressure required with steeper banks (see Fig. 5-5 for an example).

If the student is beginning to smooth out his turns, move on to climbs and climbing turns to keep up his interest.

■ **The Climb.** As stated earlier, most of your students will benefit from handling the controls in the initial climb after takeoff. If things go badly and you take the controls back with a casual: "Okay, I just wanted you to get a short stint at it. We're not really that far along and I don't want you to get tired," he'll feel that progress is being made and it will start the flight off well. In your instruction you should let the student know that while certain minimum standards are required, he can't immediately do a perfect job. You should put him in the same situation as a little guy challenging a big guy to a fight. He's expected to lose, so there's no disgrace if he is beaten. And if he wins, he's really done something.

You'll generally find that the initial pattern for the student as far as directional control in the climb is concerned is

an undercontrolling for "torque," leading you to believe he has a paralyzed right leg. After a good amount of persuasion you may find that during the second hour the other extreme tends to occur. If you realize that this overcompensation is likely to happen, you will not emphasize the "more right rudder" quite so much at the beginning.

The student should be given several transitions from climb to level flight and back to climb. There's no need at this point to emphasize extended straight and level flying; that can come a little later in this flight, or the next one, when straight and level is introduced as one of the Four Fundamentals.

The things you should watch for and correct if necessary are:

A. *The transition to climb*
   1. Make sure the nose is raised to the climb attitude *before* the power is advanced.
   2. Work on having the student detect and correct any directional (torque) problems as soon as possible. Too many students are still having this trouble when they are supposed to be ready for the private flight test.
   3. Have him check attitude *and* airspeed to confirm a proper climb setup (but don't let him stare at the airspeed indicator and neglect his outside scan). Have him use elevator trim after the first couple of climbs.
B. *Recovery from the climb*
   1. Have him ease the nose to the level position and hold it there.
   2. Keep the climb power on until the airspeed is approaching the cruise value.
   3. Then set the cruise power and trim. If the student is allowed to throttle back as soon as the nose is lowered, the transition to cruise is delayed; also, if he throttles back and sets the trim as soon as the nose is lowered, the power will have to be reset and the plane retrimmed. He should be given the proper procedure so that he won't have to be reinstructed later. It's all new to him anyway, so it's no harder for him to learn the right way than the wrong way.

CLIMBING TURNS. These should be introduced as soon as the turn and climb are established in the student's mind and he can do them with reasonable skill.

If you haven't taught him the proper rudder correction for "torque," the left turns will be skidding and ever steepening; right turn banks will be ever-shallowing.

Do some straight climbs and level turns after the climbing turns have been introduced. Don't let the student get in a rut with one particular Fundamental—review those covered earlier, too.

Fig. 4-6. (A) Have the student pick a reference off the wing tip, turn, and (B) roll out headed directly for it, to establish precise 90° turn work.

■ **The Normal Glide.** The expression "The Normal Glide" has lost popularity somewhat in the last few years since most descents and approaches are made with partial power. It's a good move to demonstrate, however, that the airplane doesn't "fall" when the power is reduced to idle. It will also demonstrate the use of trim in power-off conditions and show that (hands-off and no trim) the airplane *will pick up speed* when power is reduced. That is, you will establish that power is not used to go faster, but to control altitude.

This is the one of the Four Fundamentals that probably could cause the most concern, particularly for the timid student.

The *Pilot's Operating Handbook*s of some of the newer general aviation trainers do not recommend the automatic use of carburetor heat for approach and landing, and as an instructor, you'll have to keep the student aware of the possibility of carburetor ice. You can work it out best for your airplane, but for extended glides at idle it would be well to have the student use full heat throughout, since it will be some time before he has developed the judgment to be able to immediately spot carb ice. This is particularly true at idle in the glide, where the engine may have already iced up and the prop is merely windmilling. You'll have experience in the particular airplane and can set up your own procedure such as, "Always use full heat for descents at power settings below 1800 rpm . . ."

Warn the student that first time about the relatively strong nosing down tendency when the power is reduced to idle. Explain that it's a safety factor and that the plane wants to nose down to maintain a safe flying speed. (If you forgot to tell him this before he chopped the throttle, wait a couple of minutes, because he'll be trying to get his stomach back down where it belongs.)

One of the biggest problems, even with the advanced student, is the inability to use the trim tab properly. It seems superfluous now, but if he flies heavier airplanes it's bound to be a serious factor in his flying. However, first let him fly the airplane in the climbs, glides, and turns with the airplane *trimmed for the straight and level* cruising flight, but go into trim use as soon as feasible. He should get an idea of how power and airspeed affect the control pressures required. Later when you introduce slow flight, have him do it at first with no trim—then move on to the use of trim in this maneuver.

Demonstrate glides that are too slow and too fast. Show that power controls altitude by demonstrating the *power-off* glide at the *recommended glide speed,* then flying straight and level at the same speed using *some power,* and climbing at that airspeed using *full power.* Needless to say, the recommended glide speed may not be the exact speed for best rate of climb, but it will show the idea.

GLIDING TURN. Again, you'll combine two of the Four Fundamentals after each is understood by the student. All of this practice should be in power-off condition; you'll have time to let him practice partial power descents later. For the presolo student the two major types of descents will be (1) the power-off glide and (2) the reduction-of-power type of approach used in making landings. After solo, when he is working on the instrument flying requirements for the private certificate, he should get the power setting and airspeed recommended for a descent at 500 fpm—but not yet.

You can expect the gliding turns to be of the "slipping" type until he gets used to the lack of slipstream past the rudder.

■ **Straight and Level.** As many instructors have found, you may have trouble convincing the student that straight and level flying is one of the most important Fundamentals be-

cause when he is a pilot on his own, probably 90 percent of his flying time will be straight and level. You then, of course, prove that you are a liar—at least in the presolo phase—by making sure that over 90 percent of his flying is *not* straight and level. In the later stages of presolo work, on some of the lessons shooting takeoffs and landings, his experience of straight and level flying may consist of a few seconds on the downwind leg. He should have some breaks from the takeoff and landing routine or he will get stale. You can work on this Fundamental while flying to and from the practice area during the presolo phase.

The whole secret of straight and level, as you know, is to get power and trim set up and *to do as little as possible* after that. In smooth air, with a properly rigged airplane, you should not have to *do anything;* even in turbulent air, corrections should be held to a minimum. You'll find that with proper instruction the average student can do a pretty fair job of flying straight and level in smooth air, but his performance may suffer badly on the next flight when the air is rough. You'll have to convince him to "ride with the airplane" (you can't very well expect him to ride without it, at 4000 ft AGL). At any rate, have the student take hands *and* feet from the controls to show that his own manipulations have contributed to the problem and made the airplane jump around more. If you think you may have trouble proving your point, keep *your* feet on the rudders while ostentatiously keeping your hands off the wheel or stick. This may seem as though you are cheating, but it is used to illustrate the point of smaller or no control movements to recover from bumps. He'll feel better about it if he sees that it works and so won't fight the controls so much. He'll still use them as necessary but won't overcontrol or make corrections when they aren't needed.

About that rudder you "secretly" used to prove a point: In much of your instruction you may have to exaggerate both errors and certain techniques. For example when you show him a turn that was started with too much aileron, make sure that it is steep enough and slipping enough so that he gets the idea (without shaking him up). If you make a gentle, shallow turn with an imperceptible (to him) slip, you aren't showing the problem. Granted that with any experience at all he won't make such a poor (slipping) turn as the one you demonstrated, but you've shown him what you definitely don't want. This is cheating a little, but it's "good" cheating. Bad cheating is when you help him by staying on the controls all through the presolo period and/or always tell him when to make that final turn, etc. He won't know whether he has been flying, or can fly, the airplane or not. And think about it; when you've been doing this all along and see this student of yours up on that first solo—and you suddenly realize you've been flying the airplane for him—it's a very disconcerting discovery.

Even if you don't teach trim *at first* in the other three Fundamentals, you must teach it for straight and level flight from the beginning. Straight and level trim (at cruise) is the starting point of feeling the control pressures for speed changes. You went over this on the initial level-off of the flight but should review it here.

Stay with the student on his straight and level work. Expect him to hold his altitude within 50 ft of the assigned altitude *from the beginning.* It's more work for you, but too many instructors keep quiet and say to themselves, "Oh heck, it's only his first (or second, or third) flight; what's a couple of hundred feet?" You'll have to clamp down before the private flight test and he'll think that you've suddenly turned into a grinch. Why not be a grinch from the beginning? This doesn't mean that you'll use shouting or abuse, but if a student thinks you are tough and exacting because you want him to be a good, precise pilot, then wear the title proudly. Tighten your altitude requirements as he flies more.

Some instructors use a *phugoid oscillation* to demon-

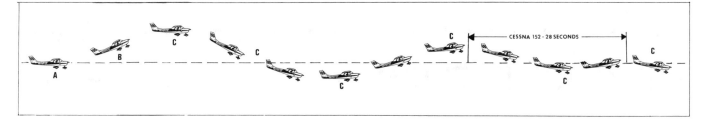

Fig. 4-7. A demonstration of an airplane's longitudinal dynamic stability (for Pete's sake, don't call it that or call it a phugoid oscillation, either).
(A) Establish trimmed cruising flight (85 K here).
(B) *Ease* the nose up to 15–20 K below cruise and release the control wheel.
(C) Allow the airplane to go through several oscillations to return level cruise flight. Use the rudder pedals to keep the wings level. The time from peak to peak of the oscillations is 28 sec for this airplane.

strate that the airplane is longitudinally stable both statically and dynamically. The "phugoid" is a long-period pitch-oscillation (about 30 sec from peak to peak for most current light trainers) with each movement through the level-flight pitch attitude smaller and smaller until the airplane is flying straight and level at cruise. This can build student confidence if done under the right circumstances.

For one trainer, which normally cruises at about 85 K IAS, the procedure (after checking for other aircraft) is to set the power, trim the airplane for cruise, and ease the nose up gently until the IAS is 65 K. You then release the back pressure, allowing the airplane to assume its own flight path, which consists of the nose gently easing up and down with decreasing airspeed variations until level flight is reassumed. You can count on 4 to 6 oscillations to get back to cruise.

Keep the wings level with the rudder during the process since you don't have a hand on the control wheel. You can stop the process anytime you choose by putting the nose back in the level-flight attitude.

A couple of pieces of advice, however:

First, if your trainer does not have a positively stable phugoid (and you certainly should check this out before demonstrating it to a student) don't give this demo. Trainers aren't required by the FARs to have a stable phugoid mode.

Second, if you think it might be counterproductive with a particular student by creating anxiety or queasiness, then don't do it. Fig. 4-7 shows a couple of the decreasing oscillations.

After he has demonstrated and has practiced all of the Four Fundamentals, give him exercises with definite requirements. Here's one possibility starting at 4000 AGL:
1. Straight and level cruise—1 minute—± 50 ft (trim)
2. Climb to 4500 ft—a straight ahead (trim)
3. Straight and level cruise—1 minute (trim)
4. Power-off glide straight ahead to 4000 ft (trim)
5. Straight and level cruise at 4000 ft (trim)
6. 90° climbing turns left and right to 5000 ft (trim)
7. Straight and level cruise of less than 1 minute (trim)
8. 90° gliding turns (power at idle) left and right to 4000 ft (trim)

See Fig. 5-6 for a slight variation of this. If he starts looking tired you'd better head back to the airport. It's quite possible that you would not be able to get all of the effects of controls plus the Four Fundamentals, *plus* the exercises just mentioned, in the first flight; but it's best to at least introduce the Four Fundamentals.

Give the student short breaks occasionally during this flight. Take over and let him relax while you show him landmarks and the practice area.

The ideal first flight would be to cover *effects of controls* and the *Four Fundamentals* with a comparatively fresh and

eager student who can be vectored and directed back down to the airport traffic pattern (he's doing it all). Under your direction he uses the landing checklist, applies carburetor heat on the downwind leg, reduces power (or you can handle the power and flaps for him), and makes the descents and turns. After he turns final *you'll take over* to land the airplane, unless you crave a great deal of excitement. He'll have a chance to use the practical aspects of the instruction he received at altitude.

After you've landed the airplane and everything is under control, let him taxi the airplane most of the way back to the hangar under your close supervision while you handle the radio as necessary. Have him do the shutdown using a checklist, and you can explain the idea of the postflight check. Make sure the airplane is secured—chocked, tied down, or hangared as dictated by wind or weather conditions.

Informally review the flight and get his reaction to the idea of flying. Fill out his logbook and give him the address of a nearby AME (aviation medical examiner). Set up an appointment for the next flight, issue him a *Pilot's Operating Handbook* for your trainer, and assign him reading material. The student should have his own FAR Parts 1, 61, and 91 and a copy of the *Practical Test Standards*, plus, of course, *The Student Pilot's Flight Manual*.

SUMMARY

Various problems, such as airsickness, may arise on this first flight. (If you *must* get airsick, let the student do the flying until you feel better.) Seriously, for some students this is a definite problem for the first, or first few, flights. It can be a sign of trouble ahead if he gets airsick without a "good" reason—a series of snap rolls or turbulent air. (The chapters on aerobatics go into detail on this problem.) Aside from the obvious problems, it's hard to judge how well the student will progress based on that first flight. Occasionally, however, you'll find one so outstanding (or so below average) in his reaction to flight that you'll have a fair idea of his flying future.

You should always, on this and future flights, encourage the student to ask questions during the pre- and postflight briefing. You should ask *him* questions to check on his progress.

Don't overwork him but try to cover, or at least introduce, all of the Four Fundamentals. This will allow him to make good use of the trip back to the airport and the traffic pattern. You can be sure he'll "casually" mention to his friends in talking about the first lesson that *he* flew it back to the field and *he* flew the traffic pattern and approach.

He is now a member of the group.

THE SECOND FLIGHT

PREFLIGHT DISCUSSION

The second preflight discussion should review the preflight check, taxiing, the takeoff, and the Four Fundamentals. The local traffic pattern should be reviewed and traffic patterns in general should be covered at the chalkboard before the flight.

■ **Traffic Pattern.** The student should be shown a chalkboard diagram of the traffic pattern with explanations of the various legs, plus traffic pattern exit and entry.

The following approach is suggested, using *what, why,* and *how:*

What — The traffic pattern (for a particular airport) is a standard procedure in use by airplanes for takeoffs and landings and for arrivals and departures.

Why — The pattern is used to establish an orderly flow of traffic, so that pilots are (generally) able to predict the actions of other airplanes taking off or landing at the airport.

How — Discuss a diagram of the traffic pattern, covering the various legs, altitudes, and entry and exit procedures to be used.

Emphasize alertness in the traffic pattern, because the airport is a central point of traffic. Too many pilots spend all of the pattern time looking toward the airport; this is especially true on the base leg. Look to the right on a left base leg (and vice versa); someone may be making a long straight-in approach.

■ **Taxiing.** Review any problems the student may have had taxiing the airplane during the first lesson. Briefly introduce the idea of control positions used in taxiing in various wind conditions, if you feel he's ready for it.

■ **The Four Fundamentals.** Note that this flight is basically a review and repeat of the first flight as far as the maneuvers are concerned, but also indicate to the student that he will feel more at ease, and the maneuvers and his orientation of the practice area will be much better.

You should review the Four Fundamentals as covered in the first flight, and if all of the Fundamentals could not be covered, discuss them in detail before *this* preflight session is ended. This is the time to clear up any misconceptions concerning turns, climbs, descents, and straight and level, including the need for trim (elevator *and* rudder, if both are available). Emphasize that beginning with this flight he will be expected to use trim with all changes in airspeed and attitude without your telling him.

■ **Rectangular Course.** Depending on the student's progress on the first flight, you might discuss the rectangular course briefly at the blackboard so that, if time (and his progress) permits, he may get an introduction to the maneuver during this flight.

Always work it so that if the student is making unexpectedly good progress you can move on to the next step of learning. It's discouraging to the above-average student for you to continue to have him practicing basics when he can see that he's ready for the next step. However, don't place too much emphasis on the rectangular course or other advanced maneuvers when you introduce them early. Indicate that there is an outside chance of doing the maneuver this lesson, but actually it is a maneuver usually done on the third or fourth flight. You will be following the idea as given in the first lesson during the climbout. If he doesn't get to do a rectangular course, well, that's usually reserved for later lessons; if he does get to practice the maneuver, he's that much better off as far as his morale — and progress — are concerned.

PREFLIGHT CHECK

The preflight check should be thorough (as always), and explanations given in the first lesson should be repeated as necessary. After this second experience, the student should be able to perform the check himself under your *direct* supervision. This means that you are making the walk-around with him, letting him do the actual check of the plane while you observe on the spot, not from the airport office. Some instructors figure they've done their part by going through the routine once and after that send the student out to "look the airplane over" while they stay in the office.

Following through with a student on a preflight can be very dull, particularly if it's the last flight of the day, but on the other hand, to depend completely on the inspection of an airplane by a low-time student could lead to an anything-but-dull flight if something is neglected because of his inexperience and your inattention (Fig. 5-1).

Here's an idea to consider: When (or if) things are in a temporary lull between flights, go out and wash a windshield, wash dust off a wing, or even wash the airplane. Cracks in the cowling, bad rivets, bad baffles (and baffle rubber flaps) can be found under leisure conditions, whereas they might be overlooked in a hurried preflight inspection.

■ **Points on Checking the Fuel.** At some time during the presolo training get a bottle or the clear container used to check for water in the fuel, and fill it with fuel. Go to a faucet and run a couple (or more) drops of water into the fuel and let the student see what the mixture of fluids looks like. You will find after this demonstration that he will check for water more carefully now that he knows what to look for. Too many pilots in their preflight check fill the container with fuel and quickly throw the contents away without really looking. Have the student hold the fuel up to the light so that water or trash can be seen more easily. If water or trash is found, the checking container should be carefully cleaned and the process repeated until no trace of foreign substance is seen. It's a good idea to check the fuel after every refueling; it's possible that there is water in the fuel storage tanks at the airport.

As you well know, 100LL fuel is blue (well, *light* blue), and if you hold a container of water up against the sky, it will be blue. Hold the fuel checker up against the sky to check for water or other contaminants, but then hold it against the side of the white portion of the fuselage to make sure it *is* blue (if you airplane is to use 100LL).

By draining fuel where it can't be checked, you are taking a chance that water or other debris may still be in the system. In some extreme cases you might have to drain several gallons of fuel before getting clean fuel. If you didn't know this, you may feel satisfied with the quick draining (but no check) of a small amount of fuel.

The manufacturers, in their race to make things more convenient for the pilot, have come up with such ideas as draining fuel from the cockpit (the clear plastic tube through which the pilot should be able to see the fuel—and water—is not always in the best light). If you have a real friend, he can lie on his stomach under the plane and observe (or catch) the flow as it comes out.

Another type allows the pilot to reach into the cowling and pull a handle, draining the main sump. The stream is usually so directed that it's sometimes possible to filter the fuel/water through your left trouser leg and observe the result on top of your left shoe. In order to really check both of these types mentioned, they should either be drained on a clean area where the fuel may be examined (if you are alone) or caught by another person in a clear container, since an impossibly long reach may be necessary to "pull and check."

It's a good idea to establish the habit of draining the wing tank drains *first* and *then* the main fuel strainer, particularly in high-wings. Doing the reverse procedure might pull water from the wing tanks down into the lines between tanks and strainer, and it would be there at a bad time (such as takeoff). Check your trainer's system and manufacturer's recommendations.

Encourage your students to open all cowlings possible when they run the preflight check. Just reaching into the oil flap to check the oil doesn't do it as far as finding other problems is concerned. Granted, some of the planes today would require two men working with screwdrivers for 30 min to get the cowling open, but there are too many pilots of airplanes with easy-opening cowlings who haven't been instructed to take advantage of it. Any student (or private or commercial) pilot who had an accident could come back with, "My instructor never told me to open the cowling, so I didn't. How was I to know that there was a bad oil leak pending?"

In cold weather, make sure that the student checks the breather line for ice that may have formed from engine water vapor during the previous flight(s). If the breather line is sealed by ice, the pressures in the crankcase could cause damage to the engine.

STARTING

This lesson the student should start the airplane (under your supervision) using the proper checklist.

Too many pilots start the airplane without checking about who or what might be in the vicinity of the propeller. A shout of "Clear!" or "Clear prop!" could save the life of a child or an adult who may be bent down in front where you can't see them. Allow some time after hollering for people to get out of the way before starting.

One of the bad habits students get into is to start the engine without having a hand on the throttle. In the usual two-place, side-by-side trainer the starter is on the left and the throttle is controlled by the right hand (from the left seat). Since most people are right-handed, the tendency is for the student to use the starter with his right while his left hand lies

Fig. 5-1. Here is Joshua Filstrap, flight instructor, supervising the preflight inspection of his second-hour student. Filstrap is continuing with his long-time experiments on the effects of 1 g on the human when acting transversely to the long axis of that body.

idly in his lap, and the throttle (which invariably is opened more than it should be) is unattended. Of course, you're there to pull the throttle back when the engine blasts up, but if he doesn't get out of this bad habit before solo, he might give himself—and various spectators—a thrill as the engine starts.

Some trainers have vernier throttles, and you should be sure the student has an understanding of not only how the throttle works—push for more power, etc.—but also how to use the vernier release to pull the throttle back in a hurry.

It's best to have the student hold the brakes for starting rather than depend on the parking brake. The parking brake is not always the most dependable aid, and it's better for the student to use it only as a temporary brake when the engine is not running. (Chocks or tie-downs are better for parking for more than 5 min.)

You should be in the airplane for all presolo student engine starts. If you tell the student to go ahead and start the airplane—you'll be out in a minute—and he roars into another airplane, you're both in trouble. He hasn't been endorsed for solo operation of the airplane.

TAXIING

You shouldn't allow the student to repeat maneuvering, as shown back in Fig. 4-2, but if he did well on the first flight you should let him take more of the taxi responsibilities. Your requirements should be tighter and you should correct his smaller errors that were not mentioned in the earlier lesson. One of the major taxi problems the new student has is trying to turn the airplane from a stationary condition. He will usually try to start the turn at the *same time* power is applied to start rolling. This tends to "cock" the nosewheel and no progress is made. You may have to demonstrate the proper technique a couple of times.

His coordination of throttle, rudder pedals, and brakes should start falling into place this second flight, and if he is having unusual difficulty in taxiing, it might be good to spend some extra time concentrating in this area. You may find that with some students special taxi sessions are necessary.

It's best at this stage to allow the student to taxi with as few distractions as possible. He should be able to keep his gaze outside the cockpit, as should *always* be the case in taxiing; and, again, unless the wind is strong or he is extra sharp, it's best to leave any involved discussion of flight control position until later flights.

One tendency for the student, particularly one with a great deal of driving experience, is to come up to the place of stopping at too great a speed (it's fine for a car, but not for an airplane). The airplane's brakes aren't designed for this type of use, and there's always the problem of the wind increasing any nosing-over tendency.

You should again watch for a tendency to ride the brakes. This is one of the most common errors in taxiing—too much power and too much brakes.

One thing lacking in taxi instruction these days shows up when the average private pilot is being directed to a parking spot or a gas pump—he doesn't know how to follow the taxi director's signals, and what follows can be either humorous or dangerous. Too many pilots come roaring up to the gas pit totally ignoring the poor guy who's trying to show them where to turn, who's rapidly scuttling backward to avoid having the prop cut all the buttons off his mackinaw. Usually the line serviceman just gives up and retreats to the office; after the pilot has shut down the engine, he will venture out to push the airplane to the proper spot. Make sure that *your* students know what taxi signals mean (Fig. 5-2).

You might mention that on sunny days when taxiing a high-wing trainer near another high-wing trainer, watching the wing tip shadows of the two gives a good indication of the amount of clearance. It's more accurate than trying to look out along the wing and judge the gap. Of course, for cloudy days, early and late day sun positions, and low-wing trainers, this doesn't work.

PRETAKEOFF CHECK

The checklist should always be used for the pretakeoff check. You can use it as a reference as you point out each item to be checked. Let him check the magneto and carburetor heat.

■ **The Magneto Check.** You may have a preference as to which magneto should be checked first, but the student should be taught to go back to the BOTH position after checking each magneto to get a double check on the drop (and pickup). Mention to the student that if he should inadvertently turn the switch OFF, to avoid a backfire, the throttle should be closed before turning it back on. Students sometimes turn off the switch during the mag check—and try to turn it back on immediately, *before* you can get the throttle closed (with a resulting backfire). Tell them that there will be plenty of time to close the throttle and turn the switch on again before the engine stops.

■ **Carburetor Heat.** The carburetor heat check should not be just a matter of pulling the control to see if a drop in rpm occurs, but the heat should be left on at least 10 sec (or more, depending on the airplane) to see if ice is present in the carburetor. If after the initial drop the rpm picks up, with the heat still ON, there was ice in the system and you and the student are forewarned.

One possible problem is that if the carb heat control is hidden from your view by the control wheel, you may find that you will tend to forget to push it back in after this more extended test. Some instructors continue with the rest of the pretakeoff check leaving the carburetor heat ON, but this would increase any tendency to inadvertently leave it that way. Also, the air going to the engine is unfiltered when carb heat is on. In a situation of bad carburetor icing you should leave the heat ON until the takeoff roll is started.

Pull the throttle back to idle with the carburetor heat ON (if the heat is to be used throughout the approach and landing). If the engine doesn't want to idle properly (or quits), now is the time to find out, rather than on final when you suddenly need power and the engine has quit (you've been fooled by the windmilling prop to believe that the engine is still in action). If your *POH* doesn't recommend carb heat on approach and landing, check the idle with the heat OFF.

Complete the pretakeoff check using the recommended checklist.

Check such items as the suction gauge and other systems of the airplane *even though these may be of greater interest to the more advanced pilot.* This is not to say that the marker beacon and other IFR items should be checked for a primary flight, but *systems*—electrical and vacuum—should be introduced from the beginning.

After the check is completed, mention that the student should take a good look at the traffic pattern, particularly the base and final legs. At an airport with a Unicom the common tendency is to expect *all* airplanes to call in when approaching the field, but you know that this doesn't always happen since some airplanes don't have radios and some pilots with radios don't bother to call in. It's even a good move to instill in the student the idea to check before pulling out on the runway even when cleared for takeoff by the *tower*. Everybody has complete faith in the tower, but people up there can also goof,

SIGNALMAN DIRECTS TOWING

SIGNALMAN'S POSITION

FLAGMAN DIRECTS PILOT TO SIGNALMAN IF TRAFFIC CONDITIONS REQUIRE

ALL CLEAR (O.K.)

START ENGINE

POINT TO ENGINE TO BE STARTED

STOP

LEFT TURN

PULL CHOCKS

COME AHEAD

RIGHT TURN

INSERT CHOCKS

EMERGENCY STOP

NIGHT OPERATION

SLOW DOWN

CUT ENGINES

Fig. 5-2. Taxi signals.

although it's not as likely to result in a problem as in the no-communications or Unicom situation.

THE TAKEOFF

Unless crosswind conditions are bad, definitely let him steer the airplane during the takeoff roll, while you take care of ailerons, elevators, and throttle this time. Naturally he will have been briefed on the left-turning tendency on the roll (you'll still have to tell him "right rudder" from time to time throughout the roll). The average student—the vast majority of students, in fact—tends to underrate the amount of right rudder needed to keep the airplane straight on takeoff that first, and maybe second, time. Some may comment after the takeoff that it was "hard to use enough right rudder to get it straightened back out again—it felt funny." Your initial personal reaction might be to say that it wouldn't feel nearly so

funny as going across that ditch (fence, etc.) off the left side of the runway, but you should merely state that he is to use whatever rudder is necessary to keep the airplane moving straight.

With students who have a particularly hard problem with this, perhaps a couple of takeoff runs (without lift-off) up or down a deserted runway at the end of the period might help. You will handle the throttle and other controls, letting him keep the airplane straight with the rudders.

To avoid mechanical "walking" of the rudder on takeoff, this exercise should be done with you varying the power throughout the roll. This will teach him to use rudder as necessary with the change of power effects. This may be combined with lifting the tail (tailwheel type) or raising the nosewheel to get experience in these aspects of the takeoff roll. However, most students will need very little extra practice in keeping the airplane straight, particularly with tricycle-gear types. More about this in Chap. 10.

THE CLIMBOUT

Here is the time to start pinning down good habit patterns. What you teach here concerning the climb can stick with him the rest of his flying career.

It should be emphasized that he should *look around* as he climbs (or anytime). In the climb the nose is higher than the path of the airplane, compared with straight and level flight, and will obscure more of the area ahead. Shallow climbing turns should be made as the airplane climbs to the altitude and area of practice. Explain that the turns should be shallow so that the climb rate won't suffer. These clearing turns and looking around are necessary because, while the sky is not as crowded as indicated by the newspapers and television, it only takes one other plane to make a midair collision, and a midair will spoil your day.

This second flight is the flight in which you will require definite performance limitations from the student in all areas.

After the specific prechosen practice altitude is reached, the proper level-off procedures will be used.

Remind him again that the following three items (Fig. 5-3) are to be completed for the *level-off*:

1. Ease the nose over to the proper level-flight attitude.

2. Reduce the power to cruise setting as the airspeed reaches expected cruise. (Trim.)

3. As the speed increases ease off the right rudder or change rudder trim, as applicable.

Note that in entering a climb, the items mentioned above are normally in the same order (nose attitude, power, rudder, trim). *To enter the climb:*

1. Ease the nose up to the proper attitude.

2. Add power as the airspeed approaches the climb value. (Trim.)

3. As the speed decreases be prepared to use right rudder.

In the climb entry, steps 2 and 3 will be nearly simultaneous or may on some airplanes be reversed (Fig. 5-4).

Discuss any deviations from climb airspeed, poor correction for "torque," neglect of proper clearing turns, or proper bank. As far as you (and the student) are concerned, there is to be *no* deviation from the proper climb airspeed (75 or 80, etc.). Some *Pilot's Operating Handbook*s say the climb speed is "70–80 K," but you should hold the student to a specific airspeed; you can later explain the need for changing the best rate of climb speed with altitude, but not yet. You will have to remind him that he should get the proper nose attitude and *then* check the airspeed. The tendency is to chase the airspeed—usually moving the nose the wrong way at that—on the first couple of lessons, and if you aren't firm about establishing good habits in this regard, it could give him problems all through his flying career.

You might note the "rule of threes" in flight instruction. Nearly every maneuver can be initiated or completed in three steps, as was just done for the climb entry and level-off. Setting up your own rules of threes for maneuvers can help you in both ground and in-flight discussions.

STRAIGHT AND LEVEL

Two or 3 min of straight and level with a 180° level turn, followed by another couple of minutes of straight and level, should allow him to get another look at the practice area and

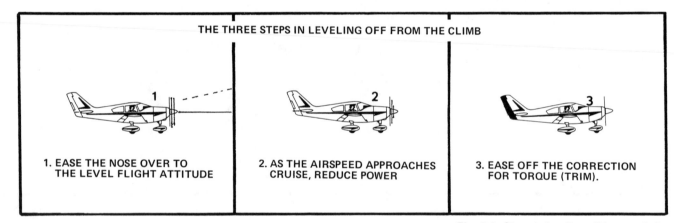

THE THREE STEPS IN LEVELING OFF FROM THE CLIMB

1. EASE THE NOSE OVER TO THE LEVEL FLIGHT ATTITUDE

2. AS THE AIRSPEED APPROACHES CRUISE, REDUCE POWER

3. EASE OFF THE CORRECTION FOR TORQUE (TRIM).

Fig. 5-3. Steps in leveling off from a climb.

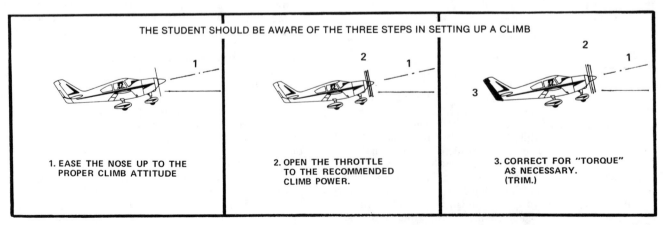

THE STUDENT SHOULD BE AWARE OF THE THREE STEPS IN SETTING UP A CLIMB

1. EASE THE NOSE UP TO THE PROPER CLIMB ATTITUDE

2. OPEN THE THROTTLE TO THE RECOMMENDED CLIMB POWER.

3. CORRECT FOR "TORQUE" AS NECESSARY. (TRIM.)

Fig. 5-4. Steps in setting up a climb.

be ready to get on with the exercises. The student should do all flying at first using outside references, except for checking the altimeter. This book does *not* recommend the integrated method of instruction for the student the first few flights, because in too many cases the student establishes the habit of staring at the attitude indicator when he should be checking the airplane's attitude by the real horizon (and looking for other airplanes). (Student pilots tend to become fascinated by the attitude indicator.) This is not to question the value of VR/IR training (and showing the similarities in controlling the airplane)—but not yet. A good time to start the concept is shortly after solo, and certainly before the solo cross-country.

TURN EXERCISES

The second flight should make use of the idea of making turns to definite headings, or of definite changes in headings. Of course, if the student needs work on the mechanics of entering and recovering from turns or altitude control during turns, you'll want to work this out before setting up the more rigid roll-out requirements.

Use medium banks of 20–30° for ironing out problems of the turn, showing the student the bank angle by reference to the natural horizon, *not* the attitude indicator. Some private and commercial pilots still can't judge within 10° of a particular bank without using the attitude indicator.

The first "precise" turns should also be of medium bank, and turns of 90°, 180°, and 360° should be made, maintaining a constant altitude. Fig. 5-5 shows another exercise for possible use during this period of instruction. (He should be able to satisfactorily perform the medium banks before this one is attempted.)

Talk to the student during the process. You should have already briefed him on the exercise.

■ **Overbanking.** "Overbanking tendency" should not be mentioned to the student when he is first working on turns. The zealous student (if made aware it exists) will usually correct for it even when it is no problem, and mentioning it introduces a factor that may distract him from the turn as a whole. If he is told about it he may be correcting so much for overbanking that the bank is too shallow or the point of roll-out is forgotten. This is another item for the timid soul to cope with and he may have visions of the airplane rolling over on its back—with him inside.

Strangely enough the average student, not wanting to bank more steeply than required, will take care of the problem automatically. If an overbanking tendency is a noticeable problem in the type of trainer you are using, mention it only casually as a factor in the turn.

The best method during this second flight is to offer ad-

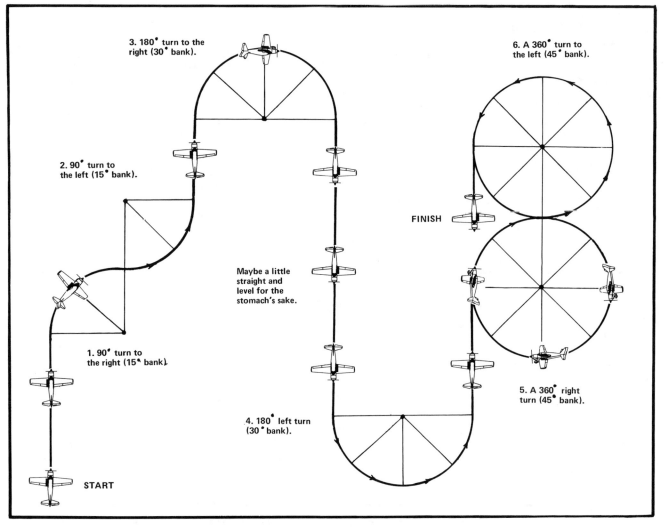

Fig. 5-5. An exercise for setting up predetermined banks and degrees of turn. The student may have a little trouble with the 45° banks at first, but this will be a good introduction to the 720° power turns.

vice such as, "Shallow out your bank slightly, keep it at 30° . . ."

In any maneuver, don't introduce details so early that they could confuse the student.

■ **Altitude Problems.** You can pretty well predict that the first 45°-banked turns will bring in some altitude problems. The student won't use enough back pressure at the beginning of the first turn or two and then won't relax enough back pressure on the roll-out, so that the 360° turn is (1) a diving turn, followed by (2) possibly a level turn, followed by (3) a climbing roll-out. Items 1 and 3 may cancel each other out and the airplane will end up at the same altitude it started. Needless to say, this should be corrected so that a *constant* altitude is maintained throughout.

If you use the exercise in Fig. 5-5, you should first have the student only work with the 15° and 30° bank (90° and 180° turns) and have a brief period of straight and level flying between each turn. You can then work up to the full exercise with no hesitation between turns.

CLIMBS, DESCENTS, AND TURNS

Another exercise useful during the second (or following) flights could be that as shown in Fig. 5-6.

The point of the exercise is to use all four of the Four Fundamentals and iron out any problems on transitions to and from climbs or descents.

You should not have to demonstrate the exercise since it is a combination of things the student has done before.

■ **Climb Exercises.** The student should concentrate on keeping a straight course during the climb, using outside references with occasional glances at the ball and heading indicator. Watch for an unconscious tendency to hold a wing low to maintain a straight course rather than hold the proper correction for "torque." The transition to and from climb, and the climb itself, are integral parts of the exercise; don't think in terms of "getting on with it" to the loss of effective instruction.

The problems you should look for in the climb were covered in Chap. 4.

If your student is having a particular problem with any part of the climb, you may want especially to have him practice this several times before moving on to the rest of the exercise.

■ **Introducing the Stall.** Depending on how it's going, you may want to demonstrate the effects of climbing at an airspeed above and below that recommended and give the student his first introduction to the stall. *You* should be the one to do the flying if a stall is to be approached; make sure that it is a "gentle" entry and recovery.

While discussing the effects of too slow an airspeed on the climb, you might demonstrate that the slower the airspeed the more the climb is affected—until you get the buffet and a gentle break. You can note, "That was a stall and by relaxing the back pressure the airplane is back flying normally *as it always wants to fly.*" A casual approach here could save a lot of trouble later when stalls are to be practiced. Too many people flying airplanes today are afraid of the common everyday 1-g stall, and it often goes back to the instructor who taught them. In fact, some aerobatic instructors have noted that their students can get acclimated to loops and rolls but are *afraid of the 1-g stall* until later in the course.

It will be your job on the ground to clamp down on other students' descriptions of "how rough stalls are." Your attitude with regard to stalls will be very important in how they are accepted by the student.

If your required recovery from a stall is violent, with an abrupt forward motion of the wheel or stick and a resulting violent pitch down on the nose, then either you aren't teaching stalls correctly or you are using the wrong airplane for primary training.

The chances are good that after the easy introduction to the stall the student will breathe a sigh of relief. Don't *you* be a contributor to the fallacy that stalls should be a stomach-wrenching maneuver—too many potentially good pilots have quit flying because of that. Later, you can cover the idea that stalls can be dangerous at low altitudes, but wait until he's seen that *he* can recover from them.

If you use this approach to the introduction of stalls, be sure to discuss the stall briefly after the flight, giving the general theory and covering the recovery techniques. *Note that stalls may be done power-on or power-off.*

As far as the normal climb is concerned, the proper transition techniques should be emphasized as necessary at each entry and level-off.

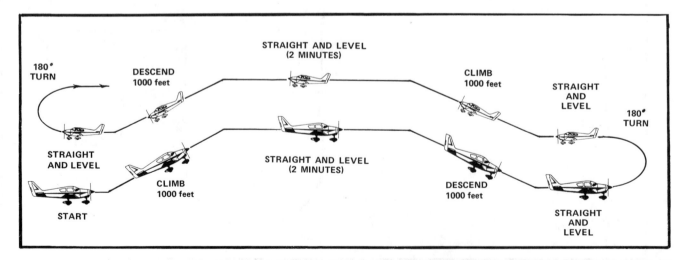

Fig. 5-6. An exercise using all the Four Fundamentals. The first climbs and descents should be for 1000 ft; if you decide that the student needs work on transitions or trim, you can cut them to 500 ft to get more action.

■ **Straight and Level Part of the Exercise.** The straight and level portion of an exercise should now be given full weight and not just be used as a breather for the student. Trim should be used for all parts of the exercise. The period of time for each straight and level part should be at least 2 min to allow the student to work out any trim problems. Be firm on altitude and heading requirements and have him correct for any variations.

The student is likely to gain altitude if he is slightly tense. (He is subconsciously pulling away from the ground.) Use one of the straight and level periods to demonstrate the inherent stability of the airplane. A couple of minutes of hands-off flying will raise his confidence. (One instructor issues his students a stick of chewing gum before the flight and uses the "chew-rate" as a tension indicator—the faster the rate, the greater the tension.)

As indicated in Chap. 4, more than 90 percent of the average private pilot's flight time is straight and level, so cover the proper techniques and require precision in straight and level.

■ **Descents.** These descents should be made at idle, using carburetor heat as suggested by the manufacturer. You'll discover on level-offs that the average student is timid with the throttle. Bring out the fact that about 50 ft is needed as a lead in leveling off the average light trainer or, as a general rule of thumb, lead about 10 percent of the rate of descent.

Fine, you tell the student this and he nods his understanding—and proceeds to descend right on through the altitude (Fig. 5-7).

Stress that he should get the power on without hesitation but without ramming the throttle open, because there may be times when definite power application is needed to stop a rate of descent—*now*.

If you're using, for example, 2300 rpm for straight and level flight during the presolo practice period, the student may spend a lot of time trying to get *exactly* 2300 rpm while the airplane is roaming all over the sky. Tell the student to firmly increase the power to the *general* amount required to stop the rate of descent and then make a more precise setting as necessary.

Again, you should note that the rudder is suffering from lack of slipstream, compared with the climb or cruise, so that a definite effort at coordination must be made when first practicing descents.

If the student's progress permits, during this flight you should set up an exercise using different flap settings in a constant-airspeed, power-off descent. Even if you find that the second flight is too soon to introduce the following exercise, you should include it before leaving the high work for the wind drift correction maneuvers:

1. Using the normal approach speed for your trainer (for instance, 70 K), have the student set up a wings-level, clean, power-off descent. Have plenty of altitude and do this in smooth air if possible. Indicate that he is to maintain a constant heading throughout the exercise.

2. Explain that you will add one notch of flaps—or about one-third flaps, if they are electrical—and that he will have to lower the nose to maintain 70 K.

3. After he has stabilized at the new attitude—and at 70 K—add another notch (or another one-third), and have him readjust the pitch to maintain a constant airspeed. After the airplane has stabilized again, increase the flaps to the full setting, and have him adjust the pitch to maintain the 70 K.

4. Reverse the procedure, upping the flaps in increments and requiring him to maintain a constant 70 K by raising the nose with each change. (Tell him before you change the setting each time.) Climb back up to the original altitude and repeat as you feel necessary.

5. *After he is proficient* at the exercise you may want to have him set up a descent, and you put in a flap setting without telling him beforehand the amount you will use. He will

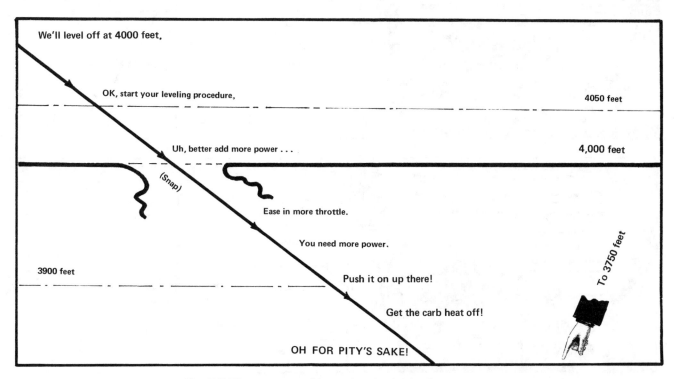

We'll level off at 4000 feet.

OK, start your leveling procedure.   4050 feet

Uh, better add more power . . .   4,000 feet

(Snap)

Ease in more throttle.

You need more power.

3900 feet

Push it on up there!   To 3750 feet

Get the carb heat off!

OH FOR PITY'S SAKE!

Fig. 5-7. The average student tends to be timid at first in applying power for leveling off. Note that no bad language was used.

enjoy the challenge of changing attitude to maintain the correct airspeed. If he isn't ready for the "unwarned" flap changing, he'll probably hit a plateau of learning.

Point out the attitudes required to maintain 70 K at the various flap settings. If he is progressing well with the exercise steps 1 through 4 you might have him set up the attitudes with the airspeed covered. After he has established what he considers the proper attitude for each flap setting, uncover the airspeed to see how he's doing. He will take a great interest in how he's doing, and you are also teaching him the idea of attitude as a means of airspeed control. As a next step you can have him make the different flap settings and fly the airplane. Fig. 5-8 shows the idea of flap setting and the attitude required to maintain a particular airspeed in a power-off descent.

Following are some of the most *common errors* made during descents:

1. Directional problems. He may be concentrating so heavily on the attitude and airspeed that bank or yaw complications arise.

2. Too low an airspeed, or airspeed decay each time that additional flaps are set. The common tendency for the student is to be slow in lowering the nose.

3. Allowing the nose to rise after the correct pitch attitude was set up. (You'll find that the major problem will be a too-slow airspeed.)

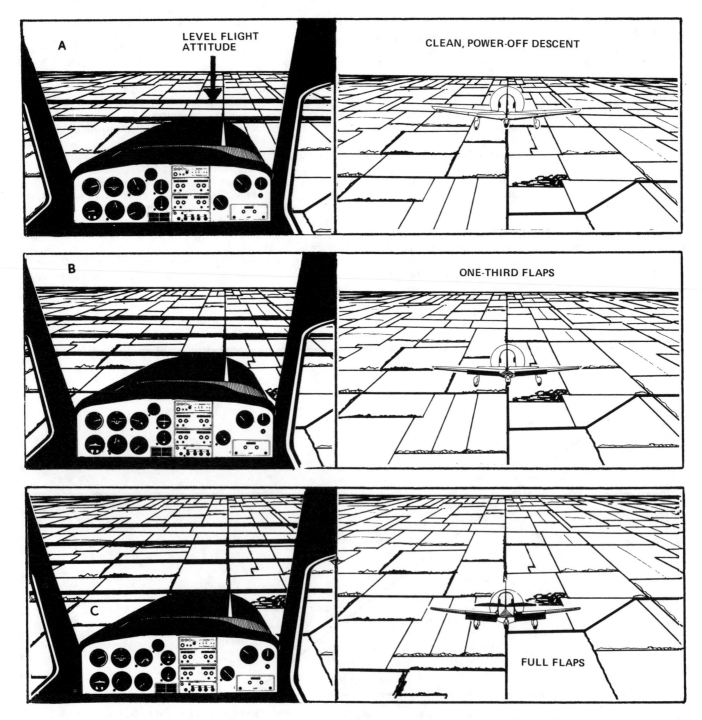

Fig. 5-8. Attitude changes with flap settings (at constant airspeed).

4. Keeping the nose *too low* as flaps are retracted, allowing the airspeed to pick up.

5. Fixation on the airspeed indicator.

6. In the cases where he is setting the flaps himself, neglecting to keep up with the airplane—having pitch and heading problems.

As a separate exercise, you could have him practice a simulated 180° side approach including use of flaps, reducing power continually until the throttle is at idle after the second 90° turn. It would be well for you to at least demonstrate this at altitude before returning to the airport, since he should do most of the pattern and approach at the end of this flight.

■ **Turns.** He may not be ready for the 45°-banked turns, so these may be left for the third lesson. Don't do a great number of turns now without some sort of break; let him practice climbs and descents and straight and level. By the end of this flight he should be able to turn the airplane consistently to within 10° of a preselected heading and in the shallower turns (30° bank or less) should be able to maintain his altitude within ±50 ft (with an occasional sashay 100 ft off). *You will be shooting for a zero error in heading and altitude.* The mentioned limits are maximum limits for smooth air.

His roll-ins and roll-outs should be coordinated with no excessive slipping or skidding.

Don't tire him out with too many of the exercises; work it so that he is still interested when the flight is over.

RETURN TO THE AIRPORT

Under your direction the student should return to the airport and descend to traffic pattern altitude. As you fly back, review the traffic pattern and describe the route he must fly in order to enter traffic for the particular runway you will use. Emphasize that he should be down to the pattern altitude before starting the 45° entry leg. Letting down in the traffic pattern can be a good way to have a midair collision.

He should use the Unicom, if available, far enough out so that he can start planning for the pattern entry. Remind him that Unicom is an *advisory* service and cannot "clear" him to land. There may be airplanes in the area without radios or airplanes whose pilots don't have them turned on, so he'll always have to keep a sharp lookout.

Complete as many of the checklist items as feasible before starting the entry leg. Items such as fuel on best tank, mixture rich, electric-driven pump on, and brakes checked should be completed early so that attention may be directed outside to look for other airplanes—and check the wind and traffic on the runway. You may want to revise your trainer's checklist to get the following items first:

1. *Fuel*—If you find yourself on the downwind leg and still on a tank that is one-quarter full (and you know this is correct) while the other has one-half, don't switch tanks even though the usual recommendation is to use the fuller tank for landing. You might switch to the full tank and discover that it isn't working when power is needed on final. Take a look at the fuel situation while you are returning to the field (but still far enough out to give the "new" tank a chance to prove itself) and switch as necessary. Some trainers have only two fuel selections, OFF and ON; so there's no problem.

2. *Brakes*—When an airplane lands and the pilot suddenly discovers as he is running out of runway that he has no brakes or only one brake, it can be extremely interesting, to say the least. Teach your students to check the brake pedal pressure before landing; this doesn't always guarantee good braking, but if a pedal (or pedals) goes to the "floorboard" he has some warning and can plan for a longer runway, or can go

to another airport where longer runways and better maintenance are available. Teach him how to pump the brake pedals to build up braking effectiveness if needed.

This follows the concept mentioned earlier in the book that a good pilot keeps surprises to a minimum. A careful preflight check might show up an in-flight problem still several flying hours away. You should continually remind your students that a good flight is the result of thorough preflight planning and, in the case of a brake problem, a successful and safe landing can be assured by his actions while he is still several miles from the airport.

■ **Traffic Pattern.** The student should make the entry (you will tell him when to turn parallel to the runway on the downwind leg). Have him use a prechosen power setting on the downwind leg. Some instructors use 100–200 rpm less on the entry and downwind leg than is used out in the practice area; this keeps the speed down and gives more time to get set up for the landing. This practice may be useful later when the student is flying faster aircraft where it is necessary to slow the airplane below the gear and/or flap extension speed. The airspeed should be well under control during the entry and pattern itself.

On the downwind leg the carburetor heat, if required, should be ON long enough to assure that heat is being developed to clear out any ice in the system before the power is reduced. You should teach the "gradual reduction of power" approach from the beginning. A power setting of about 1600 rpm is good as an initial figure for the beginning of the approach for most of the trainers in current use. The power should be continually reduced to idle on final (Fig. 5-9).

■ **Flaps Again.** Using the flaps (or more accurately, *when* to start the student using the flaps) is the cause of some controversy among instructors. Some do not teach the use of flaps until some time after the student has soloed; these advocates are usually old-timers who did a great deal of instructing before trainers had flaps and haven't gotten into the use of them themselves.

If your trainer has flaps, *you should teach your students the use of them from the beginning.* Flaps are part of the airplane's control system. Nearly all trainers have them, all of the advanced airplanes have them, and their use should be standard operation to the student.

If the student is taught to not use flaps during the presolo period, two things will probably occur:

1. Since the airplane will have different landing attitudes, flaps-down or no-flaps, he will probably have problems landing when the use of flaps is introduced at a later date and his progress toward the private certificate could be noticeably retarded.

2. Because of these possible problems, he may go back to not using flaps during his solo periods and might require extra time getting used to them for the practical test (and he *will* be expected to use flaps on approaches and landings on the practical test, although a no-flaps approach could be requested).

As you get more experience in the right seat you'll see how important are the habits established in the first few hours of instruction. You may remember that some of your own current attitudes were established well before you soloed.

It's been said that the student has to learn 1000 new items when he starts to fly; okay, make it 1001 and teach him to use flaps from the beginning.

Some instructors advocate the use of full flaps on every approach for every airplane, but local conditions may not make this practicable. You may have to use the amount of flaps suitable for the particular condition. For instance, the Cessna 150, which is a widely used trainer, can have a flap

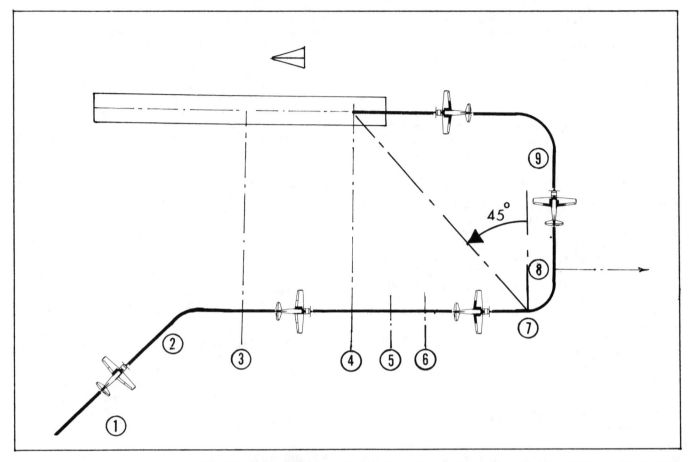

Fig. 5-9. (1) The student should have the fuel tank, fuel pump, and mixture items on the checklist completed before turning on the entry leg. (2) The turn to parallel the runway is made. (You'll have to coach him on this the first few times.) (3) The carb heat is checked, or left ON, as recommended by the manufacturer. (4) Power is reduced opposite the point of intended landing. (5) The nose is kept level to aid in slowing to flap speed. (6) Flaps are set. (7) The turn to base is made. The turn should be started when the selected point of touchdown is 45° behind the wing. (8) A check is made for possible straight-in traffic. (9) At the proper point the turn to final is started.

deflection of up to 40°. The stall speed CAS (no-flaps, max certificated weight) is 55 mph; 20° of flaps lowers this 6 mph to 49 mph. The last 20° (to 40°) only lowers the stall speed 1 mph (to 48 mph), but may not give good go-around characteristics under certain conditions. These days, the primary purpose of flaps is to decrease the landing speed (the rate of descent is steeper too, but with the approaches-to-airport obstruction requirements this isn't as important as it once was, except for short-field obstacle clearance). A suggested "normal" approach flap setting of 20° could be used for this trainer. Other settings could be used for special procedures, such as shown in Fig. 5-10:

1. *Normal landing* — 20° flaps, landing at 49 mph; easy go-around.

2. *Short- or soft-field landing* — 40° flaps, landing at 48 mph, taking advantage of the higher drag and 1 mph lower touchdown speed for a special situation. Higher drag means a steeper approach and more rapid slow down on the landing roll. A go-around is expected to be a little more complicated.

3. *Gusty air or crosswind* — 20° *or less,* to ensure that gusts don't cause problems of deceleration with a following slow acceleration. In *gusty* air the airplane's speed at touchdown may have to be great enough to assure aileron and rudder control to recover from gust effects.

You should take a good look at the airplane you'll be using and decide on the best combination for the airplane,

local conditions, and the student's training process (and it could be that you'll advocate the use of full flaps for *all* approaches).

Notice in the example of the Cessna 150 that the student has an approach setting of 20°, which is a good all-around setting for normal situations, but he can be taught variations (and what flaps are all about) for specific purposes. He will

=Power Off= **STALLING SPEEDS** MPH ═ CAS				
	ANGLE OF BANK			
┌Gross Weight┐ └─ 1600 lbs. ─┘ **CONDITION**	**0°**	**20°**	**40°**	**60°**
Flaps UP	55	57	63	78
Flaps 20°	49	51	56	70
Flaps 40°	48	49	54	67

Fig. 5-10. Flap effects on the stall speed in straight and turning flight. (*Cessna Aircraft Company*)

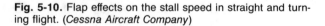

understand better the function of flaps on airplanes he flies after he gets the private certificate.

A suggested method of flap use for the C-150 during the presolo period is to throttle back; as speed permits, set the final landing flaps *with one move;* and, as the approach speed is reached, trim the airplane. There is no need for the student to set 5°, then 10°, 15°, and finally 20°, which would require continual pitch changes and diversion of his attention from the runway and approach. Use flaps, but don't force the student to make a production of it on landing approaches. Explain, however, that more complex airplanes may require setting flaps in increments and *other* trainers may require a couple of steps to get to the final setting.

The example used is the C-150, but the same approach could be made with the C-152, Tomahawk, Skipper, and other trainers (using knots instead of miles).

THE APPROACH

It is suggested, then, that the student be taught a one- or two-step-flap-setting, single-airspeed approach. There is no advantage to be gained by paving the approach start at, say, 80, then 75 K on base, and 70 on final. Why not start at 70 K, if that is the recommended speed "over the fence." Too many times the student starts at 80 and crosses the threshold at 80 because he hasn't had time to get organized. Perhaps several seconds can be saved during an approach if a faster initial approach speed is used, but the student needs extra time during the earlier stages of his landing training.

It's a little early to worry about his traffic pattern on the second flight, but there is one problem nearly every student (99.9%) has in turning onto the base leg—tending to make that first turn too shallow. In a high-wing airplane the runway is hidden from view, and when the airplane is rolled out of that first 90° of turn, the student may discover that he's crossing the center line of the runway (Fig. 5-11).

The first bank should be 20–30° (no more than 30°). This is not critical because the airplane has altitude and (usually) more speed. Better to have a moderately steep bank up there at the first 90° turn than to have a moderately (or more) steep bank turning final.

Let him fly the airplane down until just before the round-out is begun (and tell him well beforehand that you'll take

over and land it). You'll probably have to remind him to throttle back and when to turn, but he'll be doing more of the approach than he did on the first flight.

LANDING

You should explain that the landing is nothing more than a stall, the power-off version of what you demonstrated with power in the too-steep climb. This should help dispel fears about the stall (i.e., it's a normal thing, done at least once every flight). Don't drop it in, or he'll think the stall *is* a very violent maneuver.

You keep control of the airplane until it has slowed to taxi speed. Tell him the landing roll requires the full attention of the pilot; opening windows, upping flaps, pushing off carb heat, opening cowl flaps, and *retracting gear* (which is sometimes the result of confusing one control for another at a busy time) should only be done after the airplane is slowed down and the landing phase is over. Well, the gear shouldn't be retracted anytime when the wheels are on the ground. As will be further covered in Chap. 10, more experienced pilots are usually the culprits in this sort of practice.

TAXIING AND SHUTDOWN

He should taxi the airplane back to the parking spot (if you think it's not too crowded) and shut down the engine. Too many checklists for shutdown have the actual shutting down of the engine far down the list after many other less important items. *This is no time to have your head down in the cockpit reading—in a congested area of people and airplanes with that meat cleaver whirling out in front.* There have been cases of the instructor killing the engine, bypassing several "important" items, just in time to prevent a child from getting hit by the propeller—while the student was still faithfully following the checklist. Remember, the student will later be solo and could inadvertently let the plane ease into another plane while he's reading, if the engine is not secured.

Some instructors bring the rpm back to idle and turn the ignition OFF and back ON to check if the switch is properly grounding out the magnetoes. Be sure that the engine is not at a higher rpm or a backfire could occur, as noted during the

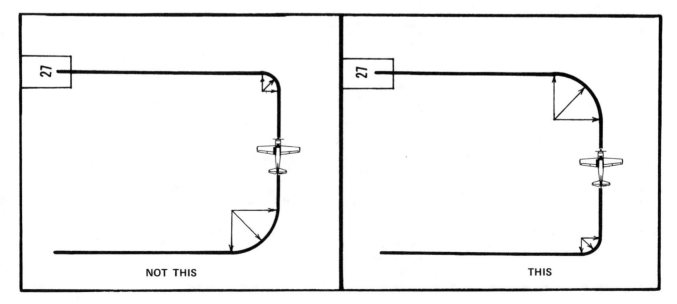

Fig. 5-11. A comparison of two turns to base and final.

pretakeoff check discussion earlier in this chapter. After you've done the ignition check, run the engine up to 1500 rpm for about 20 sec to help clear the spark plugs of possible lead fouling, then shut it down with the mixture control.

Set up your procedures so that the engine is shut down as soon as possible after reaching the parking spot. If electrical requirements for your particular airplane are such that the equipment must be shut down with the engine running, then get this out of the way as soon as possible.

POSTFLIGHT CHECK

You and the student should make a postflight check of the airplane, noting and reporting any discrepancies. He may think of questions that slipped by him on the preflight.

POSTFLIGHT DISCUSSION

Review the flight, bringing out any points you did not have time to discuss in the air.

Here would be a good time to begin a discussion of stalls: *what* they are, *why* they are practiced, and *how* recoveries are made. Emphasize that the airplane *wants* to recover. Don't overdo it, however, because too much attention could defeat the idea of a "casual" approach.

Set up a schedule for the next flight and remind him to get his physical if he hasn't already set up an appointment to do so. Fill out his logbook, and if he has been doing a good job with the Four Fundamentals give him further reading assignments in preparation for the next lesson.

SUMMARY

This second lesson can be more important than the first. If the student has had little or no flying experience, much of the first lesson may have been only an orientation and some of the instruction will not have soaked in. The second flight is basically a repeat of the first, and the student will begin to understand more. This is probably the point at which he will really start establishing habit patterns. The first lesson may seem like a blur to him, but it wasn't lost—it was the beginning of a repetition pattern. He will now start to feel that he is flying the airplane and will begin to see his own mistakes.

How long should this flight be? It should be like a good speech, long enough to cover the subject, but short enough to be interesting. You should leave him enthusiastic and looking forward to the next flight. Less than 45 min might mean that you couldn't complete all of the work you felt was needed; over an hour could tire him out and learning could suffer. If in doubt, though, keep it on the short side. It's better for the student to say to himself, "Why did we come back so soon?" than "We should have come back sooner."

It may be that the student needs more work on the Four Fundamentals, and the *third* flight could be a partial repeat of this one. If he is beginning to have good control of the airplane and knows where he is, the confidence building and elementary precision maneuvers may be introduced.

In the case of high work and low work combined in one flight (720° power turns *and* wind drift correction maneuvers), it's best to do the high work first. This is true particularly in the earlier stages, since the climb to altitude will give him a chance to get accustomed to flying again without the pressing requirements he would have going into a rectangular course just out of the traffic pattern. You want to move on with the training process, of course, but it's better to start each flight easy with the presolo student.

Fig. 6-1. A lesson plan for 720° power turns.

LESSON PLAN

LESSON —	Performing 720° turns of 45° and 60° banks, using climb power and maintaining a constant altitude.
ELEMENTS —	The effects of increased bank on required back pressure to maintain altitude. Power as an aid in maintaining altitude. Coordination required in rolling into and from a steep bank.
SCHEDULE —	Preflight instruction . . . . :10 Instructor demonstration . . . . :05 Student practice . . . . :15 Postflight critique . . . . :10
EQUIPMENT —	Blackboard and model airplane.
INSTRUCTOR'S ACTIONS —	Explain the objectives using the blackboard and model plane. Flight demonstration of 720° power turn using a 45° bank.
STUDENT'S ACTIONS —	Discuss the maneuver and resolve any questions. Perform the 720° power turn using 45° banks until the maneuver is satisfactory, then 60° banks may be performed.
EVALUATION —	The performance is satisfactory when the student can perform the 45° banked turns, holding altitude within ±100 feet and rolling out within 15° of the selected heading reference.

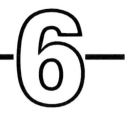

HIGH WORK—THE CONFIDENCE BUILDING MANEUVERS

720° POWER TURNS

■ **Preparation.** Fig. 6-1 (on the opposite page) is a sample lesson plan.

■ **Explanation.** The 720° power turn is used as a confidence building maneuver and demonstrates that the airplane is turned by the same method for all degrees of steepness in banks. (You might want to mention, however, that it is impossible to continually maintain a constant altitude in a 90° bank.)

Power is used as an aid in maintaining altitude and to give the student practice in coordinating power and flight controls on the roll-in and roll-out. Explain that this particular maneuver is really one of the Four Fundamentals he has done before.

The 720° power turn fits into the overall goal by extending the student's experience so that he can get more performance from the airplane.

What—The 720° power turn is a maneuver consisting of two complete circles (720°) with the airplane maintaining a constant altitude at a 45° (or steeper) bank. Added power is used throughout and is coordinated with the flight controls on roll-in and roll-out.

Why—The maneuver is good for building confidence and for developing planning, coordination, and orientation abilities.

How—An easily recognized reference on the horizon is picked and the airplane is rolled into a 45° (later 60°) banked turn. Added power is applied as the bank increases. The coordinated turn should be established by the time the plane has turned 45°. (You'll find that in most cases, it will be established well before this.) You may, on the first turn, help the student keep up with the reference point by saying "one turn" as the nose passes it. The higher rate of turn usually surprises him, and this plus the more radical attitude could keep him too busy to watch for it.

You should explain how the bank may be shallowed slightly to stop a temporary loss of altitude, but also note that on the private flight test he'll be expected to stay within certain limits of bank.

Indicate that because of added power and the slower speed caused by the greater back pressure required in this steeper banked turn, he can expect "torque" problems. The usual tendency is to skid in a left turn and slip in a right turn. It's possible that he may have to hold right rudder for turns in both directions, but the usual need is to use it only as "bottom" rudder in a right turn.

■ **Demonstration.** One 720° turn should be enough of a demonstration. You may even prefer to do only 360° of the turn (reminding the student, however, that he is expected to go around twice). The 360° turn allows a demonstration of the roll-in and roll-out, but may not provide enough time to demonstrate the shallowing or steepening of the bank as an aid to altitude control. You can make up your mind on this for a particular student.

As you demonstrate the maneuver, explain what you are doing. If you hit your own wake turbulence, explain what happened. If the maneuver should turn sour, admit it and do it again; he has never seen the maneuver before and won't know what's right.

Show him the procedure of adding power *smoothly* as the turn is entered and retarding it smoothly on the roll-out.

■ **Practice.** During the first couple of 45°-banked 720° power turns you will usually have to remind him to steepen the bank. Even though he has done 45° banks in the turn exercises, he turned only 360°. The second 360° of this maneuver is where he may have trouble with a shallowing bank and loss of orientation.

Have him check the pitch and bank attitudes (visual reference) and altimeter. He should be moving his eyes constantly between these items throughout the 720° of turn.

These steeper-banked turns will exaggerate any nose position problems caused by his sitting to the left of the airplane center line. If he tends to climb in right turns and lose altitude in left ones, you can figure that this is at least part of his trouble; this is why the altimeter is an important part of his scan. Until he gets experience enough to know the required nose reference, the altimeter will be a good aid in telling him he's doing right or wrong. Let him know when he has the proper pitch and bank attitudes and is maintaining altitude.

You may have to talk him around the first couple of 720s while telling him to steepen or shallow the bank or to increase or decrease the back pressure. Expect on his first turn that he probably won't use enough back pressure and altitude will be lost. You may even have to require him to break off the maneuver and start again. You could demonstrate again that shallowing the bank *before* increasing the back pressure is the proper way to stop altitude loss. Keep an eye on the tachometer if he gets "wound up"; be prepared to reduce throttle if he doesn't catch it.

This maneuver can be a little active for some of your students, so ease into it and take a break between each 720. This is intended as an early confidence building exercise so you shouldn't expect perfect performance. However, don't leave 720s if he's losing control of the airplane; this wouldn't help his confidence at all (Fig. 6-2).

You can expect some, or all, of the following *common errors* during the student's first session of 720° power turns:

1. Anticipation of required back pressure. The nose is brought up before the bank is established and the airplane gains altitude initially.

2. Poor throttle use in the entry. Abrupt opening of

the throttle, or neglecting to add power at all.

3. Keeping the bank too shallow, and not reaching the required bank of 45° (or 60° for later 720s).

4. When altitude is being lost, trying to raise the nose by back pressure without shallowing the bank.

5. "Losing" the checkpoint; loss of orientation.

6. Coordination problems. The usual difficulty is slipping, particularly in a right turn.

7. "Fixation," tending to stare at the altimeter or over the nose, neglecting the other aspects of the maneuver unless you remind him to keep looking around.

8. Failure to relax back pressure during the roll-out and so gaining altitude. (Sometimes the altitude loss in error 4 is exactly compensated for by error 8, which looks good when the maneuver is complete.)

9. Forgetting to reduce power *in coordination* with the roll-out.

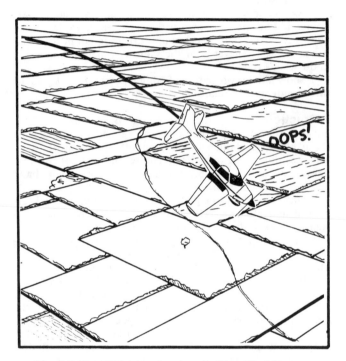

Fig. 6-2. The 720° power turn is a confidence building maneuver. Try to keep it that way.

■ **Evaluation and Review.** The student should be more at ease in the Four Fundamentals and other areas of his flying after introduction to this maneuver. After a couple of practice 720s he should show an improvement in his general confidence and orientation ability. A 3- to 4-hr student should be able to remain within ± 100 ft of the starting altitude during the maneuver—*under your direction.*

You can check the vertical speed indicator as your aid to see the trends of altitude loss or gain. You will be watching his reactions and performance, as well as looking out for other airplanes and taking a quick glance at the VSI, which can help you catch up to what's about to happen. You won't mention the use of that instrument to him this early, because he can get "fixated" and chase it.

After the flight, review 720° power turns with him, bringing out points that were mentioned briefly as you and he were doing the maneuver.

SLOW FLIGHT—MANEUVERING AT MINIMUM CONTROLLABLE AIRSPEEDS

■ **Background.** Your student may have heard "If the airplane is pushed to fly too slowly it will stall," which is true, but how slow is "slow"? He should fly the airplane in that area close to the stall; it will boost his confidence when he discovers that the airplane can still fly with the airspeed well below that of cruise. He should have had a brief demonstration of lower speeds in the introduction of effects of controls, but this practice period will allow him to spend enough time in the slow-flight or minimum-controllable-airspeed regime to become more comfortable with it. (It's also called "flight at critically slow airspeeds" but "slow flight" is most often used.)

After he's had some slow-flight experience you may find that this is a good time to introduce stalls (gentle ones). If you were able to ease into them during earlier climb practice, this could be the point to go into them more thoroughly.

■ **Explanation**

THE POWER-REQUIRED CURVE. The chalkboard will be the major aid during the preflight discussion. Briefly discuss the drag curve, indicating that the airplane is the victim of two types of drag. Discuss induced and parasite drag (Fig. 6-3A); the total drag on the airplane is a combination of the two types.

Because of the characteristic U-shaped total drag curve, the horsepower required to fly the airplane at a constant altitude varies in the same general way as the drag. It requires less and less horsepower to maintain a constant altitude as the airplane slows from cruise—until a certain airspeed is reached. This airspeed naturally varies in different airplane types, but it also decreases with the weight decrease of a particular airplane. *Higher altitudes (and/or higher weights) require more rpm for a given indicated airspeed.*

Draw on the chalkboard the curve shown in Fig. 6-3B, but instead of using horsepower, plot *airspeed versus rpm* for *your* particular airplane. You don't have to demonstrate a number of points on the curve; three airspeeds will be enough to show the idea of power (rpm) required to maintain a constant altitude as the speed changes. You should have flown a number of points yourself to set up a general curve for your airplane. Use a medium altitude to get the numbers (6000 ft MSL if terrain permits).

In your discussion start at point *1*, which you should pick at the "normal" cruise airspeed used in the practice area. You'll use the rpm required at the three points you worked out before starting slow-flight instruction. Assume that 2350 rpm is required at point *1*. Explain that less rpm is required to maintain altitude as the airspeed is slowed to *2*. At point *2* the minimum rpm is required to maintain a constant altitude (say, 2000 rpm). This would be the general airspeed to set up if *maximum endurance* were needed.

Because of *induced drag* which rises sharply at lower speeds (higher angles of attack) as the airplane is slowed to speeds *below* point *2*, more and more rpm is required to maintain a constant altitude. This is the "backside of the power curve." If the power required (2350 rpm at point *1*) was used at the airspeed of point *2*, the airplane would have excess rpm available and would climb.

At point *3* the power (rpm) required to maintain a constant altitude just above the stall is the same as that required to fly at 95 K for this example. Your numbers may not work out this close, but you can show him that almost as much (or more) rpm is required at a low airspeed as at a higher one. If you fly and establish a number of points from an above-cruise airspeed down to just at the stall, and you draw a curve like

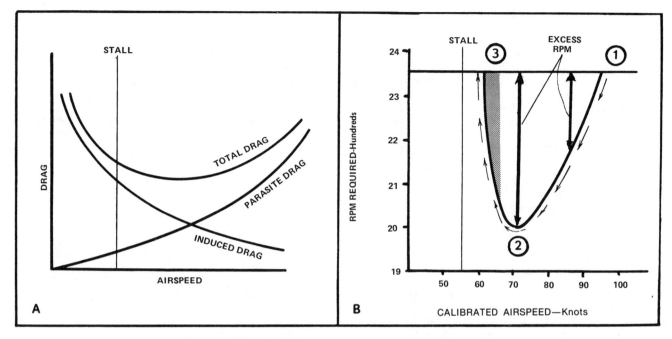

Fig. 6-3. (A) The drag existing and (B) the power (rpm) required to maintain a constant altitude at various airspeeds. The gray area shows the slow-flight regime.

that in Fig. 6-4, you can pick out "pairs" of airspeeds requiring the same amount of power and set up your own program.

The numbers you get are only exactly correct for that density altitude and weight at which the airplane was flying, but you'll be surprised how close they'll be for other weights and altitudes (within reason). Be sure to tell the student that there may be slight differences from the numbers you give him during the briefing and the ones actually required in flight, so he won't wonder what went wrong between talking and doing.

You may not want to go into such detail on the power curve with some of your students, but for those with enough technical background to understand, it would be wise to shoot down some wild theories that even some experienced pilots

believe. (One of these theories is that you *must* reduce power on the back side of the power curve if you want to go faster.) The truth, of course, is that you can maintain a constant altitude with less and less power as the airspeed increases from *3* back to *2* in Fig. 6-3B.

Okay, stop for a minute. Here's where the old controversy about power/elevator or what controls what should be discussed.

It's fully realized that when you do something with the throttle, you have to do something with the elevator (or trim), or vice versa. One affects the other, but the question is that of *primary* controls of airspeed and altitude.

Assume here that the airplane is *right side up (and the*

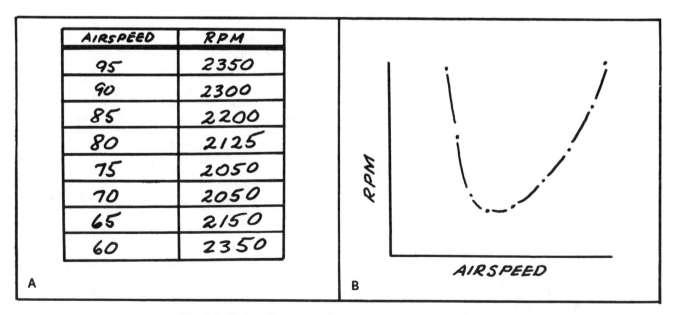

AIRSPEED	RPM
95	2350
90	2300
85	2200
80	2125
75	2050
70	2050
65	2150
60	2350

Fig. 6-4. Finding the rpm required to maintain altitude at various airspeeds. After you've gotten the data (A) in flight, plot the points and draw your curve (B).

wings are level, to simplify matters). For purposes of discussion, think of elevator and power as being two separate entities:

1. Elevator or stabilator controls airspeed.
2. Throttle controls altitude.

By exerting forward pressure on the wheel or stick the airplane will go faster. By exerting back pressure on the wheel or stick the airplane will go slower. This is true whether right above the stall or at the red line. (It is tempting to put *always* in those sentences.)

For a given (constant) airspeed:

Addition of power will cause
    a. an initial climb if in level flight,
    b. an increase in rate of climb if climbing, or
    c. a lesser rate of descent, if descending.
A reduction of power in a climb will result in
    d. a lesser rate of climb,
    e. level flight,
    f. a descent, or
    g. a greater rate of descent.

These reactions are based on the altitude status of the airplane, that is, whether it was flying level, climbing, or descending at that airspeed when the power change occurred. Note that the power change required a *change in elevator or trim in order to maintain that airspeed,* so you can't do something with one factor without affecting the other.

One school of thought *(not this book)* indicates that if power is variable (and available), elevators control altitude and power controls airspeed. If power is fixed or in transit, the elevator controls speed (some say).

The average student does not have the experience and the time to sort this all out when he sees a cow that has just wandered onto the runway as he's crossing the fence on final.

Suppose that the airplane is well on the back side of the power curve, just about to touch down (the airspeed is just above the stall) when that cow or a car moves onto the runway. After he opens the throttle all the way he has to move the wheel or stick *forward* to climb. If he keeps the wheel back at the near-stall condition, the lack of *excess* horsepower will not allow him to climb. The best rate of climb speed, for example, is 80 K; he has to push on the wheel to get that value.

Assume, on the other hand, he is roaring in on final at 120 K and sees he will run out of runway because he's going too fast. There's no cow, he's just too fast on final. This time he has to exert *back pressure* from that airspeed, to climb at the best rate of climb speed of 80 K, and of course, adds full power to get all the climb he can.

He eventually attains a steady-state climb at 80 K in both examples, but the student had to push in one, and pull in the other to go "up."

So you would say this to the student: "Elevators (stick or wheel) control airspeed and you exert forward pressure (or trim) to go faster. If you want a particular airspeed and the ASI indicates below it (or less than desired), you would put forward pressure on the wheel or stick to get it." (It probably would be better to talk in terms of attitude plus power equals performance, but the just-soloed student pilot on a go-around doesn't have the finesse to set the exact pitch attitude and also make a go-around. So you give him an airspeed to go to and he "pushes" or "pulls" as necessary.) This thinking works on both the front or the back side of the power curve, whereas if he thinks of elevator controlling climb (altitude) it's confusing because (again) sometimes he has to "push" the wheel forward to climb and at other times he "pulls back" to climb.

The U.S. Navy should know well which control should control what, since carrier approaches require precise airspeed (or rather angle of attack) and power control, and that organization teaches that the stick or wheel controls angle of attack (or *airspeed,* if your airplane doesn't have such an angle-of-attack indicator). Power (or thrust) is used to control the angle and/or rate of descent on approach.

The argument against the power-for-altitude and elevator-for-airspeed concepts often uses the idea that it wouldn't work for jets. The navy pilots don't know this and it would be better not to tell them at this late date.

A couple of other examples:

Suppose that you and a friend are flying cross-country in a current four-place lighter airplane. Because of some fictitious and exaggerated company or club requirements, he is totally in charge of the throttle (but cannot touch the yoke or elevator trim) and you are in charge of elevator and elevator trim (but cannot touch the throttle).

The airplane has plenty of altitude and is trimmed with power set in level, cruising flight. After a couple of hours of this you are tired of seeing 120 K on the ASI and ask him to increase the airspeed indication by use of the throttle *only* (with no worry about losing or gaining altitude). He will have to *reduce power or close the throttle to increase airspeed* but will lose altitude—but this is no immediate problem. To decrease the airspeed, the throttle is opened, the nose moves upward, and *altitude is gained.*

In this example, the throttle had to be used backward to control the airspeed if that was its assignment, but it directly controlled altitude in both cases (*more* power, *more* altitude, and vice versa). Some airplanes have small pitch changes with power and might not give such an immediate pitch response.

The elevator, or stabilator, should be the item to control airspeed because it is the most effective if not the most logical procedure since, as said before, you "push" to increase airspeed and "pull" to decrease airspeed no matter where you are in the airplane's speed spectrum or envelope. (Assuming, again, that the airplane is right side up.) For instance, you are again in charge of the elevator/stabilator system and your friend has the power controls; you can completely control the airspeed from stall to the red line, no matter what he does with the power. He can go from full power to idle but the effect is negligible on your airspeed control. You can stall it and a few seconds later be diving to the red line and then back to the stall and his use of the throttle is an exercise in futility.

On the other hand, he has control over whether you can set up a steady-state climb of the airplane or maintain any particular altitude. No matter what you do with the elevator/stabilator, if he doesn't want the airplane to climb (disregarding any temporary "zooms" you might do) he'll keep the power off and the airplane must lose altitude over the long term.

Look at Fig. 6-3B again. You could be flying straight and level at point *1* and, without touching the throttle, ease the nose up and move the airspeed back to point *2.* Note that by pulling the wheel back, the airplane did set up a rate of climb since at any point between points *1* and *2* there is excess rpm and in this example excess thrust horsepower exists, allowing a climb. And also as shown in Fig. 6-3B, point *2* is the place to be for the best climb, at least for the rules established here.

Next, get the student set up at point *3,* the slow-flight speed, just above the stall, carrying the right amount of power to maintain level flight.

There you are, not gaining or losing an inch of altitude, flying with the nose cocked up and carrying plenty of power and then you tell the student about that imaginary ridge just ahead. He must climb the airplane over it and so "pushes" the nose gently over (*don't change power*). You can see that the airplane is starting to climb as indicated by the vertical speed indicator and altimeter.

The power is constant and, just as in the earlier "rule," by using elevator alone the airplane is made to climb and does better as point *2* is reached. So, the elevators in both cases make the airplane climb—except that the pilot "pulled" the first time and "pushed" the second time to get the same climb rate at point 2. Does the student push or pull to climb? He doesn't know.

The job of the wheel or stick is to move the airplane back and forth along the airspeed line in Fig. 6-3B and the power is set as necessary to get altitude performance. If you wanted to fly straight and level at point 2 (about 72 K) you would carry 2000 rpm and *slowly* go on about your flight. Of if you wanted to set up close to a maximum endurance situation you could use a rule of thumb for some types of airplanes of 1.3 times the stall speed (see *The Advanced Pilot's Flight Manual*) and get 1.3 × 55 = 72 KCAS (again). You'd set this airspeed up with the elevator (trim) and use whatever power necessary to maintain altitude. You'd lean the mixture, also.

Looking at a practical point for your demonstrations, assume that points 1 and 3 require 2350 rpm and point 2 requires 2000 rpm as indicated on the tachometer. You'll have to take into account windmilling effects on a fixed-pitch prop when doing this exercise. For instance, to get 2350 rpm at point 1 does not require the throttle to be as far forward as to get 2350 rpm back at point 3. You may have to have the throttle nearly open at the lower airspeed to get the same rpm. This windmilling effect can puzzle the student at first. If he lowers the nose to go from 3 back up to point 1 without touching the throttle he may find that as he reaches 95 K the rpm may be 2450 or more and the airplane wants to climb if he holds that airspeed. If you set 2350 rpm at 1 and, not changing power, slow up to 3 you'll find that the power is too low (2250 rpm) to maintain altitude and you'll have to push the throttle farther forward to get 2350 rpm again.

The main thing is to choose your airspeed with the stabilator or elevator and set the power to get the altitude effects desired.

What—Slow flight, or flying at minimum controllable airspeeds, is flying the airplane at a speed close enough to the stall that if power is suddenly reduced to idle, immediate indications of a stall result. As a general explanation, it is assumed that the airplane will be flown within 5 K of the stall for a particular power setting and configuration.

Why—Slow flight is demonstrated to and practiced by the student to show the characteristics of the airplane at speeds near the stall. It is used to show him one end of the range of flight limitations for his airplane, so that confidence in his ability to fly in that area will increase.

How—The airplane is slowed from cruise by throttling back well below the power required to maintain altitude at slow flight. (Set up your own rpm requirements.) A constant altitude is maintained as the nose is slowly raised. As the proper airspeed is approached, the power is increased to that required.

Discuss the practical aspects of slow flight, including control sloppiness and how he will use sight, sound, and feel to fly the airplane. Indicate that he will fly the airplane in the slow-flight regime at first without using trim; later he will use trim in all slow-flight exercises. Remind him of "torque" complications.

Brief him that he will do the Four Fundamentals in the slow-flight regime but will keep climbs to a minimum because of possible engine overheating at the low speed. Note that at first you will want very shallow-banked turns (10–15°) because of the increase in stall speed as the bank increases, but tell him you will require 20–30° banks as he gains experience.

■ **Demonstration.** When you introduce the student to slow flight, your demonstration will be more thorough than when

you review it with him after a layoff. You might set up the following sequence:

1. Throttle back to well below the power setting required to maintain altitude at the selected airspeed and make the transition to level slow flight, adding power as necessary. Indicate that you are correcting for "torque" and raising the nose at the proper rate to maintain altitude as the airspeed decreases from cruise. The airspeed should be held *constant* in the slow-flight regime.

2. Fly at a constant airspeed, heading, and altitude.

3. Turn 180° in each direction, using a shallow bank of 10–15°. Add power as necessary to maintain a constant altitude, telling him why you are doing it.

4. Fly straight and level for a few seconds to reestablish, if necessary, the original airspeed and power setting.

5. Climb straight ahead at the slow-flight airspeed and normal climb power for 1 min. Emphasize the need for "torque" correction (and heading control). Indicate that the climb is not efficient because the airspeed is well below that for best rate of climb, and repeat that such a climb should be kept of short duration because of engine cooling problems.

6. Level off and fly straight and level at slow flight for at least 2 min to let the engine temperatures restabilize.

7. Set up a power-off descent straight ahead using the slow-flight airspeed. (Note: Some airplanes may have comparatively *limited up-elevator* travel so that they may not be capable of flying at as low an indicated airspeed power-off as they can power-on. If this is the situation for the airplane you are using, you can either increase the airspeed used for the other slow-flight exercises, or use a higher speed for this part. Tell the student why.)

8. Level off at a predetermined altitude and established straight slow flight for a few seconds.

9. Make the transition to normal cruise maintaining a constant heading and altitude. Discuss your use of trim in steps 1 through 9.

You might do better to demonstrate and let him practice only one part at a time to keep his interest.

■ **Practice.** Have the student make the transition to slow flight and fly straight and level for several minutes to get the idea—without his use of trim at first. Correct any airspeed heading or altitude problems as they occur. After he's settling down with straight and level you may add items 3 through 9 as he is ready for them and have him use trim each time. (Don't expect him to remember the sequences, but you'll call out the maneuver as you want it flown.)

Common errors made in slow flight include the following:

1. Gaining altitude in the transition from cruise to slow flight. (Raising the nose too quickly.)

2. "Torque" (heading) problems during the straight and level slow flight.

3. Timidity in adding or decreasing power if altitude problems occur.

4. Letting the airspeed get too high. (It's interesting to note that very seldom does the student let the airspeed get too low during the introduction to slow flight.)

5. Heading problems in the straight climb.

6. Letting the airspeed increase in the level-off from the climb.

7. Too high an airspeed in the descent.

8. Losing altitude in the transition from slow flight to cruise. (Lowering the nose too quickly.)

Fig. 6-5 shows a couple of these errors.

SLOW FLIGHT—LANDING CONFIGURATION. At first you will

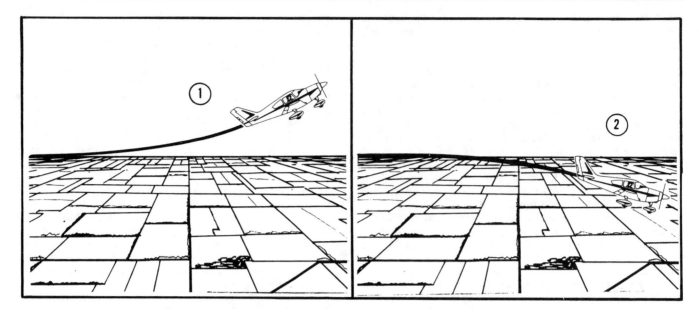

Fig. 6-5. The average student tends to gain altitude in slowing up and lose altitude when increasing airspeed back to cruise (slow flight transitions).

want to keep things simple and so should have the student fly the airplane in the clean configuration. If his progress is good you should have him work on slow flight with recommended landing flaps (and gear down if applicable). He should practice this before solo, but you probably will want to wait until the postsolo period before pushing for real precision.

Use a lower airspeed for these exercises, since the stall speed will be lowered with the use of flaps. Don't have him practice the climb exercise in the dirty configuration—that's asking a little too much of the airplane. You might demonstrate for a *very short while* the folly of trying to climb at a very low airspeed with garbage hanging out.

■ **Evaluation.** The student will be making satisfactory progress in flying at minimum maneuverable speeds in the introductory lesson if at the end of the period he can keep the airspeed within ±5 K and heading within 10° of that required in straight and level trimmed flight in smooth air and if he makes corrections in the proper direction. His confidence in the airplane should be increased after this session. You'll decide whether he needs another short go at slow flight.

■ **Review.** Review *specific* items or problems *in flight* immediately after a segment is completed. Review specific items in more detail and cover general problems concerning a whole maneuver after getting back on the ground. The term "slow flight" has been used for purposes of brevity. Be sure that the student understands that this may be also described as "maneuvering at minimum controllable airspeeds."

SIMPLE STALLS

A good way to introduce stalls, if you haven't already done so earlier, is to tell the student that you will slow it up even more and show him the stall during a demonstration of straight and level slow flight. Make your entry and recovery as easy and smooth as possible. Don't add power for recovery during the demonstration(s); let him see that the airplane can get out of a stalled condition without a great deal of hoopla. Tell him how you recovered—by relaxing the back pressure—and that the wings were leveled by coordinated controls *after* the back pressure was relaxed. During one of the later straight

and level slow-flight segments of the lesson have him ease the nose up and gentle-stall the airplane and recover. If this works well you should also have him stall it in the other phases of slow flight.

It's best to have the student directly participate in flying the airplane as soon as possible in the various maneuvers if there is no possibility of a sudden fright that could cause a setback or a loss of confidence.

It was recommended in Chap. 4 that the student be given responsibilities *from the beginning* (taxiing, climbout, etc.), if possible. However, he should be *eased* into stalls. As will be noted later, you will work to continually increase his knowledge and skill in this area of flight. All too often the student is introduced to stalls too abruptly (which may frighten him), and his knowledge and skill in stalls and associated maneuvers (spins, deep stall conditions) remain the same or increase very little throughout his flying career because he wants to avoid going any more deeply into them. It's as if the instructor who is introducing and teaching stalls has only one level of knowledge himself—too high for the beginning student and too low for the more advanced (preprivate) student. Fig. 6-6A shows an exaggerated example of two types of stall instruction.

As the student progresses toward the private certificate he should have increasing knowledge and skills concerning stalls, but you should insure that even as he loses fear he maintains respect for stalls and spins.

Fig. 6-6B shows desired, and not so desired, learning curves for *general* skills and knowledge in handling the airplane. The curves are both *classic and exaggerated* but are intended to show that unlike stalls, which may introduce a fear element if improperly introduced, it is beneficial to introduce the general handling skills of the airplane as soon as feasible for all students.

The extended dashed lines show that at some point of experience two students of equal ability would have the same general knowledge and skills; but this does not always work out for stalls, which may have a fear element for certain students.

MIXTURE CONTROL

Some instructors have their students fly a full-rich at all times during the preprivate work. One reason is that "it's a lot

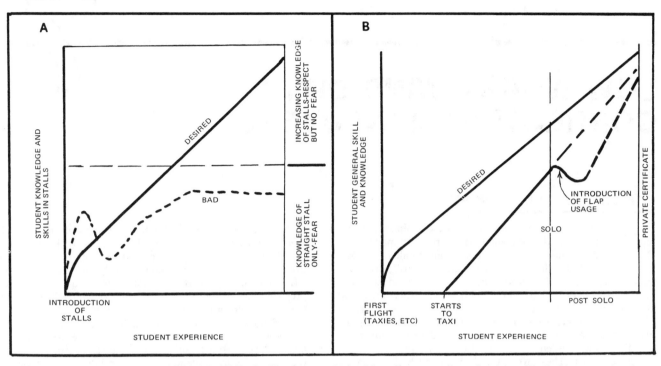

Fig. 6-6. (A) Desired (and not so desired) learning curves for stall training. Too sudden or too harsh an introduction of stalls may cause plateaus and dips. (B) Learning curve comparisons of the student's general skill and knowledge.

simpler" to just tell them to leave the mixture control in full-rich. Others may mistakenly believe that no leaning is to be done below 3000 ft density altitude at *any* power setting. This is a misconception. Engine manufacturers recommend that the mixture be leaned at cruise, at 75 percent power or below, at any altitude. Generally, the unsupercharged engine can be judiciously leaned in the climb above 3000 ft (but check the *Pilot's Operating Handbook* or *Engine Manual* for your airplane for any special leaning mandates for a particular engine). In fact, as you're climbing to 8000 or 9000 ft a *full*-rich mixture means a power loss and a rough-running engine. Teaching the use of the mixture control, like the use of flaps, will be your own decision. But the sooner the student pilot views mixture control use as part of his normal operations, the better his later transitions will be. Sure, you say, what if he's out on a solo flight and lets down, forgetting that he has leaned the mixture? He'll have had the problem explained and will know to richen it as he descends.

SUMMARY

As the title of the chapter implies, these are maneuvers that will not only increase the student's ability to fly the airplane, but will also increase his confidence in flight. An *easy* initiation with ever-increasing skill requirements (move him on as he can take it) is the way to handle these and other maneuvers.

Make sure that you don't *decrease* his confidence by the way you instruct the confidence building maneuvers.

LOW WORK—WIND DRIFT CORRECTION MANEUVERS

TRANSITION TO LOW WORK

Parts of at least two lessons in the presolo phase of training should be devoted to the wind drift correction maneuvers. Don't spend an entire period on them, but break it up, perhaps with half the flight spent on high work followed by a simple simulated high-altitude forced landing demonstration. Show the student how to pick a field, set up procedures, and glide down to set up a pattern for it. You can either continue the approach down to a safe lower altitude or break off the problem at the altitude used for the wind drift maneuvers.

For some students the introduction to low work can cause problems. The lower altitude with the higher relative speeds may cause a minor onset of "ground shyness." You will find in the rectangular course and other such maneuvers that most students have a tendency to climb, but they will work out of this with experience. As mentioned in Chap. 5, it's a good idea to do some high work at the beginning of the flight before moving down to the low maneuvers, because this allows students to get flight oriented again or "warmed up." If the low work is done immediately after leaving the traffic pattern they generally have trouble catching up with the air-

plane. So, for the presolo flights requiring both high and low work, the best procedure is to do the high work first.

These low maneuvers are used to show the idea of wind drift correction, to teach students to fly the airplane in a specific flight path with their attention focused outside, and to prepare them to fly the traffic pattern (which will require precise maneuvering even lower—to the runway).

If possible, students should have their introduction to low work in smooth air; turbulence can make them tense. If it looks like rough air can't be avoided, tell them about it beforehand and don't spend a long period doing the maneuvers. (Climb up to smooth air for a break and do some more high work.) It's easy to say that smooth air is needed for the first low work, but their schedule and/or yours may make it necessary for the flight to be made at a time of day or a season when the air is just naturally rough (you can't wait 6 months for smooth air or meet them at dawn). This book talks in terms of what should ideally be done in the way of flight training, but local situations frequently require changes. For instance, finding a field for a rectangular course or a long

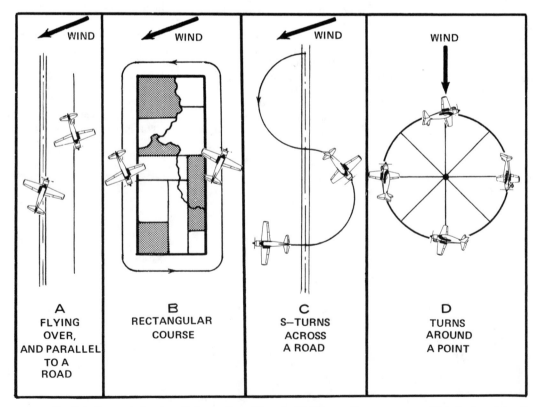

Fig. 7-1. Suggested order of introduction of the wind drift correction maneuvers. Note that each leads to the introduction of the next.

straight stretch of road for S-turns is no problem in Kansas or Iowa, but these references may be practically nonexistent in eastern Tennessee or western North Carolina. Some suggestions for alternate actions will be made later throughout this chapter and in the rest of the book, as applicable.

WIND DRIFT CORRECTION MANEUVERS

These maneuvers should be introduced in order of increasing complexity (Fig. 7-1).

■ **Introduction to Crabbing.** Have the student fly the airplane directly over a straight stretch of road or railroad track as a first encounter with wind drift correction.

In explaining these maneuvers before the flight(s), you may want to use the analogy of a boat crossing a stream, or just go right into the idea that the *airplane* is affected just like a *balloon* (see Fig. 10-5 in *The Student Pilot's Flight Manual*) and that both aircraft are affected equally by the movement of the air mass. Be sure to shoot down any possible misconceptions about holding any control pressures (rudder or aileron) to correct for the wind. If you choose, and depending on the student, you may mention that at only one time is the airplane not flying in a balanced condition when correcting for wind drift (in the wing-down approach and landing in a crosswind and in some cases immediately after lift-off in a strong crosswind).

Fig. 7-2 shows how the student can be introduced to the crab concept and to the idea of maneuvering the airplane to set up a particular requirement. Note that the turns used for reversing are upwind *and* downwind. You might point out as the turns are being made that there is a need for steepening or shallowing the bank as necessary in the roll-out over the road. This will probably result in your first use of body English with a particular student; he won't be rolling out soon enough (or too soon), and you'll just have to sit there with your hands off the wheel.

The next step should be to have him fly parallel to the road, both on the upwind and downwind sides (Fig. 7-3). The odds are against the wind being perpendicular to the road, but try to pick one that at least has a good crosswind component. Have the student fly parallel to the road at the same distance from it that he is to fly from the boundaries of the rectangular course.

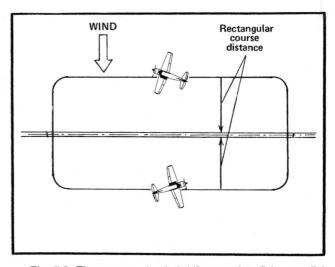

Fig. 7-3. The next step in wind drift correction, flying parallel to a road or railroad.

During these elementary exercises, emphasize maintaining a constant altitude. While these maneuvers are used as an introduction to the traffic pattern concept, you may want to do them at a lower altitude, say 500 or 600 ft rather than the usual pattern altitudes of 800 or 1000 ft, so that errors may be more easily seen. This lower altitude may help prevent ground shyness in the lower parts of the traffic pattern, such as on base or final. (Maintain legal obstacle clearance.)

Fig. 7-2. Flying along a road or railroad to introduce the idea of correcting for wind effects. The long stretch of straight flight gives the student a chance to see the effects of crab and to ease into low-altitude flying. Have him make the turns alternating on the upwind and downwind sides of the road.

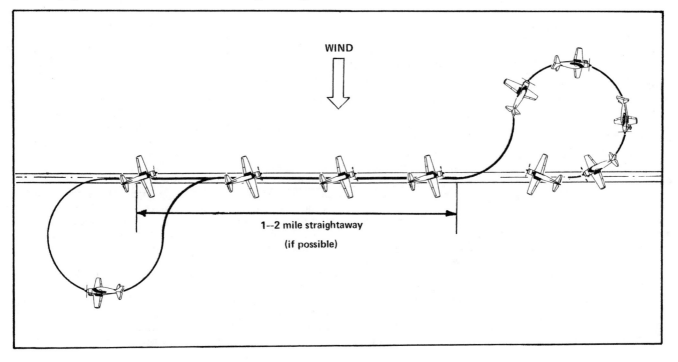

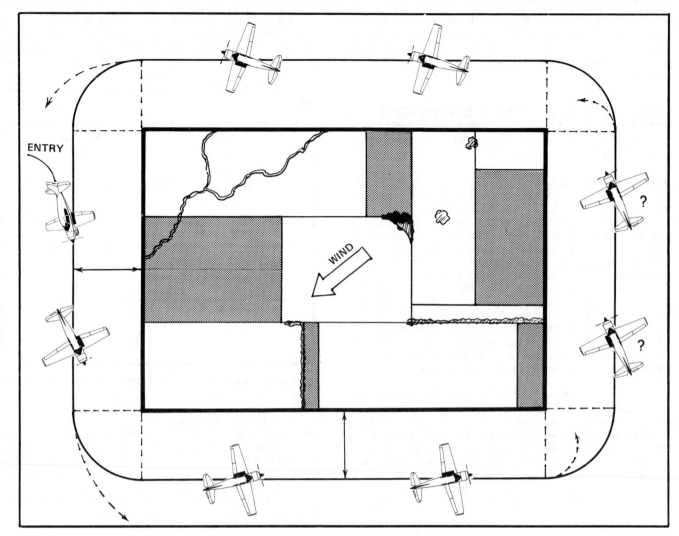

Fig. 7-4. The rectangular course. The explanation of the turns can allow an introduction of S-turns across a road and turns around a point. While it is "traditional" to have the airplane enter downwind at a 45° angle, you should also use other entries.

RECTANGULAR COURSE

This should be the next maneuver introduced. It combines elements covered earlier (crab and turn) but has stiffer requirements of performance. Try to pick the field or group of fields making up the rectangular course so that there is a crosswind component on all sides (Fig. 7-4).

■ Explanation

WHAT, WHY, HOW. In explaining the maneuver at the chalkboard, stress starting the turn as the airplane gets to the corner each time and varying the bank in order to remain the same distance from the boundary after the turn. On the chalkboard exaggerate the effects of a turn that's too shallow or too steep. Try drawing the first two sides with the airplanes crabbed in the proper direction, but on the third, "accidentally" show the planes crabbing in the wrong direction (Fig. 7-4). If the student has the crab idea well in mind, he'll catch it right away, and it gives you a chance to give him a buildup if you think it's needed. If he doesn't catch it, correct it yourself and tell him that one of the main points of the maneuver is to teach the pilot the importance of being aware of the wind and its effects on *every* flight. There will be local flights during which no navigation is required, and flights with no stalls

(except for the landing); but *every flight* will be affected by the wind or its absence.

As noted earlier, in some parts of the country the rectangular course is practically impossible because of the terrain. Fig. 7-5 shows one road setup that can be used to cover the rectangular course idea if good square fields aren't available.

You can use powerline or pipeline right-of-ways also for the introductory wind drift maneuver and for the S-turns. Many instructors have a particular reference they always use for these maneuvers; after the student has been to the place the instructor may (on following flights) have him locate it and let down for the rectangular course or other low work, after the high work is completed. Or, you may locate it for the student and have him glide down to set up an approach pattern to it.

■ Demonstration.

It would be best for you to demonstrate a minimum of two sides and both an upwind and a downwind turn to show how the turn is played to assure the proper distance from the field. Fly the airplane at such a distance that you will be looking down at the boundary of the field at about a 45° angle. This will give a distance approximately equal to

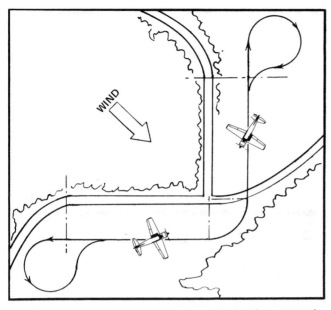

Fig. 7-5. A practice maneuver for introducing the rectangular course if no good square fields are available. Repeat the pattern as necessary.

the altitude above the surface. This could be slightly too close for some wide side-by-side airplanes when the student is flying the pattern to the right, so you might have to widen it out.

■ **Practice.** After the earlier practice with crabbing, students shouldn't have much trouble with flying the sides of the pattern, but the turns might give them a little trouble. Be sure they know where the corners are (and this is an argument for flying all four legs during the demonstration). Some students get so involved in a particular leg that they would blissfully fly on and up into the setting sun (Fig. 7-6)—unless you remind them that *that* was the corner.

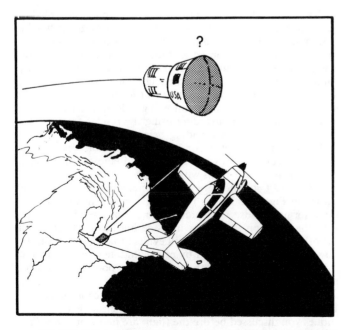

Fig. 7-6. It is a common tendency for students to gain altitude while doing the rectangular course or S-turns across a road.

Emphasize as they fly the maneuver that *the point to initiate the turn is when the airplane's position is at the corner of the rectangular course, not when the wing is pointed at it (if the plane is in a crab)*. If there is no crosswind component on that particular leg and the airplane heading and track are the same, the wing will be pointed down the next side at the point at which the turn is to be started.

Check the student on any unconscious slipping or cross-controlling going on during the maneuver. Again, the initial tendency of the student is to gain altitude. If it's turbulent, any problems with altitude and heading will be compounded. He or she should have experience flying the low work in rough air, but not the first time if it can be avoided.

A good round figure for the maximum number of circuits of the field(s) would be *three each way*. This would pin down the idea of the maneuver and still not tire the student. It can be very boring for you both if you continue to grind around, particularly in rough air. This could be the place where the student starts getting a little queasy. Go back up to smooth air if this occurs.

> *Common errors* during the rectangular course include the following:
> 1. Poor crab correction; moving into or away from the field.
> 2. Gaining altitude. A few will surprise you and lose altitude.
> 3. "Losing" the corner and neglecting to turn at the proper time.
> 4. Poor bank control in the turns; not compensating for the wind effects.
> 5. Becoming "fixated" on the field; neglecting to look around.
> 6. Jerky turns. (Turns were good at altitude.) Expect a slight loss in coordination at first because of the student's divided attention.

■ **Evaluation and Review.** By the end of his first period of rectangular courses, the student should be well oriented in the area and able to return to the prechosen rectangular course if you fly him away from it or if he has to leave it to set up turns in the opposite direction.

You can check his orientation at the end of the maneuvers (and this can be done for the other phases of low work as well) by telling him to "take you home" at the lower altitude. This probably would be difficult for him to accomplish and you can make a point that you are, for instance, northwest of the airport, and that a heading of 135° (southeast) will get you back to more familiar territory. Don't let the situation go so far that he starts getting really shook up; use humor to make a point. Help him get back to an area where familiar landmarks come in sight. Let him know that from now on he will have to start keeping up with his position relative to the airport or other local landmarks. Again, don't leave him confused more than a minute or two because he could decide that flying is not for him, when a little easier approach to the problem on your part could help him establish a good habit.

After you've felt he's had enough of wind drift correction maneuvers, you might demonstrate a simple low-altitude emergency. You handle the throttle and fly the airplane as you talk to him. This will be covered in more detail in Chap. 8.

On the ground you should review the rectangular course and cover any mistakes he made. Assure that he has no misconceptions about what he is doing both in the turn and in the straightaway parts of the maneuver.

This would be an excellent time to briefly introduce the S-turn on the chalkboard as a direct follow-up to what was learned in the rectangular course turns.

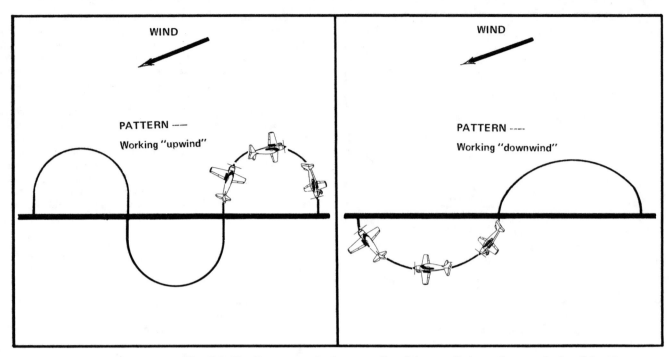

Fig. 7-7. The S-turns may be "squeezed" or "elongated," depending on whether flying the pattern upwind or downwind.

S-TURNS ACROSS A ROAD

This maneuver is considered slightly more complex than the rectangular course because of a lack of definite reference points. For this reason you may not want to introduce the idea of the crab in the turn at first since it is difficult to see.

Your problem here is to get the student to fly the pattern so that the airplane crosses the road with the wings parallel (and at a prechosen altitude) each time.

He won't pay much attention to how far down the road he's gone in doing the maneuver, but he will see it more as a matter of time involved in the maneuver. If there is a component of wind along the road, the semicircles may be "squeezed" or "elongated," depending on which direction you are flying (Fig. 7-7), but he will not realize this at first. The maneuver can still be very helpful even under these conditions.

■ **Preparation.** A chalkboard is the best aid, but a model can be used as well. In drawing the maneuver, introduce it with the idea that the wind is perpendicular to the road.

■ **Explanation.** Here, as in many situations in your instruction, you may exaggerate in a couple of spots.

Draw the maneuver with three semicircles as symmetrically as you can, so that the student can see the maneuver as a whole (Fig. 7-8).

What — This maneuver consists of a series of symmetrical semicircles across a road or some other straight geographical reference.

Why — S-turns are done for the same reasons as the rectangular course:
1. Wind drift correction practice (in this case, a correction while turning).
2. Flying the airplane by references outside the cockpit.
3. Practice for low-altitude precision flying such as the traffic pattern.
4. Confidence building by flying at a low altitude.

How — Your chalkboard talk and flight demonstration should cover this in detail. (The steepness of bank is proportional to the groundspeed.)

■ **Demonstration.** This one will normally require demonstrating, but for most students flying two semicircles is enough to show the idea. Talk it up as you do it. Here, again, the temptation is to bring up the crab idea in the turn. If there is a very strong wind and you can make an obvious point with a *particular student,* then mention it briefly and follow up with a better coverage when you get back on the ground.

■ **Practice.** This is the point at which you will probably enter the true world of flight instructors' body English—the first time you teach S-turns across the road, or more accurately, the first time you let a student do S-turns. The body English used by you in the earlier drift corrections is nothing compared to this. He is not turning enough (or too much), and since you can't grab the controls (or shouldn't, no matter how great the temptation), you will get down after the flight with leg cramps from avoiding using the rudder for him and shoulder cramps from avoiding using the ailerons for him. Well, that's the way it is for a flight instructor. You will do even more contortions when takeoffs and landings are started in earnest, but the S-turns will seem plenty fatiguing for now.

The most common error you'll encounter with S-turns is shown in Fig. 7-9. This is particularly prevalent in stronger-than-normal winds. The student is more or less able to cope with the turns on the downwind side but has problems with the other half. However, it is still evident that the turns on the downwind side are affected by errors initiated on the upwind side. You'll also run into the problems of poor bank control resulting in a poor pattern.

Sometimes flying S-turns in light wind conditions does more harm than good unless you make it clear that the techniques you discussed before the flight are based on a *need* for wind correction. The puzzled student is mechanically doing what he was told on the ground, but the maneuver is going to

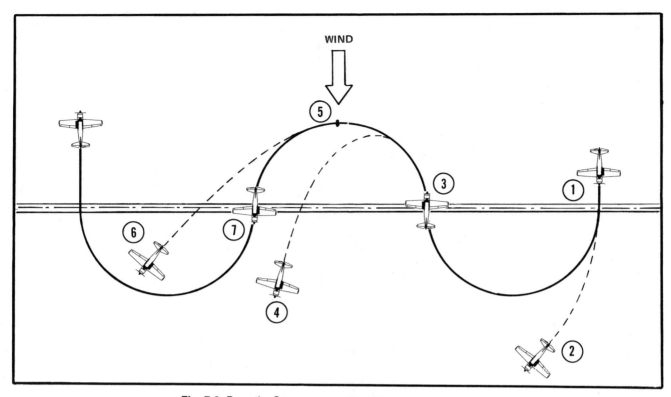

Fig. 7-8. Draw the S-turns as symmetrically as possible and explain that the goal is to fly a series of equal-size semicircles, correcting for the wind in the turn. Follow the manuever, starting at point *1,* and explain what happens if too shallow a banked turn is made (point *2*). The airplane should recross the road with the wings level and parallel to the road (point *3*). Exaggerate the effects of too steep a bank (point *4*). Note the effects of continuing a shallow bank at point *5* and how the plane would follow the path (point *6*), unless the bank is steepened to end up at point *7*. The third "loop" is a repeat of the first.

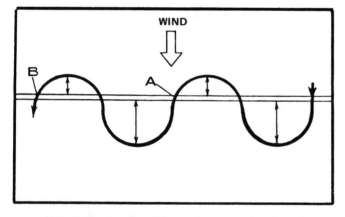

Fig. 7-9. One problem students have is poor wind correction of the upwind half of the maneuver. This, of course, can affect the downwind semicircles as well, since the tendency is to *not* cross the road with wings parallel to it at points *A* and *B*.

pot anyway. On some flights you may have the rectangular course and S-turns as pure ground reference maneuvers without a need for a discussion of wind correction.

Here are some *common errors* made in doing S-turns:

1. Poor pattern control, most likely originating on the upwind side.

2. Altitude problems, with the usual tendency being to climb.

3. Mechanically flying the pattern. Some students make bank changes mechanically but get confused about where the bank should be steepest. This usually requires some review on the ground.

4. Similar to the problem in 3, the student forgets the wind direction. This is more of a problem in light wind conditions, naturally.

■ **Evaluation and Review.** By the end of his first encounter with S-turns the student should be able to at least remain oriented and make corrections in the proper direction in moderate wind conditions and relatively smooth air. His altitude should be generally held within ±100 ft of that selected for the maneuver.

Review the maneuver briefly on the ground, and if time permits at that point you might introduce turns around a point if the student is ready for it.

TURNS AROUND A POINT

■ **Explanation.** Unless the student is above average you will probably not introduce this maneuver until after solo, but it's presented here as a natural next step in wind drift correction maneuvers. You might well use it before solo with some students who would enjoy the challenge.

In introducing this maneuver, you should review S-turns using only the two semicircles (Fig. 7-10).

It's best to enter the maneuver downwind, and it should be practiced with turns in both directions although you may want the student to make the first turns to the left until he gets the idea.

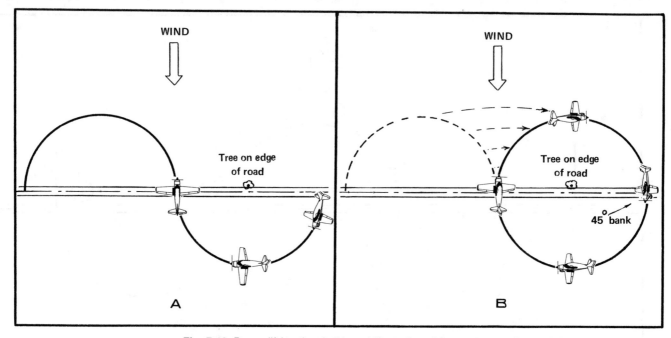

Fig. 7-10. By modifying the chalkboard illustration of S-turns across the road (A), you may move on into discussion of turns around a point (B). By "turning over" one of the semicircles of the S-turn, the pattern is established for the turns around a point. Tell the purists that the line for the S-turns was the north side of the road, in case they want to argue about the "offset" position of the example—or maybe you can use an oil spot in the center of the road.

Students sometimes have trouble with this maneuver because they can't seem to convince themselves that the steepest bank is required when the airplane is headed directly downwind, as indicated in Fig. 7-11A.

You may also introduce turns around a point as the four corner-turns of the rectangular course put together to make a 360° circle (see Fig. 7-4).

You might explain the maneuver by having the student suppose an airplane is flying tangent (wings level) to the circle at positions *A, B, C,* and *D* in Fig. 7-11A. At *A*, the airplane is moving at its true airspeed *plus* the wind speed. This results in the highest speed relative to the reference point of any position around the circle. At *C*, its relative speed is at the lowest, because it is moving into a headwind. The arrow (vector) behind each airplane gives a picture of the relative speed at the four positions. The dashed lines show the comparative distances the airplane could be "away" from the circle for any given interval of time, if no turn were made. The rate of turn, then, must be greatest at *A* if it is to "follow the circle." Since this is a coordinated maneuver, the bank must be greatest at that position (and the bank is shallowest at *C*). The banks at *B* and *D* will be comparable.

THE VECTOR APPROACH TO THE MANEUVER. Fig. 7-11B shows wind triangles for the airplane at eight points around the circle. As noted, the airplane's true airspeed is 100 K and the wind is from the top of the illustration (north) at 40 K. The solid thin arrow represents *groundspeed, or the airplane's path and speed with respect to the ground.* This groundspeed vector varies in size (speed) because of the wind, but at each of the eight points shown (or at *any* point on the circle) it must be tangent to the circle—or the plane would not be following the prescribed path. This is the first consideration.

The true airspeed is indicated by the dashed arrows and is always 100 K for this problem. (Disregard slowing the airplane in the steeper banks.) However, the 100 K true airspeed vector must be "pointed" in such a manner that the result of

the plane's heading, plus the wind, makes the airplane's path tangent to the circle at any position. The wind is a constant velocity and direction.

The circle can be thought of as "an infinite number of short, straight lines," and the airplane is flying "one leg" of a rectangular course for each one. In order to do this, the plane must be crabbed to fly the line tangent to the circle. The reason for the bank is to get the proper heading for the next "leg." The bank at point *1* must be steepest because the airplane is approaching the next "leg" at the greatest rate. At point *5* the opposite is true.

Of course, practically speaking, you fly the airplane, maintaining the proper distance from the point by looking at it. But given the true airspeed and wind, a navigation computer could be used to work out the required headings for each of the eight points given here. For instance, the course at point *1* is 180°. At point *2* the course would be 135°. At point *3* it would be 090°, and so on, and the wind correction angle could be found. The required banks at each point for the radius of the circle to be flown could be obtained mathematically using a turn equation similar to that given in Chap. 19. The maneuver could be theoretically flown "under the hood" once the airplane is established in the maneuver. The proper banks would be matched to the various required headings found for the eight (or more) points around the circle.

■ **Demonstration.** In demonstrating and having the student practice this maneuver, the idea of using a particular number of turns is purposely ignored. It's good for training to continue to circle the point until the student sees what is going on (but don't let him get queasy). Emphasize later as he is preparing for the *flight test,* that a certain number of turns may be required and he should know his heading of entry. He'll be graded on planning as well as other factors—and going around the point several extra times is no way to "make a point."

You should go around no more than three times on your

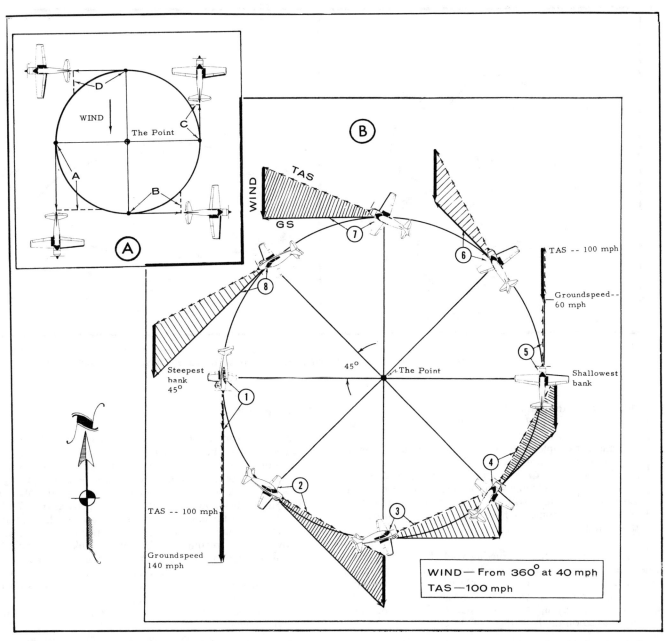

Fig. 7-11. Methods of explaining the theory of turns around a point (see *The Student Pilot's Flight Manual*). (A) Draw this figure on the blackboard first to prove the point that the steepest bank is required when the airplane is flying directly downwind. (B) This figure, or a variation of it, can be drawn on the board to introduce the concept of the crab at various points on the circle.

first demonstration while you are discussing the requirement for changing banks.

■ **Practice.** Let the student "take over" while you are still in the turn pattern so that he or she can see the proper radius of the turn. Generally, you can expect that the student will tend to let the airplane drift out away from the point so that the radius is too large for a 45° bank at the steepest point.

This is a maneuver in which it is easy for the student to become disoriented as to the direction back to the airport, and you might use the occasion to stress again the importance of knowing where you are at all times. This is also a good time to demonstrate a simulated forced landing (see Chap. 8).

It is mentioned in *The Student Pilot's Flight Manual* that a student can be doing wind drift maneuvers very well, but

when you give him a simulated forced landing he has no idea from which direction the wind is blowing. The human mind is a wonderful thing; it starts working the instant you are born and doesn't stop until you have a forced landing or go up for a flight test!

Common errors will produce several "variations" of turns around the point (Fig. 7-12):

1. *The Amoeba.* As you know, the Amoeba is a shapeless single-celled animal, and some of the turns will resemble it. The student, who isn't exactly sure of what's required of him but hopes to please, varies the bank angle—and radius of turn—in an attempt to stay within a reasonable distance of the point. If you see this type of maneuvering you will know that he doesn't understand

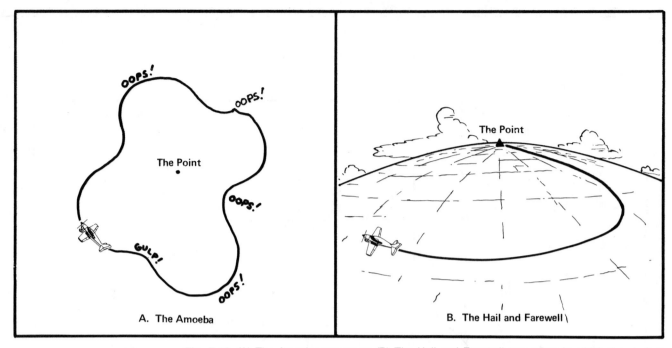

Fig. 7-12. (A) The Amoeba maneuver. (B) The Hail and Farewell maneuver, in which the student departs the point on a circular cross-country. These are variations you'll encounter while instructing turns around a point.

the theory of the maneuver.

2. *The Hail and Farewell.* In this one the student gradually eases away from the point until you suspect that another couple of turns would result in the reference sinking below the horizon.

3. *The Roller Coaster.* In this one the student is definitely having problems with altitude. When the Roller Coaster is combined with the Amoeba, even instructors with cast-iron stomachs have been known to throw up their hands.

When you first start instructing you'll be naturally alert whenever a student is flying the airplane, but as you build experience you may tend to get complacent and let your mind wander. Remember that *you* are responsible for the airplane and student, so keep up with what's going on even though it's rough to do so on the fifth flight of the day.

■ **Evaluation and Review.** If you introduce the turns around a point before solo—and again, except for an above-average student, you're probably better off not to—don't expect perfection, because you will still need to get him sharp for this maneuver on the private flight test.

On the ground you should review the need for crab in the maneuver, particularly if the wind is strong. ("Remember how you were looking back over your shoulder at the point when on the upwind side of the circle?")

He won't be ready for traffic pattern work until he can fly the low-work maneuvers maintaining altitude within 50 ft (closer is preferable) and has good control of the airplane.

AROUND-PYLON EIGHTS

This maneuver has been around a long time and was required on flight tests before turns around a point were used.

■ **Preparation.** As with other ground reference maneuvers, a chalkboard and model are the best teaching aids.

■ **Explanation**

What—The around-pylon eights (eights-around-pylons) maneuver may be considered a variation of two-turns-around-a-point patterns "back to back." The term "variation" was used because there are steep eights requiring a brief period of straight flight, as well as the shallow eights, which approximate two tangent circles (Fig. 7-13).

In the turning portions of the eights, the airplane should maintain a constant distance from the pylon, with the pilot correcting for drift with bank and crab. As noted in Chap. 19 about *on*-pylon eights, the steep versions of this maneuver are easier than the shallow ones because the turning portions are shorter and the student doesn't have so much of a chance to get disorganized; he can use the straightaways to get back in the vicinity of the pylons again.

Your students who have been flying *amoeba* and *hail and farewell* turns around a point will also exhibit their peculiar skills and abilities in this maneuver, and an "amoeba eight" can be an impressive thing to ride through. This type of problem is more prevalent in the shallow eights but some will manage to do a quick version in the steep ones.

Why—The around-pylon eight is more advanced than the other wind drift correction maneuvers discussed so far, and will require more planning, coordination, and alertness on the part of the student. He will have to fly the proper wind correction around one pylon while keeping the position of the other in mind, all the time maintaining a constant altitude. I can also be a good air in showing a cocky student that he still needs to work.

How—Two pylons or references are selected about three-fourths of a mile apart (for trainers of 85- to 105-K cruise) and at the same elevation. Crossroads make good references because the pilot can have an estimate of the position of the intersection even when it's obscured by the wing. The line between the pylons should be perpendicular to the wind and the maneuver entered downwind. Use

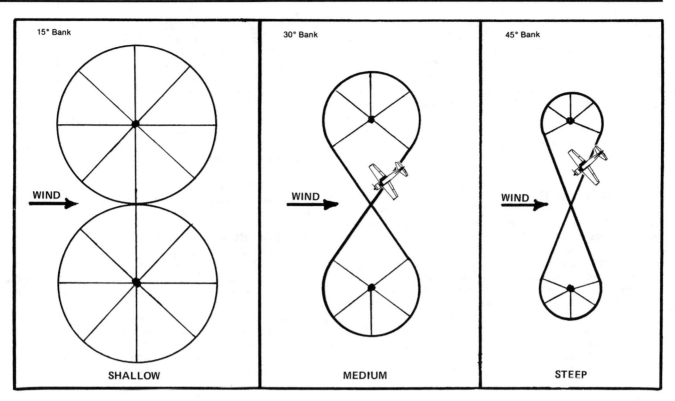

Fig. 7-13. Around-pylon eights (shallow, medium, and steep).

the same altitude as for the turns around a point (Fig. 7-14).

The principles of drift correction in the turn are the same as for S-turns and turns around a point; the steepest bank is required at the point of highest groundspeed, etc.

■ **Demonstration.** One entry and a couple of full eights of each of the types (steep, medium, or shallow) as it's introduced should be a sufficient demonstration for the average student who's had a good preflight briefing on the maneuver.

You should demonstrate pylon-picking (additional details on this art and comparisons of around- and on-pylon eights are given in Chap. 19) and enter the maneuver, pointing out other references to be used in keeping the pylons in mind. Fly a couple of full eights and then let the student take over during the process and continue the pattern. After he sees the idea and is coping, have him leave the maneuver and reenter using the same pylons or new ones he's picked under your supervision. The main thing is that this first time, he needs to understand what you want in the way of picking pylons and entering the maneuver and flying it before he's required to do it

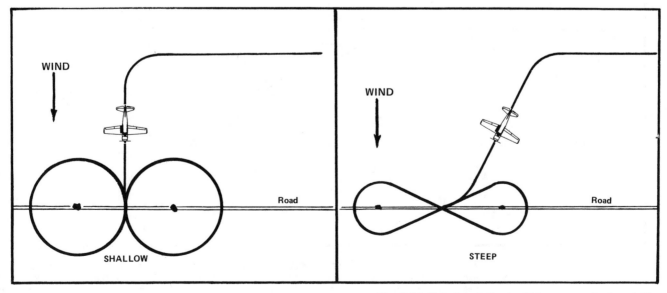

Fig. 7-14. Entering shallow and steep around-pylon eights.

himself. He makes his transition to flying it during the process that first time. In later sessions you should expect him to pick pylons and do the maneuver without your help.

■ **Practice.** You'll see the usual errors of altitude control and poor wind drift, and probably the student will tend to overconcentrate on one pylon and "lose" the other—which was no problem for turns around a point. If he seems to be seeing the principle, probably five or six complete eights would be sufficient for this first session. Make sure that he understands what's required for steep, medium, and shallow eights before sending him out for solo practice.

■ **Evaluation and Review.** The way he handles this maneuver is a good indication of a pilot's ability to divide his attention between flying the airplane and keeping up with outside references. You'll find people who do turns around a point very well but aren't able to handle that extra pylon and its turn. This maneuver can be an early indication of a student's reaction to a situation where things start "piling up" on him, such as in an emergency situation. Keep a close eye on the student who is unable, after a reasonable amount of dual and solo practice, to keep up with the airplane. Be prepared to work extra with him in later flights as more complicated requirements are encountered.

After each dual, review the student's problems with the eight and have him work on these in the solo sessions.

ADDITIONAL GROUND REFERENCE MANEUVERS

You may have to improvise maneuvers, because as mentioned earlier, your local situation may not allow the classical S-turns and rectangular course.

Aside from that, you may want to use other maneuvers to develop the student's skills in this area. These additional ones are useful if he is having ground reference problems and you don't want to wear him down to a nub with S-turns and rectangular courses.

■ **Eights across a Road.** This maneuver is a good one for establishing an unconscious correction for wind and is shown in Fig. 7-15. The straight flight legs should be of equal length.

■ **Eights along a Road.** Fig. 7-16 shows eights along a road. You can first demonstrate the maneuver by picking a road parallel to the wind and making the turns of constant bank, requiring a straightaway to get back at the point of beginning of each loop. You would then demonstrate how the bank is varied as necessary.

Eights along a road are more difficult than S-turns because of the more precise maneuvering required to get back to the reference point at the proper ground track angle. It can be interesting for an advanced student, particularly in a strong wind. You may use a crossroad as a center reference.

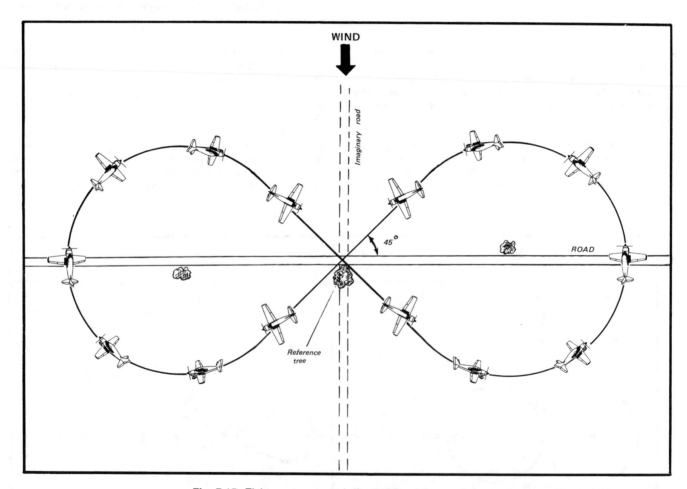

Fig. 7-15. Eights across a road. If possible, pick a road perpendicular to the wind. Trees or other references may be used to keep the pattern intact. If a crossroad is not available, the student should visualize one at the position shown so that he will cross the road at the proper angle. (A fencerow may serve the purpose.)

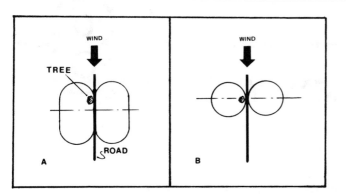

Fig. 7-16. Eights along a road. (A) A turn of a constant bank requires some straightaway flying to get back to the reference (tree). (B) The bank is varied as necessary to maintain constant radii. (See *The Student Pilot's Flight Manual.*)

■ **The Diamond Eight.** This maneuver is a good one to combine the requirements of the *rectangular course* and *S-turns* without overdoing either idea. It's basically a ground track maneuver that has the student flying a prechosen path without (again) conscious wind correction (Fig. 7-17). It should be done after rectangular course and S-turns practice.

SUMMARY

As you instruct more, you'll find yourself using the local topography for an original approach in teaching a general concept of ground track maneuvers. Don't hold back on new ideas. As long as they are safe, many different ideas can be used to good advantage for the student. Don't let books (including this one) or other aids get you hidebound in any phase of flight instruction.

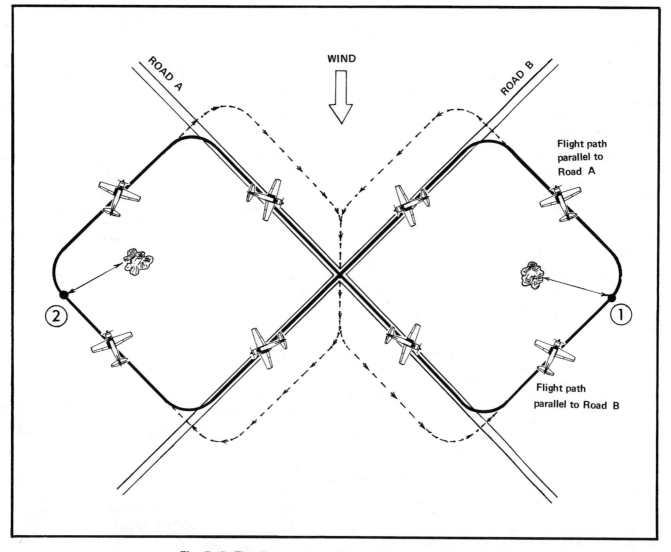

Fig. 7-17. The diamond eight. This maneuver requires better orientation and recognition of drift because part of the flight path will be some distance from references (such as the portions of the diamond flown parallel to the road). The student will probably have to use imaginary points such as *1* and *2 in relation to* some reference such as a tree or fencerow intersection to start his turn. This serves as a good introduction to Search and Rescue procedures where a target is initially sighted in a large wooded area and its position must be kept located by reference to other items (such as taller trees) that are not directly on the site. This requires continual monitoring of the ground so that the target is not lost.

A variation of this maneuver is indicated by the dashed lines, in which the airplane is flown parallel to the road.

INTRODUCING EMERGENCY SITUATIONS

8

SIMULATED EMERGENCY OR FORCED LANDINGS

In Chap. 7 it was indicated that your demonstration of a simulated emergency landing is a good way to make the transition from high work down to the wind drift correction maneuvers. It's always best to talk about it during the preflight briefing, but if a good opportunity arises in flight you should briefly tell the student what you are going to do, then continue your explanation as you demonstrate it. Point out the various types of fields available (good and bad, and why) as you demonstrate the procedure.

Make a point that the recommended glide speed is the major factor in knowing where the airplane is going, and since the student has been doing normal glides himself, the situation appears more normal. One of the most frequently-asked questions concerns what would happen if the engine actually stopped. Even people who've been doing power-at-idle descents sometimes have the impression that the small amount of power available when the engine is at idle is a big factor in maintaining control. Indicate that the glide distance would be less with a dead engine because of a windmilling or stopped prop, but that would be the only effect.

After you've demonstrated a simulated emergency landing in that first transition flight or a later one, let him practice an approach with you taking over at a safe altitude. At the end of the same flight you may give him a simulated engine failure at a high altitude over the airport, coaching him and then taking over for the landing. However, this practice is usually best reserved for postsolo dual periods when he can carry it to conclusion and actually land the airplane on the runway. There are some disadvantages in practicing at the airport, the main one being that it is a "canned problem," and making an approach to a familiar runway doesn't require as much judgment as picking a strange field. However, one confidence building factor in using the runway is *knowing* that he made it. (Insurance and legal and other factors make it hazardous to complete the exercise by landing in a farmer's field.) Keep other traffic informed about your actions on Unicom.

Many accidents have been caused by poor fuel management—a wrecked airplane is found with the selector on an empty tank, while plenty of fuel is in the other(s).

Your checklist for elementary high-altitude emergency landings will probably be something like the following:

1. Establish the recommended glide speed (max distance).

2. Pull carburetor heat to full ON as the airplane is being slowed up to the recommended glide speed. If the stoppage is sudden it's most likely caused by a tank running dry. If the loss of power takes a longer time, carburetor ice is likely the culprit. If ice appears to be the problem, the carburetor heat should be applied *immediately* so that residual heat can clear out the ice. You'll want to have him analyze the situation if possible, but he must do *all* of the remedial steps listed here.

On cold winter days during high work you can make a graphic point about how carburetor heat needs power to be effective. Tell the student that the cabin heat for your trainer is obtained by the same method as the carburetor heat (if it is); and while it is comfortable *now,* wait a minute. Retard the throttle to idle, using carburetor heat as recommended, leaving the cabin heat ON as before. The student will very soon get the idea about how the effectiveness of such a system drops when power is not being developed; his chilling feet will put the point across. Don't carry this demonstration to extremes, as the engine could get *too* cool.

3. Switch tanks. Some trainers have only ON/OFF selections but the student should be taught that the fuel selector should be one of the first, if not *the* first, items to be considered in the event of an engine failure. He may not be able to switch to another tank in a particular training airplane, but he should at least consider it for future purposes.

4. Pick a field and check the wind direction. Tell him *how* you are finding the wind (smoke, trees, waves, etc.). Later when you give him simulated emergency landings he will pick a field and point it out to you.

Some instructors, after checking at high altitudes directly over the airport (solo) the reactions of a *specific* airplane to having the fuel selector turned OFF, have done this surreptitiously on dual flights with postsolo students to set up a "realistic" situation. This is very realistic training, but if a vapor lock occurs, the selector breaks off in the OFF position, or other such problems pop up, the simulated emergency could become a real one (Fig. 8-1). *Don't do it.*

5. Electric fuel boost ON, if so equipped.

6. Check the mixture. You should tell students that later when they are flying at higher altitudes and have the mixture well leaned, a descent to lower altitudes can cause the engine to run rough and even stop if they forget to richen the mixture.

Fig. 8-1. "Stand by and I'll see if I can't get that li'l ol' fuel selector handle back on the shaft." (More about this type of problem in Chap. 17.)

7. If the engine doesn't restart, continue the approach to the prechosen field, watching for power lines and other obstructions. You would secure fuel, ignition, and electrical switches before contact in an actual situation.

Be very cautious when you are demonstrating, or letting the student practice, simulated emergency landings. It's possible that nearly as many deaths and injuries have resulted from simulated emergencies as from the real thing.

You will likely have special considerations that apply in the area over which you are flying (Alaskan Tundra, Death Valley, or the Rockies), and you should bring these out along with the general approach to emergency landings.

The first simulated emergency landing you will demonstrate that will take you *below 500 ft should be during the low work.* When you are flying around a rectangular course or doing S-turns across a road (tell him what you are going to do), pull the carburetor heat and close the throttle. *You* should have a field picked out and be in a position to set up an approach to it *before closing the throttle.* In other words, you will deliberately set up a situation to show *how it should be done.* No more than 90° of turn should be required to make the field. You should go through the procedure, describing your moves, and recover at a safe altitude. You may get engrossed in talking it up during the procedure and suddenly find that you are lower than you meant to be. Also, you'll find that under some conditions the engine may not take power as quickly as you thought, and you get ever lower. Remember that even in sparsely populated areas you are not to get closer than 500 ft to any person, vessel, vehicle, or structure.

You will probably have prechosen fields in the practice area picked out for your emergency-landing demonstrations (where you will know where all the obstacles are).

When the student gets his first one or two of these low-altitude emergencies, close the throttle at a point in the rectangular course or S-turns from which he can easily make the field you have in mind. Let him make the simple approach with the *understanding that you will be responsible for taking over at a particular altitude and climbing out.* You should discuss or demonstrate to him the use of flaps in hitting the field.

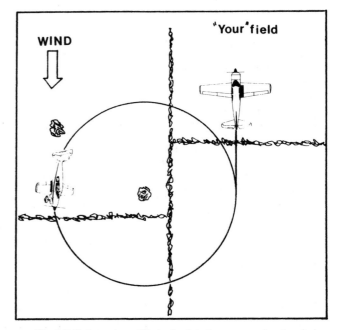

Fig. 8-2. Set up a low-altitude simulated emergency landing during a wind drift correction maneuver.

If he chooses the wrong field, or misses his approach to the field you picked, take over and *climb the airplane to the same altitude and position it was when you closed the throttle. You then demonstrate the proper approach, explaining why* you picked that field and *how* you are making the approach. Although you will have him go through the motions—checking fuel, boost pump, etc.—of an emergency in the simulated high-altitude emergency, his *first real* effort at judging how to hit a field should be from a low altitude because it's easier for him to judge from there (Fig. 8-2).

Later you can have him bring it down from a higher altitude, using the idea shown in Fig. 8-3. (Chap. 19 goes into more detail about the use of "keys" in the forced landing.)

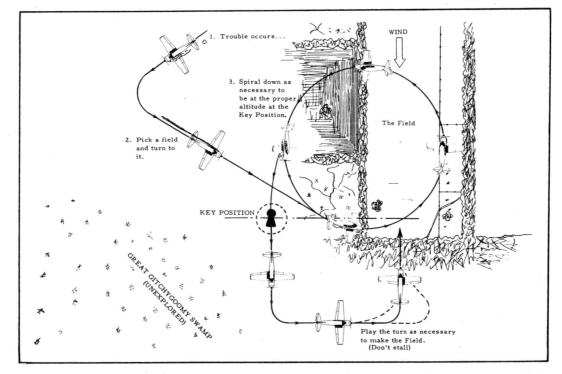

Fig. 8-3. A more complex simulated emergency landing. (See *The Student Pilot's Flight Manual.*)

Following are some *common errors* in high-altitude emergency landings:

1. A too-fast glide speed.

2. Poor field picking, choosing a plowed field over a firm pasture because of inexperience in noting field colors and tints. You'll also find that student pilots have an affinity for choosing fields with a power line or fence running across them.

3. Unaware of, or not checking wind direction and approximate speed on the ground.

4. Losing the field picked originally. This is why you should have him indicate which field he has chosen early in the exercise; if he has a tendency for disorientation you can see and work on this problem.

5. Neglecting the cockpit check; getting so involved in the glide that possible causes of the engine failure aren't considered.

6. Letting the airplane drift away from the field so that, without a reminder for wind correction from you, he would not make the field.

7. Too low to make the chosen field. This is usually a result of error 6. You might bring up the point that he can shoot for the middle of the field initially and slip or use flaps to hit this end of it, if necessary.

8. Too high at the final turn. If wind drift was no problem (see error 6), the average student will tend to turn in too soon, to "make sure he doesn't undershoot." Judicious flap use may save this situation, but tell him that he mustn't get too distracted from keeping the airplane under control.

9. Getting wrapped up in the final turn—with possibly dangerous attitudes.

10. Later when using flaps on the approach during a simulated emergency landing, he will tend to let the airspeed decay as he neglects to lower the nose each time the flaps are lowered more.

Here are some *common errors* in low-altitude forced landings:

1. Slow reaction to the emergency, which is more noticeable at this low altitude. (Actual unexpected engine failures are always followed by varying periods of shocked disbelief.)

2. Picking the wrong field.

3. Not knowing the wind direction. Students who were flying a wind drift correction maneuver when the simulated forced landing was given sometimes get blank minds concerning the wind when the throttle is closed.

4. Poor airspeed control. Usually the airspeed is too high.

5. Overshooting the field, that is, being too high.

ELECTRICAL PROBLEMS

Before the student solos you should discuss his moves in the event he smells or sees smoke from the electrical system components.

This is a good time to go over the circuit breakers with him. He'll have a "need-to-know" interest now. Before, during the introduction of the cockpit controls and systems and during the earlier flights, the circuit breakers were of only passing interest; but now things will start falling into place as he sees the *why* of their use.

If your airplane has the type of circuit breakers (or fuses) that can be pulled, later as he nears the solo cross-country stage, you should distract his attention and pull different ones at different times and have him scout out the problem. At that

point you should also show him where the spare fuses are and how to replace them.

One common origin of an electrical fire is the installation wiring of the radios. The electronic bodies themselves rarely are the culprit, but if one of the radios (wiring) is giving trouble he'll have to shut it off.

When he is in the pattern of a busy tower-controlled airport, the loss of a radio is of more consequence than if he were alone in the pattern of a small airport. If you are operating out of a busy airport, you might tell him that the procedure would be to let the tower know that he has an electrical fire and will shut off the equipment and continue his pattern and landing—to a full stop, naturally. Depending on the amount of smell and smoke, he may be better off keeping the radios in action as long as possible if traffic is a critical factor.

For the first-solo or local-solo student flying from an uncrowded airport, you'd probably be better off just telling him to secure the master switch(es) if he suspects an electrical fire, and to return and land. You might even "simulate" an electrical fire on one of your later presolo duals and have him turn the master switch off temporarily in flight, if the *Pilot's Operating Handbook* allows it. Even though it sounds ridiculous to *practice* turning off a single switch in flight (it seems that you could just talk about it on the ground), it will help ease his way in an actual situation if he has done it before. Besides, in an actual situation in his excitement he might think you said "turn off the mag switch," rather than "the master switch." If in practice he turns the master switch off himself, he'll remember which one he needs to hit in an actual situation. It will also prove once and for all that the plane will keep flying without electrical power.

If your student has a good technical background you might go into more detail so that he can eliminate the problem and still have the use of the rest of the electrical components.

Following is a possible procedure in the event of an *electrical fire:*

1. Master switch OFF.

2. All radios OFF and system circuit breakers or fuses pulled. Leave the alternator switches ON.

3. After the smoke and smell have gone, turn the master switch back ON. If the problem reoccurs, the alternator switches should be turned OFF and any electrical components used very sparingly since the battery is carrying the electrical load on its stored capacity.

4. If, when the master switch is turned back ON (all radios OFF and circuit breakers/fuses out, but the alternator switches are ON) there's *no* problem, turn on the components individually until the problem reoccurs. That last item was the culprit and it should be isolated. The student should be taught to return to the airport as soon as practicable after this and *to fly a normal traffic pattern and approach* with no rushing. As discussed in Chap. 1, accidents have occurred because of excessive reactions to a comparatively minor problem.

While the master-switch-OFF-and-return-to-the-airport response is fine for local-solo flights, you should have an emergency checklist available for solo cross-countries. In fact, it should be available in the airplane at all times (for *your* use, too). You might type out a procedure somewhat like the one just given, but applicable to *your* airplane. It should be a part of, or attached to, the normal checklist so that he will be used to seeing it and know where it is. (If the flaps are electrical, they will be out of action. Discuss this.)

DIVERTING TO OTHER AIRPORTS

This will not be a real factor in the presolo phase, but if there are one or more airports reasonably near the practice area of your airport, point it out to the student during one of the early practice sessions. If it is larger or has more runways than yours, you might even use it in that first hour of concentrated takeoffs and landings to more easily introduce him to the idea with fewer distractions than at your home base. It's also common practice in some areas for the instructor to take the student to a nearby airport for his first solo if crosswinds or traffic are causing problems at the home airport. This doesn't mean that the instructor has the student fly over there and immediately solo at the strange airport; there should be a transition period of dual pattern work there.

In flying to another airport, whether in a presolo period or after solo when you plan to turn him loose solo in the practice area, point out landmarks along the route. If the home airport is sometimes fogged in or has sudden bad crosswinds and the "auxiliary" airport has to be used for landing, be sure to tell him. The thing that should be made clear is that you may fly over to pick him up when, say, the crosswind at the home base is too much for a pilot of *his experience.* (*Or* the home airport may be fogged in and you have to drive over to get him.)

If you have another available airport, you should take him into it at least once before you put him in a situation where he has to go into it solo. If he's been there before, his mind will be more at ease if he has to go there in what to him is an emergency situation.

This problem is most critical in the period when he is first going out solo to the practice area and has not gone cross-country, and so does not have experience operating into and out of strange airports.

Don't forget to make an entry in the student's logbook if you find him competent to make takeoffs and landings at other airports within 25 NM from your home base (FAR 61).

One thing you should assure him of is that if he decides to go to the other airport, you won't jump all over him about it. (You'll probably be in contact on Unicom anyway.) It's very important for the student to understand that he will not get in trouble with you if he chooses the safe way, even though you think he could have made it to the home airport. If he feels pressure, conscious or unconscious, from you to come home or to *not* take it around on an approach that doesn't "feel right," he might push too hard to do what he thinks would please you and have an accident.

If that nearby airport is bigger and better than yours, don't become too dependent on it for shooting takeoffs and landings. The student will still have to fly solo in and out of your smaller airport—unless you plan to transfer all operations to the other airport. Use the other one when needed, but plan on his first and subsequent solo flights to be out of your airport under normal conditions. If he does all of his takeoffs and landings and first solo on that 5000-by-150-ft runway, he may have a setback when he starts getting checked out on your 2500-by-50-ft strip. There will be more detail about this in later chapters.

SUMMARY

Simulated emergencies, like the various other maneuvers, should be introduced in a comparatively simple way with more complex problems following as the student's experience increases. A few years back there was a tendency in aviation to try to tone down the idea that anything bad could happen in an airplane, and some said that flying had no worrisome problems and could be done as easily as driving a car. ("Drive into the air.") This attitude may be considered acceptable when you are talking to nonpilots, but the person taking flight training will have to be taught the bad along with the good. Interestingly enough, after he's had practice at emergency situations, the average student is usually relieved and feels much better about the subject. An emergency isn't nearly so bad as he'd heard or imagined, and he learns that he can even get himself and the airplane down in one piece if he does as he's been taught.

As you give unexpected emergencies later in the presolo stage you can get an idea how the student would react in a real situation, which will have a definite effect on when you solo him. Mechanical "perfection" of takeoffs, landings, and traffic patterns are not the only items to be considered in preparing the student for his first solo.

Note that there are basically two kinds of emergency situations: (1) immediate, such as an engine failure on takeoff (or less immediate, such as an engine failure at altitude) and (2) deferred, for example, a landing gear problem when fuel is available to work with it.

During simulated emergency practice you would continue training the student in the decision-making process, teaching him that overreacting to a fairly minor problem could cause a worse situation. For example, if a door opens right after takeoff or a seat belt or shoulder harness hangs out, banging against the fuselage, overreacting could cause the pilot to put the airplane down gear-up on or past the runway or to spin out of a tight turn while trying to get back to land the airplane. You should describe and discuss these problems, whether you demonstrate the problem at altitude or not. (Check the *Pilot's Operating Handbook* of your airplane for suggestions.)

One subject that should be mentioned is that it's considerate to avoid using fields containing livestock in your simulated emergencies (even though in an actual situation you could land to one side away from them), because the noise of opening the throttle to break off the exercise could cause panic in the herd. (The worst thing you could do would be to climb out low and noisy over a mink farm.) You can get irate calls from beef and chicken farmers and shoot the airport's public relations program. Also try to avoid giving simulated emergencies to fields that would require passing close over, or close by, a house. Remember that *you* are responsible for safe operations during simulated emergencies and that you will be expected to abide by FAR 91.119 *Minimum Safe Altitudes; General.*

More complex emergencies will be covered as they pertain to later phases of training.

Be sure it's understood that the student is not to practice simulated emergencies solo.

ELEMENTARY STALLS

BACKGROUND

You'll find that even with some experienced pilots, stalls remain a source of anxiety. Stalls should be respected but not necessarily feared — it depends on the situation.

For instance, the aerodynamic stall might be compared with an automobile skid. (Sure, the problem in the car is rotation around the vertical axis and the airplane has a pitch problem, but in both cases, for recovery the "vehicle" has to be pointed in the direction it's traveling.)

Compare an airplane stalling at 6000 ft AGL with an automobile skidding on ice on a mall parking lot where there are no utility poles or other cars. In either case a normal recovery can be done without danger to vehicles or occupants. On the other hand, a stall occurring on the turn to final, or a car skidding on a busy highway with oncoming traffic, can be fatal. But in both examples, the *aerodynamics/physics* of the problems (stalls and skids) are exactly the same. The danger is created by *where* the exercise occurs.

Yet as an instructor you'll run into certificated and experienced pilots whose reactions to stalls at 6000 ft are unnatural. They are tense, overcontrol the recovery, and show a definite fear of what's happening. This usually is the result of an ongoing problem; they were not taught proper stall recoveries and/or don't know what the airplane is doing. If a student sees his instructor literally "climbing the walls," leaning away from that dropping wing at the stall break, he'll get a deep-seated fear of stalls and *The Unknown Horrors That Must Follow.* Aspiring flight instructors should get more spin training than is currently being given.

There's no point, of course, in giving the student stalls that would not apply to the type of flying he will be doing before and shortly after solo. To have him highly proficient in accelerated stalls at this stage is not as important as being sure he is aware of possible problems in the climbout after takeoff (takeoff and departure stall) or in a low and slow turn to final (approach to landing stall). However, you should demonstrate a simple accelerated stall-type situation to back up the point you make in the initial discussion that *stalls are a function of angle of attack, not airspeed.* The best route to take in the first full-fledged practice session of stalls is to discuss them as an aid to the landing, let the student work on this aspect for a while, and then go into the elementary turning stalls. The interesting thing about this technique is that the student is usually so wrapped up in practicing "landings" at altitude that his fear of stalls decreases. He is learning to unconsciously make good recoveries, using both elevators and power. You'll teach him to respect stalls and know how to avoid them at low altitudes and how best to recover when he doesn't avoid them. Work to keep him from having an unreasoning fear of them, which could cause him to clutch up in a tight situation and get into worse trouble.

While you won't demonstrate or have him practice stalls from extreme attitudes, you should make sure that the student carries the maneuver to the point that the airplane is completely stalled unless you *deliberately* want him to practice recoveries from an approach-to-the-stall-only situation. It appears that many private pilots are afraid to get the airplane into a fully stalled condition — they've either never been there (maybe their instructor didn't like stalls, either) or were there and got the wrong idea. When you are instructing some private pilots for higher certificates and the requirements for stalls come up, note their reaction. They'll bring the nose up, and shortly after the stall warner goes off or at the first sign of a buffet, they'll lower the nose and in relief say something like "Well, there it was!"

Always have plenty of altitude when practicing stalls, not only for the obvious safety reasons, but also to put the student more at ease. An altitude of 1500 ft above the ground should be the minimum for recovery, and 3000 ft is even better for the beginning student. *Always clear the area for straight-ahead stalls and keep looking around in the turning stalls.*

STALLS AS AN AID TO LANDING PRACTICE

■ **Preparation.** A chalkboard and model are necessary and you might have him read the references in the Bibliography.

■ **Explanation**

STALLS IN GENERAL. You most likely eased into stalls in one of the earlier presolo lessons during climbing, glides, and slow flight, so the student will be generally aware of what happens in the airplane when the nose is too high for the power being used (or available). However, at this point he thinks of the stalls as only resulting from too high a nose position. Your explanation here should emphasize that *the stall is a function of angle of attack only and that the airplane can be stalled at any airspeed, attitude, and/or power setting.* Too many pilots have stalled on the final turn or on the final itself because this idea was not firmly fixed in their minds. The airplane had a nose-low attitude, yet a stall accident occurred. One failing with current stall training is that even though stalls are practiced from "real" situations (climbouts and approaches are set up and stall recoveries practiced), the attitude is usually exaggerated and easily recognized. This is obviously not the case when a pilot inadvertently stalls on climbout or approach. After the student has practiced a few of these stalls, demonstrate at altitude how that extra back pressure in a final turn can cause problems even though the nose is low.

What — A stall is a condition in which the angle of attack is too great to allow the air to maintain the smooth flow for required lift.

The stall has nothing to do with the power being carried. The airplane engine doesn't stop as in the case of a car stalling — a notion the student may have gotten from hearing the word. The power being carried *does* have an effect on the stall speed; because at high power settings a small vertical component of thrust is present at the nose-high attitude as set up in the deliberate, practice stall. If an airplane had a thrust-to-weight ratio of 1 (the thrust used is equal to the weight), it could hang motionless on the propeller—or act as a helicopter.

The elevators stall or unstall the airplane. Abuse of this control will cause the stall in *any* attitude or power setting, and proper use of the elevators will recover from a stall at any airspeed or attitude. If you also teach aerobatics, you know that one of the problems in the loop is stalling the airplane after it is over the top and the nose is pointed straight down. The student unconsciously feels that since the nose is pointed straight down, he can pull back briskly with no fear of stalling. What happens is that his application of back pressure is too much for the airspeed available (or in more technically correct terms, a high angle of attack already exists) and the stall break occurs with the airplane rolling as much as 90° from the original position. He is puzzled since he unconsciously figures that it's impossible to stall an airplane when it's pointed down.

You may demonstrate an accelerated stall or two, but as noted earlier, don't push that stall too hard at this time. It could induce added anxiety and slow the training process. (He'll probably still be thinking about those accelerated stalls later when you are talking about and demonstrating stalls as applied to landings.) You might draw Fig. 9-1 on the chalkboard and discuss coefficient of lift and angle of attack and what happens at the stall. The dashed line shows that some airplanes like delta wing fighters don't have a definite stall break but have a prodigious sink rate in that flight regime.

Since the average training airplane at present does not have an angle of attack indicator, the airspeed indicator is the instrument that *sometimes* warns of an impending stall. Tell the student that the indicated airspeed at the stall depends on the airplane's weight, the acceleration forces present, and the instrument and position errors—plus parallax and other factors. In short, the airspeed indicator is *not* a good aid in finding the stall; but as the inveterate gambler said, "It's the only game in town," for some older airplanes.

Most of the newly manufactured airplanes have a stall warning system that, unlike the airspeed indicator, warns when the airplane approaches a *critical angle of attack,* which is what the stall is all about. Fig. 9-2 shows one type of indicator and you should explain how it works. One thing the student can't understand when first looking at the tab on the wing is how the airplane could get to an angle of attack great enough for the tab to move *forward* and make contact to give the warning by a light in the cockpit or a horn.

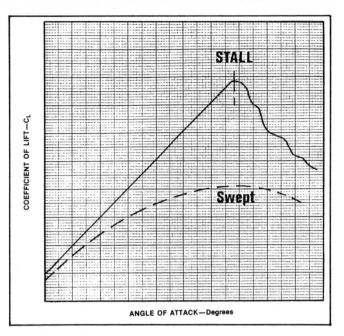

Fig. 9-1. Coefficient of lift versus angle of attack in straight- and swept-wing airplanes.

Another type of stall warning system uses a reed in an opening in the leading edge of the wing, and it changes its tone to higher and higher pitch as the stall is approached. If it starts playing a hymn, you could be in a little trouble.

The pilot depends on *sight* (as mentioned in discussing the loop, sight is not always the best aid since the airplane can be stalled at any attitude), *sound* (for some airplanes the sound of the stall warner is about the *only* sound aid), and *feel* (the controls feel mushy and buffeting of the airplane may be felt, but this sense is less valuable with airplanes with boosted elevator control—unless a stickshaker warning system is installed).

The stall recovery can be summed up in one sentence: *To recover from a stall, point the airplane the way it's going, and this is done with the elevator, or stabilator.*

You might note that an airplane can be temporarily stalled by vertical gusts, but the primary cause of accidental stalls is the pilot's mishandling of the elevator controls.

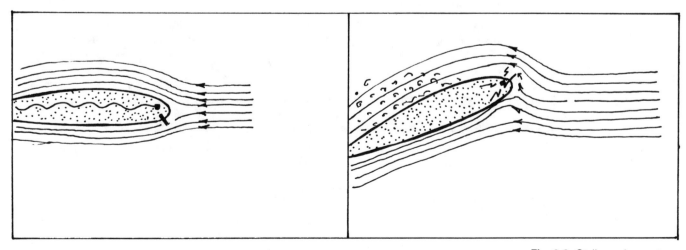

Fig. 9-2. Stall warning system.

Why — Stalls are practiced so that the pilot may recognize their onset and so be able to avoid them and effect the most efficient recovery should an inadvertent stall occur.

The student will be taught stalls in different attitudes and power settings so that he will become familiar with the different situations in which stalls may occur.

How — Any normal stall is entered by an *increase of back pressure* regardless of the airplane's attitude. The stall is recovered by relaxing back pressure, and the only effect of added power is to reduce the altitude loss during the recovery. The student should think of stall recovery actions as the simultaneous forward movement of both hands (throttle and wheel) to expedite the return to un-stalled flight. You might tell him that you will practice stalls and recoveries without power, to show the idea that the recovery from the stall is done by the elevators and not the throttle.

■ **Demonstration.** In the normal landing, whether in a tailwheel or a nosewheel type, the airplane should be stalled or very close to it at touchdown. You will practice a number of "approaches" and "landings" at altitude before seriously attacking the problems at the airport. (Fig. 9-3).

You might explain the procedure by saying the airplane will be set up in the *landing configuration,* power at idle, at a safe altitude. The student will set up the recommended approach speed. As the altitude is approached you will read off the numbers — 100 ft, 50 ft, etc. — and at about 20 ft above the "ground," you'll say, "Start your round-out." The student is to start looking along the left side of the cowling and gradually ease the nose up. (More about this later.) Have him use trim for the glide. The usual error is for him to raise the nose too fast initially, which would cause him to balloon spectacularly if close to the ground; but this is not noticeable at this altitude. You will have to tell him the rate of change of pitch to the proper landing attitude; and you may find yourself saying, "Not so fast," and then shortly adding, "Now bring it up some more." You may be doing this until he gets the idea that the rate of aft elevator movement is inversely proportional to his airspeed. As the airplane slows, the elevators must be deflected a greater amount to get the desired result. Make it a point to indicate the proper nose position for touchdown ("That's it, now pin it there by continuing back pressure"). He is to recover from the stall by relaxing back pressure and adding full power simultaneously. Depending on his experience and the altitude situation (it's best to practice initially up

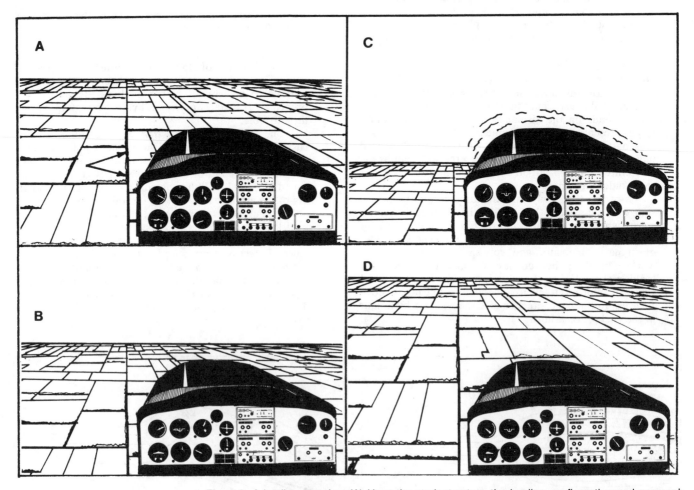

Fig. 9-3. A landing exercise. (A) Have the student set up the landing configuration and approach speed. Line up over a straight stretch of road or railroad or a reference on the horizon. (B) He should make the transition to the landing attitude (about halfway here), looking along the left cowling. Remind him to watch his lineup and to keep the wings level during the transition. Caution him about too rapid an application of back pressure initially. (C) After the proper landing attitude is reached, he is to pin the nose at that position by continued and increased back pressure. He will check the bank attitude of the airplane from the corners of his vision. (D) When the stall breaks, the landing exercise is complete. Don't worry about slight direction change with the break (theoretically, he's already landed). Have him recover from the stall and either set up another exercise immediately or clean up the airplane and climb.

in smooth air), you may have him clean up the airplane for a climb—or you may have him close the throttle and start immediately on another exercise. It's important that you tell him about the varying rates of elevator application. You are *simulating* the landing situation as closely as possible, and if you allow him to start the round-out with rapid elevator action, he will transfer this practice to the actual landing. One of your jobs during all this discussion and demonstration and practice is to make sure that no midair collision occurs, so keep looking around.

Mention also that although he is looking out the left side of the airplane, he must be aware of the wing attitude out of the corners of his eyes and must keep the wings level. If an interstate highway or other long straight reference is available he should also keep the airplane directionally straight during the "landing" process.

■ **Practice.** It may take a number of these "approaches and landings" before you feel that he is ready to begin the actual landings. *These are not just repetitive exercises. You will have to be instructing constantly throughout each one.*

Here are the *common errors* made while a student is learning landing exercise stalls:

1. Improper approach speed, usually too slow when the landing-flaps setting is used. The student tends to keep the nose up at the no-flaps attitude.

2. Raising the nose too fast during the early part of the transition.

3. Staring up blindly over the nose during the exercise. You cannot see directly over the nose of many airplanes in the landing attitude, even with flaps being used; you should break him of this habit before starting actual landing practice. This high-altitude simulation of landings will allow you to watch him during the process to check where he is looking, a practice that is not recommended during an actual landing or at least not in the early phases of landings. This is a good place to stop such habits as trying to stare over the nose.

4. Lateral control problems, not keeping the wings level during the approach and landing. Warn him that while it makes little difference at altitude, during the landing a wing being down can cause the airplane to "drift" off a narrow runway or even on a wide runway, and that such drift can cause directional control problems at touchdown.

5. Letting the nose ease back down after the landing attitude has been established, but before the stall has occurred.

6. When the stall break occurs, not relaxing enough back pressure when power is applied; a secondary stall and some lateral control problems arise.

■ **Evaluation and Review.** If he is able to ease the nose up and pin it at the landing attitude, is looking out at the proper place, and has the airplane under reasonable lateral control, you should be able to let him shoot a couple of full-stop landings (with you ready to take over if necessary) at the end of this dual period. You will likely give him a couple of short periods of this practice, both before and after the other stalls have been explored.

OTHER STRAIGHT-AHEAD STALLS

The student should get practice in other straight stalls before moving on to the elementary turning stalls. After he has had a chance to practice a few simulated landings at altitude and has gotten used to them, you should give him straight stalls (power-on and power-off) to work on the idea of lateral and directional control in slightly more exaggerated pitch attitudes. He should be taught that coordinated use of ailerons and rudder is the best method for maintaining lateral control. As will be covered in a later discussion of spins, cross-controlling at the stall break can result in lateral and directional control problems. Coordinated use of ailerons and rudder, especially after the stall recovery is initiated, is the most effective means of maintaining or regaining lateral control for later model airplanes. It's possible to raise a wing by pushing the opposite rudder for most airplanes, but for those with full tip tanks (inertia problem) and small rudders (low control effectiveness), such action is not effective.

As an overall practice it would be better not to try to raise a wing (using ailerons and rudder) in any stall until the nose is lowered to the horizon or below in the recovery. While most airplanes will respond reasonably well to the lateral controls even in the stall, the main idea is to decrease the angle of attack; then the wings may be leveled. Don't let your students get into the habit of worrying about the wing attitude and neglecting the major requirement of relaxing back pressure to recover from the stall.

If there has been one other failing in stall instruction (in addition to the timid approach to the subject) it's that of a too-brisk recovery. When this happens the nose is pushed too low at a rate that's too fast, "to make sure the stall is broken" and the student's stomach is left up there a hundred feet or so above the airplane for a while. This type of recovery can induce fear of stalls and is inefficient as well. *The proper stall recovery is that technique requiring the least loss of altitude, even if the airplane is near stalling for a brief period during the recovery.* If the airplane is near the ground when the stall occurs, the nose should not be shoved over hard for obvious reasons; and in addition the recovery might even be delayed by this method, compared with a less-athletic move. Suppose a pilot has been taught to shove the nose over *well* below the horizon for all of his stall recoveries; this has worked well for his high-altitude practice sessions because altitude loss was of little significance up there. Then comes the day he gets into a stall condition on final for some reason or another, and following his training he shoves the nose well down. He then sees how close the ground is and quickly pulls the nose up. The airspeed is still at the stall value just after he put the nose down, and his abrupt reaction will ensure that the airplane will be fully stalled—if it wasn't before. If, instead, he had lowered the nose at a less-steep angle his recovery would have been made with less loss of altitude and no complete stall break. Fig. 9-4 gives an exaggerated comparison of the two methods.

Your philosophy of teaching stall recoveries should be to begrudge every degree of nose-down pitch below level-flight attitude. This is *not* to say that you will have the student always stop the nose in the level attitude as the recovery is made—you'll encounter a lot of secondary stalls that way. It means that you should teach the student that the nose should not be lowered any more than necessary to recover from the stall. In a stall where the break occurred at a high nose-up attitude, it's impossible to recover at a level-flight attitude; you'll get another stall. *The nose attitude below the horizon necessary to recover from a stall is proportional to the nose attitude above the horizon when the stall break occurred.* The higher the nose at the break, the lower the nose must be for recovery and *the more time and altitude required for recovery from the break position.* Fig. 9-5 shows some examples.

You may want the student to practice approaches to stalls ("partial" or imminent stalls), normal stalls, and the complete or exaggeratedly nose-high stalls both *power-on* and *power-off.* Fig. 9-5 is an exaggerated comparison of the three types. These are useful training exercises, but you may not want to

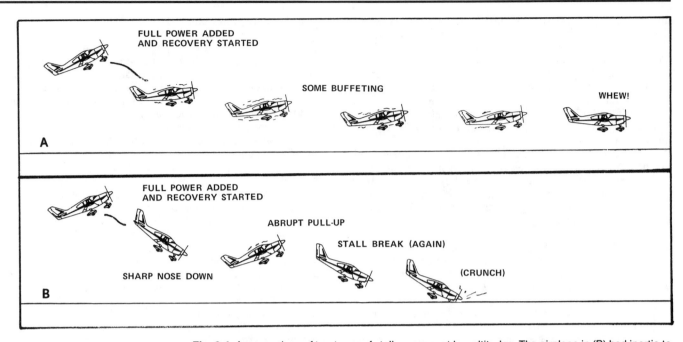

Fig. 9-4. A comparison of two types of stall recovery at low altitudes. The airplane in (B) had inertia to overcome; it was moving downward rapidly during the first recovery. When up-elevator was applied, the airplane rotated but the downward motion assured too high an angle of attack—and a complete stall.

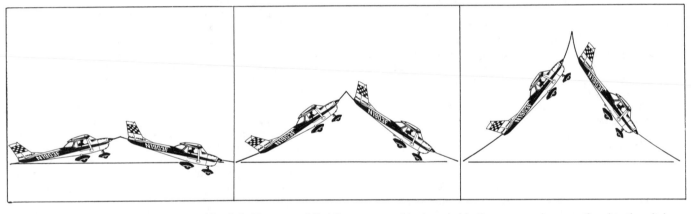

Fig. 9-5. The amount that the nose must be lowered in the recovery is proportional to the pitch-up attitude at the break.

give them to all of your students. Full power is applied for recovery in each case.

1. *Partial, approach to, or imminent stall* (power-off, clean). These three terms are often used interchangeably, though "imminent" is losing popularity at this writing. This one is little more than easing the nose up, and when the indications of an approaching stall are seen, heard, and felt, the nose is lowered and full power is applied.

These partial, approaches to, or imminent stalls should also be practiced later in turns, both power-on, clean, and in the *landing configuration* as if an approach is being made and the airplane is near a stalled condition. Teach the student that he is to recover from this *near*-stalled condition with the least loss of altitude (which is also the case for a situation where the break has occurred). Power should always be increased to the maximum *automatically* unless you ask him to demonstrate a power-off recovery. Such power-off recovery demonstrations should be kept to a minimum to avoid establishing a habit of not using power for recoveries in actual situations. The stall

break should never be allowed to occur in this one. This is a good exercise for people who are dubious about stalls, and it can lead into the normal stall.

2. *Normal stall* (power-off, clean). In this one, the nose is eased up to about 15° above the horizon and pinned there by continued application of back pressure. *The recovery is made when either the break occurs or the wheel is full aft against the stop.* This is the definition of the stall, since limited elevators on some airplanes may preclude a definite break at lower pitch attitudes.

3. *Complete or full stall* (power-off, clean). This is a good exaggerated stall to show that the nose must be lowered more for recovery if it is well up above the horizon when the break occurs. Ease the nose up to about 30° above the horizon and pin it there by continued back pressure. After the break occurs, hold the wheel full back until the nose crosses the horizon, and then release the back pressure as full power is applied. Some of the more timid souls you'll be flying with won't consider this their favorite introduction to stalls, so save

it for later with them. This stall is a good exercise in lateral control effects but isn't a realistic approach to the subject.

Airplanes of earlier times lost control effectiveness in alphabetical order (A-E-R) as the stall was approached. The *a*ilerons were the first to become ineffective, followed by the *e*levators (you had to keep moving the stick or wheel back as the airspeed decayed—which, of course, is required with the latest models) with the *r*udder being effective in most cases throughout the stall and recovery (which is not always the case with some later-model airplanes).

One of the required maneuvers in those earlier airplanes was the *rudder exercise stall,* in which the airplane's nose was smoothly brought up to 45° above the horizon and pinned there by increasing back-stick or -wheel pressure. The stick or wheel was kept full back (but centered laterally) and any wing dropping tendency was corrected by opposite rudder. (And too much or too long application of corrective rudder could result in the *other* wing dropping, to the pilot's surprise and sometimes dismay.) Some airplanes were so laterally unstable that rapid movements of rudder were necessary to keep the wings level at, and after, the stall break—hence the name *rudder exercise stall.* Use of the aileron to raise a wing in the stall was generally counterproductive and the down aileron could drag that wing back and cause a spin in that direction.

Another problem was that a new private pilot, after going through numerous rudder exercise stalls and being very sharp at them, might 5 years later get into a stall condition at a low altitude. Did he remember to use rudder-only to keep the wings level? Probably not, since he had been using aileron (and rudder) to raise a wing in the hundreds of hours of normal flight since his stall training (and very few private pilots go back for interim stall training). The FAA requires, as part of the manufacturer's certification process, that the airplane be laterally controllable during the stall recovery (within certain limitations) with *normal use of controls* (FAR 23). It makes more sense this way since the pilot's reaction to a wingdrop in this regime would be to use normal control (opposite aileron and rudder) to raise the wing.

■ **Power Effects in the Stall.** The stalls just described were power-off. They can also be done with power, which means good news and bad news.

The good news is that because of the power already being used, the stall recovery (even if no more power is added) is quicker and has less loss of altitude than would be the case in the power-off condition. The stall speed is less and the airplane is flying again at a lower speed during the recovery.

Now for the bad news: The nose must be raised higher above the horizon for the same type of stall, and this could be uncomfortable for some students. Also, the airplane is definitely less laterally stable since "torque" effects are present and the student will have more trouble with a wing dropping and heading changes. A case in point is that up until the stall break in the power-on stall the pilot must cope with the slipstream effect and hold right rudder to maintain heading. As the stall breaks, the airplane may temporarily move down out of the slipstream, and since he is still holding a good amount of right rudder (which isn't needed anymore), he could in extreme cases be introduced to the idea of right-spin entries.

ELEMENTARY TURNING STALLS

■ **Takeoff and Departure Stalls.** This stall should be demonstrated and practiced before solo because it simulates a situation that could occur during the student's solo practice and is not violent enough to give him great concern if he has already had the stalls covered in this chapter.

As was noted, very few accidental stalls occur from an exaggerated pitch attitude; the pilot can see that trouble is pending in such a position. Stall/spin accidents happen from less-radical attitudes, when the pilot is caught by surprise. With this in mind, don't make the pitch attitudes in these and approach to landing stalls too exaggerated.

What—The takeoff and departure stalls simulate a condition in which the pilot immediately after takeoff pulls the nose inadvertently (or deliberately, to show off) into a steep climb or climbing turn.

Why—This maneuver is used to teach the student how to recognize and recover from a dangerous stall situation close to the ground. The main idea is to teach him to recognize the impending stall and avoid getting into such a predicament.

How—While this is a turning stall and no formal clearing procedure is necessary, the student should keep checking outside the cockpit to look for other aircraft. The straight-ahead version of this maneuver should be preceded by clearing turns, since the nose will remain in one spot on the horizon and will obscure forward vision in that area as the stall is approached.

Set up the situation as follows:

1. Throttle back and slow the airplane to just above normal lift-off speed. The stall should be practiced using 0° flaps as well as the flaps required for short- and soft-field takeoffs.

2. Add climb power as the nose is raised; a bank of 15–20° is established in a prechosen direction. Have the student tell you beforehand which way he plans to turn so that you can help watch for other airplanes in that direction.

3. The nose is raised to a higher-than-normal climb attitude and the airplane is stalled in the turn.

4. The recovery is to release back pressure *first* and then level the wings with coordinated controls. Expect that the "higher" wing on most training airplanes will drop immediately following the stall break, although some few airplanes may fool you and the "lower" wing will drop, particularly if the ball is kept centered.

Before teaching stalls you should be familiar with the reaction of your airplane to various stall conditions; it's pretty embarrassing to grandly announce to a student that the airplane will now do thus-and-so and then find it does something else. One thing, too, you'll find in instructing is that an airplane, even a particular one you've been flying for many hours, sometimes develops a mind of its own and in your stall demonstration decides to drop that "other wing" for a change. This usually happens after a good discussion on the ground about how your airplane reacts, compared with others, and you've confidently described just what will happen. You start demonstrating the stall, and it doesn't happen that way—and there you are with egg on your face. On the next flight everything works fine. Gremlins.

Fig. 9-6 shows what *usually* happens during a climbing turn stall. The ball will generally move away from the wing that is going to drop (a slip occurs, tending to blanket the higher wing) and the procedure will usually be in this order:

1. The stall break occurs in the banked attitude.

2. The nose drops directly down; slipping occurs and can be seen by the ball moving to the "lower" side of the instrument.

3. The higher wing, now "away from the ball," drops.

4. Recovery is made by relaxing back pressure and leveling the wings *after* the nose is at or below the horizon.

While there are exceptions, you will find that normally 2 happens before 3; the break occurs first, followed immediately by the slip, and then the wing drops. Check this for the airplane you are using.

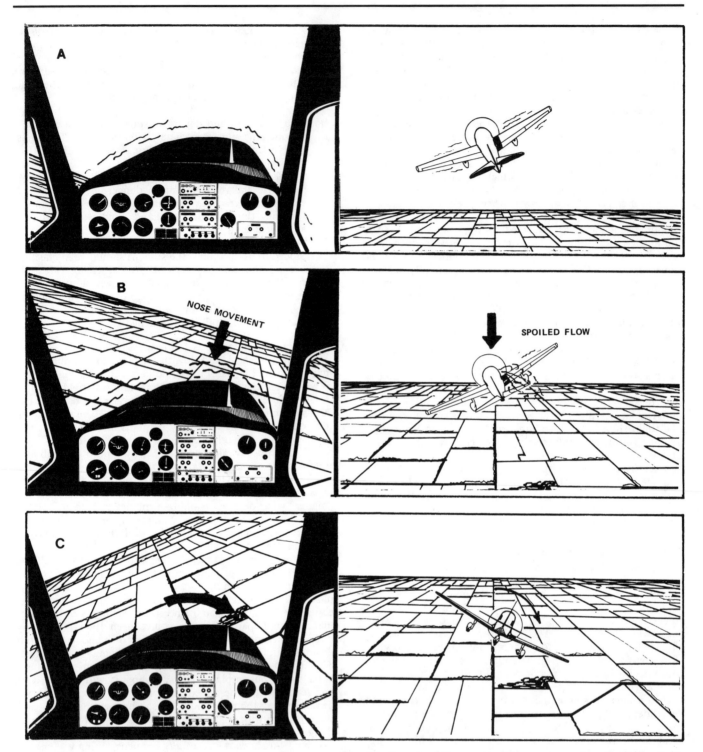

Fig. 9-6. Reactions of many airplanes in a takeoff and departure stall.

You can expect the following *common errors* when the student gets his first shot at takeoff and departure stalls:

1. Letting the bank get too steep. The nose lowers and the airplane mushes and buffets in a tight spiral when the back pressure is increased.

2. Raising the nose too high. An exaggerated, unrealistic situation is set up.

3. Not raising the nose high enough to attain the stall. For the timid student, the normal climb nose position (or a little lower) is just fine, thank you.

4. Trying to raise a wing after the break, but before the relaxation of back pressure is initiated.

5. Not relaxing back pressure after the stall break (usually a sign of extra tenseness).

6. Not using coordinated controls in the wing-leveling process, after the nose is lowered.

7. Failure to apply more power, if available, in the recovery.

■ Approach to Landing Stalls

FLAPS AND STALLS. These stalls are less obvious to the student because of the lower nose attitude at the break. Fig. 9-7 shows attitudes required for three different stalls.

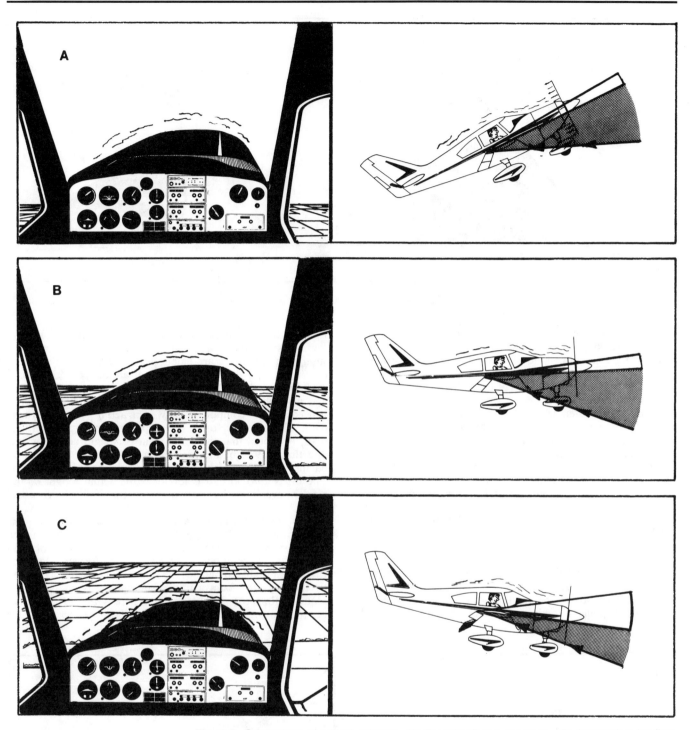

Fig. 9-7. Comparison of critical attitudes. (A) For takeoff and departure. (B) Approach-to-landing stalls, clean. (C) With flaps.

A point you should make with your more technically minded presolo students and certainly with all advanced students is that the airplane will stall at a lower angle of attack (and hence a lower nose attitude) with flaps than without.

Fig. 9-8 compares the angles of attack and coefficients of lift for a particular wing, clean and with flaps. Note that although the wing with flaps has a higher maximum coefficient of lift at the stall (and hence the airplane would stall at a slower indicated or calibrated airspeed), the stall angle of attack and pitch attitude are much lower for a given power setting. Compare Figs. 9-7B and 9-7C for a practical example. Later, this same principle will be used to show why the inside

wing in a cross-control stall is the first to go, even though the down aileron on that side should be "holding it up."

The example wing has a nonsymmetrical airfoil (as do most nonaerobatic airplanes today). Hence the zero lift angle of attack will be negative even in the clean condition.

Suppose the plain wing (no flaps) will stall at an angle of attack of 15°, and at a particular power setting it will stall at an *attitude* of 15° nose-up as well. The pilot in this simplified situation sees that the airplane stalls when the nose is 15° above the horizon. He moves the nose from level flight up 15° and gets a stall. He now "knows" that as long as he keeps the nose below a pitch of 15° above the horizon at that power

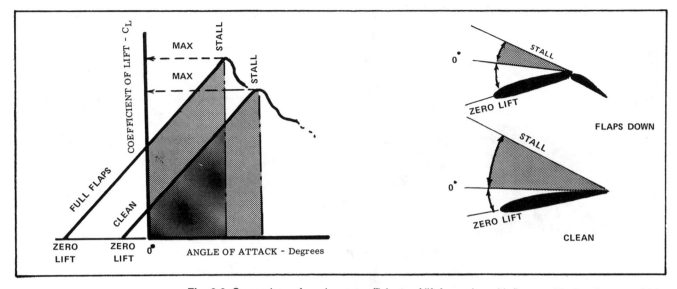

Fig. 9-8. Comparison of maximum coefficients of lift for a wing with flaps and in the clean condition.

setting he doesn't need to look at the airspeed to avoid a stall.

He is making an approach at that same power setting with *full flaps* and unconsciously eases back on the wheel and suddenly gets a stall well before he thinks he's in trouble, because he was thinking in terms of the stall *attitude in the no-flaps condition.*

One fallacy often accepted about flaps is that they "give more lift" than would be found for the same unflapped wing throughout the approach and landing. Disregard the tail-down force in the following discussion.

In the landing the airplane's lift should theoretically just equal its weight at touchdown. If at that instant the lift is greater than weight, the airplane will accelerate upward. If, on the other hand, lift is much less than weight at the touch-down, the airplane will be accelerating downward; and de-pending on how great the difference in the two forces the result can be anything from a "solid" landing to structural failure of the gear.

As far as flaps giving *more* lift on the approach, the actual situation is just the opposite. The mechanics of flight require that for a state of equilibrium to exist in a straight glide (the airplane is maintaining a constant airspeed and

flight path), the summation of the forces must be zero and the same must stand for the moments acting on the airplane. As-suming the moments are balanced, take a look at the forces existing in a power-off, *no-flaps* constant-airspeed glide for a 2000-lb airplane (Fig. 9-9A). The forces are measured parallel and perpendicular to the flight path. The aerodynamic drag just equals the component of weight acting along the flight path. Checking the forces acting perpendicular to the flight path, it's noted that the component of weight acting in that direction is 1987 lb; and for equilibrium to exist, this must be balanced by lift, which acts perpendicular to the flight path. The lift is 1987 lb and is less than the actual weight of the airplane even though the requirements for equilibrium are sat-isfied.

In Fig. 9-9B, the airplane is using *full flaps* in the power-off approach at the same airspeed, and the path is much steeper since the nose must be pointed down farther. This is to assure that the component of weight acting *forward* along the flight path is able to balance the sharply increased drag caused by the flaps (or from a pilot's standpoint, to maintain the required airspeed). The lift required here is *less* (1970 lb) than required in the no-flap approach, *but both will have to have*

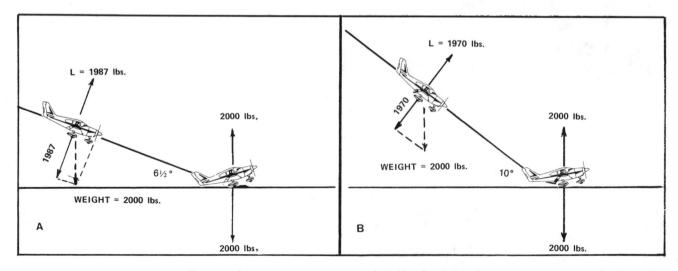

Fig. 9-9. Lift required during an approach and landing for an airplane—clean and using full flaps. The angles are exaggerated.

2000 lb at touchdown. This means the airplane with flaps has "farther to go" to get that lift, and the rotation for landing will be more definite (and pull more g's), even in normal conditions. If the "flapped" airplane is very slow with a steep angle of descent, the rotation to get the desired lift may result in a stall and drop-in.

So, it boils down to the fact that the new student may be fooled by the pitch attitudes when he starts practicing flaps-down, power-off stalls. But getting back to practical matters:

What—The approach-to-landing stall is a stall set up in a simulated approach condition from straight or turning flight.

Why—It is used to show the student how such stalls can occur and how to recover from them should they occur, and most important, how to *avoid* getting into such a stall situation.

How—There are two methods of setting up the stall.

The first is to close the throttle and simulate a landing at altitude, deliberately raising the nose to the landing attitude and pinning it there until the stall breaks. The recovery should be at the instant of break, and full power is applied simultaneously with the relaxation of back pressure. Generally, unlike the landing practice at altitude, *don't* have the student wait until the nose has dropped before initiating recovery; you'll want to have him get the habit of recovering as soon as he realizes the airplane is on its way to stalling or has stalled.

After the recovery is initiated, he is to level the wings with *coordinated* use of controls unless the airplane specifically requires only rudder action. This brings up a point you may have already considered: At low airspeeds adverse aileron yaw is more noticeable than at cruise and *more* rudder must be used. In recovering from the stall and leveling the wings, it would appear that most of the work has been done by the rudder, since that control was deflected more than the ailerons. The fact is that the rudder requires more effort because: (1) the adverse aileron yaw is stronger at the speed close to the stall, requiring more rudder, and (2) the rudder is less effective because of the lack of dynamic pressure and *requires more deflection to get a particular response.* The effects are additive. You wouldn't recover with the rudder alone in most airplanes, but use whatever amount is necessary to coordinate with the ailerons.

The second type of approach to landing stalls is more realistic and follows the idea that the nose is raised only enough to produce a stall—which may be about level-flight attitude in some airplanes (power-off) with full flaps being used.

Have the student set up an approach speed (power-off) and configuration and establish a 20–30° banked turn. He should ease the nose up to the lowest position at which a stall will occur, as demonstrated by you earlier. At the first indications of a stall he should relax back pressure (or move the wheel forward briskly as required), add full power, and level the wings with coordinated controls. The flaps should be retracted in the manner you have worked out as most efficient for your airplane. Generally, you'll find that quickly getting rid of flaps as soon as the nose is lowered is *not* the best way to recover from a stall, because the higher stall speed in the clean condition may offset the effect of increased power and put the airplane back into another stall. As he stalls the airplane, you should remind the student to check the altimeter at the beginning and end of the recovery (or you may do it), so that he can see the altitude loss involved.

After recovery and cleanup you might have him climb 300 ft above the recovery altitude to simulate a climbout after a near-stall situation. You could have him climb a certain altitude at the max angle of climb speed (say, up 150 ft) and then complete the 300 ft at the recommended max rate of climb speed. You might then have him level off and establish cruise or set up another approach to landing stall, as appropriate.

The approach to landing stalls should be practiced clean and with varying flap settings up to full flaps, so that he can see the different nose attitudes as the stall occurs.

Common errors made doing approach to landing stalls include the following:

1. Too steeply banked, the airplane mushes in a turn rather than stalls.

2. Nose not high enough for flap setting; more mushing.

3. Nose too high in the second (realistic) version of the stall, making an exaggerated demonstration of the maneuver.

4. Letting the wings become level during the approach to the turning stall.

5. Poor throttle handling during the recovery; abrupt or overly timid adding of power.

6. Lowering the nose too quickly or too far in the recovery and losing an excessive amount of altitude.

7. Coordination problems during the wing-leveling process. Usually too little rudder is used and adverse yaw occurs.

8. Poor poststall procedures. Flap retraction is neglected or poorly sequenced.

SUMMARY

Be sure that the student is able to recover from any of the realistic situation stalls covered in this chapter before you start to shoot takeoffs and landings in earnest, because he could encounter such problems during these periods—and after solo. If he should get in a fairly tight situation on one of your dual periods of shooting landings, you could tell him he was getting close to the situation he had simulated earlier, and you will have some basis for discussion. You may want to take a break from takeoff and landing practice during one period and review these stalls and do other high work.

SHOOTING TAKEOFFS AND LANDINGS—AND SOLO

BACKGROUND

After the student is proficient in stall recoveries and slow flight, understands wind drift effects, elementary emergencies, and other basics, you should start full periods of takeoffs and landings. He's been making the takeoffs since the second or third flight and should have a pretty good idea of landings from your demonstrations and perhaps from his "practicing" at the end of flights under your supervision.

Talking a student through his first takeoff and landing session requires a well-coordinated tongue and eye—and hands that can spring from a position of apparently complete relaxation (they aren't relaxed) to the wheel and throttle. *Never* pull your feet back away from the rudder pedals (and brakes) during the student's takeoffs and landings (this rule also applies any time you are giving dual takeoffs and landings and are pilot in command, no matter what the other pilot's credentials are). Make sure that your feet aren't touching the rudder pedals but are so close that you can use them without delay. Directional control problems on the takeoff and landing roll usually happen more quickly than pitch or bank problems, because the airplane configuration may tend to aggravate the heading problem (tailwheel airplanes with full flaps blanketing the rudder, etc.). When you first start instructing, the idea that you might relax at any time during a student's takeoff and landing practice will seem preposterous—you won't be able to relax an hour after you've finished flying on some days. But as you get more experience, it's possible that you might get complacent, tired, or distracted and could space off, particularly if the student has been doing well during that flight period. *The first full lessons of the student's takeoff and landing practice are critical, both from a learning and a safety standpoint.* Later he *may* be good enough to take off or land the airplane without complications (don't count on it) if you're woolgathering, but not the first period or two.

Some of the information in this chapter, plus your own, may be given before the early flights so that the student will be able to absorb the landing demonstrations, and practice, at the end of those flights. However, you should wait until the takeoffs and landings begin in earnest before going into all the detail given here.

The student should have his student pilot certificate (and medical, if he is getting them separately) before you start full periods of landings.

Crosswinds and gusty air conditions should be avoided during the first couple of periods of takeoffs and landings, or at least until he has a good grasp of the principles. After that, introduce him to crosswind takeoffs and landings as necessary for him to cope with local conditions. Some instructors prefer to stay away from crosswind takeoff and landing instruction until after solo; but if you feel he might need those skills during the first solo, you should work on this area after the normal takeoff and landing procedures are well in hand.

And along this same line, some syllabuses include the instructing of short- and soft-field takeoffs and landings, including maximum angle climbs, before solo. This tends to confuse the student about the performance that is to be expected and could set back his understanding of normal takeoffs and landings. Ascertain that he has the basic ideas well established before introducing variations—this is true for any maneuver. After solo and certainly before the solo cross-country he should be taught short- and soft-field takeoffs and landings, but not yet.

As mentioned in Chap. 4, you should make sure that the student's seat is high enough for him to see out properly for the landing practice. On occasion, students have spent hours of takeoff and landing practice without "finding the ground" (in fact, without even *seeing* the ground in the latter stages of the landing), and the addition of extra seat height or a cushion saved the situation.

THE NORMAL TAKEOFF

■ **Preparation.** You will have discussed the takeoff earlier but should review the process again before the first full session of takeoffs and landings. A chalkboard and model are the main aids.

■ **Explanation.** The *what* and *why* of takeoffs are self-explanatory, and your instruction and review will consist of *how* to make the takeoff. The following discussion assumes that you are using a nosewheel airplane unless otherwise noted.

How—After the pretakeoff check is complete and there is no traffic problem (that is, after tower clearance is obtained, and it is visually noted as clear at a controlled *or* an uncontrolled field), the airplane is lined up with the center line of the runway. The heading indicator should be checked and reset after the airplane is lined up. Taxi forward 5 or 6 ft to straighten the nosewheel.

Advise that as power is applied the student will be expected to correct for "torque" as necessary to keep the nosewheel on the runway center line, unless, of course, you are flying from a grass field. And if you are flying from an old concrete war surplus field, have him concentrate on a particular tar seam. The main thing is for you to have him stay in the middle of the runway, no matter how wide it is. (A deviation from this rule is in the case of suspected wake turbulence on takeoff, which should be covered later in the training.) Since it's possible that a tire might blow or a brake problem could arise, he'll have more room for corrective action on either side. Of course, you may be operating from a strip so narrow that if you're not in the center, you're not on the strip. As one instructor said about landing on a strip of that type, "In the last part of the landing when the nose is up, if you can see the strip you're not going to land on it." An exception to the center-line takeoff and landing is on a runway with center-line lighting. This could be a little rough on the nosewheels of some smaller airplanes, but stay *close* to the middle of the runway. Usually such runways are very wide, anyway.

The normal takeoff is one in which the airplane is helped in the lift-off with gentle back pressure or trim. It is neither stalled off nor held on to an excessive airspeed before lift-off, so a proper safety margin exists in normal wind and obstacle conditions.

As power is applied the pilot keeps it straight; and depending on the airplane, as the controls become firm or a specific airspeed is reached, gentle back pressure is applied to ease the nosewheel off and assume the proper attitude. You should advise during your explanation that when the nosewheel is raised, the student will have to depend on the rudder (and aerodynamic pressure) for directional control and should be prepared to use more right rudder.

After the airplane lifts off, the back pressure is relaxed slightly to avoid a nose-high attitude that would delay getting the max rate of climb speed. He will be expected to correct for "torque" on the climbout and to keep the airplane's track lined up with the runway center line.

If you are instructing in a *tailwheel type* you may want to discuss the takeoff as being in three phases: (1) the initial tail-down roll, (2) the tail-up roll, and (3) the lift-off. Emphasize the need for an increase in rudder deflection as the tail is raised, and spell out that the tail should not be raised too abruptly or gyroscopic effects will cause the nose to take a quick swing to the left with potentially interesting results.

■ **Demonstration.** Your demonstration of a takeoff in the first and possibly the second lesson may be the last time you'll have to do it for some students, but for most you will have to show the proper technique a couple of times again when you start concentrating on takeoffs and landings.

Talk it up as you demonstrate, mentioning any problems *you* had and reviewing the technique during the climbout after takeoff. If there is a particular problem that should be seen again, after explaining what you plan to do, stay in the pattern, land, and do another takeoff.

To repeat, the demonstrations and the student's initial practicing should be done with no crosswind problems. He should not have to cope with added aileron or rudder work at this early stage.

Show him that the elevators are kept in the neutral position or slightly aft. Note that the tendency of most students is to relax pressure so that the control wheel goes all the way forward. This leads to a possibility of "wheelbarrowing," or having too much weight on the nosewheel, causing directional problems during the takeoff run. You may have to remind him several times to hold the wheel back in the proper position. For some airplanes, the hole through the wheel shaft used for the control lock may be used as a reference for the control wheel position on the takeoff roll; later he will not need such mechanical aids (Fig. 10-1). He can also look back at the elevators to check that they are at the proper position and confirm the control wheel position before starting the takeoff run.

■ **Practice.** As pointed out in Chap. 4, the student should start making takeoffs as soon as possible after starting to fly, with his first takeoff effort being with use of the rudder only. After that, in good wind conditions, he should control the

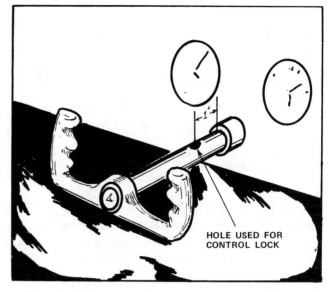

Fig. 10-1. Mechanical references may be used at first to aid the student in setting up the proper elevator (or stabilator) position for takeoff. In this example, the hole for the control lock should be about 1 in. out from the panel. You probably have your own ideas about this.

elevators as well. When he starts making takeoffs himself, you may see a renewed tendency to use the control wheel to steer the airplane on the ground.

In the tailwheel airplane, you can expect a certain amount of "rudder walking" as he attempts to "average out" any pending directional problems. This tendency is usually less pronounced in the tricycle-gear airplane because of the firmness of the rudder pedals when the nosewheel is on the ground.

If you think the student is mechanically correcting for "torque" or if he is having directional control problems of any kind, a special session of high-speed taxi runs up and down the runway (during a light- or no-traffic period) will help a great deal, as mentioned back in Chap. 5. You handle the throttle and have him control direction with the rudder. As he becomes more at ease in the exercise, you should vary the power (resulting in a variation in the left-turning tendency) unexpectedly so that he can break any mechanical habits. He will enjoy the action as he becomes more at ease, and his directional control problems will usually be eliminated after a short session of this. During the exercise you should open the throttle slowly to full power, close it as speed is built up, then open it again fairly quickly, and so forth. Don't run off the end of the runway.

A few people tend to open the throttle abruptly, a practice that will also occur later in some students' experience. However, the usual throttle handling problem at first is to be too timid with power application on the takeoff run, which can use too much runway. You may have to remind students several times to "open the throttle all the way."

Remember the advice in Chap. 2 about not using "colorful" expressions. This is particularly important during takeoffs and landings when instant understanding may be critical to safety. "Give me 100 percent!" may be your old jet pilots' phrase, but to a new pilot, it's totally meaningless. Another expression sometimes used by older instructors when they want full power is "Throw a (coal) scuttle on!" The legitimate command is, "Open the throttle to full power," or later, "Open it all the way," but you must make sure he understands you all the time.

Emphasize that it is much better to stop small directional

Fig. 10-2. It's best to correct immediately for minor heading problems rather than wait until something drastic occurs.

deviations immediately than to wait and have to make larger corrections (Fig. 10-2).

You'll discover that to most students—and you may have felt this way yourself—the *landing* is much more important than the takeoff so they may tend not to work as hard in smoothing out the takeoff procedure.

Generally, you'll find that after you've worked with take-offs in the several flights preceding the full-scale takeoff and landing practice, the student will require only minor corrections and advice from you in the last couple of hours before solo.

■ Evaluation and Review. As the student approaches solo, in normal wind conditions he should be able to consistently take off with only minor deviations in the takeoff path and make required corrections without prompting. Your evaluation of his progress will depend on this.

Review his takeoff(s) during the postflight briefing in the early lessons. It's sometimes hard to remember the details of a takeoff at the beginning of a maneuver-filled flight, but in any postflight review you should have in your mind the step-by-step moves from the preflight check to the shutdown and cover them in that order. Make a couple of notes in flight, if necessary. Explain what you are doing. Don't jump around in your review of any flight; you might miss something. You'll find that some students' early takeoffs are so memorable that not only can you remember them at the end of *that* flight, but several months later the details are still burned into your mind.

The student should be taught that, like the altitude above him, the runway behind him is useless, and you should insist on starting at the beginning of the usable runway. The use of intersection takeoffs, even at large airports, should not be encouraged in the early part of flight training (or later, either).

THE NORMAL LANDING—TRICYCLE GEAR

■ Preparation. A large percentage of accidents occurs between the touchdown and the slowing to taxi speed (this is not news to most pilots), but landing the airplane as slowly as possible without dropping it in will shorten that critical period. The airplane's kinetic energy is a function of its mass (weight) and the velocity squared (true airspeed in the air, and ground velocity after touchdown). Since the velocity is *squared,* any excess of airspeed at touchdown will have a great effect on the inertia of the airplane. Its landing roll will

be increased and—should it be deflected from its path by poor pilot technique, a gust, or other factors—it will be in a more dangerous situation.

The equation for kinetic energy is $KE = \frac{1}{2}MV^2$, and basically it means that if an airplane touches down at a speed 10 percent higher than normal, the landing roll will be 21 percent longer or $(1.1) = 1.21$. If the airplane lands at a 20 percent higher speed, the landing roll will be increased 44 percent over normal or $(1.2)^2 = 1.44$. This is assuming the same airplane weight and wind and runway conditions.

Suppose an airplane stalls at 60 K in the landing configuration at a particular weight. The pilot, who puts it on at 72 K (and this happens all too often), will give himself a 44 percent longer landing roll and, possibly, a 44 percent greater chance of an accident in the operation from touchdown to taxi: $(72/60)^2 = (1.2)^2 = 1.44$.

The nosewheel airplane has decreased the incidence of ground loop accidents, but too many pilots depend on it to cover for them if they goof on the landing. Because they feel they can get away with it, some land the airplane on all three wheels at a higher-than-normal speed—and then bounce. Directional control problems can arise during the landing roll too by wheelbarrowing (putting too much weight on the nose-wheel, making it extra sensitive to inputs from the pilot), as noted a while back.

The nosewheel airplane should be landed at the same attitude as a comparable tailwheel airplane. Compare a Cessna 180 and 182: both have the same airfoil, flap settings, and landing speeds. The 182 would probably forgive a nose-low landing attitude error that in the tailwheel-equipped 180 would set off a series of impressive bounces. However, the 182 should be landed in the *same attitude* as required by the 180 so that the minimum touchdown speed would occur for the conditions existing.

Watching private and commercial pilots who fly out of your airport you'll see that this tendency over a period of time is to start landing with the nose at a lower and lower attitude at touchdown. You may have to talk to them about this; they may get away with it for months or years on the 5000-ft runways they've been using, but when they go into a 2500-ft strange field, "bad luck" catches up with them. *The common tendency of a pilot is to be high on an approach to a strange field.* This tendency, plus the fact that the field is much shorter than he is used to, plus excess landing speed, can add up to running out of runway.

Your philosophy in instructing landings should be to train the student from the beginning always to land as if the runway were only about 2500 ft long, even though he may be landing on a 10,000-ft runway. Of course, there will be times at a controlled field when the tower or traffic may require that the airplane be landed long, but he can adapt to that situation better than to the reverse.

■ Explanation

MECHANICS OF THE LANDING. *Mechanics* may not be the best word, since the student should not be taught to land the airplane mechanically; but in addition to talking about *where* on the runway he's going to land, you'll certainly have to teach him *how* to actually go through the mechanics of the process—and there's the rub.

Okay, so the idea of the landing is to have the airplane touch down at the lowest speed consistent with the wind (gusty or crosswind) conditions and with the center line of the airplane parallel to (and preferably, on) the runway center line. The pilot is apparently trying to keep the airplane from landing when he's using the proper technique.

Discuss the landing at the chalkboard (Fig. 10-3), then review the procedure several times as you demonstrate, approaching and landing the model on a table as you explain.

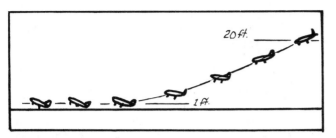

Fig. 10-3. Sometimes a landing sequence drawn on the chalk-board can help the student visualize the requirements of the landing.

Fig. 10-4 shows one method of setting up a landing. Fig. 10-5 gives a variation of a landing technique. You might want to go over this at the board. This one is the recommended method of landing.

When you explain the landing, tell the student that from a *normal power-off glide* (you can point out that some types of landings will require the use of power all the way to the ground, but that this will come later), start easing the nose up, or breaking the glide, at about 20 ft above the ground (it depends on the airplane size and rate of descent but 20 ft is good for the average light trainer). About 19 of that 20 ft is used in the transition from the glide attitude to the landing attitude. Theoretically, the nose could be raised at a predictable rate and would be at the proper landing attitude just as the 1-ft height is reached. An outside factor such as a thermal might require stopping or slowing the backward motion of the control wheel temporarily, or a downdraft might require a faster rate of rearward movement of the wheel or stick *to*

match the airplane's attitude to its height above the runway. Also, the pilot may have to correct for under- or overcontrolling the elevators.

A plot of elevator position versus airspeed during the landing in *ideal conditions and perfect techniques* might be as shown in Fig. 10-6. The required elevator or stabilator deflection would increase at an accelerated rate as the airspeed decreased, particularly during the final stage of maintaining the attitude while the airplane settles.

WHERE TO LOOK. The student should be taught to look to both sides as well as directly over the nose during the approach, but since he is sitting on the left side in a side-by-side airplane and would favor that side in a tandem trainer, he should be taught to look out the left side during the transition from glide to landing. Some instructors argue that the student should be told to look *over* the nose (as well as out both sides) as the landing is made because many of the trainers today have low nose attitudes at touchdown, particularly if flaps are used. However, there are still many airplanes from which it is not possible to see the runway over the nose in the proper landing attitude, so the technique of looking at the ground along the left side of the nose will be valid for *all* airplanes.

As he breaks the glide, the pilot should move his sight to alongside the nose and visualize the wing attitude out of the corner of his eyes as he continues to ease back on the wheel. This will be difficult for the student at first and your continued attention to this will be needed.

It's been said that the student should be told to look the same distance ahead in an airplane as he does in a car. But car drivers look well ahead for obstacles, and the judging of *height* is not involved. Following this advice, the student will look too far ahead and tend to fly into the ground. Tell him he

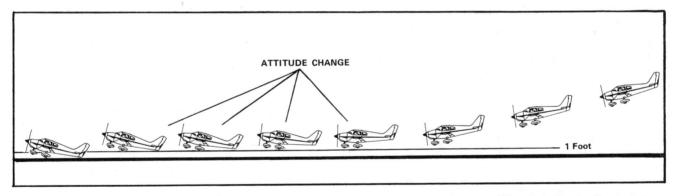

Fig. 10-4. One landing technique.

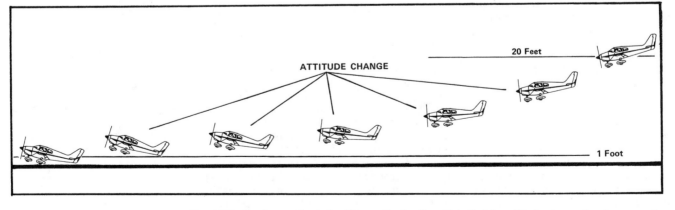

Fig. 10-5. Another method of setting up the landing. Note that the attitude change has been greater during the "19 ft of descent," compared with Fig. 10-4.

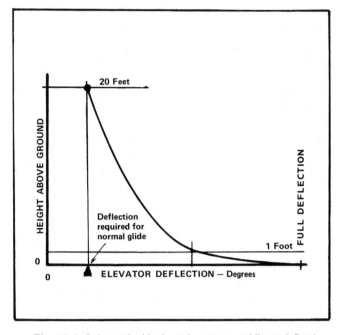

Fig. 10-6. A theoretical look at elevator or stabilator deflection as the airplane descends in a normal landing. Note that the last foot of height requires a large portion of the total travel during the landing process.

should look far enough ahead so the ground is not blurred and that *this will vary with airspeed and height.* Taking only airspeed into consideration would lead to the belief that as the airplane slows, the gaze should be brought back closer to the pilot since the relative motion would be *less due to the speed decrease.* On the other hand, at the beginning of the round-out, the pilot's eyes are about 20 ft above the ground (the height judgment depends on how far the pilot's eyes are above the runway, not the landing gear height above the surface), and are perhaps 5 ft above the ground at touchdown. Looking strictly at speed effects at first, however, and assuming an approach speed of 70 K and a touchdown of 45 K (a slightly wider-than-normal range of airspeeds admittedly, but it makes for easier arithmetic), the pilot would theoretically have his sighting area about two-thirds as far ahead at the touchdown at *50 K,* as was the case at the beginning of the transition at *70 K.*

Apparent relative motion also depends on the distance between the eye and the moving object. (The runway is "moving," not the airplane.) It would appear from a simple arithmetical analysis that the pilot should look 4 times as far *ahead* when his eyes are 5 ft above the runway at touchdown as when they were 20 ft above the runway at the transition. *It is not that simple, of course.* If this arithmetic were correct for actual conditions, the correction for height change would well overcome the correction for the *airspeed* change, and the pilot would have to look *farther ahead as he got closer to the ground!* A quick check indicates that the area to scan would be about 2⅔ as far ahead at touchdown as at 20 ft—which is certainly not correct, but height *does* have an effect.

The idea here is to *not* tell the student to "keep moving your eyes closer to the airplane as you slow up during the landing." He may dutifully and mechanically move his sighting area back so close that the ground is blurred—and the tendency would be to think that his height above the ground is *less* than it really is—and so he holds it off too high and drops it in. *There is a definite correlation between apparent relative motion and judgment of height (or distance) if the size of the reference is known.* This becomes apparent in shooting down-

wind landings; the new student will tend to level off too high because (1) he's trying to "slow" the airplane (the high ground speed is fooling him), and (2) the faster-than-normal movement of the runway tends to make him believe that he is lower than he actually is, so he gets the nose up to the landing attitude too soon. (Crunch.)

Your method of teaching landings, after the student knows where to look, should follow the *attitude versus height* idea during the transition and touchdown. He will start the round-out at, say, 20 ft at a certain attitude, use (about) 19 of that 20 ft making the transition to the landing attitude, and then maintain that attitude as the airplane continues to settle. The attitude versus height idea rightfully ignores ground-speed, as could be seen in an exaggerated example of a steady wind of 70 K at 20 ft with a calibrated change to a velocity of 45 K at the ground. The example airplane used earlier would come straight down, even though it's flying very well; the pilot would still make the same attitude-height *transition* as for a no-wind condition.

Fig. 10-7 shows the example just mentioned. (The airplane here starts the transition apparently higher than 20 ft but spacing of the illustration was a problem.)

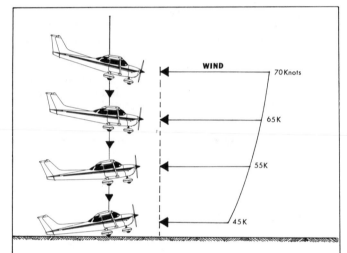

Fig. 10-7. One concept of attitude versus height. This approach to the subject tends to eliminate the student's fixation on his speed relative to the runway. (He should have already confirmed that he is landing upwind or downwind and will have done the right thing—land or go around—before getting to this point of the landing.) In this fictitious situation the airplane is over the landing point at 70 K on final and will touch down at 45 K. The wind velocity helpfully changes as the airplane slows up during the landing process, so that it is moving straight down, with attitude changing from the approach to the landing attitude as the height decreases.

Before starting intensive landing practice you should show the student the general area of the runway at which he should be looking—how far ahead, how far to the left, etc. You should have him sitting in the airplane with the seat adjusted normally *and the airplane in the landing attitude as would be found using the normal landing flap setting.* For tricycle-gear airplanes this would mean that someone may be required to push down on the tail (if placards and other notices don't preclude this) to get the landing attitude, while you walk out to the area ahead and slightly to his left (Fig. 10-8).

You might have the student ease the wheel back a couple of times, simulating a landing as the attitude is held—assum-

Fig. 10-8. Have the student sit in the airplane, with it in the landing attitude, as you show him the *general* area to scan. You may have to sit in the airplane first to check where you actually do look during the landing. In extreme cases, you may put a "window" outline of masking tape on the windshield to show the student where to look during the landing practice.

ing that the man holding the tail down can carry on. *You* should watch your trainer being landed by other instructors to find the proper attitude to simulate. You can see how high the nosewheel is "up" at touchdown (one-half diameter or one diameter, etc.) with normal landing flaps.

The average student will try to look over the nose at first during the landings, and the earlier simulated landings at altitude should have given you a chance to work on this problem.

If you have a long runway, when things are quiet you can, with some trainers (the Cessna 150 and 152 come to mind) taxi down the runway with the nosewheel off the ground in the landing attitude. You handle the throttle and elevators and the student keeps the airplane rolling straight by looking at the runway along the left side of the nose, as would be the case just at touchdown. He *has* to look at the left side to keep it straight, since he can't look over the nose. (Don't get so engrossed that *you* let the airplane run off the end of the runway or hit that cow that wandered onto it.)

You will have to learn to look out the right side in the side-by-side trainer when you start instructing even though it's awkward at first.

■ **Demonstration.** You'll demonstrate the landing on the

first few flights and talk it up as you do so. One of life's more embarrassing moments, after going into precise detail as to how it should be done, is bouncing or dropping it in on your demonstration (Fig. 10-9). It's not considered cricket to mutter about a "downdraft" or "gust."

Assuming all is going normally, as you land try to describe what you are doing. After landing safely, explain, for instance, why you had to stop easing back on the wheel for a second (the rate of descent had stopped). Or you may have to explain that the rate of easing back increased to adjust the airplane's attitude to its height above the ground. Thermals or downdrafts may affect the airplane's height, so the attitude must be held or changed more slowly until the proper combination of height and attitude is again attained. Each height must have its attitude and vice versa.

■ **Practice.** The student is eager to get on with practice of takeoffs and landings, because at his stage of experience this seems to be what flying is really all about. He is also seeing other near-solo students shooting takeoffs and landings and may start getting restless.

You, too, will look forward to starting work in the traffic pattern, and because of possible pressure from the student and within yourself, you may tend to start full periods of takeoffs and landings before he's ready. This is a mistake that nearly every new instructor tends to make because he feels that a particular problem "can be worked out in the pattern." Unfortunately, when the problem is not worked out, takeoffs and landings may have to be stopped and a review of high work and fundamentals has to be done. This is bad for morale because the student realizes there's been a setback. In addition, other students may make comments such as, "I thought you were shooting landings; how come you're back up there doing high work again?" So, if in doubt, it's better to wait a little longer before starting full-scale takeoff and landing practice.

One of the problems to expect is that initially he will use asymmetric force in easing back on the wheel, which causes the lowering of a wing and "drifting" sideways during the round-out and touchdown. A crosswind may add a confusing factor to the situation. If you suspect that he might not be coming straight back on the wheel during the landing (and you can usually feel and see the slip), you might have him sit in the airplane after shutdown and simulate a landing while you watch for any aileron deflection as he moves the wheel back. As you gain experience, you can also glance quickly at your wheel as he is landing to see if he is using ailerons when they aren't required.

While there are individualists who will drop the *right* wing while landing a side-by-side trainer, the usual procedure is to apply *left* aileron as the wheel is pulled back. This hap-

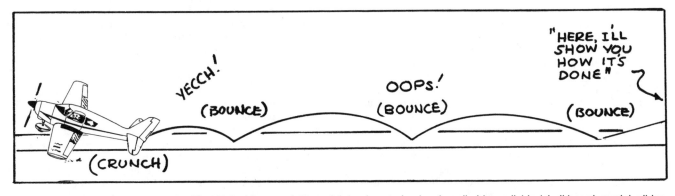

Fig. 10-9. After carefully explaining how to land, using all aids available (chalkboard, model, slides, etc.), you demonstrate how a landing should be made—it happens.

pens because the wheel is held on the left side by one hand only and the weight of the arm will tend to turn it that way. However, it depends on the vertical position of the wheel in relation to the pilot; if it is in a very low position, you may expect a pulling *upward* on the left side of the wheel as it is pulled back and the *right* wing is lowered. In a stick-controlled airplane the right hand controlling the stick may pull it in that direction, lowering the right wing during the landing.

As practice continues you will have to indicate that the wings must be kept level (as seen from the corners of the pilot's eyes) and that *a greater deflection of the ailerons and rudder is necessary to raise a wing as the speed decreases in the landing process.*

The average student will tend to concentrate on the pitch attitude and ignore heading and bank attitudes. He'll probably have to be reminded a number of times about this—and you may have to make a "save" as he is making a perfect landing as far as pitch is concerned but touches down headed off the runway for rough terrain.

If he is having problems "finding the ground," or finds it too readily (thud), check to see that he is still scanning the proper area ahead. His scan may be too close to the airplane (giving a tendency to level off too high), too far ahead (flying into the ground), or trying to look over the nose (and not seeing much of anything but the sky). If he's looking over the nose he may either (1) tend to land nose-low and bounce because easing the nose up to the proper attitude would put it in a position so that he couldn't see the runway directly ahead, or (2) mechanically ease back while staring up over the nose and hoping everything will come out all right. A student may even alternate procedures 1 and 2, which clouds the issue, but if he is having problems with landings and his use of the elevators is reasonably consistent, the odds are great that he isn't "seeing" the ground. Poor sunglasses could be causing his problems, so look into this.

One method of introducing the actual landing process is for you to fly the airplane along the runway several times starting at the round-out height and decreasing altitude and airspeed at each progressive pass. This will let him see the attitude/height combinations. (Use power to maintain a constant height each pass.) You may then let him practice until he is landing the airplane. However, this is normally reserved for situations involving height-judging problems. Another procedure is to have him make *power-off* approaches with landing flaps, starting at the maximum flap-operating speed and decreasing this in about 5-K increments each time. He will try to maintain an altitude of about 1–2 ft each time by proper use of the control wheel, gradually easing it back to maintain a constant height. The airplane shouldn't be landed during the exercises and *you* will take it around as the end of the runway approaches each time. This is a good exercise if he has been tending to overcontrol the elevators during the actual landing practice.

You may expect the average student not to make outstanding progress during the first full period or so, but if he has had a good background of stalls and is looking at the proper area during the landing, you'll find that at some point he will "find the ground" and have no major problems afterward. Tell him when he's got the idea of the landing. "That's it, exactly," will help him see what is to be done on subsequent landings.

If he "greases it on" accidentally by a fast, level-attitude landing (tricycle-gear), you pull it off again, if runway permits, to show him that the airplane *was not through flying.* It's best not to take this action until he's had some time in the pattern and is beginning to relax a little. If he touches down too fast in the tailwheel type, the airplane will usually let him know of his blunder by either bouncing or lifting off again as he tries to ease the tail down. But don't be slack about the

proper attitude for the nosewheel airplane, even if he's getting away with landing at an improper attitude.

The student should be taught to keep his hand on the throttle at all times in the pattern except as necessary to use carburetor heat, flaps, trim, and other items of the checklist. He should definitely keep his hand on the throttle throughout the landing process, since he may need power immediately to help ease a drop-in or recover from a bounce.

One of the best training techniques you can use is to have the student go to a spot near the end of the runway and observe landings. If he picks the proper position (not too close to the runway) he can see the landing from the breaking of the glide to the touchdown. If possible, you should go out there with him the first time to explain which approaches and landings were good and which were not, so that he develops a standard for his own later observations. Observing is especially helpful for students who are having problems visualizing the proper pitch attitude at touchdown.

Don't get complacent with a student who has apparently gotten the landing idea exactly right from the beginning. Sometimes it's a matter of lucky combinations, and you may find that by relaxing you're setting yourself up for an unpleasant surprise.

GROUND SHYNESS. As you continue to shoot landings with a particular student you may notice that he will make a normal approach, start the round-out, then have a tendency to "put it on." He doesn't try to keep the airplane from landing, but apparently wants to get the process over with and the airplane on the ground as soon as possible. This action can be one manifestation of ground shyness. The student does not like to be in a close-to-a-stall condition close to the ground, and won't bring the nose up. He is uncomfortable during the transition. You'll probably see added signs of nervousness on final as the round-out point is approached.

You may be able to get out of this situation by talking him through the last part of the landing several times, letting him see that the situation is not as he feared.

In more extreme situations you may have to go back to flying the airplane several times at the landing attitude a few feet over the runway to let him get used to the idea. He should also fly the airplane this way (while you handle the throttle) until he works his way out of the difficulty.

You will have people who do very well in the simulated landings up at altitude, consistently holding the proper landing attitude until the wheel is full aft and the stall occurs. But when they start the actual landings they begin to hesitate in bringing the wheel back, and start "putting it on." (And this problem can even crop up after a couple of periods of reasonable progress in landings.)

In some circumstances where such a loss of confidence has occurred, instructors have taken presolo students up and demonstrated and let them practice spins (in the proper airplane), with good results. Before you make such a move, be sure that you know your student; such practice *could* cause an even greater loss of confidence.

■ **Evaluation and Review.** As the student gets more takeoffs and landings you should see an increase in confidence and improvement in his handling of the airplane in the landing. After a couple of full periods of dual he should be "finding the ground" with only minor problems of overcontrol of the elevators (which he sees), and should be at least making corrections, even if they are not precise yet. After he is able to judge the attitude-height relationship, you may expect that he will still have some problems with wing attitude and/or heading during the round-out and touchdown, but these usually can be worked out without too much trouble.

After each session, review the landings and go into more

detail on specific problems that you only mentioned briefly during the flight.

The approach, landing, and landing roll are usually the most critical phases of the flight, and in your evaluation of the student, you should consider and correct immediately any problems. He may be very sloppy at altitude and get away with it without damage, but not in the landing.

SHOOTING TAKEOFFS AND LANDINGS

■ **The Traffic Pattern.** Briefly go over the traffic pattern for shooting takeoffs and landings, since this will probably be his introduction to the "closed pattern" (Fig. 10-10).

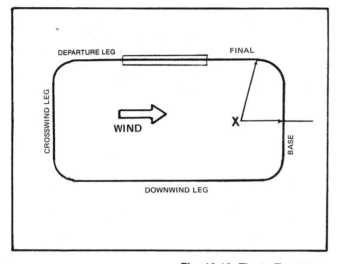

Fig. 10-10. The traffic pattern.

You should particularly indicate the area of the pattern shown at X, because for the first time he will be required to judge when to make the turn (and how steep to make it) in order to line up with the center line of the runway. One common error at this point is turning too soon and angling toward the runway. Go over each of the legs for a left-hand pattern in order.

DEPARTURE LEG. Remind the student to check his alignment with the runway a couple of times as he climbs out and to make corrections as necessary. This is difficult in airplanes without a rear window, but he should be made aware of the requirement and keep his climbout path as straight with the runway as possible. Point out that he can spot a reference soon after he gets off and can track to it as an aid in staying lined up.

He should have a definite altitude to use for that first turn, and for most traffic patterns the turn is made 200 ft below the downwind leg altitude used. In calm to moderate wind conditions, this will allow the airplane to reach the pattern altitude of 800 (or 1000) ft well before it is halfway along the downwind leg. Faster airplanes should use a higher pattern.

Before making the turn he should clear the area to the left and left rear. You may want him to give you a "Clear left," after checking. Too many students get into the habit of looking at the wing tip area in the direction to be turned but not behind them. Fig. 10-11 shows that if another airplane is in that area (abeam), a midair collision is less likely than would be the case for an overtaking airplane in a more rearward position. The student should check 180° to the side to which he is turning—other airplanes may be ahead, and coming your way. While you are emphasizing the need for a special check in the direction of turn, you should add that he must look in *all* directions in the traffic pattern. It's possible that some other airplane could be making a straight in approach, or even a right-hand approach against traffic. And while you would be in the right, this knowledge probably wouldn't be of much use after a midair collision (Fig. 10-12).

CROSSWIND LEG. He should be instructed to keep a sharp watch for traffic entering the downwind leg. Too many people only look back at the runway. The subject of correcting for a drift on this leg should be covered, and you might demonstrate the proper technique the first time around. In a strong wind, poor drift correction on this leg could affect the rest of the pattern. The downwind is rushed and/or stretched; the base turn may be late, with a parallelogram pattern, as shown in the exaggerated example in Fig. 10-13.

You'll have to show him when to turn during the first few patterns. Direct his attention to the relative position and distance of the runway at the proper point of turn. He should

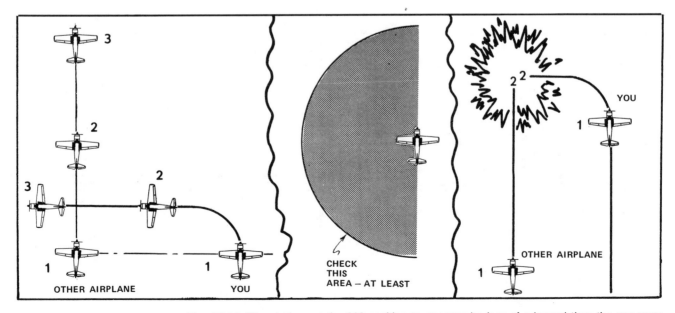

Fig. 10-11. The airplane at the 90° position to you may be less of a hazard than the one more rearward and overtaking. The student should be taught to check 180°.

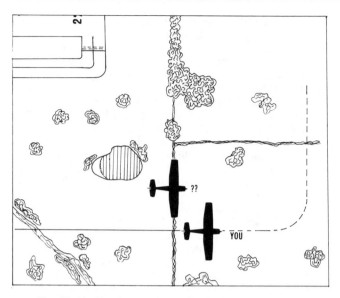

Fig. 10-12. Check your airplane's shadow when in the pattern. There's little that starts adrenaline moving faster than to see *two* shadows close to or at the same point in the pattern.

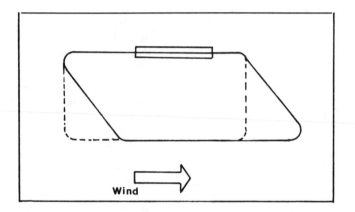

Fig. 10-13. Poor drift correction on the crosswind leg can affect the rest of the pattern.

base his point of turn on *how the runway looks.* (While *you* won't do it, a few instructors have had their students use landmarks such as barns or lakes for references as to where to make this turn and the turn onto base. This may help the student solo a little sooner using a particular runway at the home field, but it won't help him if another runway is used, or at another airport.)

He will depend on you to tell him when the airplane is at the proper distance from the runway on the downwind leg at the beginning, and after a few corrections by you, he should soon settle down and be able to judge the proper distance from the runway.

DOWNWIND LEG. Most traffic patterns require a continual climb from takeoff to the required pattern altitude (no level turns), and normally the student should be able to level off shortly after making the turn onto the downwind leg. This should give a few seconds of straight and level flight, allowing him to trim the airplane at a speed *below* the max speed for flap extension, and you should have him use a specific (lower-than-cruise) power setting for the downwind leg. If he lets the airspeed build up, it may cause an extended downwind leg and an unusual pattern.

As he turns on the downwind leg, require a check for airplanes that may be entering the pattern. At any time in the pattern, but particularly on downwind and base legs, assure that the student checks the ground for shadows of other airplanes that may be hidden in his blind spot(s) (see Fig. 10-12).

One of the difficulties that may arise on this leg is his allowing the airplane to move in to or out from the runway, even with no crosswind component existing. It's tough for the student to recognize that the airplane may be pointed a few degrees off the required heading; tell him immediately if he's easing in or out.

Military flight training procedures often use references on the wing, or the wing tip itself on low-wing airplanes, as a check for proper distance from the runway. For high-wing airplanes with struts, a reference can be made on the strut (Fig. 10-14).

The use of this method of checking for the proper distance from the runway may seem mechanical, particularly the use of tape to mark the reference, but it's mentioned as a possible aid in some more extreme student problems in this

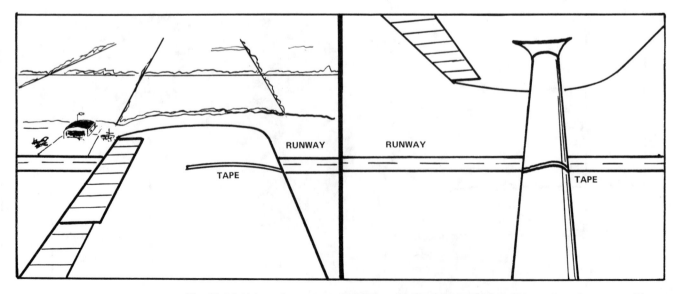

Fig. 10-14. Using references on the airplane to check for the proper distance from the runway. These references can change with students' heights. (The wings must be level, also.)

part of the pattern. Naturally, the airplane references are only good for a particular airplane *when the wings are level,* so you will mainly be working to teach him to judge by looking at the runway perspective. If in doubt, it's better to be slightly wide on the downwind leg than too close—it gives more time during the approach.

If the airplane has simple systems as most fixed-gear trainers do, you may have him go over in his mind a short checklist such as carburetor heat, mixture, fuel, boost pump (if needed), brakes, and flaps; or he can actually check the five or six items. You might make up a checklist and post it on the instrument panel for use in the pattern. Too many of the current checklists are hard to read and require the pilot to keep his head in the cockpit too long. Some trainers already have such a standard posted checklist, and it should be used.

In nearly all of the older trainers and some of the newer ones, carburetor heat is recommended for use throughout the approach. Many new airplanes, however, require use of carburetor heat only in times of suspected, or actual, carburetor ice. Fig. 10-15 shows a possible checklist route you might take in the pattern with these newer airplanes.

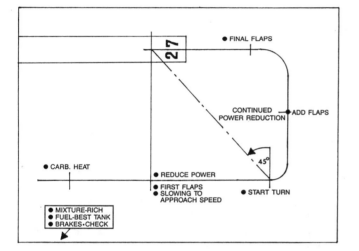

Fig. 10-15. Use of carburetor heat and suggested approach procedures.

The carburetor heat in these new airplanes should be pulled full ON about halfway down the runway and left full ON until time to reduce power to start the descent. If there is no evidence of carburetor icing the carburetor heat should be pushed off at that point and not used further during the rest of the approach. If there is evidence of icing or it is suspected that icing might occur, the heat should be left full ON. The student does not have the experience to be able to use partial heat, so he should be taught to use full heat or none. You, of course, will establish procedures for the airplane being used and the climate in which you fly.

Getting back to the pattern, as the airplane reaches a position opposite the point of intended landing (remind him that you won't be trying to land right on the end), throttle back to about 1600 rpm, and apply recommended landing flaps. A power setting to 1600 rpm works well for the first power reduction for the approach for most current trainers. Set up and trim for the recommended approach speed and turn on base.

One question that might come up is: Suppose you are following another airplane for landing—how much interval do you give him before turning base? A rule of thumb— assuming that the other airplane isn't going to have to taxi back up the hot runway after landing, but can get off the

runway after slowing down—is to start your turn when, on its final, the other airplane has passed behind your wing (Fig. 10-16).

Fig. 10-16. A good interval will result between airplanes of comparable approach speeds if the turn onto base is started just after the airplane on final passes behind your wing.

THE BASE LEG. As noted in Chap. 5 and as shown in Fig. 10-15, the turn onto base should be slightly steeper than the turn onto final, because it's best to get established on base reasonably soon. A long, shallow-banked turn could mean that the runway was out of sight to the pilot in a high-wing airplane, and he could be over to the runway center-line extension before he realized it. Besides, if any comparatively steep banks are to be made, it's best that they are made at a higher altitude.

Check his power reduction and drift correction. As on the crosswind leg, the student will tend to scan only one area, the runway, while on base. As soon as the turn onto base is completed and the wings are level, he should get into the habit of checking to his right for possible straight-in traffic. It's also very interesting to be on a left base and discover that someone else is making a *right* base to the same runway, so he should check straight ahead, as well.

Most beginners will make a number of approaches before they are able to judge the turn to final well enough to do the roll-out reasonably lined up. *You* can run into this problem also if you've been instructing in a J-3 Cub and suddenly switch to a Bonanza or Centurion; the greater radius of turn at the higher approach speeds will make you tend to go past the runway that first approach. The opposite will occur when you go back to the J-3 after flying faster airplanes; you'll tend to turn too soon. Chap. 17 covers this idea further.

Note the approach speed and advise when the student starts deviating from it. Emphasize that a proper pitch attitude will result in the desired airspeed.

It's your job to note the wind and remind him of a tailwind or headwind component on base and how it will affect his point of turn to final. You might indicate that if there is a direct crosswind on a runway and traffic permits, it is always better to pick the runway that would give him a *headwind* component on base (Fig. 10-17).

Student and other low-time pilots don't like this because in a standard left-hand pattern, it means a *right* crosswind on approach and landing, which for some is a more difficult correction.

Also indicate sometime during the presolo takeoff and

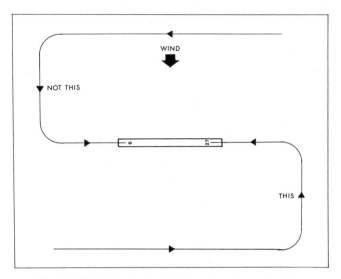

Fig. 10-17. Tell the student that in a direct crosswind where he has a choice of runways, it's safer to land on the runway heading giving a headwind component on base.

landing periods that if he overshoots the wind line (an imaginary extension of the runway center line), it's a better procedure to add power, shallow the turn, and climb out parallel to the runways for another try rather than rack the airplane up in a dangerous attitude. Go-around procedures will be covered later. It's better to start the final turn slightly early and have to shallow out than to be late and make an extra steep (often skidding) turn.

FINAL. The student should make a pattern large enough to assure sufficient final leg length to get stabilized on airspeed and the approach path. You will have to watch him closely the first few finals; the average student will tend to delay his track corrections until he is so close to the runway that radical maneuvering is required to avoid landing off to one side. He will have to be taught to make *small* corrections early here, too.

One of the big difficulties he will have is maintaining a constant airspeed on final. Usually, he does a reasonable job on base; but after that last turn, when the airplane is headed at the runway, the common tendency is to let the airspeed get too high. This is not to say that you won't have some students who will get too slow on final, which gets your attention, and others who are too slow on one final and too fast on the next one; but the biggest tendency is to have extra speed there. This is more of a problem if flaps are not used.

You may use the illustration that the purpose of the traffic pattern, particularly the base and final, is to put the airplane at a "window" at the approach end of the runway at the proper airspeed (Fig. 10-18).

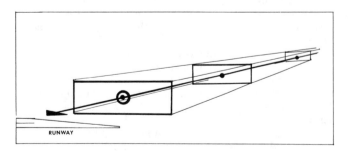

Fig. 10-18. The base and final legs are adjusted to put the airplane through a "window" at the end of the runway at the proper airspeed.

By setting up the proper attitude and adjusting power in the pattern, the airplane should fly through the final "window" with the power at idle and start the landing transition immediately after. Note in Fig. 10-18 that the airplane would pass through a series of windows on a perfect approach (smooth air and continual and smooth reduction of power throughout); but a perfect approach doesn't often happen, so he will try to make corrections to hit that final window at the proper airspeed. You can "adjust" that window to fit your particular situation; a very short runway will require a lower window, etc.

Because you are interested in having him establish a consistent approach pattern, you won't push this idea too hard. However, when you stop to think about it, the airplane can fly any type of traffic pattern imaginable. As long as it is at the proper airspeed when it passes through that last window, the landing *should* be successful. The student won't be allowed to do any radical maneuvering, but he will make corrections to get back on the proper approach path and have the correct airspeed at that point (Fig. 10-19).

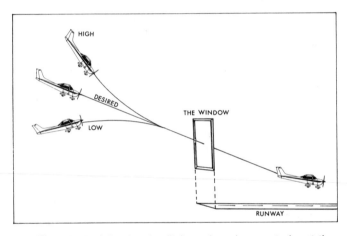

Fig. 10-19. Adjusting the flight path and power to be at the proper spot over the end of the runway. (The positions are exaggerated.)

Many students and private pilots later become so used to the mechanical repetition of a standard pattern that if they were put on the downwind leg a couple of hundred feet higher or lower than usual, they would have a problem hitting the runway. The chances are that they would be *high and/or fast in both instances*. On the higher-than-usual downwind leg the tendency would be to fly the same pattern and carry the same power as usual, so that the window they would hit would be on the "third or fourth floor." On the low downwind leg situation, they would also tend to fly the same pattern and power, and would finally discover that they won't make it that way and so have to use a great deal of power to get to the runway. In this situation, nine times out of ten the airplane would pass through the window at 10 K or more above the proper airspeed.

Before he gets his private certificate you should have him make approaches from high downwind and base legs (requiring idle approaches) and low downwind and base legs (requiring extra power all the way to the runway) to check that he's not mechanical. But for now, push a standard pattern.

SOME ADDITIONAL POINTS. In the latter stages of presolo work, you should cover up the altimeter during the climbout and have the student level off on the downwind leg at what he estimates is the proper pattern altitude. Uncover the altimeter to check. It is interesting to him to see how close he has estimated it, and after several tries he'll see that he can get reasonably close. Students are soon able to judge the lateral distance from the runway reasonably well but will tend to rely on the altimeter for height information. How far the runway is *below* them on the downwind leg doesn't always register unless it's brought to their attention.

There have been cases, after a front has passed or after a rapid and large degree of local pressure change, of pilots at uncontrolled airports setting the altimeter to "field elevation" by moving the hands of the altimeter the wrong way in the pretakeoff check. (They don't check the Kollsman setting and see how far off it is; an altimeter setting check with the tower would have straightened out matters immediately at a controlled field.) If the pressure change is ½ in. (500-ft change in the indication), when he gets in the airplane he may correct it the wrong way and set it 1000 ft higher or lower than the field elevation. He doesn't notice it and takes off for a local solo or a solo cross-country. Everything is fine until the time comes for landing. He may fly a pattern that is ridiculously high or dangerously low (he won't fly it into the ground but is puzzled at how low he is—and he's not even down to "pattern altitude" yet). It can make for a confusing situation and cause people to get well shook up before they get back on the ground. If he has a good idea of what the runway should look like from the proper downwind altitude, he may call Unicom and get the altimeter setting before things go too far.

One answer to the just mentioned problem is to train him to check both the 100-ft and 1000-ft hands when setting the altimeter to what he thinks is the field elevation.

From time to time check the student on his preflight and prestart checklist. You might turn the fuel selector off before the start of a dual flight and see if he really checks (or is used to the idea of the fuel selector being left on all the time). One instructor noted that some students were getting rather lackadaisical in their preflight check and, for instance, disdained to bend over and look under the belly. Well before the flight the instructor would go out and put a strip of masking tape on the belly. After a student's check, the instructor would ask him if he had checked "the whole airplane." The answer would be "yes," so the instructor would reach under and pull off the tape, which had written on it, "an 11-in. rip in the skin." The students got the idea. Put a point across this way without getting sarcastic and it will be remembered. Your flight schedule may be too tight to do a lot of sticking of masking tape on airplanes, but you will have to stop unsafe tendencies, either verbally or by demonstration.

■ **The Takeoff and Landing Practice.** There are several things to consider when shooting takeoffs and landings with the presolo student.

Again, the first period(s) of shooting takeoffs and landings should be done under smooth-air conditions, if possible. This means scheduling for late afternoon and early morning flights, and this isn't always easy since your other students and advanced trainees may have those periods as the only time they can fly because of work schedules. But smooth air is best for letting the student find out what is required. Rough or gusty air during the first couple of full periods of this work can mean absolutely no progress in learning about landings for some students. In fact, it could cause a major setback in morale as well. *The student might require another extra hour of work just to get him back to where he was before the starting of the rough-air landing practice.*

During the first few sessions you should normally make full-stop landings and taxi back for another takeoff. Touch and go practice obviously gives more landings in a given time, but for the first few instruction periods you should stick to full-stop landings so that the student will get experience in the critical phase between touchdown and turning off the runway. As he approaches solo, *only* full-stop landings should be done.

It may not be a factor for students or low-time private pilots, but after a while you may ride with a pilot working on a higher certificate or rating in his own airplane. You'll find that he is so much at ease—too much at ease, in fact—during the landing roll that he opens windows, opens cowl flaps, ups the flaps, and generally relaxes now that the "flight is over." He may get away with this *most* of the time in a tricycle-gear airplane in favorable wind conditions, but in a tailwheel airplane things could get completely out of hand. If he shows this diversion of attention with you, stay very alert, since you may have to take over to avert a catastrophe. After the airplane is stopped, tell him about the hazards of such a habit. (Don't discuss it during the landing roll.)

While the tailwheel airplane is touchier during the landing roll, it's quite possible to have directional problems in a tricycle-gear if the student relaxes back pressure too much. To repeat, wheelbarrowing results when a greater-than-usual amount of weight is on the nosewheel, making its steering effectiveness much higher than normal. The steering is ultrasensitive in this situation, particularly in airplanes with a more direct rudder-pedal-to-nosewheel linkage, and the pilot is caught by surprise as a comparatively minor correction results in the airplane darting off the runway to an accompaniment of screaming tires. In the normal landing roll, the wheel or stick of the tailwheel airplane should *be kept full aft throughout the roll to ensure the best tailwheel steering.* In the nosewheel airplane, the wheel should be full back at the beginning of the roll and may be only relaxed slightly as the roll continues, to help the nosewheel become more effective. Under no condition should the control wheel be relaxed past the neutral position in the tricycle-gear airplane; it should be kept well aft.

If the student has the ground roll well in hand and the only problem he's having is finding the ground, touch and go landings may be used to get in more patterns. You should still intersperse the touch and go landings with an occasional full stop so that in working on one problem you don't let him get rusty on landing rolls—and introduce a "new one."

If he is only having difficulty with the round-out, you can make two or three landings per approach if the runway is long. Have it thoroughly understood beforehand what you are going to do. On a particular pattern he is to make the approach and land, and immediately after touchdown you will say, "I got it," add power, and ease the airplane up to nearly round-out height. He'll be ready to take over when you tell him (you'll close the throttle again) and make a second landing. This can be repeated if the runway is long enough, but don't get so engrossed in the landings that you run out of runway. Affirm who is to make that last takeoff! You probably won't take the airplane up as high as the round-out height or get back up to the approach speed each time, and because of the runway distance required you will settle for a lower height and lower airspeed. One advantage of this procedure is that the student may correct a mistake immediately in the next landing, rather than having to fly a full pattern before getting the chance again. This is also useful if you think he is mechanically breaking the glide at the proper height but then really isn't sure of his height during the landing. By setting up the lower-height, lower-airspeed, and higher-nose-attitude situation for landing for him, you can usually break any mechanical habits. You should know your airplane, airport, and student before using this procedure (Fig. 10-20).

Before taking off each time in the normal one-landing-per-pattern practice, you can have your own system of confirming that he has read the checklist correctly. Set up a systematic check, starting at a particular item and always following the same procedure. Fig. 10-21 shows one method

Fig. 10-20. If the runway is long enough, you may add power, ease the airplane back up, close the throttle, and let the student land it a couple of extra times.

of "checking the checklist" quickly. This procedure for a particular airplane uses a "question mark" to ask if the airplane is ready to take off again. In that example, the fuel selector, trim, flaps, mixture, and carburetor heat are quickly checked before the throttle is opened for takeoff.

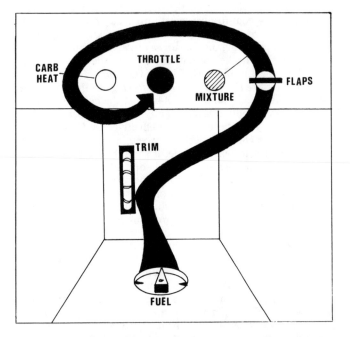

Fig. 10-21. The "question mark," a quick check for a particular airplane. You may set up a procedure applicable to your own airplane to back up the student's use of the checklist.

■ **Takeoff and Landing Problems.** When thinking of that first student you'll solo, the biggest concern you'll have will be about his problems during the landing, such as dropping it in or bouncing. After you've instructed awhile, on the student's first solo you'll be watching his pattern and approach more closely, because you'll realize this will definitely affect the landing. With more experience, you'll be able to tell much earlier in the pattern whether he might run into kinks on the landing. If he turned on the downwind leg too soon (his crosswind leg was too short), you can see possible complications, since he will probably be too high and have to go around; or he may dive for the runway, gaining too much speed and having touchdown problems (ballooning, etc.), as well as facing the possibility of running off the end of the runway. *Again, if in doubt, lean toward a slightly wider, more leisurely pattern in training your students.* Later you should tighten it up as necessary.

TOO HIGH IN THE APPROACH. The presolo student should not be expected to deliberately S-turn or fly past the runway center line if he is too high on any of his approaches. While it might save a not-very-high approach, it could also get the airplane in a wrapped-up and dangerous final turn. Set up your own requirements for student go-arounds and stick to them. Your initial thought will be to take over on an approach and make a gentle S-turn to get the airplane to the "window," but you should make him go around if he is too high.

How high is too high? You may establish a reference by the runway—such as a taxiway, hangar, wind tee, or other object—and tell the student that if he is not firmly on the ground as he passes that point, he must open the throttle and take it around. The usual requirement is that the airplane should be on the ground within the first third of the runway, but this may vary with local conditions. While such references are helpful, don't encourage too many mechanical aids. If you are flying out of an airport with a 12,000-ft runway, the first third, or 4000 ft, would allow *far too much latitude for error* and would show up when he tried to land on a 3000-ft runway on that solo cross-country. You might use a taxiway turnoff 500 to 1000 ft from the approach end of the 12,000-ft runway as your reference (Fig. 10-22). It's too easy to get lax about the required point of touchdown on a long runway.

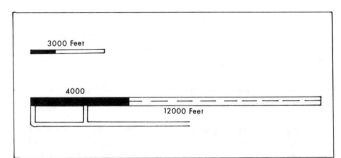

Fig. 10-22. The expression "first third of the runway" can mean different things.

There are several reasons for the student being too high to land the airplane safely (*common errors*):

1. Making the first turn after takeoff at too high an altitude, resulting in a *too-high downwind leg*, because other factors follow as well.

2. Making the turn to the downwind leg too soon; the downwind leg is too close to the runway.

3. Being distracted and continuing to climb well above the required downwind leg altitude. This can happen even if the first turn was made at the proper altitude and the turn onto the downwind leg was properly posi-

tioned. The student is behind the airplane, or the airplane is ahead of the student.

4. Failing to reduce power enough initially as the approach is begun. You will have to remind various students about this during your instructing career. It also could be that by the time he gets through concentrating on setting the power *exactly* as he thinks you want it, the airplane has moved well past the point of the turn on base and may be *too low* in the approach. This requires that power be added during the latter part to make the runway and he gets high and/or fast.

5. Making the turn to base too soon. In this situation he will be above his approach altitude all the way around, and in addition he will not have a long enough final to get set up for the landing.

6. Being slow to use the flaps, or using less than the required approach setting. It is sometimes puzzling to the student when he has set the flaps (not enough, however) and is still too high and fast, all the way to the landing point. He thinks he has two (or three, as required) notches of flaps, but has set only one (or two). Since he subconsciously, or consciously, remembers doing something about the flaps at the beginning of the approach, the idea of an improper flap setting doesn't occur to him. Have him ascertain, as he sets the flaps on each approach, that the proper amount is being used. With some electric flaps and indicators, you will run into flap jugglers who overshoot the number of degrees required and then take off too much, then add again, and so on. These types, if allowed a free hand, would still be peering at the indicator and working on the problem as the airplane flew into the ground. If you have to cope with a student who tries to set the initial approach power of, say, 1600 rpm *exactly* and then juggles the flaps for an *exact* setting, your patience will be strained indeed.

7. Failing to continue reducing power as the approach is continued. *Students in the early stages of shooting takeoffs and landings tend to get fixated on one phase; things are moving too fast.* This fixation is usually the cause of most of his problems; the airplane is at a new position, or going on about its business, while he is still considering and trying to cope with what happened several seconds ago. As he gets more practice things will start to slow down for him, so if he is having problems at first keeping up with the airplane, assure him things will get better. Some pilots continue to carry power during the landing roll. This is a sign of tension.

8. Initiating the final turn too soon. He may "cut across" and end up too high and also be in a poor lineup position.

THE GO-AROUND. If the student doesn't make an approach requiring a go-around (he will probably make quite a few) during the periods of takeoffs and landings, you should set up an approach that is too high and *demonstrate* the procedure. Talk about it on the ground before you fly and be certain that he knows the steps. The position to emphasize is that *once he has decided that a go-around is required he is to do it,* even if he decides later that he might have made it. He should be taught that indecisiveness is a big factor in takeoff and landing accidents. If a pilot adds power for the go-around, then decides he can land, chops the power, etc., he might end up in a position where he can neither take it around nor get down and stop on the remaining runway—and an accident occurs. Making up his mind can cost height and runway distance (he's still getting lower, which is bad for the go-around, and is overflying good runway, which is bad for a landing).

Your policy should be that if the student decides he's too high and starts to add power, he should continue on with the go-around, adding full power and cleaning up the airplane.

In the preflight briefing you might describe the procedure in this way:

1. The decision is reached that a go-around is needed.

2. Full throttle.

3. Carburetor heat OFF, if used in the approach.

4. Retract the flaps to one-half of approach setting (or one notch less than used in the approach, as applicable).

5. Adjust the nose position to maintain altitude or have a slight climb. Trim.

6. Flaps to zero, as airspeed and airplane's reactions permit. Trim. One method of flap retraction might be, after power is increased, to initially bring the flaps to one-half the approach value, then halving this until 10° or less is left, then upping them to zero. For most trainers, this would mean no more than three adjustments. Again, don't let him juggle to get *exactly* half of the approach flap setting, etc. An approximation is fine, because his concentrating on the flap indicator might cause him to fly into obstructions.

7. Climb out for a new pattern, flying the downwind leg out wider this time. Correct the factors that may have caused the requirement for a go-around.

Your attitude about his go-arounds is of paramount importance. If he takes it around during the presolo training when you think he could have landed and you show this by tone of voice or facial expression (or, bluntly say so), he will become uncertain and probably will be so worried about your opinion that he will try to land when he shouldn't on a solo flight. When the student says, "I'm too high," and starts around, some instructors might grab the controls, violently slip the airplane, land it, and act disgusted that he should have "such bad judgment." It is very tempting on the last approach of a period to take over and land the airplane from a high (to him) position. Such a procedure would hurt his development of judgment, even if you say, "That was right, but . . ."

If he takes it around, you might say something like, "*If in any doubt* about the landing, take it around." You want to help him establish judgment criteria without having him worried about your displeasure if he leans toward the safe side.

So, as he builds up dual time he'll have to develop the proper sense of whether to break off the approach or not, and he is not to depend on help from you. At some time during the presolo period you should make an approach deliberately too high (but not telling him) and then let him have the controls "to complete the approach." You should have him at the stage of saying, "We're too high; I'm going around," and then proceeding to do it without looking questioningly at you to get an opinion. If after things are under control he looks at you and shakes his head over *your* approach, you're making real progress. When he does start making his own decisions, "good" or "exactly right," from you will be a form of praise that will help establish his criteria for required go-arounds.

What about his actions in the pattern following the broken-off approach? If the unusual situation has rattled him badly you are forewarned of possible problems when he encounters unforeseen trouble. If he has no difficulties, this will help you pin down his nearness to soloing.

During the climbout ask him if he knew the reason he was too high, and give him a chance to analyze the problem and decide what he should do during this new pattern to avoid it. It's better for him to be slightly low on the next approach, requiring more power than usual, than to repeat the same mistake. If he overcompensates (within reason) for the original error it shows that he's thinking, anyway.

The student should be encouraged to consider that he might have to make a go-around *early* in the approach (within reason). If he is obviously too high but tries to take it on in anyway, requiring you to take over, he's certainly not ready for solo.

When the student pilot is getting close to solo and he makes a mistake in judgment like the one just mentioned, it will take several successful go-arounds initiated by him before you will start feeling comfortable about letting him go. You'll want to be sure that he's well aware of what is required and *will do it*. You will find that during the initial part of the approach, if another airplane is taxiing onto the runway for takeoff, the average student will start making noises or motions about "taking it around." Depending on the circumstances, such as the possibility of wake turbulence, the position of your airplane in the approach, and other factors, you'd best have him continue the approach and decide that if that other airplane is not well off the ground when he completes the final turn he should take it around. He should be prepared for that possibility and decide *where* to do it and, if conditions require it, do it at that point; *but he should not take it around as soon as he sees the airplane*. Most students are intrigued by another airplane on the ground near or on the runway; they watch it so carefully that sometimes the pattern is pulled in too close and they are too high. In this case a go-around becomes necessary, and the other airplane's proximity to the runway had nothing physically to do with it. To generalize a little, at first the student will tend to take it around *too soon* if another airplane is involved, and he'll take it around *too late* if the pattern is being messed up.

The student should be taught that being too high is not the only reason for adding power and breaking off the approach. Following are some other reasons:

1. Overshooting the runway center line, requiring a steep, racked-up turn to get back to line up with the runway. In this position he should shallow out the turn, add full power, clean up the airplane, and climb out parallel and to the right of the runway. As he climbs he may again get back over the center line or an extension of it, but to try to get back too quickly might compound his problems, especially during the cleanup process. This will be covered also in Chap. 11 in the section on advanced stalls.

2. Another airplane whose pilot is taking his time taking off or clearing the runway after landing. Even though the landing airplane has the right-of-way, a rendezvous with another airplane just over the runway isn't a good idea. Spell out that if the other airplane is taking off, the student on his go-around should fly well to the right of the runway so that he can keep the other guy in sight during the climbout.

3. Possibilities of wake turbulence on final and/or during the landing. The sight of an airliner taking off or landing during your downwind leg in calm air could be a real factor to consider in deciding whether to go around or not.

4. He feels that the approach is going wrong. He's uncomfortable about it but can't pin down any one thing. It's better to go around and get set up better than to risk an accident. The student should understand that the man in the airplane has a better idea of how *he* feels, or how the airplane is reacting, than the hangar fliers on the ground.

FORWARD SLIPS AS AN AID TO LANDING. It's too bad that a full pattern is wasted because the airplane was just a bit too high, even with flaps, to get on the runway safely, when the student could have slipped it and made the difference. However, there are a couple of points to consider.

The two main reasons you should avoid letting the student put the flap-equipped airplane into a forward slip in the presolo period are (1) he does not have the skill and experience to safely slip the airplane without added instruction and (2) airplanes with flaps extended may have bad reactions to excessive cross-controlling and an accident could result (or at best he could scare himself on one of his early solo flights).

Some airplanes have placards forbidding deliberate slips with flaps extended. These same placards mention nothing about the use of a sideslip in a crosswind landing. No doubt it is assumed that the deliberate forward slip to lose altitude would sometimes be much steeper than the sideslip as used for wind drift correction since the airplane has a demonstrated limit by the manufacturer for a crosswind component for takeoff and landing.

For the airplane not equipped with flaps, the forward slip is the only means of steepening the glide path, and for the older, slower airplane such as a T-craft, J-3, or Champion you may decide that the airplane is forgiving enough and the slip is a valuable aid, so you teach it before solo. In the preflight discussion you should tell him the difference between a forward slip and a sideslip, and why each is used.

If it comes to a strict yes or no decision as to whether slips should be taught before solo in a normal situation, you should lean toward the NO side, even in an airplane that doesn't have flaps. If you are teaching from a short field with obstructions and this is normal for your flight training, you may have to introduce them early. If they aren't actually required as a matter of routine, it's suggested that you wait until after solo to teach them. Before the student's solo cross-country he should have practiced slips and be able to do them safely. You can show him how the forward slip can be used in a *no-flap* approach for flap-equipped trainers, and certainly the ability to do a good slip is a mandatory requirement for the pilot of an airplane not having flaps. So the only question would be *when* to let the student do deliberate forward slips, and it's normally best not to have him do them before solo.

The sideslip as used in crosswind landings will be covered later in this chapter.

If you decide that he should be able to slip the airplane, the demonstration and practice should be at altitude so that the student will get used to the uncomfortable crossed-control requirement.

Line up with a straight stretch of road, railroad, or transmission line right-of-way and demonstrate setting up the forward slip. Use slightly less than cruise power for practice at first, explaining that in a real situation requiring a slip the throttle should be fully closed since power will tend to cancel the effects of the slip. The reason for using power here is to decrease the amount of altitude loss and allow more practice.

As he becomes more proficient, decrease the power until he is making a forward slip along the road in a closed-throttle descent. In some cases you may not go through the full procedure of gradually decreasing the power settings, but you may have him practice power-off slips as soon as he has practiced a couple with power. For certain students you may have to move gradually so they will get used to the forces and psychological factors acting on them in a steep, slipping descent. You'll notice that some people are quite uncomfortable at first and try to lean well to the high side of the airplane.

The student should do a number of power-off slips over the reference—each lower than the last, until you have him break it off at a safe altitude—and then climb back to repeat the procedure (Fig. 10-23).

In the side-by-side trainer he will always want to slip to the left, since he can see better that way, but have him get proficient in slipping both directions. Indicate that if there is a slight crosswind and he is high on the approach and needs to make a forward slip, he should always slip into the wind, because the forward slip then can be converted into a sideslip for the crosswind landing.

As to the method of entering the forward slip, some airplanes have enough adverse yaw at low speeds so that when the aileron is applied the nose yaws briskly in the opposite direction of its own accord, and opposite rudder is only needed to hold it at the proper position. Others require that

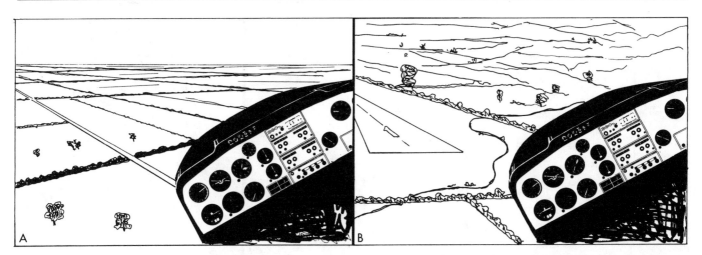

Fig. 10-23. (A) Have the student practice forward slips at altitude using partial power at first; then as he feels more comfortable and develops his technique, have him practice them power-off, losing altitude. He's to break it off at a safe altitude and climb to repeat the procedure as necessary. (B) Try to work it so that he will get a chance to practice one or two at the airport at the end of that period.

opposite rudder be used as the aileron is applied, to help move the nose over to the proper place.

The direction of bank controls the slip; if the left wing is down, the slip is a left one. Some students tend to get confused about which wing should be down for a left slip; they think the variation of nose position (right, in this example) indicates the slip is to the right.

The amount of bank controls the rate of descent, and the rudder should be considered as a "lock," keeping the nose at the proper point for the degree of slip desired.

Common errors you'll encounter when you introduce the forward slip are these:

1. *The "flat slip."* The student feels uncomfortable in a slipping bank and unconsciously allows the wings to ease up to, or near, the level position while the nose is still yawed considerably from the flight path. He is holding just enough aileron to balance the banking tendency created by the use of opposite rudder. Usually the nose is too high, and the nose will tend to move back to the original preslip heading as the airspeed decays and the rudder loses effectiveness. However, a dangerous situation could exist for airplanes with very effective rudders, because the airspeed is low and the controls are crossed (Fig. 10-24).

2. *The "turning slip."* The student may either get too slow and the rudder loses effectiveness, or he may unconsciously relax rudder pressure so that the airplane ends up in a slipping turn in the direction of bank. When this latter happens (relaxing of rudder pressure), he will usually come back in strongly with top rudder alone to correct his heading. The result of that may be a flat slip as the wings react to the yaw and tend to become level.

Even experienced instructors sometimes get edgy at some of the slip efforts of their students close to the ground. When you teach slips, convey firmly that the student is *never* to forward slip below the point of breaking the glide. He should recover from the slip and have plenty of time to be straightened out and concentrate on the landing. Low-time pilots, and sometimes high-time instructors, get fooled by the rate of sink and the airplane is on the ground before they are ready. If slipping all the way down to (and through) the round-out isn't going to prevent the airplane from landing too long, *it's best to take it around.* The exaggerated slipping could cause more problems for the inexperienced pilot than running off the far end of the runway.

You should warn that the airspeed indicator is not reliable in the slip and can indicate excessively high or low values, depending on the direction of slip. The pilot should pay more

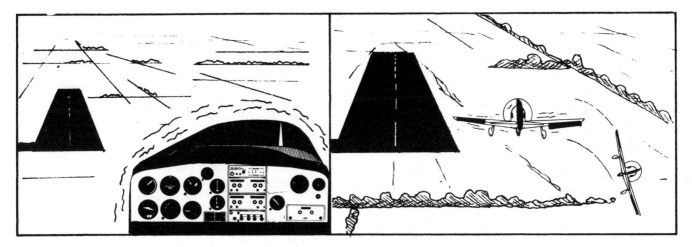

Fig. 10-24. The "flat slip." It could lead to problems.

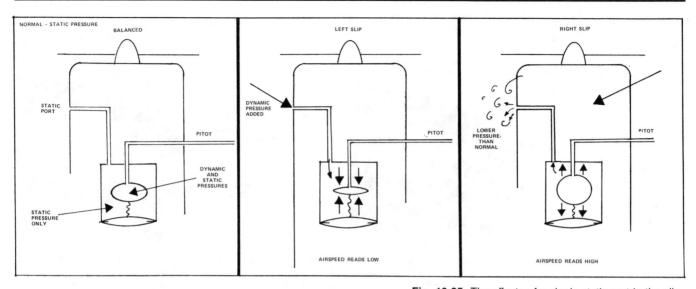

Fig. 10-25. The effects of a single static port in the slip.

attention to the airplane's pitch attitude than the numbers on the ASI. A discussion of *why* this happens can also help him better understand the airspeed system (Fig. 10-25).

The Y-type system, which has static points on both sides of the fuselage, is more accurate; or more to the point, its smaller errors are approximately equal in a slip in either direction (see Fig. 2-6).

If the student believes in the airspeed religiously (Fig. 10-25), he may stick the nose down to get the "correct" airspeed in the left slip and be much too fast (on recovery he may float on down the runway wondering why the slip didn't work), or he may get too slow in the right slip and drop in.

You should caution the student about slipping *when the fuel level is low.* When an airplane has two separate wing tanks, it's possible in the slip, if the "lower" tank has been selected (the left tank in a left slip), for the fuel to be unported, which means that power would not be immediately available if needed. The delay in recovering power at a low altitude, even after the wings are leveled, could result in an accident.

AND EVEN MORE ABOUT FLAPS. There's always the question about the student's use of added flaps if the airplane is too high on the approach. In the example of the training airplane in Chap. 5, it was suggested that 20° rather than 40° (full flaps) be used, because the higher setting only decreased the landing speed 1 mph below the lower and could offer complications if a go-around became necessary and there was a delay in flap retraction.

Suppose during your presolo instruction you have been using 20° of flaps as standard. Since you will likely leave the teaching of slips until after solo (and may never have him slip it with flaps because of the airplane's restrictions), what about having him use full (40°) flaps if he thinks he's high on final? Again, it's best that you don't throw in added distractions because (1) the added rate of descent with the extra flaps may be greater than he's been used to, (2) he may not have had enough experience with full-flap usage and so flies the old pitch attitude, getting too slow and sinking too fast for him to recognize and readily recover from, (3) the extra flaps may be a crutch that he starts depending on and so does not make needed corrections of pattern and airspeed (feeling that if he's too high, what the heck, the use of more flaps will save the situation), and (4) if the full flaps still aren't enough and he still has to take it around, things can be a little more complicated, as mentioned at the beginning of this section. If he has

been using full flaps for landing in all of his presolo training, fine, but you should try to keep things stable until he has more experience.

TOO LOW IN THE APPROACH. The too-low approach can usually be salvaged and a go-around avoided. Like the final that's too high, this problem often starts well back in the pattern at the turn onto the downwind leg. If that turn is delayed so that the downwind leg is extra wide, the usual result is a long base leg and a low final turn.

The student should be taught to add power as soon as he sees that the airplane will be low on the approach. It's better that he have the power at idle on the last part of the final, so he won't be dividing his attention between throttle handling and the landing process. When he first discovers he is low, enough power should be added at that time to ensure that he can make the runway and have the throttle closed in the last part of the approach. Low-time pilots tend to add a small amount of power and need it all through final (and round-out) to make the runway. In addition, in such situations he may have to add *full* throttle at the last part of final to make it; you'll feel as though the airplane is being carried up to the threshold by its bootstraps, in a cocked-up-attitude, low-air-speed, full-power situation. Gusty air or misjudgments could cause the airplane to land short in this instance. Fig. 10-26 shows the wrong and right way for the student to correct for too low a position in the approach.

You'll have to make an effort to see that he doesn't get into the habit of adding so much power that the too-low approach turns into a too-high or too-fast approach, requiring a go-around.

Suppose he adds power, as suggested in Fig. 10-26B, closes the throttle, and discovers that he is still too low. He should again add enough power so that the last part of the final and landing can be at idle. You'll encounter a few of these stair-step finals, but after the average student has a couple of hours of traffic pattern work he won't have this much difficulty in judging the approach.

As he gains experience he can make corrections closer to the "wrong" way in Fig. 10-26A, but in the presolo and early postsolo stages he should be taught the "right" way.

Some *common errors* a student may make while trying to correct for a too-low approach are these:

1. Failure to recognize that the airplane is going to be low. This will change with experience, but it is a prob-

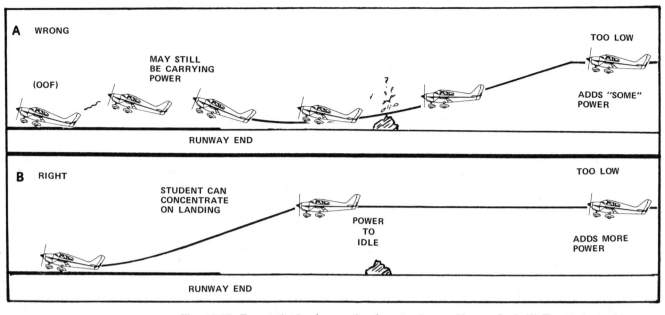

Fig. 10-26. Two methods of correcting for a too-low position on final. (A) The student adds *some* power and has to carry it all the way to the ground in order to make the runway. His attention is diverted and the landing usually suffers. In addition, the more shallow approach, in some unusual instances, could cause him to have problems with obstructions. (B) By adding *more* power at the beginning, it may be reduced to idle in the last part of the approach so that the student can concentrate solely on the landing itself. Hopefully, your airport has lower obstacles on final than shown here.

lem with the low-time pilot, who may allow the airplane to get critically low before deciding that something has to be done.

2. Timidity in applying power. The student adds just enough power to drag the airplane up to the runway and the issue seems in doubt for a while. If he does this on one of his first solo landings, it will get your attention as much as anything he can do, so break the habit early.

3. Too rough on the throttle; abrupt opening and closing of the throttle, resulting in gaining too much airspeed or altitude after power is applied and having pitch problems when the power is chopped off.

4. Trying to stretch the approach without adding power. As one instructor said to a student about the use of power in the normal approach: "In making the runway, you can use alternately full throttle or idle, but I'd rather you didn't do it that way." If the student is taught to make consistent patterns and approaches, he will soon develop the ability to eyeball his relative position and altitude and make small corrections as needed rather than waiting until a large correction is required.

BOUNCE RECOVERIES. Discuss on the ground the procedures for recovering from a bounce, using a model to demonstrate the process. Point out that the recovery procedure varies with the amount of bounce, and the reaction of the airplane is mostly due to aerodynamics and the pilot's reactions rather than the shock absorbers and rubber tires.

The tailwheel-type airplane is more subject to bouncing, because if the nose attitude is too low the main wheels strike the ground first, moving the nose up. The reaction is as if the pilot had jerked back on the stick; and the angle of attack, and lift, is suddenly increased. In addition, the pilot usually sees the problem at the last second and pulls back, trying to avoid flying into the ground; and his reaction time delay is usually just about right to *add* to the bounce reaction. At the apex of the maneuver, the nose starts dropping again. The pilot, feeling that a bounce recovery effort should be initiated,

eases the nose down, again out of phase, and a hard landing (and probably more bounces) results.

There is always the question, in tricycle-gear and tailwheel airplanes, as to how much power should be used in a bounce recovery. A general answer is that power should only be used if the pilot believes that the airplane has gotten slow and is "dropping in" from the peak of the bounce. Students often make things worse by adding power when it's not needed. The airplane bounces slightly and is just settling back on the ground when the student "saves" the situation by *adding power and leaving the power on*. The airplane, which would have remained on the ground if left alone, is pulled back up into the air, power may or may not be reduced, and the process is repeated.

The tricycle-gear type of skip or bounce may be the result of a too-fast, level (or perhaps the nosewheel touches first) landing. Unless the student adds to the reaction by pulling back on the wheel, it usually will be minor, but if he doesn't correct the problem the result will be an extended session of "heel and toe" reactions from the airplane, hitting on the nosewheel, then the main gear, and so on.

The cause of this problem is that the student does not try to *hold the airplane off after the first skip or bounce*. Make him hold the airplane off in the proper attitude until it's through flying. The first bounce can be excused as a temporary lapse of attention; the second and following bounces should be charged to carelessness and lack of skill. *The idea of the normal landing is to keep the airplane from landing as long as possible, and this should be done even if it's the second landing in an approach.*

DROPPING IT IN. The drop-in can be caused by three situations: (1) leveling off too high because of the pilot's not knowing "where the ground is" or not having established his scan during the landing, (2) immediately following a bounce, the pilot tensing up and not checking his height as he should and so "re-landing" a few feet off the ground, or (3) ballooning. After the student's training has progressed to knowing where the ground is, ballooning is the biggest culprit causing a drop-in. He may be normally making good landings but occasionally bounce or level off too high and land a little "solidly" but ballooning, whether caused by gusts or thermal effects, or by his own actions (too much airspeed, or too fast an application of back pressure, or both), most often initiates the drop-in.

The student does not understand how the airspeed can decay so rapidly at the end of a ballooning process. He may unconsciously think that the airplane is getting something for nothing when it's gaining height with no power being added, but the swift loss of airspeed and subsequent high rate of descent can remind a pilot that he got a "bonus" and now must pay the Piper (or Beech, or Cessna, etc.).

Throughout the landing instruction the student must be reminded that there will be times he will have to temporarily stop increasing back pressure and times when he must move the wheel back at a greater-than-usual rate. It should be made clear that he may relax the back pressure slightly but should never push forward on the wheel during the landing because of the comparatively low airspeed. Low airspeed plus an imposed low angle of attack can result in a high rate of sink in a poor attitude for contacting the ground (nosewheel first).

So, you'll show him that ballooning, which gives him an idea that the airplane has plenty of excess energy and apparently doesn't want to come down, is a *temporary* situation and he must make corrections at the beginning of the process in order to forestall problems at the end. He should stop the rearward movement of the wheel or relax the back pressure, and he should be prepared for a requirement of faster elevator action and possibly the use of power in the last part of the landing as the airplane starts to settle again. Ballooning is more apt to occur if flaps are not used, because there is a tendency to have a too-high airspeed. However, if the airplane does balloon with flaps, expect an even higher rate of airspeed decay at the peak.

In summary, nearly all students have early problems with power use in a landing recovery situation. They are usually out of phase, adding too little power too late, and leaving the power on after there's no further need for it. This often further complicates the matter by dragging the airplane back into the air right on the edge of a stall, and the student then decides to close the throttle. You should demonstrate the best method of power use in the bounce and drop-in, reminding him of the tendency to leave the power on and also noting that power is not necessary for all recovery situations.

ABORTED TAKEOFFS. Aborting a takeoff can be an even more critical test of decision making than is the go-around. Except in critical situations (the airplane is very low and slow and the decision was made late), the go-around is pretty cut and dried if the proper power and cleanup procedures are used. An early (but wrong) decision to go around costs another trip in the pattern and takes time but lean to the safe side. Aborting a takeoff, however, can mean airplane damage and injuries, if done late. The problem is that the low-time pilot may not have the experience to catch problems during the early part of the run.

You should bring to the student's attention the normal rpm (and feel of the engine) as any takeoff run starts. If the rpm is lower than normal at the beginning, the throttle should be closed and the airplane slowed and taxied back for a static rpm check. It could be that on a very hot day the high-density altitude and full rich mixture could cut down on the rpm, compared with that cool day 3 weeks ago when he flew last. Maybe a little judicial leaning should be done before takeoff, but the main thing is for the student to realize that the airplane's response was different and should be analyzed. One last quick glance at oil pressure and rpm is always in order as the takeoff starts.

Unlatched doors and seat belts hanging out usually make their presence known just as the airplane is at the lift-off airspeed. The door pops open, and the immediate response is usually one of three things: (1) the student turns his entire attention to the noise and buffeting (staring at the door in disbelief while the airplane is free to go its own way), (2) he jerks the throttle back to idle immediately (airborne or not) with no thought as to how much runway is left or, if airborne, how close those trees are, or (3) he closes the throttle, decides that won't work, opens it, and maybe then closes it partway.

You may not want to demonstrate opening a door or letting part of your safety harness hang out to start the banging (it sounds like the engine is coming apart) on takeoff, but you might demonstrate at altitude, depending on the airplane. The least you should do is discuss the problem in the quiet of the briefing room, describing the sounds and feel of an open door or safety belt or harness hanging out—and what he should do about it as follows:

1. He should go over the procedure to be followed for the takeoff problem in his mind, on the ground. The problem is that in too many cases the pilot is taken totally by surprise and hasn't even considered what a safety belt in the slipstream sounds like or doesn't know the sounds and feel of an open door. He may be *sure* in the actual event that the belt must be an engine problem since he hasn't thought of other possibilities. Or he may be unaware that the airplane will fly with an open door and so may put it down immediately without thinking.

2. Most important, he should maintain control of the airplane.

3. He should check his relationship with the remaining runway and obstacles. One rule of thumb for short-field takeoffs is to set up a "point of abortion" consisting of one-half of the expected (or computed) takeoff run distance. If at that halfway point the airplane does not have 70 percent of its lift-off airspeed, the takeoff should be aborted. (The airplane is not accelerating properly due to runway surface conditions or the engine not developing the expected horsepower.) While the student won't compute the exact point of no return on each normal takeoff, he should have an idea of expected distances required for lift-off. If the distance is paced off, consider each step to be 2½ ft, rather than 3 ft.

Basically, on each takeoff the pilot should be aware of the possibility of aborting and should be prepared to act decisively. If he makes the decision to abort, the throttle should be *fully* closed, and braking should be done with the control wheel or stick in the full aft position.

Under safe conditions you might have the student abort one takeoff and go through the stopping procedure during the presolo takeoff and landing phase.

ENGINE FAILURE IN THE PATTERN. At some time during the takeoff and landing sessions, you should explain and demonstrate the procedure following an engine failure during climbout after takeoff. This is the most critical place for a power loss because the airplane is in the climb attitude (which is a power-off stall attitude, as well), the altitude is low, and since airports attract all kinds of commercial activity in the immediate area, there will likely be a used car lot or a shopping center right off the end of the runway.

Explain the procedure on the ground, emphasizing the need to get the nose down fast after the power failure because of the rapid decay of airspeed in the climb attitude. He should then pick his spot for the landing, make shallow turns to avoid obstacles, and try to touch down in as normal a manner as possible. He should cut off the mags, master switches, and fuel if possible before touching down, but he shouldn't sacrifice control of the airplane for this (and there won't be much time).

There is a continuing controversy about making a 180° turn and attempting to land back on the runway after an engine failure on the upwind leg. Experienced pilots might be able to make a 180° turn if they are near the point of first normal turn. Some have successfully done it, but others have had stall/spin accidents trying it. You'll get arguments that the airplane can make a 180° gliding turn in, say, 300 ft of altitude, and it may be possible. This information is usually based on a preplanned situation with the airplane established in a power-off glide when the exercise begins. It doesn't take into account the surprise factor in an actual emergency, the utter incredulity that this could be happening to a person of your caliber. While the disbelief is reviewed the airplane is slowing up, and when the decision is made to turn back to the airport the airspeed is critically low. It's an old aviation belief that you should always be prepared for an engine failure on takeoff, but it always seems to happen at the one time your attention is elsewhere. Although the chances of an engine failure on takeoff are extremely small, the student should have the procedure demonstrated and the throttle should be closed unexpectedly to simulate the emergency at least once during the practice sessions. If his response is slow you should explain the procedure again, telling him that you will close the throttle on the next climbout and he is to get the nose down *immediately*. It's best to do this simulated engine failure reasonably early in the takeoff and landing sessions so he can be somewhat used to the idea. A number of routine takeoffs and landings after he encounters the simulated problem will restore his confidence. He should know what to do, but the worst thing you could do would be to *initiate* this training in the same session in which he is to solo. You may want to review it that session, but hit him with it cold then—no. It may be too much on his mind when he solos.

Fig. 10-27 illustrates the point, particularly for airports with long runways, that the student will tend to fly a pattern with a final leg that gives plenty of straightaway before the airplane gets over the runway, certainly a desirable condition in a normal approach. But in an emergency or simulated emergency, he may think "normally" and he'll either make a dangerously low turn onto final or won't make it to the runway. You should at least talk about this, if not demonstrate it.

Naturally, you will have to plan your simulated emergencies with respect to other traffic, and you also must keep the tower or Unicom informed of your plans if there might be traffic conflicts.

Have the student practice several *power-off* approaches from the point where power is normally *reduced,* so that he can see the requirement for a tighter pattern.

It's interesting that you can shoot a number of power-at-idle approaches with a student, not mentioning anything about emergencies, then close the throttle in the pattern just as he is about to do so normally, and he forgets all. At your statement of "emergency landing," the student will sometimes look *away* from the airport to find a place to land, even though he's been shooting consistent power-off landings to the runway from exactly the same position.

You may simulate an engine failure on base so that he can see the need to "cut across," as opposed to the pattern used when power is being gradually retarded.

Your particular student's reactions or the airport situation may be such that it would be a waste of time before solo to give a large number of simulated emergencies from many points around the pattern. Besides, an overabundance of simulated emergencies could cause the student to decide that flying is extremely dangerous. Some of the more exotic pattern emergencies can, however, help get a cocky student back on course.

LOSS OF THE AIRSPEED INDICATOR. The student should fly several dual patterns with the airspeed indicator covered to show him that power plus attitude equals performance; that is, keeping the nose in the proper pitch position combined with required power means the airspeed will take care of itself.

You should uncover the airspeed briefly during the climbout, downwind leg, and approach to let him confirm that the attitude and power combination is giving him the airspeed he wants. If the pitch attitude is off and you suspect that the airspeed is high or low, have him correct the pitch attitude, then wait a few seconds for the airspeed to stabilize before you uncover the instrument. It's better to have the right combination of pitch and airspeed when he checks, so that he will see the proper attitude. There is only one attitude for each condition, and you should impress him with this idea.

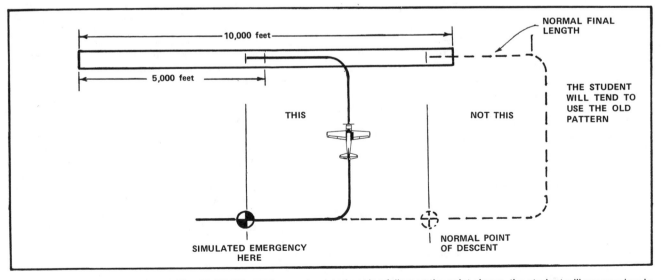

Fig. 10-27. When given a simulated engine failure at the point shown, the student will unconsciously back away from turning inside the airport boundary, even though there is plenty of runway left for landing.

In *your* early flights as an instructor you may want to "cover" the airspeed indicator in such a way that he can't see it but *you* can. Since you are responsible for the safety of the flight, you can use this to help you keep abreast of the situation in an unfamiliar airplane. After you've instructed in a particular model for a while you will have the required pitch attitudes well in mind, and it's more impressive if the airspeed instrument is completely covered.

Your question might be, "Am I trying to impress the student or teach him?" Both. *Also, if you are good at flying the airplane, he will be more apt to listen to your instruction and to emulate you—you'll have his attention.* And looking at it from a practical standpoint, if your flying is impressive, he will talk you up among potential students and you'll get more business. However, as stated in Chap. 2, the most skillful pilot around has to teach well or the student won't get much, if anything, from a lesson.

If feasible, have the student later shoot one or two patterns leaving the airspeed indicator covered throughout (no checking).

■ **Crosswind Takeoffs and Landings.** It's been indicated that it's best if the presolo takeoff and landing practice can be done without encountering crosswinds, but your local situation might be such that a crosswind *nearly always* exists. In that case, the student should be given presolo crosswind training, *but not until he has had a chance to practice takeoffs and landings in calm or straight wind conditions* and "find the ground" without extra distractions. If possible, he should have several periods of normal pattern work before starting crosswind takeoffs and landings since to many students the first crosswind landings are like trying to simultaneously pat your head and rub your stomach. He will tend to concentrate on either the crosswind correction or the landing, to the loss of the neglected procedure. He will usually forget about the crosswind correction until the jolt of a drifting touchdown and the sound of tortured tires jars him. So, even though you might need to teach him crosswind takeoffs and landings before solo, don't introduce them until he understands the normal pattern work.

CROSSWIND TAKEOFFS. Assuming that there is a requirement for presolo crosswind training and the student is proficient in normal takeoffs and landings, you should brief him on the techniques during the preflight session and demonstrate at least one crosswind takeoff and landing before letting him practice.

You might review with him that the airplane on the ground has two main reactions to a crosswind, and these reactions depend on the strength of the crosswind component. Explain headwind and crosswind components of a particular wind at the chalkboard (Fig. 10-28).

Weathercocking, as shown in Fig. 10-28, is much more of a factor with tailwheel airplanes but still will have some effect on the takeoffs (and landings) of tricycle-gear airplanes. Directional control is more difficult to deal with on takeoff in left crosswinds, since the left-turning tendency caused by "torque" is aided and abetted by weathercocking. For some airplanes very strong left crosswinds, may make it impossible to maintain a perfectly straight path on takeoff, but the student shouldn't have to worry about that at this point. You should also mention that a right crosswind tends to cancel the "torque" effects.

Emphasize that the airplane should not be forced off in a crosswind; the risk would be great that the airplane would touch down again, drifting. This is true for both tricycle-gear and tailwheel airplanes. It should be held on the ground longer to a higher airspeed, then brought off with a firm, but not abrupt, back pressure. A turn is then made into the wind

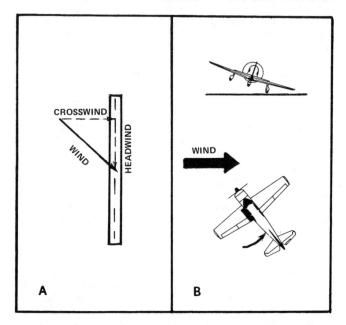

Fig. 10-28. (A) Headwind and crosswind components. Explain that the *angle* that the wind makes with the runway, as well as its velocity, decides the headwind and crosswind components. (B) The crosswind component acts on the airplane in two ways: it tends to cause the airplane to "lean" away from *and* weathercock into the wind.

to the proper crab angle, the wings leveled, and proper "torque" correction held to maintain the heading.

Teach the use of full aileron into the wind as the takeoff run is started as well as the requirement to ease it off as the airspeed picks up. You will have to demonstrate such a mechanical method at first, but as he gains experience he will set up the aileron use initially corresponding to the need for correction. He will always have a slight amount of aileron deflection at lift-off and can use this to establish the turn into the wind to set up the crab.

Usually the student has comparatively little problem keeping the airplane straight. Heading deviations can be fairly easy to spot, up to and including running off the runway. Developing feel for the amount of aileron deflection required is another matter, however. You'll encounter people who want to hold full aileron all the way (and tend to make interesting two-point takeoffs—with the upwind wing and upwind wheel touching) and those who let the aileron-deflection change several times during the early crosswind practice sessions.

Another crosswind takeoff procedure involves using enough aileron into the wind so that on lift-off the downwind wheel leaves the ground first and the airplane is banked into the wind. Enough opposite rudder is held so that the initial few feet of the climb is in a wing-down attitude, keeping the center line of the airplane parallel to its path along the runway. The advantage is that if the airplane touches again, there will be no, or nearly no, side loads on the landing gear. (The same principle is used in the wing-down crosswind landing.) After the airplane is definitely airborne, the controls are coordinated to turn it into the wind for a balanced crab climbout. A valid argument (for some airplanes) against lifting the downwind wheel off first is that if you have so much airspeed that the ailerons are that effective the airplane should have been airborne long ago.

Common errors you might encounter in *crosswind takeoff* practice include these:

1. Easing off of the aileron deflection too much as the takeoff roll begins.

2. Mechanical use of the ailerons, not feeling the need for varying the deflection. The student may take some time to work this out. A clue that this is a problem is when he abruptly remembers that aileron is needed, or not needed, and moves accordingly.

3. Poor directional control, particularly in a left crosswind. (In a right crosswind, he may tend to mechanically use the same amount of "torque" correction as used for normal conditions, and the wind helps him turn to the right.)

4. Trying to force the airplane off and setting up a skipping, drifting situation.

5. Holding too much aileron in the latter stages of the takeoff so that the airplane banks too steeply into the wind as the wheels leave the ground.

6. Poor drift correction on climbout.

CROSSWIND LANDINGS. The most comfortable, as well as an easy, crosswind approach and landing for the more experienced pilot is a crab approach and a wing-down landing. The crab approach doesn't require slipping, which can make the passengers or pilot uncomfortable, and the wing-down landing requires less judgment than holding the crab throughout and kicking the airplane straight with the rudder just before touchdown.

Wing-down method. Many instructors use the *wing-down (sideslip) method for the approach and landing* for students because there is often a problem in the transition from the crab to the wing-down procedure. As you have students make the transition, you'll be amazed at how many kinds of cross-controlling techniques exist.

Before the flight you should compare the sideslip and forward slip using a model and the chalkboard, and emphasize that the center line of the airplane must be parallel to the path it is traveling at touchdown. Hopefully, the airplane's center line and flight path will also be parallel to and on the *center line* of the runway at touchdown (taking into consideration the earlier discussion in this chapter about center-line lighting on some runways). Impress on him that it's better to make minor corrections all the way down on final than to have to make radical maneuverings at the end of the approach and during the landing. This is common sense, of course, but you may have students who have drifted well off the center line of the runway as the round-out is started, and luckily after some wild maneuvering managed to grease it on the center line with the proper drift correction at the last split second. It's hard to argue with that result since it boils down to the fact that the thing that counts is touching down properly, no matter what was done on final. The counterpoint is that it is rare that radical maneuvering is followed by a good landing.

Fig. 10-29 shows a way to describe the difference between a sideslip and forward slip, using a model and a pencil. Hold the model and pencil as shown, with the pencil representing the airplane's flight path. Indicate to the student that for a given angle of bank in a constant slip (no turn occurring), the flight path will be a certain angle from the longitudinal axis (angle *A* in Fig. 10-29). The airplane does *not* know where it's headed; the pilot's job is to establish that. Aerodynamically, the slips are the same as far as the airplane is concerned.

Pick a portion of a table or desk top as the "runway" and explain the difference in slips as followed (assume no wind or a wind right down the runway for your discussion):

1. *Sideslip.* In this example the airplane is high and to the left of the runway center line. The airplane heading is parallel to the center line but a sideslip is used to lose altitude *and move sideways* with respect to a chosen reference (the runway). Fig. 10-30 shows a vector analysis of a sideslip for crosswind approach and landing. The bank is exaggerated.

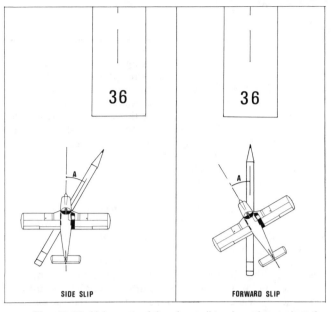

Fig. 10-29. Using a model and pencil to show the student the difference between a sideslip and a forward slip. (The airplane doesn't know the difference since angle *A* is the same in both kinds of slips.) Be sure that you are aware of any manufacturer's restrictions against slipping with flaps down.

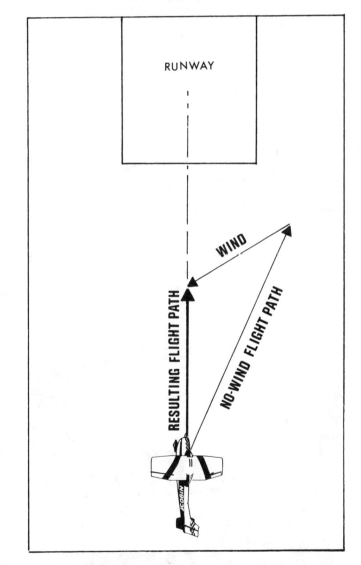

Fig. 10-30. A vectorial look at the sideslip as a method of crosswind correction on approach and landing. (The bank is, uh, exaggerated a little.)

2. *Forward slip.* Without moving the pencil relative to the model, turn the model so that the pencil is pointing the airplane's flight path down the center line of the "runway." Emphasize that in this example the banks are the same, hence the airplane's flight path will be at the same angle from the longitudinal axis in both cases. In the forward slip, the airplane's longitudinal axis is not pointed at the runway reference but had to be turned so that its flight path was that required. You might add here that doing a forward slip all the way to touchdown is indeed a bad habit.

The basic technique for the full wing-down crosswind approach and landing is to roll the airplane out on the runway heading on final and lower the upwind wing and apply opposite rudder as necessary to keep the nose aligned (or parallel) to the runway center line.

The word "parallel" is used because there are two methods of checking the required amount of drift correction. The first is to roll out lined up with the upwind side of the runway (wings level) headed parallel to the center line. The airplane is allowed to drift to the center line and the pilot gets an estimate of the amount of drift correction required. It's hard for the student to tie in that initial drift with the amount the wing should be down, particularly on a narrow runway situation. He often doesn't have the wings level at the point of roll-out on the final, so that an erroneous input to his personal computer occurs (Fig. 10-31).

Another procedure is to roll out on final lined up with the

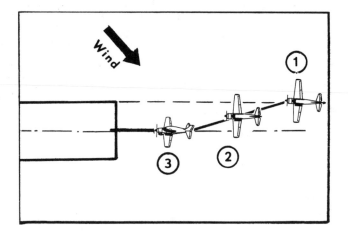

Fig. 10-31. One method of setting up the proper drift correction on final.

center line and immediately set up an estimated correction.

Unlike the crosswind takeoff, which can only be practiced on the ground, the wing-down approach and landing can be practiced at altitude. Some instructors set up the problem at 4000 ft above the surface over a road, but this isn't always a realistic situation because the drift may not be evident or may be exaggerated and the student gets a false picture.

A good exercise at altitude is to use an integrated flight situation. The flight instruments may be put to good use in teaching him to *cross control,* a practice you've had him avoid throughout the earlier part of his flying (assuming he hasn't been introduced to forward slips). He'll have to "unlearn" something, and this can better be done at a comfortable altitude. You might mention that coordination does not always mean using the ailerons and rudder in the same direction but that they are used in the proper proportions to make the airplane do what is required.

Pick a heading on the heading indicator (preferably one of your airport's runway headings) and have the student set up a partial-power descent at the recommended approach speed and flap setting. He should bank the airplane 15° and maintain that heading by use of the opposite rudder (Fig. 10-32).

Have him practice several of these "approaches" correcting for "wind" from both directions. You may even have him "land" the airplane at altitude several times, letting him see the need for added aileron and rudder deflection to maintain 15° of bank as the airplane is slowed to the stall. You should mention that in an actual situation the 15° of bank used during the approach will have to be increased (and opposite rudder added) as the airplane is slowed in the landing because the crosswind component will be comparatively greater at the lower airspeed.

In using this technique, which calls for more monitoring of the instruments than is done for a landing, be prepared to tell him to keep his eyes out of the cockpit when you get back down to an actual touchdown. It is a good way to introduce him to the idea of crossed controls. You may also give him some "go-arounds" at altitude.

The next step in working on crosswind approaches and landings is to have some extra-long final approaches at the airport so that he can set up the drift correction and see the requirements. The average final is too short to allow him to be properly introduced to the idea, so you might even set up a few as long as 3 or 4 mi in an *actual* crosswind. (Don't try to demonstrate or have him practice crosswind techniques when there is no crosswind; it will confuse him.)

Fig. 10-32. The use of the flight instruments at altitude to introduce the idea of the wing-down approach (15° shown here).

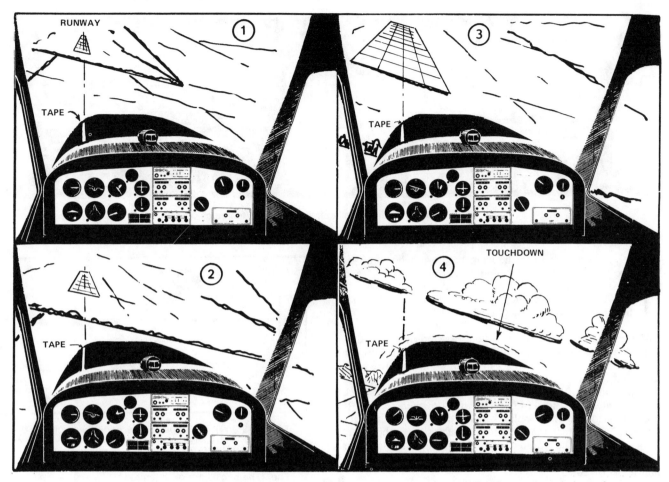

Fig. 10-33. Using masking tape and a long approach to establish the proper techniques for a cross-wind landing.

A strip of tape on the cowling directly in front of the student's line of sight and parallel to the airplane's center line is a good training aid for keeping lined up if there isn't already a hinge line or other reference on the nose directly in front of him (Fig. 10-33). (Be careful of what kind of tape you use; some tapes take paint off when they are removed.)

In this practice, you should set the student up on the long final (recommended landing flaps) and handle the throttle yourself. Make the throttle handling smooth so that he won't be distracted as he handles the flight controls throughout the approach. You'll probably be continually telling him that he is over- or undercorrecting for drift. You'll also have to remind him to keep correcting during the round-out and landing. If traffic permits you should practice several of these long approaches from both directions, so that both left and right corrections may be done. As is the case for most phases of flying, correcting for a *left* crosswind is easier for students. A right wing-down correction during the landing is more difficult from a coordination standpoint, and also it's usually harder for him to see the ground in that attitude.

One common problem is the turning of the nose away from the airplane's path and turning the sideslip into a forward slip. This is particularly true in a left slip (side-by-side airplane). He subconsciously would like to get the nose farther around so that he can see better. In the right slip, he won't have quite as much tendency to drag the nose around with the rudder; it will be in his way. He could not get away with a true forward slip, since the airplane would drift off the runway center line, but he may do a half-side, half-forward slip. He may do this using 15–20° of bank to maintain the required flight path, which would, for instance, only require 10° in a true sideslip in this wind condition. Besides, he shouldn't land in a forward slip—it's hard on the landing gear (Fig. 10-34).

The tape on the cowling is useful in breaking this habit and as a reference to use in keeping the airplane center line parallel to the runway. Later you may remove it as he begins to visualize the proper attitude of the airplane in the wing-down approach.

Common errors to be expected in this type of approach and landing are these:

1. Letting the airplane turn when the wing is lowered.

2. *Undercorrecting* for wind drift. This is much more common than overcorrecting.

3. Not having the nose lined up with the runway, setting up a forward slip type of approach.

4. Working at the drift correction. Constantly changing the bank (and rudder correction) when it isn't needed—a sign that he is mechanically trying to do the job and isn't recognizing the drift.

5. Letting the bank shallow out during the landing, or not increasing the bank as needed to correct for the crosswind at the slower airspeed. A drifting touchdown results.

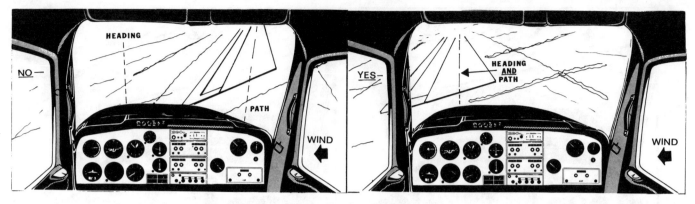

Fig. 10-34. The student will at first tend to use too much opposite rudder and turn the sideslip into a forward slip.

6. Trying to raise the wing before touchdown. Assure the student that he should land on one wheel and that no problems will be encountered.

7. Getting so wrapped up in the drift correction that the landing process is neglected.

He should understand that the crosswind may vary during the approach, and he is to fly the airplane to follow the required path. To expect to set in a certain amount of bank and hold it constant all the way to the ground isn't realistic. He may have drifted off and had to turn slightly to get back on the center line and set up a new wing-down correction. If the airplane has drifted well downwind of the center line, it's better to make a turn than to set up a much steeper slip to get back on the line; the required slip may be dangerously steep or may not be effective enough to make the correction before the landing (Fig. 10-35).

There may be times when he gets so behind in a crosswind approach that a go-around is the best move. Make it clear again that he should not try to salvage a dangerous situation when close to the ground; take it around.

Crab approaches. More advanced students should be shown the *crab approach and wing-down landing,* since this is the best combination of passenger comfort and easy landing (Fig. 10-36).

Point out that the crosswind component will normally decrease slightly because of ground friction as the airplane gets close to the ground. However, it usually does not decrease

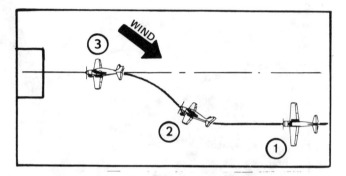

Fig. 10-35. (1) The airplane has drifted well off the desired flight path on final. (2) A shallow turn will get back to the center line and set up the proper wing-down attitude, as indicated by (3). This is better than trying to slip it back over to the center line. (The angles are exaggerated.)

enough to cancel out the need for added deflection of the controls as the wing-down landing is completed. The airplane's decrease in airspeed (and decrease in control effectiveness) usually is a bigger factor than the decrease in crosswind velocity, except for obstacles, such as trees, alongside the runway.

In other words, pilots who think in terms of positive crosswind correction throughout the approach and landing may be surprised at a strange airport where obstacles such as

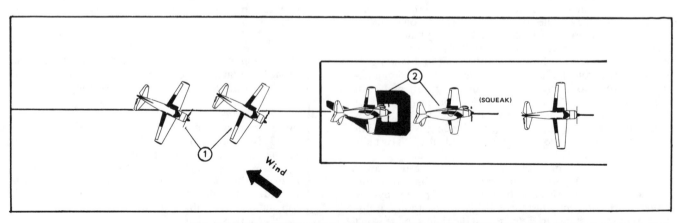

Fig. 10-36. The crab approach and wing-down landing. (1) Setting up the proper crab angle on final and maintaining it until the point of round-out is reached. (2) Then straightening the airplane and making a wing-down landing. In the crab method, as in the wing-down correction, the required corrections will vary as the wind varies in the approach. (The crab and bank angles are exaggerated.)

trees or hangars are comparatively near the edge of the runway. If there is suddenly no further need for a correction, the airplane may move toward the upwind side of the runway before the pilot realizes what is happening (Fig. 10-37).

Fig. 10-37. Warn the student (or private pilot) that he may "lose" the crosswind close to the ground because of obstacles. Also bring up the point that these obstacles may cause turbulence and downdrafts.

While at first they may have some trouble with the transition from the crab to the slip as the landing is started, most people will prefer this method over the long slipping final. You may demonstrate both methods and see which the individual student is able to grasp most easily and safely, and go on from there.

You should briefly discuss the crab approach *and* crab correction during the landing, but you shouldn't push this idea in the earlier stages of his experience. He will see that "kicking" the airplane straight at the last second requires better judgment than he has with his limited experience. For airplanes of large wingspan, such as jet liners, in strong crosswinds the wing-down landing *could* cause the wing to contact the runway. Jet airliners or jet bombers won't be his worry for some time, but the idea should be brought up as general background.

When taking a full look at the crosswind approach and landing, you should also discuss the combination crab and wing-down method for strong crosswinds. The crab *or* wing-down landing procedures each have limits of crosswind correction, and by combining the two the pilot may almost double the crosswind limits that can be handled, but this is for later use. If the crosswind is strong enough to require a student or low-time private pilot to use the combination method, he'd better use another runway, or airport.

In talking about crosswind landings, discuss the maximum crosswind demonstrated values as given in the *Pilot's Operating Handbook* or other sources. As a rule of thumb, most airplanes have a demonstrated crosswind component of 20 percent of the maximum certificated weight stall speed (CAS) with recommended landing flaps.

You'll probably see the need to give at least some instruction in light crosswinds before solo, but you should try to solo him that first time in a straight wind and nonturbulent conditions.

SOLO

■ Getting Ready

THE STUDENT CERTIFICATE AND MEDICAL AND OTHER REQUIREMENTS. Confirm that the student has his medical and student certificate well before the day for solo. He could be as hot as a firecracker, with wind and traffic conditions perfect, but because of paperwork problems it's a no-go situation. Even a day's delay can sometimes mean extra dual because he's allowed to slip over a peak of performance. It's possible too, that if he has a medical problem, a delay of several weeks could occur, so you should get him on the way with a medical and student certificate immediately after that first flight. It's also possible that he may have a diabetic problem or another physical ailment that would preclude the issuance of *any* kind of a pilot's certificate—and he could waste his money by flying 8 or 10 hours of dual. It could also hurt a lot if he has started to love flying.

He must have passed a written test on FARs, operation of the airplane being used, and general safety practices, before solo. You may use a true-or-false exam, but if you want to get your students started early on the technique of taking the private written you should make up a multiple choice test for them. Some instructors require a minimum passing grade of 80 percent, but depending on the toughness of the questions, you may use 70 percent as a minimum, as is the case for FAA writtens.

You won't find an actual sample test here because you should do the research and writing yourself. You'll learn more about your airplane and the student's knowledge requirements if you dig it out for the particular airplane and conditions. With that practice you can then make tests for other airplanes and other conditions. However, you should cover the following general areas, and it's best to use an "open book" test so the student can become familiar with the *POH* for the airplane he's flying (plus other publications). He should know where the *POH* is kept in the airplane.

Here are some suggestions for the areas to cover:

I. THE AIRPLANE

    A. *Fuel system.*
        1. Usable fuel (gallons).
        2. Weight of fuel per gallon.
        3. Expected average fuel consumption (gph) for cruise at the power setting used in the practice area.
        4. Minimum octane fuel to be used.
        5. Color of fuel used.
        6. Electrical auxiliary fuel pump operations (if applicable); when and why it is used.
        7. Preflight fuel check items and order of checking them.

    B. *Oil system.*
        1. Brand and viscosity number of oil currently in use in the airplane.
        2. Total capacity and minimum operational volume (quarts).

    C. *Electrical system.*
        1. Equipment and instruments (flaps, oil tempera-

ture, fuel gauges, etc.) that are electrically controlled.
2. Effects of turning off the master (electrical) switch on the running of the engine in flight (none).
3. Steps to be used in a suspected, or real, electrical fire.

D. *Instruments.*
1. The instrument(s) that would be affected if the static port was totally closed.
2. Location of the static port.
3. Significance of each of the following airspeed markings:
   a. Green arc.
   b. White arc.
   c. Yellow arc.
   d. Red radial line.
4. The max rate (V_Y) and max angle (V_X) climb airspeeds. (Note that these values change with density-altitude but have the student give the numbers you've been using.)
5. The number of seconds for the oil pressure to come up to normal in moderate ambient temperatures.
6. The normal fuel pressure (if applicable), in psi (green arc).
7. Steps to follow if the low-voltage light comes on in flight. Discuss several options with the student: If in the traffic pattern continue the approach and land, terminating the flight; if some distance from the airport, recycle the master (electrical) switch, etc.

II. TRAFFIC PATTERN AND PRACTICE AREA OPERATIONS

A. *Pattern.*
1. Have the student sketch the traffic pattern for the home airport for all runways. (If some patterns are right hand, show this.) Show the altitudes (MSL) for the first turn after takeoff and the downwind leg(s) for a normal approach.
2. Have the student sketch the approximate practice area showing towns and other landmarks that would help pin down the boundaries.
3. Review airports in the area that can be used as alternates should weather or other factors preclude the return to the home base. (You may have already taken the student to shoot landings at one or more of these airports, but you should review here.)

B. *Discuss the following situation.* You are solo and doing a takeoff and departure stall in a right-climbing turn at the break and the airplane rolls abruptly to the left. The nose is down and the airplane is rotating. The wheel is full back but is not bringing the nose up. You should immediately do which of the following?
1. Hold the control wheel full back and make sure the engine is at full power to try to stop the altitude loss.
2. Close the throttle, apply full rudder opposite to the rotation, and immediately move the control wheel forward. (This is the correct answer for a spin recovery. Briefly discuss spins.)
3. Apply rudder into (with) the rotation, maintain full power and stop the nose from getting farther down with full-back wheel.

III. AIRSPACE AND REGULATIONS

A. *Airspace.* You should be sure that the student is aware of airspace that would be a factor in his solo operations in the pattern and practice area. You should cover local airways and Class B, C, D, E, and G airspace; prohibited and restricted areas; plus MOAs and Military Training Routes, *but only* as they would affect him at this stage of his local training. To ask about operations in Class B airspace 200 miles away does nothing but clutter up his mind with trivia.

B. *Regulations.*
1. Your questions on regulations Parts 61 and 91 should also be applicable to the current situation. For instance, the requirement of 30 min minimum fuel reserve for VFR flight (FAR 91.151) has probably been superseded by your, or the school's, hour or more fuel reserve requirement.
2. Apply your questions to what is required at this point in his training, both general and local. For instance, there may be local restrictions such as "Don't fly below 3000 ft AGL over the Veteran's Hospital" and other rules that would only apply to your operations.
3. While it's not required, you might later want to make up a written test for the student to take before going on that first solo cross-country, with general questions on FARs and airspace and other questions on the trip itself, with emphasis on the destination airport facilities, traffic patterns, etc., plus alternate plans for weather and other factors. You might not want to bother with tests for the following solo cross-countries if you are satisfied with his judgment on the first one.
4. You should keep the presolo written test and any others on file for future reference.
5. To repeat: The point of making up your own test(s) rather than getting a canned quiz from commercial publishers is twofold: First, *you* know more about your local operations and requirements than somebody sitting at a desk a thousand miles away. Second, making up such a test will add to *your* knowledge of the requirements for a safe postsolo period for the student and will also help you gain experience in this area for your ground school instructing.

IS HE READY? Assuming the paperwork is okay, you'll still have the problem that faces all new flight instructors: When is a student ready to solo? As you gain more experience you'll develop a feeling for the right time, but now you'll want to have some criteria for that first student solo.

1. *He makes corrections without your prompting him.* This is the best single sign that he is capable of going on his own. His corrections don't have to be perfect, but he sees the need and is making them in the right direction. Sure, he's bounced a little on a landing or two but moves to correct the problem. *You* might have made a smoother correction than he did, but the issue was never in doubt. *He is aware of the airplane and its position relative to the airport* (or runway, etc.) *at all times.*

2. *You do not have to instruct him in basics in the last couple of periods before solo.* You should be able to keep your comments to a minimum unless you need to straighten

him out on a minor point. During the last five or six dual takeoffs and landings before solo you should only make comments for small corrections or say something like "That was a good touchdown; you have the landing attitude just right." You are riding as a check pilot the last dual session before solo, to see if he's ready; the last dual flight should require a minimum of instructions.

3. *He is generally relaxed as he flies the pattern.* (Nobody is completely relaxed.) If he flies the pattern without unnecessary comment and isn't jerky on the controls, you can figure that he is becoming at ease in the airplane. Many mutterings and much chatter as the pattern is being flown is generally a sign of excessive nervousness. And the student who is trying to please *you* rather than fly the pattern for himself is not ready for solo.

4. *He is consistent in the pattern and in flying the airplane.* This is very important in determining whether the student is ready for solo. But consistency is *not* mechanical flying. One method of checking for mechanical flying is (if wind and traffic permit) to change runways during one of his later takeoff and landing sessions to check his reaction.

A student who is erratic, excellent on one pattern and landing and bad on another, is *not* ready for solo.

■ **The Solo.** The student has his medical and student certificate, has passed your written test, and is in your opinion ready for solo as far as flying the airplane is concerned.

So, the first solo should be done at a quiet time when the wind is steady down the runway or calm and other traffic is very light or nonexistent. (If you've been shooting landings at a busy airport, he'll be used to heavy traffic and this won't be a problem.) Generally, the first solo should be as free of "outside" distractions as possible. If you are flying from a controlled field, take the student to the tower well before solo, introduce him to the people there, and mention that he is working on solo. He will feel better about his first (and following) solo flight because he will be more at ease with the tower people. He'll feel *in the group* (Chap. 2) and that they will "look out for him." Most student pilots regard the tower at first as a faceless, overly authoritative system that is "against them." (Remember?) He'll do a better job on that first solo if he's met some of the people he's been communicating with.

Early morning, shortly after sunrise, is a good time for solo, particularly for *your* first few solo students. Some argue that later afternoon is as good because conditions are calm then, but there will be likely heavier traffic, and if problems do arise while he's solo, there could be the added factor of encroaching darkness.

Don't solo any student that first time unless you have shot at least three takeoffs and landings with him immediately preceding the solo. In other words, he did a very good job on the dual flight late yesterday afternoon, and he was ready for solo except for the fact that it was too close to darkness. Should you meet him at the airport first thing this morning and send him out solo immediately? NO. You should allow him to warm up with you again before turning him loose. In addition, you don't know what the effect of an overnight wait (or longer, in some cases) may have had. Some students can run into a slump overnight. The transition to solo is easier from a dual situation; it makes for better continuity in training.

Most instructors don't tell the student when he will solo. Few would dare tell a student that he will solo "in 2½ more hours," because there are too many variables to consider. Forecasting the date and time tends to tense up most students and could set them back, which means that you can't safely stick to your promise, causing them to lose some faith in you as an instructor. Individual reactions vary so widely under

different conditions that it's practically impossible to predict the exact time of solo. You can say to yourself, "At the rate he's going, I can solo him after four more takeoffs and landings," but the key is *"At the rate he's going."* His rate may change so that you can solo him after one or two more patterns, or you may discover a problem that requires much more dual.

When you've decided that all factors are right for his solo, make the transition as casual as possible. After he has landed and has taxied back, you should ease out of the airplane, fastening your seat belt so it doesn't get in his way, and tell him that you want him to shoot three full-stop landings. After the first one he is to taxi back, and you will probably discuss any errors he may have made on the first circuit before you wave him on. Or you just might wave him on to taxi back for another takeoff. If he makes minor errors in that first pattern and landing but is safe, you should grin and *wave him on for another.* This helps his confidence a great deal. Don't stop him and start working on minor details; it's distracting and could cause his performance to suffer.

Be sure that the student is told sometime before he solos that the airplane will get off the ground and climb better without your weight.

One good psychological move for some students is to stop the airplane and say that the takeoff pattern and landing were good (excellent, etc.) and that you'll be in the office while he shoots three more full stops just like it—and then you get out. Always have the solo student monitor Unicom if your airport is so equipped (no tower).

Never have the student shoot touch-and-go's on the first solo. You may need to stop him if he suddenly starts doing something radical and dangerous, or if new situations arise (such as the sudden arrival of a 100-plane fly-in, or sudden weather changes). There would be nothing worse than to see your student do something extremely dangerous in his first pattern and know that he is going to shoot two or more additional patterns before you will be able to talk to him, either personally or on the Unicom. "Wait," you say, "I wouldn't solo him if there were any doubt." True, but your judgment has been based on his actions when *you are with him.* While the number of incidents of student difficulties on that first solo is extremely low, you won't honestly know how he'll react to solo until he does it, and the steps just mentioned are a little extra insurance. Your judgment of students will improve with experience, but you should still have an ace in the hole.

After a while you'll find the best technique for your airport and situation for soloing a student the first time (and will vary it with students). You might stand at a position on the side of the runway as a reference to decide on a go-around—if he isn't on the ground by the time he passes you, he's to take it around. Or, using Unicom, a casual comment like "That looked good from in here; do it just like that again," will help his morale. You might say something like "That was okay; have that nose just a tad higher next time and it will be just right." You aren't going to go into the instructing business on Unicom (you did your instructing in the airplane), but a few comments may ease his uncertainty. If he's doing well and you want to pin down his solo work, you might tell him over the Unicom to shoot a couple of extra patterns before having him come in.

Always congratulate him after the first solo. While you may only have a slight feeling of relief, it's a high point for the student and he's relieved, too. Have a debriefing, but keep details to a minimum.

A point you should consider is the signing of his student pilot certificate (Fig. 10-38). *Legally, he should be signed off before he goes solo.* You could endorse the student certificate before the flight you expect to solo him (or even the day before) so that there is no interruption of the training process, but there could be problems if the student isn't ready to solo when you think he might be or a long siege of bad weather sets in and you have signed the certificate. It's possible that he might move away in the meanwhile, or go to another airport to rent an airplane (a few flight operators don't give dual checkouts before renting an airplane!) and have an accident. Even though you had not permitted him to fly because he may have had some new problems—legally, you are responsible for his solo actions after signing, even if he goes to another airport and solos without your knowledge. Endorse only when he's ready to go.

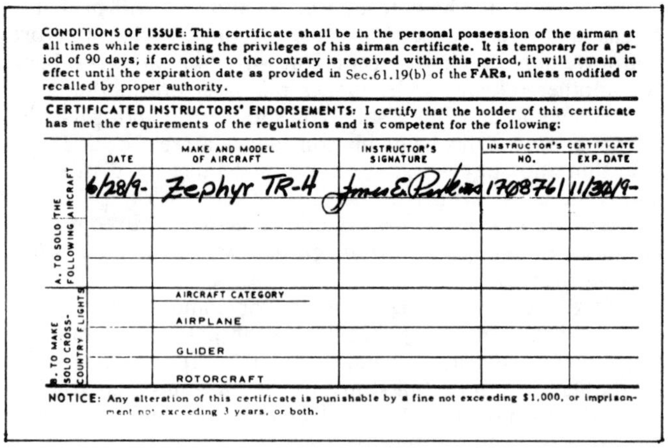
CONDITIONS OF ISSUE: This certificate shall be in the personal possession of the airman at all times while exercising the privileges of his airman certificate. It is temporary for a period of 90 days; if no notice to the contrary is received within this period, it will remain in effect until the expiration date as provided in Sec.61.19(b) of the FARs, unless modified or recalled by proper authority.

CERTIFICATED INSTRUCTORS' ENDORSEMENTS: I certify that the holder of this certificate has met the requirements of the regulations and is competent for the following:

	DATE	MAKE AND MODEL OF AIRCRAFT	INSTRUCTOR'S SIGNATURE	INSTRUCTOR'S CERTIFICATE NO.	INSTRUCTOR'S CERTIFICATE EXP. DATE
A. TO SOLO THE FOLLOWING AIRCRAFT	6/28/9-	Zephyr TR-4	[signature]	1708761	11/30/9-
B. TO MAKE SOLO CROSS-COUNTRY FLIGHTS		AIRCRAFT CATEGORY			
		AIRPLANE			
		GLIDER			
		ROTORCRAFT			

NOTICE: Any alteration of this certificate is punishable by a fine not exceeding $1,000, or imprisonment not exceeding 3 years, or both.

Fig. 10-38. Signing the student off for solo.

It's hoped that a Regulation change will be worked so that the flight instructor will not be legally required to sign the student certificate before the *first* solo if he releases the student and is personally supervising the solo. This would make a better transition from dual to solo. After the first solo flight is completed, the student certificate could be signed and any *following* solo flights without a signed certificate would be a violation of the FARs. But as it now stands, the *Regulations call for the certificate to be signed before any, repeat, any solo flights are made by the student,* so to be legal you'd better do it.

■ **Immediate Postsolo.** You should schedule him to fly again as soon as possible after the first solo to ensure that he doesn't get stale. There is a natural tendency for a letdown after solo since this has been his goal since the first flight. You'll have to convince him that the first solo is one of *three* primary goals and that the next two steps (preparing for solo cross-country and preparing for the private flight test) are equally, if not more, important. His real goal is to become a safe pilot, and these particular well-defined points in his training are merely way stations to that end.

Keep him busy after solo; introduce new material after he has had a couple of flights in the pattern.

You should ride a couple of circuits (or more) on the flight following the first solo and then let him shoot solo takeoffs and landings—all full-stop. These flights are to help build his confidence in solo flight and should be done when wind and traffic conditions are good, but not necessarily as perfect as for the first solo. You will be gradually letting him get into more complex situations as his skill improves.

To get the most from his periods of solo in the pattern you might set up a table as shown in Fig. 10-39. This way he will be getting the confidence-building solo practice and still have the benefit of your postflight briefing. Sure, you may be very busy and unable to do this, but it would be better if instructors could be paid *dual rates* for the student's first couple of solo periods in the pattern. You could concentrate on helping him and still get paid for it.

You can review the flight with him after he's on the ground and may see some weaknesses in specific areas as shown by the table. In this sample, the student is tending to turn too soon onto final and is angling. This is a new habit he has developed, and while it's not dangerous, it should be corrected on the next dual. Certainly any *dangerous* actions should call for a dual flight immediately. Terminate the solo session right then if necessary.

It would be best to have a minimum of two supervised solo sessions in the pattern so that he is comfortable flying by himself and you have a chance to check any problems.

The conscientious instructor always is a little tense on the first solo and during solo cross-countries. No matter how many students he's soloed, he doesn't take it as a light matter.

PATTERN NUMBER	TAKEOFF RUN	LIFT-OFF	CLIMB-OUT	X-WIND LEG	DOWNWIND LEG	BASE	FINAL	ROUND-OUT	LANDING ROLL
1	OK	PULLED OFF	DRIFT	OK	WIDE	ANGLING	GOOD	GOOD	OK
2	OK	OK	OK	OK	OK	OK	ANGLING	OK	OK
3	LINE UP	OK	DRIFT	OK	OK	OK	TURN TOO SOON—ANGLING	OK	OK (Brakes ?)
4	OK	OK	OK	OK	WIDE	ANGLING	OK	OK	OK
5	OK	HELD ON TOO LONG	OK	OK	OK	OK	OK	OK	SKIPPING
6	OK	OK	OK	OK	OK	OK	ANGLING	TOO-NOSE LOW ON TOUCH DOWN	OK

Fig. 10-39. A homemade table of your comments while watching the student's solo pattern work is basically a memory jogger for your postflight briefing.

PRESOLO SYLLABUS

Following is a syllabus suggested for use in the presolo portion of a student's training. Like the actual instruction, this syllabus is heavier with details at the beginning, gradually tapering off as the student approaches solo. As noted in the text, in the last few flights before solo, the flight instructor is acting more or less as a check pilot, with occasional comments or instruction, as needed. By the tenth flight, most of the necessary instruction has been given and only reviews and reminders are necessary as the student gains confidence and develops the skills for taking the airplane around the airport by himself. Remember, too, that a syllabus may be modified for a particular student (see the Summary).

■ First Flight

GROUND INSTRUCTION. Discuss the following in the office or classroom before flying. The amount of detail for the first flight will depend on the student's background. Some points may be reserved for second or third flights.

— 1. *Effects of Controls.*
—   a. *The elevators.*
—   (1) Pitch control.
—   (2) Control angle of attack and airspeed.
—   (3) Use of elevator trim in various phases of flight; the importance of trimming to relieve pressure of the control.
—   (4) Change of elevator effectiveness with airspeed. (Briefly note that slow flight will be covered later in the course.)
—   b. *The ailerons.*
—   (1) The roll control.
—   (2) Why and how an airplane turns; brief discussion of horizontal lift vector.
—   (3) Adverse yaw; why rudder is used with the ailerons in starting or completing turns.
—   c. *The rudder.*
—   (1) Yaw control.
—   (2) Use in turns.
—   (3) Use in slips. (Indicate that this will be covered in more detail later.)
—   (4) Correction for the left-turning forces and moments in a climb or slow flight.
—   (5) Primary use for steering on ground for some airplanes; nosewheel or tailwheel steering in taxiing.
—   (6) Adverse yaw. (Repeat the discussion of adverse yaw in the aileron section, to emphasize.)
—   (7) Rudder tab or rudder trim use.
— 2. *Four Fundamentals: Turns, Climbs, Straight and Level, and Descents.*
—   a. *The turn.*
—   (1) Ailerons and rudder.
—   (2) Back pressure in the turn.
—   (3) Three aspects of the turn (ailerons, rudder, and elevator).

—   (4) Load factors in the turn.
—   (5) Rolling out of the turn; need for relaxing back pressure.
—   b. *The climb.*
—   (1) Ease nose up to climb attitude.
—   (2) Increase to climb power.
—   (3) Correct for "torque."
—   (4) Ensure that climb *attitude* is for V_Y; check with instruments; trim off pressures.
—   c. *Straight and level (from climb).*
—   (1) Ease nose over to level flight, maintain climb power.
—   (2) Let speed build up and ease off rudder.
—   (3) Adjust power cruise setting.
—   (4) Assume straight and level *attitude;* check with instruments; relieve pressures with trim.
—   d. *The descent (glide).*
—   (1) Carburetor heat ON or check for presence of carburetor ice, as recommended by the manufacturer.
—   (2) Smoothly close throttle.
—   (3) Hold nose in level flight *attitude* until airspeed reaches recommended value.
—   (4) Assume glide *attitude;* check with airspeed indicator.
—   (5) Trim.
—   e. *Straight and level (from glide).*
—   (1) Lead altitude by about 50 ft for most trainers (depends on rate of descent).
—   (2) Add cruise power. Add enough power to stop descent, then adjust as necessary. Stop nose from rising with forward pressure.
—   (3) Carburetor heat OFF (if applied).
—   (4) Maintain level flight *attitude;* check with instruments; relieve pressures with trim. (The secret of straight and level flight is proper power setting and trim.)
—   f. *Climbing turns.*
—   (1) Differences in left and right climbing turns ("torque" effects).
—   (2) Why banks are kept shallow.
—   g. *Descending (gliding) turns.*
—   (1) Airspeed close to that of climb but the rudder comparatively weak because of lack of slipstream (coordination problems).
—   (2) Keep banks shallow at this point of training.

FLIGHT INSTRUCTION. Discuss the following at the airplane or in flight, as applicable.
— 1. *Preflight check.* Discuss controls (what controls yaw, pitch, and roll) and *why* each item is checked on the airplane.
— 2. *Cockpit familiarization.* Get the student comfortable and check seat adjustment to assure his ability to see out and move all controls. Discuss sight-picture.

___ 3. *Instruments and controls* (including trim tabs). Go through the Four Fundamentals briefly, with the student moving and watching the flight controls to simulate entry and recovery from each.

___ 4. *Instructor starts engine.* Use a checklist and explain the procedure. Instructor uses the radio.

___ 5. *Instructor demonstrates taxi technique* with explanations.

___ 6. *Pretakeoff check* (using checklist). Explain *why* for each item.

___ 7. *Takeoff.* Instructor makes takeoff, with explanation.

___ 8. *Max rate climb attitude.* Demonstrate and explain.

___ 9. *Level-off procedure.* Demonstrate use of trim and leaning of mixture, noting that a full explanation will come later.

___10. *Effects of controls.* Student uses rudder, ailerons, and elevator controls singly and at high and low airspeeds.

___11. *Turns.* Demonstrate medium and shallow turns with aileron, rudder, and elevator. Student practices. (Say "Clear right; clear left"; use outside references.)

___12. *Climb entry.* Demonstrate and identify sight-picture *attitude.* Student practices.

___13. *Straight and level from a climb.* Demonstrate and identify sight-picture *attitude.*

___14. *Student climb straight ahead and level off.*

___15. *Student turn while climbing* (with proper rudder), then level off.

___16. *Normal glide.* Demonstrate and identify sight-picture *attitude.* Student practice setting up glide, then leveling off at a prechosen altitude.

___17. *Return to airport. Instructor uses the radio. With student observing goes through the checklist.* Student makes turns in pattern. Instructor handles power and takes over on final. Instructor uses and explains postlanding and shutdown checklist.

___Evaluation
___Critique
___Review
Assigned Reading_____

Comments_____

_____

_____

Instructor_____

Date_____ Ground School Time_____ Flight Time_____

■ Second Flight

GROUND INSTRUCTION

___ 1. *Review the Four Fundamentals.*

___ 2. *Discuss the local traffic pattern* and traffic patterns in general. Introduce the wind indicators (wind sock, tetrahedron, wind tee).

___ 3. *Explain taxiing* in various wind conditions (establish forward motion before turning). Do not overemphasize crosswind taxiing and control positions at this point.

___ 4. *Discuss the use of trim.*

___ 5. *Discuss the chance of doing some wind drift correction practice,* including rectangular course, on this flight. Introduce briefly the concept of crabbing and the rectangular course.

___ 6. *Discuss the radio procedures* to be used on this flight.

___ 7. *Posttakeoff checklist.*
___   a. Climb straight ahead or as directed by ATC.
___   b. At pattern altitude or as locally recommended, make 45° left turn to exit airport traffic pattern (controlled field).
___   c. Turn off electric fuel pump (if so equipped) and check to make sure the engine-driven pump is providing fuel pressure. Give brief explanation of electric fuel pump use.

___ 8. *Landing checklist.*
___   a. Descend to pattern altitude at least 2 mi out.
___   b. Before entering the pattern:
___   (1) Check brake pedal(s) pressures.
___   (2) Check fuel on fullest tank. Turn on electric fuel pump (if so equipped). For Cessna C172 and C182, check fuel selector on BOTH; for C150/152 check that fuel selector is seated properly ON.
___   (3) Undercarriage coming down if not already welded in place.
___   (4) Mixture RICH. Give brief discussion of mixture control.
___   (5) Make traffic call.

___ 9. *Enter downwind from 45° angle if possible.* When established on downwind:
___   a. Pull carb heat ON; leave ON for 10 sec to clear out any ice that has formed. Either leave ON or OFF as required by manufacturer. Briefly discuss *why* carburetor heat may be needed.
___   b. Report position to tower or Unicom, as applicable.
___   c. Perform G-U-M-P check:
___   (1) *G*as on fullest tank. If tanks were switched earlier and there is plenty of fuel in the one being used, leave the selector alone. C172/C182 selector on BOTH.
___   (2) *U*ndercarriage. Check DOWN indications if retractable.
___   (3) *M*ixture rich.
___   (4) *P*ump (electric fuel pump) ON, if so equipped.

—10. *Abeam landing point,* reduce power to approximately 1600 rpm. Maintain altitude and let speed dissipate.

—11. *When airspeed below* V_{FE}, within white arc, lower 2 notches of flaps (20° on C172/C152 and 25° on PA-28-161 Warrior).

—12. *Stablize airspeed* at recommended approach speed. Trim for no pressure required. Set prop for go-around as recommended by the *Pilot Operating Handbook,* if airplane has a constant-speed propeller.

—13. *Turn base and final* as directed by instructor. Explain reference 45° behind wing to start turn onto base. Set final landing flaps on base.

—14. *Take over on final and land airplane,* explaining procedure.

FLIGHT INSTRUCTION (at the airplane or in flight, as applicable)

— 1. *Preflight.*

— 2. *Student starts airplane and taxis to run-up position* under instructor's supervision.

— 3. *Pretakeoff check.*

— 4. *Check traffic.* Student uses radio under supervision.

— 5. *Student takes off* with help from instructor.

— 6. *Assume climb attitude,* check with airspeed indicator, trim.

— 7. *Explain scanning pattern* in checking for other traffic. Use definite climbing turns to clear area.

— 8. *Level off,* using three steps.

— 9. *Turns, left and right* (15–30° bank). Student practices.

—10. *Glides.* Student practices.

—11. *Level off from the glide.* Emphasize effects of power and nose-up trim on pitch-up during recovery from the glide.

—12. *Exercise.*
— a. Climb 1000 ft.
— b. Straight and level 2 min.
— c. Descend 1000 ft.
— d. Straight and level 2 min, then 180° turn.
— e. Repeat sequence.

—13. *Demonstrate approach attitude* with/without flaps, as applicable. (Refer to the *Pilot Operating Handbook* for the airspeeds at various flap settings.)

—14. *Student uses radio* under supervision and enters traffic pattern, performs landing checklist, handles throttle and flaps with instructor's help. Instructor takes over on short final, lands airplane while describing the process.

—15. *Student taxis to ramp* and shuts down under supervision. (Check mag ground wires for proper operation by turning ignition OFF, then ON at idle.) Student then runs up engine for a few seconds and shuts down the engine with the mixture. Student uses postlanding and shutdown checklist.

—Evaluation
—Critique
—Review
Assigned Reading_____

Comments_____

_____

_____

Instructor_____

Date_____ Ground School Time_____ Flight Time_____

■ Third Flight

GROUND INSTRUCTION

—1. *Review flights* 1 and 2 as necessary.

—2. *Resolve questions* relative to assigned reading material.

—3. *Discuss 45°-banked power turns* (360° and 720° turns).
— a. Discuss forces acting on airplane.
— b. Discuss elevator, aileron, rudder, and power requirements.
— c. Pick good reference on horizon.
— d. Roll into 45°-banked turn (coordinated); then neutralize ailerons and rudder.
— e. Add power as bank increases.
— f. Add back pressure as bank increases.
— g. Continually check pitch and bank attitudes and check with altimeter.
— h. Roll out on original reference on horizon after one and two complete turns, as planned. As bank decreases, ease off back pressure and extra power.
— i. Note correction for "torque" in left and right turns.

—4. *Explain Power-required/Drag curve* in simplified terms on chalkboard.
— a. Induced drag (brief, nontechnical explanation).
— b. Parasite drag (brief, nontechnical explanation).

—5. *Discuss slow flight* (flight at minimum controllable airspeed).
— a. Discuss aileron and rudder control effectiveness at low airspeeds.
— b. Explain the relation of stall speed to angle of bank.
— c. With airspeed close to stall at straight and level flight attitude, discuss the necessity for making all turns very shallow.
— d. Airspeed control with pitch (elevators); altitude control with power (throttle).

—   e. Pitch changes necessary with addition or retraction of flaps during slow flight.

—6. *Review* (if covered earlier) *or introduce the concepts of wind correction.*

—7. *Discuss procedures for flying over and parallel to a road.*

—8. *Review rectangular course.*

—9. *Introduce S-turns across a road.*

FLIGHT INSTRUCTION

— 1. *Student performs preflight, run-up, and takeoff* with instructor's supervision, uses radio.

— 2. *After takeoff, execute posttakeoff checklist* at a safe altitude.

— 3. *Student assumes V_Y attitude with altitude indicator covered.* Check with airspeed indicator.

— 4. *Demonstrate 45°-banked power turn* for 360°.

— 5. *Student practices* both right and left. If time permits, go on to 720° turns.

— 6. *Demonstrate slow flight,* with shallow turns left and right.

— 7. *Student practices.*

— 8. *Student establishes glide* and levels off at approximately 600–900 ft above the ground.

— 9. *Student flies over and parallel to a road.*

—10. *Demonstrate rectangular course.*

—11. *Student practices.*

—12. If time permits, *demonstrate S-turns across a road.*

—13. *Student practices.*

—14. *Listen to ATIS, student calls approach,* return to airport (controlled field). Student calls Unicom or uses radio procedures as required (uncontrolled field).

—15. *Student performs prelanding checklist,* enters traffic normal pattern or as instructed by ATC, performs GUMP check, assumes approach attitude, and flies airplane to short final.

—16. *Instructor lands airplane,* again describing the procedure during the process.

—17. *Student retracts flaps after clearing runway* (being careful to verify that, as applicable, he's moving the flap, not the gear, handle).

—18. *Student taxis to ramp* and performs shutdown checklist.

__Evaluation
__Critique
__Review
Assigned Reading_____

Comments_____

_____

_____

_____

Instructor_____

Date_____ Ground School Time_____ Flight Time_____

■ Fourth Flight

GROUND INSTRUCTION: Introduction to Stalls

— 1. *Review* as necessary.

— 2. *Resolve questions* relating to assigned reading material, etc.

— 3. *Review use of mixture control.*

— 4. *Review carburetion,* carb ice, and carb heat.

— 5. *Discuss the following items relative to aerodynamics:*

—   a. The Four Forces.

—   b. Airfoils (wings and propellers).

—   c. Chord.

—   d. Angle of incidence.

—   e. Relative wind.

—   f. Angle of attack.

—   g. Lift as it relates to angle of attack and airspeed: $L = C_L \, S \, \frac{\rho}{2} \, V^2$

—   h. Critical angle of attack.

—   i. The stall as a function of angle of attack, not airspeed.

—   j. The elevators as they relate to stall entry and recovery. To recover from a stall, point the airplane in the direction it's already going. Do this with the elevators.

—   k. Power is used for only one reason during stall recovery: *to minimize altitude loss.*

—   l. To recover from the stall, simultaneously move wheel forward just enough to break the stall (lower angle of attack) and add full power.

—   m. The stall as an aid to the landing.

— 6. *Review concepts of wind correction.*

— 7. *Review rectangular course and S-turns* as necessary, citing common errors.

— 8. *Discuss turns about a point* (optional).

— 9. *Review the traffic pattern.*

—10. *Discuss the normal landing:*

—   a. Establish a stabilized-approach gradually reducing power, and with elevator pressures trimmed off. Close throttle when runway is "made."

—   b. On final, at a height of approximately 20 ft (about hangar height), start the transition from approach attitude to landing attitude. *Gradually* ease the wheel back. *The transition must be gradual.*

—   c. About 19 of the 20 ft are used in the transition. The airplane is slowing up during this period.

—   d. By the time the airplane is about 1 ft above the ground, it should be in the landing attitude—nose high. The pilot will not be able to see directly over the nose in most airplanes. Look out along the left side of the nose to line up with the runway. *Don't try to look over the nose.* If the pilot can see over the nose, some airplanes won't have the proper landing attitude and will fly into the ground.

—   e. Once in the landing attitude (about 1 ft above the ground):

    (1) Hold that attitude. Try to keep the airplane from landing. As airplane approaches the stall, increase back pressure on the wheel as necessary.

    (2) Keep eyes focused about 10° left of the nose, and far enough ahead so the ground is not blurred. Don't stare at one spot. Scan the ground.

    (3) Keep the plane lined up with the runway and have wings level at touchdown (assume no crosswind).

    (4) During ground roll, continue to hold back pressure. Let the nosewheel come down by itself.

    (5) Brake as necessary. If the runway is long enough, the pilot may want to save brakes and turn off at a taxiway farther down the runway. Be sure the airplane is going slow enough when turning off the runway.

FLIGHT INSTRUCTION

— 1. *Student performs preflight,* taxi, run-up, takeoff, and posttakeoff checklist. Student makes climbout at V_Y, and handles radio communication under supervision.

— 2. *Student reviews 45°-banked power turns* 360° left and right.

— 3. *Student reviews slow flight without flaps.*

— 4. While in slow flight, *instructor demonstrates simple stall and recovery* without addition of power.

— 5. *Instructor sets it up again for student to practice.*

— 6. *Student establishes approach attitude* with landing flaps. Instructor demonstrates full-stall landing at altitude.

— 7. *Student practices,* then cleans up airplane for climb.

— 8. *Student climbs at V_Y* and levels off at safe altitude. Instructor demonstrates departure stall and recovery.

— 9. *Student practices departure stall.*

—10. *Demonstrate approach to landing stall* (straight flight) and recovery.

—11. *Student practices.*

—12. *Student executes descending turn(s)* to an assigned heading and altitude for low work.

—13. *Demonstrate turns about a point* (optional).

—14. *Student practices.*

—15. *Start return to airport;* listen to ATIS; check heading indicator with magnetic compass, call approach control (controlled airports). Call Unicom for airport advisory (noncontrolled airports).

—16. *Return to airport.* Student uses checklist, handles controls, using gradual reduction of power during the approach, and "lands" the airplane with help and advice of instructor.

—17. *Student retracts flaps after clearing runway,* being careful to verify that his hand is on the flap handle—not the gear handle.

—18. *Student taxis to ramp* and performs shutdown checklist.

—Evaluation
—Critique
—Review
Assigned Reading_____

Comments_____

_____

_____

Instructor_____

Date____ Ground School Time____ Flight Time____

■ Fifth Flight

GROUND INSTRUCTION

—1. *Review* as necessary.

—2. *Resolve questions* relative to assigned reading material, etc.

—3. *Discuss wake turbulence* and avoidance procedures.

—4. *Discuss collision avoidance* procedures.

—5. *Briefly discuss maximum performance climbs,* V_X and V_Y.

—6. *Introduce emergency procedures.* (Use checklist.)

—   a. Simulated engine failure at altitude.

    (1) Establish best glide *attitude* and check with airspeed indicator. Trim to maintain.

    (2) Carb heat ON as best glide is being set up.

    (3) Mixture RICH.

    (4) Electric fuel pump ON, and switch tanks (if so equipped).

    (5) Pick a landing area and turn toward it.

    (6) Note wind direction.

    (7) When switching tanks, wait a few seconds to let the pump get fuel from new tank.

    (8) The propeller will be windmilling; so if the fault is remedied (carb ice or a dry fuel tank), the engine will start again. Do not try to use the starter.

    (9) If engine starts, climb while circling the field until sure the problem is corrected before departing for further practice or return to the airport.

    (10) *If the engine will not restart,* the following must be done before landing (Emphasize that these are all secondary to

the primary need to keep the airplane under control at the best glide speed, and make a safe landing.):

— a. Declare an emergency on frequency 121.5, or if in contact with Approach Control or Unicom (uncontrolled airports), stay on that frequency.

— b. Set transponder to 7700.

— c. Mixture to IDLE CUTOFF.

— d. Fuel selector to OFF position.

— e. Use flaps as necessary to land on chosen spot. It is best to use full flaps and 1.3 V_{so}, but the first requirement is to make the field. Try to land as slowly as possible.

— f. After flaps have been extended and committed to landing, turn OFF master and ignition switches.

— g. Unlatch doors prior to touchdown.

— h. Land nose-high, tail-low. Touch down as slowly as possible.

— i. REMEMBER, THE FIRST RESPONSIBILITY IS TO CONTROL AND FLY THE AIRPLANE TO A SAFE LANDING. NOTE THAT WITH LOW-ALTITUDE EMERGENCIES, LINE-OF-SIGHT FACTOR (AND DISTANCE), AND TIME LIMITS MAY PRECLUDE (a) AND (b).

—7. *Review* the normal landing process. Transition from approach attitude to landing attitude must be done carefully to avoid ballooning or flying into the ground.

FLIGHT INSTRUCTION

— 1. *Student performs all pretakeoff functions* and executes the takeoff with appropriate checklists.

— 2. *Student levels off,* then performs shallow (15°-bank) and medium (30°-bank) turns.

— 3. *Student does stall series* (approach and full), in both straight and turning flight.

— 4. *Student climbs at* V_Y, establishes pitch *attitude,* with airspeed indicator (ASI) covered in such a way that the student cannot see it but the instructor can check it.

— 5. *Simulated power failure,* as follows:

— a. Establish best glide *airspeed.*

— b. Pull carb heat immediately to take advantage of engine's residual heat.

— c. Electric fuel pump ON, switch tanks (if so equipped). Use the emergency checklist discussed under "GROUND INSTRUCTION."

— d. Pick a landing area. Check wind direction.

— 6. *Student glides toward landing spot.* Instructor takes over at proper safe altitude.

— 7. *Set up rectangular course and/or S-turns across a road.*

— 8. *Return to airport.* Student handles the radio, flies the pattern approach, and lands the airplane with the instructor's advice and help.

— 9. *Instructor may elect to have student taxi back* to shoot one or more patterns and full-stop landings before returning to the ramp.

—10. *Student handles radio,* flaps (flaps up after slowing down and turning off the runway), and taxis to the ramp. Student shuts down the airplane and uses postshutdown checklist.

—Evaluation

—Critique

—Review

Assigned Reading_____

Comments_____

_____

_____

Instructor_____

Date_____ Ground School Time_____ Flight Time_____

■ Sixth Flight

GROUND INSTRUCTION

—1. *Review* as necessary.

—2. *Resolve questions* from assigned reading, etc.

—3. *Review normal takeoff.*

— a. Check for traffic. Do not blindly trust tower at a controlled airport.

— b. Line up with center line of runway.

— c. Smoothly apply full power; keep hand on throttle; extra right rudder may be needed.

— d. Check engine instruments.

— e. As controls become firm, use back pressure to obtain proper angle of attack on takeoff. (Note that larger airplanes may use "numbers" for takeoff.)

— f. Let plane fly itself off the runway. Be ready to add right rudder as necessary, as nosewheel lifts off runway.

— g. Assume V_Y attitude (unless V_X is needed); check with ASI.

— h. Climb out straight ahead; hand on throttle; continue correcting for torque as necessary.

—4. *Review traffic pattern* (including departures and entries); prelanding checklist, both initial and following power reduction; flap settings; establishing approach attitude; and negotiating the approach down to the point to transition to a normal landing.

—5. *Review the landing process* from breaking the glide to touchdown and roll-out to taxi speed.

—6. *Discuss the power-off approach.*

—7. *Discuss the go-around* maneuver.

— a. Add full power to stop the descent. Forward pressure on wheel or stick is needed to keep nose from pitching up. Carburetor heat OFF if used. Trim.

— b. When there is a positive rate of climb, start getting the flaps up in 10° increments (or by notches, as applicable). Trim.

— c. Climb out straight ahead. Trim.

—8. *Discuss recoveries from bad situations* during the landing.

— a. Flying into the ground: Bounce and ease nose over. Add power (if bounce is estimated 5 ft or more). If bounce is higher, or if runway is short, add full power and go around.

— b. Dropping it in. Add full power as airplane descends to lessen impact effects.

FLIGHT INSTRUCTION

—1. *Preflight.*

—2. *Normal takeoff.*

—3. *Practice full-stop landings* followed by taxi and normal takeoffs.

—4. *Practice touch-and-go's* if appropriate. Instructor may have student practice several power-off approaches, explaining differences in downwind abeam positions.

__Evaluation
__Critique
__Review
Assigned Reading_____

Comments_____

_____

_____

Instructor_____

Date_____ Ground School Time_____ Flight Time_____

■ Seventh Flight

GROUND INSTRUCTION

— 1. *Review* as necessary.

— 2. *Resolve* any questions relative to assigned reading, etc.

— 3. *Review traffic pattern,* approach, and normal landing procedures.

— 4. *Review recoveries* from bad situations.

— a. Flying into the ground (bouncing).

— b. Dropping it in.

— c. Too low in the approach.

— d. Too high in the approach.

— 5. *Discuss emergencies* on takeoff.

— a. Power failure while on runway.

— b. Power failure after lift-off. Discuss the hazards of turning back.

— c. Partial power failure after lift-off.

— d. Emphasize never to be afraid to abort a takeoff if you don't feel good about the conditions.

— 6. *The crosswind takeoff.* (Optional.)

— a. Taxiing in a strong crosswind.

— b. Maintaining directional control in a strong crosswind. (Holding full ailerons into the wind and decreasing deflection as airspeed increases.)

— 7. *The crosswind landing.* (Optional.)

— a. Crabbing as a drift correction maneuver in the pattern.

— b. The sideslip as a drift correction maneuver on final and through touchdown.

— c. The crab approach and sideslip correction during the landing process.

— 8. *Introduce the effects of gusty air and wind gradients.*

— 9. *Discuss loss of directional control.* Causal factors and recovery procedures.

—10. *Cover wake turbulence avoidance* procedures as necessary.

—11. *Review scanning* and collision avoidance procedures.

FLIGHT INSTRUCTION

—1. *Preflight.*

—2. *Normal takeoffs and landings.*

— a. Instructor checks climb at V_Y and approach attitudes with ASI covered.

— b. Intersperse full-stop and touch-and-go landings, as appropriate for the training situation.

—3. *Simulate engine failure* on climbout after takeoff and demonstrate procedure (get nose down immediately and make only shallow turns, depending on altitude).

__Evaluation
__Critique
__Review
Assigned Reading_____

Comments_____

_____

_____

Instructor_____

Date_____ Ground School Time_____ Flight Time_____

■ Eighth Flight

GROUND INSTRUCTION

——1. *Review* as necessary.

——2. *Resolve* questions relative to assigned reading material.

——3. *Discuss* simulated engine failure at altitude
——   a. Establishing best glide speed.
——   b. Picking a field.
——   c. Restoring power if possible.
——   d. Pattern procedures at field.
——   e. Using voice or transponder procedures.
——   f. Immediate prelanding steps.
——   g. Postlanding procedures.

——4. *Introduce the forward slip* as a means to lose altitude without picking up airspeed. (Does the manufacturer allow slips with flaps?) (Optional.)

——5. *Note the sideslip* as a means to correct for wind drift in a crosswind landing.

——6. *Discuss landings* with various flap settings for particular situations, but don't overemphasize.

——7. *Review procedures* for being too high or too low on final, and for bounce or high level-off on landing.

——8. *Discuss pattern altitude,* entry procedures, and radio communication requirements at a nearby airport where you will be practicing.

——9. *Cover the need to make traffic patterns,* approaches, and landings applicable to varying environments of different airports.

FLIGHT INSTRUCTION

——1. *Execute preflight,* run-up, and normal or crosswind takeoff. It's best to avoid crosswind work if the student does not have a complete grasp of normal takeoffs and landings.

——2. *Fly to a nearby airport* (if available) to practice takeoffs and landings.

——3. *En route to the practice airport, review:*
——   a. Climb attitudes at V_Y and V_x.
——   b. Leveling-off procedures.

——4. If time permits, *practice 45°-banked turns.*

——5. *Make touch-and-go and/or full-stop landings* at the nearby airport.

——6. *Demonstrate procedures* for too-high or too-low approaches, bounce recoveries, and throttle use if drop-in is imminent or occurring.

——7. *Set up a too-high and/or too-low approach* and turn the airplane over to the student to check his reaction and procedures.

——8. *Return to your home airport* and shoot two or three landings (full-stop) as time and/or student fatigue permits.

___Evaluation
___Critique
___Review
Assigned Reading_____

Comments_____

_____

_____

Instructor_____

Date_____ Ground School Time_____ Flight Time_____

■ Ninth Flight

GROUND INSTRUCTION

——1. *Review previous flights* with emphasis on problem areas. Depending on the student's progress, go over the emergencies on takeoff and in the pattern. The emergencies should be reviewed a flight or two *before* solo but not emphasized in the briefing just before the solo flight, to avoid creating anxiety.

——2. *Resolve questions* concerning the traffic pattern, such as altitudes, departures, and entries.

FLIGHT INSTRUCTION. The last few flights should not require that you say much, as noted in the text, but you should monitor the student's action. Here is where you will get the sense of when he is ready to solo. When you first begin instructing, it is a good idea to have an experienced instructor ride with the student as he approaches solo and give you a report on his progress.

It's best to shoot full-stop-and-taxi-back landings as the student approaches solo. This will make the transition to solo easier than if he has been shooting touch-and-go's and at the point of solo has to change his routine.

——1. *Shoot full-stop landings.*

——2. On at least one landing, *set up a bounce condition* for the student to recover from (if the student hasn't already done it inadvertently during the previous takeoff and landing sessions). Also check his response to an imminent or actual drop-in.

——3. *Depart and reenter the traffic pattern* at least once during the period.

——4. Some instructors *have the student depart the pattern* and go to the practice area for stalls or slow flight to break up the continuous landing practice (students can lose progress by too much concentration on pattern work).

—————————————————————

__Evaluation
__Critique
__Review
Assigned Reading_____

Comments_____

_____

_____

Instructor_____

Date____ Ground School Time____ Flight Time____

—————————————————————

■ **Tenth Flight and Following.** As noted earlier, the tenth and following flights until solo are essentially a review of the previous instruction, with the instructor doing less flying and less talking as the student becomes more proficient. Expect most students to solo between the 12th and 20th hours, depending on consistency of schedule, airplane type, and airport environment (traffic density, tower control, crosswinds, length, and width of runways and other factors). The instructor should not keep the student senselessly shooting takeoffs and landings if no progress is being made. (The frustrated student will be set further back under this policy.) Stop and analyze the situation. There is a need for repeated takeoff and landing practice, not only to ensure that the student can make safe takeoffs and landings but also to see his reactions to flight and to establish good habit patterns.

If the student has a plateau that seems to be unbreakable, talk to one of the more experienced instructors and if necessary have this instructor fly with the student. A situation of no progress can be pretty hard on the nerves of the new instructor, who doesn't have enough experience, in many cases, to solve the dilemma and hence suffers a loss of confidence. This is particularly unnerving if it happens with the first student and the instructor should, as indicated, at least discuss the problem with a more experienced instructor, who may be able to resolve the problem with a couple of quick suggestions. It takes experience to analyze all of the problems presolo students have.

■ **Syllabus Summary.** The term "Flights" has been used in the Syllabus, but perhaps "unit" would be a better heading because, for instance, on the third flight the student still needs work on the Fundamentals of the first and second flights. Or you'll find that a timid student shouldn't be introduced to emergency procedures as early as the fifth flight; he's still worried about banking over 10° at that point. Certainly you would not want to get into crosswind takeoffs and landings if the student doesn't understand the no-wind or wind-down-the-runway techniques. Under certain conditions you may hold off until after solo to introduce crosswind procedures, or slips as noted earlier in the chapter.

While the suggested order of introduction is listed here, you will modify it for a particular student and/or your local requirements.

3 FIRST SOLO TO THE PRIVATE CERTIFICATE

THE POSTSOLO VISUAL MANEUVERS

BACKGROUND

After the student has had two or three solo periods in the pattern, you should fly a dual with him. Do this not only to introduce new maneuvers but also to see if he has started developing bad habits that couldn't be seen from the ground.

Many students quit flying shortly after solo because the instructor has allowed them to drift and mark time. The solo was a high point, and a letdown sometimes occurs afterward. Students generally do some sightseeing, but without definite short-term objectives they soon lose sight of the goal of getting that private certificate and becoming a good pilot. "Flying around" is enjoyable, but not for long at $30–$45 an hour. You will have to keep him busy and interested after solo. One way to do this is to alternate dual and solo flights. Try to work it so that there is always something new to demonstrate on a dual session.

Shortly after solo, you should lay out the rest of his training up to the private flight test—not in detail, but cover enough ground that he can see where he's going.

Start tightening up your requirements on all phases during this postsolo work. Watch his basic flying and continue to instruct on areas of weakness. Some instructors are lax on the Four Fundamentals during the latter part of training, not realizing that many problems of advanced maneuvers can be traced back to the basics.

STALL REVIEW

■ **Elementary Stalls.** A review of elementary stalls will be needed, because the student may not have done these as airwork for several flight hours.

At this time, it would be a good idea to briefly discuss the spin entry, spin, and spin recovery, since the student will soon be out practicing various stalls solo (and it seems that the takeoff and departure stall is the favorite situation for a solo student to get accidentally introduced to the spin entry—and recovery). Some instructors used to shy away from demonstrating or even discussing spins, but the current regulations require that the private pilot certificate applicant shall have the aeronautical knowledge of "stall awareness, spin entry, spins and spin recovery for airplanes" (FAR 61.105).

Whether or not you plan to actually teach the student how to recover from spins (assuming you have a proper airplane and parachutes), this is a good time to discuss (1) the incipient spin and (2) the spin itself, with emphasis on what the student will see as the airplane proceeds through the stall and how to stop the problem. Spins are covered in detail in Chap. 21 and you may want to review that chapter yourself before introducing the subject.

Demonstrate and have the student practice stalls straight ahead, using cruise power and idle, both clean and with flaps. Cover the procedures required for straight-ahead takeoff and departure stalls and approach-to-landing stalls. He will be practicing various stalls solo, so confirm that the recoveries are safe. You should require the recovery from some of the power-off stalls without adding power, so that the difference in altitude loss can be seen. Here you will also establish procedures for use on the flight test. Keep him clearing the area and looking around.

Also demonstrate and have him practice the elementary turning stalls, noting any problems and correcting them. Fig. 11-1 shows an exercise that he can use on his solo sessions.

■ **Slow Flight.** On the ground, review the principles of slow flight when you discuss the stalls. Fig. 6-4 shows an rpm versus IAS for a particular trainer at a constant altitude, and you could have the student fly one for your airplane.

Have the student perform slow flight or flight at minimum controllable airspeeds using outside references in cruise and landing configurations. After the student has been introduced to hood work, have him practice slow flight using the instruments only. He can get particularly good practice in instrument work in the transition from cruise to slow flight (and flying slow flight under the hood) and back to cruise. This is good for scan development.

ADVANCED STALLS

These stalls should be introduced before cross-country work, but not necessarily immediately following the elementary stall review.

These advanced stalls are rougher on the student, with attitudes and forces more radical than he's had before. You want him to respect stalls, but he shouldn't clutch in them. With some students at first you may only *demonstrate* the cross-control stall and keep it as easy as possible. Show them what to avoid as far as the airplane's attitude and the control uses are concerned, and don't let the airplane get into an extreme attitude.

■ **Cross-Control Stalls—Skidded Turns.** All student (and private) pilots should be aware of what can happen if the controls are abused in the turn onto final (or anytime). Aviation is still plagued with stall/spin accidents and one of the problems is that many pilots have never been in a deep-stall condition; they were shown approach-to-landing stalls as required for the flight test, but they didn't know what could occur in other attitudes and conditions.

PREPARATION. A model is your best preflight discussion aid, but a chalkboard will also be helpful. You might want to draw Fig. 11-5 for some of your more technically minded students.

EXPLANATION. This cross-control stall is one in which *inside* rudder and opposite aileron are being used as the stall is entered. (Cover *what*, *why*, and *how*.)

You should explain that steeply banked turns should be

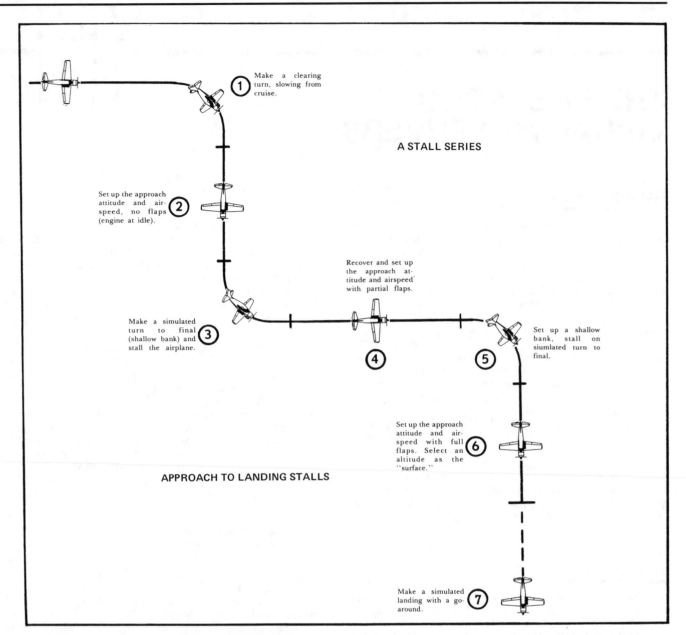

A STALL SERIES

① Make a clearing turn, slowing from cruise.

② Set up the approach attitude and airspeed, no flaps (engine at idle).

③ Make a simulated turn to final (shallow bank) and stall the airplane.

④ Recover and set up the approach attitude and airspeed with partial flaps.

⑤ Set up a shallow bank, stall on siumlated turn to final.

⑥ Set up the approach attitude and airspeed with full flaps. Select an altitude as the "surface."

⑦ Make a simulated landing with a go-around.

APPROACH TO LANDING STALLS

Fig. 11-1. Stall exercises allow the student to practice a number of stalls in logical sequence.

avoided close to the ground, but you should add that steep turns close to the ground, if the airplane has sufficient airspeed, are better than the flat, cross-controlled turn you are about to describe.

The problem of this stall usually arises when the airplane is about to overshoot the runway center line while on base. Maybe the pilot has been told not to bank steeply at a low altitude (good advice, generally), or maybe he just feels uncomfortable about such an attitude. He "cheats" by adding rudder to hasten the turn—which requires opposite aileron if the bank is going to be kept shallow.

The result is a vicious circle. The nose will tend to drop because of the airplane's bank (and inside rudder), shown in Fig. 11-2.

In extreme conditions the end result can be the wheel turned all the way to the right and pulled back, and the left rudder pedal moved to the stop. The airplane won't usually allow things to go this far but will whip over toward the low wing as the stall occurs. (It's nearly impossible to stop the airplane in a less-than-vertical bank.)

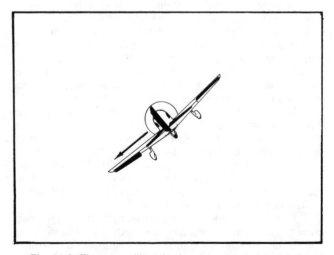

Fig. 11-2. The nose will tend to lower because of the bank and skidding condition.

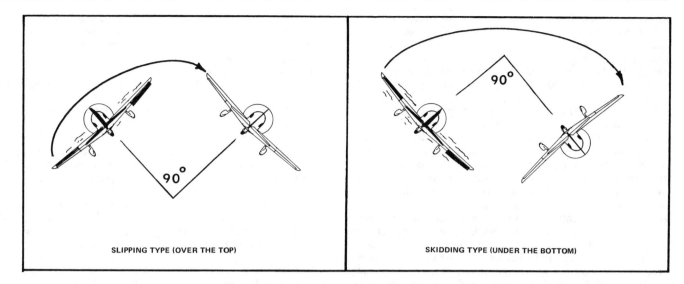

Fig. 11-3. A comparison of attitudes after the stall break (90° of roll from the original bank).

One complication is that the forces working on the pilot in this inside-rudder situation aren't particularly uncomfortable. As indicated in Chap. 4 in the introduction to the turn, the average student feels more comfortable in a skidding turn than a slipping one, because this is the type of force felt every time he turns a corner in a car. The situation causing this "natural" feeling is more dangerous than a slipping-type turn because many airplanes resist a slipping stall more than a skidding one. There is usually more buffeting and early warning in the slip-type (top rudder) stall approach since the airplane will tend to go "over the top" and the pilot has more time to recover before radical attitudes are reached (Fig. 11-3).

Another contributing factor to the setting up of an under-the-bottom stall condition is a tailwind component on base. This tends to make the airplane overshoot the wind line, and the pilot may get the false impression by the groundspeed that his airspeed is higher than it actually is. The overall problem is considerably aggravated by that tailwind; as stated in Chap. 10, teach your students that if the wind is directly across the runway and traffic permits they should use a pattern that would give a *headwind* on the base leg. Fig. 11-4 shows a view of a cross-control stall accident as seen from above.

Interestingly enough, you may run into pilots who say that the inside or lower wing can't stall first since the down aileron on that side is acting as a flap, and everybody "knows" that flaps hold off the stall.

In Chap. 9 it was indicated that flaps increase the maximum coefficient of lift, C_L, but that the stall will occur at a lower angle of attack and lower nose attitude than for the unflapped wing. The situation applies here, too. The outside wing with its up-aileron can be considered to be "twisted" to a lower angle of incidence and has farther to go than usual before the stall angle of attack is reached. Fig. 11-5 shows comparisons of the three wing configurations.

The student might look at Fig. 11-5 and say that it proves the point that the inside wing can't stall first: look at the comparative lift. That is only the *coefficient* of lift; the inside wing will be moving more slowly than the outside one because of the rudder-induced skid. Lift depends on velocity as well as coefficient of lift; with a strong inside rudder its velocity may be so low that it has less actual lift than the outside wing in spite of the higher lift coefficient. This can set up a roll toward the low wing, reaching its critical angle of attack well before the outside wing considers the idea of stalling. You might ex-

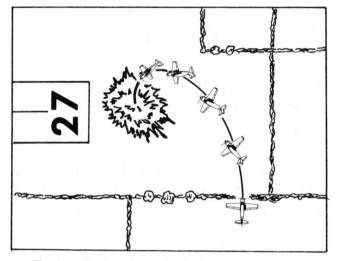

Fig. 11-4. Top view of an under-the-bottom stall accident in the turn onto final.

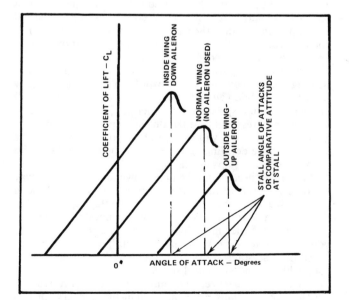

Fig. 11-5. A comparison of the coefficients of lift and angles of attack for three wing conditions.

plain this by describing a wings-level, cross-control turn.

Impress on the student that if he thinks he's getting into a situation that requires a tight final turn, he's better off to shallow out and break off the approach. He can climb out parallel to the runway and set up a new pattern. This was mentioned in Chap. 10, but hit it again when you introduce these stalls.

An example of the factors indicated in Fig. 11-5 could be evident when a student is doing a takeoff and departure stall to the right. When the airplane rolls to the left after the break, he tries to stop it with full aileron and maybe not enough rudder, so two things are happening at that point: (1) the angle of attack of that dropping (left) wing is increasing because the relative wind has changed — there is an upward component as well as a rearward component of relative wind — and (2) the application of right aileron (left aileron down) moves the C_L versus angle-of-attack curve to the left, as shown by Fig. 11-5, and things meet in the middle. The roll to the left is aggravated, and a left spin may result. The recovery procedure would be to neutralize the ailerons, apply opposite rudder followed immediately by forward motion of the wheel or stick, and then pull out of the ensuing dive.

DEMONSTRATION. You might warn that there will be some slipping effect in the recovery because the speed is low as the roll-out is made and the top rudder could cause a slip, if overdone. There will likely be some scrambling around over in the left seat when the stall breaks, so warn him about the attitude *before* you do it. As always, recover from these practice stalls at least 1500 ft above the surface.

Note the altitudes at stall entry and recovery. When you are setting up the demonstration mention that the first turn after takeoff is another point where people tend to make flat turns.

Clear the area.

Carry a slight amount of power, maybe 1500 rpm, since this not only will help get a better stall entry but will also set up the partial-power conditions that might be encountered at the turn onto final. Don't use flaps for this first one.

Set up a shallow turn, gradually increasing inside rudder and the opposite aileron. Increase the back pressure to keep the nose up — in some airplanes you may have to exaggerate the nose-up attitude to get the stall.

As the break occurs, neutralize the ailerons. The nose attitude will be low and the airspeed will pick up quickly when the back pressure is released. The recovery is stopping the roll with opposite rudder, then simultaneously applying aileron and rudder to level the wings as the nose is brought back up to level. Full power is added at the start of the recovery.

You would not use flaps at first because you are demonstrating the maneuver so that he will have a pattern to follow (hopefully). While *you* will be able to recover without exceeding the maximum speed for flap extension, he might not, and unnecessary stress could be put on the flaps if he lets matters get out of hand. You will have to know your student and the airplane before deciding at a later date whether he should do the stall with the flaps down.

If you do decide that the flaps-down approach is the way to go because the attitudes and reactions are more realistic in that configuration, be sure that *you* get the flaps up as soon as the stall breaks; for electrical flaps, you're not likely to exceed V_{FE} by the time they are fully retracted. The chances are that you'll feel better if he keeps the flaps up if you allow him to practice these stalls solo — and you may decide against any solo deliberate cross-control stalls.

In a real occurrence, on a final turn, it would probably be best to leave the flaps down throughout the recovery for two reasons: (1) upping the flaps could distract attention from more vital things such as adding full power and using the

primary controls in recovery, and (2) the down-flaps make for a lower stall speed and the airplane can be back in a non-stalled condition (and some semblance of control) more quickly than would be the case in the clean condition. In a real cross-control stall on final, or at any point close to the ground, slightly exceeding the listed flaps-down airspeed is the least of the pilot's worries.

You should check your airplane (solo) on the best methods of recovery from stalls and other performance problems, and it may be that upping the flaps is easy and aids in the recovery process so well that you might recommend it as part of the recovery. *However, it's one thing to be ready for the practice stall and another to get caught unexpectedly.* The pilot probably would not get the flaps up, anyway.

Your demonstration of this maneuver, like other demonstrations, can sometimes go awry, so grin and do it again, if necessary. You might make the comment in this circumstance that *intentional* stalls are sometimes hard to get into.

PRACTICE. The usual student reaction to his own first cross-control stall is to sit there more or less rigid while the airplane makes its own recovery — at a high airspeed. It will usually take a couple of tries before he gets the idea. Don't have him do more than a couple during any particular flight, since it's a fairly violent exercise.

EVALUATION AND REVIEW. When he is able to initiate the recovery without delay and understands the recovery technique, you should move on from this one. In your discussion of spins later you will note that this stall could go into an under-the-bottom spin entry. Review the procedure and recovery after the flight.

■ **Accelerated Stalls.** This maneuver is used to show the idea that the stall is a function of angle of attack rather than airspeed. If your trainer has an accelerometer you can *show* your more advanced students what is only theory in most presentations.

A good demonstration to show that stalls *are* a function of angle of attack is to approach a stall by slowly increasing the pitch attitude and noting the airspeed at which the stall-warner first sounds. (In one trainer this will be about 45 K.) Then set up a 60°-banked turn, increasing the back pressure more rapidly until the stall-warner sounds. *Its first sound will be at the same angle of attack,* but you will note that the airspeed is 60 K or higher. The stall speed has increased because the g's being pulled make the airplane "weigh" more.

EXPLANATION. Before the flight, discuss the idea of the stall being a function of angle of attack rather than airspeed. Define the fact that the load factor goes up as the square of the airspeed at the stall. The normal stall speed occurs at 1 g, and if the airplane is stalled at *twice* the normal stall speed, a load factor of *four* is imposed; at *three* times the stall speed, *nine* g's result. Set up the exercise so that the airplane is never stalled at an airspeed greater than 1.5 times the stall speed as indicated by the bottom of the green arc (which would result in a load factor of 2.25 at gross weight — and more g's at lighter weights).

The rule "1.5 times the bottom of the green arc" for setting up an accelerated stall refers to the *calibrated* stall speed. Airplanes that were manufactured before 1976 are marked in CAS and the 1.5 factor will be correct. But if the stall speed is marked in IAS, which means that there is a large error at stall and a lesser error at the 1.5 value, you should convert the value at the bottom of the green arc to calibrated airspeed (using the *POH* correction table), multiply that by 1.5, and then reconvert to indicated airspeed. You'd remember this final number for use with your accelerated stalls. This is the procedure for all rules of thumb using the bottom of the green arc (IAS).

The airplane can be stalled in any attitude and airspeed but may suffer structural damage at higher airspeeds.

DEMONSTRATION. There are two basic ways to demonstrate accelerated stalls.

One way is to set up a 45°-banked turn and gradually decrease power while trying to maintain altitude with continually increasing back pressure. This makes a good, comparatively gentle introduction to the accelerated stall. However, with some airplanes there is buffeting and pitching but no sharp indication of the higher-speed stall.

The second way to demonstrate accelerated stalls is to set up a 45° bank with cruise or climb power at a prechosen safe airspeed and "tighten the turn" by rapid back pressure on the wheel, as if trying to turn inside of another airplane in a dogfight. This is a more abrupt action, but it usually results in a definite break. Your ROTC Flight Instruction Program students will enjoy your demonstration (and their practice) of what can happen when you try to pull too tight a turn in a hassle. With some of your more timid students, you might be better off to stick to the first demonstration until you see their reactions to the accelerated stall. Don't overdo the demo with any of them.

PRACTICE. The student will probably be too cautious at first in his application of back pressure in the accelerated stall when using either of the types of entry techniques described above. Have him repeat the procedure until the entry and recovery moves are well in mind. It's interesting that while a conscious attempt to get an accelerated stall may not be successful at first because of insufficient and/or not brisk enough back pressure, an actual accelerated stall can occur because the pilot doesn't realize that he's using too much back pressure—he's absorbed in something else. It depends on the airplane, of course, as to its reactions to the accelerated stall entry; but a common instructor error is to not push for a complete stall and a definite break, whether demonstrating or having the student practice the stall (and that's often also the lack in normal 1-g stalls). Don't let him practice more than three or four of these at a sitting—one of you might get queasy (Fig. 11-6).

EVALUATION AND REVIEW. If the student gets a complete stall, understands the principle of the stall and recovery, and has shown an increase in confidence in the stall situation, you can move on. The main idea, of course, is for him to understand the circumstances that could cause the stall.

After each of these sessions, briefly review the accelerated stall. The student may nod his head and agree that the stall is a function of angle of attack and not airspeed, but it may take several emphasized repetitious times before he really believes it. This is particularly true of those students transferring from that flight school across the field who haven't had the earlier benefit of *your* good training techniques.

OTHER ADVANCED STALLS

■ **Trim Tab Stall.** Discuss and demonstrate this stall, and if you feel it's warranted, have him go through the process a couple of times. Set up a power-off approach using recommended landing flaps, and roll the trim to the full nose-up setting. Continue to slow the airplane and then apply full power as the airspeed decreases to about 5 K above the expected stall speed. You are simulating a sudden go-around. Demonstrate that forward pressure is required to keep the

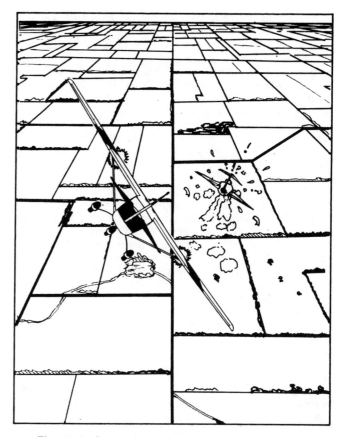

Fig. 11-6. Some of your students will see themselves as fighter aces during the accelerated stall practice.

nose from rising to a critical pitch attitude. The nose should be held to a position *no higher* than the climb attitude as the trim is corrected. For some higher-power airplanes this could require both hands while climbing the airplane to a safe altitude. Don't have him close the throttle as the stall is approached; at low altitudes this could be fatal. Recovery may require a quick release of one hand to get to the trim and then back to the wheel, repeating these motions until the trim is normal. Work out the best clean-up procedure for flaps for your airplane in this situation. It is unlikely that the trainer you are using will need such action (the nose can be easily held down even with full power and full nose-up trim), but the student should have an idea of what could happen. On the other hand, see Fig. 11-7.

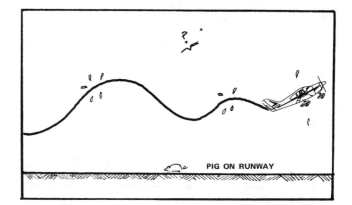

Fig. 11-7. Adding full power with the trim tab in the full nose-up position can be an interesting experience in some airplanes.

■ **Top-Rudder Stall.** This is a cross-control stall, but it is more likely to be encountered in a slip. It's entered as a slip (power-off) in either direction, and top rudder is increasingly applied as the aileron is increased in the direction of the slip. (You may have to keep the nose at an unusually high attitude.) The airplane should roll toward the higher wing when the break occurs; and the recovery technique is to relax back pressure and add full power, leveling the wings and then raising the flaps as a full recovery is made. As in the skidding-type, cross-control stall recovery, leaving the flaps down may help in an actual situation (such as a stall in a slip on final), getting the airplane back into an unstalled condition at a lower airspeed.

There are arguments that it's impossible to stall some airplanes in the manner just described (a power-off slip). Admittedly, some airplanes do tend to hang on and shudder, reluctant to go over the top; others (particularly tapered-wings) will break very nicely. You'll find it frustrating at times demonstrating "how the airplane will go over the top when you stall it this way." You'll have better success, if not as practical a demonstration, if you introduce the idea in a power-on, clean situation.

Even if the airplane doesn't give a clean roll over the top in the approach configuration, you should note that such abuse of the controls in a slip can result in a much higher rate of sink than normal and might catch the pilot unprepared. Pilots have been known to slip into the ground.

Some airplanes have a tendency to suddenly tuck the nose down in extreme slip conditions; the wing isn't stalled, but the horizontal tail is blanketed and the tail-down force is suddenly decreased. This could be an unpleasant surprise on final.

■ **Summary of Advanced Stalls.** You don't have to work on these stalls (except for the accelerated ones) until the student does a slick job of entry and recovery, but make him aware that an airplane can be stalled in some unusual attitudes. Help him see that certain situations should be avoided, but if he does get caught, there are ways that he has had some experience with to help him get out of it.

The *rudder exercise stall* was discussed in Chap. 9, and if you didn't mention it at that phase of training you might now talk about the effects of raising (or trying to raise) a wing with rudder alone. You should also describe the secondary stall (a new stall resulting from an improper stall recovery—usually a result of trying to get the nose back up to the level-flight attitude too soon).

SHORT-FIELD OPERATIONS

■ **Background.** The student should have practice in short-field takeoffs and landings before the solo cross-country.

Short-field takeoffs and landings can be hazardous for a student pilot because his judgment of just when short-field techniques are required isn't developed yet. Don't give him the impression that a short-field procedure *must* be used when conditions aren't good enough for a normal takeoff or landing. For example, if he has a choice, it's better to go on to another airport rather than try to set it down on that 1500-ft ag pilots' strip. Or if he has landed on a short strip and conditions have changed (the dirt strip is now muddy after that heavy rain that ran him in there and/or the wind has died), he should not take off to continue the solo cross-country until he's talked to you.

Unless the student has been given plenty of dual practice in short-field work, he may actually lose performance by using short-field techniques in an actual situation. He might have had better results by using normal procedures and perhaps converting to the max angle climb speed after takeoff. Attempting to rush the takeoff will in most cases delay the process, and in some instances this could result in a dangerous situation that could have been avoided using normal procedures. He can make a mess of things if the proper special techniques aren't used.

■ **Short-Field Takeoff**
EXPLANATION

What — The short-field takeoff of interest here assumes that a 50-ft obstacle must be cleared shortly after lift-off. Make it clear that the airplane configuration and pilot technique must be such that the 50-ft obstacle *is* cleared. The student often thinks only in terms of getting airborne in the shortest distance and tends to ignore the climb part of the exercise. You will see students rush the airplane off the ground and stagger along for a considerable distance just above the surface—and you're glad the 50-ft obstacle actually doesn't exist near your runway (Fig. 11-8).

Suppose your trainer uses *some* flaps for the soft-field takeoff and none for the short-field takeoff and obstacle clearance. You can make the point that while the flaps help the airplane break free from the ground and so are used for soft-field takeoffs, down-flaps often hurt the second phase.

You might compare the requirements of short-field and soft-field takeoffs. *Short-field* takeoffs must be the best combination of takeoff *and* climb. The takeoff surface is assumed to be firm and smooth. In *soft-field* takeoffs, techniques are used to break the airplane free from the surface as soon as possible (which may take a long distance in very-soft-field conditions). Flaps are normally used to get weight off the wheels at low airspeeds so that the high ground-drag is decreased as efficiently as possible. Once the airplane is airborne a normal climb is estab-

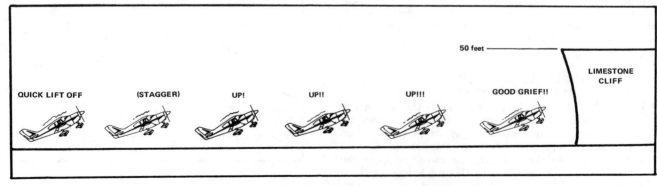

Fig. 11-8. The student tends to think in terms of a quick left-off and often ignores the climb requirements. In this case he may be taking that limestone cliff for granite.

lished. There is assumed to be no obstacle in practice soft-field takeoffs.

For a short-field takeoff you could theoretically work out some combination of flap usage for your trainer where flaps are used until lift-off and then are retracted to get a better climb to clear the obstacle. But it could also cause a decrease in performance as the flaps are juggled (the airplane might settle back in as the flaps are upped). You might cover this combination in your explanation of the maneuver, but you should indicate that it's always more practical to set the manufacturer's recommended flaps before the takeoff run and be able to concentrate on the takeoff and climb.

The maneuver is assumed to be complete when the obstacle is cleared. (You may use a height of 100 ft above the runway when simulation is used.)

Most students, and some experienced pilots, are confused about the difference between maximum angle and max rate of climb. The definitions are these:

Max angle of climb—The greatest amount of vertical distance per unit of horizontal distance.

Max rate of climb—The greatest amount of vertical distance per unit of time.

An exaggerated curve showing rate of climb versus airspeed can be drawn on the chalkboard and the two different types of climb shown (Fig. 11-9).

In plotting the points on your chalkboard graph, note that these rates of climb are for a fictitious airplane at a particular density-altitude and weight. You should also draw in the rate of climb curve for a higher altitude (smaller curve).

You can make your point better if the curve is stretched out slightly; this makes the difference between the two types of climb more obvious. Use figures for velocity and rate of climb that are reasonably close to those of your airplane.

The max *angle* of climb, then, will allow the airplane to clear obstacles that the max rate would not, and that's

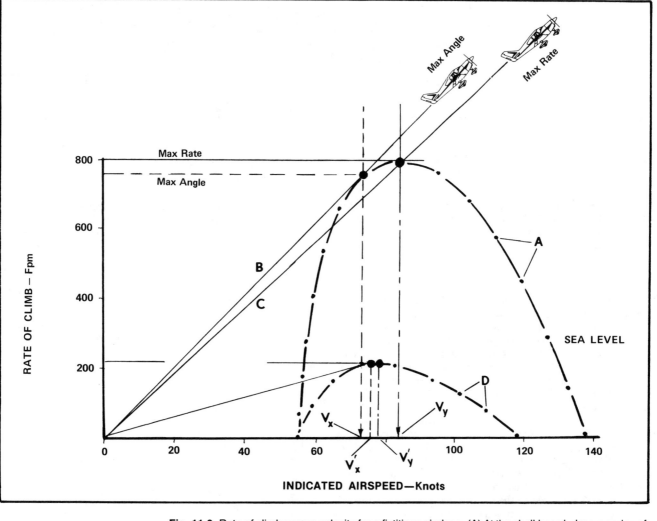

Fig. 11-9. Rate of climb versus velocity for a fictitious airplane. (A) At the chalkboard, draw a series of dots as shown to indicate the rates of climb at various airspeeds at sea level. (Rate of climb zero at stall speed and at the top level-flight speed.) Connect the points to form a curve. Point V_y will be the airspeed for the max rate of climb. (B) Draw a line from the origin, *0*, tangent to the curve. Airspeed V_x is that required for max *angle* of climb. (C) Draw a line from the origin, *0*, to the peak of the curve and compare angles. Even though line *C* shows a greater *rate* of climb, line *B* would have a greater *angle* of climb. You might draw the curve for a higher altitude *D* and show how the airspeeds vary with altitude; the required IAS for max angle increases and for max rate decreases. V'_x and V'_y are the speeds at altitude. Your drawing doesn't have to be completely accurate but should show the principle.

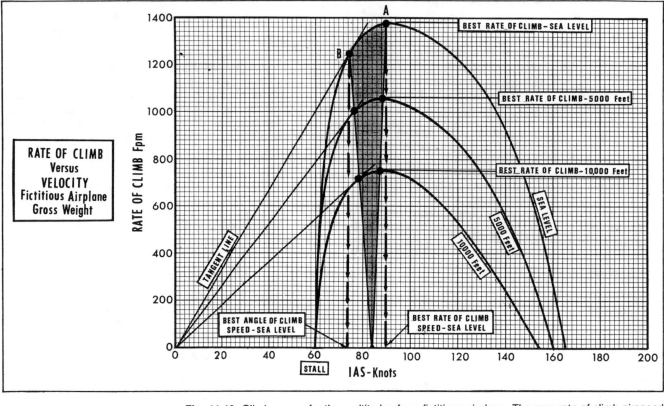

Fig. 11-10. Climb curves for three altitudes for a fictitious airplane. The max rate of climb airspeed (*A*) decreases with altitude and the max angle airspeed (*B*) increases with altitude. (See *The Advanced Pilot's Flight Manual* for more on this.)

why the short-field takeoff and max angle climb are used.

Fig. 11-10 shows the climb curves for three altitudes, with the shaded area showing that V_Y *decreases* with altitude and V_X *increases* with altitude and the two airspeeds are equal (theoretically anyway) at the absolute ceiling of the airplane.

Okay, let's back up a little. The curves of Figs. 11-9 and 11-10 were developed from the clean airplane (in some cases, both *gear* and *flaps* are up), and the resulting airspeeds for max angle and max rate climbs resulted. A *Pilot's Operating Handbook* may give a recommended airspeed for *obstacle clearance* immediately after takeoff that is lower than the airspeed cited for max angle climb cited elsewhere in the book. Your students may need an explanation of *why* this is the case.

Normally the airplane must clear the obstacle shortly after takeoff and the distance data includes the climb to 50 ft with the landing gear down (even in a retractable, since the gear would be still down or in transit) and the recommended flap setting for the best obstacle clearance. Gear and flaps create *parasite drag,* and a lower airspeed is required to decrease this. The rate of climb curve in Fig. 11-9 is moved to the left and has decreased values of climb in the dirty condition, but *you are getting the best possible results for that configuration.*

Fig. 11-11 gives a comparison of the sea level climb curves in the cleanest configuration and with the gear and recommended takeoff flaps down between lift-off and obstacle-clearance height.

A *fixed-gear airplane* without flaps (or if flaps are not recommended for the short-field takeoff) would have the same obstacle clearance and max angle climb airspeeds.

How — The airplane is taxied to the extreme end of the takeoff area. This may seem elementary, but since the average lower-power trainer will be climbing at about a 1-to-10 ratio (1 ft up per 10 ft forward), the use of every additional 10 ft of takeoff run will mean an extra foot of height at a particular point on the climbout. Needless to say, a headwind on takeoff is a much desired commodity if all other things (such as obstacles at each end) are equal.

A good pretakeoff check is made to make sure that the engine is developing max power for the density-alti-

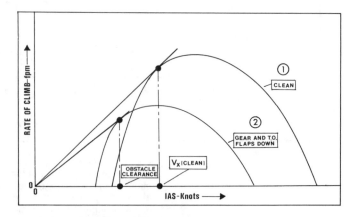

Fig. 11-11. Sea level rate of climb curves for a fictitious airplane clean (*1*) and with the gear down and recommended takeoff (*T.O.*) obstacle clearance flaps (*2*). Note that because of the characteristics of the two curves, the "dirty" condition has a lower rate of climb and a lower best angle airspeed.

tude at the takeoff area. This may include leaning of the mixture at higher elevations.

There is some argument concerning the advisability of the holding of brakes, running up to full power before releasing them, to get a shorter takeoff run. Doing this might be a small advantage for reciprocating engines with constant-speed propellers, but for the average trainer, it's best to open the throttle smoothly and let the airplane accelerate normally. You'll find that students using the "hold brakes and full power—then release" technique sometimes have difficulty keeping the airplane straight during the first few yards of the run, and some performance may be lost. In addition, the noise and vibration of the full power-and-brakes method (the student is holding brakes so hard that his calves are quivering) plus your shouting instructions over the noise isn't a help. So, while you are teaching a special technique, don't add factors that could adversely affect the procedure.

The most common error is trying to pull the airplane off, both in practice and in an actual situation. It takes "cool" to sit there as the obstacle gets closer and not pull the airplane off abruptly.

When the airplane is ready, it should be flown off and the obstacle-clearance climb speed attained and maintained. The student, not used to the more radical pitch position of this type of climb, will tend to keep the nose down and pick up too much airspeed. This is the most common error in the lift-off and initial part of the climb.

Expect directional control problems the first few takeoff runs and climbouts. The airplane is slower than usual and "torque" will be a greater factor than in the normal takeoff he's used to flying. After a height of 100 ft above the runway is reached, he is to assume a normal climb and clean up the airplane. Allow him to wipe his sweaty palms after a safe altitude is reached.

DEMONSTRATION. A good exercise to compare the normal

and max performance takeoff and climb is to fly a short-field takeoff while explaining what you are doing (particularly reminding the student of maintaining the max angle of climb speed after lift-off until the obstacle is cleared) and then letting the student complete the pattern with a normal landing so that he can practice a short-field takeoff or two. Then, after he has the technique clear in his mind, you may compare the altitude above the runway in normal and short-field takeoff procedure. You'll make a takeoff and climb of each type, using a ground reference (lining it up with the leading edge of the wing or other reference on the airplane) and having the student check the indicated altitude as you say "mark" when the references line up each time (Fig. 11-12).

You should demonstrate the comparisons in the same wind and weather conditions, probably immediately following one procedure with the other. One thing you'll learn after doing this is that some airplanes have very little difference between max and normal performances. Also, it will become evident that if the wind changes between the two comparisons, the normal phase might be better than the max performance effort. This can be embarrassing (Fig. 11-13).

PRACTICE. You will probably ride through a half-dozen of his takeoffs before you'll feel that he is safe for solo practice.

> Some of the *common errors* you'll see are these:
> 1. Abrupt power application at the beginning of the takeoff.
> 2. Trying to pull the airplane off too soon. (This is usually just after you've chided him for keeping it on the ground too long.)
> 3. Picking up excessive airspeed after lift-off; not holding to the max angle of climb speed.

EVALUATION AND REVIEW. Confirm that he understands the principles of the maximum performance takeoff and realizes that he will have to fly by the numbers given by the manufacturer; he cannot find the airspeeds by "feel." Note that it took

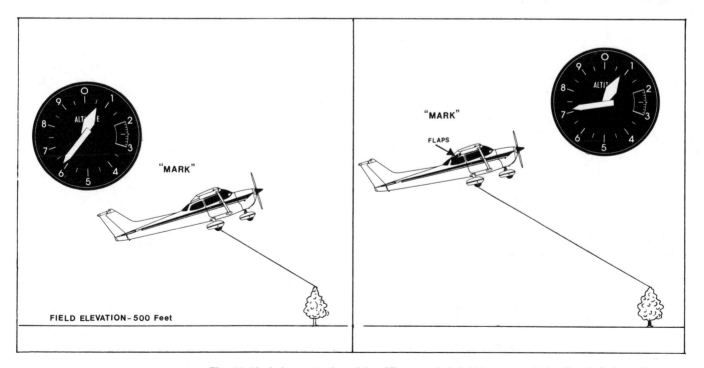

Fig. 11-12. A demonstration of the differences in height in a normal takeoff and climb and in a max performance takeoff and climb, at a particular reference.

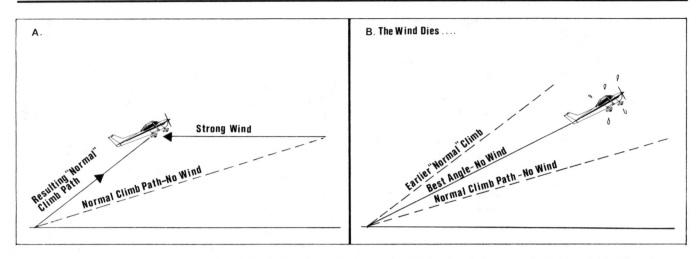

Fig. 11-13. (A) You demonstrate the normal takeoff and climb and check the height at the reference. (B) Now you're going to show the student how to get maximum performance from the airplane—and the wind dies. (This is an exaggerated example.)

a number of climbs at different airspeeds to establish a graph and the resulting information to obtain the max angle of climb speed. Some old-time pilots say they can feel the airspeed for best angle of climb, but usually they will be too slow, giving an impression of climb by the cocked up attitude but not gaining altitude well because of high induced drag.

■ **Short-Field Landing.** This can be a critical maneuver if the pilot allows the airplane to get slow with a high rate of descent. The first solo flight, the solo cross-countries, *and the* sessions of solo short-field landings will "get your attention" more than any other phases of primary training. Only the strongest of instructors can watch one of his students practice solo short-field landings without just a little flinching at each approach. As would be expected, the approach and landing is touchier than the takeoff. A student's short-field takeoffs may be good, but extra flight time might be needed for the landing.

EXPLANATION. A model and the chalkboard will be your best aids in the preflight discussion.

What — The short-field approach and landing is, as the name indicates, a special maneuver used for a landing area shorter than "standard." It requires special techniques, lower airspeeds, and usually more flaps than normal (full flaps are to be used—there may be particular airplanes or circumstances where this isn't so, but you'll know this for your situation).

A 50-ft obstacle is assumed to exist so both the approach from the obstacle to the touchdown point *and the* landing roll must be an optimally short distance. As in the short-field takeoff, you'll find that the student tends to think only of the ground roll portion. In the landing, however, a fast and extended approach can affect the landing roll. Incidentally, it makes for less confusion if you refer to the "takeoff *run*" and the "landing *roll*"; it separates the two factors better for many people.

Why — The teaching of the short-field landing to the student pilot makes more sense than teaching him short-field *takeoffs*. For the preprivate pilot, short-field landings are taught as a safety aid. If the student gets lost on a cross-country and has to make an off-airport landing, he should have been given a procedure to give him the best chance of making a safe landing. Don't let him get the idea that the short-field takeoff is to be used after he's landed in that farmer's field and decides to fly home

when the weather is better. *If a student pilot has to make an unplanned landing away from the home airport, whether on or off an approved landing area, he is to call you before making any further moves about flying again—ALWAYS.* It's possible that after talking to him you figure it's okay for him to proceed (999 times out of 1000 you *won't,* since it's hard to judge the situation over the telephone). If he's made a precautionary landing off an airport, the airplane should stay on the ground until you or another instructor get there to fly it out; and you could decide that it's better to dismantle it and haul it back by truck.

How — A wider-than-normal pattern is used so that the proper airspeed can be set up and the airplane trimmed. If the approach is rushed, the result is a too-fast airspeed and a tendency to "put it on." It's suggested that the flap setting be completed on the base leg to (1) help control airspeed and (2) allow the pilot to give his full attention to that turn on final and the last part of the approach.

There are two types of approaches to consider, as shown in Fig. 11-14. Fig. 11-14A shows the "classic" type of short-field landing over an obstacle. The airplane is dragged up just above a stall and power is chopped and the nose lowered quickly just past the obstacle. The airplane is rotated to the landing attitude and (usually) power is added. An experienced pilot may use this type of approach without bending the airplane, but the student or private pilot could make a big impression.

Fig. 11-14B shows the type of obstacle approach you should teach, at least at first. A steady flight path is established at the recommended airspeed (a good rule of thumb for initial training, if no data is available, is 5 K lower than the normal approach speed or 20 percent above the full flaps *calibrated* stall speed, whichever is greater). Remember that turbulence on final could require a higher approach speed.

After the touchdown, the throttle should be completely closed, *the wheel or stick held full aft,* and judicious braking applied. He is not to allow the tires to skid; if there is the sound of punished tires, he's not getting the optimum braking effects and this is costing money. Be especially careful braking a tailwheel-type airplane, for obvious reasons.

The flaps should be up after landing if the most effective braking is to be obtained, but he should practice a number of short-field landings without this distracting action. He should apply brakes, leaving the flaps down in

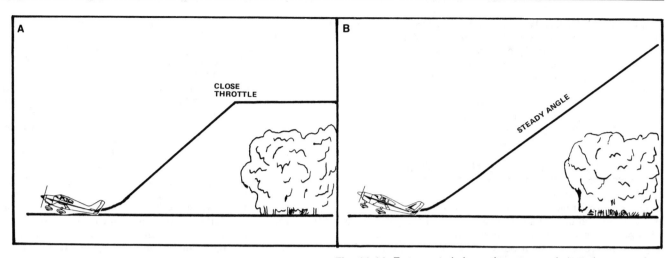

Fig. 11-14. Exaggerated views of two types of obstacle approaches.

the first few landings. In early practice there is usually more loss of efficiency caused by diverting attention to the flaps than is gained by the added braking effectiveness. *Don't* pull up the gear.

DEMONSTRATION. Some flight instructors carefully explain how they want the short-field approach and landing made, cautioning the student to maintain the proper airspeed (don't let it get too slow, use proper braking, etc.). Then on the demonstration they bring the airplane around hanging on the prop, shuddering and buffeting, slam it on, and bring it to a skidding halt on squared-off tires. You'll certainly be tempted to get the extreme performance on a short-field landing, but you'll have to demonstrate it the way you want the student to do it—safely (Fig. 11-15).

Fig. 11-15. You'll be tempted to make the world's shortest landing when demonstrating short-field techniques. Don't—unless you'd like to watch the student using the same procedure, solo.

PRACTICE. Riding through the practice of short-field landings with a student, like the practice of his first few normal landings, can be "interesting" at least. You'll be alert (absolutely) for dangerous practices and should have him do a number of patterns dual before turning him loose solo. Many instructors don't encourage the solo practice of short-field landings, but they fly dual enough to ensure that the student knows the

procedure and can do it if needed on a solo cross-country. Later, after he's built up more time and is approaching the private flight test, the instructors allow solo practice.

Here are several *common errors* to check:

1. Letting the airspeed creep up above the recommended value for the approach. Some students let it get too slow, but the usual airspeed problem is that of being too fast.

2. A pattern that's too close in and doesn't allow the approach to get stabilized.

3. Poor vertical flight path control; carrying too much power in the early part of the approach so that the throttle must be completely closed in an attempt to land on the predetermined spot. Nearly every instance in which the throttle has to be closed in the latter part of an approach will result in landing long. Because of the low-time pilot's inability to realize that he is too high until the latter part of the approach, even completely closing the throttle is often not enough to save matters and a go-around is required. You should stress that a go-around is much more important on a high short-field approach than on a high normal approach—after all, the special approach is used because the field is *short*. If the approach isn't right on, he should take it around and give himself more room in the pattern next time. There's little or no margin for error in an actual short-field approach and landing.

4. "Putting it on." If the airplane is fast, there is a tendency to put it on the ground instead of taking it around. You'll see this tendency even if the airplane *isn't* too fast. If it's landed this way, it will touch down too fast and the braking effectiveness suffers. He will try to correct for previous errors in the approach by slamming it on.

5. Poor braking technique. It was noted earlier that the tailwheel airplane can run into trouble (nose over) if brakes are applied too hard, but by misusing the brakes in the tricycle-gear airplane a nose-over can also occur. The student may start hard braking but is exerting uneven pressure (students do this) and the airplane starts turning. It's possible that the airplane can nose over around a line from the outside main wheel to the nosewheel. *You and the student should have your shoulder harness and belts tight when shooting short-field landings,* because the added braking may cause the student to move forward and further increase his pressure on the brakes for a very short period of time. Sometimes, that's all that's needed to cause trouble.

EVALUATION AND REVIEW. By the end of the first dual session of short-field work he should be making consistent, safe approaches. If you have to fly extra patterns to ensure this, do so. Don't release him until he is safe, because a superficial knowledge of short-field takeoffs and landings (you explain them and he does one or two) could result in an accident if he decides to try a few on one of his solo flights.

Review short-field techniques (explanation and dual practice) before his first solo cross-country.

SOFT-FIELD TAKEOFFS AND LANDINGS

Like the short-field takeoff, an actual *soft*-field takeoff should be discouraged for student pilots. Again, if he's made an off-airport landing in a soft area, he should call you before attempting to get off again—and most likely you would want to go to the scene and fly it out yourself. You'll show him the procedure for future use; he may get the private certificate and never get any more dual, and if he ever has to get out of a soft-field situation, he should have had some experience with the proper technique. As a private pilot, he should be encouraged to discuss the problem with more experienced pilots before taking off.

▪ Soft-Field Takeoff

PREPARATION. A model is the best aid; this one is difficult to draw on the chalkboard. Have the student read about soft-field takeoffs before the introductory flight.

EXPLANATION. The soft-field takeoff technique is used not only for soft surfaces such as mud, but also for high grass, snow, and rough fields. Drag on the wheels (rolling coefficient of drag) is the big problem, and the takeoff procedure decreases this factor as soon as possible. On a rough surface, it's best to get the weight off the landing gear to avoid damage. After the airplane is airborne, the exercise is considered ended (no obstacle).

Why—The soft-field technique is used to effect a takeoff in field conditions that would preclude a lift-off using normal procedures.

How—The run-up area is considered firm, but the takeoff area is soft (or rough, or has tall grass, etc.). After the run-up is made, ensuring that the engine is developing full power (lean the mixture if the density-altitude requires it), the recommended flaps are set, the airplane is taxied onto the takeoff area *and is kept moving* (otherwise, the maneuver could "bog down"). The throttle is opened fully as the airplane is lined up and a tail-low takeoff is made with the wheel held full back as the run continues. A reference at the far end of the runway should be used from the start of the takeoff to check direction and pitch attitude requirements.

You need to establish an angle of attack to get the weight off the main wheels as soon as possible and to keep weight off the nosewheel to avoid damaging it. Keep

the tail of the tailwheel type low, decreasing a nose-over tendency.

You might point out that the pilot should expect the rolling drag to vary; there will be softer spots of mud, drifted snow, or higher clumps of grass; and sudden deceleration forces may be encountered. This is why the tail-low attitude in the tailwheel airplane must be maintained. There is sometimes a tendency by pilots making a soft-field takeoff to let the tailwheel airplane assume a more-level attitude if a comparatively hard area is reached, and "snags" can occur if the airplane suddenly hits a soft area again.

Okay, so the run is started with the control wheel or stick in the full-aft position. In the tricycle-gear airplane the nosewheel will leave the ground, and you should warn the student that he must relax some back pressure as the nose comes up or the airplane will assume a pitch attitude that's too high, which would delay the takeoff. This is most often a factor when simulating the procedure on a hard surface runway where the airspeed picks up rapidly and the controls become effective sooner. *You* should be ready for this. It's possible in some airplanes that he could drag the tail tie-down ring, getting such a high pitch attitude that the elevators (stabilator) lose effectiveness and the nosewheel bangs down again and again as the cycle repeats itself while he continues to hold the wheel full back (Fig. 11-16).

The back pressure must be relaxed slightly on the tailwheel type so that the proper tail-low attitude is attained and held.

Spell out to the student that he can expect some directional control problems with the tricycle-gear airplane. This is because the nosewheel will be "lighter" than usual due to the full-back wheel as the run starts, and after the nose is raised he can expect stronger "torque" effects. In addition, because the nose is higher than normal on the takeoff (and will be that way at lower speeds and for comparatively longer times than he's been used to), a visibility problem could pop up for some airplanes.

After lift-off and when the airplane is definitely airborne, the flaps are smoothly retracted and the maximum rate of climb speed is maintained. It's assumed that there is no obstacle to be cleared in the demonstration and practice sessions, but in real life circumstances he may have to get out over some trees after a soft-field takeoff and would use the max angle climb technique.

DEMONSTRATION. One demonstration of a soft-field takeoff should be enough; you can make corrections and explanations as he practices the maneuver dual. Make sure that *you* don't overdo the rotation and drag the tail. Don't get so involved in talking that you let directional control problems arise.

As the airplane lifts off, make the transition to the recommended climb speed smoothly. One of the common errors seen in soft-field takeoffs is easing the nose over too quickly after getting airborne, allowing the airplane to touch again (Fig. 11-17).

SCREECH!   BANG!   SCREECH!   BANG!   SCREECH!   (ETC.)

Fig. 11-16. The heel-and-toe takeoff procedure is caused by poor elevator usage.

Fig. 11-17. Don't let the airplane touch again once it's been lifted from the quagmire.

PRACTICE. During the student's takeoffs in the tricycle-gear airplane, you may have to remind him of something that was only considered for the landing—he'll probably have to look alongside the nose when it's in the proper attitude in the take-off run. Expect some directional control problems.

He will probably not hold the wheel full back the first couple of times.

> *Common errors* a student pilot may make in soft-field takeoffs include the following:
> 1. Not keeping the airplane rolling from the run-up position; stopping for lineup (and sinking in an actual circumstance).
> 2. Forgetting to set the recommended flaps.
> 3. Neglecting to hold the wheel or stick full back at the beginning of the roll.
> 4. Holding the full-back pressure after the nose starts to rise, with the result that it is *too* high, delaying the takeoff.
> 5. Directional control problems.
> 6. Stalling the airplane off, so that it falls back and the earth smites it.
> 7. Poor transition after takeoff, touching again, or maintaining a nose-high attitude too long, delaying the climb.
> 8. Forgetting to up the flaps as the climb is entered.

EVALUATION AND REVIEW. Check for safe operation at this stage rather than the ultimate in performance. When the student is able to lift off in a shorter-than-normal run and can make a safe transition to the normal climb he is ready to practice solo soft-field takeoffs.

In reviewing the theory of the soft-field takeoff you might draw a version of Fig. 11-18 on the chalkboard for some of your more technically minded students.

The pilot who sets up the angle of attack as shown at point *A* has the best amount of lift per knot (without getting too close to the stall) and a reasonably low amount of aerodynamic drag. The pilot who pulls the nose too high and gets the angle of attack shown at point *B* not only has a higher aerodynamic drag but also has less lift (per knot) acting to decrease the weight on the wheels in the muck—he's losing both ways.

As the student approaches the private flight test you should review the procedure and have him practice several soft-field takeoffs at the home airport to pin down the procedure.

An angle-of-attack indicator would be a useful aid, if it were available.

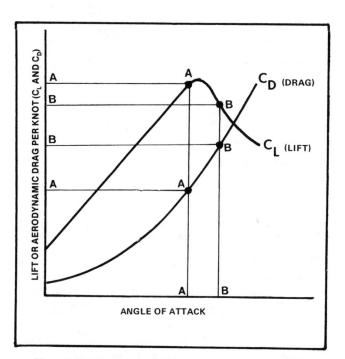

Fig. 11-18. An analysis of the best angle of attack for the takeoff run on a soft field.

■ **Soft-Field Landings.** This can be a valuable maneuver if the student gets lost on a cross-country and has to make an off-airport landing. Before he goes on the first solo cross-country you'll demonstrate and he'll practice until you're sure that he could make such a landing. As with short-field work, the ability to make a safe soft-field *landing* is more important to the student pilot than the takeoff.

PREPARATION. A model is the best ground aid, although the chalkboard can be used to advantage.

EXPLANATION

What—The soft-field approach and landing is one in which the airplane touches down at the lowest speed possible on a surface that is known, or suspected, to be muddy (or have a high rolling resistance due to other reasons, such as snow or tall grass) or rough.

Why — If a normal approach and landing is made, a strong nosing-over tendency may exist at touchdown as the airplane suddenly decelerates from a comparatively high landing speed. The *point* of landing can be better picked with a short-field approach (using power all the way around); usually the landing area has varying degrees of softness and a controlled approach makes it possible to land on the best portion (Fig. 11-19). Or drifted snow might make it advisable to pick a particular area.

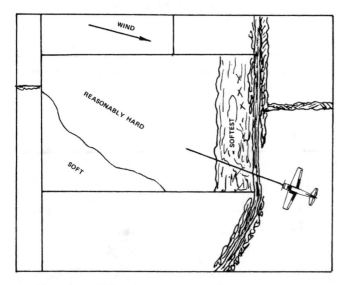

Fig. 11-19. A short-field approach is the best way to start a soft-field landing.

You might note the types of fields in your area and the comparative hardness of their surfaces, such as these examples:

1. *Rich brown* — Plowed or bare field. Very soft if wet, very rough if dry.

2. *Rich green* — Usually a sign of much moisture; may be immature grain or tall grass in a wet area.

3. *Brown-green* — The color of pastures, which are usually much firmer than 1 and 2. The presence of livestock is a giveaway. Needless to say, you'll remind the student to land away from the livestock (and to watch where he steps after getting out of the airplane).

If there is standing water in areas of a field he can pretty well assume that the rest of the field is soft. Even if the "dry" portion of the field might support the landing, it's possible that he could overrun the firmer ground and go into the water, with probable damage. If there is a choice, the pilot should pick a field some distance away from *that* one.

So it's good to discuss local crops and how they look from the air, and during the dual sessions in the area you can take a short break from maneuvers and point out some examples.

How — It's assumed that it has become necessary to make an off-airport landing, or a landing on a turf or grass strip in unusually wet conditions. First, the student (or any pilot) should *drag the area* (discussed in a following section, see Fig. 11-20) to check field conditions and obstructions. Emphasize that this should be done any time the student is unsure of a landing surface, airport or not.

After the field has been chosen and checked, a wide pattern is set up. The approach speed should be that of the short-field landing to ensure that the touchdown is

slow. Use full flaps unless the *Pilot's Operating Handbook* or other reliable sources say otherwise.

While in actual situations it isn't always possible, the airplane should touch down *with power* so that the stall speed is lowered. If you can fly the airplane up to the point of touchdown gradually using more and more power trying to hold it off as the airspeed decays, you'll get the slowest touchdown speed. The ideal, of course, would be using full power as the airplane stalls and touches the prechosen point. The throttle would then be closed.

Talking about flaps — should they be retracted immediately after touching down? Mud, mixed with gravel perhaps, would be thrown up and could damage the flaps, particularly on a low-wing airplane (or slush, ice, and other debris could be present) so there is an argument *for* raising them. Also, any loss of aerodynamic drag resulting from upping the flaps would be more than offset by the very high rolling resistance when the wheels "sink in."

Some arguments against retracting the flaps are these. (1) The constant flaps will let the weight be placed on the landing gear more gradually than would be the case if they were suddenly retracted, and there would likely be less of a chance of nosing over because of suddenly added deceleration forces. (2) The student or low-time pilot will probably relax the back pressure as he divides his attention by retracting the flaps, and this could be an added contribution to a nosing-over tendency (or having a nosewheel dig in and break off). (3) When the student pilot lands on a soft or rough field it is likely to be an emergency situation, and minor damage to the flaps wouldn't be as bad as the other possibilities of more damage plus the possibility of injuries by nosing over.

So, generally, students should be taught to leave the flaps down during the landing roll. However, if you are instructing in the north where operations are routine in slush and mud, you may as a matter of regular training have your students up the flaps to protect them after touchdown. You might also have your more experienced trainees (commercial, flight instructor, etc.) use the idea of retracting the flaps to avoid damage if they are flying into a soft-field area (don't pull up the landing gear).

You may be wondering why we went all around Robin Hood's barn to come up with the idea that students should leave the flaps down throughout the soft-field landing roll. The point is that some of your brighter students may ask, "Why not bring up the flaps after touchdown? It would save them from damage." You would answer, "That's a good point and here's another one for retracting the flaps (you state the position), *but* the advantages are outweighed by the disadvantages such as . . . ," and you give the reasons for keeping them down. Be prepared to defend your position in *any* of your statements or demonstrations. If you know and give both the pros and cons, the student will remember better, as well. When you are reviewing a maneuver or ground school subject for yourself, try to think of questions the student(s) might ask about that subject. Always push the idea of *why* it is done this way. Your answer of, "I don't know why we do it this way — it's the way I was taught," or, "Do it this way because I have more flying time than you, and I say do it this way," is not likely to win the respect and admiration of your students. When you are preparing for a ground school session or flight, work as if the student (who knows absolutely nothing beforehand about the subject you are about to discuss) has brought his two brothers, one of them is the world's sharpest aeronautical engineer and the other a very experienced pilot and instructor. You will have to (1) make sure the subject is covered in such a manner that the student can readily understand it, (2) make it technically accurate so the engineer doesn't jump all over you, and (3) satisfy the sharp pilot with your description of the flight technique.

Basically, it would be a hard job to state categorically that the addition of power would always help prevent (or tend to cause) a nose-over in a soft-field landing roll. It would depend on the power (thrust) available and the elevator effectiveness in the slipstream.

Summed up, the *student* should leave the flaps down and the throttle closed after touchdown.

DEMONSTRATION. You may have to make more than one demonstration of the touchdown and ground roll procedure to ensure that he will keep that wheel full back. With some students, you will have to repeat demonstrations after they've tried a few landings themselves.

PRACTICE. Here you will again experience the fact that an airplane can be dropped in by the student pilot. Generally, you can look for a tendency to level off and to hold the airplane off with power at too great a height. He'll hold the airplane off and increase the power so as to land as slowly as possible, as he's been told but will tend to be a foot or so too high. Watch for it. Put your word in when you first see the problem coming on. If the airplane is stalled too high, *with full power,* you won't have much of a chance for added recovery efforts. Depending on the student's skill, you might at first have him land with power off and then later introduce power use. The throttle should be closed after touching down. Expect that if power is used at the touchdown, either as a part of the drill or added to save a drop-in, the student will tend to leave it on too long, as in the presolo landing sessions.

Be ready for directional control problems on the roll-out. This is a possibility, particularly if he's dropped it in and has landed on one wheel (and adds full power too late, roaring toward the edge of the runway).

Common errors made on soft-field landings are these:

1. Too tight a pattern. This doesn't give the student time to get set up and results in poor airspeed control.

2. Poor airspeed control in the approach; usually too fast, but watch for the extra slow, cocked-up (with full flaps!) condition where a great deal of power and altitude might be needed to get out of a bind. Make your corrections *early* if he's headed for this.

3. Leveling off too high. In his desire to touch down as slowly as possible, he may drop it in.

4. Failure to hold the control wheel full back after landing.

5. Poor power control at, and after, the touchdown.

6. Directional control problems, particularly in the tailwheel type.

EVALUATION AND REVIEW. Before letting him practice solo soft-field landings you'll work out any of his dangerous tendencies. You may decide that if his soft-field work is good enough at the end of the dual sessions, no solo practice is warranted.

If there is a layoff between the practice sessions and the cross-country phase, you'd better review the procedure in the classroom and fly a couple of dual soft- and short-field patterns with him.

OTHER TAKEOFF AND LANDING PROCEDURES. On one of the dual flights in the postsolo period you should have the student simulate taking off at a high-altitude airport. You'll handle the throttle and restrict the rpm (or manifold pressure) for his takeoff and climb. Also, if you are using a four-place airplane for training you should have the student take off and land it at or near max certificated weight. Do this also for your people checking out in high-performance airplanes.

■ **Dragging the Area.** Before attempting to land on an unknown area (including some airports), take a good look at the surface and obstructions. To set up an approach and landing on a suspicious-looking field without a low pass could result in an unpleasant surprise. Certainly, you wouldn't send a student on a solo cross-country to an airport with dangerous obstructions or a poor surface, but he should have the knowledge to cope with such circumstances if he needs to go to an alternate airport or make an off-airport landing.

As with any training technique, if he has done it once, a lot (but not all) of the pressure will be off in an actual situation. You might explain the procedure this way:

If the pilot has a choice he should stay as *high* (up to 4000 ft above the terrain) *as the ceiling permits* when looking for a suitable field. This gives a look at a greater number of possible landing areas. If the search is deliberately conducted too low, some good fields may be missed. If he is *too* high, time and fuel may be wasted as he descends to check fields that looked good from high altitude but turn out to be pretty grim when he gets down a little closer. Basically, an altitude of between 2000 and 4000 ft is best for the initial look. If you (or the student) get much lower or higher than these outside limits, the spotting of a good field may be delayed.

Once a suitable landing site has been spotted from altitude, the pilot should set up a pattern and make a low approach.

While the length of the landing area is of great importance, in most cases a field that is too short would have been rejected from altitude, so the surface and obstructions are of utmost interest in the dragging process. Fig. 11-20 shows how a pattern might be set up, and you may want to make a sketch of this type on the blackboard.

No more than half flaps are to be used by the inexperienced pilot during the flyby; you may decide that he is not to use any flaps during that pass. Flaps would allow slower airspeed and a longer look but could cause problems for the student if he gets too slow and suddenly sights an obstacle requiring a quick climb and go-around. Once the pilot has decided the landing can be made, full flaps are used for the actual approach and landing.

The student should be encouraged to land on the longest length of the field, taking the wind into consideration. The example in Fig. 11-20 shows that a landing diagonally across the field takes advantage of both wind and field length.

The student may tend to get *low* during the checkout pass, as he is awed by this version of legal buzzing. Watch for this and remind him that he must keep an eye out ahead to avoid flying into some object (Fig. 11-21).

Emphasize that he may check several fields before getting one that is satisfactory. All other factors equal, a field should be picked near a road or houses, but this should not be an overriding factor. You might also mention that he may be faced with two very different pressure situations. One type occurs when the fuel is low and the weather is getting bad enough that he feels that he's "boxed-in," so he may pick an unsuitable field even if there are better fields nearby. He checks a field and finds it to be very rough or with ditches but feels he has to get down immediately anyway. In some circumstances, depending on the rate of deterioration of the weather, a crash landing with aircraft damage but little or no injury to the occupant(s) is better than getting in a full IFR situation where the chances of a fatal accident could be high. It's hard to be able to cover all of the possibilities of weather and terrain,

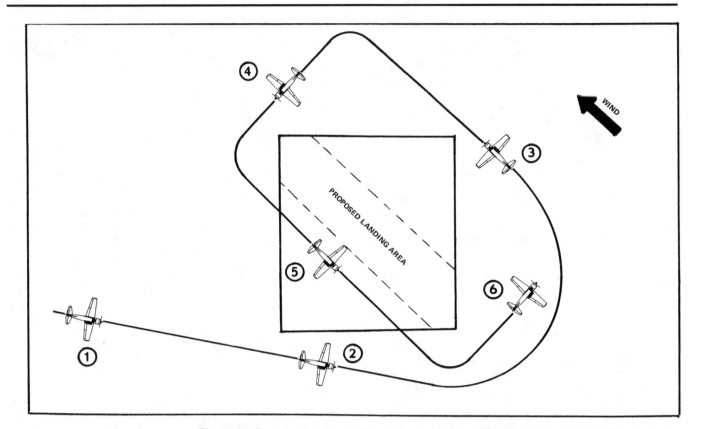

Fig. 11-20. Dragging the area. (1) The pilot sees what appears to be a suitable field. (2) He looks the field over from pattern altitude as he approaches the downwind leg. (3) He flies a wide downwind leg for the approach. (4) A power approach is set up and the airplane is descended to about 100 ft above the surface (if a previous sighting of higher obstacles earlier hasn't precluded this minimum). No more than half flaps are used in the approach. (5) The airplane is flown at a speed about 1.3 times that indicated by the bottom of the green arc (calibrated airspeed marking). (6) If the surface and approach path seem satisfactory, a wide pattern with a short-field approach and soft-field landing is made. The low pass can be repeated if the first one didn't give a good picture of the situation.

Fig. 11-21. Inform the student that he must not only check the surface but also keep an eye ahead for possible obstructions.

and that's one reason flight instructing is a job with never-ending inputs.

The other situation occurs because of "wishful thinking." The pilot keeps rejecting fields (from altitude or after the first pass) because he doesn't want to make an off-airport landing (understandably) and hopes the problem will "go away." He could procrastinate enough that there's no choice left when fuel starvation occurs.

One more thing: He's on a solo cross-country and is overdue; the weather didn't follow the forecast and has gone to pot. There's no question about the fact that you are getting really concerned (and maybe the chief instructor starts asking questions). It's a bad feeling. If he's well overdue, you call the other airports that he was to fly to and discover that he hasn't shown up at one (or either) of them. You're just about ready to call Search and Rescue (SAR) when you get a collect call. (It sure takes a long time to get to a telephone.) He says, "Mr. Jones, this is Jack." Your reaction during the next few seconds can affect his outlook on flying and your image as a professional flight instructor. *Always,* repeat, *always,* ask if he is all right—*first.* Airplanes can be repaired or replaced, and the fact that *he* is making the call himself should immediately ease the pressure considerably. Keep your cool. This is no time for recriminations; that can come later, if needed, in a debriefing when emotions have cooled. Right now you need to get such things done as take care of the airplane if it has been damaged in an off-airport landing (or even if it hasn't been damaged). Plan to call the closest airport or Civil Air Patrol unit to get someone to guard the airplane from souvenir hunters, and make arrangements for the student's return. You'll have your hands full taking care of all the details in such a situation. The student will be shook up and will depend on you to straighten matters out. (*He* sure didn't want to land where he did and start all of this pandemonium.)

This last point will be touched on again in Chap. 14.

TEACHING EMERGENCY FLYING BY REFERENCE TO INSTRUMENTS

BACKGROUND

When teaching the student pilot to fly the airplane by reference to instruments, emphasize that the training is for use if he *needs* it to get out of a sticky situation and that he is not to start thinking he can push on into bad weather. Your own attitude toward this training will help determine whether he keeps it only as a hole card and makes a 180° turn when necessary or uses it to "fly on a little farther."

This chapter will deal with instrument training as required for the private certificate; more detailed information on instrument training is given in Part 6. (Check Part 6 before instructing instruments to your first few preprivate trainees; there may be some ideas and techniques that you could apply here.)

Whether to teach students using the integrated method (combined visual and instrument references) from the first flight is the question all instructors have to face. Some prefer to use the VR-IR method from the beginning, while others wait until after solo to introduce flying by reference to instruments. Your own preference, the local situation, and the individual students will decide your course of action. You may decide that a particular presolo student is concentrating too much on the instrument, and so you stop work on flying strictly by instrument references until after solo. Or if you didn't start out with the integrated method, you may start as soon as you see that the student can handle it and still keep looking around at other times. It's a good move for the student who is sharp but needs more time in the airplane to develop airmanship before solo; he may be getting a little bored with rectangular courses, etc., and will benefit from a change of pace.

The majority of instructors do *not* teach any hood work until after solo, and this book agrees with that approach. However, the following information is set up so that the order of presentation can be used at any point in the training when you decide IR work should be instituted. You may, for instance, give the student VR-IR work on the Four Fundamentals before solo and then work on specific exercises after solo. Or you may not introduce it until *after* solo. Only in an unusual situation would you give enough hood work before solo to cover five of the six basic maneuvers listed later in this section (straight and level flying is the exception).

What if you're in the air and decide that the student is ready to go from VR to IR flying and there's no hood? Somebody has "borrowed" it; any time you plan a session of IR flight you should check for it. While IR work can continue without a hood, it's better to get him used to it. You'll find a sectional or radio navigation chart and a pair of earphones can work as a satisfactory hood, at least for short periods (Fig. 12-1).

One other problem often discovered after the airplane is airborne is a lack of a cover for the various instruments.

Fig. 12-1. A chart and a pair of old earphones can make a practical (if slightly uncomfortable) temporary hood.

You'll be wanting to have the student or instrument trainee fly the airplane with one or more of the flight instruments covered and so should have material available for covering. If airplanes have a plastic face on the panel (the instruments are recessed), a piece of paper can be readily slipped in (Fig. 12-2). A folded dollar bill (even a five or ten) is a good size to be slipped in to cover individual instruments. It goes without saying that you shouldn't use the airworthiness or registration certificates, even though when torn in half each document would take care of two instruments. For other instrument panels, masking tape may be needed to keep the paper over the instrument. So make sure a hood is available in the trainer, and have a small notebook and tape as needed to cover instruments. There also are suction devices sold commercially that can be used to cover separate instruments; check aircraft equipment catalogs.

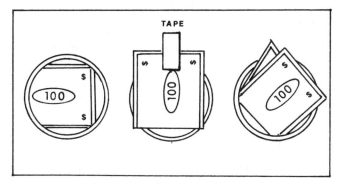

Fig. 12-2. Three ways to cover flight instruments. Of course, a smaller-denomination bill may be used.

The term *turn indicator* will be used in this book to describe the turn and slip *or* turn coordinator, as applicable.

THE FOUR FUNDAMENTALS – AND INSTRUMENT FLYING

Assuming that the airplanes you're using for primary training have a full panel of flight instruments (and the vast majority do), you will teach him "attitude instrument flying." This is basically the same as "attitude plus power equals performance." Make the point that the *rate* instruments indicate rates as they do because of this fact.

Under the hood the airplane is flown exactly as when outside references are used, but your order of introduction of the IR Four Fundamentals will not be the same if instrument training is not taught from the beginning. The suggested order of introducing the student to strictly *visual* flying in Chap. 4 was (1) turns, (2) climbs, (3) descents (and climbing and descending turns), and (4) straight and level—the idea being that this is the best sequence for keeping the student's interest, because in the first three items "things are happening." Your sequence in reintroducing the Four Fundamentals for *IR* work might be: (1) straight and level, (2) climbs (straight), (3) descents (straight), and (4) turns (level)—then climbing turns and descending turns.

■ **Straight and Level.** The reason straight and level is listed *first* in the order is the same reason it was listed *last* in the VR training: it is a static maneuver. Here it allows the student to get accustomed to flying by instruments and developing his scan. Small corrections can be made for altitude and heading, but things aren't continually changing as in a climb or descent or turns. If you compare the first three Fundamentals listed in the approach to instrument training as given in Part 6 of this book, you'll note that they introduce the pilot to *pitch* control. In those first three items pitch will be the primary interest and bank will be secondary. (Have him make corrections to stay on heading, of course.)

The fourth Fundamental will introduce him to bank control as a means of deliberately changing heading. The student should be expected to maintain altitude reasonably well since he has already had the pitch practice, but you won't overdo your corrections as far as altitude is concerned the *first* turn or two.

When you introduce flying by reference to instruments, first have the student fly the airplane without using the hood so that he can see the correlation between the airplane's attitude and the instrument indications. Call his attention to various instruments as each indicates what the airplane is doing.

You might introduce straight and level flight (no hood) and note how the attitude indicator shows even a slight change in the pitch of the nose. Be sure to stress this idea; many pilots never get proficient in using the attitude indicator for pitch control because they think it's hard to read. They haven't been taught to use it right. Note the wing attitude as well but don't call for an exact heading at first. (Remind him to raise a wing if he gets careless.) After he has settled down and is responding properly with the controls, you may use the hood. It's a good idea for you to set the airplane in straight and level flight and let the student adjust the small airplane so that it is proper for *him.* Tell him to set the top of the wings of the small airplane level with the top of the horizon line; it's easier to read this way (Fig. 12-3).

Since at this point you are not working on an instrument rating for him, most of your instruction will be full panel, but you can introduce briefly the actions of the four pitch instruments. Later you may cover the attitude indicator, altimeter, and airspeed, leaving the vertical speed indicator the only

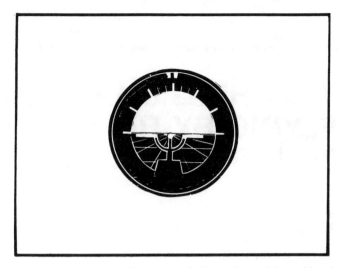

Fig. 12-3. Setting the attitude indicator in straight and level flight.

pitch instrument showing. Have him fly a couple of minutes straight and level by using the heading indicator and vertical speed, and then uncover the altimeter after the prechosen time is up and check his progress. This is a good confidence maneuver and brings the vertical speed to his attention. It's too bad that some instructors ignore the turn indicator and vertical speed indicator in emergency-type instruction and in work on the instrument rating. You could also have some students flying straight and level by using the airspeed as a pitch indicator. The main thing for you to do during this first phase of instruction will be to acclimate the student to the idea. Get his feet wet but don't let the water be too hot or too cold.

When you see that he is becoming at ease using full panel and can hold the altitude within ± 50 ft and the heading within ±10° for 4 or 5 min (with your instruction), move on to straight climbs.

■ **Straight Climbs.** Here, you should start the student out with a definite heading to be held throughout the climb. "Torque" can be an insidious factor in pulling the airplane off heading. You can detect troubles with scan if he continually lets the airplane ease off to the left; he's probably fixed on the airspeed or, less often, the attitude indicator. Don't use the hood at first.

A student may like to chase the airspeed in a climb under the hood, and this can lead to some pretty radical pitch attitudes. He's eased the nose up a little too high, and the airspeed has dropped to 60 K instead of the required 80 K. He eases the nose down and the airspeed moves to 72; he eases it farther down (students get impatient), and so it goes. The attitude indicator has been ignored throughout this—or at least the *pitch* indication has been left out of his consciousness. He sees now that the airspeed is 90+ K, so the process of slowing up starts and the airplane could approach a stall before you take over.

There is a proper pitch attitude for every climb airspeed you choose for your airplane, of course, but you should stick to one recommended airspeed and its corresponding pitch indication on the attitude indicator. Whatever the indication is, set it up and demonstrate that the proper airspeed results if climb power is being used and, conversely, that pitch attitude results if the desired airspeed is maintained. Set up one of the parameters while covering the other instrument and then uncover and show the results. Fig. 12-4 shows a typical pitch attitude and airspeed combination for a general aviation trainer.

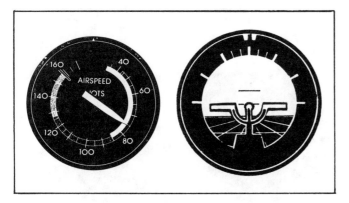

Fig. 12-4. Airspeed and attitude indicator readings for a climb (full power) in a fictitious trainer. Show the student the comparison of the two instruments in your airplane.

The point to make is that the student is not to chase the airspeed in a climb. If he is having trouble with an airspeed that's too high or too low, he is to set up the proper pitch attitude with the attitude indicator, and the airspeed will settle down of its own accord.

You may want to show a sharper student how to use the vertical speed indicator to demonstrate cause and effect. For instance, in your trainer in the altitude range you are flying (say, 4000–6000 ft MSL), you may expect a rate of climb of about 600 ft per minute at recommended climb power. In smooth air you can cover the airspeed and attitude indicators and demonstrate that by holding a rate of climb of 600 fpm the airspeed and attitude indicator are normal (uncover them after the VSI indication has stabilized). And, naturally, you can demonstrate that *that rate of climb* results if the airspeed and/or pitch is held to the proper indication. This is a good way to introduce the vertical speed indicator and bring it to his attention. You should spell out that in turbulent air the VSI indications will have to be averaged.

Also demonstrate that keeping the wings level in the climb, as per the attitude indicator, does not necessarily result in a constant heading. Many students (and more advanced pilots, too) keep the wings level nicely with the attitude indicator by using aileron, but "torque" can move the nose around to the left if the proper amount of right rudder isn't exactly held and the heading indicator is ignored.

Elevator trim is very important in maintaining the proper climb speed. Impress on the student that a properly trimmed airplane will maintain its attitude if the pilot's attention is distracted briefly to charts or radio work.

In leveling off, he should ease the nose down to the level-flight position shown by the attitude indicator and hold it there. Forward pressure should be relieved by trim. As the airspeed approaches the cruise value, the power should be reduced to cruise setting. This is the same procedure as used in leveling off from the climb under visual conditions; the attitude indicator replaces the natural horizon.

Here are some *common errors* students make doing straight climbs by reference to instruments:

1. "Overrotating," bringing the nose up too rapidly as the climb is entered. This is usually caused by the student not checking the attitude indicator as the climb is entered; he is depending on the airspeed indicator. Remind him to ease the attitude indicator to the proper attitude, and the airspeed will take care of itself. You may also have to nudge him to go to climb power.

2. Directional control problems.

3. Chasing the airspeed.

4. Poor leveling-off technique. Expect that he will let the nose creep back up during the leveling process, usually because his trimming isn't catching up with the airspeed increase (and climb power is still on).

■ **Straight Descents.** You can use a little psychology here by bringing up the idea that "500 ft per minute is considered a standard rate of descent for all instrument work." (The student then pictures himself in the left seat of a 767 on final approach.) You are bringing him further "into the group" and giving him the idea that he will go on and get an instrument rating—or ATPC—and fly IFR like a pro.

Pick a descent airspeed for your trainer that is slightly above the normal approach speed to assure good control in turbulence, but don't let it get so fast as to cause structural problems or difficulty after he breaks out of the clouds. For one trainer a speed of 80 K is a good figure between the extremes of stall and overstress, and 1900 rpm serves to set up a 500-fpm descent. "Nineteen hundred and eighty" or "nineteen and eighty" is easy to remember. If the student, or later the private pilot, gets into bad weather he may tend to panic, so in all of this instrument training your aim will be to give him something to hang onto. Hopefully, he'll remember "nineteen and eighty" as an old friend that he used with you to make a controlled descent, or conversely, he'll use the pitch attitude for a climb that you showed him. In an actual situation he'll need something to get back to, something familiar to build his confidence. The tendency of the noninstrument pilot who gets into the soup is to stab around using different power settings and airspeed to descend—with a strong leaning toward loss of control. Emphasize that prechosen and practiced numbers or airplane attitudes are what he must use. So, you'll have several reasons (one of them ease of remembering) for setting up a descent speed and power setting for a 500-fpm descent.

Make it clear that he won't usually be able to hold *exactly* a 500-fpm descent in an emergency situation, but a 400- to 600-fpm range will get him down with good control. You aren't aiming at having him settle for a less-than-safe rate of descent, but you don't want him chasing the vertical speed indicator trying to get an exact rate of descent.

Here you can demonstrate again that if the proper pitch attitude—or proper airspeed—is maintained with the right amount of power, the desired rate of descent will result. You can illustrate this by covering up one at a time the tach, airspeed, attitude indicator, or vertical speed, and show the relationships. As an example, you could cover the tachometer and set up the descent airspeed and/or pitch attitude. Add or decrease power as needed to get 500 fpm, then uncover the tachometer and see that 1900 rpm (or whatever rpm your airplane requires) is indicated there. By doing this sort of demonstration you'll increase his confidence, and he will also understand and remember better—when he has to (Fig. 12-5).

You should point out that in turbulent weather the vertical speed and the airspeed may be fluctuating, and he should set the power and try to hold the proper pitch attitude. He may see the vertical speed at 1000 fpm down, which could be caused either by a downdraft or by his unconscious lowering of the nose. He is *not* to chase the vertical speed but should check the power and attitude. If both are okay, he holds what he has. If the nose is low as shown by the attitude indicator he makes it right and doesn't worry abut the vertical speed, which will take care of itself. However, you'll find that students often become fascinated by the vertical speed indicator once they've been introduced to it and start "flying" it—with interesting results.

PROCEDURE. Have the student get organized on a cardinal

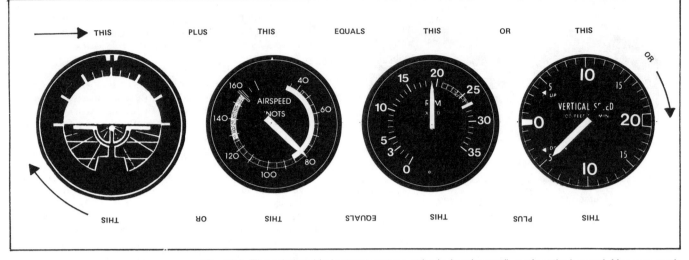

Fig. 12-5. The relationship between power, attitude (or airspeed), and vertical speed. You may work several combinations between the instruments.

heading and reduce the power to that required for a 500-fpm descent.

The nose should be held in the level-flight position until the descent airspeed is reached. The nose is lowered as necessary to maintain that figure. Power is readjusted and the airplane trimmed.

To level off, cruise power is added and the nose is brought to the level-flight attitude. Power and trim are adjusted as needed, as the cruise speed is reached. Note that the steps in the level-off procedure are the same as those of the visual.

> *Common errors* made in instrument descents are these:
> 1. Poor transition to the descent. The usual problem is to let the airspeed stay too high.
> 2. Directional control problems.
> 3. Chasing the airspeed or vertical speed.
> 4. Letting the nose rise as the power is added at level-off. (Not resetting trim fast enough.)

■ **Level Turns.** *Shallow* constant-altitude turns should be the next of the Four Fundamentals introduced. The student has been using the bank instruments in leveling the wings and maintaining straight and level flight so that it won't be completely new to him. As was the case for the straight and level and other work, the first part of the practice should be without the hood.

You might start out with a minute of straight and level flight and then direct him to set up a shallow turn in either direction. Don't have him roll out on a predetermined heading, but have him do it at your "command." Work on the maintenance of a constant altitude and a constant shallow bank. Expect him to start having minor problems with altitude in the turn, and concentrate on working this out first rather than insisting on a specific heading for roll-out. As he begins to see the pitch changes required to maintain altitude in that shallow turn of, say, 10°, start him on 15°, then 30°, and finally 45° banks (still no requirement for a specific heading), then work under the hood. Intersperse turns under the hood with straight and level hood work.

Fig. 12-6 shows a constant-altitude exercise you can use after the student has smoothed out his turns and straight and level work and is able to maintain altitude within ±50 ft. You may prefer to skip this with some students and move on to (and stay with) standard-rate turns.

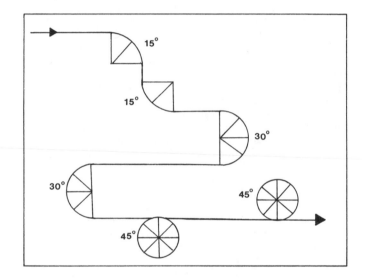

Fig. 12-6. An exercise to get the student to work out problems with altitude control in turns of various banks. A constant altitude is to be maintained.

Another exercise you may use to aid the student in use of the attitude indicator in the turn is to have him make a turn of a certain angle of bank (not over 30°) with the altimeter covered for 1 or 2 min. Have him roll out at your say-so, and uncover the altimeter. Or you may have him turn for 2 or 3 min and uncover the altimeter a couple of times during the turn to check progress. You may expect the majority of altitude problems to pop up during the roll-in and/or roll-out. Fig. 12-7 shows a comparison of pitch attitudes as shown on the attitude indicator in level cruising flight and in a 45°-banked, level turn.

STANDARD-RATE TURNS. Once the exercises such as the 15°-, 30°-, and 45°-banked turns have been completed (if you used them), the idea of the standard-rate turn should be introduced and promoted. The student should understand that he must never deliberately exceed a standard-rate turn of 3° per second when flying under actual instrument conditions; there's too much chance of a loss of control at steeper angles of bank.

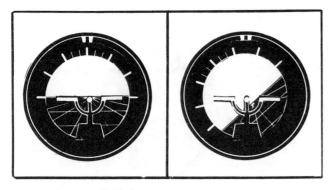

Fig. 12-7. Comparison of pitch attitudes in straight and level and in a 45°-banked level turn. You can show the student the proper attitude indicator perspective for your airplane.

Demonstrate some standard-rate turns and show the comparison of the bank angle as given by the attitude indicator and the turn indicator. The rule of thumb of dividing the airspeed (mph) by 10 and adding 5 to get the required angle of bank in degrees can be demonstrated.

Since you are likely to be working in terms of knots, in that case divide the airspeed by 10 and add one-half of the answer. For instance, at 130 K you would divide 130 by 10 to get 13, add one-half of 13 (6½) to 13, and get a required 19.5° bank necessary for a standard-rate turn (call it 20). For 160 K, you'd get 16, plus one-half of 16, to get a bank requirement of 24° (call it 25). This rule of thumb is most accurate in the range of 100–170 K. For *emergency* training such as this, in some of the faster trainers you might hold the maximum bank to 15° on the attitude indicator and forget about standard-rate turns.

After he's seen your demonstration and practiced a few standard-rate turns full panel, cover up the heading and attitude indicators and have him practice a couple of level, timed turns using the turn indicator. He should start on a cardinal heading, with you doing the timing for 180° or 360° turns. A 90° turn takes too short a time to allow the student to settle down, but you may have him do a few after he has the grasp of timed turns.

Require several 180° and 360° timed turns with the turn indicator covered, using the *attitude indicator* and the bank as established by the rule of thumb. Uncover the heading indicator after 60 or 120 sec, as applicable, and check the results.

In addition to the timed turns you might have him fly the airplane using the heading indicator for bank and heading information. With the attitude indicator and turn indicator covered, turns may be made to specific headings and the rates of turn estimated (keep it slow). You can demonstrate also that the wings can be kept level for straight flight by checking the heading indicator for signs of direction change.

In all of the level turns, timed or otherwise, watch the student's altitude control.

Common errors in level-turn practice include these:
1. A tendency to gain altitude in the first shallow turns (overcompensating for the expected altitude loss).
2. Losing altitude in the 30°- and 45°-banked turns.
3. "Losing" the required roll-out heading in the exercises requiring a specific heading.
4. Poor bank control. It's part of the scan development for the student to maintain a constant bank during the turn. Expect a tendency to shallow out of the 45°-banked turns.
5. A tendency to gain altitude when rolling out from the steeper turns. Have him move the reference pip of the attitude indicator down to the horizon bar in the roll-out.

■ **Climbing Turns and Descending Turns.** Demonstrate and have the student practice climbing and descending turns after he is able to handle the Four Fundamentals. For example, using the full panel you could start at an even altitude and descend 1000 ft and turn 360° in a 2-min timed turn and 500-fpm descent, explaining how to "catch up" or "slow up" the turn and/or descent. And you might have him do the same exercise with the heading indicator and altimeter covered. At the end of exactly 2 min have him simultaneously roll out and level off. Uncover the two instruments and see how things worked out. Review the maneuver and note what might have caused errors in heading or altitude. If he is very close, mention this also (he "held it steady throughout," etc.). If he is 20° off heading and 150 ft off altitude at the end of the exercise, remind him that this is *emergency* training and even errors like that are within safety bounds—if he should ever need to use the method. It's admitted that 360° turns aren't the best for reversing course to get out of bad weather, but it's a good training exercise anyway. Encourage him to work for more precise results in the exercise just mentioned.

You may use the same general procedures for climbing turns, also.

SIX BASIC MANEUVERS

Before the student goes on a solo cross country, he should have had practice in, and be able to safely complete, periods under the hood performing the following maneuvers or flight regimes (he may also be required to demonstrate some of them on the private flight test):
1. Straight and level flight.
2. Left and right 180° turns to within 10° of the preselected heading.
3. Shallow climbing turns to a predetermined altitude and heading.
4. Shallow descending turns to a predetermined altitude and heading.
5. Recovery from the start of a power-on spiral.
6. Recovery from the approach to a climbing stall.

These maneuvers are given in recommended order of introduction for instructing purposes. As you introduce each one, indicate how it can be actually used in flight in an emergency situation; these aren't just pure training exercises, but have purposes for survival. List them on the chalkboard in the above order and discuss them *all* before flying.

■ **Straight and Level Flight.** Straight and level flight was discussed and practiced earlier as an exercise. Here you want to bring out the idea that it is an important part of survival in an actual weather encounter.

The ability to fly for a number of minutes maintaining a constant altitude and heading is of paramount importance. You should tell the student that to get into weather and make a perfect 180° level turn—and then to wander all over hell's half-acre trying to get back to visual conditions is no way to operate. The longer the student or low-time pilot stays in instrument conditions, the more he is apt to panic. In your instructing you will see students who turn as far as 90° off heading while trying to fly straight and level. In the real circumstance, the tendency to let the airplane get out of control will be increased because of fear or tension.

Make a humorous but emphasized point about how long it takes to fly out of bad weather, so that in actual weather the student will remember and his tension will be eased (somewhat). One instructor used to say to each of his students when these specific maneuvers were introduced: "You fly into weather and are there 10 to 15 seconds before starting the 180° turn. After the turn is completed, you then fly 2 hours and 45 minutes to get back to good weather." It sure seems that way to the guy who wishes he hadn't been so careless as to get in the solid stuff.

At first you might set your own limits for his work—of ±50 ft and ±10° of the chosen altitude and heading. Always work for perfection as he gets more experience.

As part of straight and level training have the student (under the hood) intercept and track to or away from a VOR on a prechosen radial, or home to a radio beacon or a commercial broadcast station using the ADF.

The most common error here is "weaving" across the heading, overcorrecting for minor errors.

■ **180° Turns, Left and Right.** A variation of ±10° from the selected heading may be allowed on the flight test, but you should work for total accuracy with each turn.

Point out that the 180° turn is used to get out of the weather, and he's being taught to do it by reference to instruments in case he hasn't been keeping up with things and it's too late to make the turn under visual conditions.

Indicate that he may *start* the turn under visual conditions but immediately run into clouds, and he is to *stick to the gauges* to assure that control is maintained. Looking for the ground in this circumstance could also mean that he makes a 360° turn instead of a 180° turn (Fig. 12-8).

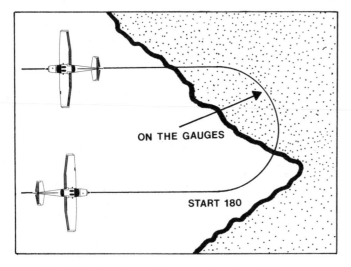

Fig. 12-8. Even though the pilot started the 180° turn before entering the clouds, he'll have to complete it by reference to the instruments.

Point out that a 180° turn is made if the airplane *hasn't turned* after visual references are lost. *The heading to which the pilot should turn is the reciprocal of the original course, and he should keep this number in mind.*

In this early stage of instrument training, don't press too hard on the detail of turning the shortest way to a heading if it's nearly 180° away. If he turns 210° to get to an escape heading instead of doing a 150° turn, that's all right. (If he turns 340° *left* to get to a prechosen heading, instead of using a 20° *right turn,* that's a different matter.) The heading indicator is a good aid in checking which way to turn to a near-reciprocal heading (Fig. 12-9).

Demonstrate and have him practice 180° turns using the turn indicator, airspeed, and altimeter (and clock) only.

At this point you also should demonstrate the reactions of the magnetic compass in a turn and how it could be used to turn to various headings if the lead and lag are known. (Keep the bank below 20°.)

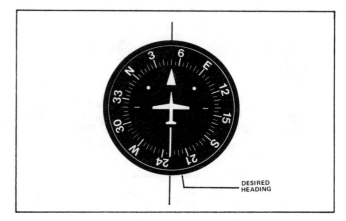

Fig. 12-9. Remind the student that the heading indicator will show the shortest way to turn.

■ **Shallow Climbing Turns to a Predetermined Altitude and Heading.** The pilot can continue turning as he climbs (maybe making several 360° turns as he moves up to that altitude), but as he climbs it would be best if the airplane is being turned to a heading that will move it back into VFR conditions. Basically, you could set up a problem such as this: The pilot flies into the bad weather and not only needs to get altitude to clear obstacles but also wants to reverse course. Or perhaps altitude is needed because he got into a power-on spiral after entering the clouds.

Before starting the exercise, indicate that a particular altitude *and* heading must be attained. In the descending turns discussed earlier, a canned problem of a 360° turn and a required descent of 1000 ft at 500 fpm was given. With luck (and skill, of course), it worked that after 2 min the airplane had made a perfect 360° turn and was exactly on the required altitude. That was a good training exercise, but you'll get more response from the student now by setting up a situation that he may actually put to use later.

The point should be made that he may get to the required altitude before the heading is reached, or vice versa, but he should continue to complete both requirements. If the altitude is reached first, the airplane should be leveled off and a level turn continued as necessary. If the turn is completed first, the wings should be leveled and a straight climb continued to the required altitude.

You can expect the student to sometimes let the item first completed get out of hand while he concentrates on the second.

Remind him to keep that scan going.

■ **Shallow Descending Turns to a Predetermined Altitude and Heading.** This maneuver has a twofold value. First, it may be needed if an approach to a stall occurs after flying into instrument conditions—the pilot has climbed well up into the murk, and now has to descend and turn to get back out (Fig. 12-10). Second, it is good practice of a radar-controlled letdown which may be necessary in actual weather conditions.

You should act as "approach control" at the end of several dual flights and direct the student on an "ASR" or "PAR" approach. (You may have been working on strictly visual maneuvers throughout that particular flight but can have him put on the hood for the return.) Also, in one of the dual flights into a controlled field later you might ask for a practice ASR approach if it's available so that the student can get used to working with the actual procedures.

Assume that he has gotten into the position shown in Fig. 12-10. He has to turn back to the prechosen heading (reciprocal of the original course) and descend to the original altitude,

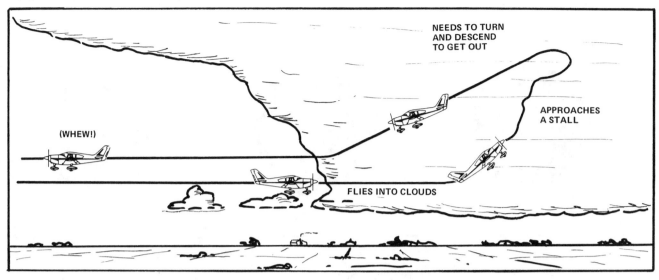

Fig. 12-10. The requirement for a descending turn to a predetermined altitude.

or slightly lower if safe, to get clear of the clouds.

He should by this time automatically set up the power and airspeed for a 500-fpm descent and should use a standard-rate turn. Indicate that in a real situation his biggest problem will be to avoid extra-steep turns and above-normal rates of descent because the pressure will be on to "get out of this mess" as soon as possible. He may have to take a couple of slow deep breaths and use the numbers that he knows will work. Like flying back out of bad weather, flying back *and* descending can seem to take a very long time.

The most likely error he'll make during this maneuver will be to overshoot the heading, followed closely by a tendency to descend well below the recovery altitude as he stops concentrating on altitude and starts concentrating strictly on heading corrections. You'll notice as you instruct preprivate instruments that you often have to remind the student not to get stagnated on one instrument. Teach him that if trim or radio frequency setting requires a division of attention, he should trim some, check the flight instruments, then go back to finishing the job, doing this procedure several times if necessary. Loss of control results when too much time is spent away from the instruments. Point out that if the airplane starts a power-on spiral, the pilot may not realize it because the forces acting on him will still be balanced.

Again emphasize that the exercise is complete when the airplane is on the prechosen heading and attitude. He may have to turn 90° (at 3° per second) and descend 1200 ft (at 500 ft per minute) so that he'll roll out on the heading and continue the straight descent, or vice versa. Like in the climbing turn to a prechosen altitude and heading, the student may forget about the heading (or altitude) once it's been reached and put too much concentration on the other requirement.

■ Recovery from the Start of a Power-on Spiral

FULL-PANEL. On the flight test the student may use the full panel of flight instruments for hooded flying if they are available, and you should certainly encourage him to do this.

The spiral is the most common result of loss of control. Nearly all general aviation airplanes are designed to be slightly (?) spirally unstable; if they are left alone in a bank, the bank will tend to become continually steeper. This is better than having too much dihedral, which would make it more spirally stable but could cause bad Dutch roll characteristics.

The attitude indicator will be the primary instrument here for discovery and recovery.

Depending on how far matters have progressed (i.e., how high the airspeed is), the wings may be *simultaneously* leveled *by reference to the attitude indicator* as the nose is brought up. This is the best method if the spiral has just started.

A rolling pull-out puts more stress on the airplane than a straight pull-out; but this depends on airspeed, rate of pull-out, and roll rate during the process. A rule of thumb for *rolling pull-outs* is not to pull more than two-thirds of the limit load factor because of the bending *and* twisting of the wings in that type of recovery. In other words, if the positive load limit is 4.4 g's, as for a utility category, in a rolling pull-out, 3 g's should not be exceeded. There's one problem, however; most airplanes you'll be instructing in won't have a g meter, but the point is that you should use caution in a high-speed rolling pull-out. The most judicious method for the student in the *start* of a power-on spiral is to bring the nose up as the wings are leveled, using the attitude indicator (Fig. 12-11).

Fig. 12-12 shows a procedure that would also be used for partial-panel recovery from a power-on spiral, although it is

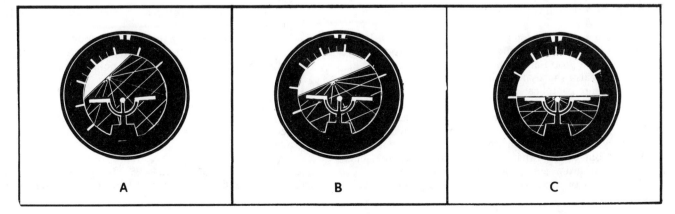

Fig. 12-11. Using the attitude indicator to roll out and pull up from a spiral simultaneously.

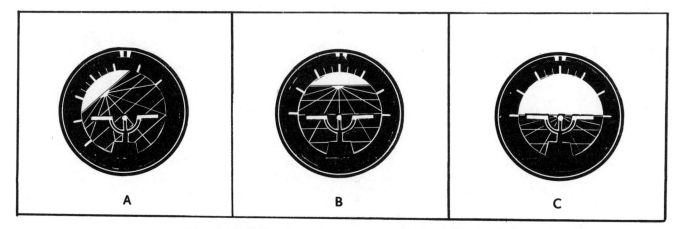

Fig. 12-12. (A) The airplane is in a nose-low, banked attitude (the airspeed is high and/or increasing). (B) The wings are leveled using the attitude indicator as a reference. (C) The nose is then raised to the level flight attitude (and the wings *kept* level). As will be shown later, these are the steps used for emergency, or partial, panel recoveries (turn indicator and airspeed), except with different references.

not as efficient a method. This procedure could naturally lead on to turn indicator and airspeed recovery practice.

The student will usually allow the nose to come up well past level unless he is reminded that the excess residual airspeed from the dive will tend to bring the nose up.

Okay, what about throttle handling as part of the recovery process? Do you close the throttle before the recovery is started? Power should certainly be retarded in the fixed-pitch prop to keep from exceeding the rpm red line. However, closing the throttle may hurt recovery as well as cause backfires and other problems (cooling or carburetor ice) in exaggerated conditions because in an actual situation he will be in cool or cold and moist clouds.

Here's an experiment for *you:* Take your trainer up and set up a spiral condition of, say, 45° bank starting at a particular altitude. Leave the power at cruise rpm. Take your hand off the wheel as it starts but keep one on the throttle. When a prechosen airspeed is reached, start a recovery. Retard power slightly. When the airplane is level, check the altitude. Don't let the airspeed get high enough to cause engine overspeeding. Climb back to the same altitude as before, set up the same bank with your hand off the wheel but the other on the throttle, and set up the same spiral condition with power at cruise and a 45° bank. When the previous airspeed is reached, close the throttle and complete the recovery to level flight. Compare the two altitudes at recovery.

That's a crude test and you may set up your own, but it shows that closing the throttle usually means a steeper angle and greater rate of descent. The recovery might have to be drawn out to see any significant difference in altitude loss. There are advantages in a not-so-abrupt power change that have to do with pitch effects and trim. You may decide that *closing* the throttle is always a good first step in a spiral recovery.

The constant-speed prop gives a good argument for leaving the power alone. Some argue that closing the throttle with a constant-speed prop flattens the blades, causing more drag, but this could also shortly result in a steeper pitch-down attitude and a more complicated power adjustment after the recovery.

Getting back to the fixed-pitch prop, the less change made in power, the less power adjustment will be required as the recovery is completed (don't let the prop overspeed, however). Find out exactly what is best for your airplane, but always lean toward a general procedure for all airplanes.

Indications of a power-on spiral as shown on the full

panel are shown in Fig. 12-13. You might explain to the student that the indications are these:

1. Airspeed high and/or increasing.

2. Attitude indicator shows left bank and nose-down attitude.

3. Altitude decreasing.

4. Turn indicator shows left turn, and ball may or may not be in center. (It will probably be to the left if the airspeed is high when the problem is discovered.)

5. The heading indicator will be showing a turn, or if the bank has exceeded limits the card may be spinning.

6. Vertical speed indicator shows a high rate of sink.

Recovery steps are these:

1. Adjust power if necessary. (Usually a good reduction is required to avoid fixed-pitch prop overspeeding.)

2. Level the wings using the attitude indicator.

3. Bring the nose up to the level-flight attitude by reference to the attitude indicator. Hold that pitch attitude keeping the wings level.

4. As the airspeed moves back down to cruise, adjust power as necessary.

5. Reset the heading indicator, if necessary.

6. After the airplane is under control, climb and turn to a prechosen altitude and heading and establish cruise.

Before setting up a spiral condition for the student, always require a base altitude and heading for him to return to. It doesn't have to be the numbers you started from but should be something that would require a climb of 1000 ft or less and a turn of up to 180°.

Common errors in the recovery from a power-on spiral are these:

1. Letting the propeller overspeed (fixed-pitch prop).

2. Overcontrolling the bank on recovery.

3. Letting the nose move up past the level-flight position. You'll have to continually point out this problem during the first few recoveries.

4. Failing to keep the wings level after the recovery; in some instances a new spiral is started.

5. The heading or altitude problems mentioned earlier in the climb and turn to the prechosen heading and altitude.

EMERGENCY PANEL (PARTIAL-PANEL) RECOVERIES. The student should be shown this method of recovery even if the

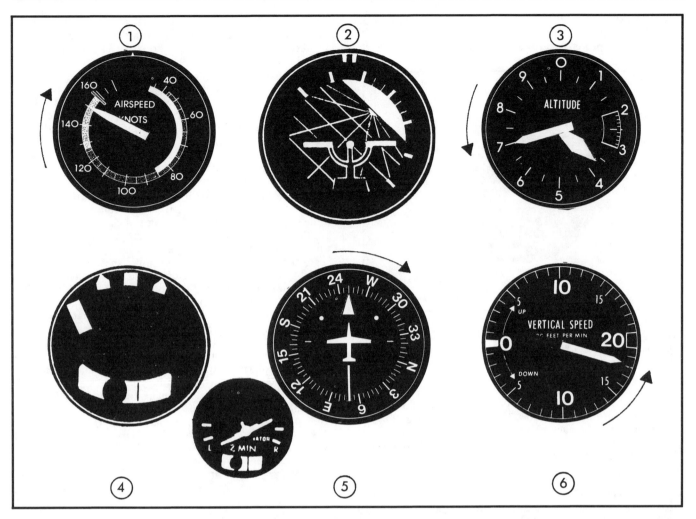

Fig. 12-13. The instrument indications in a power-on spiral. You might make a sketch on the chalkboard to show what each instrument is doing. The turn is to the left so the *card* in the heading indicator will be turning to the right.

airplane has a full panel. He might someday be caught in a circumstance where the attitude indicator has tumbled or be in an airplane that doesn't have the full panel.

Fig. 12-14 shows the instruments you might sketch on the chalkboard. The *indications* (point these out on the board during the preflight explanation) are these:

1. The airspeed is high and/or rising.
2. The altitude is showing a rapid drop.
3. The turn indicator in this example shows a turn to the left, and the ball will likely be to the left at higher speeds because of a right-yawing tendency.
4. The vertical speed indicator, if available, will show a high rate of descent.

You should point out that (1) trying to pull out of the dive without leveling the wings or (2) trying to level the wings while simultaneously pulling the nose up isn't the way it's done. The first procedure won't work because pulling back on the wheel *without* leveling or starting to level the wings makes matters worse (tightens the turn and increases the rate of descent) whether on full panel, full visual references, or partial panel—it's a matter of physics. As for indicator number 2, the simultaneous type of recovery, it's too difficult for the low-time pilot to accomplish efficiently without the full panel, so the extra step is included as shown for the attitude indicator in Fig. 12-12.

The *recovery* (a model can be used to back up your chalk-

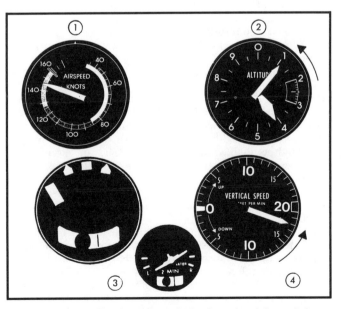

Fig. 12-14. The instrument indications (partial-panel) in a power-on spiral.

board drawings and the following explanation) steps are these:

1. Adjust power (retard) as necessary (Fig. 12-15).

Fig. 12-15. Retard the throttle to avoid overspeeding in the fixed-pitch prop, as shown here. This action with a constant-speed propeller will tend to make it go into a flatter pitch, increasing drag, but may complicate power handling later.

2. Center the needle and ball (or turn indicator) through the use of coordinated controls. (This levels the wings.)

3. Increase the back pressure to bring the nose up. (Explain that the nose will tend to come up of its own accord as the wings are leveled, so he shouldn't "haul back" on the wheel or stick for this reason, as well as structural ones).

4. Check the airspeed. Because there is no attitude indicator to see the pitch attitude, the airspeed indicator is the first reference after the pull-up begins. When the airspeed changes (stops increasing or decreases if it has been high and steady) the back pressure should be stopped. The nose at that instant is approximately level.

5. Check the altimeter; the altitude (to the nearest 100 ft) should be "flown" using the elevators. (Forward pressure may be needed at first.)

6. Keep the turn indicator centered.

7. Adjust power as necessary as the airspeed returns to cruise.

You may require that the attitude indicator be kept covered during the turn and climb to the predetermined altitude and heading after recovery. (The heading indicator may be quickly reset by the mag compass after the latter settles down.) Or, you may let him use the full panel to go to the chosen numbers. While it's doubtful that the altitude indicator would be tumbled in an actual incident, you should give him at least one climb and turn without it so at least he knows the procedure if it's ever needed.

Common errors made in partial-panel recoveries are these:

1. Overcontrolling the turn indicator. The usual problem is an overshoot of the indicator in rolling out, followed by a strong correction back, and so on. An interesting thing to watch for is that while the student is juggling back and forth trying to get the wings *exactly* level, the nose is rising (a normal tendency as you can demonstrate). So by the time the wings are leveled according to his satisfaction, he is in an approach to a stall (recovery procedures for which are covered in the next section).

2. Missing the airspeed change. The student is still staring at the turn indicator when the great event occurs. Remind him that he is not to chase the airspeed *or* the vertical speed.

3. Overcontrolling the elevators when trying to pin on a particular altitude. Sometimes he will forget which altitude he wants (for instance, the airplane is at 4200 ft and he goes from 4100 to 4300 ft trying to decide).

4. Failing to keep the turn indicator centered, ending up in another spiral, requiring another recovery.

5. Poor scan during the turn and climb to the pre-chosen heading and altitude.

Your procedure in setting up the unusual attitude will be to have the hooded student look away from the instruments while you maneuver the airplane and when you say, "You take it," he takes over and recovers. Some instructors have the student take his hands and feet off the controls and put his head down while the instructor pulls the airplane up and down and banks it in both directions, then finally lets the student have control of the airplane when confusion reigns supreme. This teaches the student to recognize the problem quickly and to start a recovery move. It's true that if he's been looking at charts, trying to find a pencil under the seat, or has other problems requiring a long period of being away from the instruments, he may return to some wild indications, but the usual problem is a slow entry into a spiral, which could have been corrected easily had he come back to the instruments every few seconds. So make the exaggerated entries, but tell him that he will get into these final positions by a comparatively long period of inattention. *If he continually makes small corrections as needed he'll never suddenly have to make a big one.*

■ Recovery from an Approach to a Climbing Stall.
This is the second of the unusual attitude recoveries that may be required on the private flight test. It could be the result of an improper recovery from a power-on spiral or an exaggerated climb attitude (he was just generally not paying attention). You'll find that some students create a great many unusual attitude recoveries by going from a diving spiral to an approach to a stall, back to a spiral, and so on. You'll have your work cut out with these people.

The chalkboard and a model are the aids for the preflight explanation.

FULL-PANEL RECOVERY. Draw a quick version of Fig. 12-16 on the chalkboard, and describe the situation. The *indications* of an approach to a climbing stall are these:

1. The airspeed is low and decreasing.

2. The attitude indicator shows a nose-up attitude. The wings may or may not be level. In this example, assume they are not.

3. The altitude is increasing, or if the stall is imminent, it may be steady.

4. The turn indicator shows a turn to the left, and the ball is to the right. This is usually caused by "torque," since it's assumed that the entry is accidental and hence no corrective right rudder is being held.

5. The heading indicator will (probably) be showing a left turn, because of uncorrected "torque" effects.

6. The vertical speed indicator will show a high rate of climb.

These clues are what might generally be expected. However, the airplane may be indicating a *right* climbing turn because the pilot is subconsciously holding right rudder or has established a climbing turn to the right. So, the student is to correct for the approaching stall whether the airplane is banked or the wings are level.

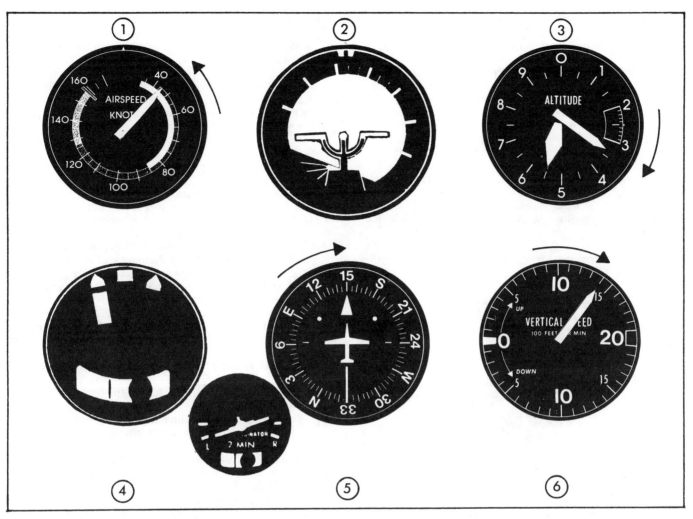

Fig. 12-16. Instrument indications in an approach to a climbing stall.

The major problem here is the low airspeed, and unlike the power-on spiral, the *pitch* correction will be the first move for the pilot to make after he takes care of the power detail. Fig. 12-17 shows how the attitude indicator could be used in a simultaneous type of recovery from the approach to a climbing stall.

One disadvantage in leveling the wings first and then lowering the nose is that a further pitch-up tendency may be present, and the recovery could be delayed by an extra step. It is a good training technique to be able to say, "In both the spiral and the approach to a stall after the power is adjusted, the first move will be to level the wings, then adjust the pitch attitude as required." The student could bring the recovery procedures back to mind, but they wouldn't result in the most efficient means of recovery from the stall approach. You might cover the following recovery steps, referring to the

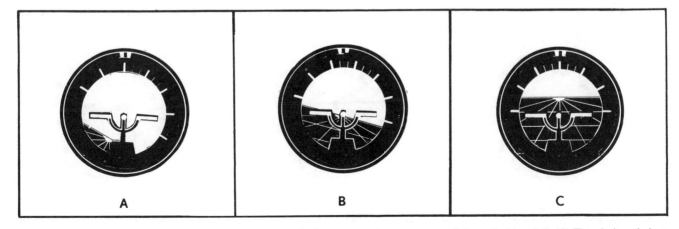

Fig. 12-17. The steps in the recovery from an approach to a climbing stall. (A) The airplane is in a steep climbing turn to the left. (B) The nose is lowered as the wings are leveled. (C) The wings are level, but the nose is lowered past the level-flight position to avoid a secondary stall.

sketch of the instruments on the chalkboard. (It's assumed the condition is entered from cruise, at cruise power.)

Your step-by-step explanation of the *recovery* might be as follows:

1. Add full power. This lowers the stall speed and gives a greater margin for recovery. Warn the student that the added power will tend to make the nose pitch up higher (Fig. 12-18).

2. As power is added the nose is simultaneously lowered and the wings leveled. The attitude should be that shown in Fig. 12-17C.

3. As the airspeed starts to increase, ease the nose back up to the level-flight attitude.

4. Adjust power to the cruise setting as the cruise airspeed is reached.

5. Reset the heading indicator as needed.

6. Descend and turn to the predetermined altitude and heading.

Fig. 12-18. Stall recovery steps.

Note that nothing was mentioned about trim during the recoveries from either the spiral or approach to a stall. It's assumed that the airplane got into these circumstances from cruise power and trim, so things will be back to normal after recovery. The airspeed and wheel forces are changing so fast that trying to use trim would be futile and could distract attention from power requirements or instrument scanning.

It's more important to have the nose down than to have the wings leveled concurrently with the pitch change. Any

added heading change caused by the bank can be taken care of later. (In VFR *stall* recoveries, the wings are leveled after the nose is lowered, but this is a recovery from an *approach* to a stall.) Fig. 12-19 shows another look at a recovery from a stall when it is done by reference to the attitude indicator.

In the "straight" recovery, as in Fig. 12-19, a respectable slip will occur; in a left bank the ball in the turn indicator will be well to the left. If the "pushover" is abrupt, the slipping feeling will be intensified and may be uncomfortable.

It's best, after taking all factors into consideration, to recover from an approach to a climbing stall by simultaneously lowering the nose and leveling the wings if the full panel is available (Fig. 12-17). The other factors were mentioned because sooner or later a student will ask about them, and you'll have to be able to tell him *why* the simultaneous method is suggested for the full panel in the approach to a stall.

Common errors students make in full-panel recoveries from an approach to a climbing stall are these:

1. Overenthusiastic application of forward pressure, usually because of improper visual stall recovery training. Of course, this happens to those students who didn't get their earlier stall training from *you*.

2. Timidity in lowering the nose. You may have to remind some students several times to move the nose down. This is often a sign of tenseness.

3. Forgetting to add power as the recovery is started.

4. Getting a secondary stall (if the first stall has been allowed to occur).

5. Poor scan technique during the descent-and-turn to the prerequired altitude and heading.

PARTIAL-PANEL RECOVERY. Discuss at the chalkboard the indications of the instruments in the approach to a climbing stall (Fig. 12-20).

The *indications* are these:

1. The airspeed is low and/or decreasing rapidly.

2. The altimeter shows an increase, or if the stall break is near, is steady.

3. The turn indicator shows (usually) a left-skidding turn.

4. The vertical speed indicator shows a high rate of climb.

The *recovery* steps are these:

1. Add full power. Ease the nose over. The wings are

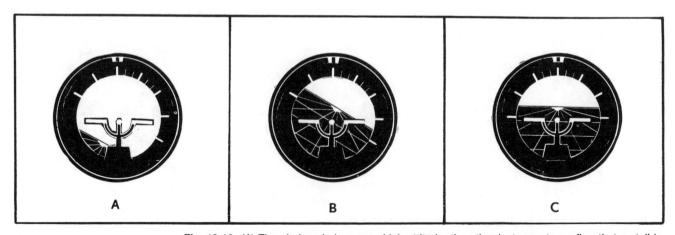

Fig. 12-19. (A) The airplane is in a nose-high attitude; the other instruments confirm that a stall is close. (B) The nose is pushed over to a below-level-flight position without leveling the wings. (C) The wings are leveled in the nose-low position and the recovery is completed.

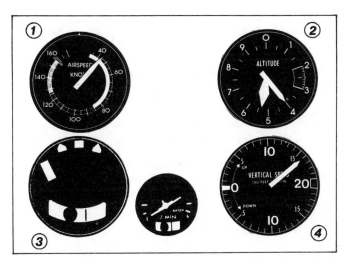

Fig. 12-20. The emergency, or partial-panel, instrument indications in an approach to a climbing stall.

then leveled by centering the turn indicator.

2. As the nose is lowered, check the airspeed; the nose is approximately level when the airspeed makes a change (starts increasing if steady, or stops decreasing). At that point, do the next step.

3. Check the altimeter and descend approximately another 100 ft from the indication at the time of the airspeed change. This is to avoid a stall.

4. Keep the turn indicator centered as the selected altitude is held.

5. Adjust power to cruise setting as the airspeed approaches cruise value.

6. Descend and turn to a prechosen altitude and heading.

You may require a partial-panel return to the prechosen numbers, as was done for the spiral recovery, or in most cases you may allow the student to get the altitude and heading indicators back before turning and descending. As was noted in the section on spiral recoveries, always encourage the student or private pilot to get the heading indicator back to work even if the altitude indicator will be out of action for the several minutes it will take to re-erect if it has spilled (no caging knob). If the airplane is in the process of stalling or has stalled, the most important thing is getting the nose down; the wings can be leveled later.

Common errors made in a partial-panel recovery are these:

1. As in the spiral recovery, overcontrolling the turn indicator.

2. Failure to note the airspeed change and lowering the nose too far. There will be a tendency for the student to lower the nose until the airspeed is well away from the stall value; this will result in too low a nose position.

3. Altitude problems. The student may forget to note a particular altitude and chase the altimeter.

4. Directional problems because of a poor scan continuation. The turn indicator is ignored as the student concentrates solely on the altitude.

5. Altitude and heading problems as the descent and turn are made back to the preselected altitude and heading.

SUMMARY

After hood work has been introduced, whether from the beginning or after solo, try to give the student some practice on every flight, if possible. This won't mean that you'll spend 15–30 min each period on instrument flying; a 4- or 5-min session under the hood while climbing out and flying to the practice area or on the flight back to the airport can give variety in the lesson, as well as extend his experience in flying without outside references. You may insert a break between high work and low work by having him go under the hood and make a 500-fpm descent to a predetermined altitude and heading, and then have him continue with the visual low work. Depending on the student, you might also have him do some of the takeoff and departure stalls and approach to landing stalls under the hood using the full panel.

Before he goes solo cross-country he should have instruction and practice intercepting and/or flying a VOR radial or homing on a radio beacon or broadcast station while under the hood (full panel).

Be sure that he is also able to demonstrate the proper procedures for contacting Approach Controls or Flight Service Stations for emergency assistance and is able to follow radar or DF heading instructions in simulated (or actual) instrument conditions.

You should tell him that the attitude indicator could tumble in extreme conditions and not to use it blindly if the other instruments disagree.

If possible—and legal—let him get a few minutes of dual with you in actual instrument conditions. Let him see the "grayness" and feel the slight turbulence often found in actual conditions. Then if he later accidentally flies into clouds, he'll have experienced the "strange" sensations (the airplane seems to be standing still, or sometimes moving extra rapidly in the murk) and won't add that problem to his already "shook up" condition. If you don't fly him in actual conditions, at least warn him of possible different sensations and emphasize that he is to fly the instruments just as he did under the hood.

Discuss vertigo briefly and again make the point that he is to ignore his own feelings and fly all of the flight instruments available.

■ **PEDPER.** For the Four Fundamentals and the Six Basic Maneuvers as introduced in this chapter, it is a good idea to set up the lessons following the general checklist, PEDPER, so that nothing is overlooked.

PREPARATION—Model, chalkboard, and hood should be available for the ground and flight instruction. Before the flight have the student read or review references concerning the maneuvers to be covered.

EXPLANATION—Cover the lesson before the flight for each segment of the instrument training to be dealt with, whether it is one of the Four Fundamentals or one or more of the Six Basic Maneuvers.

What—What are the instrument-reference maneuvers to be discussed and practiced this flight?

Why—Why are they to be done? Illustrate how these maneuvers fit into actual flight situations. Make a special effort in this regard. The student may think of practicing climbing turns to a predetermined altitude (and heading) as just a training requirement to keep him busy and make you money, unless you explain it.

How—Spell out how each maneuver is done and what the instruments will read during the process and what common errors he may expect to encounter.

DEMONSTRATION—Demonstrate each maneuver, again explaining how it works, bringing the student's attention to the instrument indications, and mentioning the problems and errors to avoid in each instance.

PRACTICE—Let the student practice each item after your demonstration and running explanation. Don't work on a particular maneuver too long; it can get boring and then learning suffers. For instance, a couple of practice descending turns to a predetermined altitude and heading will be plenty before moving on to the next item. After various exercises are introduced, have him practice them in an alternating fashion; don't have him do too many successive tries of the same one. Early periods of instrument practice can be very fatiguing so don't forget to give him a break every 10 or 15 min. A couple of minutes out from under the hood can perk him up.

EVALUATION—Is he able to translate the instrument indications into the airplane's actions? He should soon start to recognize deviations and correct for them without prompting from you. Before you send him on the first solo cross-country be sure that he can safely make a 180° turn and fly straight and level by reference to instruments. Also be sure he can recover from unusual attitudes with safety, if not precision. You'll have to extrapolate his hooded practice performance to that of an actual situation where tension moves in. If he is barely able to do the job under simulated conditions with you, his instrument work probably wouldn't be adequate if he had to save himself after flying into instrument conditions.

REVIEW—As was noted earlier, try to give the student some hood work practice each dual session. If you fly a couple (or more) hours of concentrated hood work in the early postsolo dual and then let it drop completely, he won't remember or respond as well as he would with periodic reviews.

It probably goes without saying, but you might indicate to the student that practicing under the hood while *solo* just isn't done.

CROSS-COUNTRY— GROUND INSTRUCTION

INTRODUCTION

This chapter assumes that you have no audiovisual aids available, or that you want to add a more personal approach to certain areas of cross-country instruction. Don't get lazy and depend strictly on audiovisual aids to teach the students about cross-country (or any other phase of flying); these aids (and books) are valuable but can never replace a versatile, skilled instructor. Students always pose questions that fall outside the presented material. An overhead projector and transparencies are helpful, especially when you are showing and explaining information such as in Figs. 13-5 (cruise performance) and 13-12 (weather reports, terminal and area forecasts, and charts).

In this chapter the suggestions for cross-country ground school instruction are discussed in the order that works best for the student. A quiet hour or more of ground school with one or several of your students, with no plans for immediate flying, is best for introducing cross-country work. Give them time to go over at home the material you present that first time. Several sessions would be better to cover various aspects before scheduling the period of actually planning and flying a specific flight. Remember that you will be introducing *and* reviewing parts of this material as you fly, but don't hit the student with too many basics in the preflight discussion for a particular flight; it could interfere with the details of his planning. (Charge for your ground time.)

MERIDIANS AND PARALLELS

Reference—Chap. 19, *The Student Pilot's Flight Manual* (*SPFM*).
Aids—Chalkboard and sectional chart.

As mentioned in Chap. 2, people want to concentrate on realistic problems. The following discussion of meridians and parallels may not seem to apply directly to a cross-country flight to an airport 50 mi away, but unless the student knows how his flight and the sectional chart fit into the whole navigational system he won't do as good a job. If he can "see" the relationship of latitude and longitude and how measurements are taken, he can find locations on the sectional chart with fewer errors.

Fig. 13-1 is a drawing that you might sketch on the chalkboard to show the meridians and parallels and how they apply to local circumstances. After completing the layout, roughly sketch in the location of the United States so the student can see the general position. (The 48 states portion of the United States extends from about 25° north latitude to approximately 49° north latitude and approximately from 67° west longitude to 125° west longitude.) Each degree is broken down into 60 min of latitude or longitude. Discuss UTC or Zulu time and how to convert to and from local time.

Note that U.S. latitude values increase as a person moves *north* and the longitude increases as one moves *west*. In the United States the pilot will only have to cope with north latitude and west longitude.

Break out the sectional chart and check the latitude and longitude of your airport. Have the student find the latitude and longitude of a couple of airports or other geographic points in the general area.

While on the subject of minutes of latitude and longitude, you might bring out the fact that one minute of *latitude* equals one nautical mile (6080 ft) and that this is true because the parallels *are parallel* to each other. The meridians converge on the north and south poles (refer back to the chalkboard sketch); and while the distance between them in *degrees*

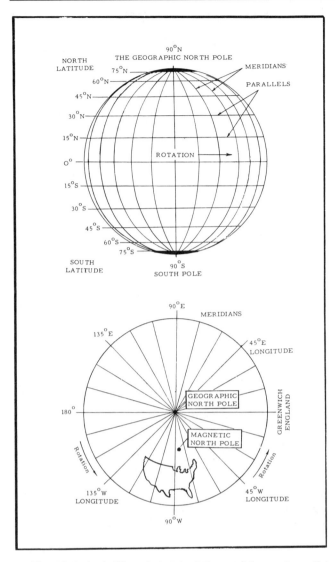

Fig. 13-1. A chalkboard sketch of the meridian and parallel idea. (See *The Student Pilot's Flight Manual.*)

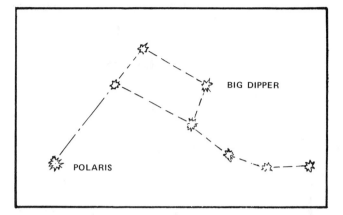

Fig. 13-2. Locating Polaris. While this may seem a digression from the subject of flying cross-country in the local area, most people are interested in the subject since they can show their friends this "new" knowledge.

or minutes stays the same with latitude, only at the equator would a minute of longitude equal one nautical mile.

As a point of interest, the North Star (Polaris) can be used in the northern hemisphere to give a good approximation of latitude. It is directly above the true north pole (astronomers, please forgive the slight error) so that it would be 90° above the horizon at that position (90° north latitude). At the equator (0° north latitude), it would be seen (?) right on the horizon; at 30° north latitude, it would be 30° above the horizon, etc. Polaris can be located by reference to the Big Dipper (Fig. 13-2).

So, the meridians converge on true north (or south), and each is considered a line to the true poles.

Assign such projects as finding a couple of local airports by giving the latitude and longitude (no names). Use the large planning chart on the wall as well as the sectional chart. Don't overdo the number of problems because, as an adult, the student will quickly tire of the process once the principle is understood.

MEASURING COURSES

Reference—Chap. 19, *SPFM.*
Aids—Plotter, pencil, blank paper, and sectional chart.

■ **True Course.** True directions and courses are measured clockwise from true north, using 360° of arc. Sketch this on the board using a known city or airport. ("Sherwood airport is south-southeast of us, so the true course would be about 160°.") Teach the student to estimate the true (*or* magnetic) course before measuring it with a plotter or protractor; otherwise he may mechanically get a number that is 180° out of phase. Require that he carefully check the course as indicated on the protractor, because the tendency is to read the "wrong" side of a major indication (he may read 145° as 135°, or 143° as 137°). At this time you might have him measure several true courses to other airports to pin down the procedure. A good move here is to also have him measure the true courses for that first triangular cross-country to show an immediate application of the theory.

■ **Magnetic Course.** Because the magnetic compass (plus the heading indicator set to the magnetic compass) is used for heading information, corrections must be made for the difference in locations of the magnetic and true north poles.

Some students will get confused by the use of east or west *variation,* thinking that their course (easterly or westerly) has something to do with whether they subtract or add variation. A sketch and discussion of the relative positions of the two poles will usually clear it up. If the variation is east it means that from the position cited the magnetic north pole is 5° east of the true north pole; the compass starts measuring from its "zero reference" 5° later (clockwise) than the zero reference of the meridian. Fig. 13-3 shows the variation idea as might be shown in an exaggerated chalkboard drawing. In Fig. 13-3A the magnetic course is 225° and in 13-3B the magnetic course is 235°.

Variation is a function of geographic position and is the same for all airplanes regardless of size. (East is least and west is best.)

After you see that the student's idea of true and magnetic courses is clear and he's had practice in locating positions and drawing courses on the sectional chart, move on to discussing magnetic compass deviation.

Deviation is a compass error found in individual airplanes; it is caused by ferrous materials in the airframe and use of electrical equipment. Its value depends on whether the engine is dead or operating and/or the amount of electronic equipment being used.

Each airplane has (or should have) a compass deviation card, and you can show the student a sample. Assume no-wind conditions, then demonstrate and have him practice un-

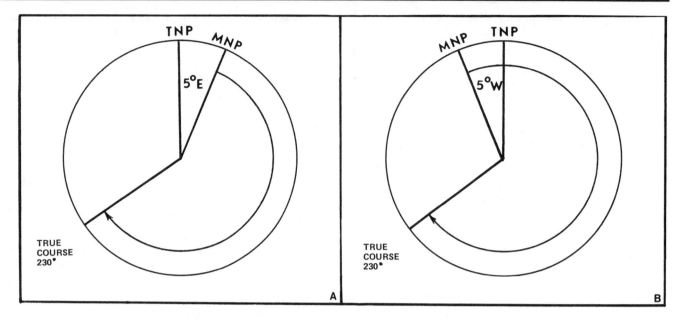

Fig. 13-3. (A) The variation is 5° E at this location. The compass starts its "counting" 5° later. (B) The same true course of 230° in an area where the variation is 5° W; the compass starts "counting" 5° earlier. (The angles are exaggerated.)

der your supervision getting several true, magnetic, and compass courses by using the sectional chart and the compass correction card. Because you are assuming that there is no wind, the compass course and heading are the same.

Any *course* (true, magnetic, or compass) plus or minus *wind* correction is the *heading* (true, magnetic, or compass). The *track* is the airplane's path over the ground; it should exactly correspond to the preplanned course. The steps taken to go from true course, or heading, to magnetic course, or heading, can be remembered by TVMDC—*True Virgins Make Dull Company.*

    T—*True* course (or heading) plus or minus

    V—*Variation,* equals

    M—*Magnetic* course (or heading) plus or minus

    D—*Deviation* equals

    C—*Compass* course, or heading.

Go through a few problems on the board to show the procedure of going from true to magnetic to compass and back the other way, and note that in going from magnetic to true, east variation is added. If this isn't clear to the student(s), you might refer to the sketch on the board (Fig. 13-3). If you don't have that much chalkboard available or have already erased it, use a piece of paper or redraw it on the board.

Assign homework. Give the student several problems of finding true, magnetic, and compass courses based on routes between airports or other points in your general area. Furnish a sectional chart, plotter, and a sample compass correction (deviation) card.

THE SECTIONAL CHART

Reference—Chap. 20, *SPFM.*

Although you used the sectional chart in the earlier lessons to show the principles of locating airports and drawing course lines, look at it now for details and the following items

that you should discuss as you and the student(s) check the chart covering the local area:

    1. *Terrain features.* This includes differing colors at different elevations; references such as cities, towns, rivers, roads, railroads; and other physical features.

    2. *Airport data.* Compare controlled and uncontrolled airports and the information available as listed on each. Note airports with Flight Service Stations, and briefly discuss functions and services available at the FSS.

    3. *Airspace.* Go over the chart, finding Class B, C, D, E, and G airspace and how each is designated (colors and markings). Make sure that the student knows the requirements and restrictions for the different airspace classifications plus restricted or prohibited areas, MOAs, and Military Training Routes. It's particularly important that he (she) knows where to look on the chart borders for times and altitudes of the various restricted areas.

If information on certain items (such as airport data) is also found in other sources, such as the *Airport/Facility Directory,* note this briefly, but save detailed discussion of those other sources until later. To mention that certain things will be covered in *detail* later is a good ploy that sets the student up for that coverage when the time comes; you are in effect laying out the short-range goals for this part of the instruction process.

Basically, your technique is to introduce material as it is needed and to review it from time to time as the student approaches the written and practical tests and the private certificate. He'll remember much better this way than if you wait until the last minute to bring up a subject and have him cram. Not only is last minute cramming bad for the basic idea of teaching what should be known about the practical aspects of flying, but it also is bad for the process of passing the written test. *Time, plus useful repetition,* are the best methods to nail down learning.

Give the student time to study the sectional chart at his leisure. He should be able to take it home and go over it and have a chance to think up questions on areas where he's uncertain (suggest to him that such questions be written down).

As part of your work with the sectional chart, if this hasn't already been done, have the student draw the first (and subsequent) cross-country triangles so that he can study the routes before you do the detailed preflight briefing of each trip. Be sure that he studies the legends and symbols as listed on the chart.

WIND TRIANGLES AND THE COMPUTER

Reference—Chap. 19, *SPFM*.
Aids—Chalkboard, yardstick, pencil, paper, plotter, a computer with a wind-side, or an electronic computer.

In bringing up the idea of wind correction for the cross-country, you might mention that the procedure is the same used in flying the legs of a rectangular course except that the line to follow is longer and is imaginary. The rectangular course has well-defined boundaries that allow the pilot to set up the proper crab angle; the legs are so short that predetermining the heading for each wouldn't be worthwhile (although it could be done).

Pilotage is the primary method of navigation for the student pilot, but the wind-side of a computer can be used to get a good estimation of heading—and pilotage will be used to confirm or correct that estimation.

Too many instructors introduce the wind functions of the computer (mechanical or electronic) without discussing the principles involved. The student is mechanically taught to insert numbers for various problems, but he isn't taught *why* it must be that way. No doubt some ground instructors would throw up their hands at the thought of drawing wind triangles in this day and age, but the basics of wind correction and resulting ground speeds *are* more readily seen if the subject is introduced this way.

Using local situations for one sample leg of the first triangular cross-country *true* course, work out the wind triangle (pick a wind with an even number of degrees and velocity, such as 240° and 20 K) on the chalkboard, using a yardstick as a guide. Use 100 K as the true airspeed for the first example. Have him work it out on paper using a protractor and measuring the true course and true airspeed and finding the true heading and ground speed. *The solving of a wind triangle, whether done on paper or on the computer, is the solving of a vector system.*

After the student has worked a problem on paper, have him work that same problem on the wind function of the computer. The number of problems you'll work this double way depends on the student, but be sure he is able to "see" the situation before dropping the paper triangle problem. Usually two or three examples are plenty for the average person. (They'll need practice in setting up the problem even after the principle is clear.) Use different winds and/or true airspeeds as necessary to keep up the interest. Then go on to examples using the computer only.

Teach that in working any of these problems the student should first have a rough picture of the course and wind in his mind: He's flying nearly due west and the wind is from the southwest, so he can expect that his correction will be to the left and the groundspeed will be less than the true airspeed (Fig. 13-4).

If the principle is considered each time, the pilot will have little problem using any computer on the market, mechanical or electronic. If after he works the same type of problem shown in Fig. 13-4 (with numbers, naturally) he comes up with a true heading of, say, 315° and a groundspeed higher than the true airspeed, he'll know he put in the wrong directions and values. As computer operators say, "Garbage in—garbage out."

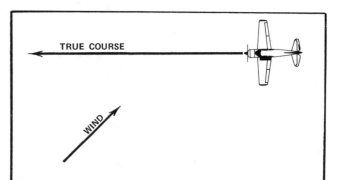

Fig. 13-4. The student (and other pilots) should visualize the situation before working a wind drift correction problem (at least at first). Here, before taking the computer in hand, he can see that a left correction will be needed and that there will be a headwind component.

After his computer handling is smoothing out, go to the *Pilot's Operating Handbook* for your trainer and get the true airspeeds for various altitudes and power settings to use in the problems. This gives you a chance to get into the use of the power setting chart as you note what rpm and fuel consumption (plus hours of endurance and miles of range) are tied in with various true airspeeds (Fig. 13-5).

CRUISE PERFORMANCE

CONDITIONS:
Recommended Lean Mixture
1600 Pounds
Flaps Up

NOTE:
Cruise speeds are shown for an airplane equipped with speed fairings which increase the speeds by approximately two knots.

ALTITUDE	RPM	20°C BELOW STANDARD TEMP			STANDARD TEMPERATURE			20°C ABOVE STANDARD TEMP		
		% BHP	KTAS	GPH	% BHP	KTAS	GPH	% BHP	KTAS	GPH
2000	2650	- - -	- - -	- - -	78	102	5.9	72	101	5.4
	2600	80	101	6.0	73	100	5.5	68	99	5.1
	2500	70	96	5.3	65	95	4.9	60	94	4.6
	2400	62	91	4.7	57	90	4.3	53	90	4.1
	2300	54	86	4.1	50	86	3.9	47	85	3.7
	2200	47	82	3.7	44	81	3.5	42	80	3.3
4000	2700	- - -	- - -	- - -	78	104	5.8	72	103	5.4
	2600	75	100	5.6	69	99	5.2	64	98	4.8
	2500	66	95	5.0	61	94	4.6	57	93	4.3
	2400	58	90	4.4	54	90	4.1	50	89	3.9
	2300	51	86	3.9	48	85	3.7	45	84	3.5
	2200	45	81	3.5	42	80	3.3	40	79	3.2
6000	2750	- - -	- - -	- - -	77	105	5.8	71	104	5.3
	2700	79	104	5.9	73	103	5.4	67	102	5.1
	2600	70	99	5.2	64	98	4.8	60	97	4.5
	2500	62	94	4.7	57	94	4.3	53	93	4.1
	2400	54	90	4.2	51	89	3.9	48	88	3.7
	2300	48	85	3.7	45	84	3.5	42	83	3.4
8000	2700	74	103	5.5	68	102	5.1	63	101	4.8
	2600	65	98	4.9	60	97	4.6	57	97	4.3
	2500	58	94	4.4	54	93	4.1	51	92	3.9
	2400	52	89	4.0	48	88	3.7	45	87	3.5
	2300	46	84	3.6	43	83	3.4	40	81	3.2
10,000	2700	69	102	5.2	64	101	4.8	59	101	4.5
	2600	61	97	4.6	57	97	4.3	53	96	4.1
	2500	55	93	4.2	51	92	3.9	48	91	3.7
	2400	49	88	3.8	45	87	3.6	43	85	3.4
12,000	2650	61	99	4.6	57	98	4.3	53	97	4.1
	2600	58	97	4.4	54	96	4.1	50	95	3.9
	2500	52	92	4.0	48	91	3.7	45	90	3.5
	2400	46	87	3.6	43	86	3.4	41	83	3.3

Fig. 13-5. Cruise chart. (*Cessna Aircraft Company*)

The next step would be to use the computer to find true airspeeds when calibrated airspeeds, pressure altitudes, and temperatures at those altitudes are given. This could lead to a discussion of density-altitude and its effects on airplane performance.

While talking about density-altitude, note that *moist* air is less dense than dry air if other factors are equal, so "wet is worse" for airplane performance. (Usually this is only a noticeable factor in the takeoff. In fact, the tendency is to overstress the effects of moisture on performance; calculations show that only an additional 3 percent would be added to the takeoff run by moisture even in very humid conditions.)

Indicate that the computer "works through" density-altitude (by the setting in of the outside air temperature and pressure altitude) in obtaining the true airspeed.

You should briefly discuss pressure and true and absolute altitudes at this point. Use these memory aids:

True altitude is the height above sea level (MSL).
*Ab*solute altitude is that altitude *ab*ove the terrain.

You could demonstrate how to find density-altitude on the computer as well as correct for nonstandard temperatures in finding indicated altitude, but don't overwork details; you'll want to concentrate on information that will be used immediately.

For practice on the computer, pick an indicated airspeed in the cruise area and use the airspeed correction table for your trainer to get the calibrated airspeed. Using a sample pressure altitude and outside air temperature at that altitude, have him find the true airspeed.

After that, set up a wind problem using that true airspeed to find the true heading and groundspeed, and have him work it out on the wind-side of the computer. The magnetic and compass headings should be found, using local variation and a copy of the airplane compass correction card. *Note that you've started putting it all together in a problem.* Repeat the process using other true courses, indicated airspeeds, altitudes, temperatures, winds, variation, and compass deviation figures. Give him some sample problems to work at home if you feel it would help matters. This would be a good time to introduce the idea of flying at odd or even + 500-ft altitudes for different magnetic courses. Interpolation between altitudes for TAS as given in the *POH* and for wind information could now be brought up.

In one or more of the problems, give him a distance to be flown and fuel consumption in gallons per hour, so that he can use the computer to find time en route and fuel used, introducing another use of it. Again, keep all of your numbers within the range of those he'll use in his cross-country work. Use specific routes and points on the sectional chart to set up the problems.

Don't forget to show him how to convert from knots to mph and vice versa. This is a good place to introduce the flight log sheet and show how to fill it out. This will sum up the previous material.

COMMUNICATIONS AND RADIO NAVIGATION AIDS

Reference—Chap. 21, *SPFM*.

Unicom will likely be a familiar item, and you've probably given the student some rudimentary work in using the VOR in earlier local flights, (and, of course, you may be instructing from a controlled airport and the student is quite familiar with the various procedures and nav-aids at this point). If not, here you'll bring up the idea of communica-

tions and radio nav in actual cross-country use. Compare LF/MF and VHF frequency bands (kHz and MHz). Discuss communications procedures.

If the first couple of cross-countries only involve the use of Unicom, then review it briefly. Stay away from a detailed discussion of local, ground, and approach control frequencies for now; tell the student that he will be going into a controlled field later and that you will discuss these frequencies and procedures at that time. Cover in detail only areas that will be immediately useful. While the operation of the Air Route Traffic Control Center is very interesting (and you'd like to let the student know that *you* know a lot about it), obviously such knowledge will be of no use at this stage and will occupy mental processes that would be better used for matters at hand.

■ **The VOR (VHF Omnidirectional Range).** Discuss the VOR and how it works, with particular attention to the 180° ambiguity. The student should note whether the TO-FROM indicator says *to* or *from* when he centers the needle and checks the OBS.

Discuss how to intercept, and track to or from, a station on a particular radial. Use the sectional chart and note VOR stations in the area, and if applicable, illustrate how the VOR can be used in one of the upcoming actual flights.

An exercise that people in ground school usually enjoy is locating themselves on the sectional chart using cross bearings you set up from VORs. You might indicate, for example, that the student(s) is over a small town with a railroad running north and south and a highway running east and west (use actual towns on the sectional). In tuning in and *identifying* one VOR, the needle is centered on 320° FROM. The same process gives a 230° *from* for a second VOR. What is the name of the town, and how does he get to the nearest airport? If you asked 10 student pilots that question, probably 9 of them would think of drawing a line from the present position to the airport and going through as lengthy a process as was done in planning the flight. The big difference is that the initial navigation and planning efforts were done on the ground, not in a bouncing airplane. You might note that it's perfectly legitimate to follow a road or railroad in getting to an airport.

Another item that should be emphasized is that VOR cross bearings only give the *area* the airplane is in. Most omni receivers aren't accurate enough to pinpoint a location, so the best method of locating oneself is to fly (and stay) over a prominent landmark and then set up cross bearings. Sometime during your instructing career be prepared for an argument with a student who has run VOR cross bearings and states, "That can't be Jonesville, the cross bearings show us 4 miles north of there." (The name on the water tank down there plainly indicates "Jonesville.")

You can go further in locating the airplane's position by VOR cross bearings by telling the student(s) that the stations are identified by Morse code symbols (then give them on the board). The student locates the VORs on the sectional by finding these symbols at the proper VOR(s) and locates himself by the radials you've given.

Illustrate how VOR cross bearings can be used to get LOPs (lines of position) for *rough* groundspeed checks. (Fig. 13-6 shows the idea.) The station used should be as near as possible to 90° to the track, at the estimated position of the airplane.

Discuss the idea of radials and how they spread with distance; a 10° bearing change may be only a couple of miles when close in, but it could 10 mi or more well out from the station. Explain that most VORs have voice facilities, and the pilot can listen to Flight Service Stations on the navigational frequency—he transmits on 122.1 MHz in that instance.

In short, discuss the VOR receiver and how it can be used

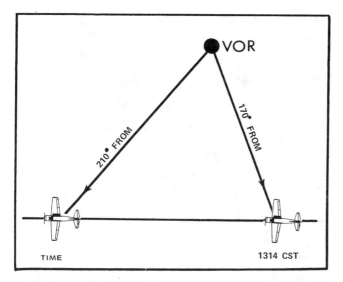

Fig. 13-6. Using the VOR to get (rough) lines of position (LOPs) and a groundspeed estimate.

by the *student* and *private* pilot. You could mention that later he will be using it for VOR approaches, but for now don't go off the main subject for discussion.

■ **The ADF.** Use of the ADF should be discussed (and covered in detail if the airplane is so equipped). Also, go over such things as tracking or homing (the student will only be expected to be able to home on a station), what sources may be used (radio beacons 190–535 kHz, and commercial broadcast stations 550–1650 kHz), and any other items pertaining to your local situation. On one of the local flights or maybe the first dual cross-country you can have him home to a nearby LF/MF facility. Or you might have him tune in and divert to a local broadcast station at the last part of the return from a dual cross-country (Fig. 13-7).

When teaching the use of the VOR or ADF, it's best to go out to the airplane on the ground and have the student set up some of the problems you discussed in ground school while

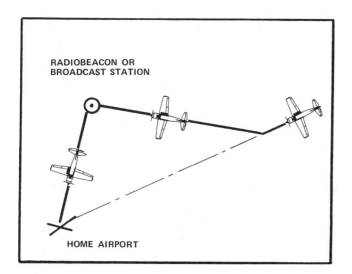

Fig. 13-7. Diverting to an LF/MF facility in the latter part of a dual cross-country is a good way to show the idea of homing without distracting the student from the overall exercise. Later you may have him divert in the middle of a leg to simulate a bad weather situation.

you supervise and comment. This allows him to be introduced to the procedure of proper tuning without the distraction of flying the airplane. You might turn on the master switch for a short period and have him tune in a local radio beacon, commercial broadcast station, or VOR that can be picked up from the ground and let him set up the receiver for it. (Don't run the battery down.)

A discussion of VHF/DF, how aid can be obtained, and where facilities are located on your sectional chart (by referring to the *Airport/Facility Directory*) should also be covered.

Summing this up, assume that your student will get lost on the first or following solo cross-countries and will need to use a VOR (or two), radio beacon, commercial broadcasting station, or VHF/DF to help locate himself. There are numerous instances every year of student and private pilots who get lost and are *unable to set up a VOR receiver to tell people on the ground their relative position (FROM – 180, etc.) from a VOR!* Don't *you* be an instructor who has to explain why his student wandered around a couple of hours calling for help on the wrong frequencies and not being able to aid his "helpers" because he didn't know how. Figure out in advance what problems the student could possibly have on a cross-country and be sure he knows what to do about them.

AIRMAN'S PUBLICATIONS

Reference—Chap. 20, *SPFM.*

Go over the publications briefly with the student, looking at the items of most interest at this point of training and tying in earlier information on Flight Service Stations, VORs, and weather aids. Show him how to look up airport and other information concerning the cross-country flights he'll be making. There's no point in listing the items here, but for your own benefit you should go through the publications before starting to instruct, and also review them briefly just before each of your students gets to the cross-country phase.

After you've gone over the publications with the student, have him go to a quiet spot and browse through them (let him take them home overnight, if you trust him). It's better to let him have a chance to look through it quietly and find items of interest, he'll remember them better.

You should note that the AIM—*Official Guide to Basic Flight Information and ATC Procedures* contains information of a more permanent nature and is good as a training manual.

You should go over briefly a current copy of an *Airport/Facility Directory* for your area of the United States and go through the information for some airports you'll be using in the cross-country training. Look at some of the information for the more complex airports also, to be sure he can use the *Directory* Legend.

Note that other information is available in the *Directory*, such as Special Notices, FSS and National Weather Service telephone numbers, aeronautical chart bulletins, and En Route Flight Advisory Service (EFAS) locations.

There are also Air Route Traffic Control Center frequencies, VOR receiver check points, parachute jumping areas, preferred IFR routes, and GADO/FSDO addresses and telephone numbers.

If the student is originally from another part of the country and you have an *Airport/Facility Directory* for that region, look up his hometown airport. He may have no idea of the facilities at the airport. He can now look at the airport "as a pilot" (one of the group). *Anytime you can include something of the student's personal interest in the training process you'll ensure his attention.* ("Did you know that your hometown airport has a 5000-foot runway? See, this abbreviation shows it right here. They have jet fuel, too.")

Class II NOTAM information should be covered and the fact noted that while it is issued every 2 wk, much can happen in between so the pilot should check with a Flight Service Station concerning strange airports before starting a cross-country—they could have rolled up the runway(s) yesterday.

FEDERAL AVIATION REGULATIONS

Parts 61 and 91 will be of particular interest, with 91 being the major item to be discussed during the cross-country phase. As the student approaches that new private certificate, Part 61, *Certification of Pilots,* will become more important and then more easily remembered.

Discuss how FARs will apply to his flying before and after he gets the private certificate. Break out the sectional chart and talk about the altitude/magnetic course rule again and the visibility and ceiling minimums he can operate in on a cross-country. Such things as cloud clearances in different types of airspace should be covered again as *they will directly apply on his cross-country trips.* Note, as applicable, the bases of transition areas and control areas (700 or 1200 ft above the surface) he will be flying through, and the limits below these heights. "Fly him" on the sectional chart through various classes of airspace and discuss the communications requirements, if any.

The main thing is that you are covering FARs that he will be using and will remember. *Don't* discuss the need for oxygen above 12,500 ft MSL, or operations above 18,000 ft MSL, or other bits of useless information for a student pilot flying a 100- or 150-HP trainer. You might mention that certain requirements exist above certain higher altitudes and that he'll get into that as he has the experience and equipment to cope with that type of operation.

What, why, and *how* still apply to your discussion of controlled airspace. If the student (or low-time private pilot) is getting back into flying after a long lay-off, you may have to compare the older types of airspace (control zones, ARSAs, etc.) with the new Class A, B, C, D, E, and G airspace; for the brand new student, refer only to the current designations because mentioning that "this used to be called an airport traffic area" would be of no benefit in helping to remember.

What (it is)—Discuss the particular types of airspace, laying out their boundaries and how they are referred to on charts and in Regulations.

Why (it is there)—Explain how air traffic and other considerations make it necessary to have certain requirements for communications and navigation equipment and weather minimums.

How (you operate in the various classes of airspace)—You could simulate a trip within or into the various airspace classes so that the student can learn the procedures involved.

The Federal Aviation Regulations can be boring, or they can be made interesting by you. Droning the FARs at your students, starting from Paragraph 1, Part 1 on through Part 91, can ensure that they won't be listening very long (Fig. 13-8). Keep it practical, humorous (or humerus, if he or she is a medical student or doctor), and therefore interesting. Encourage the idea of having a copy of the pertinent FARs available for reference. *Nobody* knows them all.

Fig. 13-8. The effect on your audience of covering the FARs step by step as they are printed (not as they are used). Use a local or practical situation that could apply to the students' flight experiences.

KNOWING THE AIRPLANE

Reference—Chap. 23, *SPFM.*

Before the student goes cross-country solo, he should have a better idea of the airplane's systems and weight and balance information. He will need to know such things for the written and oral part of the flight test, as well as for practical application later as a private pilot. This is a good place to talk about weight and balance, or fuel to be used, because for the first time he can see a need for such information. (What type and viscosity of oil does he add if needed on one of the cross-countries?)

■ **Systems.** The student should know where the battery is, *minimum* octane fuel and usable fuel in the airplane, viscosity and type of oil, alternate brands of oil, and other such information. He doesn't need to memorize tire pressures, but he should know that such numbers are found in the *Pilot's Operating Handbook.* He should have been issued a *POH* for his own use during flight training—you may sell him one or have copies for lending. A brief review of that book would be in order so that he has a good idea of where to find information. Go over in your mind what he might need to know about the airplane in order to accomplish a safe trip, and bring out those points.

In Chap. 8, emergency use of fuses and circuit breakers was discussed. Before you send him on a solo cross-country, review this information.

If he is at another airport and has a problem with a system, he is to call you before leaving, even if a competent mechanic has looked at it; you are the person who has signed that he can safely complete the flight and you should know of anything that happens out of the routine. Maybe the (only) radio has gone out of action and will need repairs back at the home field, but the weather now isn't as good as it was and you'd rather he had that piece of equipment for the return trip. Or he might have smelled smoke from electrical sources on the way over to the other airport and after close examination no reason was found. Would you want him to take off again without your talking to him and a mechanic and/or flight instructor there? Perhaps when he checked the brake pedal pressures before landing, one (or both) pedals had to be pumped to keep up pressure. He got down okay. The question is, should he continue the round robin to that other airport and home having to pump the brakes before and during the approaches?

To list dozens of items that "could go wrong on the cross-country" wouldn't be your approach to training, but the above problems are examples of what can happen and you should review the various items to consider about your particular airplane systems. In some cases you might mention the most critical items (rough-running engine, brakes, etc.).

You won't let the student go (or he's been trained by you so that he won't go) if the airplane isn't up to snuff before starting the solo cross-countries, and he should feel that there will be no repercussions if he calls you from another airport about a new development concerning the airplane systems.

■ **Weight and Balance.** Weight and balance should be covered in this phase of training because cross-country flying will be the first time that extra people and/or baggage is a factor. (Very few people carry baggage on a local flight.)

In working on weight and balance in the classroom, emphasize the idea that the student will be using this information after he gets the private certificate and flies on those vacation trips. (*That's* been in his mind all through training.) Also emphasize that consideration of weight and balance will be his responsibility as a private pilot and will be an integral part of his cross-country planning from now on.

Many students who lack a technical background are apt to look at weight and balance as a subject only understood by engineers. You have to make a special effort to show that the basics are simple (see Chap. 2).

It's suggested that a simple system, shown in Fig. 13-9A be used first; 13-9B shows a slightly more complex arrangement but the principle is the same. Indicate that *each moment is first considered separately,* and then the moments acting in the same direction are added to get the total moment for that side. The system in B is in equilibrium, and this is confirmed by comparing the moments. On the "minus" side, the moment is

Minus
(1) 35 lb × 20 in. = 700 lb-in.
Plus
(1) 20 lb × 10 in. = 200 lb-in.
(2) 10 lb × 50 in. = 500 lb-in.
700 lb-in.

Fig. 13-9C shows what appears to be a very complicated system, but it's as "simple" as the first one; the principle is exactly the same, although it takes a little longer to complete. You can set up the problem in 13-9C on the board and ask what amount of weight would be required at 20 in., as shown, to balance the system. On the right side the total moments will be: 4200 + 7200 + 6600 = 18,000 lb-in. What weight will be required at 20 in. on the other side to give an equal moment? (The answer is 900 lb.)

You may have a student who figures out a "shortcut" for a problem like 13-9B. The suggestion will likely arise in class or in a one-to-one discussion that you could average the distance of the two weights on the right side from the datum and use this as a multiplier for the two weights. Well, the "average" distance of the two weights is 15 in. and their total is 40 lb, for a total of 600 lb-in.—no deal. You make the point that weight and balance problems are to be worked step by step to avoid errors. In your ground school, you'll find that the major cause of error in weight and balance problems is simple arithmetic.

Discuss the datum and how it can be established at any point but is usually an even distance well ahead of a prominent point on the airplane (such as the leading edge of the

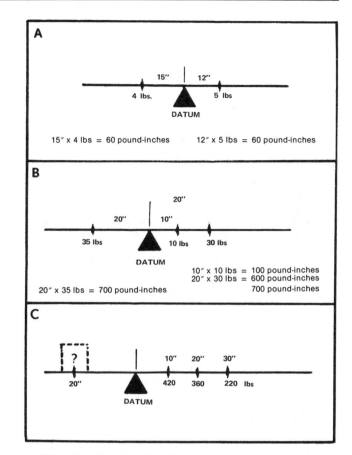

Fig. 13-9. Introduce the idea of moments simply and then gradually increase the complexity of the computations. Always emphasize that no matter how large the values used or the number of items to be considered, the principle is the same.

wing at the root). This allows for all of the moments to be acting in the same direction. Other airplanes use the leading edge—at the root—as the datum, which usually means that the oil is ahead of the datum and its moment is given a minus value. One error often made is to forget that all of the "minus" moments must be subtracted from the *sum* of the "plus" moments to get a final answer.

Fig. 13-10 shows the idea of a mirror-image airplane that has all of its weight located in precisely one spot and is exactly balancing the airplane being loaded.

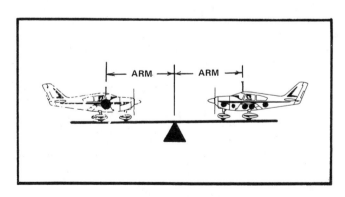

Fig. 13-10. The imaginary airplane with its mass found at exactly one point is balancing "your" airplane, which has weight at various points—you can find its center of gravity by working out the moments created by your weights. This approach often helps those people who have problems seeing why the datum can be well outside of the airplane.

The datum can be located anywhere and could be 5000 in. ahead of the wing. The center of gravity could be at a position of 5089 in. instead of 89. The datum is set up by the manufacturer so that a constant reference is available for all measurements. The limits of CG might be 5084–5092 in., or 84–92 in., aft of datum. Or the datum could be behind the airplane and the problem worked in another way.

Use the sample problem in the *Pilot's Operating Handbook* or Airplane Flight Manual for your trainer (using the actual basic empty weight). Cover the steps slowly on the board. Work a couple of problems with normal answers and then set up one with the total weight below maximum but the CG outside of the allowable limits; too many students think only of staying within the weight limits and believe that this will take care of any situation. After it's been determined that weight must be moved (or removed), what is the effect on the airplane's range for the flight? An extra fuel stop could be necessary if a particular problem shows that fuel has to be removed to meet weight limits. Baggage may have to be removed, also.

Baggage compartments are limited for *two* reasons: (1) CG limits and (2) stress. This could lead to a discussion of airplane categories and limit load factors. The limit load factor for a normal category airplane is 3.8, and a 200-lb-limit baggage compartment can support 3.8 by 200, or 760 lb without damage. Pulling more than 3.8 g's with 200 lb, or overloading the baggage compartment more than 200 lb and pulling exactly 3.8 g's, would overload the structure.

It will probably be tempting, at this stage, to go into the effects of gusts on fixed-weight components such as engines and baggage, but too much information may be coming too fast for the student pilot to handle. A student with a technical background may want more of such information.

SOME MORE POINTS ABOUT GROUND INSTRUCTION. When you start ground instruction in subjects that require formulas or the use of figures such as solving a weight and balance problem, go over the numbers before you stand up and give the students the word. Ground (and flight) instructors have been caught in simple arithmetic errors by their students and have almost come unglued trying to justify some (wrong) conclusions.

Use notes in teaching ground school. As with a flight, the planning done beforehand can make or break your ground school sessions. Notes don't have to be elaborate (in fact, too elaborate notes are as bad as none at all since they often can't be read). You could talk about weight and balance for one full 50-min period using the following notes:

1. *Background.* Need for proper weight and balance for stability and performance reasons.
2. *Basics are simple.* Seesaw-analogy. (Here you could sketch the items in Fig. 13-9, including the numbers, so that they can be drawn on the chalkboard as needed.)
3. *Weight and balance terminology.*

Reference datum	An imaginary vertical plane from which all horizontal distances are measured for balance purposes.
Station	A location along the airplane fuselage, usually given in terms of distance in inches from the reference datum.
Arm	The horizontal distance from the reference datum to the CG of an item.
Moment	The product of the weight of an item multiplied by its arm. (Moment divided by a constant is used to simplify balance calculations by reducing the number of digits.)
Center of gravity (CG)	The point at which an airplane would balance if suspended. Its distance from the reference datum is found by dividing the total moment by the total weight of the airplane.
CG arm	The arm obtained by adding the airplane's individual moments and dividing the sum by the total weight.
CG limits	The extreme CG locations within which the airplane must be operated at a given weight.
Usable fuel	Fuel available for flight planning.
Unusable fuel	Fuel remaining after a runout test has been completed in accordance with governmental regulations.
Standard empty weight	Weight of a standard airplane including unusable fuel, full operating fluids, and full oil.
Basic empty weight	Standard empty weight plus optional equipment.
Payload	Weight of occupants, cargo, and baggage.
Useful load	Difference between takeoff weight, or ramp weight if applicable, and basic empty weight.
Maximum ramp weight	Maximum weight approved for ground maneuver. (It includes weight of start, taxi, and run-up fuel.)
Maximum takeoff weight	Maximum weight approved for the start of the takeoff run.
Maximum landing weight	Maximum weight approved for landing touchdown.
Maximum zero fuel weight	Maximum weight exclusive of usable fuel.

4. *Sample problems for airplane in use* (put in your own figures).
   a. *Basic empty weight and moment.*
   b. *Fuel* (6 lb per gallon, oil 7.5 lb per gallon).
   c. *People* (actual weights).
   d. *Baggage.* Limits (1) stress, (2) stability and control problems aft CG.
   e. *Use of a loading graph. Sum up moments.* Students use *Pilot's Operating Handbook* and follow.
   f. *Two types of CG envelopes.* (1) Weight versus moment or (2) weight versus CG position. (Note that either can be converted to the other. The principle is the same.)
5. *Airplane categories and load factors.*
   a. *Normal, utility, and acrobatic—positive and negative.*
   b. *Maneuvering speeds.*
   c. *Fixed-weight components (baggage compartment).* Ultimate load factors.

It's better to have notes that include more information than you think is needed for that period than to be caught short. By including information on airplane categories with the weight and balance notes, you might be able to go on and cover the added material if the session has gone more quickly than anticipated. While you'll use the notes as a reference, you may expand the information as the need arises. The following material might be added to the notes above (match the following numbered items to the numbers used earlier):

1. *Background.* Stability and control details, tail-down force, slipstream and downwash effect on tail-down force.
2. *Basics are simple.* Levers and automobile jacks and other such aids.
3. *Weight and balance terminology.*
4. *Sample problems for airplane in use.*
   a. *Basic empty weight and moment.* How the empty

weight and CG are found by weighing. The procedure.
Empty weight as weighed and standard and basic empty
weight. What is included in the two empty weights
(paint may add 10–15 lb to a small airplane's empty
weight).

  b. *Fuel.* Octanes, fuel colors.
  c. *People.* Use actual weights of people in class—if they'll
give this information. ("George and Joe here are going
on a fishing trip carrying this much baggage. . . .") Get
the people in the class involved. ("My mother-in-law
weighs 300 pounds; where can we put her?")
  d. *Baggage.* Use specific examples of items that could be
carried.
  e. *Use of a loading graph.* Note why 100 lb in the baggage
compartment can affect the center of gravity more than
150 lb in the front seat. The slope of the lines in the
loading graph shows the relative effect of the weight.
Use some numbers to make your points.
  f. *Two types of CG envelopes.* Have actual examples of
each type (Piper and Cessna) to show that both use the
same principle. If time allows, convert one system to
the other.
5. *Airplane categories and load factors.* This could be expanded into one or two full periods for the advanced pilot.
You could cover such items as *why* the maneuvering speed
decreases with a decrease in weight; how fixed-weight components are affected; gusts, and the factors that affect the
airplane's reaction to gusts (wing lift slope, airspeed, wing
loading, and gust velocity); maneuvering and gust envelopes.

It's likely that in a private pilot course you wouldn't go
into too much detail, particularly in item 5; this is just a
sample of how you could expand notes to cover more area if
the need arises. As you get experience in more formal ground
school, you'll find that a few words of reminder on your notes
will be enough for a 15-min coverage of the subject.

Getting back to the weight and balance class, have the
students work some problems on a piece of paper after you've
shown a couple of examples. (You write the requirements on
the board.) Move around the class (if you are conducting a
class for several people) and quietly check the progress and
answers. Some of the people will be slow and you shouldn't
stick there long enough to draw attention to their plight, but
make a special effort in your review on the chalkboard to
cover the weaknesses you've noted.

■ *Pilot's Operating Handbook*

PERFORMANCE AND SPECIFICATIONS. You can review the *Pilot's Operating Handbook* with the student(s) and make it the
driest subject possible, or you can really teach something as
you move through it. Fig. 13-11 may look like just a bunch of
numbers, but the information can be used to cover not only
the specifics but also some basic principles of airplane performance:

SPEED—With these specifications, 109 K is the max level-
flight speed, and for unsupercharged airplanes this will
be found at sea level (density-altitude) where the engine is
developing the maximum power.
    Cruise speed of 75 percent power at 7000 ft is 106 K;
this is the fastest cruise speed because 7000 ft is the highest altitude where this percentage of power can be maintained (and resulting true airspeed benefits occur).
CRUISE—You'll make a point that *22.5 gal* (of usable fuel) is
mentioned here, but the total capacity for standard tanks
is 26 gal. You can ease into the idea that max range is
found at CAS of about 1.5 times the airspeed given at the
bottom of the green arc, after *that* value has been con-

PERFORMANCE - SPECIFICATIONS

```
SPEED:
    Maximum at Sea Level . . . . . . . . . . . . . . . . 109 KNOTS
    Cruise, 75% Power at 7000 Ft . . . . . . . . . . . 106 KNOTS
CRUISE: Recommended Lean Mixture with fuel allowance for
        engine start, taxi, takeoff, climb and 45 minutes
        reserve at 45% power.
    75% Power at 7000 Ft . . . . . . . . .  Range   340 NM
        22.5 Gallons Usable Fuel           Time    3.3 HRS
    75% Power at 7000 Ft . . . . . . . . .  Range   580 NM
        35 Gallons Usable Fuel             Time    5.5 HRS
    Maximum Range at 10,000 Ft . . . . . .  Range   420 NM
        22.5 Gallons Usable Fuel           Time    4.9 HRS
    Maximum Range at 10,000 Ft . . . . . .  Range   735 NM
        35 Gallons Usable Fuel             Time    8.5 HRS
RATE OF CLIMB AT SEA LEVEL . . . . . . . . . . . . 670 FPM
SERVICE CEILING . . . . . . . . . . . . . . . . . 14,000 FT
TAKEOFF PERFORMANCE:
    Ground Roll . . . . . . . . . . . . . . . . . . 735 FT
    Total Distance Over 50-Ft Obstacle . . . . . . 1385 FT
LANDING PERFORMANCE:
    Ground Roll . . . . . . . . . . . . . . . . . . 445 FT
    Total Distance Over 50-Ft Obstacle . . . . . . 1075 FT
STALL SPEED (CAS):
    Flaps Up, Power Off . . . . . . . . . . . . . . 48 KNOTS
    Flaps Down, Power Off . . . . . . . . . . . . . 42 KNOTS
MAXIMUM WEIGHT . . . . . . . . . . . . . . . . . . 1600 LBS
STANDARD EMPTY WEIGHT:
    Commuter . . . . . . . . . . . . . . . . . . . 1111 LBS
    Commuter II . . . . . . . . . . . . . . . . . . 1129 LBS
MAXIMUM USEFUL LOAD:
    Commuter . . . . . . . . . . . . . . . . . . . 489 LBS
    Commuter II . . . . . . . . . . . . . . . . . . 471 LBS
BAGGAGE ALLOWANCE . . . . . . . . . . . . . . . . 120 LBS
WING LOADING: Pounds/Sq Ft . . . . . . . . . . . . 10.0
POWER LOADING: Pounds/HP . . . . . . . . . . . . . 16.0
FUEL CAPACITY: Total
    Standard Tanks . . . . . . . . . . . . . . . . 26 GAL.
    Long Range Tanks . . . . . . . . . . . . . . . 38 GAL.
OIL CAPACITY . . . . . . . . . . . . . . . . . . . 6 QTS
ENGINE: Teledyne Continental . . . . . . . . . . . O-200-A
        100 BHP at 2750 RPM
PROPELLER: Fixed Pitch, Diameter . . . . . . . . . 69 IN.
```

Fig. 13-11. Performance and specifications for a trainer.

verted to CAS. (Later in the *POH*, V_{s_1} will be found to be
48 K.) This figure of 1.5 is good for gross weight and is
reduced by less weight. A student who needed to get the
max range on one of his cross-countries could set up an
airspeed of 1.3–1.5 times the number on the bottom of the
green arc (again converted to CAS if necessary) for a
better stretch. This could also lead to the discussion of
max endurance, as time in class allows (Fig. 23-15,
SPFM).
RATE OF CLIMB AT SEA LEVEL—670 ft per minute is the
maximum rate of climb he could get there at *max certificated weight,* because the rate of climb drops with an
increase in altitude (density-altitude). You could bring up
the point that as *weight* decreases, the climb rate increases, so that at 4000 ft and a *much lower* weight, the
airplane could climb at the sea level rate.
SERVICE CEILING—This is the altitude at which the airplane would only climb 100 ft per minute. *The rate of
climb drops off in a straight line so that the two points of
climb (sea level and 670 fpm, and 14,000 ft and 100 fpm)
may be joined, and intermediate rates of climb may be
found.*
TAKEOFF—Note that the lift-off distance is about *one-half*
of the *total* distance to clear the 50-ft obstacle. This is a
rough rule of thumb for most trainers (Figs. 17-10 and
17-13, *SPFM*).

LANDING—The landing roll is slightly less than *one-half* of the *total* distance from the 50-ft obstacle to the stopping point, again a rough rule of thumb for most trainers (Fig. 17-10, *SPFM*).

STALL SPEED (CAS)—You can bring up the idea of stall speeds at IAS and CAS and how the CG can affect stall speed. (Forward CG means a higher stall speed for the majority of airplanes.)

MAXIMUM WEIGHT, STANDARD EMPTY WEIGHTS, AND MAXIMUM USEFUL LOADS—You can review these and bring up any points missed earlier.

BAGGAGE—120 lb maximum, and this airplane has two possible arrangements for that weight. (Not indicated in Fig. 13-11 but shown later in the *Pilot's Operating Handbook* of this sample airplane.) Repeat that these limits are not to be exceeded.

WING LOADING—If the information were needed you could divide the weight (1600 lb) by this figure (10.0) and get the wing area of this trainer.

POWER LOADING—If you were selling a particular model airplane you could compare its power loading with its competitors. The lower the power loading, the more "pizazz" an airplane has. Power loading = pounds of airplane weight per total horsepower at max gross weight.

FUEL CAPACITY—Fuel capacity does not mean all of the fuel is available for use. Note that the information under the range heading only mentions 22.5 and 35 gal, respectively, for standard and long-range tanks.

OIL CAPACITY—This could be a good place to mention what type of oil is used in your trainer and talk about required viscosities, straight mineral oils, and additive oils.

ENGINE—What does O-200-A mean? Use some other engine designations as well (IO-360-C, etc.).

PROPELLER—In addition to the length given, you might use this as an opening to discuss simple propeller theory and care and checking of propellers.

Moving on through the *POH*, use the various sections listed (Electrical System, Fuel System, etc.) as topics for your notes. Write notes or reminders on the margins of the pages of the *POH* you use in class and in pre- and postflight discussions.

The operating limitations and performance sections are good for using items and numbers as applied to the particular airplane the student will actually use and also to give the principles for *all* airplanes. A *POH* should be kept permanently in each airplane, even if the Regulations don't specifically require it.

Try to give each student access to a *POH* for the airplane. They can buy it or you can lend copies for the duration of their preprivate flight training.

■ **Airplane Papers.** Discuss the *Certificate of Registration* and when it should be removed or changed. Have one available for examination firsthand. A new one is needed if the airplane is sold.

The *Airworthiness Certificate* should also be checked by the student and you should emphasize that this certificate has to be *displayed* in the airplane so that pilot and passengers can see it. The *Airworthiness Certificate* is valid as long as the aircraft is maintained in accordance with the FARs and, *unlike* the *Certificate of Registration,* stays with the aircraft regardless of changes in ownership.

The airframe and engine logbooks should be brought out and *annual* and *100-hr* inspections discussed. Who may do either inspection, and where is the information noted?

Weight and balance data and the *equipment list* for the airplane should be covered and examined. You might make a copy of such information for each student to keep and study at his own leisure.

If *Major Alteration and Repair Forms* are a part of your life, discuss them. Again, obtain *copies* of actual forms for the student, if possible.

The *Aircraft Radio Station License* should be discussed, with the special note that it pertains to *transmitters* (including DME and transponder) and is included on the license, which is kept *on board* the airplane. This is Federal Communications Commission business rather than FAA, but it would be bad to get in trouble with them, too.

So, when you start instructing, find out what papers are necessary for the airplane you are using and cover them in class (formally or informally) in logical order. Some airplanes below 6000 lb max certificated weight have a formal Airplane Flight Manual type of arrangement; others may use placards and markings in lieu of an AFM. Take the time to sit down and make notes for laying out a logical way of presenting the material to the student.

If you use actual *airworthiness* and *registration certificates* in the classroom, don't forget to put them back in the airplane before you fly.

■ **Weather Services.**
Reference—Chap. 22, *SPFM*.

Now you should introduce the student to the subject of weather; he'll be using it. First, you'll start off with weather theory of air masses and the earth's general circulation pattern; he should get a couple of hours of this to have a good background . . . WAIT A MINUTE! What good is knowing that the weather is caused by a MTW (Maritime Tropical Warm) air mass when there's nothing he can do about it? Your job will be to show the student pilot what *services* are available and how he may best use them, not to give a lot of meteorological theory, though this can be covered in some detail later.

So cover what is available to him at Weather Service Offices and Flight Service Stations. Fig. 13-12 shows a system sometimes used to help train students about weather services in a day-to-day manner. If this information is posted the student may read it in his spare time. The new student should be in such an environment from the beginning, even without specific pressure to learn, because he will see the more advanced students using it and he will likely have a head start in reading hourly reports and other weather information.

It may seem like a lot of trouble to set up such an arrangement, but you may be starting your own flight school someday and it could be a valuable aid.

■ **Knowing the Available Information.** The student should have an idea about frontal movement and the weather normally associated with warm and cold fronts and about the circulation around high and low pressures. He should also know that weather in the United States moves generally west to east, so conditions west of his position may be *local* conditions in an hour, or several hours, depending on how far away the reporting station is. If 40 mi *west* it's not a fit night out for man nor beast, the same conditions will prevail here within a short enough time that that cross-country flight should be canceled.

Get enough copies of the various reports and forecasts so that each student may have one. Also get old winds aloft and other charts from the nearest Weather Service Office—they'll usually be glad to send you the material after it has been in their file a specified number of days. The FSS or weather people will also furnish a number of copies of guides for translating various reports and forecasts, or you may order them from the U.S. Government Printing Office.

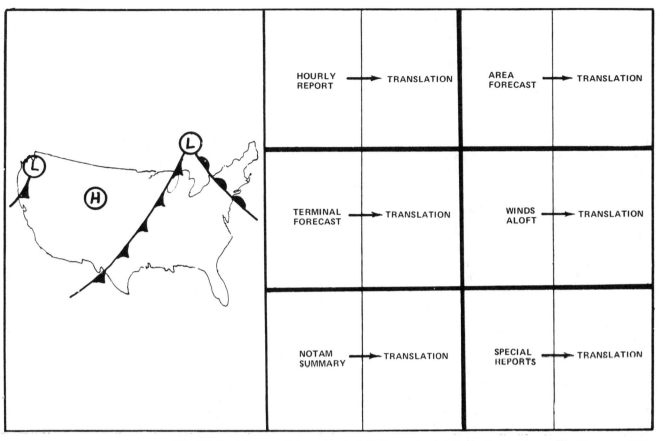

Fig. 13-12. A weather information bulletin board for the flight office or classroom. The map of the United States has a clear plastic cover and the weather systems can be drawn in with grease pencils daily, using either teletype or other weather information sources. You may have students do this while waiting for a flight. On the right are teletype reports and forecasts plus translations. By using copies of actual information (translating them to plain language) and replacing them from time to time, you'll be more up to date than your textbooks.

If possible, the student should have a chance to visit both a Weather Service Office and a Flight Service Station. You should plan one or more of the dual cross-country flights so this can be done.

Work at the chalkboard, writing out sample copies of hourly reports and terminal forecasts and having the student(s) translate, or better yet, as noted earlier, have samples on transparencies that can be shown by the overhead projector so that all can see it as you read. Writing an *Area Forecast* on the board is asking for writer's cramp, so you should read your copy (and have the students check theirs) aloud. The student must be able to decipher *Routine (Hourly) Weather Reports* (METARs), *Terminal Aerodrome Forecasts* (TAFs), *Area Forecasts, Winds Aloft Forecasts,* and applicable NOTAMs.

You should review the weather charts available to pilots such as the Surface Analysis, Weather Depiction, Radar Summary, and Prognostic Charts. Depending on the student, you might move on to the more complex forms, but too much information can stun some people.

The temperature and dewpoint don't give any problems to a student who can think in terms of Celsius.

The wind is given in true directions if printed (the tower or FSS *talk* about it in magnetic terms) and knots. A "2815G28" can cause mild consternation until you explain about gusts.

The altimeter setting is straightforward and usually gives no problems, but the "Remarks" following the information

might require a little extra instruction on your part.

The widespread advent of TV weather channels, the direct user access terminal (DUAT), and other computer weather "publishing" brings the information instantly. It may be that the flight school where you are instructing has such equipment, making the overlay/grease-pencil map crude and outdated. But for smaller schools the national Weather Service or FSS specialists, via a toll-free telephone number, can give a quick rundown on highs, lows, and fronts and precipitation areas that can be put on the overlay.

You may want to teach your students the "older" shorthand for cloud cover so that they can copy telephone briefings more quickly. Those symbols can be written faster than OVC, BKN, and SCT:

⊕ — Overcast
◍ — Broken
◌ — Scattered

Have the students read a couple of TAFs and know when the forecasts are issued and what information is carried. Check with the Weather Service Office for the latest information. When *Area Forecasts* are covered, use a sample and go over each part, noting specific points that you think might cause problems in understanding the information. Punctuation can cause trouble for some—a comma at a different place in a sentence could give it a widely different meaning.

Compare *Area Forecasts, Terminal Forecasts* (TAFs), and *Routine (Hourly) Weather Reports* (METARs) for the same period. (Work it so that your samples and students' copies of these sources cover the same date and time period.) An hourly report of *actual* weather may be compared with what the forecasts *said it would be.* The same idea goes for *Winds Aloft Forecasts* and other data. This way, you'll not only be teaching the students to decipher the information but also compare the sources.

Winds Aloft Forecasts are, as the name says, *forecasts,* and you should cover the forecast periods and time of validity. Most of the information is easily read after a brief explanation, but you might explain how aloft winds above 100 K velocity are written (50 is added to the direction and 100 subtracted from the velocity); 781061 would be a wind from 280° true (78 − 50 = 28) at 110 K and the temperature is −61°C. This kind of wind and temperature might be found at 39,000 ft for a particular station.

You can expand your discussion of weather principles just as was done in the sessions on the *POH.* For instance, in teaching how to decipher a *Winds Aloft Forecast* you can stress that the winds at higher altitudes are from the west and compare the temperatures at various stations at these altitudes.

SIGMETs, Convective SIGMETs, and *AIRMETs* should be covered.

The student pilot should be taught the value of pilot reports (PIREPs) in conducting a safe flight. The pilot should use the information and also make reports of his own.

Flight Watch (EFAS) is the direct in-flight weather information service. A frequency of 122.0 MHz is used now, but it's assumed that additional frequencies will be required as the use of the service becomes widespread.

In short, take time to find out what weather facilities are available in your local situation, such as Automatic Weather Observation Stations, telephone numbers for briefing (1-800-WX BRIEF, or others) to be sure that the student can take advantage of all sources of information.

Some of your students may have computers and DUAT; encourage their use of this service. (You may want to use it yourself.)

The AIM—*Official Guide to Basic Flight Information and ATC Procedures* has the weather services available, and you should review this information during the ground school sessions on weather.

Remember that times and presentations of weather services may change—you'll have to stay on top of this.

SUMMARY

The material covered in this chapter represents a number of hours of ground school, and much of it should be covered as the student flies, after he's had the introductory material. Your teaching will give background, as well as specific flight information. To teach the cross-country ground school material in a big lump would cause him to forget too much of it.

You may also introduce the use of electronic navigation computers with at least some of your students.

This chapter contains the suggested order of introduction of material for the average student, but you will naturally vary the details with individuals and circumstances.

The following syllabus might help you in laying out the ground school portion of the cross-country, particularly with your first few students.

■ Cross-Country Flight

GROUND INSTRUCTION

— 1. *Types of navigation.*
—   a. Pilotage.
—   b. Dead (deduced) reckoning.
—   c. Celestial navigation (briefly).
—   d. Radio navigation.
— 2. *Meridians and parallels.*
—   a. Meridians run North/South and converge at the True North and South poles.
—   (1) Prime meridian runs through Greenwich, England.
—   (2) Lines of longitude run north and south, from the prime meridian (through Greenwich) and are measured east and west.
—   (3) Time zones/Universal Coordinated Time.
—   b. Parallels (lines) run East/West.
—   (1) Prime parallel runs through the equator.
—   (2) Divisions measured north and south of the equator.
— 3. *Degrees of latitude.*
—   a. 60 minutes = 1 degree.
—   b. 1 minute = 1 nautical mile or 1.15 statute miles.
— 4. *True North.* Measuring distances clockwise from True North. Using the protractor.
— 5. *Magnetic North Pole in Canada.*
— 6. *Magnetic variation* = difference between Magnetic North and True North. Does not necessarily mean the geometric angle between the two poles.
— 7. *Isogonic lines* = lines of equal magnetic variation.
— 8. *East is least; West is best.*
— 9. *Compass deviation.*
—   a. Swinging or correcting a compass.
—   b. TVMDC.
—10. *The plotter.*
—   a. Measuring the true course on the mid-meridian.
—   b. Mileage scales.
—11. *The wind triangle.*
—   a. The vector system.
—   b. Wind directions (winds aloft) given in True directions and in knots.
—   c. Selecting VFR cruising altitudes.
—   (1) 0°–179° = odd thousands plus 500 ft.
—   (2) 180°–359° = even thousands plus 500 ft.
—   (3) Selecting altitude with most favorable winds.
—   d. Drawing True North/South line for reference.
—   e. Using a plotter to measure the True course line through some point on the North/South line. Draw this line of indefinite length.
—   f. Using the nautical mile scale on the plotter,

draw wind vector in the proper direction and length from some point on the course line.

— g. From the wind arrow, swing a line the length on the true airspeed (again, use the nautical mile scale on your plotter), until it hits the True course line.

— h. Discuss the meaning of each line in the triangle.

—12. *Using the E6B* and other computers.

— a. Finding true airspeed.

— b. Checking fuel consumption and speed.

— c. Finding density altitude.

— d. Finding groundspeed and estimated times of arrival.

— e. Finding fuel consumption.

— f. Correcting indicated or pressure altitude.

—13. *Converting* the True Course to True Heading to Compass Heading and back.

—14. *The sectional chart.*

— a. Aeronautical symbols.

— b. The chart's coverage.

— c. Topographical symbols.

— d. Radio aids to navigation and airspace information.

— e. Data on obstruction symbols.

— f. List of ATC frequencies.

— g. Prohibited, Restricted, Warning, and Alert areas.

— h. Cover the A, B, C, D, E, and G classes of airspace.

—15. *Aeronautical Information Manual* (AIM)—*Basic Flight Information and ATC Procedures.*

—16. *Airport/Facility Directory.*

—17. *Notices to Airmen.* L, D, and the FDC publication.

—18. Using the flight computer to solve wind triangle problems.

—19. *The airplane.*

— a. Systems.

— b. Weight and balance.

— c. Emergency procedures.

—20. *Weather.*

— a. Hourly reports.

— b. Forecasts.

— c. Weather charts.

— d. Winds aloft.

— c. In-flight advisories.

—21. *Navigation Aids and Communications.*

a. VOR.

b. NDB.

c. LORAN (if applicable).

d. Global Positioning System (GPS) (if applicable).

e. Unicom frequencies.

f. FSS common frequencies.

g. Area tower frequencies (list on a card).

h. Emergency communications procedures.

—22. *Filing flight plans.*

— a. Advantages.

— b. Procedures for opening and closing a VFR flight plan.

— c. Use of transponder or VHF/DF steer in emergency situations.

CROSS-COUNTRY—
FLIGHT INSTRUCTION

INTRODUCTION

Chap. 13 presented the ground school subjects the student should be taught before and during the cross-country flight training. This chapter covers the actual preparation for, and flying of, dual and solo cross-country flights with some suggestions for your instruction.

Sending a student on a solo cross-country requires consideration of more factors than is realized. It's not just a matter of seeing that he can navigate a particular route; weather and time of day (the sun's position), for example, will have a definite effect on whether a triangular cross-country would be flown clockwise or counterclockwise. And every flight will be unique because of wind, visibility, and sun position (plus other things that crop up). Assume for the purposes of this chapter that the flying so far has been from a small, noncontrolled airport, unless otherwise noted.

THE FIRST DUAL CROSS-COUNTRY

■ **Choosing the Route.** Fig. 14-1 shows an example of a first dual cross-country. The trip forms an almost perfect equilateral triangle, with the legs only about 29 NM each. In this cross-country, the student will get set up on his pilotage and get the experience of landing at two strange airports without building up too much flight time.

SOME IDEAS TO CONSIDER. The minimum requirement for cross-country experience for the private pilot aspirant is 3 hr of dual and 10 hr of solo flight. Each of the solo cross-countries must include a landing at an airport more than *50 NM* from the point of departure; and one must have landings at three points, each of which is more than *100 NM* from each of the other two points. In addition, the student will have to make three solo takeoffs and landings to a full stop at an airport with an operating control tower.

You'll probably use the minimum 3 hr of dual for the first dual cross-country and a dual to an airport with a control tower (assuming that you aren't instructing from such an airport). The "long" triangular solo cross-country with 100-NM legs would take nearly 4 hr for some trainers, so this would leave about 6 hr solo time for other phases of cross-country you want to cover. You'll find that this leaves little time to give across-the-board practice in pilotage and the use of the VOR and/or ADF and other cross-country work.

You'll want to give the greatest variation of experience that can be safely done in the time allowed. Remember that 3 hr dual and 10 hr solo are the *minimums* required; you might base your private course on more time in either, or both, categories. Few people get their private certificates at exactly the 40-hr minimum anyway (the national average is 60+ hr). Don't forget, also, that one of the dual cross-country flights must be of 2-hr duration (FAR 61.107).

Another requirement is 3 hr of night dual including 10 takeoffs and landings if the new private pilot expects night flying privileges. You might use part of this time for a short dual round robin to teach him cross-country flying at night, using both pilotage and radio navigation. For best training, the student should prepare for it just like a day flight—making a flight log, filing a flight plan, etc.

You may decide to set up the first dual with a landing more than 50 NM away so that the student can still *fly the same route by himself* and get credit toward the required 10 hr of solo cross-country time. Fig. 14-1 is included only to demonstrate some points in picking airports for the first and subsequent cross-countries.

Note that Fayetteville is a typical general aviation airport with a radio beacon, but Shelbyville has a VOR on the airport. You could have the student track into or away from the VOR, if it works out that he is ready for it on this trip. The main purpose of this first dual is to introduce the idea of *pilotage,* but a couple of minutes of tracking on the way to and from the VOR would be worthwhile for most students. More about this later. You can work with the ADF if your trainer is so equipped.

It's suggested that the triangular cross-country be clockwise in direction in this example, that is, from Sewanee to Fayetteville to Shelbyville and return. Why? Always consider the sun position in setting up a cross-country training flight; it's better for the student to fly into strange territory with the sun in a favorable position. Note that on the leg to Fayetteville only in the very late afternoon would the sun be ahead, and you likely would not plan a dual *or* solo cross-country training flight that late in the day. At *any* time of the day, the leg from Fayetteville to Shelbyville will have the sun in a position behind the airplane for best visibility. In returning to Sewanee, if the trip was started very early in the morning, the sun would be in the least advantageous position, but the student is flying back to territory that he has flown over many times. By this point in training he'll be familiar with the country within a 10-mi radius of the home airport.

One good reason for choosing airports to the west of Sewanee *is not as evident from the chart as it should be,* but Franklin County airport is on the western edge of the Cumberland Plateau. The terrain immediately west of the airport drops sharply about 900 ft and the trip would be made over farmland with good checkpoints and brackets (compare the elevations of Franklin County and Winchester airports). The territory south, east, and northeast of the Sewanee airport is rugged woodland with only scattered checkpoints—not very good for training or the student's frame of mind, if he is the timid type. So, when you set up your own first cross-country routes, take into consideration terrain, checkpoints, brackets, and the position of the sun at various times of the day.

Look at Fig. 14-1 again: A triangular cross-country could be made from Sewanee to McMinnville (Warren County airport, north of Sewanee) to Shelbyville and return (leg 2). It's suggested that for best sun positions the trip be made in the order just mentioned (again, it depends on the time of day).

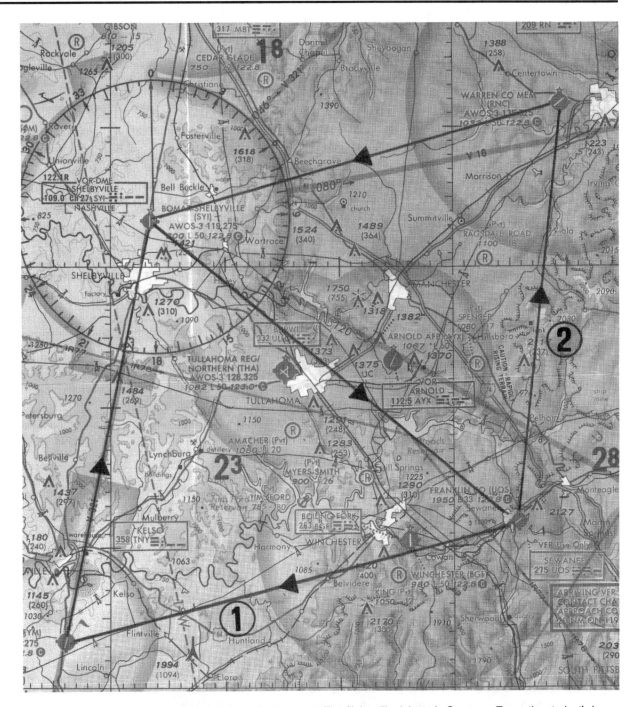

Fig. 14-1. A first cross-country route. The flight will originate in Sewanee, Tenn., the student's home base. The trip shown here would not meet the 50-NM minimum requirement for solo cross-country and is used only to bring out some points. You should have a dual trip that could be reflown solo for credit toward the private requirements.

The legs would be approximately the same length as the earlier route discussed, but the checkpoints wouldn't be so numerous or well defined.

In addition to the "sun factor" for the routes of the two trips, note that in both cases Shelbyville is the second of the two strange airports where the pilot will land—and it's the one with the VOR. This means that the student would have flown *at least one* leg by pilotage before you decide whether to introduce the idea of using the VOR going to Shelbyville. If he's having trouble with pilotage on that first leg, you would prob-

ably skip VOR work and might introduce it on a second dual over the same route.

The McMinnville-Shelbyville route might be good for the second solo cross-country (except for the 50-NM requirement, but it's only an example). The student is introduced to one strange airport (Warren County) but is going into Shelbyville solo for the second time. This procedure would be good for the person who is safe for solo cross-country but who needs his confidence built up (and you don't want to send him too far afield solo just yet).

The first solo cross-country, like the first solo flight, is mainly a confidence-building effort and should follow the carefully thought out dual cross-country; plan it so that he doesn't get too far from the home airport, but then set up the following solo trips with increasing distances and more complex requirements.

This writer had phone interviews with FAA inspectors at 19 FAA FSDOs to get information for this edition on the weaknesses of CFI candidates during the practical test. Basic aerodynamics and practical pilotage navigation were the two items cited most often. In some cases the CFI applicant had problems with basic navigation, using a sectional chart and measuring true courses or distances. They had begun to rely too heavily on VORs, Flight Following, and other methods of navigation.

Be sure that *your* students are well grounded in basic pilotage and use of the charts before they move on to navigation by electronic aids. (One transient pilot landed at our local airport and asked, seriously, whether he was in Alabama, Georgia, or North Carolina. The LORAN C had failed and he never bothered to carry sectionals, so he landed after spotting an airport. He was not in any of the three states mentioned — he was in *Tennessee*.)

Working with the basic drawing of lines on a sectional chart doesn't mean that the student ignores the use of electronic nav-aids as he progresses, but in many cases after you move on to radio navigation, you'll have trouble going *back* (moving backwards in *his* mind anyway) to the drawing of course lines. You'll run into people who have plenty of money and buy a well-equipped airplane to use in learning to fly (rare, but it *does* happen); they want to use that expensive equipment and forget the "basic kid stuff." You'll sometimes have to insist that their navigation training starts with basics.

■ Preflight Briefing

WINDS AND WEATHER. Allow the student to have an overnight look at the cross-country and do some of his planning at home, if possible, this first time. He should draw and measure the course lines and fill out a flight log (which you will check), leaving out the groundspeed and time information but including prominent checkpoints about 10 mi apart on all three legs. You can give him a quick pre-preflight briefing about this the day before the flight so that he'll know the procedure. Check the course lines and true courses on the chart to confirm his protractor accuracy.

Call in and copy the winds and weather (you might remind him that this is the first move in any flight; if the weather is too bad at the destination, all that careful planning will be wasted). Let the student listen on an extension phone and copy the weather and winds. *Your* copying is necessary to ensure that his information is correct and to straighten out any errors he made in writing it down. Of course, if a computer or teletype service is available you would use this, but he should also have some supervised practice in writing down verbal weather information. Some may be able to get the weather on their home computers.

Let the student decide on an altitude for each leg, taking the VFR required altitude/course rule into consideration. Suppose he decides on 4500 ft MSL for the leg from Sewanee to Fayetteville (or *your* equivalent flight). If the nearest winds forecast shows that at 3000 MSL the wind is from 250° at 15 K and at 6000 from 280° at 20 K — he should interpolate and use 265° and 17 K in his computations. Always have the student work wind computations for his trips even though you know, and he'll find out, that it doesn't usually come out so neatly on the flight. It's good practice for the written and flight test and, in addition, teaches him to consider all factors before flying. Later, when he moves up to the airlines or goes to the military, the ability to be sharp with a computer will be

to his advantage. *Remind him that he will probably have to make corrections in heading during the flight by pilotage and, like using the VOR, he isn't to blindly follow numbers.*

He should work out the magnetic and compass headings, time en route, and fuel required for each leg. Allow fuel for climbs, although in many cases the extra fuel used in the climb is almost balanced by the lower fuel consumption of the letdown. One rule of thumb to try with your airplane is of adding two-thirds of a minute for every 1000 ft to be climbed, after working out the en route time using *cruise* groundspeed figures. You might come up with your own rule. Better yet, use the information for taxi climb and cruise from the *POH*.

Have the student estimate the total fuel to be used on every flight. Fill the airplane before leaving, refill it after getting home, and check his figures (naturally taking into account any fuel bought at another airport). It may seem a bit of trouble, but you can remind the line attendant as you get out of the airplane that you want to know the amount of fuel it took. The students will enjoy seeing how close their estimates come, and you'll soon have a close estimate for dual trips that you've flown. If you're able to estimate it before leaving and the results are very close, the students will realize that you know what you're talking about and will listen when you talk about other flight factors.

While the student is doing his paperwork you might do other things and check his work after he's finished, correcting and explaining any errors. *In fact,* as long as he is a student pilot under your supervision, you should oversee and closely check his pre-cross-country planning for *every* flight. Too many flight instructors offer the student (more or less) carte blanche after the first dual cross-country, and he is not supervised or checked for each flight as the FARs require. FAR 61.93 states that you may sign off a student for repeated solo cross-country flights under "stipulated conditions" over a course not more than 50 NM, if you've given him instruction in both directions over the route, including takeoffs and landings at the airports to be used. This could encourage a student not to go on for the private certificate, since he can use the airplane on a chosen route when he wants to. You'll run into "professional student pilots" who may have 300 or 400 hr and are still flying on a student certificate, usually because they don't have the discipline or ability to pass a written and/or flight test.

Getting back to the first dual cross-country briefing: *Go over the flight, using the sectional chart, and note outstanding checkpoints.* Here again you'll be directing the student's *perception,* as discussed in Chap. 2. The average student can pick out outstanding terrain features such as towns, railroads, and major highways; but things such as pipelines, streams, and rivers are often overlooked until you bring his attention to them. Teach him to recognize towns by the railroad or road system around them.

STRANGE AIRPORTS. If the student hasn't been to the airport(s) before, draw a sketch on the board showing runways, taxiways, the location of the wind indicators, and the flight operations where you'll be stopping (Fig. 14-2).

In your briefing for the solo cross-country to these same airports, you might review this information, particularly if there has been a layoff. This type of information should also be covered in the following solo cross-country flights to other airports. Booklets of airport layouts have been published and you can make good use of that information.

While you're talking about the airports, note that the common tendency is to be too high when landing on a strange area. He should know this and plan accordingly. Compare the runway widths and lengths with the home airport.

The *Airport/Facility Directory* is a very important part of the preflight briefing; the student should have reviewed the

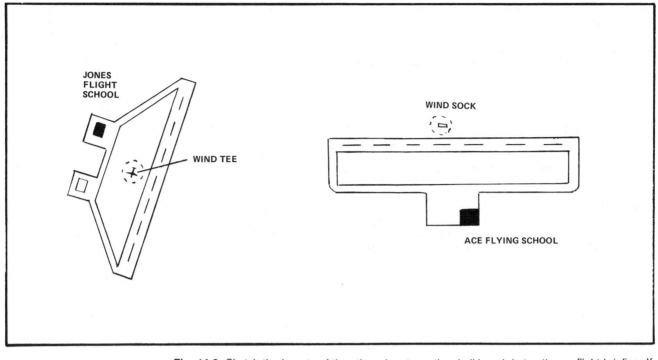

Fig. 14-2. Sketch the layouts of the other airports on the chalkboard during the preflight briefing. If you are doing enough cross-country business you might reproduce copies of the layouts to give to each student sometime before the first dual X-C.

information and have an *A/FD* with him on the flight so that if he has to divert he won't be operating in the blind at that new airport.

Discuss how you would enter traffic for various runways at the strange airports and the use of Unicom, if available. Note that the heading indicator is a good aid in finding the proper runway on a multirunway airport (Fig. 14-3).

GEMINI. Embry-Riddle Aeronautical University has used a Gemini (The Twins) program in dual instruction. One student

flies the airplane with the instructor, while the other student observes from the rear seat (assuming that a three- or four-place airplane is being used). This is particularly valuable during the cross-country phase (practicing stalls with this setup wouldn't be so good), and you could approach the triangular cross-country two ways. One way is to have the students alternate navigating and flying: Jones flies the first leg with Smith navigating; Smith flies the second leg with Jones navigating. A disadvantage to this is that neither student will have the experience of "commanding" the airplane for the full trip and

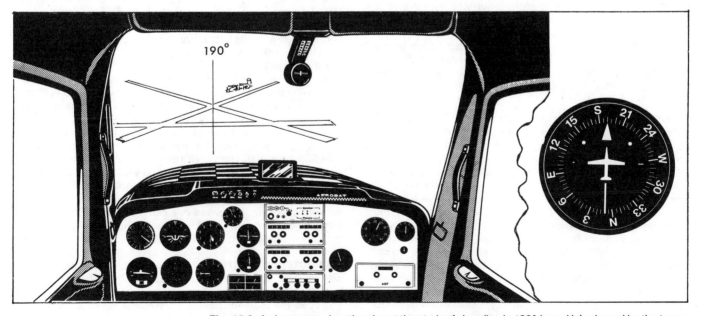

Fig. 14-3. As he approaches the airport the student's heading is 190° here. He's cleared by the tower to enter right downwind for runway 33. He can check the heading indicator and compare it with the various runway directions to help pick the proper one. Student pilots often have trouble getting set up for the proper runway. So be sure to discuss and demonstrate the procedure.

might have a gap in knowledge that could cause a problem on the solo trip. Another approach would be to make the trip twice, having each student act as "pilot in command" for a whole cross-country, while the other navigates.

Advantages of this second approach to training are obvious. Both students will have made the cross-country twice and been into the strange airports twice, once as a pilot and the other as a navigator/observer, and so they will be more at ease on the solo.

Don't allow the navigator/observer to sightsee. He's just as responsible for navigation and keeping up with checkpoints as the student in the front seat. Ask him questions as you fly. *Each* student, pilot or navigator, on *each* flight is responsible for his own preflight planning; and after the return, the times, fuel used, and estimates should be compared. Each will learn by the other's mistakes; or course you should avoid deliberately embarrassing a person when using this arrangement.

Usually you'll find that the person who navigates the first round will do better as pilot in command, when he gets his chance at it, than the student who flew the first trip. He's made the route before and has profited by the mistakes of the other.

This approach is *not* as valuable if the student will take another airplane (say, a *two*-place trainer) on the solo cross-country because on the dual he won't have been using the fuel system or radios of the airplane he'll take solo.

FLIGHT PLAN. (Throughout this book the phrase "Depending on the student, you'll do thus and so . . . " is used and since that is the "ultimate truth" as far as training goes, here goes again.)

Depending on the student, you should encourage the use of all facilities available to the pilot as early in training as possible, so that he will take advantage of them and be at ease with their use when he goes out on his own. *So, again, have the student file a flight plan for each flight from the first dual X-C if you think that one additional item wouldn't be too much for him.* With some of your slower students you may want to concentrate strictly on flying the airplane from point to point at first. You will save radio work or flight plan filing for that second (or third) required dual, before they're allowed to go on the first solo cross-country. You'll be able to decide about this after seeing the students' actions during the ground school and preflight planning session. Have each student keep his paperwork after the flight so that he can take it home and review the trip that night and on later occasions as well. (Encourage him to keep a copy of the flight plan and flight log on file.)

■ **The Flight.** The preflight inspection of the airplane should be thorough as always. Expect a tendency by the student to rush the check a bit, especially when a flight plan has been filed; he thinks that something catastrophic will happen if his flight plan isn't activated exactly on the dot as estimated. Don't let him overlook any inspection items because he is thinking about the trip.

A clipboard is handy for keeping charts, flight logs, and other paperwork, plus pencils, computers, and protractors. A cross-country checklist on the back of the clipboard would be a help to assure that all required equipment is being taken, including the latest *Airport/Facility Directory* and a reliable timepiece. It delays the trip to discover after leveling off that the charts are back in the airport office. You might keep a permanent spare set, with the training routes marked, in the airplane for "emergencies" such as this; but remind the student that he won't always be so lucky and should check the items before leaving. Again, file a flight plan every time, if possible.

Okay, you are now leaving on what the newspapers always call a "routine training flight." Notice that *all* aerial disasters occur on a "routine flight" or "routine training mission," at least as the news media state it. If a disaster occurs, it could hardly have been a routine flight. (Since accidents only happen on routine flights, a great step for safety would be to call each flight a "special flight," which naturally would eliminate all problems.) Anyway, on the cross-country, as on all flights, be sure that the student's preflight work will result in "routine training flights" that *stay* that way.

PRETAKEOFF AND CLIMBOUT PROCEDURES. After starting the airplane, set the heading indicator to the magnetic compass. Check the heading indicator for drift after the airplane is taxied to the run-up area. However, indicate to the student that the deviation may be different for different headings and this might be a factor if the heading in the run-up area is different from that at which the heading indicator was originally set.

The pretakeoff check should be thorough with special attention to electrical and vacuum system operations. The radio should be checked with Unicom if possible, or if there isn't a communications facility on the field, the student can set up the Unicom frequency and listen for other planes transmitting to ground stations.

The heading indicator should be checked and set with the runway heading or the magnetic compass again just before starting the takeoff run.

The time off should be noted for fuel consumption check purposes.

There is always the decision of whether to have the student leave the traffic pattern and go on his way or to require him to fly back over the center of the airport. For the first couple of flights, at least, he should fly back over the center of the airport. It's best for initial training purposes to have him start where the line drawn on the map starts—the center of the home airport. He will normally be close to the cruising altitude by the time he gets back over the airport (and so will have a good check of groundspeed, without considering a lot of climb to be computed), and can get a time check at that point. On solo cross-countries he can call in to the Unicom (if available) at that point and say, "En route at 4500 at 1024 CST." Remind him that the other method of getting started will usually have him slightly to one side of the course line but will be used after he gets more experience.

If the trainer you are using has fuel tanks that can be selected separately, you should be sure that the student is at ease with switching tanks and knows the proper procedure if he runs a tank dry. For most airplanes it's best *not* to deliberately run a tank dry, and for student pilots, the *rule* is *never* to deliberately run a tank dry. You should describe the sounds, feelings, and moves to make if he accidentally runs one out.

A good suggestion for the student (or anybody) on a cross-country is to change to the other tank shortly after he levels off and has cruise established. If that new tank presents problems he can switch back to the one used for takeoff and return to the airport. If he uses all the fuel in that first tank without checking the second one early, he might find too late that the second doesn't work—just when he's over bad terrain.

He should activate his flight plan.

EN ROUTE INSTRUCTION. Point out checkpoints and generally oversee his flying and navigating, particularly the first leg. With the average student, on the first dual cross-country you'll do the most instructing on the first leg of a triangular cross-country and by the third leg will only be required to make a few comments and suggestions. With some students you'll be working the entire trip.

Basically, your main instruction here will be to "direct his perception" so that he can relate what is on the sectional chart to what is on the ground. You'll point out specific references and checkpoints and teach him to estimate his distance from those points. On the dual cross-country flights you'll be teaching orientation by calling his attention to such things as a bridge over a creek, small towns, etc. Show him that villages may have the same symbols on the chart but vary in size from a store and a couple of houses to several stores and service stations, plus 20 or 30 houses. Before the flight he'll have been briefed on the landmarks on the chart and now can point out and compare them. You can't expect unlimited visibility on your dual or his solo flights, but if you plan on a successful *first dual* you should have a *minimum of 5 mi* for good pilotage training.

Check his cruise control procedures and remind him to keep looking around for other airplanes.

MAGNETIC COMPASS. Before letting the student go solo you should have him fly one leg of a trip using the magnetic compass only. He may be having problems using the heading indicator, and you'd want to work this out first, maybe even requiring a second dual before turning him loose. But he should be able to operate the navigation and communication equipment and to use the magnetic compass as the *only* heading reference. One problem that sometimes comes up is that the student may set the heading indicator to the runway heading at the home airport and doesn't compare it with the magnetic compass *or* set it to the runway headings at the other airports. These pilots may visit strange and exotic places as they blindly follow the indications of the heading indicator which, as a result of normal gyro drift and maybe a couple of hard landings at the strange airports, is pointing 45° or more off the actual heading. *The heading indicator should be set to the magnetic compass* (when it is settled down) at least once each leg of the cross-country and every 15 min on long flights. Emphasize this on every leg of every dual cross-country flight you make.

You'll play each leg and trip by ear, noting new items or special situations en route and at the other airports.

THE STRANGE AIRPORT. As you approach the airport, have the student call Unicom to get wind and traffic information. If no Unicom, FSS, or control tower is available, he can circle the airport *above* the traffic pattern altitude to check wind and traffic and then descend to report entering traffic (and/or downwind leg), giving his position and what runway he plans to use at "Houston County Airport," on 122.9 MHz. This is a good operating practice so that other airplanes in the immediate area will know what's going on. He should also report on final on 122.9 at an airport without the facilities just mentioned. (Sure, you do this yourself, but the student may not have been given the word.) You may have been requiring this at the home airport all along so it may be routine for him already.

As he enters traffic, point out the wind indicator, taxiways, and other items covered in the preflight briefing. Emphasize that he should keep a sharp eye out for other airplanes. Remind him that pilots tend to be high on an approach to a strange airport. A slightly wider downwind leg is a good idea; it gives him more time in the pattern to see the layout and also to avoid the too-high final. You might have him shoot a couple of dual takeoffs and landings to a full stop and taxi back at each of the strange airports if you think that would help him be more at ease during the solo trip.

Have him taxi in to the tie-down or parking area under your or the line attendant's direction. Secure the airplane and fuel up as necessary. If you plan on having him fuel up there on the solo cross-country to give a little extra margin that isn't needed on the dual, fuel up so that he'll be familiar with the

procedure. This first dual, you'll want to cover details he'll face when he does it solo. You won't be doing this on later dual cross-countries, because he'll need to be more and more on his own. However, that first solo X-C should be well rehearsed on the dual since there may be some trepidation on the student's part.

Stop for a soft drink, and introduce the student to the people at the airports; mention that he will be coming back solo. This will make him feel better about coming alone, and it is particularly important for your more timid students. If you've filed a flight plan, include in it time for an extra 30 min per stop, and remind the student that he'll have to take any ground time into consideration when he files on a solo trip.

You'll have some comments and suggestions during the flight and should have a general review after getting home. Check the actual times over the checkpoints and the fuel consumption.

■ **Evaluation and Review.** You'll decide after the flight whether he is ready to go solo or not. Your evaluation will be close to that used for the first solo. Does he make corrections as needed without prompting from you? If he gets off course badly, and doesn't realize it, and you have to tell him *how* to get back, he's certainly not ready to go by himself. How did he react to the strange airports? Was he overly confused about traffic patterns and taxiing? Can he use the sectional chart and radio with accuracy?

If he had problems on the first leg but did a good job on the last two, you can feel that he's ready to go. You'll be more interested in the trend in his proficiency than in requiring a perfect flight all the way.

If in doubt, fly the trip again dual. This, of course, can be bad for morale, but it's better to be on the safe side; and like in presolo work, you'll find that some of your slow starters do very well as they get experience.

It would be a good idea to have the student actually work with VHF/DF either on the dual cross-country or on a local flight, if the facility is reasonably close, before letting him go, so that he will be at ease on the first and following solo cross-countries. You should have discussed the procedure for this and the VOR and ADF orientation during the ground sessions, but it's best to have him do one himself.

A large percentage of training airplanes now have transponders, including altitude encoders and if your airplane(s) is so equipped, you should have introduced and demonstrated its use during the pre- and early postsolo periods.

The student should be well checked out in transponder operation before going solo cross-country in an airplane with one. You can sit with him in the airplane on the ramp (engine dead and master switch OFF) and have him set in various codes under your direction, explaining that 7700 is to be used for an emergency but otherwise left alone. Explain the use of the DIM rheostat, the IDENT button, and the STANDBY, ON, ALTITUDE, and TEST, as applicable. Be sure that you have the transponder back on 1200 after all this, because the guy who turns it on for the next flight might not notice and take off squawking "7700" or some other attention-getting code.

THE FIRST SOLO CROSS-COUNTRY

The first time you send a student on his first solo cross-country you'll find that time can pass very slowly.

After you've decided that he is proficient enough to go alone, sign his student certificate (Fig. 14-4).

Brief him just as was done for the dual. Check his paperwork (flight log, etc.) and oversee his weather check and flight plan filing, if indicated. Review the airport layouts.

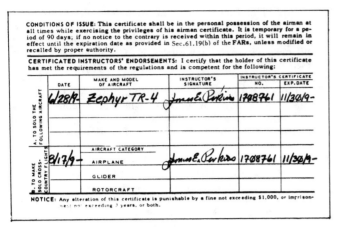

Fig. 14-4. Sign the student certificate when the student is ready to go solo cross-country.

You might run the dual in the morning, and if he's not too tired, have him fly the solo in the afternoon following the same directions of flight (check the sun position). This is a good plan if it looks like bad weather could move in during the next day or so.

■ **Winds and Weather.** Carefully consider the weather and wind conditions before you let him go; 3 mi is legal for a student pilot's (daytime) VFR flight, but if you send out a solo student on a 50-NM—or more—cross-country with only 3 mi of visibility, you (and he) can expect some problems. Of course, you can't set hard and fast rules for all combinations of routes and student ability, but you should think in terms of a minimum of 10 mi visibility for the first solo. (Remember 5 mi was a suggested minimum for the first dual.) You may lower the solo visibility minimums for later cross-countries for individual students.

The surface winds are a factor to be considered, too. It's good to send a student out on solo X-C in the morning, maybe getting started shortly after sunup. The wind will usually be calm, and he'll have the maximum number of hours of daylight if he needs it (let's hope he won't). In this situation, while you're briefing him, *keep a check on the local wind.* You may get the future surface wind in the terminal forecast, but that information may not be geographically close enough to be accurate for the airports he will be using.

Local factors such as mountains and other terrain features could cause radical variations from other airports. For instance, as noted earlier, the Sewanee (Franklin County) Airport is close to the edge of a plateau, and the single runway is aligned so that much of the time a crosswind exists. A student departing from such a situation early in the morning could find on his return later in the day that the crosswind is too strong for a safe landing. In an early morning briefing, check the *rate of increase* of the wind velocity. If it was calm when you started the briefing and only 30 min later is hitting 15 K and gusting, you'd better figure that it will get much higher as the day moves on. Your decision will be whether to let him go (there's no problem of his getting off) or delaying the flight.

You have another option, though. In Chap. 8 the idea of using an alternate airport was discussed. This was an airport nearby where the student had landed during some part of the pre- and immediate postsolo work, one he's familiar with. For example, Winchester Airport is 9 mi west-southwest of Sewanee (Fig. 14-1) and is in the flatland. The winds there are usually lower in velocity and closer to being lined up with the runway. If you knew that the wind was strictly a home airport problem, you could brief him on the possibility that he may have to land at Winchester (as an example) on the return flight, in which case you'd come down and pick him up. This

way you might salvage a flight (it may be the best visibility that you've had for weeks) and the student is prepared for an alternate plan of action. Advise him to contact the home airport's Unicom on the return leg while he's some distance out and get a check on the wind. If your airport doesn't have a Unicom, he can look the situation over and decide what to do. Keep it casual so that he won't be worrying throughout the trip about the wind at home. Naturally your situation may not be as clear-cut as that for Sewanee and Winchester, but you might be able to get in additional solo cross-countries by looking at some simple alternate moves.

■ **Reserves**
DAYLIGHT RESERVE. The first solo cross-country should be started soon enough that it's completed at least an hour before sunset, based on the most pessimistic estimate of time required. This is a good margin for any solo cross-countries, but you may decrease it slightly for the more advanced—and sharper—students. However, it's bad enough to be sweating a slightly overdue student without the added factor of impending nightfall pushing at you. If in doubt, require an extra-early start.

The student should have *at least one hour* of dual instruction at night (including a short trip to a nearby airport you have flown into during the day) before the first solo cross-country is flown.

FUEL RESERVE. Most of the trainers in use today have a 4-hr or more endurance at normal cruise power. Require a 2-hr fuel reserve for student solo cross-countries. This means that on the long cross-country (the 100-NM legs) he must fuel at least once, if not twice, during the trip. It wouldn't be a bad idea to have him refuel at both stops to get a rest break and more experience, plus having an added safety factor.

It would seem that refueling every stop in a solo triangular flight of 50-NM legs wouldn't be necessary, but you should look at each situation and decide.

For Pete's sake, no student pilot should be sent on a solo cross-country without having the maximum amount of fuel (based on weight and balance requirements) *on board.* This may seem elementary, but students have been sent out without full tanks. ("Go ahead. It's only flown about 30 min and I don't want to gas it.") Thirty (or 15, or even 5) min worth of fuel can mean a great deal in some situations. You're not only training him for specific cross-country flights but are establishing his attitudes for later when he'll be carrying passengers and on his own.

■ **Logbook.** Have the student take his logbook to get it signed at each stop.

Before he goes, you'll endorse it: "Briefed and approved for a solo cross-country from Homeville to Landville, to Windville and return, in a Cessna 150 this date, June 1, 19—, Lee V. Yellowhammer 442723CFI." You might not use the same words (possibly your name *isn't* Lee V. Yellowhammer), but you are naming the route, the date, and the airplane you're approving. Do this for each solo trip after you've briefed him.

Before each solo remind the student that if he is delayed, or thinks he will be delayed, for 30 min or more, he's to call the home airport.

You won't be sitting in the office all the time he's gone because you will naturally have other flying to do. But other people in your operations should be aware of where he's going and when he's expected to return.

If possible, you should be available for immediate debriefing. Have him review the trip and you can make comments and suggestions. Check the fuel consumption and actual en route times.

In Chap. 11 it was emphasized that if a student comes dragging back late or calls from another airport, it's sometimes hard to keep your cool, but keep your cool you must. It's a good rule to not start worrying until 30 min past his ETA home, allowing a reasonable time for ground visiting. You'll find with experience that student pilots tend to kill more time than you expect. They may circle the strange airport twice to check it out, or have trouble starting the engine there. Also, there are interesting people to talk to and antique airplanes and/or homebuilts in the hangar, so they take more time than they plan to.

SUGGESTED CROSS-COUNTRY TRAINING

Following are some ideas for setting up and flying your cross-country syllabus. It's assumed that you are flying from an uncontrolled airport. The first dual and first solo have just been discussed but are included here again for continuity.

■ **First Dual.** Preflight brief, and fly to two other airports. Decide whether the student is ready for solo X-C to these airports or needs another trip. Again, if possible have the student file a flight plan on all cross-country flights and also check his fuel consumption on completion.

■ **First Solo.** After deciding that he is ready for a solo flight, sign his student certificate. Brief and check his paperwork and weather information. Review the dual trip. Sign his logbook for this flight for this date.

■ **Second Solo.** Brief him and check his preflight work for a triangular trip to two airports he hasn't flown to previously. Draw the airport layouts on the chalkboard and discuss any facets of the trip that might be unusual or that he hasn't experienced before. Sign his logbook for this particular trip. Fig. 14-5 shows a sample first "new" trip, with landings at two airports. Debrief after the flight.

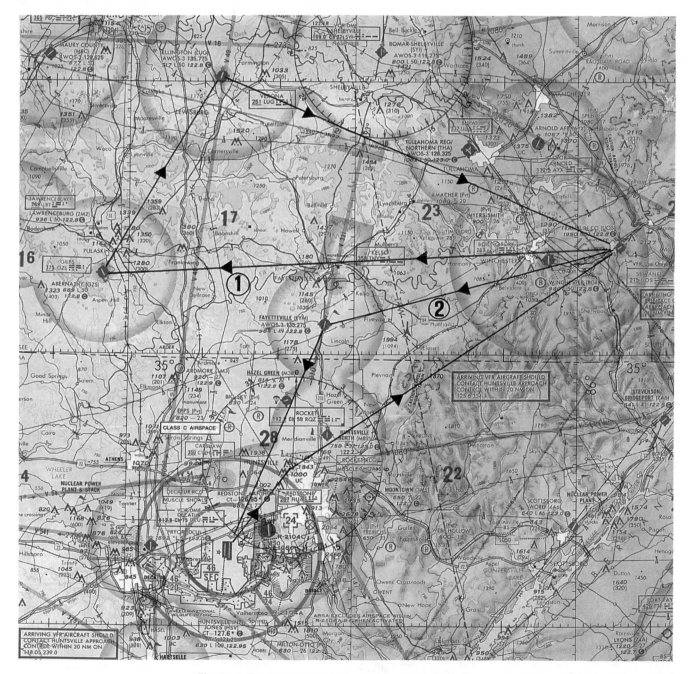

Fig. 14-5. Two sample trips to give an idea of the student's second solo (1) and your dual (2) to a small airport and then to a controlled airport.

■ **Second Dual.** While you could let him go on more solo flights before flying another dual, it is a good idea to fly cross-country with him after the second solo for two reasons: (1) to check his progress to see that he is not developing bad habits and (2) to introduce new material as early as feasible in his cross-country training.

This dual should be to a controlled airport using pilotage and radio nav as best suited to the situation (Fig. 14-5). Brief on the communications procedures and the layout of the airport. Review light signals—as out-of-date as this procedure may seem, it's good to have a copy of the signals permanently in the airplane where he can use it as necessary (Fig. 14-6).

Color and Type of Signal	On the Ground	In Flight
STEADY GREEN	Cleared for take-off	Cleared to land
FLASHING GREEN	Cleared to taxi	Return for landing (to be followed by steady green at proper time)
STEADY RED	Stop	Give way to other aircraft and continue circling
FLASHING RED	Taxi clear of landing area (runway) in use	Airport unsafe—do not land
FLASHING WHITE	Return to starting point on airport	
ALTERNATING RED & GREEN	General Warning Signal—Exercise Extreme Caution	

Fig. 14-6. The light signals could be typed up, covered with celluloid, and taped somewhere in the cockpit (or kept in the map compartment) for reference when needed.

BRIEFING. During the preflight briefing, rehearse going into the controlled airport following step-by-step the moves in dealing with Automatic Terminal Information Service (if available), approach control, local control, and ground control. Cover the departure from the big airport from listening to ATIS and talking to clearance delivery before taxi, then to ground control, local control, and departure control, as applicable.

This would be a good time to talk about wake turbulence again. Tell him to avoid it by asking for another runway, circling, or delaying his takeoff, as the case may be. The student pilot may be slightly intimidated by the tower and will land—or take off—when he is in doubt about wake turbulence. Remind him *he* is responsible for the airplane. Review on the chalkboard the techniques used to avoid wake turbulence and cover the best moves to make if it is encountered. Advisory Circular No. 90-23D covers its actions, and you can use it or its replacement as a guide to your discussion. This AC should be available to each student. (If you've been flying from a controlled airport all along, you would go to a controlled airport unfamiliar to the student—with more complicated requirements and more traffic, if possible.) The AIM—*Basic Flight Information and ATC Procedures* has a good section on wake turbulence, also. Review collision avoidance procedures.

THE FLIGHT. Visit the tower at the big airport and introduce the student to the controllers. Mention that he'll be coming back solo.

When asking for takeoff clearance, tell local control that you'll be staying in the pattern for some full-stop landings. Shoot several to let the student get more experience in the controlled environment and working with the tower. The pilot certification requirements say that before the private certificate is issued, the student must have had three solo takeoffs and landings to a full stop at an airport with an operating control tower. On his solo trip to this airport he can fulfill this requirement, but it's best that he have some dual practice if flying at a big airport is new to him.

If you can, also plan a stop at an airport with only a Flight Service Station so that the student can get experience in using Airport Advisory Service (AAS) and seeing the functions of the FSS. If the big airport has a Flight Service Station and/or a Weather Service Office, a visit there would be valuable. He should see the layout in the Weather Service Office so he'll know what services are available.

On the return home, require him to start diverting to another airport or set up a dragging-the-area situation. Or you may have the student put on the hood and turn and climb, or descend, to get out of simulated bad weather.

After the flight, review the trip and cover any points of weakness. He will probably have questions.

■ **Third Solo.** Brief him for a solo trip to the controlled airport visited earlier. As always, check his weather and wind information and navigation computations. Review the airport layout and communications procedures. Remind him to shoot a minimum of three takeoffs and landings to a full stop. Review wake turbulence avoidance.

Sign the logbook for this trip (and be sure he takes it with him).

See that the light signals are available.

Have him file a flight plan.

Make sure he has charts, the flight log, the A/FD, and other necessary paperwork.

Debrief and review the trip after his return.

■ **Fourth Solo.** If he's progressing suitably, the next and probably final flight should be the triangular flight with 100-NM legs. Pick new airports that give the best combinations of experience, perhaps one with an FSS and the other with a control tower. Brief, as always; check that he's taking the proper charts, A/FD, etc.; sign his logbook; and establish an ETA for his return. Allow plenty of time to complete the trip before dark. He should file a flight plan for all legs. Make sure he has money for fuel or can charge it.

Debrief and have him review the flight.

SUMMARY

The trips just suggested may not exactly fit in with your geographical and training situation; they are intended to demonstrate some general points. The 3 hr of dual and 10 hr of solo cross-country are considered minimums by this instructor, and you may exceed these as needed for an individual student. For instance, if a student gets lost or has other problems on a solo cross-country, sending him out immediately on another solo would *not* be a good procedure. You'd want to fly cross-country with him to work out his problems and put both of you at ease before he goes out solo again.

The student will also be flying local dual and solo flights, working on maneuvers during the cross-country phase. There will be days of conditions suitable only for local operations and you'll want to get on with his training.

On occasion, a student may ask to make a solo cross-country on which he will spend the night at another town and return the next day. As an overall policy, overnights like this should be avoided because of possible weather problems. Of course, if weather is forecast to be outstanding for the next week, if the RON location is reasonably close, and if he calls before leaving the other airport, you might consider an overnight but should coordinate with an instructor at the other airport. Sometimes maintenance problems arise (fouled plugs, flat oleos, ignition lead problems, etc.) after an airplane has "cooled down" overnight, so there may be the problem of your having to go get the airplane.

In the cross-country training you should provide as many combinations of trips and airports as safely feasible. Your approach should be to brief and supervise carefully as he starts each *new* type of trip, and gradually extend his boundaries and lessen your grip on his actions. The first solo should be a repeat of the first dual—to small airports in uncomplicated situations. He should then fly solo to uncontrolled airports new to him and farther away, with the number of trips depending on the solo time required.

The introduction to flying to a controlled airport follows the same idea. You'll fly a dual "rehearsal" and he will make the same trip solo. You can then send him solo to a different controlled field farther away, and so on. You'll set up a program so that you'll feel that he can carry passengers safely to either type of airport before you sign the recommendation for his flight test.

The average private pilot spends many more hours flying cross-country than doing turns around a point. While he should be safe in doing such maneuvers as turns around a point, don't let the cross-country training suffer while you are trying to attain perfection in specific maneuvers.

Some instructors like to have the student get the Knowledge Test out of the way as soon as possible, but for the best results wait until at least after that first solo cross-country when he has actually used the information. His chances of passing will be much better.

His attitude toward preflight planning and weather checking (and not going, if he's in doubt) will be based on what he's been taught. You'll be doing his future passengers a great service by establishing good preflight habits. Many cross-country accidents can be traced back to a lack of preflight preparation and the ignoring of available weather and other information.

TRANSITION FROM STUDENT TO PRIVATE PILOT

BACKGROUND

This chapter is aimed at "pulling it all together" and getting the student ready to be a private pilot. Sure, your main job has been to teach him to be a safe pilot, but you'll also have to consider the practical aspects of passing the Knowledge and Practical Tests (as per FAA requirements), or he won't be able to use all that good safety training as a private pilot.

KNOWLEDGE TEST

Some of your students may not have taken a knowledge test for 20 years and may have forgotten not only how to study but also how to take a knowledge or practical test.

You'll find that individuals with less education will usually fear the Knowledge Test more than the Practical Test, and those with college degrees tend to breeze through the Knowledge Test and worry about the Practical Test. Naturally, there are exceptions to this "rule," but you'll be surprised how often it applies.

Because some of your students don't have experience in the mechanics of taking tests, you'll have to work with them on this requirement. They may have the knowledge but are unable to write it down, even in a multiple choice test, so you'll run into people who can answer oral questions on the subjects required, but whose encounter with the FAA Knowledge Test is nothing less than a disaster.

These people need sample tests, plus added work on your part, strictly in learning how to take the tests.

Having taken FAA Knowledge Tests yourself, you're aware of the controversy concerning whether the questions are "tricky." To persons not used to such tests, the answers *can* seem ambiguous.

During his flight training the student didn't get, or use, all of his knowledge at once. On the Knowledge Test, he will be expected to be able to produce any or all of his theoretical knowledge at one sitting, and this is difficult for some.

If he has been taught correctly, using integrated ground and flight training from the beginning, the preparation for the test will be a brief review of theory and some sample questions to get him oriented to taking the test.

All Knowledge Tests are now given at computer testing centers and the grade is available in a few minutes after completion. You should have Knowledge Test guides for the areas in which you are instructing, such as *The Recreational Pilot and Private Pilot Knowledge Test Guide* AC 61-117 plus Test Guides for Commercial, Instrument, etc., as required.

After the student has taken the test and gotten his grade slip back, go over the areas in which he was weak, as indicated by the coded references on the slip (A09, A22, B08, etc.). Sometimes you'll run into the situation where the student is shown to be weak in a certain area and, after questioning, you find that he is very sharp there. It's possible that he made a mistake in selecting the answer(s), but you should approach the subject from different directions to be sure of what he knows.

A high grade on the Knowledge Test makes a good impression on the check pilot. A *98* rather than a *70* shows that the student at least did his homework. It doesn't show how well he can handle the airplane, or what his judgment is like, of course—that will be seen on the Practical Test. Professional people will usually make the best grades on the Knowledge Test, as will women; both groups generally are more faithful about hitting the books.

Students sometimes fail the Knowledge Test the first (or second) time. If you think that a mistake has been made in scoring, you can write the FAA Aeronautical Center and have them *re*check it. (It's standard for all failed tests to be checked by a human, just in case the computer goofed.) This *will* happen: You have a student who knows the subject and has answered your questions during the review, and he brings a grade slip to you with a big fat *60* on it. The chances are that your next move will be to have him take a number of sample tests, just to get used to the process.

Okay, he's failed the Knowledge Test, so what's your best next move? If it's his first failure you can give him further instruction and, when he's ready, sign a statement to the effect that you find him competent to now pass the test, or he can wait 30 days to take it again. Subsequent failures mean a mandatory 30-day wait, so some of your students with test problems can be getting pretty far along in age following a number of failures.

So, again, try to make sure that he is ready to take the Knowledge Test and *pass it the first time;* for students with lower education levels, prepare to spend some more time on just how to take the test. Both you and the student will start losing confidence with each failure.

PRACTICAL TEST

Appendix A is a sample application for an airman certificate and/or rating. The first few times you fill it out, better have one of the more experienced instructors look it over, since it would be embarrassing to have a student sent back because of an oversight in the application.

■ **Recommendations.** More important than the paperwork involved is whether you think the new private pilot is ready to go out on his own; he can fly months without getting any sort of dual. Just ask yourself this question when you start to sign a recommendation: "Would I let this person fly my family?" Of course, he isn't as skilled or as experienced as you are, but the main thing is, does he realize this and would he stay on the ground if he has doubts about weather or other factors? If he has average or above average skills in handling the airplane, your decision will be based on your knowledge of his judgement, based on the hours spent with him.

You'll have people come up and say, "I've been flying with my brother-in-law, who is an ex–military pilot and holds a commercial certificate. We figure he can teach me the maneuvers and we could get you to ride and recommend me for the flight test." As you know, flight instructing is not just a matter of "demonstrating maneuvers" or being a good pilot; it requires the ability to teach and to instill judgment. You don't know what kind of judgment either the speaker or his brother-in-law has. They do have a certain amount of gall to ask you to lay your certificate on the line for a couple of bucks worth of flying. (And if you think you aren't laying your certificate on the line when you sign a recommendation, just have your students flunk several flight tests in a row.)

Maybe with experience you'll be able to fly one flight of private maneuvers with a person and read him well enough to recommend him, but in the beginning you should use a rule of thumb of five dual flights or 5 hr of dual, whichever is longer, with a student before signing the paper. Admittedly, this is tough on a student who has more than enough flight time for a private certificate before he transfers to your operation, but you can check his logbook and discuss his background with his former instructors if you want to ignore the minimum dual just mentioned.

In your early experience in instruction you might want to get the opinions of more experienced instructors by having them ride a "prerecommendation ride check." Many flight schools routinely use the procedure of such cross-check rides during the training. Bad habits that are overlooked by one instructor may be seen by another; sometimes you're so involved in the overall preparation of your student that you don't pick up problems as early as another would.

So your decision on recommending a student pilot will be mostly based on his headwork and maturity in flying. If you've had problems with his actions on solo flights (he likes to fly *low*), you should have talked to him and, failing to see improvement, discussed his problem with your Flight Standards District Office people. The thing to consider is that he might get worse on his own. Of course, if he is a real menace you would have made sure he was grounded; the types with marginal judgment will cause you the most worry.

Assuming that you've recommended him and he's going off to take the test at another airport, better go over the checklist as given in the *Practical Test Standards,* FAA-S-8081-14 (or the latest issuance).

Another bad scene, in addition to having the airplane needing more controls for the flight test (FAR 61.45), would be to not have the Airworthiness and/or Registration Certificates in the Airplane.

■ **The Flight.** If the airplane has seating available the instructor is allowed to ride on the Practical Test and (quietly) observe. Generally speaking, while this is educational for you, your presence could put added pressure on the student at a time when he doesn't need it. You will also find that you are using body English you've never used before as he's not *quite* steep enough in that bank or needs to start the roll-out *now* (Fig. 15-1).

Fig. 15-1. Riding as an observer on your student's check flight can be mentally and physically exhausting.

Particular circumstances might warrant your riding shotgun, but as a practical rule it's better not to ride with the student on his private Practical Test. For more advanced checks (such as an instrument or flight instructor's ride) you might discuss the possibility of going along with the individual involved, but don't put extra pressure on him.

You may want to fly to the other airport and do a couple of landings and takeoffs with the student before the check ride if he's unfamiliar with it, and then listen to the oral part of the exam. The examiner will likely have general records of accidents and incidents and will quiz and talk to the student on subjects where weaknesses have been found, as shown by statistics. No, you aren't listening so you can get the next guy past his oral, but you will learn perhaps that you need to cover certain areas of, say, the FARs better than you have before.

CHECKITIS (YOURS). The check ride, like the first solo cross-country, can take a very long time, whether you're waiting in the other airport's office or at your home airport. (You watched the takeoff and thought it was lousy, and of course, this one time your sharp crosswind corrector let the airplane drift over into the next state on the climbout.) They've been gone an hour now (well, 30 min) and time creeps by on leaden feet. Finally, you hear or see them enter the pattern and casually ease over to where you can see better.

Your ground help is required to help him make that proper abeam distance on the downwind leg; he seems to be drifting in to the runway. It looks like he's high, and hot, and overshot, and you're sure that landing must have loosened the fillings in the check pilot's teeth.

If they taxi in after that one landing, you know that things were so bad that *it's all over*.

If they shoot another landing, you know that things were so bad that *it's all over*—the check pilot is working on the Fundamentals that have entirely slipped the student's mind, and you become mentally prepared to take a check ride yourself.

They taxi in and shut down and sit out there in the airplane and *talk*—the check pilot must be asking, "Did your instructor really want you to recover from a stall that way? He told you *what?!?*"

After several weeks, they get out and you casually (?) saunter out to meet them. There is a ritual involved at this point; the instructor is bound by unwritten law *not* to come right out and ask if the student passed. This would indicate anxiety and lack of self-confidence. On the other hand, the check pilot is bound by the same unwritten law to keep the results a secret from you as long as possible. You are allowed to ask, "How did it go?" to the student, but students aren't true judges of the results of a flight test—they tend to be pessimistic, so this is not a criterion. You may ask the check pilot, "How did it go?" and if he is canny and experienced he will answer with a noncommittal statement or shrug. This talent is developed by experience and is much admired.

When you get into the airport office, the game ends and the straight story comes out. He passed and did a good job, so you relax—*until the next time*—and a temporary airman certificate is typed out (Fig. 15-2).

The last few paragraphs were, of course, slightly exaggerated, and after you've sent a few students up for checks, you will find that you can confidently go on about your business, instructing or whatever, because you know where you stand as an instructor and the student stands as a pilot. There'll be an occasional failure because a student had an off day, but you'll take care of your, and the student's, checkitis by making sure he's ready.

Fig. 15-3 is a notice of disapproval of application (a "pink slip"), and the odds are that you may see a few of these during your instructing career, but keep them to a minimum. Your first pink slip will cause a temporary drop in your morale and confidence, but use it to learn something and improve your instructing. Talk over the details with the check pilot so that you can avoid the problem with other students.

Fig. 15-2. A temporary private pilot certificate.

Fig. 15-3. The pink slip, a notice of disapproval. Some instructors' reaction to a pink slip is close to that of the sea captain who got the "black spot" handed to him in *Treasure Island*. There may be instructors who've never had a student flunk a flight test, but they're hard to find.

POSTCHECK POINTS

After you've congratulated the student—er—private pilot, you might review a few last points:

1. You'll be available for help later in his flying career and can give him the biennial flight review.

2. He owes it to his passengers to do the best preflight planning and use in-flight safety procedures. They have (more or less) blind faith in his ability.

3. When taking nonpilots for a flight, particularly for the first flight, he should make the flight as smooth as possible, with shallow turns and continual explanations of what is going on.

4. His actions will be watched by students and less-experienced pilots throughout his career, and his influence can be good or bad.

5. If, by his careless or reckless action, he has an accident, he may be sure that the communications media will spread the word—usually inaccurately. ("He was on a *routine flight* and the airplane exploded in midair. . . .") He owes aviation something, too, and the fact that he's by himself in his own airplane doesn't mean that it isn't going to affect other pilots.

6. *You'll* keep an eye on him if he's flying from your airport, and be ready to clamp down on problems as they arise.

4 ADVANCED VFR INSTRUCTION

NIGHT FLYING INSTRUCTION

THE FIRST FLIGHT

■ **Preparation.** The most important piece of equipment for night flying (besides the airplane) is a dependable flashlight. Part of the preflight check should make sure that it's available for *that* flight. It's a bad feeling to be en route on a cross-country and find out you've forgotten it (Fig. 16-1). Sure, the airplane's electrical system has been operating well and you figure that the instrument lights will be available, but it makes for an uneasy feeling for the rest of the flight. As soon as you realize you've forgotten it, the instrument lights will flicker from time to time.

Fig. 16-1. Always confirm that *somebody* brings a working flashlight. Instruments are hard to read by matchlight.

Check the flashlight before the flight; if the light is not as strong as it could be, put in new batteries. An extra bulb stashed in the map compartment is a good idea, too.

While red lighting is usually reserved for night fighters and night attack or other military operations, it's a good idea to cut out a piece of red cellophane to fit under the lens to give you a red light. For many civil night flying activities, white lighting is used for the cabin and instruments because the abundance of outside white lights would destroy a carefully nurtured night vision, anyway. If you were going to fly off a carrier in nearly total darkness it would be worthwhile to spend 30 min beforehand in a red-lighted room getting night adapted.

Briefly discuss the eye's rods and cones and how the student must scan differently at night than in the daytime (there's a blind spot in the center of his scan because the day cells, the cones, aren't being used as much). He must look slightly to one side of where he thinks the object is. Also note that red light is considered best for preflight and flight dark adaptation.

Commercial "cold light" capsules sealed in foil packages (trade name Cyalume) can be kept in the cabin for emergency. If no flashlight is available, one of these may be removed from the foil and used as per directions. A cool green light will last a minimum of 8 hr and maps or other material may be read easily in the light. Since the total endurance of most airplanes is not much more than 6 hr, this length of illumination would more than cover the time required to get the airplane to a safe port. There is no danger of heat or fire, so the light may be stored out of sight for various times during its illuminating period.

■ **Explanation.** You should have a briefing session to tell the student exactly what you are going to do and what he can expect in night flying.

You might mention that the operation will seem a little strange at first (distances are harder to judge), but things will soon fall into place. Some of your students may be overly concerned about flying at night and you'll have to spend extra time reassuring them. "Automatic rough" should be mentioned—the pilot's imagination can make the engine sound rough after dark. Discuss *aircraft lighting* and *airport and obstruction lighting*. The AIM—*Official Guide to Basic Flight Information and ATC Procedures* manual has a section on these latter two subjects; have students read it before the first night flight.

Assuming that you'll be giving several sessions in order to meet the experience requirements, the first flight should be on the order of familiarization with the new condition and putting the student at ease. You might consider flying a short trip to a couple of towns that are several miles away but are easily seen shortly after takeoff.

Naturally it won't always be practical, but for your more timid students try to fly that first time when the weather is well above average. Starting off with bright moonlight at night would be better for them, if it can be done. Make a point to mention before the first flight that most people enjoy night flying very much once the initial strangeness has gone. Other airplanes and towns are more easily seen, which is an advantage over day flying.

Night flying, like all new phases of flying, should be well covered on the ground, and a suggestion is that the *first* ground session be completed at such a time that the preflight check, pretakeoff check, and takeoff can be done by the student before dark. It seems to work better for most new people if the transition to flying in the dark is gradual. This is also a good idea for the first full takeoff and landing session. After the first flight or so of each category (airwork or pattern work), you'll normally have the flights start in darkness.

Night flying should not be done until the student has had under-the-hood practice flying the airplane. In your later discussions note that it's possible that instrument conditions could be suddenly encountered at night and he would use the procedures described in Chap. 12 to get back out of it. (Indications of getting into instrument conditions could be a gradual, or sudden, disappearance of lights on the ground and "haloes" around the wing lights.) While in instrument conditions, the rotating beacon or strobe should be turned off to avoid getting vertigo; it can be turned back on after the airplane is out of the bad weather.

In each of your explanation periods you should compare what the student will be doing and seeing at night with the same maneuvers or flight regimes already flown in the daytime. The night dual should repeat flying already experienced in day flying; don't decide to introduce some exotic maneuver during a night session. While you might create an interesting psychedelic effect by spinning the airplane and watching the lights on the ground, the student would likely take a dim view of the procedure.

In a later ground session discuss the problem of taking off or flying in a sparsely lighted area and how it's possible to get a fixation on a single light on the ground near the horizon, with possible altitude loss as the airplane is "flown to it."

■ **Demonstration.** Depending on the student's experience, he should do as much of the work as possible from the beginning. You'll have to be even more alert during ground and pattern operations than in the daytime (if such a thing is possible). As always, keep demonstrations to a minimum, although you might shoot that first pattern and landing to show him what's required. More about that later.

■ **Practice.** The first night flight should be for familiarization; the student should fly the airplane as much as possible but not be given any demanding tasks until he's at ease. It's assumed here that he's been flying the same make and model airplane in his day work.

1. During the *preflight check,* stress that the aircraft lighting must be checked. Sometimes pilots start a flight in daylight hours and know, or suspect, that the flight could terminate in darkness but neglect to check lighting.

A flashlight with a white light is best for the preflight check, and you should emphasize that even more care is necessary at night to ensure that gas caps aren't left off or the cowling left unfastened. In the daytime the pilot might glance back and catch such a problem, but at night it could well be overlooked.

2. *Starting.* The position lights should be turned on just before starting so that people near the airplane are forewarned of the propeller action. If you are on a well-lighted ramp this would not be as vital, but even then, people walking or standing nearby might not see the shadowy figures (you and the student) in the airplane. If other airplanes are running up in the area, a "Clear!" would not be heard. It's startling to be walking across a darkened ramp with airplanes parked around and suddenly see the position lights of an apparently empty airplane come on. It makes the pedestrian a little more cautious about walking close to other parked airplanes. Make sure that *all other* electrically operated equipment is off for the start.

3. *Taxiing.* The sea of blue taxi lights can be confusing for the new pilot at a big airport, but if he's been flying out of it all along, he'll have a head start. The bad thing is to land at a large strange airport for the first time at night and have to grope your way to a well-hidden flying service. Ground control will be a great aid, but taxiing on a strange field can still be confusing.

Taxiing at night should generally be slower than in the daytime, and the airplane taxi or landing light should be used. For some airplanes it's best to leave all unnecessary avionics equipment off to ease the load on the battery during the comparatively low rpm (and thus comparatively low generator/alternator output) during taxiing.

Don't let the student put his head down in the cockpit while taxiing. It seems that this is the time the cockpit or instrument lighting needs to be set *just so* or other "vital" adjustments have to be made in the cabin. (Such actions are often the result of nervous tension about the flight.)

Turning from one taxiway to another or to the runway may result in the airplane turning into an area not lighted by the taxi light. Airplanes with the taxi light on the nosewheel don't usually have this problem (Fig. 16-2).

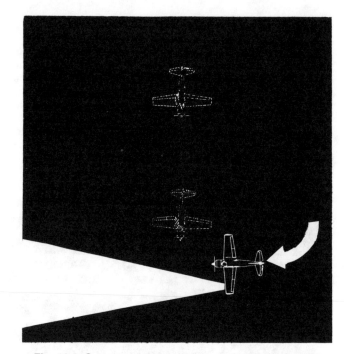

Fig. 16-2. Some taxi light installations leave blind areas in the turn.

4. *Pretakeoff check.* Usually the cabin lighting makes trim indicators, fuel selectors, and other instruments hard to see, so a flashlight should be used. You'll probably have to remind the student to turn off the taxi light while the check is being made. Advise him to take more time in the check than during the daytime. You may have to use a higher rpm while doing these trim and fuel checks than in the daytime, because of the added electrical load. Check *your* airplane for this.

5. *Takeoff.* Let him make the first takeoff, if possible, using the taxi and/or landing light. Warn that the airplane will *seem* to be moving faster on the takeoff run than it is and he shouldn't pull it off prematurely. (It will appear to be moving faster during the landing round-out, also.) After the airplane is well airborne have him turn off the landing light and adjust the instrument lighting. He may have to rely on instruments during climbouts in sparsely populated areas with few ground lights.

6. *Airwork.* The student will soon discover that on darker nights with no well-defined horizon he will have to rely more on instruments for attitude information than in the daytime. On your short cross-country and other night trips you'll likely have to remind him about heading and altitude corrections.

A good length for this flight is 45 min. More than that

would probably tire the student, since he'll be slightly tense to start with.

Generally, in addition to the short cross-country you should have him do some easy climb, turn, and descent exercises. Don't do stalls or slow flight this first time. It may seem that if he is going to make one landing (or more) at the end of the flight he should try one or two simulated ones at altitude. The arguments against it are (1) it could tense him up if he lets the attitude get out of hand due to lack of good references and (2) it's assumed that he has a number of landings in daylight in this same airplane and simulating them at altitude in the dark won't compare with using the landing lights on the actual landings. (The landing technique using a landing light is just like that of day landing.) In later flights he should do the stalls he's done in daylight, slow flight, and other maneuvers.

7. *Pattern.* You make the first pattern and landing. If he makes the first one without knowing what's needed, your verbal corrections may add a sense of urgency and tense him up. Let him relax and observe that first one. He's probably been worrying about the landing(s) throughout the flight.

After you've landed it, have him taxi it back and do one or more full-stop landings himself. Call for a slightly wider, more leisurely pattern. It will help his confidence if *he* has landed the airplane; he won't be worrying about it during the time before the next night session. It's best if there is no crosswind to cope with this first session, but he should get some night practice in those conditions later.

■ **Evaluation and Review.** After the flight, go over what you did and answer any questions he might have. Spend a little more time than usual in the postflight briefing to see that any anxiety is dispelled. His actions during this flight will direct your plans for the next one. If he has done well you might have him plan a cross-country to another lighted airport, doing flight log work and other planning just as for a day flight.

FOLLOWING FLIGHTS

Your decision as to whether the student should fly solo at night will hinge on his ability and your insurance. Generally, the preprivate student should fly dual only.

The 3 hr of night dual required for the private certificate will vary with individuals, from 3 hr of practically all straight and level flight to an almost full coverage of the private maneuvers, including ground reference maneuvers on bright nights. (Short- and soft-field takeoffs and landings would be better left for daylight.)

For most students it would be well to shoot one pattern (or more) without the landing and/or taxi lights used in takeoff and landing. With some students, however, you could go further and have them fly a pattern or two without using cockpit lighting, but this hinges on *your* experience and their reaction to night flying.

SUMMARY

You'll have to be more alert during night instruction, but as you gain experience you'll find it an enjoyable change from day work. (After a few hours of it you'll let a little of your weight down.)

Most of your night instructing will be the introductory type for preprivate students, and you'll soon work out the best general syllabus for your area and airplane type. When instructing these people, work as if they would need the information if inadvertently caught out after dark. You can do a lot *now* to avoid a panicky situation later by showing the student that he can cope with it. If people have "been there" before, they aren't so shook up if they accidentally get caught in such a bind. After all, that's the whole point of any flight instruction—you "take them there" under safe and controlled conditions so that they can handle it by themselves later.

INSTRUCTING IN COMPLEX SINGLE-ENGINE AIRPLANES

INTRODUCTION

Commercial applicants must have instruction in airplanes with flaps, retractable landing gear, and controllable-pitch propellers. Since you will be giving dual in these types you'll need to consider what to do for the most effective instruction.

The people you'll be teaching will obviously be more experienced than your student pilots, so you may tend to relax. *Don't.* These airplanes aren't as forgiving in stalls, high sinkrate approaches, etc., as the lighter trainers. They are also more complex, and you might overlook something (Fig. 17-1). (A favorite for many trainees and their instructors is to forget to open the cowl flaps after landing and then taxi back to the hangar or takeoff spot with a hot engine. Of course, the low point of your instructing career would be to get so involved in talking that you let the trainee land the airplane *gear up*.)

It's assumed here, at first, that you are already qualified in the machine but have not demonstrated it or checked anybody out in an airplane of this type, so this chapter will give ideas on *procedures in instructing. The Advanced Pilot's Flight Manual* and other books may be used for details on operation of advanced models and types. More about instructing in "strange" airplanes later in the chapter.

If you haven't flown the airplane recently, you'd better study the *Pilot's Operating Handbook,* sit in the cockpit awhile, and maybe fly it again before the dual starts.

The suggestions here are general ones based on a type of airplane rather than any specific model, so there may be individual requirements you'll have to consider as you fly. It's assumed in each section that the trainee has had no previous

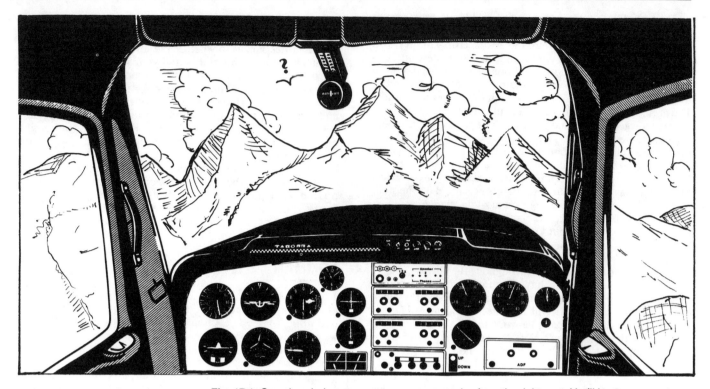

Fig. 17-1. Complex airplanes seem even more complex from the right seat. You'll have more systems to monitor and will need to keep up with the trainee *and* the airplane.

experience in the airplane type discussed but is familiar with flap usage. (If he isn't, you might review Chap. 10 with him).

Make up a multiple choice or true-false written test for each of the complex airplanes your school uses and make sure the new pilot passes it before taking the airplane out by himself. You may have to set up a special quiz for the new owner of an airplane that is not used by your school.

Remember that these airplanes will have more complex fuel and oil systems. Some may be fuel-injected, a new concept to most of the trainees. You'll also have to be sure that there are no misconceptions about these systems that could lead to incidents or accidents.

AIRPLANES WITH CONTROLLABLE-PITCH PROPS (FIXED GEAR)

Reference—Chap. 12, *The Advanced Pilot's Flight Manual.*

This is likely to be the first type of complex airplane you'll be instructing in. If you plan to set up a commercial course, you might consider using a system of a couple of hours dual and some solo in a fixed-gear Skylane or Cherokee Six before moving on to a retractable, because retractables also have a controllable-pitch prop. Hitting former Cessna 150 or Cherokee 140 pilots with too much at once could cause learning problems. Economics might not allow for the use of two types of complex airplanes, and your training procedure would be modified. That will be discussed later.

Assume for simplicity's sake that you are assigned to check out a pilot in an airplane with a controllable-pitch prop and fixed gear, so for now the only new techniques he'll have to learn are the prop and manifold pressure procedures. How do you go about it?

■ **Ground Instruction.** One thing instructors sometimes overlook is that the airplane itself is new to the person who's

stepping up to a controllable-pitch prop. Since manufacturers often move controls and instruments from place to place from model to model, you should give him a chance to sit in the cockpit to see where the engine controls and instruments are. Be sure he pins down the positions of the manifold pressure gauge and tachometer, so that on the first flight or two he won't have an added factor of confusion when setting power. Taking it step by step, you should cover all items discussed below.

MANIFOLD PRESSURE AND RPM. A ground session on the theory and use of manifold pressure/rpm combinations before the flight would be good, preferably the day before so that the trainee can have a chance to absorb the information.

1. *Manifold pressure.* Be sure he knows how and where it's measured. It is the absolute pressure of the fuel-air mixture in the intake manifold (*m*anifold *a*bsolute *p*ressure hereon called "map") and is measured in inches of mercury. When the engine isn't operating it measures about 30 in. Hg (29.92 to be exact) at a sea level airport on a standard day. When the engine starts, the engine acts as a "suction pump" and the manifold pressure drops to less than 10 in. at idle. As the throttle is opened the manifold pressure increases until the maximum of slightly less than ambient pressure (due to slight loss of efficiency) is reached—this is for an unsupercharged engine with no ram effects. Explain the brake mean effective pressure and manifold pressure relationship.

Discuss the principle of the supercharger and throttle-handling in using the supercharged engine (overboosting possibilities).

2. *Rpm.* Expound on the variable-pitch prop and how it is more efficient by getting a better lift-to-drag ratio at more speeds. The fixed-pitch prop is only at a peak of efficiency at one airspeed. Cover geometric and effective pitches.

The constant-speed propeller is the most useful and popular of variable-pitch propellers. Discuss the propeller of the airplane to be flown, noting such things as oil pressure/counterweight operations, which increase or decrease the pitch, or oil pressure/air pressure propeller operation. Explain that there are high- and low-pitch-limit stops and note the reason why the propeller isn't controllable below certain values of rpm and/or manifold pressure. (When the throttle is closed the propeller can't maintain the preset rpm but moves back down to an idle value.) To *change power,* move the propeller forward first when power is increased. It will help the pilot to remember that the prop control will be ahead more than the throttle. (To reduce power, throttle back first; to increase power, move prop control forward first.) Also cover the effects of losing oil (or air) pressure for the various types of constant-speed propellers. You might briefly go into the feathering of propellers, even though it's not a factor for single-engine airplanes.

SETTING POWER. One of the questions most asked by people starting to fly an airplane with a constant-speed prop is, "Why are there several manifold pressure/rpm combinations for a particular power being used?" Fig. 17-2 gives a good illustration of this.

You might use such a table to show, for instance, several combinations of map and rpm for a given amount of power, as underlined for 65 percent power at points *A, B, C,* and *D* in Fig. 17-2.

The combination of "high" manifold pressure and "low" rpm gives the same power as "low" manifold pressure and "high" rpm (*A*). Which should he normally use? If the best efficiency is desired (he's using 65 percent power at 7000 ft, has a good tailwind and wants to save an extra fuel stop), he'd be better off to use 23.5 in. and 2100 rpm (*B*), rather than 20.7 in. and 2400 rpm (*C*). Fig. 17-2 demonstrates that high rpm means higher friction losses; on the other hand, you'll notice that it is necessary to use higher rpm at higher altitudes to keep 65 percent because the engine is running out of manifold

pressure. At 11,000 ft (*D*), 2400 rpm is needed to develop 65 percent power.

The new pilot has heard from other pilots around the airport in various hangar flying sessions statements such as, "I always use 23 and 23, that's all you have to remember," or "Never have the manifold pressure more than *one* inch above the rpm in hundreds" (for example, 24 inches is as high as can be used with 2300 rpm). Looking at Fig. 17-2 at 75 percent power at sea level, at 2200 rpm the recommended map is 26.9 in., a difference of 4.9; or note that at 65 percent power at sea level, 25.3 in. is used for 2100 rpm, a difference of 4.3 in. While you probably won't ever be flying at sea level, you can nevertheless use this example to show the new guy that the manufacturer's numbers are to be used rather than those given by various "experts" sitting in front of the airport office. While we're on the subject, you'll often have to contend with bum information that gets to your students from such sources. You'll find that a talk with some of these people may be necessary to cut off this problem.

Getting back to power settings, Fig. 17-3 shows a power setting table for one airplane, using a different format. Note that here, too, several map/rpm combinations are indicated for a particular percentage of power.

Look at the cruise performance at 10,000 ft in Fig. 17-3, in the standard temperature column. Using 63 percent power at 2450 rpm requires 19 in. of manifold pressure and has a fuel consumption of 11.7 gph. At 2300 rpm, 63 percent power requires 20 in. of manifold pressure. You should be completely familiar with the charts for *your* airplane.

For a given rpm, the manifold pressure required to get a specific percentage of power decreases with altitude. In Fig. 17-2 at *75 percent power using 2300 rpm,* you'll note that at sea level 25.8 in. are required; and at 6000 ft density-altitude only 24.1 in. are needed at that rpm, or a drop of 0.3 in. of *map* per 1000 ft. This is because of the lower temperature at

---

Power Setting Table - Lycoming Model IO-540-D, 260 HP Engine

Press. Alt. 1000 Feet	Std. Alt. Temp. °F	143 HP - 55% Rated Approx. Fuel 11.4 GPH RPM AND MAN. PRESS. 2100	2200	2300	2400	169 HP - 65% Rated Approx. Fuel 12.7 GPH RPM AND MAN. PRESS. 2100	2200	2300	2400	195 HP - 75% Rated Approx. Fuel 14.1 GPH RPM AND MAN. PRESS. 2200	2300	2400	2500
SL	59	22.3	21.5	20.7	19.8	25.3	24.1	23.2	22.2	26.9	25.8	24.8	24.0
1	55	22.1	21.3	20.5	19.6	25.1	23.9	22.9	22.0	26.6	25.5	24.5	23.7
2	52	21.9	21.0	20.3	19.4	24.8	23.6	22.7	21.8	26.3	25.3	24.3	23.5
3	48	21.7	20.8	20.0	19.2	24.5	23.4	22.5	21.6	26.0	25.0	24.0	23.2
4	45	21.4	20.6	19.8	19.0	24.2	23.1	22.2	21.4	25.7	24.7	23.8	22.9
5	41	21.2	20.3	19.6	18.8	24.0	22.9	22.0	21.1	25.4	24.4	23.5	22.7
6	38	21.0	20.1	19.4	18.6	23.7	22.6	21.7	20.9	-	24.1	23.3	22.4
7	34	20.7	19.9	19.1	18.4	23.5	22.4	21.5	20.7	-	-	23.0	22.2
8	31	20.5	19.6	18.9	18.2	-	22.1	21.2	20.5				21.9
9	27	20.3	19.4	18.7	18.0	-	21.9	21.0	20.3				
10	23	20.0	19.2	18.5	17.7	-	-	20.7	20.0				
11	19	19.8	18.9	18.2	17.5				19.8				
12	16	19.6	18.7	18.0	17.3								
13	12	-	18.5	17.8	17.1								
14	9	-	-	17.5	16.9								
15	5	-	-	17.3	16.7								

To maintain constant power, correct manifold pressure approximately 0.17" Hg. for each 10° F variation in induction air temperature from standard altitude temperature. Add manifold pressure for air temperature above standard; subtract for temperatures below standard.

Fig. 17-2. Power setting table for a Lycoming IO 540-D, 260 horsepower engine. (*Piper Aircraft Corp.*)

altitude (a more dense mixture for the same manifold pressure being used) and less back pressure in the exhaust system.

If you are flying several airplanes and find power setting charts missing from them, you might make up a chart for those airplanes (Fig. 17-4). The abbreviated chart could be made wallet-sized for easy carrying. If you work for an air-

craft dealer (and most instructors do at some time during their careers), you'll be expected to demonstrate various airplanes and such an aid would be valuable. You could memorize the manifold pressure required at sea level at 2300 rpm for 65 and 75 percent power (23.2 and 25.8 respectively in Fig. 17-2) and subtract 0.3 in. (or 0.25 as applicable) per 1000 ft. Along this same line, you might memorize the *rpm* required for your *fixed-pitch prop trainers* for 65 percent and 75 percent and *add* 25 rpm per 1000 ft to maintain that percentage of power. These thumb rules are not intended to replace the actual numbers in the *Pilot's Operating Handbook*.

■ Flight Instruction

PREFLIGHT AND COCKPIT CHECK. You'll need to point out items during the preflight check that the new pilot hasn't had to cope with before, such as cowl flaps, more fuel tanks, and generally more complex equipment. Check the propeller and follow up on the ground instruction on its makeup and operation.

During the engine compartment check, point out where the manifold pressure is measured. If the engine is the fuel injection type, you could indicate the parts of that system you're able to see.

He should have had a chance to sit in the cockpit and study the *Pilot's Operating Handbook* before this flight. Once you're in the airplane you might review any items you, or he, think need to be discussed again. Rehearse the moves for using the throttle and propeller controls (and cowl flaps) during takeoff, climb, cruise, descent, and landing, to make sure he's aware of what's required. Usually, this will be his first introduction to a vernier-type control and this should be discussed briefly.

Be sure that he knows where the circuit breakers and/or fuses are and what systems are affected by an electrical failure. (Some of the older variable-pitch props are electrically controlled.)

Cover the procedure in the event of a "runaway" prop on takeoff, climb, or in cruise. Discuss the probable causes of such a problem.

CRUISE PERFORMANCE
PRESSURE ALTITUDE 10,000 FEET

CONDITIONS:
Recommended Lean Mixture
2950 Pounds
Cowl Flaps Closed

RPM	MP	20°C BELOW STANDARD TEMP -25°C % BHP	KTAS	GPH	STANDARD TEMPERATURE -5°C % BHP	KTAS	GPH	20°C ABOVE STANDARD TEMP 15°C % BHP	KTAS	GPH
2450	20	70	142	12.9	67	142	12.4	65	143	12.0
	19	66	138	12.1	63	138	11.7	61	138	11.4
	18	62	134	11.4	59	134	11.0	57	133	10.7
	17	57	129	10.7	55	128	10.4	53	127	10.0
2400	20	69	140	12.6	66	141	12.2	64	141	11.8
	19	64	137	11.9	62	137	11.5	60	137	11.2
	18	60	132	11.2	58	132	10.9	56	132	10.5
	17	56	127	10.5	54	127	10.2	52	125	9.9
2300	20	66	138	12.1	63	138	11.7	61	138	11.3
	19	62	134	11.5	59	134	11.1	57	133	10.7
	18	58	129	10.8	55	129	10.4	53	128	10.1
	17	53	124	10.1	51	123	9.8	49	121	9.5
2200	20	62	134	11.4	59	134	11.1	57	133	10.7
	19	58	129	10.8	55	129	10.4	54	128	10.1
	18	54	125	10.1	52	123	9.8	50	122	9.5
	17	50	119	9.5	48	117	9.2	46	115	8.9
	16	46	112	8.8	44	109	8.5	42	106	8.3

Fig. 17-3. Power setting table.

POWER SETTINGS	65 Per Cent				75 Per Cent			
RPM & GPH	2200	2300	2400	GPH Each	2200	2300	2400	GPH Each
O-540 (250 HP) - M.A.P.-S.L.(-0.25"/1000')	23.3	22.6	22.0	12.3	25.8	25.1	24.3	14.0
IO-320 (160 HP) - M.A.P.-S.L.(-0.30"/1000')	24.2	23.3	22.7	8.0	26.5	25.6	24.9	9.0
IO-720 (400 HP) - M.A.P.-S.L.(-0.30"/1000')	25.1	23.9	22.9	17.5	28.0	26.4	25.3	23.3
O-360 (180 HP) - M.A.P.-S.L.(-0.25"/1000')	22.7	22.1	21.5	8.8	25.1	24.5	23.9	10.0
O-320 (160 HP) - M.A.P.-S.L.(-0.30"/1000')	23.6	22.8	22.1	8.0	25.9	25.2	24.3	9.0

Subtract 0.25" or 0.30" M.A.P. per 1000 feet from sea level value as noted.

Example: For O-540 at 75% and 2300 rpm at 5000 feet, M.A.P. required = 25.1 - (5 × .25) = 23.85".

Fig. 17-4. Combined abbreviated power chart for several airplanes. Subtract 0.25″ or 0.30″ per 1000 ft from the sea level value, as noted.

STARTING. The checklist is much more important for more complex airplanes; even a comparatively short step up to a controllable-pitch prop means vital new checks are needed. For instance, trying to make a go-around with the prop in high pitch (low rpm) could cause consternation and dismay.

For most carburetor-equipped airplanes, starting is basically the same as for the fixed-pitch types. Prime or use the electric pump procedures as given by the *Pilot's Operating Handbook*. If the trainee has been flying only fixed-pitch, high-wing trainers, the use of the electric boost pump must be introduced for starting, takeoff, landing, and as an emergency aid, if you're using a low-wing airplane for this transitional training.

It's possible that the bigger airplane has fuel injection as mentioned in the section on the preflight check, and new procedures may have to be introduced for all phases of the flight.

PRETAKEOFF CHECK. The major additions to the pretakeoff check of this airplane will be exercising of the propeller and setting of cowl flaps (if equipped).

Watch for too high a map-to-rpm ratio the first couple of times a new pilot exercises the prop; he'll tend to leave the pitch control back too long. Emphasize that the propeller system can be checked by noting its *rate* of response in rpm change. As he gains experience he'll be able to judge the propeller pitch control response before it's changed 200 rpm.

For operations in normal outside air temperatures (50–80°F), three cycles of the prop control are usually enough to ensure that everything is okay. For extra-cold mornings you may have the trainee wait until late in the pretakeoff check so that the oil is warmed up before he exercises the prop, and he may need to cycle it several times to make sure warm oil is in the prop system. Warn him that he can expect poor rpm control after takeoff on cold days if this is ignored. Flying in extra-cold conditions at cruise could also require a couple of (gentle!) rpm cycles from time to time to get warm oil in the prop cylinder, so that the prop doesn't "hunt" or continually vary rpm.

It's also good to give the prop an extra cycle or two during the pretakeoff check if the airplane hasn't been flown for a couple of days.

While the *Pilot's Operating Handbook* numbers will certainly take care of this, you might mention that checking the mags when the engine is in the constant-speed rpm range won't tell much because the blades will tend to flatten out to pick up rpm losses caused by mag or plug problems.

Complete the checklist as recommended by the manufacturer.

TAKEOFF AND CLIMBOUT. After takeoff, when a safe altitude is reached, coach him about the change to climb power. (Throttle back first, *then* prop, as you told him in the ground sessions.) Speak out that first time; after that expect him to do it right, but be prepared for a mix-up in map-rpm use.

When the trainee flew the fixed-pitch prop trainer, the chances are that full throttle was used from takeoff to level-off and he didn't worry about it during the climb. Here, you'll have to remind him about the need to open the throttle to maintain the required climb manifold pressure as altitude is gained.

AIRWORK. Note that the technique for leveling off is the same as for simpler airplanes; leave the power at the climb setting until the cruise airspeed is approached, and then set power and trim.

Have him fly at various cruise combinations of map and rpm to get practice in control and the use of the power chart. You'll find that some people can take a very long time setting power to the nearest .01 in. and 5 rpm. Teach them to be realistic about it (parallax and inaccurate gauges can make such puttering meaningless).

Let him get the feel of the airplane by flying at cruise *and* slow flight speeds. Later in the flight, steep turns as well as stalls in various flap configurations and power settings can show him the characteristics of the airplane.

A graphic demonstration of how the pitch of the propeller affects the glide ratio of an airplane is shown in Fig. 17-5. This demonstration might be held for later in the checkout program. In the event of a complete power failure, the glide ratio is increased noticeably by pulling the prop control back to the full-high-pitch (low rpm) position.

Set up a power-off glide in smooth air at a particular altitude at, say, 100 K (or the recommended airspeed) with the prop control at climb setting. Let the glide stabilize and read the rate of descent. Climb back up to the same altitude and set up a power-off glide at the same airspeed, but with the propeller control full aft. Note the stabilized rate of descent and compare it to the other. You could convert feet per minute to miles per hour, or vice versa, using the numbers for your airplane and get the glide ratios (use TAS for the airspeed numbers). The faster the glide speed, the greater the effect of pulling the prop control back.

Several simulated approaches should be made at altitude, using all procedures (airspeed, flap settings, prop control, and map settings).

Better figure on an hour of airwork that first flight for the person who hasn't flown an airplane of this type before

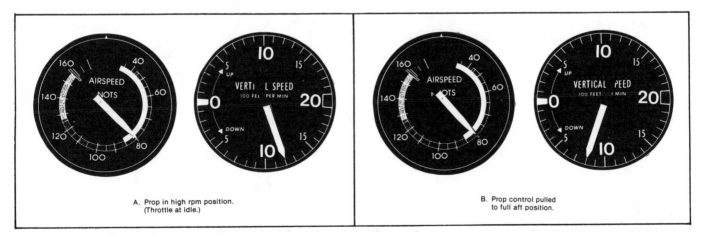

A. Prop in high rpm position.
(Throttle at idle.)

B. Prop control pulled
to full aft position.

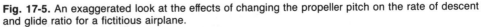

Fig. 17-5. An exaggerated look at the effects of changing the propeller pitch on the rate of descent and glide ratio for a fictitious airplane.

moving down for actual pattern work. With some people you may require several airwork sessions before starting takeoffs and landings.

TAKEOFFS AND LANDINGS. Most likely this airplane will have higher approach and landing speeds than the one(s) he's been flying, so it's suggested that at least for the first few landings a wider-than-normal pattern be used. The airplane will tend to be ahead of him because of the faster airspeeds and the extra cockpit chores required.

Use the manufacturer's recommendation on when to set the prop control and at what setting; some recommend a less-than-full low pitch for the approach.

Fig. 17-6 shows what you may expect on the turn onto final the first approach or two in the faster airplane. True, he could roll into the final turn at the point used for the slower airplanes and make the bank steeper, but he's got enough distractions right now.

For a given bank, the radius of turn of an airplane is a function of the square of the airspeed (V^2). An airplane that is approaching 10 percent faster than another will have a turn radius of $(1.1)^2 = 1.21$, or 21 percent greater. If its approach speed is 20 percent higher, $(1.2)^2 = 1.44$, or 44 percent greater radius of turn. Mention this turn radius effect before the first approach to save a little confusion.

Some airplanes in this heavier group have comparatively heavy elevator forces, and the trainee should be forewarned of the need to establish trim and power combinations during approach and landing.

Give the new pilot a chance to make a number of *full-stop* landings before throwing any go-arounds or other problems at him. Have him taxi back each time and take his time in going over the checklist before each takeoff. Probably the most common error flight instructors make in giving instruction in advanced airplanes (including multiengine) is going into emergency procedures and unusual conditions too soon. There's a strong tendency to do this because the instructor is already familiar with the airplane and just flying around normally can get pretty dull; emergency procedure practice can mean some action. Remember, though, that flying a faster, more complex airplane is anything but dull for the low-time pilot.

When you see that the airplane is firmly under control and he's ahead of it all the time, you might introduce go-

arounds, warning him well in advance that first time and then checking him later on one of the approaches without prior warning.

Review any problems after the flight(s).

He should also be checked out on short- and soft-field takeoffs and landings and crosswind techniques. You might also have him make a couple of no-flap landings.

After you see that he is safe in the airplane, have him fly solo in a fairly quiet period, getting in some airwork and full-stop landings before taking passengers. The solo work will build up his confidence and allow him to work out any minor problems without pressure.

If you are checking out a student pilot in a more complex airplane (a fairly rare but not unknown occurrence) don't forget to add this one on his student certificate (see Fig. 10-38).

AIRPLANES WITH RETRACTABLE LANDING GEAR

Reference—Chap. 14, *The Advanced Pilot's Flight Manual.*

An engineer for one of the major general aviation manufacturers was once asked where the landing gear handle should be located in the cockpit. The engineer, who had just completed a study of landing gear accidents, replied: "On the *far* right side of the panel—in a steel safe with a complicated combination." His point was that a large number of gear problems were caused by busy little fingers getting hold of a too conveniently placed gear handle at the wrong time. There are times when it would be good to be able to get the gear up quickly (an engine failure on climbout, for example), but most of the time the landing gear handle should be inconveniently located, where thought is necessary to operate it.

Ideally, the new pilot should have experience using the constant-speed prop before taking on an airplane with retractable gear *and* a constant-speed prop, but again, practically (and economically) speaking you'll probably use the same airplane for the complex systems introduction. One technique, in order not to hit a trainee with too much material at once, is for him to operate the landing gear under your supervision during the first few flights. However, don't discuss the gear

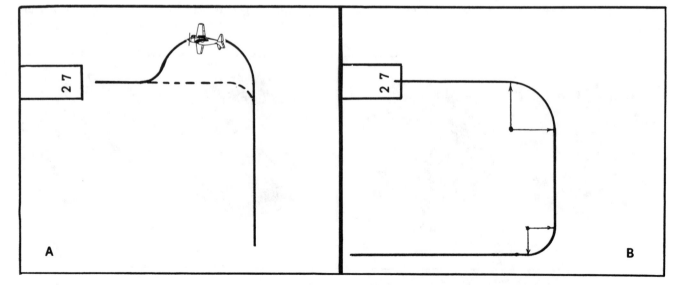

Fig. 17-6. (A) The new pilot tending to start his final turn at the same point as used for the slower airplanes he's been flying. (B) Using the idea of different banks, as introduced back in Fig. 5-11.

operation or systems. This way he can concentrate on using the propeller and get used to flying the airplane without continually thinking about the landing gear and its possible problems. Most new pilots spend the better part of the airwork for the first couple of flights reminding themselves to put the gear down when they get back to the airport and so they aren't getting the word on other items. When you feel he's ready for the next step, you can get down to serious talk about the landing gear and its system(s) and cover details of normal and emergency operation.

■ **Ground Instruction.** A quiet period discussing the system and looking at the retracting/extending mechanism of the airplane can help the flying part of the lesson. The new pilot should have some time sitting in the cockpit to see where all the handles are, particularly if he hasn't been flying this airplane during the propeller checkout.

In the classroom, you should discuss the landing gear system (hydraulic or electric), the servicing required, and any special items to consider in operating the gear normally.

Cover the maximum landing gear operating speeds. For instance, V_{LO} is the maximum gear *operating* (transitioning) speed.

V_{LE} is the maximum speed at which the airplane may fly with the gear already extended and, when applicable to a particular airplane, will be higher than V_{LO}. In other words, the airplane is allowed to go faster with the gear down and locked than during the extension or retraction process. The Navy F4U-5N Corsair used 270 K as the maximum speed at which the gear would extend fully, but once it was extended and locked, the gear could be used as a dive brake up to 350 K. These speeds will be noted again in the next chapter.

Discuss the safety switch on one of the main oleos and how it can be fooled if a gust extends the oleo during a takeoff run (the pilot has the gear handle in the *up* position).

Emphasize that the landing gear is not to be retracted until the airplane can no longer land on the runway ahead. Too many people whip up the gear and then settle back on the partially retracted gear.

Naturally, you'll discuss the emergency extension of the gear (which is probably all he's been thinking about since he decided to take the training). It would be good to have the airplane up on jacks in the hangar so that the emergency gear extension might be used. For electrically actuated gear this isn't too much of a problem if a mechanic familiar with the system is available, but if an emergency air bottle is used in a hydraulic system, complete purging of the lines is required—no small item of maintenance in most instances. Maybe if you're checking out an air taxi pilot for your company the boss might think it worthwhile, but usually the first time a pilot gets to see the emergency extension system used, it's for real. So, practically speaking, you'll probably just discuss and simulate the procedure. Probably, the best all-around advice you can give about emergency extension would be to (1) keep cool, (2) get up out of the pattern with plenty of altitude, (3) take his time, and (4) keep the airplane as slow as possible (without stalling or diverting attention from the procedure during the extension).

Spell out any specific idiosyncrasies of the gear of the airplane you're using. Discuss the warning horn (and the approximate throttle setting at which it sounds) and the gear lights.

■ **Flight Instruction**

PREFLIGHT CHECK. During the first check, you'll be overseeing his general inspection, but you should point out in detail the gear-actuating rods and how the gear is held up or down (it may have an overcenter down-lock or another type of down-lock and this could be good to know if gear problems

occur later). Point out the safety switch and see that it's properly adjusted. Let him get a good look up in the wheel wells.

THE FLIGHT. Emphasize that the first thing done in the airplane is checking that the gear handle or switch is in the proper position for a gear-down and locked condition and the gear-down light(s) is indicating that it's down and locked.

Take your time in going through the pretakeoff check, covering all items. Stay with the trainee, but don't ride the controls during the takeoff run. He can retract the gear after takeoff, set power, and make the climbout and level-off.

After he's done the Four Fundamentals, stalls, and slow flight you might have him set up a couple of simulated patterns at altitude, setting gear, flaps, and prop for an approach. As a general recommendation, the first hour might be straight airwork, with perhaps three landings at the end of the period. Assure him before the flight that there will be no tricks or deliberate emergency situations, so he can relax and concentrate on the airplane.

On the second flight, you should show him how the gear lights work and how if a light isn't ON it can be checked by swapping one of the working bulbs. Also, remind him that in bright sunlight the down-lights may not be seen on some airplanes if the navigation lights were left ON from last night's flying.

In the older Comanche models the pilot has *three* methods of checking that the gear is down: (1) the down-lights, (2) the position of the emergency handle, and (3) no horn blowing. It could be that the horn *isn't* working and the gear *isn't* down, if that's your only check. One sure check is to throttle back so that the horn is blowing before the gear handle (or switch) is put down. The gear extension will stop the horn. This is better than putting the gear down at higher manifold pressure and then closing the throttle to check that the horn *isn't* blowing.

Later in the training, pull circuit breakers or loosen bulbs to check his alertness. Insist on a full check every time he puts the gear down. Too many people flip the handle down and don't make all possible checks. If, for example, you've pulled circuit breakers and he puts the switch down and continues his approach (you have to draw his attention to the problem), he's definitely not ready to take the airplane out as pilot in command.

Let him know that the time he's most apt to forget to lower the gear is after a go-around or when he's distracted by other traffic. *He must establish the habit of making another gear check on final* (Fig. 17-7).

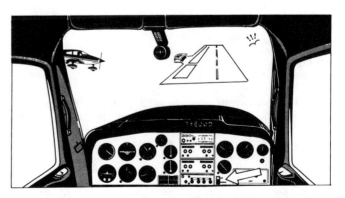

Fig. 17-7. A close look at the gear indication shows that the "minor distraction" could cause troubles later.

PATTERN, APPROACH, AND LANDING. Although warned about it beforehand, it seems that each new pilot has to learn for himself that *airplanes with retracted gear are very, very hard to slow down.* Unless you are continually on him during the pattern entry (and have him start slowing down to V_{LE} some distance out), the first pattern will consist of a high-speed flyby of the airport—at traffic altitude.

Give him plenty of time to get the gear down and complete other checks required in the pattern. Since the retractable-gear airplane doesn't usually have the gentle stall characteristics of the fixed-gear trainers you usually instruct in, keep alert. Remember that things will be moving faster for both of you, and any delay between your deciding something that should be done and his doing it could result in a tight fix. You'll find that when you first make that transition from instructing in slow trainers to higher-performance airplanes, *you* will tend to be behind the airplane, even though as pilot in command in it, you've had no problem at all. The thing is, when you are the pilot and something has to be done, you do it. As an instructor, when you see that something has to be done (say, on final), you tell the trainee and he then does it (very slowly, it seems) and then you may have to do it anyway—and a lot of airspeed and altitude have been lost while these processes have been worked out.

Your first few hours of instructing in higher-performance airplanes will be almost as strange as when you first started with the slow trainers. Always double-check his actions (Fig. 17-8).

Fig. 17-8. That step off the trailing edge of a low-wing airplane when you have just let a trainee land gear-up is like walking off the edge of the Grand Canyon.

If you are instructing in one of the "self-extending-gear" airplanes, treat it as an older retractable-gear type. You should explain how the system works but require that it be treated as a standby.

Several times during the course have him simulate putting down the gear by emergency means. Make sure that an emergency gear extension checklist is *always* in the airplane. Don't depend on the inexperienced pilot to remember the procedures while under pressure of a real emergency. By this same reasoning, be sure that *you use the checklist* when simulating a gear extension. He will feel easier about using it himself.

POSTFLIGHT. After the pilot is doing a good job with the airplane but hasn't been turned loose alone, discuss the landing procedure if one or all of the gear legs can't be extended. Review the steps on final when the runway is made (fuel OFF, mixture IDLE CUT-OFF, ignition OFF, and master switch OFF). Tell him the prop will be windmilling and to expect it to be damaged. (No stopping the engine at altitude and stopping

the prop to "save it"; he could miss the runway.) Discuss runway foaming procedures.

SUBSEQUENT FLIGHTS. Remember that he will be flying the airplane under all conditions, so work with him on short- and soft-field takeoffs and landings and crosswind procedures. If he plans to fly at night, you should give him a good checkout in this.

SET UP A SYLLABUS. This chapter was written to give you a base for setting up a syllabus for introducing complex single-engine airplanes and checking students out in them. Read the chapter again and sit down with a *Pilot's Operating Handbook* so that you can lay out your own flight and ground school syllabus.

No attempt is made to set up a fixed program here for two reasons: (1) the airplane you are using may have different requirements than other complex airplanes and (2) you'll understand and remember your own system more easily if *you* lay it out. It's assumed, as noted earlier, that you already know the basics of flying the particular airplane and using its systems. The syllabus will provide ideas on the steps required to check someone out to be safe in it.

As you review the *Pilot's Operating Handbook,* it is wise to remember how *you* felt the first time you started to fly a complex single-engine airplane. What was good or bad about the procedures as given to you? Did things move too fast or did you feel that too much time was spent on just flying the airplane around? Were important things skipped that you had to find out for yourself later? As explained in Chap. 2, if in doubt about the rate of the student's progress, keep your teaching on the simple side when introducing new areas of flying. You can always speed up the input of new material if the trainee can handle it. This slower route is better than going too fast and then having to backtrack and review. If you have a reliable, thorough program for checking people out in complex airplanes the word will get around and you will have plenty of students.

If someone approaches you for dual in their new Zephyr Six (which you haven't flown) at least get a *Pilot's Operating Handbook* for that airplane a couple of days ahead of time and go over the normal and emergency procedures and know the various systems. It's even better if you have a chance to fly it alone after studying the *Pilot's Operating Handbook.* People will know that you are more professional if you approach it this way. As you get experience in various complex airplanes you'll see that there are fundamental similarities between airplanes that have electrically operated landing gear or that have a Bendix fuel injection system, and so forth. The main thing is to be prepared, because the trainee is depending on you as the expert on the airplane.

These points will be brought out again in the Summary of Chap. 18.

You should make up a written test on equipment for each type of advanced airplane in which you are checking out pilots. Make it open- or closed-book, as your prefer—or part open-book and part closed-book. The main thing is to be sure that the trainee has a good understanding of both normal and emergency procedures. You should go over the test with him afterwards to correct any misconceptions. It's a good idea to keep the final test results on file in case you need them later.

These tests should cover propellers and landing gear, plus specific items (such as fuel systems and tankage) that would apply to complex airplanes. Make all of the questions practical, based on what the pilot can do *in flight* to correct problems. Don't ask a question such as, "What is the exact purpose of the internal frammis valve in the hydraulic pump?" It would be better to ask for a list of the steps required to lower the landing gear manually, or to ask what procedure is used with this airplane if the propeller has a runaway rpm. As noted earlier, no sample tests have been set up in this book. *You* will learn more about the airplane if you study the *Pilot's Operating Handbook* and make up your own tests, using

questions from each section of the *POH*. Figure on a 50-question, open- or closed-book test, spread over the *POH*, with special attention to emergency procedures (but don't neglect other sections).

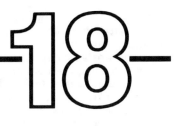

INSTRUCTING IN THE LIGHT TWIN

GROUND INSTRUCTION—FIRST FLIGHT

■ **Normal Procedures.** It would be best to cover only the normal operations in the first session of ground school and leave emergency procedures for a second (or third) time. You'll be throwing a lot of information the student's way; keep it on a "need to know" basis.

You should discuss the airspeeds to be used for normal rotation, lift-off, transition, climb, and cruise climb. Use the *Pilot's Operating Handbook* for such things as normal takeoff distances at various pressure altitudes and temperatures (and note that in most cases the airplane is kept on the ground until V_{MC} is exceeded). This chapter gives suggestions on how to *teach* multiengine flying; the *Pilot's Operating Handbook* will be the information source for *flying* any particular airplane. Assume that the twins being discussed in this chapter have reciprocating engines (no turbines).

You should review power setting procedures for cruise and tips for synchronization of the propellers.

You should also cover approach speeds and flap settings.

■ **Systems.** The twin has more complex systems, but the principles are the same as those for the single-engine airplanes, covered in the last chapter. Probably the most difficult area for the new twin pilot is the introduction of the crossfeed idea (or maybe your twin doesn't use it and has a cross-selector system). If you draw a simple schematic of the fuel system on the chalkboard before having the trainee look at the *Pilot's Operating Handbook*, it will make things much clearer. After he understands the principle you can go over the more complex diagram with him.

Note: The recognizable airplanes in this chapter are used only to illustrate the forces and moments acting on a twin. The manufacturer is the final source of information for a particular airplane.

Appreciation is expressed to Les Berven, noted industry and FAA test pilot, who let me include here information derived from his lectures on multiengine aerodynamics at the University of Tennessee Space Institute in 1978. His analysis of the factors involved in that area are the best I've heard or read on the subject.

Capt. M. R. Byington, Jr., USN, retired, furnished further actual data on single-engine performance for three different airplanes he and his colleagues ran at Embry-Riddle Aeronautical University, Daytona Beach, Fla.

It's possible, though extremely unlikely, that this will be the new pilot's first encounter with the constant-speed propeller or retractable gear, and you may have to go into basics on these systems. Even for the pilot with some constant-speed prop or retractable-gear experience a review, and certainly a discussion of feathering and unfeathering systems and operations should be covered fully.

The big question for this first ground and flight session is how much to cover. If the trainee has had no previous twin experience, you should cover the basics of flying the airplane normally so that he will have a chance to relax. In this case, the ground and flight sessions should be enough to allow the trainee to fly the airplane as a "complex single," using the engine controls as one and becoming at ease with the idea of that bigger, hotter airplane.

Your tendency will be to cover too much in ground school this first time. Save the complexities of the electrical and/or hydraulic systems for a later session. You would like to pass on your knowledge of the systems, but the trainee's memory bank is likely to break down and he could forget something really needed to help him on that first flight.

If scheduling allows, before the first ground session let him have a *Pilot's Operating Handbook* (but warn him to concentrate on normal operations, which of course ensures that he will turn directly to engine-out procedures) and have him sit in the airplane finding controls. He should "fly" the airplane through normal procedures. (Indicate to him that retracting the landing gear on the ground could cause him personal and financial injury.) A good procedure would be to give the first ground school one day and allow a day or so for studying and cockpit familiarization.

FLIGHT INSTRUCTION—FIRST FLIGHT

Allow plenty of time for that first flight in the twin; a 2-hr total period would probably be the minimum required to get an hour in the air.

Taking the activities of that first flight in order, here are some points you should consider.

■ **Preflight Inspection.** This will take more time than the single-engine airplane (big news) and will be used to review the systems discussed in ground school and seen at the airplane earlier.

Set up the most logical preflight check, which may not always mean that the airplane is inspected item by item as you come to it. You may want the trainee to check the cockpit and then get all of the fuel check items out of the way before starting to look at the physical condition of the airplane. He should also make sure that the gear handle is in the proper position and the gear-down light(s) is on.

New items he'll have to check (besides an added engine) could be (1) an anti- or deicer system on each propeller, (2) deicer boots or an anti-icing system, (3) the reservoir of the anti-ice system(s), (4) more radio antennas than the student has had to check before (point out and identify each), and (5) more fuel tanks, plus more drains (tank, sump, and crossfeed) than he believed could exist.

To keep this first flight as uncomplicated as possible, the extras should be noted (items 1 through 4) but not covered in detail. Indicate that they will be covered on later flights.

Even if the cowlings are difficult to remove, you should do this at least once before sending him out on his own—but not this first and probably not the second session. He'll need at some time to look at the various components (hydraulic pump, one generator only, etc.). If it can be reasonably done, the cowlings should be opened every flight for a check of the engines.

If your airplane is of the type requiring the fuel tanks to be "stuck" (fuel measured with a stick), you'll have to check him out on this procedure. Note that sharp dipsticks and rubber fuel cells don't always mix well.

One point to remember: It seems the bigger and more complex the airplane, the more the tendency to make only a perfunctory preflight check. You'll see this with the pilots of the bigger airplanes at your airport. Have a checklist of items that need to be covered in the preflight—and use it; the trainee will feel like a B-52 commander with such a procedure.

■ **Starting—The First Flight.** Show him which engine to start first and the reason for it if not already covered in ground school. Discuss hot and cold starting procedures. Let him start both engines under your supervision.

The usual monitoring of oil pressures after start, and other poststart procedures, should be followed, with the use of a checklist.

■ **Taxiing.** Probably most of the twins you'll be using for instruction will be the nosewheel type, which makes taxiing easier. Have the trainee use both throttles as one control in taxiing; variable throttle usage wouldn't make that much difference in directional control anyway and could confuse him at first.

Tailwheel-equipped twins can sometimes give the new pilot problems, particularly if he consciously tries to use variable throttles to aid in steering. For long, straight taxiing, it's a good move to lock the tailwheel and remind him to use the throttles as one.

Watch for general problems found in taxiing, such as moving too fast, using power and holding brakes, and taxiing too slowly (rare). If the airplane is appreciably larger than the airplanes he's been flying and the gear is wider apart (he's been flying a small, high-wing airplane with narrow gear), keep alert for his getting a wheel off the hard surface into the mud, particularly when making a 90° turn from taxiway to runway or another taxiway; he'll tend to cut across more than he realizes (Fig. 18-1).

The larger wing span could cause problems if he tries to taxi through too narrow a gap (he's made it just fine in the trainer). Don't ride the controls, but stay with him, particularly in tight places.

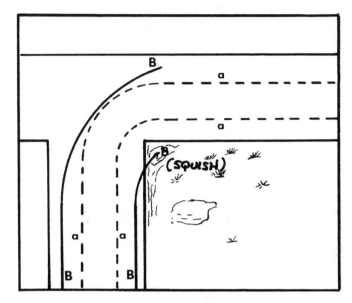

Fig. 18-1. Watch for a tendency to cut corners and get a wheel off the pavement; *a* shows the tracks of the smaller plane he's been using, compared with the twin tracks shown by *B*.

■ **Pretakeoff Check.** The first pretakeoff check will take a good while as you show and review the items discussed before the flight. Use a checklist to check mags, manifold heat, and prop operation, including feathering. Caution the student about leaving the prop control in the feather position too long.

Take your time and see that he understands the steps.

■ **Takeoff and Climb.** The new pilot's first attempt at prop synchronization when he pulls the power back to climb after takeoff is usually a frustrating experience for everybody in the airplane. He gets the gear up fine, but the power change may tend to get out of hand. First, he may repeat a mistake made on the single-engine airplane and try to decrease the rpm before throttling back (which isn't done in your twin), and finally after a little confusion he pulls back the manifold pressure and *then* starts to synchronize the props.

Watch for too low an rpm–manifold pressure combination, as there is sometimes a tendency to seesaw the prop controls back in an attempt to get things even. Remind him that sound is a good aid in synchronizing the engines. One (usually the left) prop control should be set to the recommended rpm and the other matched to it by sound, if an indicator is not available on the tachometer.

Keep your cool, and remember how your first few twin power changes went.

■ **Airwork—Normal Operations.** A short VFR pilotage cross-country that doesn't require preflight planning (the destination in sight or nearly so) is a good way for the student to get at ease. He'll have a chance to fly the airplane and find things in the cockpit without being rushed. You can also talk to him about various items of equipment and procedures. The airplane will probably be much faster than any he's flown so he'll need plenty of elbow room for a while. You might figure on possibly two hour-long flights with one landing at the end of each before doing any concentrated normal takeoff and landing practice.

In too many cases, the new pilot's flight training time is cut to a minimum. He is introduced to emergency procedures too quickly and the "training" is over in 3 or 4 hr. The trainee may be "lucky" enough to pass the multiengine rating check and then be on his own to load up the airplane with people and baggage and go to faraway places.

You might set up your curriculum, as a minimum, on the basis of 5 hr of ground school and 10 hr dual before sending someone up for a multiengine rating. True, you will run into people who've had twin experience some time earlier or who are especially talented, and you can cut their time down, but it's better to have planned for a longer curriculum than needed than to come up short when the trainee needs more work. ("Let's go out and repeat what we did the last time. Uh, what did we do last time?") The *Pilot's Operating Handbook* is a good guide for setting up a twin-engine syllabus.

If he's having trouble with synchronization you might ease one prop out of kilter a few times and let him get practice getting things together. A little listening to unsynchronized engines can go a long way, both for you and the people on the ground (Fig. 18-2). (World War II air novels and stories used to make much of people on the ground recognizing a Japanese plane "by the unsynchronized engines"; what effect it had on Japanese defeat has never been discussed in the literature.) Some newer models have a propeller synchronizer, and you should explain its use *after* the new pilot has the manual procedure well down.

Fig. 18-2. Sometimes the new pilot's attempts at synchronization leave a bit to be desired as far as the instructor is concerned. (*The Advanced Pilot's Flight Manual*)

After he is able to maintain straight and level flight with competence, you might introduce at cruise power a series of turns of 15° banks (90° turns), 30° banks (180° turns), and 45° banks (360° turns). Climbs, descents, and turns should be combined basically as discussed in Chap. 5.

Slow Flight and Stalls—The two extremes in an approach to multiengine training are (1) don't make waves—don't give slow flight or stalls or simulated landings even at a high altitude and don't actually feather an engine at *any* time—and (2) make "practice" approaches and landings, or single-engine stalls, with a prop feathered or pull the mixture or turn the fuel off at a bad spot on takeoff. The first course doesn't give the proper training, and the second is the reason for many fatalities in twin training. A big reason for accidents or near accidents is that the instructor finds himself in a situation out of his experience. *The more flying time you have in the airplane in which you are instructing, the safer you'll be, at least if you stay alert.* Time was when a brand-new instructor could spend 4 or 5 hr getting his multiengine rating in a docile twin, and then immediately start instructing in the Tornado Multi-Bomb. The requirement of 5 hr minimum as pilot in com-

mand in type and model should help this problem, but that amount of time is barely enough.

Laying it out as far as normal twin-engine operations are concerned, have him practice slow flight and stalls with *symmetrical* power (and power-off stalls) at a safe altitude of at least 5000 ft above the surface (or higher if the manufacturer so recommends). Recoveries from stalls should be made with and without power. You should avoid getting into radical attitudes but should demonstrate and let him practice stalls in normally expected situations. Some of the bigger twins may be placarded against certain (or all) stalls, but give stall practice unless prohibited by the manufacturer.

You might demonstrate the procedure for recovery from a stall in a simulated engine-out condition (use zero thrust—don't do this with a feathered propeller) by throttling back the working engine to stop rolling tendencies. *Stay alert.*

Have the new pilot do slow flight in various configurations so that he can recognize the approach of a stall.

After he's settled down to flying the airplane by visual references, you should have him do hood work to make sure (assuming he isn't instrument rated) that he could make a 180° turn on instruments or fly the airplane well enough to get out of a sticky weather situation safely. If he has an instrument rating, set up a short IFR flight and/or shoot the various types of approaches as applicable to the comm/nav equipment on board.

■ **Approaches and Landings.** As was the case in introducing the single-engine retractable-gear airplane, give the trainee plenty of time and space on the first few traffic patterns. If it seems in takeoff and landing practice that he's getting behind the airplane, leave and reenter the pattern each time.

You'll have to (1) check for other aircraft (his head will be down in the cockpit much of the time those first few patterns except for the actual takeoff and landing), (2) watch *him* to make sure he doesn't pull or push the wrong knob(s) at the wrong time, and (3) watch his actions, analyze them, and keep up with your instruction—no small task.

Most twins have a light and horn gear-warning system whereby the horn only blows if *both* throttles have been retarded below a certain manifold pressure value (for some, it's 14 in.). You should use the same technique mentioned earlier for the single-engine retractable-gear, throttling back and letting the safe gear turn off the horn. You might advise the trainee to explain to his passengers before setting off the horn.

Generally the new pilot should avoid power-off approaches in the light twin. The angle and rates of descent are high for some models and could set up an accident. It could be, of course, that the knowledge of how to set up a dead stick approach might be a good thing in case *both* engines failed, but since the chance of that is so slim, it's better to recommend a gradual-reduction-of-power technique in an approach.

Be prepared for his allowing the airplane to cross the wind line on the turn onto final.

As far as the use of flaps is concerned, there's no need to make a big production of it here, and a maximum of two moves to get the amount required for final and landing is recommended. There may be exceptions for certain twins and certainly you should follow the manufacturer's recommendations, but too many people make too much of a production of flying a twin-engine airplane in normal conditions.

When you start the normal takeoff and landing practice, take an occasional break for some airwork (good experience in making transitions to and from the pattern).

It's very easy to get wrapped up in a landing postmortem discussion and forget to open the cowl flaps before taxiing back. You'd catch the problem before takeoff if you used a checklist, but during lengthy ground operations the insulation on the plug wires and other wires may get very hot (plus the engines themselves may overheat).

Work with the pilot on short- and soft-field takeoffs and landings and no-flap landings, plus crosswind techniques as used in a twin.

As in primary instructing, after a while your biggest enemy will be complacency. The trainee has been doing a good job and you're working on smoothing out minor problems in the pattern, and it's easy to become too relaxed and start woolgathering.

■ **Review.** Review the flight and assign reading in the *Pilot's Operating Handbook,* with particular emphasis on normal operations and performance charts. The trainee should spend time in the cabin on the ground, if possible, simulating a flight from start to finish.

GROUND INSTRUCTION—SECOND FLIGHT

This should be a further look into the systems of the airplane and normal performance. You should clear up any misunderstandings of procedures that may have happened on the first flight.

The new pilot may have questions on the following areas of the *Pilot's Operating Handbook:*

1. *Airspeed calibration*—normal and alternate static sources.
2. *Altimeter correction*—normal and alternate static sources.
3. *Stall speeds.*
4. *Normal takeoff distance*—various altitudes and airplane weights.
5. *Rates of climb*—cruise climb.
6. *Normal landing distances.*

The point is that you are still in the process of pinning down normal operations (the idea of airspeed and altitude corrections with this type airplane may be new to the trainee). You might go over a couple of examples of takeoff and landing distances, so that he can work at home with the charts.

■ **Introduction to Single-Engine Operations.** This, of course, is what he has been waiting for, all too often at the cost of studying normal procedures. You'll meet people with extreme opinions concerning loss of an engine: (1) some say there will be no sweat at all because the manufacturers have ensured that the airplane will easily fly with one out, or (2) others believe the loss of an engine under any condition is a total catastrophe, based on the stories already told to them in the airport office.

You'll have to show them the practicalities of engine-out flight, with the understanding from the beginning that it is better in some situations to put the airplane into a landing area off the runway than to attempt to take it around. This is the point in training when you can instill a good philosophy of multiengine operations. As noted earlier, the first few hours of flying—and the first hour or so in a new area of flying—are the most important of all in setting up good attitudes and habit patterns. If you just give the trainee the mechanical steps in flying the airplane, he's not apt to pay much attention later to setting up particular habit patterns. He'll be thinking, "Yes, that's very interesting and all, but I've been flying the airplane for 5 (or 10) hours and am doing fine."

AIRSPEEDS. You should discuss the following stall speeds and airspeed indicator markings (FAR 23.49 and FAR 23.1545). Note that airspeeds are marked as IAS on the airspeed indicator in airplanes manufactured as 1976 models or later.

V_{S_O} —The stall speed, if obtainable, or the minimum steady speed, in knots (CAS), at which the airplane is controllable with—
1. Engines idling, throttles closed or at not more than the power necessary for zero thrust at a speed not more than 110 percent of the stalling speed;
2. Propellers in takeoff position;
3. Landing gear extended;
4. Wing flaps in the landing position;
5. Cowl flaps closed;
6. Center of gravity in the most unfavorable position within the allowable landing range; and
7. Usually max landing weight or max certificated weight, if it is higher. This is marked at the bottom of the white arc.

V_{S_1} —This one is normally considered to be flaps and gear up, at max certificated weight, and with the propellers in the takeoff position with the power set as in V_{S_O}. (Bottom of the green arc.)

V_X —This is the full climb power, max angle of climb speed at gross weight at sea level (no marking).

V_Y —Best rate of climb at sea level and gross weight with both engines going (no marking).

V_{XSE} —Single-engine best angle of climb speed.

V_{YSE} —Single-engine best rate of climb speed. This is indicated by a blue sector extending from V_{YSE} speed at sea level to the V_{YSE} speed at an altitude of 5000 ft (if the single-engine best rate at that altitude is less than 100 fpm) or to the V_{YSE} speed at the nearest 1000-ft altitude (at or above 5000 ft at which the single-engine best rate is 100 fpm or more). Each side of the sector must be labeled to show the altitude for the corresponding V_{YSE}. V_{YSE} is indicated by a blue line (a single speed) in earlier models.

V_{NO} —Top of the green arc. The maximum structural cruising speed. Do not exceed this except in smooth air, and then only with caution.

V_{NE} —Red radial line. Never exceed this airspeed at any time.

V_{LE} —Landing gear extended speed (max). (Not indicated on the ASI.)

V_{LO} —Maximum gear operating speed. (Not indicated on the ASI.) These two gear speeds may be the same, but if there is a difference, V_{LE} is normally the higher because once the gear is down and locked it's stronger than during the transition period. However, double-check the *Pilot's Operating Handbook* for the final story.

V_{FE} —Max flaps extended speed. Top of the white arc.

V_A —Maneuvering speed. This is not shown on the airspeed indicator because it decreases with the square root of the weight decrease.

V_{MC} —The minimum control speed while airborne is marked with a red radial and is defined by FAR 23.149, as follows:

(a) V_{MC} is the calibrated airspeed at which, when the critical engine is suddenly made inoperative, it is possible to recover control of the airplane with that engine still inoperative and maintain straight flight either with zero yaw or, at the option of the applicant, with an angle of bank of not more than 5°. The method used to simulate critical engine failure must represent the most critical mode of power plant failure with respect to controllability expected in service.

(b) For reciprocating engine-powered airplanes, V_{MC} may not exceed 1.2 V_{S_1} (where V_{S_1} is determined at the maximum takeoff weight) with—

(1) Takeoff or maximum available power on the engines;

(2) The most unfavorable center of gravity;

(3) The airplane trimmed for takeoff;

(4) The maximum sea level takeoff weight (or any lesser weight necessary to show V_{MC});

(5) Flaps in the takeoff position;

(6) Landing gear retracted;

(7) Cowl flaps in the normal takeoff position;

(8) The propeller of the inoperative engine

(i) Windmilling;

(ii) In the most probable position for the specific design of the propeller control; or

(iii) Feathered, if the airplane has an automatic feathering device; and

(9) The airplane airborne and the ground effect negligible.

(c) *Deleted* (for turbine powered airplanes).

(d) At V_{MC}, the rudder pedal force required to maintain control may not exceed 150 pounds, and it may not be necessary to reduce power or thrust of the operative engines. During recovery, the airplane may not assume any dangerous attitude, and it must be possible to prevent a heading change of more than 20 degrees.

In addition to those requirements, FAR 23,147 lays out the requirements for directional and lateral control as follows:

(a) For each multiengine airplane, it must be possible to make turns with 15° of bank both towards and away from an inoperative engine, from a steady climb at 1.4 V_{s_1} or V_Y with—

(1) One engine inoperative and its propeller in the minimum drag position;

(2) the remaining engines at not more than maximum continuous power;

(3) The rearmost allowable center of gravity;

(4) The landing gear (i) retracted, and (ii) extended;

(5) The flaps in the most favorable climb position; and

(6) Maximum weight.

(b) For each multiengine airplane, it must be possible, while holding the wings level within 5°, to make sudden changes in heading safely in both directions. This must be shown at 1.4 V_{s_1} or V_Y with heading changes up to 15° with the—

(1) Critical engine inoperative and its propeller in the minimum drag position;

(2) Remaining engine(s) at maximum continuous power;

(3) Landing gear (i) retracted, and (ii) extended;

(4) Flaps in the most favorable climb position; and

(5) Center of gravity at its rearmost allowable position.

Fig. 18-3 shows some principles of rudder effectiveness in directional control with one engine out and in *wings-level flight*. (The reason for that emphasis will be covered shortly.) You might sketch it on the chalkboard during this ground school session.

Basically when only one engine is operating, the pilot is balancing the yawing moment caused by the asymmetrical thrust by setting up a moment of his own with the rudder. As the lift of the wing is a function of area, coefficient of lift, and

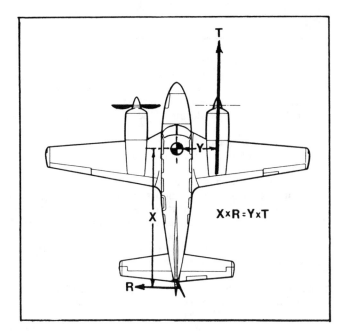

Fig. 18-3. Moments in engine-out, wings-level flight.

dynamic pressure ($\varrho/2\ V^2$), so is the "lift" of the rudder. When the rudder is fully deflected, at its maximum coefficient of lift, its yawing power will depend on the dynamic pressure $\varrho/2\ V^2$ (which is calibrated airspeed as per the airspeed indicator). As the dynamic pressure is decreased (the airplane slowed), there will be some airspeed below which the rudder moment will no longer balance the thrust moment—and directional control is lost.

As far as rudder deflection is concerned, FAR 23 sets a limiting force to be used on the rudder pedal when establishing V_{MC} for a particular airplane, so the rudder would not necessarily be fully deflected at V_{MC} as set by the manufacturer.

Note that the center of gravity is required to be at the most unfavorable position (usually most rearward). The airplane moves around its CG, and the moments are measured using that as a reference. A rearward CG would not affect the thrust moment but would shorten the arm to the center of the rudder "lift," which would mean that a higher dynamic pressure (CAS) would be required to balance the engine-out yaw. Fig. 18-4 shows an exaggerated view of the effects of a rearward CG.

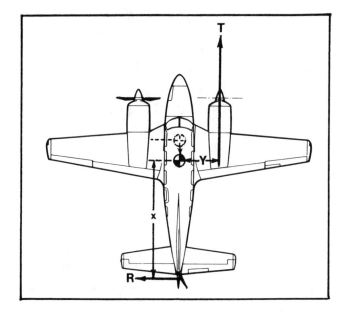

Fig. 18-4. Moving the center of gravity rearward means a higher V_{MC}.

Generally, the CG range of most light twins is short enough so that the effect on the V_{MC} is relatively small, but it is a factor that should be discussed. (Many people would consider the rear CG of their twin only as a factor for stability, not realizing that it could affect the controllability with one engine out.)

Strongly emphasize the fact that maintaining V_{MC} does not guarantee climb. In fact, it could mean a well-developed sink rate for some airplanes.

Earlier it was emphasized that the first look at V_{MC} and directional control by the rudder would be in wings-level flight. Note in FAR 23.149 that the applicant (manufacturer) can have the option of banking not more than 5°. It doesn't indicate whether the bank is toward the good or bad engine, but you would soon realize that the bank should be toward the working (good) engine.

Take another look at Fig. 18-3. Suppose you are determined to keep the wings level while flying the airplane on one engine but find you are running out of rudder at a higher airspeed than you'd figured on.

If you bank into the operating engine, the airplane is given a component of weight acting toward that side, which creates a sideslip. The rudder and fin are now more effective in fighting the yawing force of that engine. Fig. 18-5 shows the idea. (The bank angle of 5° is exaggerated to make a better point.)

It's been found in some airplanes that V_{MC} is raised up to 20 K when the wings are kept level and the pilot tries to maintain a constant heading with the ball centered.

You should coach the trainee to bank the airplane 5° into the operating engine immediately upon determining which one has failed. This, plus rudder, avoids the loss of control that often occurs just after an engine failure.

One consideration often mentioned when V_{MC} is brought up is *weight*. How does a higher weight affect V_{MC}? (Assume that the CG hasn't moved.) Common sense would tell you that the lighter the weight, the lower the V_{MC}, but that isn't the way it works as far as the 5° bank is concerned; the greater the weight, the larger the component of weight acting toward the operating engine and the greater the sideslip (and more effective rudder and fin to help fight the turning into the dead engine). This added weight helps *control* and the performance is not the best. (Be careful of stalling the airplane.) Fig. 18-6 shows an exaggerated example.

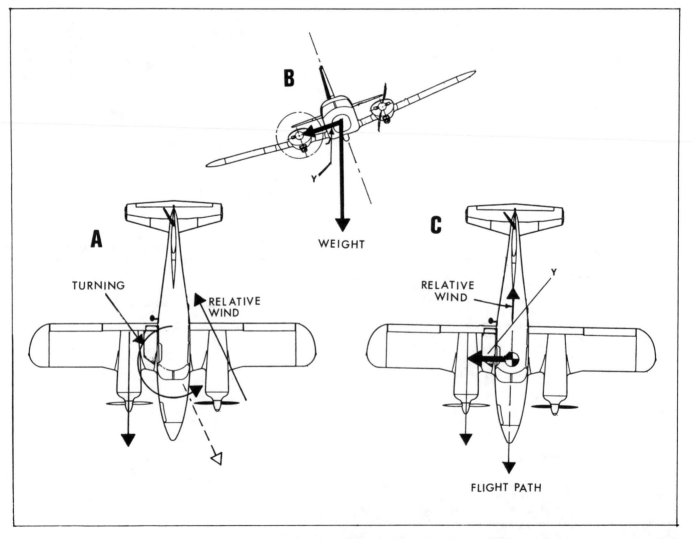

Fig. 18-5. (A) Wings level, the airplane is "translating" along the path shown by the dashed line. This results in the rudder being much less effective because of the relative wind direction and the airplane turns into the dead engine. (B) By banking into the good engine, a component of weight is acting along the wing (the Y-axis). This is roughly the same idea as used in correcting for a crosswind with a slip. (C) The fin and rudder are more effective because of the sideslip.

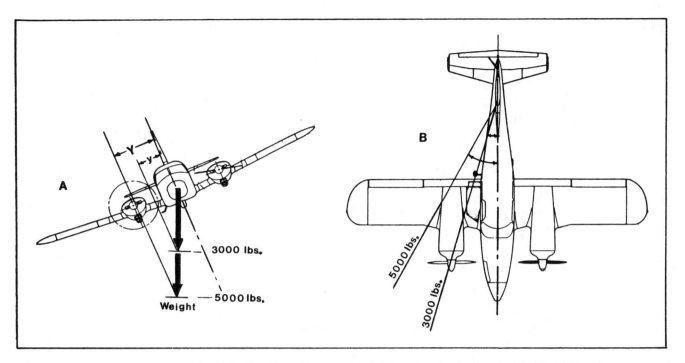

Fig. 18-6. The effect of increased weight lowering the V_{MC} (constant bank). (A) The wing component of weight and sideslip effect is greater at 5000 lb (Y) than at 3000 lb (y). (B) The angle of sideslip is greater at the higher weight. (The angles of bank and sideslip are exaggerated.) Note that added weight hurts climb performance, as does too much sideslip.

Manufacturers' pilots demonstrating certain model light twins have demonstrated the directional control (not climb performance) of their twin by picking a heading at a medium altitude, putting both feet on the floor (off the rudder pedals) and feathering the propeller of the noncritical engine. As the propeller feathered they would bank the airplane 15–20° into the operating engine and maintain heading as the airplane was slowed to a steady IAS 10–15 mph below the published V_{MC} speed. The heading could be maintained (feet on the floor) and after a suitable interval the pilot would restart the engine, gradually easing off the bank as power was eased back into the engine. The rudder pedals were not used during the demonstration and the heading throughout was generally held within 5° of that selected at the beginning of the exercise. In addition to the importance of the 15° + bank into the good engine for *directional control,* the airplane used was a lighter model of a particular twin and had a low stall speed. The pilots flew it every day in demonstrations of this sort (including taking off at light weights with the noncritical engine at zero thrust).

Another factor should be mentioned. For airplanes with normally aspirated engines, V_{MC} *decreases* with altitude, meaning that directional control can be maintained at a lower CAS (and IAS) than at sea level. The reason for this is that power (and thrust) is lost with altitude; the thrust moment (Fig. 18-3) becomes less, hence lessening the requirement for dynamic pressure (airspeed) on the rudder, which can still be deflected the same amount as at sea level. This was another reason why the demonstration could be made by the factory pilot. *You* will not have had the experience, at least at first, and will avoid getting into a critical corner for the trainee's as well as your own sake. Besides, *your* twin might not buy that procedure. The point is that banking 5° into the good engine can help directional control. (But, again, performance is not enhanced if the bank is overdone, and the bank will have to be shallowed to perhaps one-half the *control* value if you want best single-engine climb *performance*.)

Fig. 18-7 shows how a lowered V_{MC} could cause problems for the unwary multiengine instructor. Note that up to a certain altitude, as the airplane is slowed below the sea level V_{MC} (90 K here) with one engine windmilling, it will start to yaw as the pilot loses rudder power when he decreases the airspeed.

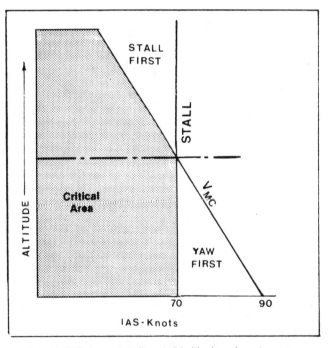

Fig. 18-7. Power-on stall speed (critical engine at max power available and the other feathered) and V_{MC} versus density-altitude for a normally respirated engine. Note that the indicated airspeed at stall (at a given weight) does not change with altitude.

This is reasonably safe until he does this experiment at a particular altitude. At this point, shown by the horizontal dashed line, V_{MC} has decreased to the stall value and the stall will break just as the pilot loses directional control. Above that altitude the airplane will stall first, creating a hazardous situation, as the operating engine creates a yawing, rolling moment in the stall, which could set up a spin.

The problem is that for several light twins, at certain weights and CG locations, the critical altitude is about 3000 ft density-altitude, often considered a good place for practice in certain parts of the United States.

If you are instructing and the trainee stalls the airplane with one propeller windmilling or feathered, close the throttle on that operating engine, use full rudder against the roll, and neutralize the ailerons as you *immediately use brisk forward movement of the wheel.* A developed spin in a twin is usually unrecoverable. Don't let it get started.

A temporary lapse of attention on your part while demonstrating V_{MC}, or any low and/or slow condition of flight could let the trainee put you in a bad spot. *The accident rate for engine-out "routine training" is high and probably exceeds that of actual incidents.* Too many new multiengine instructors have been caught short by making "practice" approaches with a propeller feathered or by deliberate or accidental stalls during engine-out practice (Fig. 18-8).

Fig. 18-8. Practicing stalls with one propeller feathered or an engine at idle could be a "good" way to find yourself in a spin. (More about twins and spins in Chap. 21.)

Some flight schools may suggest that an observer, or observers, be taken along on some of the multiengine training flights but you should not do this, particularly on V_{MC} demonstration flights. If things get out of hand and the airplane starts to spin, that weight in the back seat(s) could push an already bad situation into an unrecoverable flat spin. Chap. 21 covers in more detail the effects of CG and the mass distribution of multiengine airplanes on spin characteristics.

THE CRITICAL ENGINE. Fig. 18-9 shows the thrust patterns of the propellers of two engines at higher angles of attack, as would be the case in slow flight. If one of the engines was out and the other was producing power, the airplane would be flying more slowly and at a higher angle of attack; there would be a difference in turning tendencies because of the difference in arms to the center of thrust. The newer CR (counterrotating) twins do not have a critical engine. This doesn't mean that they don't yaw, but the loss of either engine causes an *equal* number of problems.

So far only the yaw forces created by the operating engine have been covered. You'd want to explain the induced flow difference between the two wings with one engine wind-

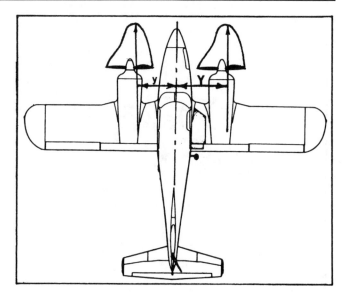

Fig. 18-9. The thrust pattern of a twin at high angles of attack. Asymmetric disk loading or P factor results in the left engine being the critical engine for those that have propellers that rotate clockwise (most U.S. engines). The CR (counter-rotating) models of twins have eliminated the idea of a "critical engine." The loss of either would have equal effects.

milling. Fig. 18-10 shows that a strong rolling moment is produced by the slipstream behind the operating engine, compared with spoiled flow behind a windmilling or even feathered propeller. The windmilling case is the worse of the last two conditions, naturally.

Some other forces to be considered in a single-engine situation are these:

1. With the wing down and a strong sideslip occurring, the fact that the rudder and fin are *above* the CG produces a force working against the bank.

2. Dihedral is also working against the bank and trying to level the wings.

3. The ailerons being held to maintain the bank can result in adverse yaw that tends to turn the airplane into the dead engine.

V_{SSE} is the *safe single-engine speed* established by the manufacturer and noted in the *Pilot's Operating Handbook* as the minimum speed at which to perform intentional engine cuts. The purpose is to decrease the number of accidents caused by loss of control during single-engine demonstrations.

To demonstrate V_{MC}, for instance, at a safe altitude smoothly reduce the power on one engine at or above V_{SSE}, setting the power on the operating engine at max continuous power. Reduce the airspeed slowly (1 K per second) until directional control can no longer be maintained or until you get the first indication of a stall. Then reduce power on the operating engine, drop the nose, and accelerate through V_{MC} to V_{YSE} (while re-adding power) for best single-engine climb rate.

Fatal accidents have occurred when a pilot, returning to the airport with a feathered propeller, *abruptly closed* the throttle on the operating engine, resulting in that propeller going into full low pitch. The sudden drag increase as the blade angles flatten (plus the fact that the rudder may be fully deflected toward that side) can cause a totally unexpected roll in the "wrong direction." When you are flying with an engine at idle or feathered at any altitude in a twin, don't *suddenly* close the throttle on that operating engine.

You might indicate to the trainee that he will see the term V_{MCA}, meaning the minimum control speed airborne. The term

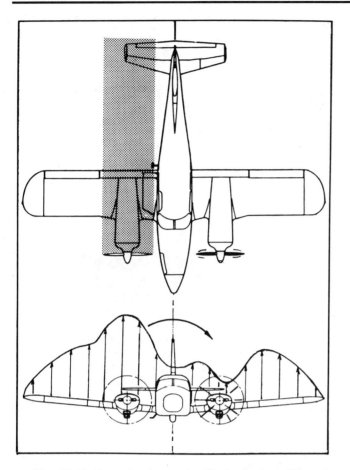

Fig. 18-10. Example of spanwise lift distribution with maximum available power on one engine and the other windmilling. With flaps down under these conditions—and only one of them in the slipstream—the spanwise lift and rolling and yawing characteristics could be even more critical.

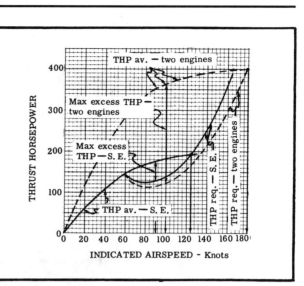

Fig. 18-11. Power required and power available for twin- and single-engine operations. (*The Advanced Pilot's Flight Manual*)

V_{MCG} (sometimes used) is the minimum control speed on the ground. (FAR 25 covers this in more detail.)

SINGLE-ENGINE PERFORMANCE. One of the commonest misconceptions of the pilot new to twins is that the performance on one engine will be approximately one-half that of both engines. As Fig. 18-11 shows, a minimum of about 30–35 percent of the *total* (100 percent) power available is needed to keep the airplane in the air, so if only 50 percent power is available, the excess power for climb is only 15 percent, instead of 65–70 percent.

Fig. 18-11 is for an *optimum* condition at sea level; on a hot day at a high elevation and/or in turbulence, the single-engine performance might be such that the airplane could not even maintain altitude.

Most inexperienced twin pilots with an engine out will try to climb in a tight situation, even at the cost of getting below V_{MC}; watch for this in your training sessions. If directional (and lateral) control is being lost, it would be better to throttle back the good engine and settle in under some control (wings level) than to dig in a wing tip and cartwheel. Below V_{MC} with one out, the rolling tendency is uncontrollable.

It seems that some training courses for twins consist strictly of seeing how fast the new pilot can recognize and feather the ailing engine; the poor guy gets emergency after emergency throughout his training and, by the end of the course, will feather one engine if the mag drop is too much on the pretakeoff check. The point is that there will be times when he shouldn't feather. For example, on a takeoff run where there is plenty of runway ahead, he should chop the throttle and land gear-down ahead.

Suppose the airplane has lifted off, the runway is gone behind, and one of the engines *loses* power. The airplane is heavily loaded and things will be close. Since the engine controls will be in the max power condition right after takeoff, the primary object is to clean up the gear and flaps. After this is done the pilot pulls the throttles back one at a time and finds that, say, the left engine has a lesser amount of response to throttle change but still is delivering noticeable power. Even that little amount of power might help clear obstacles that would be hit if the engine were feathered. To exaggerate, if that ailing engine is giving a ≠1 horsepower, it's helping. The problem, naturally, for some airplanes is whether in keeping the engine going it might be impossible to feather the prop later. However, this point should be discussed anyway so that the new pilot doesn't develop an automatic response of always feathering when an engine starts running rough.

If you draw Fig. 18-11 on the chalkboard, it's okay if you don't draw all the small grid lines.

The whole point of getting performance out of the airplane on one engine is to make sure you are getting the necessary power available from that engine, the airplane is aerodynamically clean (cowl flaps closed on the inoperative engine, gear and flaps up), and the proper airspeeds flown.

You can try an experiment in flight to determine the best rate of climb or the least rate of sink on one engine:

1. Tape a 3-ft yaw string on the nose where it can be seen from both front seats. (Don't use tape that could peel the paint when you take it off.) Only the front end of the string should be attached to the nose, leaving the rest free to follow the airflow (Fig. 18-12).

2. Pick an altitude to climb through, say, 4000 ft MSL (if possible).

3. Get about 200 ft below this altitude, set up climb power on one engine and zero thrust on the other, and establish V_{YSE} for that altitude. The airplane should be in the clean configuration.

During this first climb, try to keep the wings level and the ball centered. Don't pay any attention to the yaw string. As the airplane passes the chosen altitude, mark the time on the sweep second hand or a stopwatch and climb for 2 min. Or you could time the climb for 400 ft, for example.

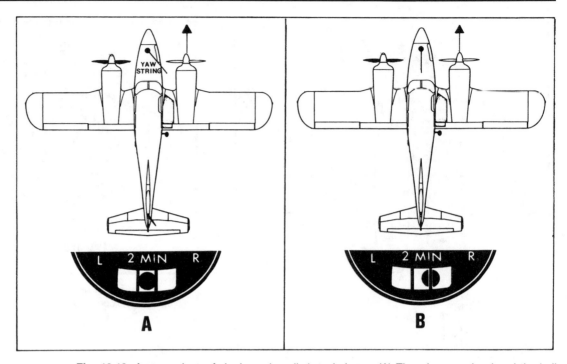

Fig. 18-12. A comparison of single-engine climb techniques. (A) The wings are level and the ball centered. (B) The yaw string is straight with a bank of one-half ball width into the good engine.

After writing the numbers down, ease back down to the 200 ft below the initial reference altitude and set up the climb again at V_{YSE}. It's best not to touch the power after setting it for the first climb, although you might check it for possible throttle creep. In other words, you'd want to have conditions exactly the same for both climbs. The second time, bank into the good engine about one-half ball width (it depends on the airplane, so you may have to experiment) and keep the yaw string centered during the climb. Use the same timing procedure as for the first climb and compare. It's most likely that with the yaw string centered, one-half or one ball-width bank to the good engine will result in a measurably higher rate of climb, but you might check it before demonstrating it to a trainee. Fig. 18-12 roughly shows the idea.

A yaw string on the nose in the conditions where the bank is overdone would show a definite sideslip existing, which is good for *control* (up to a certain sideslip angle), but the drag would be high and performance would be bad. A zero sideslip condition, shown in Fig. 18-12B, would mean low drag and better performance. A possible situation might be that initially you'd need the best control to get things organized and then go to zero sideslip to get climb performance to clear obstacles. Your airplane may require slightly less than one-half ball-width bank into the operating engine to get zero sideslip, but you would find this out in the experiment just discussed.

There is the possibility of losing an engine above the airplane's single-engine ceiling, and you may need to stretch your descent because of the terrain. It's easier to "drift down" to the single-engine ceiling than to climb up to it (Fig. 18-13). Use max continuous power on the operating engine and hold V_{YSE}.

The *Pilot's Operating Handbook* has *Accelerate-Stop* and *Accelerate-Go* charts, and you should be familiar with them. You don't have to memorize the whole chart, but you should have at least an idea of the approximate distances required for each action should an engine fail on takeoff.

Accelerate-Stop Distance numbers are what the average new pilot would consider as reasonable. Maybe 4000 ft is required at sea level at 85°F and at gross weight, but 7000 ft

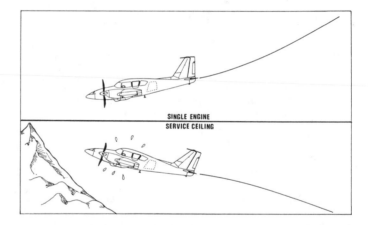

Fig. 18-13. As common sense tells you, it's easier to stretch it going down to the single-engine service ceiling than to climb up to it.

of runway is required at 8000 ft pressure altitude and 68°F at gross weight. (These are averages of some twins, but use your own actual information.) You might point out that the figures given are for a level, hard-surface, *dry* runway, and heavy braking at idle power is required.

The *Accelerate-Go Distance* is an eye-opener for most people. At a density-altitude near, but above, the single-engine service ceiling (the rate of climb less than 50 fpm), it could take over 16,000 ft to clear a 50-foot obstacle. There are very few places in the country where you could fly for over 3 mi without encountering something at least 50 ft high. Trying to spiral upward and avoid that length of straightaway could result in no climb at all.

Along this same general line, the *Rate of Climb–Balked Landing* chart should be of interest.

In your demonstrations you'll want to set up accurate zero-thrust conditions; Fig. 18-14 gives two sample graphs and a table for three twins.

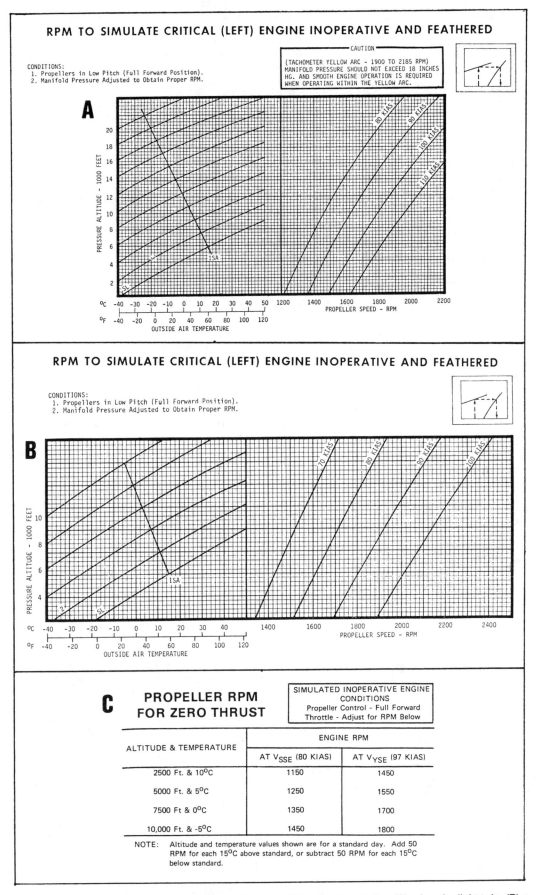

Fig. 18-14. Methods of setting up zero-thrust conditions for three twins: (A) a heavier light twin, (B) a medium, turbocharged twin, and (C) a light trainer with simplified tabular information. Note that airplanes A and B assume that the critical (left) engine is to be considered inoperative. Airplane C has counterrotating propellers, so there is no critical engine and the information is valid for either engine.

POINTS FOR LATER FLIGHTS

There's no attempt to set up a rigid syllabus because each situation is unique and, of course, your trainees will require different approaches (you may have some that are still working on normal operations at 10 or 12 hr).

Some day you may instruct in a turbocharged twin and will have to be especially careful that the trainee doesn't overboost the engines. If it's a mechanical system (which you can use or not, as you wish), you'd probably be better off teaching him normal operations (no turbochargers), emergency procedures (no turbochargers), then normal procedures (using turbochargers), and finally emergency procedures (using turbochargers).

The *Pilot's Operating Handbook* is the best source for establishing a syllabus for instructing in twins, and you should be sure that the trainee has no gaps in his knowledge that could cause problems later.

Here are some other areas in the *Pilot's Operating Handbook* that you could cover later:
—Wind component chart
—Maximum takeoff performance
—Time, fuel, and distance to climb (max and cruise climbs)
—Cruise performance charts
—Range and endurance profiles
—Holding time available (at various power settings)
—Time, fuel, and distance to descend
—Fuel flow schedule
Some other general points:

During the course you should demonstrate feathering and unfeathering and have the new pilot feather and unfeather each engine several times and maneuver at altitude (normal maneuvers, that is) to get used to the idea of seeing a prop standing still out there.

Never feather an engine for practice purposes below 3000 ft above the terrain (a flat terrain). You may decide that in your area 5000 ft (or more) is better.

The first time you feather the engine(s) of a particular airplane you've never flown before (even if you have plenty of experience in the same model and type), *do it at a safe altitude over or near a usable airport.* Individual airplanes and individual engines sometimes have their own quirks (Fig. 18-15).

Fig. 18-15. When you decide on the spur of the moment to give some "realistic practice," a recalcitrant engine can ensure that it will be very realistic.

Don't feather and unfeather each engine more than twice in a flight period, and give plenty of time between such procedures. For airplanes using the starter for unfeathering, this is particularly important because of starter overheating problems if an accumulator isn't available. Excessive and quick temperature changes are also rough on the engine.

Take care of the engine after unfeathering. Don't roar into full-power operations with it, but watch temperatures and oil pressure. A lot of cold air moves across the cylinders while it is shut down. Stress this point, and discourage the trainee from "practicing" feathering on his own.

Don't pull the mixture or turn the fuel off below 3000 ft—and particularly, don't do it on takeoff. There are instructors who advocate this, but a malfunction of a mixture control or fuel valve could result in a more dangerous situation than if an engine actually quit from unknown causes, because time would be wasted tinkering with the control instead of getting on with the feathering or other procedures. You can simulate a takeoff at altitude using a particular safe altitude as ground level and pull the mixture as you feel necessary but preferably over an airport. The trainee would go through the procedures of securing the bad engine (working foot, working engine, throttling back to confirm, feathering). One argument cited against throttling back rather than cutting mixture on a simulated engine failure on takeoff is that the trainee will see the retarded throttle and know which engine has "failed." Of course, he could see a pulled mixture control also, so the "realistic" thing to do would be to sneak the fuel off. (If you use this procedure at low altitudes you could have an interesting pattern someday.) Turning the fuel off, as noted in Chap. 8, can be a very risky business and the suggestion is *not* to do it at *any* altitude. If you pull the mixture, you can cover the two mixture controls with a chart so that he can't see which one is off.

After you see that the trainee is able to recognize and feather the propeller of the failed engine at altitude, simulate a failure on takeoff by throttling back to zero thrust while covering your actions from the trainee with a chart or other shield. He can tell you which engine has failed and then *simulate* the other checks and feathering. You would then have him fly the pattern and land. The trainee *should* have to fly close to the ground in simulated engine-out flight; he'll see that very careful planning is necessary to do the job. What seemed like a fairly reasonable rate of climb at altitude seems like nothing when there are actual obstacles to cope with; with the engine at zero thrust, you can get it back in action in a hurry if necessary.

Note that one engine may have the hydraulic pump, and the gear will have to be pumped down by hand. No matter how often you mention this, the trainee will probably forget on one of the feathered simulated approaches at altitude.

To repeat, don't make a "practice" approach to an actual landing with a feathered propeller. Something could call for an unexpected go-around and the airplane is scuttled. Even if the approach and landing are completed with no problems, it's hard on the engine mount and engine systems starting the engine and unfeathering the prop on the roll-out or after the airplane has stopped; some engines seem to want to shake out of the cowling. If a go-around should be required on one of these "practice" approaches, you could be at such a low altitude that a successful go-around could not be made. *That* would take some explaining. In fact, as a rule of thumb, better not get less than 300 ft before starting an *actual* single-engine go-around. In other words, nearly this much altitude could be lost in good conditions cleaning the airplane up and establishing the best single-engine climb speed. You might demonstrate and have him practice one or two single-engine go-arounds with the throttle at *zero thrust* to show how much altitude is lost, but by all means, take care of that idling engine on the approach and don't abuse it when power is applied. Also, emphasize that an actual single-engine go-around is extremely dangerous, and he might even consider bellying it in on the side of the runway if obstacle clearance would be a problem (if, for example, he forgot the gear and at the last minute decides to take it around on one engine).

If you can simulate *electrical and hydraulic failures,* fine; but in any event, you should review them several times during the training. It would be interesting to be making a bona fide single-engine approach and find that an electrical failure precludes the extension of the electrically operated landing gear.

SUMMARY OF CHAPTERS 17 AND 18

These flights should, like the presolo and preprivate duals, be made up of a preflight discussion, the actual flight, and a review; and if you're in doubt about which way to go in setting up any checkout flight, PEDPER is still a good guide.

PREPARATION—Have the trainee sit in the airplane; see that he has a *Pilot's Operating Handbook* and a good opportunity to go over it before the flying begins. You might review the *Pilot's Operating Handbook* if you haven't been flying that airplane every day.

EXPLANATION—This is done in the classroom and in the airplane on the ground. Don't rush it. You may use the *what, why,* and *how* approach to the new systems in the airplane.

DEMONSTRATION—Keep this to a minimum; you may want to show the trainee how to improve his synchronization of props in a multiengine airplane or simulate the *exact* steps in the emergency extension of gear, etc.

PRACTICE—Give him plenty of time in the airplane, flying it under normal conditions, before going into emergency procedures. Correct mistakes as he makes them. A single-engine approach should be treated as a normal one—no excess altitude or airspeed (twin). He must avoid getting low and slow on final, however.

EVALUATION—Is he staying ahead of the airplane and do you think he would keep calm and use the emergency procedures as necessary? Would you ride in the back seat of the airplane on an extended cross-country?

REVIEW—Review what was done on that particular flight and also go over the system(s) again, allowing him to ask any questions he might have after he's flown the airplane.

Again, noted in Chap. 17, when you are setting up a syllabus or checkout course for a more complex airplane (either single or twin), sit down and think of the information *you* needed at the new pilot's stage. Go through a flight in your mind, from the preflight check to the shutdown, and consider what the pilot does with the new systems in each phase; you'd be surprised how many things you've been doing subconsciously in flying the airplane. Think of the things that could go wrong with the system in various phases of flight and set up procedures and checklists to take care of them.

Analyze what you must teach the new pilot about the particular airplane, as well as the information that will apply to others so equipped. The whole-part-whole concept could be used in a general discussion; for instance, cover the principle of the constant-speed prop system (whole), then bring in specific numbers for specific points of flight for the airplane being used (part), and then review the principle in discussing the general emergency procedures for the system—taking a look at the overall operations of the equipment (whole). In other words, set up *your own* syllabus; it will help you learn about the airplanes.

Never sign off a pilot until you have *completed* the syllabus. For instance, you're working with a trainee on a checkout in a complex single or in a different multi-engine airplane (say, in his or her new airplane) but haven't done emergency gear extension or engine-out work yet, or maybe you feel that the individual just needs more work in that airplane. It's the weekend and he just "has" to take the airplane on that long-awaited vacation trip. You sign and may discover that the trainee decides that you aren't "needed" anymore. He later has an accident and you're in big trouble. He may turn on you (assuming he survives) for letting him go, as noted back in Chapter 2.

ADVANCED GROUND REFERENCE MANEUVERS

ON-PYLON EIGHTS

On-pylon eights are required for the commercial and flight instructor (airplane) practical tests and also are an excellent maneuver for some of your advanced students who need to build time but don't want to just be boring holes in the sky. They're particularly good for that trainee who's getting a little big for his britches.

■ **Preparation.** A chalkboard and model are needed. Review the maneuver yourself if you haven't done one lately. Assign the trainee reading on on-pylon eights (see the Bibliography).

■ **Explanation**

What—The on-pylon eights maneuver consists of flying an eight around two pylons; in the turn the pylon is kept at the same position relative to the airplane and pilot.

On-pylon eights, or eights on pylons, is one of the best maneuvers for teaching the pilot to fly the airplane automatically by reference to outside points and to control the altitude of the airplane without referring to the horizon. In olden days of flying, towers or pylons were placed in the practice area for use in this maneuver, hence the name.

In around-pylon eights, or eights around pylons, and turns around a point, the radius of the turn(s) was kept constant and the airplane flew a *specific path* in relation to a physical reference. In on-pylon eights, in the turning portion, a ground reference point is kept at the *same relative position to the wing tip.* Figure 19-1 compares around- and on-pylon eights in a light wind condition.

You might draw on the chalkboard one circle of turns around a point and compare it with one turn of an on-pylon eight in a moderate wind condition (Fig. 19-2). Compare the relative positions of the pylon to the pilot.

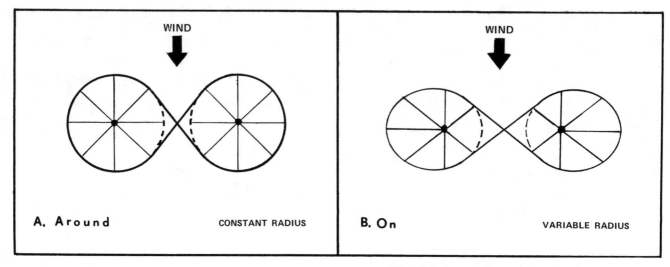

Fig. 19-1. A comparison of around- and on-pylon eights.

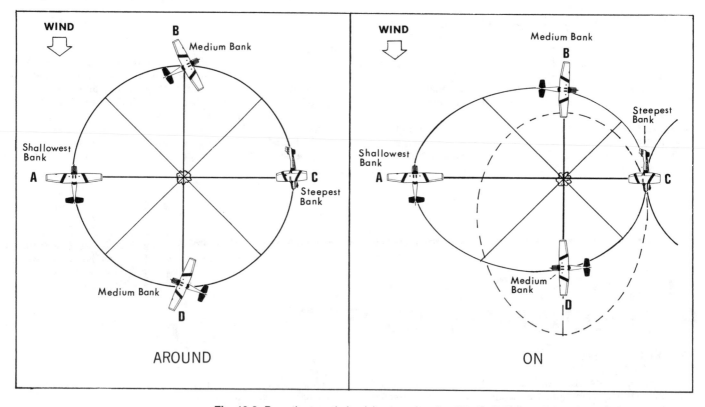

Fig. 19-2. Draw the two circles (okay, one is a "modified" ellipse) on the chalkboard side by side and trace the paths and attitudes using a model airplane. Note that in a no-wind condition both paths would theoretically be perfect circles, and banks and attitudes with reference to the pylon(s) would be constant. The path of the airplane in the on-pylon eight is not what would be predicted: the path shown by the dashed ellipse. A vectorial analysis, however, would show the path shown by the solid-line ellipse.

Comparisons of what the pilot would see at points *A* through *D* in each maneuver are in Figs. 19-3 through 19-6.

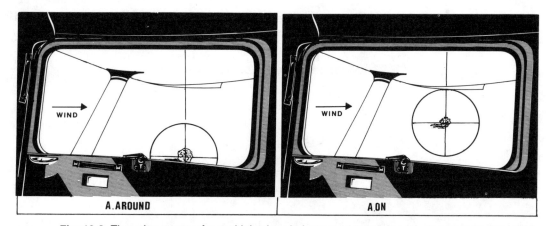

Fig. 19-3. The pylon as seen from a high-wing airplane at point *A* (flying directly upwind in Fig. 19-2).

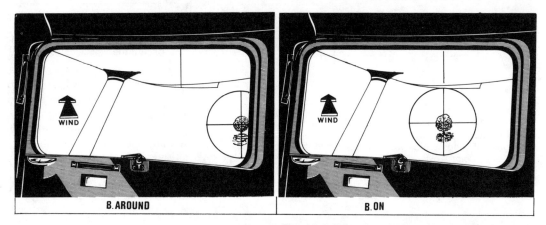

Fig. 19-4. The pylon as seen at point *B* in Fig. 19-2.

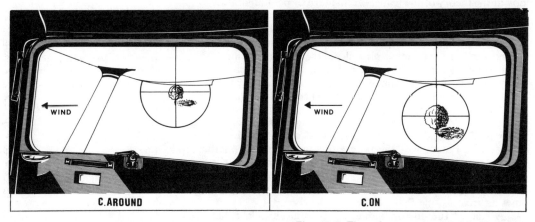

Fig. 19-5. The reference at point *C* in Fig. 19-2.

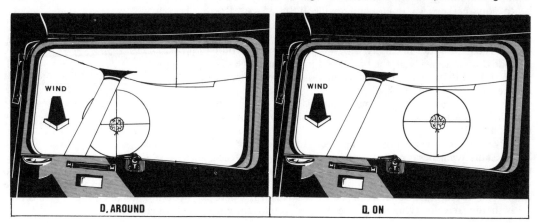

Fig. 19-6. Point *D* in Fig. 19-2.

The bank in the on-pylon eight must be changed to maintain the constant relative position as the airplane is moved into or away from the pylon (Fig. 19-7).

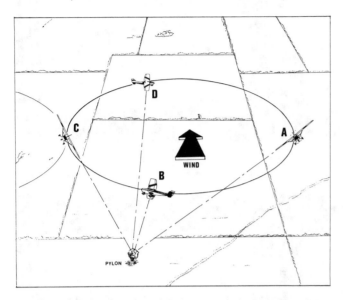

Fig. 19-7. Another view of the maneuver shown in Fig. 19-2 to illustrate that the airplane's bank must be varied to maintain a constant wing attitude with relationship to the pylon. The altitude at point *A* would be lower than at point *C* because the groundspeed is lower at *A*.

PIVOTAL ALTITUDE. After you've shown the trainee a general look at the theory of the maneuver, the idea of pivotal altitude should be brought in.

Pivotal altitude is that altitude at which the pylon may be held in a constant position and varies with the square of the groundspeed. In explaining pivotal altitude you may find it easier to introduce the idea with TAS in a no-wind situation, and *then* go into the fact that it depends on the relative velocity of the airplane to the pylon (groundspeed). It would not only vary with faster- or slower-cruising airplanes but would vary for a particular airplane when moving upwind, crosswind, or downwind. Make the point that since the pivotal altitude is a function of the *square* of the velocity, a small increase in velocity would require a noticeable increase in the pivotal altitude. For example, an airplane that is flying at 200 K in an on-pylon eight would have a pivotal altitude *four* times as great as one flying 100 K.

A rule of thumb for finding the pivotal altitude of an airplane (assuming the wind is calm, at first anyway) would be to square the TAS (mph) and divide it by 15, or PA = $(TAS)^2/15$.

For knots, $V^2/11.3$ *is the answer.*

This equation ties in closely with the fact that the pivotal altitude does not vary with the bank used in the maneuver. If you've found it for a 15°-banked or shallow on-pylon eight, it's the same for a 60°-banked maneuver (Fig. 19-8).

For your trainees with a mathematical bent you might explain it as was given in the *Civil Pilot Training Manual* (September 1941):

By geometry, the relationship between A/R and C/W will be the same regardless of bank and so will be an equal fraction (assuming that the turn is balanced), or $A/R = C/W$.

To find A (altitude) in terms of velocity (V), the law of centrifugal force must be invoked, or $C/W = V^2/Rg$, where V^2 is the linear velocity in feet per second, squared, and g is

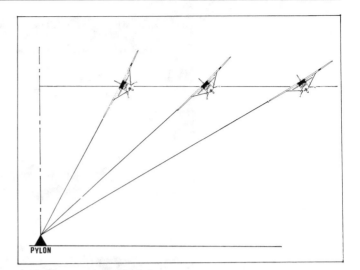

Fig. 19-8. The pivotal altitude is the same for any degree of bank if the velocities relative to the pylon are held constant.

the gravitational acceleration (32.2 ft/sec). Putting the equations together: $A/R = C/W = V^2/Rg$, or $A/R = V^2/Rg$. The Rs can be cancelled, and it's found that $A = V^2/g$ or the pivotal altitude is found by squaring the velocity, in feet per second, and dividing the answer by 32.2 to get the pivotal altitude in feet. Converting to miles per hour, further analysis would show that $A = V^2$ (mph)/15 is a good approximation for finding the pivotal altitude. Fig. 19-9 shows the geometry of the pivotal altitude.

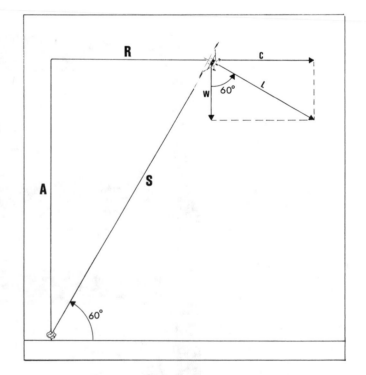

Fig. 19-9. An analysis of how the pivotal altitude is found.

Fig. 19-10 shows the effects of being at too high or too low an altitude in trying to hold a pylon. Using a 60° bank, you might sketch Fig. 19-10 on the chalkboard (assume a no-wind condition).

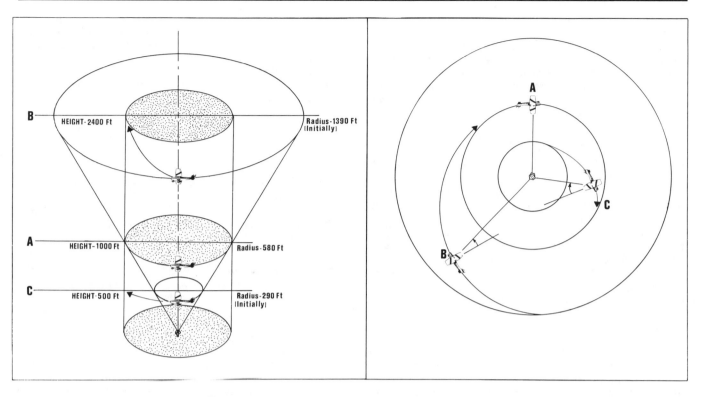

Fig. 19-10. A comparison (two different perspectives) of three airplanes "flying" a 60°-banked turn on a pylon in no-wind conditions.

For example, assume that the pivotal altitude for this airplane is 1000 ft, shown by *A* in Fig. 19-10. Solving for the *TAS* (remember this is a no-wind example) and working backward from the pivotal altitude (*PA*) equation:

$$TAS^2 = 11.3 \times PA$$
$$TAS^2 = 11,300$$
$$TAS = \sqrt{11,300} = 106\ K \text{ (actually it's 106.30146 K).}$$

There was a choice of rounding off the airspeed or the pivotal altitude to start, so *PA* was given as the fixed value here.

The true airspeed being flown around the pylon is 106 K (rounded off to the nearest knot).

The radius of that circle, right on the pivotal altitude is

$$\text{Radius} = \frac{V^2}{11.3 \tan \phi} = 580 \text{ ft}$$

(Don't write and say that your calculations, based on the above equation, gave an answer of 574.09716 ft; it's *my* book and *I* say it's 580 ft, since the rounding off created the problem in the first place.)

Airplane A then, is in a 60°-banked, balanced turn and is pulling 2 g's.

Airplane B at 2400 ft would have to have a turn radius of 1390 ft in order to follow the cone boundary, since it is 2.4

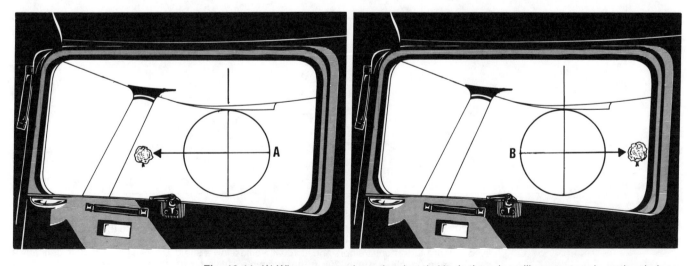

Fig. 19-11. (A) When you are above the pivotal altitude the pylon will appear to gain on the airplane. The trainee may try to hold the pylon by slipping (consciously or unconsciously). (B) When you are below the pivotal altitude, the pylon will appear to be falling behind. The trainee may skid the airplane trying to show that his pylon is in the right place, but it won't work.

times as high as airplane A. In a 60°-banked turn, it is also pulling 2 g's.

Airplane C would also be pulling 2 g's in a balanced 60° bank, and the turn radius would have to be 290 ft in order to follow the geometry of the cone.

The radius of a balanced turn is proportional to the velocity (squared) and if *all three airplanes use the same airspeed the radii of the turns will be the same* (shaded area). Assuming that Airplane A is flying the correct pivotal altitude for its airspeed, it will hold the pylon in the 60° bank indefinitely; and assuming they all start at the points as indicated in Fig. 19-11, you can see what will happen as soon as the maneuver begins.

Airplane A continues the maneuver with no problem, since it is at the proper altitude and radius of turn to point its lateral axis at the pylon.

Airplane B, maintaining a 60° bank, must turn at the *same radius as A because the airspeed is the same,* and turns inside the higher circle in which it started. As the turn radius decreases to the "proper" one for the bank, the wing (reference point) moves back, the pylon apparently moves ahead of its proper position, and the airplane turns into the pylon. (The pylon is gaining on the airplane.) To stay on the pylon in a 60° bank, the airplane must be moved *down* to fit the surface of the cone, at the proper radius for its airspeed (point *A*).

Airplane C starts off at the same relative position on the circle to the pylon as the others but must also follow the required radius for its airspeed and 60° bank, and has to leave the original smaller circle. As indicated by Fig. 19-11, its wing (reference point) will move ahead and gain on the pylon. To stay with the pylon at a 60° bank the airplane must be moved *up* to fit the cone (point *A*).

You can use this same discussion for any angle of bank.

You probably won't go into this type of detail in the explanation of on-pylon eights to many of your trainees, but you should be able to say *why* certain things are necessary in order to hold the pylon.

In your explanation of *how* to do the maneuver, note that the elevators are the primary control for holding the pylon and accomplish a double effect when used by affecting both airspeed and altitude in the desired manner. (If the airplane is high, easing the nose down will pick up airspeed, raising the pivotal altitude and lowering the actual altitude so that things meet in the middle.)

■ **Demonstration.** If the wind is calm or light you may pick a point (a crossroad is suggested) and demonstrate how the "pylon" can be held if the pivotal altitude is maintained. Show the trainee how a variation up and down from the proper altitude causes the pylon to "move" (Fig. 19-11).

PICKING PYLONS. It's suggested that the first full on-pylon eights be at 30–45° of bank and, if in doubt, pick pylons slightly farther apart than necessary so that the trainee will have time to get organized between pylons. After he sees the idea, you'll want to get started in the actual final maneuver with a 3- to 5-sec straightaway between pylons.

A short discussion at the chalkboard of how to pick pylons is a good move before the flight. You should pick a couple and have the trainee do the same, correcting his errors (too close together, too far apart, different elevation for each pylon, etc.) during the introductory flight.

You can get a good idea of the distance between pylons by using the turn radius equation given earlier. Again, ϕ is the angle of bank and several tangents are given here:

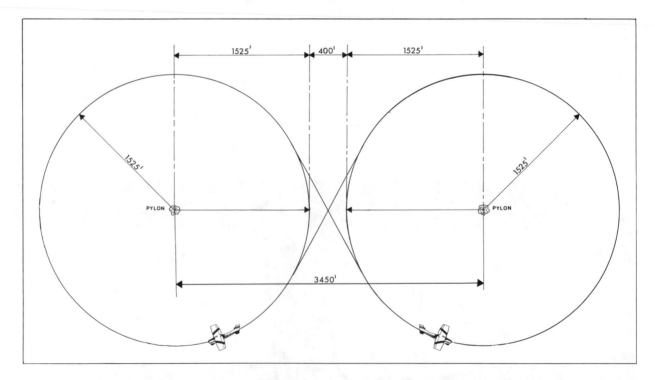

Fig. 19-12. A look at the required distance between pylons for on-pylon eights (or eights-on-pylons) for a 30° bank. Note that the distance between tangent circles would be twice the radius. A distance of 400 ft has been added between pylons to allow for a 3- to 5-sec straightaway. The 400-ft value is a fixed distance for various trainers and at 100 K would allow a straightaway flight of 1500 ft, or nearly 9 sec. Later you might add only 200 ft to the double radius for an answer closer to the 3- to 5-sec straightaway requirement. When you first introduce on-pylon eights, 3–5 sec is too short a time for the average commercial trainee to get oriented with the other pylon. (From a practical standpoint, how do you estimate 3450 ft on the ground, anyway?)

Angle of Bank	Tangent
15°	0.27
30°	0.58
40°	0.84
45°	1.0
60°	1.73

Substitute these in the equation; for instance, if you figure that the "average" bank flying around one pylon is 30° at 100 K, you would set up the following: radius = $V^2/11.3 \tan \phi$ = 1525 ft (rounded off). You would double this figure to get the distance between pylons and add 400 ft for the 3- to 5-sec straightaway for a total distance between pylons of 3450 ft or two-thirds of a mile (Fig. 19-12).

The best way to pick pylons is to fly crosswind, looking to the downwind side; when suitable pylons are sighted a turn can be made to start the maneuver.

Using crossroads for pylons (or a tree at the corner of the crossroads) helps keep track of the pylon position when it's hidden by the wing on the straightaways.

If trees are chosen for pylons, they should be well separated from others. People sometimes fly off at an angle searching for "their" trees, and the situation can get quite funny before you see the problem. Some instructors have been known to sit there and let them grope, and a pretty fair cross-country can be flown before either party breaks the silence.

FLYING THE EIGHTS. You should do at least two full eights in your demonstration. A medium-banked pylon eight is the best way to introduce the idea. You can go into steep and shallow eights after the trainee has seen and practiced the basics.

You'll have to fly the eight and talk at the same time, so keep your eyes on the pylons—and an occasional glance for other airplanes.

If there is wind the pivotal altitude will change around the turn, and you should point out and demonstrate that the airplane should be flown back to the no-wind pivotal altitude in the straightaway legs between pylons. You may sketch Fig. 19-1 on the chalkboard again after the flight.

Indicate the best ways to locate the other pylon when completing the turn. It's particularly hard for him to locate it in a side-by-side airplane when the roll-in and turn is to the right. In the first turns he should be reminded to pick some reference well out that would be in line with the pylon at the point of roll-in (Fig. 19-13).

It's better to start the roll-in fairly late, requiring a more "brisk" roll than to start too early, have to shallow out, and then roll back in (Fig. 19-14).

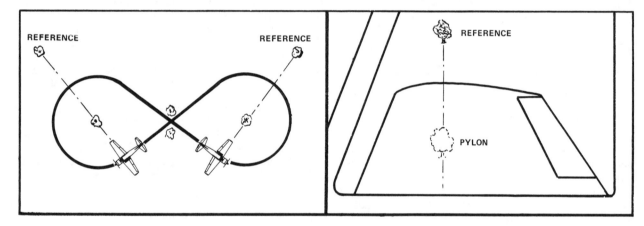

Fig. 19-13. The trainee may have trouble knowing when to roll in on the pylon, particularly in a steep eight where a right turn is required.

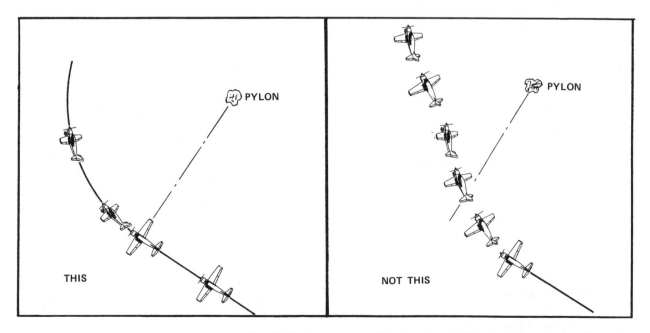

Fig. 19-14. A premature roll-in can cause a delay in getting set up on the pylon.

■ **Practice.** After you've picked the pylons and made a couple of circuits, turn it over to him for continuation of the eights. Talk him through a couple and then let him do several without too much prompting from you.

One problem the new pilot has is the tendency to check the altimeter during the turn part of the maneuver. *He is to fly the proper altitude by using the pylon;* that's the only way it can be done. By use of the elevators he will correct for pylons' apparent movement. The double effect of elevators mentioned earlier in the chapter is that their proper use gets both airspeed and the pivotal altitude together quickly. The bank must be readjusted after the pitch corrections.

The trainee will tend to concentrate on one pylon too long (Fig. 19-15). Remind him that the straightaways are good places to take a quick check for other airplanes. He'll also have to correct for drift between pylons.

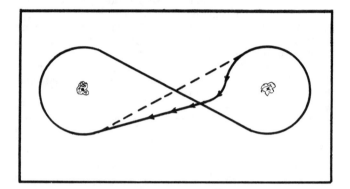

Fig. 19-15. Watch for a tendency to concentrate on one pylon past the point of roll-out.

A common error is not getting to the pivotal altitude during the straightaways; the result is that there isn't time to get set up through the next turn part either. This is particularly true of steep eights where the turn is quickly completed.

After he has the idea and is able to hold the pylon, require preselected banks such as 15° or 30° for the shallow and 45° or 60° for the steep ones. Usually the steep eights are easier if he makes a good entry; the turn is over before things can get too far out of hand. However, there is a good chance of his making the error shown in Fig. 19-15.

What's really tough for the trainee is a very shallow eight. It takes so long to get around that he has all kinds of time to get fouled up.

BACK-TO-BACK EIGHTS. Fig. 19-16 shows back-to-back pylon eights. As is indicated, a common pylon may be used for both shallow and steep eights. Save these until he has smoothed out the more elementary types.

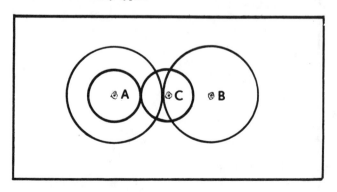

Fig. 19-16. The common pylon is *A*. *B* and *C* are the "other" pylons for shallow and steep eights respectively.

To sum up, the *common errors* that might be expected are these:

1. Poor pylon picking. The trainee takes too long or comes up with poor choices.
2. Overconcentration on one pylon.
3. "Losing" a pylon.
4. Poor wind drift in straightaways.
5. Rolling in too soon, particularly when the pylon is on the right (side-by-side airplane) and is hidden.
6. Poor bank control.
7. Too slow roll-out when back-to-back eights are being done.

■ **Evaluation and Review.** The big thing about your analysis of his progress is not whether the trainee keeps the airplane right on the spot each time but if his correcting actions are prompt and in the proper direction. Discuss accelerated stalls as they might occur in the steep eights.

Fly a dual introductory flight, giving the trainee the word; and after you see he has the principle well in mind, set up some solo practice sessions. This will allow him to work out the minor problems. Depending on his progress, you should fly a dual session periodically and check his eights as well as his other flying.

Some of your trainees might want to see what's happening as the airplane flies around one of the pylons in a moderate to strong wind. This is shown in Fig. 19-17.

Two requirements are necessary to analyze the on-pylon eight: (1) The airplane's path at any instant is tangent to the circle (or ellipse), and (2) the wing must be pointed directly at the pylon at any (and all) points of the path. By flying the airplane to maintain the pylon at the exact reference point (which also requires altitude variation as the groundspeed changes), an elliptical path is "inadvertently" flown. In other words, the pilot does not *try* to fly an elliptical path in a wind condition; it results from the two requirements. At each point shown in Fig. 19-17, a wind triangle would show that the airplane's heading and true airspeed plus the wind vector result in the path shown.

After all dual flights you should review the maneuver(s) and bring out in detail the errors or good moves he made. Have the trainee brief you on his problems during the solo sessions. Schedule for dual or solo practice as his progress dictates. Be sure to give "why" answers to any problems he had.

1080° SPIRALS

You'll have to ease into this maneuver with some of your more timid people; the idea of a 60°-banked, power-off descent doesn't grab them right away. You may have to start off with a *reduced*-power, 45°-banked descent and let them work on maintaining a constant bank before starting the variable bank spiral. This is *not* a required maneuver.

■ **Preparation.** A chalkboard and model should be available for the preflight discussion. Have the student review turns around a point and read about this maneuver (see the Bibliography) before the first dual session.

■ **Explanation**

What—The 1080° spiral is, as the name implies, a power-off spiral of three complete turns. This is the power-off (idle) version of turns around a point. The radii of the circles should be constant, and drift is corrected for by varying the bank as done in the turns around a point. Because the

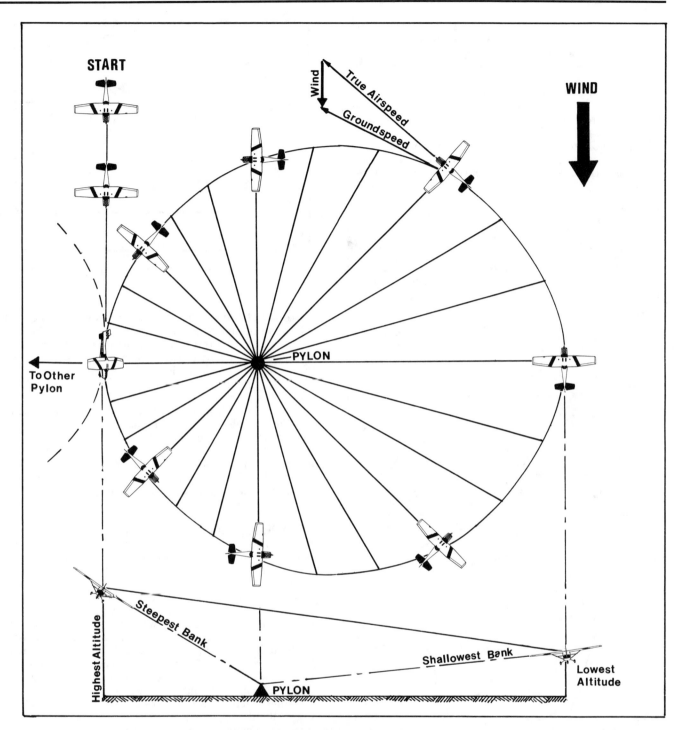

Fig. 19-17. The path of an airplane around one pylon in a moderate to strong (constant) wind. If you are mathematically inclined, or have a Navy Mark III Plotting Board, you may set up an airspeed and wind velocity and actually plot the 24 points shown. Note that the wing is to be always pointed at the pylon; if this is done with the wind, as shown, an elliptical path results, with the pylon being one of the foci of the ellipse.

maneuver must be started higher, the first turn is difficult for the average trainee.

Why — The maneuver is good for developing a skill in flying the airplane by outside references and correcting for drift in a turn. It builds confidence because the trainee finds that things don't necessarily go to pieces in a spiral (it just seems that way).

How — A reference is picked; the center of crossroads is good for orientation in the higher part of the maneuver. The spiral is normally entered directly downwind and is completed by making three 360° power-off turns maintaining a constant radius. You may later require an entry on any part of the circle. Don't get below 1000 ft above the surface, even over bare terrain. Some instructors prefer 1500 ft as the minimum altitude, but the higher the maneuver is done, the more difficult it is for the pilot to see and correct for drift.

Usually the trainee is just getting his drift correction set up well when it's time to break it off. (Because of the vertical displacement of the maneuver, varying winds could also be encountered throughout.) Fig. 19-18 shows a top view of the spiral.

Fig. 19-19 is an oblique view of the spiral. The outside lines show that the normal tendency of the new pilot is to have too wide a circle at the top of the maneuver. This is done unconsciously to help keep the reference in sight.

The use of crossroads will help keep the position of the reference in mind, particularly in a low-wing airplane and/or a right turn (Fig. 19-20). Require practice in both left and right turns.

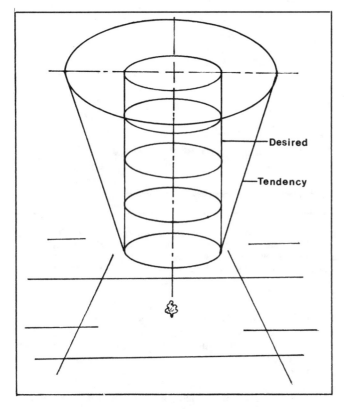

Fig. 19-18. The 1080° spiral as seen from directly above.

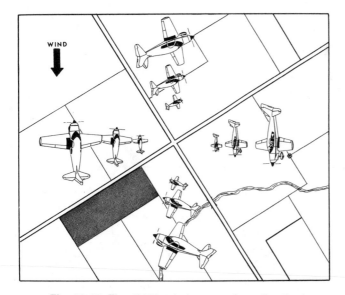

Fig. 19-19. Desired and probable paths in the first few 1080° spirals.

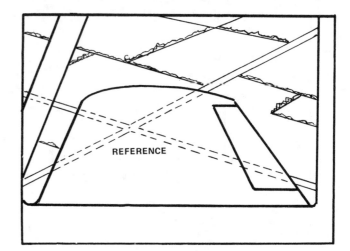

Fig. 19-20. Crossroads are good reference points.

■ **Demonstration.** The weather and your schedule might not cooperate, but if possible, the introduction of both the on-pylon eight and this maneuver should be done in smooth or maybe very slightly rough air. Certainly later these should be practiced in less-than-ideal air conditions, but as in any maneuver, the introduction should have as few outside distractions as possible.

Usually one demonstration of a 1080° spiral is enough to show him the idea. If it appears that he's having trouble getting the idea in the preflight discussion, or if he hasn't flown wind drift correction in some time, a session of dual turns around a point is in order. The turns around a point are better for introducing or reviewing the theory and practice than is the spiral, because you can make a number of turns without worrying about making a dent in the landscape.

■ **Practice.** So, if there's any doubt about his understanding the 1080s, go the turns around a point route first; that's better than having to break off work on the spirals and go back to turns around a point.

One problem most trainees have at first is not keeping up with the number of turns. You should require that they roll out within ±20° of the required heading at first and, as they get more proficient, cut this down to 5° or less. Encourage the use of landmarks for rolling in or out rather than the use of the heading indicator, which probable would "slip" in the 55–60° banked turn. The maximum angle of bank should be no more than 60°, which means that if there is a significant wind, the shallowest bank may be 30–45°.

Have him practice *constant-bank* spirals at various banks at a high altitude without using a reference. While this does not seem to apply to the final requirement, it does give good practice in attaining and maintaining a given bank in a spiral and helps the trainee loosen up. You might, for instance, have him set up a 60° bank; when he's established, call for a 45° constant bank, and then require a return to exactly 50°, 55°, 60° (it's illegal to *exceed* a 60° bank in a nonaerobatic airplane).

Common errors made in 1080° spirals are these:

1. Losing the reference in the highest (first) turn.

2. Allowing the airplane to get wrapped up—the airspeed and amount of bank increases beyond your allowable limits. Poor pitch and bank control.

3. Poor wind drift correction; forgetting the wind direction.

4. Not completing exactly 1080° of turn.

■ **Evaluation and Review.** Before allowing solo practice you would work out any unsafe tendencies, but don't keep flying with him until you think he's perfect in the maneuver. Check him out with it, then set up some solo periods so that *he* can work out the details. Follow up with a dual flight periodically.

After the first dual flight have a good review of the principles of turns around the point and the 1080° spiral. Let the trainee explain the two maneuvers at the chalkboard if you feel that this would help him see it better. Be sure he understands the maneuver before sending him out alone; if he has the wrong idea about it, he could get discouraged on the solo sessions because he can't hold the reference and he doesn't know why.

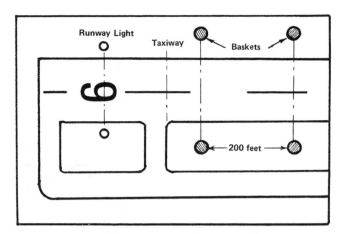

Fig. 19-22. Some references for checking the accuracy of the landings. Most people use references that already exist, but if you put out your own, make sure that they are well off the runway and/or taxiways and that they are light and easily displaced if something hits them.

ACCURACY LANDINGS

Teach the presolo student both partial-power and power-at-idle approaches (to simulate an engine malfunction); Fig. 19-21 compares the two pattern sizes.

Fig. 19-21. A straightforward engine-at-idle approach pattern as compared to the wider reduction-of-power pattern. You will use this pattern for power-off approaches rather than that shown in Fig. 19-23 with its "cutting across" or S-turning (no radical flap usage or maneuvering should be done with the student pilot. (*The Student Pilot's Flight Manual*)

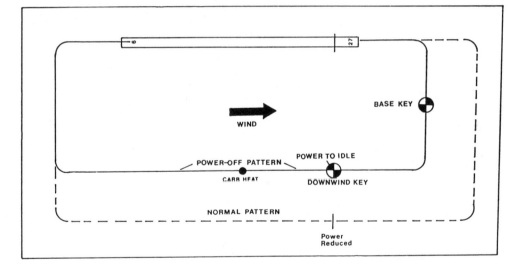

■ **180° Accuracy Landing.** The 180° approach is the best introduction to the idea; it allows enough time in the process to correct for errors, but doesn't require as radical maneuvering and close judgment as are required for some 360° overhead approaches. It is close to the type of approach the pilot has been making from the beginning.

The point of reference for touchdown should be a certain runway light or other well-defined point not on the very end of the runway. An imaginary extension of a taxiway edge might be used. If you are really going into flight training in a big way at your airport, you might place four yellow-painted, upended bushel baskets in the positions shown in Fig. 19-22 before starting each session. By using four, *you* can get an accurate check of his landings in the dual flights.

If you have only one reference line, real or imaginary, require the trainee to shoot for touching down past but within 200 ft of it.

Your *preparation* would require a chalkboard and model.

EXPLANATION. Go through the procedure step by step at the chalkboard. The use of downwind key and base key positions can be of great aid in an actual power failure. (You should point out that a dead engine and windmilling prop will hurt the glide ratio as compared to an idling engine.) The more advanced trainee should be allowed to slip and/or use flaps as necessary to land at the prechosen spot (Fig. 19-23).

The downwind key can be one point in checking how the pattern is shaping up, so that necessary corrections can be made for the base key. The base key can be the next decision point for maneuvering to make the spot.

Later have him combine spirals from a high altitude to hit the downwind key properly and complete the spot landing; however, you'll want to stay at pattern altitude and shoot landings in the pattern, as shown in Fig. 19-23, until he becomes proficient at hitting the spot. These accuracy landings are good for morale and confidence, especially for those people who've had doubts about being able to land on a precho-

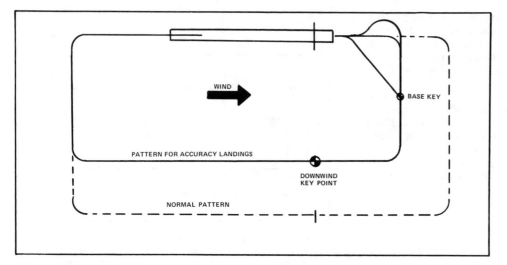

Fig. 19-23. Stay in the accuracy pattern until the trainee is competent in hitting the spot. If the airplane is a simple fixed-gear, fixed-pitch-prop type, touch-and-go's may be done; otherwise it's suggested that full-stop landings be made each time to avoid "incidents."

sen spot after a complete power failure.

The path shown from the base key in Fig. 19-23 is exaggerated, and for most cases added flaps or slipping will be better than crossing the extension of the runway center line and S-ing. Because of this, the flaps should be used in increments, starting with 1 notch, or one-fourth flaps at the downwind key, and adding flaps and/or slipping as needed. Check that your airplane may be slipped with flaps extended before advocating such a practice here. It may be that because of a late turn to base or a higher wind than expected, this initial setting would be the maximum flaps allowable if the spot is to be made. On the other hand, if it appears fairly *early* in the approach that the initial setting is too much, the flaps might be upped in increments to 0°.

The ideal approach would be to apply some flaps at the downwind key point, add more at the base key, and go to full flaps on final, touching down just over the reference line at the lowest speed possible. It is best to always have full flaps at touchdown, even if most of the approach is made with no-flaps in order to hit the spot. In some circumstances, however, last-second dabbling with the flaps could be distracting enough to cause an accident. Set a point in the approach, based on your airplane and situation, beyond which the trainee doesn't do anything else with the flaps.

If the trainee is sharp and the airplane can be slipped with flaps, a slipping turn to final works well if the airplane has a safe altitude during the turn. This is best if the airplane is high and the base leg is in too close and so the final leg will be too short to set up a slip. Fig. 19-24 shows the idea.

You'll work out the best altitudes for your airplane, but for the average two-place trainer, 800–900 ft AGL is good for a *downwind key point* and 450–500 ft works for the *base key* in a wind of about 10 percent of the approach speed—for stronger winds the height must be greater. In no-wind conditions and a perfect pattern, the base key should be at a height exactly one-half that of the downwind key. In real life this won't work because the second half of the approach will in nearly every instance be in a headwind condition, and the glide ratio will suffer with reference to the ground. You might prefer a downwind key at 1000 ft if that is the traffic pattern altitude at your airport.

The reason for setting up a pattern requiring *some* flaps initially is that the airspeed is normally better controlled with some flaps; you'll see some no-flap starts with the airplane

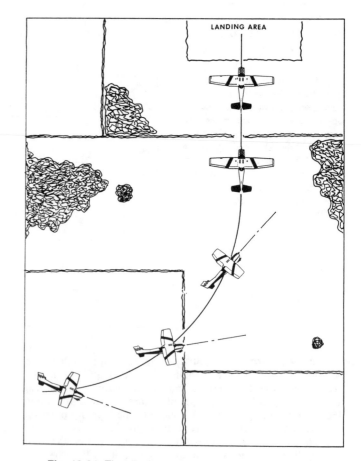

Fig. 19-24. The slipping turn can be used in emergency when the airplane is too high and too close to the runway on base. It could be hazardous in some airplanes.

moving on the descent portion of the downwind leg on past the point of no return (power-off).

DEMONSTRATION. This is always the one time, it seems, when you can't hit the runway with your hat, much less with the airplane. You have it nicely worked out on base, the approach path is wired, and suddenly the wind picks up from 5 K to 25 K and stays that way for about 30 sec right after you've turned on final. You've been talking it up all the way around and now have to add plenty of power to keep from landing on that used car lot that just appeared off the approach end of the runway. Or, when the wind has been strong and steady and goes to calm on final; full flaps and slipping won't save it. A sense of humor is useful here (Fig. 19-25).

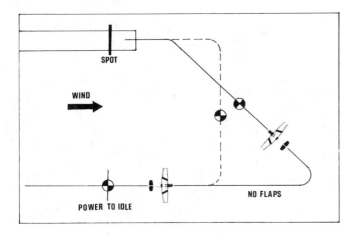

Fig. 19-25. Even cutting across and using zero flaps sometimes can't do the job if the downwind is stretched out.

Some canny flight instructors let the trainee shoot the first accuracy landing ("Go ahead and do it like I told you at the chalkboard"). They keep their mouths shut until the spot is made, or not made, and then make appropriate comments. If the trainee goofed, the instructor can then demonstrate an approach based on the information he learned riding through the first one. This shady procedure can be impressive, but sometimes the instructor misses the spot even after a free look at it.

Getting back to reality, the demonstration of one approach should be enough, since it's merely a variation of what the trainee has done many times before. If the person is sharp, you can dispense with any demonstration and let him go ahead, but *you* should be able to tell him afterward exactly *why* he had any problems—as well as make suggestions during the approach.

PRACTICE. Give enough dual on these approaches so that the trainee will be safe and will also know what to correct for during solo practice. A number of patterns may be required before he is able to put the airplane at the downwind key position, and most will tend to be too high in calm wind conditions and to undershoot if the wind is 10 K or more.

Common errors pilots make in accuracy landings are these:

1. Not starting the turn to base soon enough, the result being that the pilot is continually behind the airplane.

2. Too short a final leg—another extreme (some people alternate errors 1 and 2 in a practice session). The final turn is low and the distraction of checking the spot could result in a dangerously close approach to a stall.

3. Too fast an airspeed on final so that the airplane floats well past the spot.

4. Putting the airplane on the ground. This is usually the result of error 3. In an actual engine-out landing

it *is* better to put the airplane on the ground than to float into a barrier, such as trees, at the far end of the landing surface. The airplane will decelerate faster on the ground, and every foot-per-second velocity decrease means a better chance of walking away from an obstacle encounter. In the real situation he will have to make the best of his position since no go-around would be possible. However, even though you would advocate "putting it on" in an emergency, you'll want him to practice touching down slowly at a preplanned point if possible. If, in an actual emergency, he is floating, *then* the airplane would be put on the ground.

5. Getting so engrossed in the spot that the landing itself is neglected.

EVALUATION. See that the trainee is making safe approaches and is analyzing his errors correctly. If he is continually low (or high) he is making the same mistakes and your evaluation would be that more dual and analysis on your part is in order. It doesn't do any good to send a student (or advanced pilot) out to practice solo while he is still having *basic* problems with a maneuver; he'll just pin down the bad habits further.

REVIEW. After the dual, review errors and cover the principles of hitting a spot. If you can work it, watch some of his solo approaches and write comments, as shown back in Fig. 10-37. Discuss any problems.

■ **360° Overhead Approach.** Fig. 19-26 shows the makeup of a 360° overhead approach. This one, like the 180° side approach, may be entered from a spiral, with wind corrections used to assure hitting the upwind key at the proper point and height.

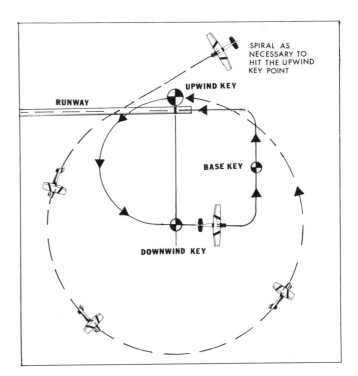

Fig. 19-26. A 360° overhead approach. For most light trainers the upwind key is 1000–1200 ft above the surface. The military 360° overhead approach is usually a full circle, without a developed base leg. It's suggested, however, that you use two 90° turns with a definite base leg as shown here. The trainee may "cut across" or S-turn to final if the situation is safe to do so.

It was noted earlier that the height of the airplane at the *base* key is in direct proportion to the wind velocity; however, it is best to have the *upwind* and *downwind* keys at their same respective heights and each time to use the same pattern between those two. The actions required to do this will better show the wind's velocity and the required height at the base key.

Expect some problems at first, particularly in a strong wind, with the pattern. If a spiral is used to get down to the pattern, the pilot is forewarned of such a problem.

You can expect the errors mentioned in the 180° side approach in the last part of the maneuver.

SUMMARY

These maneuvers are covered in more detail than might seem warranted by current flight test requirements. This chapter is included because the maneuvers are very useful for training and building confidence and also because information on them is gradually being lost over the years. As was noted in the section on the on-pylon eights, the best reference for this type of maneuver is a *Civil Pilot Training Manual* published in September 1941. Later manuals (FAA and commercially printed) have tended with each new edition to gradually reduce discussion of these types of maneuvers.

INTRODUCING AEROBATICS

BACKGROUND

Part 5 is intended for the aerobatic trainee (how to fly the maneuvers) as well as the aerobatic instructor (how to instruct aerobatics).

Don't feel that you can go out and spend a couple of hours learning to do some aerobatic maneuvers and then start teaching them. You should have at least 20 hr of experience in aerobatics (10 hr dual and 10 hr solo) before attempting to teach even *elementary* aerobatics. You not only must be able to demonstrate the maneuver but also will have to analyze the trainee's errors, and this takes a little longer to learn than you might expect.

You'll find that in aerobatic instruction, as in other areas of instruction, you'll learn a great deal more about aerobatics, and people, over the years.

Aerobatic instruction is more physically demanding than the other types of flying, but after you've gotten involved you'll find that teaching straight primary and basic maneuvers seems very tame and maybe even a little bit boring at times (but don't get complacent). The two types of instruction can be alternated, to the benefit of both.

Always keep in mind that you will need plenty of altitude (1) for safety's sake, and (2) because the trainee will feel more comfortable and the learning process will be improved. Also,

INTRODUCTION TO AEROBATIC INSTRUCTION

always make sure that your safety belts are well snugged up when teaching aerobatics; the trainee might get the airplane in such a position that you'd be hanging in there (loosely speaking).

When you get involved with aerobatics there are some new paragraphs of the Federal Aviation Regulations you'll have to know about:

FAR 91.303. *Acrobatic flight.*

No person may operate an aircraft in acrobatic flight—

(a) Over any congested area of a city, town, or settlement;

(b) Over an open air assembly of persons;

(c) Within the lateral boundaries of the surface areas of Class B, Class C, Class D, or Class E airspace designated for an airport;

(d) Within 4 nautical miles of the center line of any Federal airway;

(e) Below an altitude of 1,500 feet above the surface; or

(f) When flight visibility is less than 3 statute miles. For the purposes of this section, aerobatic flight means an intentional maneuver involving an abrupt change in an aircraft's attitude, an abnormal attitude, or abnormal accleration, not necessary for normal flight.

FAR 91.307. *Parachutes and parachuting.*

(a) No pilot of a civil aircraft may allow a parachute that is available for emergency use to be carried in that aircraft unless it is an approved type and—

(1) If a chair type (canopy in back), it has been packed by an appropriately rated parachute rigger within the preceding 120 days; or

(2) If any other type, it has been packed by a certificated and appropriately rated parachute rigger—

(i) Within the preceding 120 days, if its canopy, shrouds, and harness are composed exclusively of nylon, rayon, or other similar synthetic fiber or materials that are substantially resistant to damage from mold, mildew, or other rotting agents propagated in a moist environment; or

(ii) Within the preceding 60 days, if any part of the parachute is composed of silk, pongee, or other natural fiber, or materials not specified in subdivision (i) of this paragraph.

(b) Except in an emergency, no pilot in command may allow, and no person may make, a parachute jump from an aircraft within the United States except in accordance with Part 105.

(c) Unless each occupant of the aircraft is wearing an approved parachute, no pilot of a civil aircraft, carrying any person (other than a crew member) may execute any intentional maneuver that exceeds—

(1) A bank of 60 degrees relative to the horizon; or

(2) A nose-up or nose-down attitude of 30 degrees relative to the horizon.

(d) Paragraph (c). of this section does not apply to—

(1) Flight tests for pilot certification or rating; or

(2) Spins and other flight maneuvers required by the regulations for any certificate or rating when given by—

(i) A certificated flight instructor; or

(ii) An airline transport pilot instructing in accordance with FAR 61.169 of this chapter.

(e) For the purposes of this section, "approved parachute" means—

(1) A parachute manufactured under a type certificate or a technical standard order (C-23 series); or

(2) A personnel-carrying military parachute identified by an NAF, AAF, or an AN drawing number, an AAF order number, or any other military designation or specification number.

■ **More about Parachutes.** When practicing "acrobatics" (aerobatics) with a trainee, you'll both have to wear parachutes that have been repacked within the period noted in FAR 91. You should get checked out by a certified rigger on their upkeep and use. If possible, each parachute should have its own bag and should be stored in a warm (not hot) dry place when not in use. From time to time parachutes should be left out of the bags overnight in a safe dry place so that the harness and external portion of the packs may air out, particularly in the summer, when perspiration is a problem.

You should check out each trainee on the wearing of parachutes and procedures to use in case of trouble. This should also include an explanation of jettisonable doors and any other methods of egress.

Don't lay the parachutes on the ground; they could get damp (or stepped on, which could possibly bend the rip cord pins, making opening difficult or impossible). It would be disappointing to bail out and discover that the rip cord refused to budge.

Make sure that the parachute harness is tight and well fitting so that no excessive opening jolt might be encountered.

Nordiss Himmshau, an aerobatic instructor in the Midwest, was so conscientious about this with an attractive blonde student in shorts, that he set a new world record of 55.6 minutes for aiding a person into a parachute. His time would have been longer except for the untimely arrival of his wife at the airport. Nordiss was out of the hospital and teaching aerobatics (with trainees selected by his wife) within two weeks, however, so no permanent damage was done. It was found later that his advice to male trainees had always been, "Here, fathead, go figure out how to put this on and I'll come out to the plane when you're ready to go."

■ **The Aerobatic Trainer.** There are several good commercial aerobatic trainers on the market today and the plane you use will depend on the local situation (dealerships, local preference, etc.). The illustrations in Part 5 (Chaps. 20–23) use a "composite" side-by-side trainer with a *wheel* control. You may prefer a tandem trainer with stick controls. There seems to be a feeling, particularly among older (and military) pilots that the stick is better than the wheel for aerobatics. However, if you are teaching aerobatics to a recently trained pilot (who has never flown with a stick), there's no transition for him with a wheel-equipped trainer, and the aerobatic training can be readily transferred back to his own airplane, which is apt to have a wheel control. (This is not to say that he is to go back to his nonaerobatic plane and run through the repertoire, but if he does happen to get inverted because of wake turbulence, he can more easily transfer his training to an actual situation.)

On the other hand, there are advantages in teaching in a tailwheel airplane with a stick control. Most recently trained pilots have not flown that combination and would like to have the experience.

Impress emphatically upon the trainee that a great deal of research and engineering went into the aerobatic trainer you're using and he may get into real trouble by doing aerobatics in an airplane certificated only in the normal or utility category. At best he could get an incident report filed on him; at worst he might make a brief, but exciting, descent to the ground.

If at any time the airspeed red line or the limit load factor is exceeded, stop any further aerobatics and return to the airport for a thorough postflight inspection by you and/or a certified mechanic. Damage could have occurred in areas of the airframe that you cannot see from the cockpit, and further stresses could cause structural failure. Remember, *the limitations given for your airplane are for a new one right off the line, not one that has been flown for some time by inexperienced pilots.* While the red line gives a certain factor of safety, you still might have "pulled something loose" during your sashay into that area, so look the plane over before proceeding with aerobatics.

The verbal descriptions of the maneuvers make reference to throttling back as the airspeed increases to avoid propeller overspeeding, so a *fixed-pitch prop is assumed for these discussions.* Naturally, this would not be a problem with a constant-speed propeller, and many ex-military trainers so equipped use a *constant* throttle setting throughout various maneuvers.

It's suggested that you put a NO SMOKING sign on the instrument panel. The smell of stale cigarette or cigar smoke that has permeated the seat covers and headliner can be the last straw that pushes one of your aerobatic trainees over the line to physical illness. Usually during an aerobatic flight nobody has time to smoke anyway, but other people using the airplane for local sightseeing or cross-country flight may light up.

■ **The Aerobatic Course.** For most pilots, aerobatic training is only practical as a means of improving the other areas of their flying. There's no aerobatic rating at this writing, and the expense can discourage a potential student.

Your primary course should therefore be aimed at the pilot who can justify the cost by showing that it is going to make him a better and safer pilot.

An initial syllabus of 5 hr of dual and 5–8 hr of ground school is suggested—both the dual and ground school should cover not only aerobatic maneuvers, but also airplane stress, stability and control, and recovery from unusual attitudes (set up by the instructor—or wake turbulence). At the end of Part 5 is a sample primary syllabus.

INSTRUCTING AEROBATICS

Your approach to teaching aerobatics will naturally vary with each student. *You should not normally teach aerobatics to anyone holding less than a private certificate.* This is not to say that in an enlightened pilot training course you should never make a preprivate introduction to aerobatics, but with present-day training requirements, it would be best for the student pilot to learn the fundamentals of flying before getting too fancy. Besides, if the student has taken an aerobatic course and can do aileron rolls, loops, and snap rolls—and his flight instructor can't—it's bad for the training process.

If the trainee is a private pilot working on a commercial certificate, you would probably make use of chandelles and lazy eights as an introduction to aerobatics. This helps justify the course, also. If he is already a commercial pilot or flight instructor familiar with these maneuvers, you would only use them for a brief period in getting used to the particular airplane.

(This writer has instructed aerobatics to people with zero flight time, in a special experiment in which they were mechanically doing loops and rolls in 45 min in the air, and to pilots with 8500 hr but no previous aerobatic experience.)

During the first preflight instruction, you should assure the trainee that sarcasm or abuse won't be used, but there are bound to be times when you will need to raise your voice to galvanize him into immediate action. Things are happening fast in aerobatics; (1) pitch attitudes, (2) headings, or (3) banks are constantly and *rapidly* changing (sometimes all three at once) and often the trainee will fixate on, say, watching the wing tip in a loop, when he should be back to using the nose as a reference. By raising your voice you get him to move on. *Be sure that he understands that it is part of the instruction process and that you are still cool.*

There are some things you should have a good knowledge of before you start instructing aerobatics:

■ **Physical Condition.** The trainee should be in good physical condition; if he has passed an FAA physical, he's in good enough shape to do aerobatic flying. However, if he has a virus or some other temporary problem, he won't be at the peak of mental or physical vigor, no matter what his normal condition. If the trainee has a cold or a sinus problem, you'd better schedule another day because you will be climbing and letting down pretty fast in some of the maneuvers. Talk with him about this before flying.

It's best to introduce aerobatic flying by limiting your demonstration of a maneuver to a minimum amount of time, allowing the trainee to do the majority of the flying. Each maneuver will be explained prior to demonstration, so he will know exactly what to expect. This can make quite a bit of difference in his physical reaction to aerobatics.

Be sure to explain before the first flight that he should let you know *immediately* any time the aerobatics he's doing don't feel like the greatest thing on earth. In other words, at a particular part of the flight, he might lose some of his enthusiasm, often a sign of impending queasiness even if he hasn't felt any physical symptoms yet. When the trainee seems to get a little quieter and there are no more expressions of delight (or he's quit grinning), you'd better ease off for a while. A short period of straight and level flying, or even returning to the airport, should be done NOW. If you continue, the process will accelerate rapidly from seeming loss of interest to queasiness, to nausea, and finally to the use of a bag or carton. Once that occurs, the chances are practically out of sight of his coming back to flying aerobatics with you, or anybody else.

TRAINEE RESPONSES. Fig. 20-1 is a look at some trainee responses to physical problems in that first aerobatic flight (and later, too).

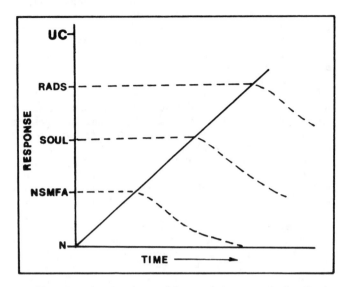

Fig. 20-1. A trainee's possible reactions to aerobatics. Don't push too hard, particularly on the first couple of flights.

N—Normal. The trainee is smiling from ear to ear. He (she) has discovered what airplanes are all about. No more flying a machine (one that was meant for three-dimensional action) like a little old lady driving a golf cart.

NSMFA—Not So Much Fun Anymore. You are elated by his earlier response and have had him do twelve (12!) more aileron rolls (or any other maneuver). You ask, "Want to do some more aileron rolls?" He nods weakly with a forced smile. Maybe he says "Naagh," or shrugs his shoulders. He doesn't quite have the courage to say, "Let's go home," but he hopes you will read between the lines. *Be warned.*

SOUL—Sweat On Upper Lip. Assuming you have ignored the warning signal and persist in making this indeed a memorable flight (and it's certainly headed that way), you will encounter the next symptom. You have moved him on to the next level of misery. (If the trainee who has a mustache, you may miss it.)

RADS—Rapid And Determined Swallowing. If you are such an unfeeling klutz that the first two symptoms have been missed, you may be warned by the third symptom. He is barely staying ahead of the game, so to speak. The trainee's voluntary actions are staying only a little ahead of his involuntary actions.

UC—Upchuck. This is the final step that tells even the most out-to-lunch instructor that he's gone too far. And another potential aerobatic pilot has been lost.

If any one sentence could cover aerobatic instruction, it's "Take it easy the first couple of flights." Don't feel you have to give him the maximum number of maneuvers in a given period, because it would be bad for the trainee physically. It can also set back the learning process by not giving him enough time between maneuvers to absorb what he's done. As a standard practice, *2 minutes should be a minimum time between maneuvers or a series of maneuvers* (such as a loop immediately followed by an aileron roll). Use this time for getting back some altitude.

Your philosophy of the instruction of aerobatics should be the same as the other phases: If possible, end the period on an optimistic or enthusiastic note. If the trainee has problems you can either (1) fly straight and level until he feels better and let him do, say, *one* more aileron roll so that the period is ended (hopefully) with a sense of some accomplishment, or (2) return to the airport. If (1) is possible, that's fine; it's good psychology to have him fly one more maneuver and then end the flight. However, for most people it may take 30 min or more (even on the ground) for them to feel completely normal again and too quick a return to aerobatics could be bad.

Returning to the airport (2) might not be the best psychological move as far as immediate goals are concerned, but it is less risky than having further problems that flight. This, of course, could leave him concerned about what could happen the next flight, unless *you make it clear that this happens to a number of people and can be overcome by experience.* It's up to you, the instructor, to provide the rate of gaining that experience. The fact that the trainee has problems at first does not mean that he cannot get over it and later be able to do a number of consecutive maneuvers without any adverse reactions at all.

Okay, so he's having problems during the flight and you've decided that the aerobatics session is over and you want to return to the airport. Should you return *immediately?* Not necessarily. Probably you've been flying up at a good altitude in smooth air (where it also is cooler). In the summer it can be quite turbulent (and hot) at lower altitudes and he may not be ready for that. Maintain altitude and turn the ventilators (yours included) in his direction. Cool air helps. *You* fly the airplane *smoothly* and give him a chance to settle down as you fly in the general direction of the airport. Sometimes chewing gum can help his return to normalcy, and you should have some available in the plane. When he's feeling better, ease down and land with no radical maneuvering. You can ask him if he would like to make the approach and landing, but usually he'd rather not. It usually helps for the trainee to fly the airplane after he's started his recovery, but let him make the decision. After he's starting to perk up, an "I know the feeling" from you can help a lot, since he will realize that you are now teaching aerobatics and have obviously overcome any problem. (If you haven't "known the feeling"—and you're a rare bird—just don't say anything except to assure him that it happens to a lot of people who later overcome it.)

On the way back to the airport keep discussion to a minimum, whether you or he is flying the airplane. People don't like to be chipped at when they are nauseated.

The trainee (particularly a male) will have suffered some loss of ego by admitting that he's getting queasy; he feels he couldn't "take it" and that's a rough blow to masculine pride, so you should let him know that since it happens with other people there's no point in your saying anything about it to the hangar fliers. If you want to discuss his particular problem with another instructor in private, that's fine, but never put him on the spot in front of the other trainees or the hangar fliers.

A light meal is usually okay for aerobatics: A heavy rich meal or an empty stomach are usually both bad. If you are going to start flying the next morning after an afternoon or evening briefing, tell him that a heavy, greasy breakfast is not a good idea until he knows his reactions. It's interesting to note that some people have less trouble at different times of day. You'll find that with a particular person you may have to cut the flight short in the morning, but he'll have no problem in the afternoon or vice versa. You might try a change of schedule if other things haven't worked. You can, for instance, fly him in the afternoons until he is well acclimated and then proceed to schedule him anytime it's convenient. Coffee can be a big contributor to nausea problems, so check on this. After he has flown a couple of flights with no problems his diet can be less restrictive.

Your aerobatic trainees will be of three main types:

1. This trainee could start with the syllabus and go right on through with absolutely no sign of physical discomfort at any time.

2. The more average person is not able to immediately "take" a rapid series of snap rolls but with an *easy* introduction to aerobatics may initially have minor problems, or none at all. Some problems may be the result of a one-time thing such as partying the night before. Tell him partying and aerobatics don't mix.

3. This third group consists of people who, although they are certificated pilots, have motion sickness tendencies so pronounced that anything more violent or lasting than one 720° power turn is more than they can handle without getting sick. You will meet members of this small group who can do everyday cross-country-type flying (many of them do it very well—and they could be making their living this way), but will never be able to hack it aerobatically. If the trainee has continual problems he'll probably never be able to do it. In other words, if you are easy in your approach (smooth air, etc.), take plenty of time between maneuvers, and he's still having problems after a maximum of four or five aileron rolls (and this is the fifth flight), you might figure—and tell him privately—that he should consider forgetting it. As you get more experience you'll see some courageous people in the third group who keep on trying but they won't always be successful. A good sign is that he starts feeling better quickly after straight and level flight is resumed. A sign of trouble ahead is if the queasiness or nausea continues well after landing.

One last thing, and this unpleasant subject will be left behind: You should have "comfort bags" handy (out of the envelope and ready for instant use), but out of sight—the power of suggestion, you know. There's nothing like trying to get the plastic sack out of the envelope and opened when it's needed *now*. You might prefer to get the medium-size plastic bags available at grocery stores, which may be easier for the trainee to handle. (One aerobatic instructor called these larger bags "banquet size" and suggested that for a more obnoxious trainee the bag could be used to fully cover the head with a string tied around the neck to secure it. Fortunately, this has not become a widespread practice.)

You'll find that hazy days with little or no horizon can induce nausea during aerobatics. The trainee has to search for references quickly and the rapid eye movement can cause problems. You'll find people who had no problem at all 2 days earlier, right after the cold front moved through and the visibility was great, who get queasy when the heavy haze moves in again.

If you can, keep the cockpit temperature down, even if it's a little uncomfortably cool. This is easy enough in the winter, but those hot August days can cause problems.

For most lower-performance aerobatic trainers, a ceiling of 5000 ft and visibility of 5 mi should be considered the minimum weather criteria. If you are instructing in high-performance trainers (turboprop or jet), this minimum should be increased proportionally.

Enough said about "problems."

■ **Acceleration or "g" Forces.** The trainee may have known about acceleration forces, or g's, for some time, but still it is valuable to review the information (Fig. 20-2).

If you are standing still, sitting in a chair, or flaked out in the hammock, you have the force of 1 g acting on you. That is, you have the acceleration of normal (1) gravity being exerted on your body. For some people, this force of 1 g is nearly too much and they lie around a lot.

In a constant altitude, properly coordinated, 60°-banked turn, you and the airplane will be under an acceleration force of 2 positive g's.

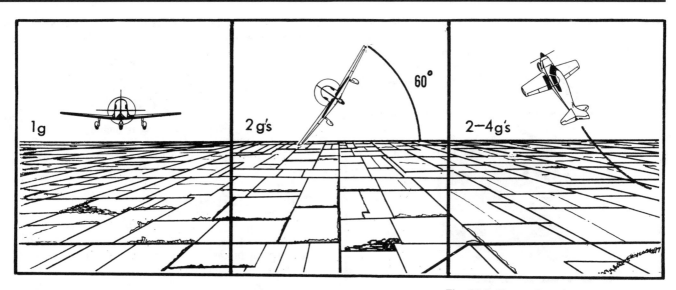

Fig. 20-2. The g effects in turns and pull-ups.

The 2 g's experienced in the 60°-banked turn are positive; that is, they are working on the pilot and the airplane from top to bottom. The 1 g experienced every day is also positive. However, a person who hangs around on tree limbs by his legs (with his head down) has 1 negative g working on him because gravity is working from his feet to his head. When you and the airplane are flying upside down in steady level flight you have 1 negative g working on you.

If you are pulling 3 g's in the airplane (say, in doing a loop), every part of your body weighs 3 times its normal weight, and your blood will tend to move down into the legs and feet.

At high accelerations, the draining of blood from the head can cause a "grayout" (loss of vision), or "blackout" (unconsciousness), at 6 g's or higher for extended times. However, the maneuvers you'll do in this introductory course (Immelmanns or Cuban eights) will not exceed 4 g's, and you will probably not experience any of these severe effects.

If you are an average pilot, 2 or 3 g's will give you the sense of being firmly pushed in your seat and your hands and feet will feel heavy if you try to move them. At 4 g's you have good evidence that acceleration forces are at work. Things get a little gray and your cheeks may feel like they are sagging. This is particularly true if you are just riding through the maneuver. The person doing the flying doesn't notice the effects of acceleration forces like a passenger does. The pilot is concentrating on the maneuver and has tensed up a bit; the other occupant, having nothing to do but watch the changing of directions of flight, may start to gray out. The pilot will likely have no effects at all, which brings up a good point: You can raise your tolerance by tensing up your muscles, particularly your stomach muscles. Navy dive-bomber pilots used to yell as they pulled out of a dive; this tensed their muscles nicely, but probably did nothing for the gunner's peace of mind, since he was facing backwards and couldn't see what was going on (Fig. 20-3).

Negative g's are harder on you than positive g's, and your tolerance will be less. At extended −2 or 3 g's your face feels full and a headache is in the offing. You are well aware that the seat belt and harness are there (since you are pushing against them). At −4 g's the brain gets all the blood it needs (and then some) and if that number of g's is extended in time you may become quite uncomfortable and the small blood vessels in the eye may rupture. It will be extremely unlikely that you will see less than zero g, or perhaps −1 g, on your accelerometer in an introductory aerobatic course.

Fig. 20-3. Yelling during pull-outs can be disconcerting.

THE ACCELEROMETER. The accelerometer measures acceleration forces ("up" or "down") parallel to the vertical (Z) axis of the airplane. Since you are sitting upright in the airplane, it measures the positive and negative g forces on you as well. The accelerometer you'll probably be using is a spring balance measuring a single weight under varying forces (Fig. 20-4).

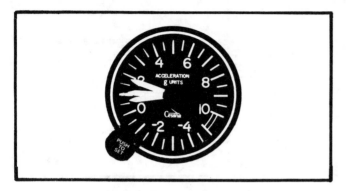

Fig. 20-4. The accelerometer. (*Cessna Aircraft Company*)

You might explain it as follows: The instrument has three hands. The "top" hand (or the one closest to you) moves with the forces being applied at that instant (negative or positive g's, as applicable). Under +g forces it will move the "middle" hand and leave it at the maximum +g's pulled in a particular maneuver. With g forces of less than 1 g, the "bottom" or third hand is moved and remains at the maximum negative reading. After the airplane is returned to 1 g (normal) flight, the "top" hand moves back to the 1-g indication and the other hands remain at the maximum positive- and negative-g indications so that you will have a record of the range of forces encountered during a particular maneuver. For instance, you could say, "I pulled a maximum of 4 positive g's and 1 negative g during that series of maneuvers and I see I'm now pulling 2 g's in this 60°-banked turn." There is a reset button on the instrument. Push it to return the two reference hands back to the normal 1-g reading for another maneuver series.

When the accelerometer is indicating +4 g's, this means you've added +3 g's to the normal 1-g reading. If a −2g's is indicated on the instrument, you've gone from +1 g to 0 and then to −2, or −3 g's on the negative side of the normal condition.

One thing you'll notice after you have used the instrument is that even in fairly easy pull-ups, such as in entering a normal climb, the accelerometer indicates a value greater than +1 g. It may be a very little change and hard to read, but it's there—anytime you change direction of flight "up" or "down."

An interesting note on the effects of g forces on the pilot is that time has a lot to do with it. For instance, you may pull 6 or 8 g's by jumping off a chair, but since the time that the g force is being applied is so short, the blood doesn't start to move downward before the force is removed—hence no grayout or blackout. Usually it takes about 3 sec for things to get started, so a snap roll with a +3-g force probably won't even let you realize that you've been under acceleration forces. On the other hand, in a loop where 3 g's are being exerted for a longer period, the "relaxed" pilot of the airplane (who didn't have any trouble with the +3-g snap roll) may decide that it's getting a little foggy all of a sudden. The airplane is not affected this way since 3 g's are 3 g's, no matter how long they last.

Turbulence can add to the load factor. A +3-g maneuver could end up as a +4- or +5-g maneuver if turbulence is encountered at that point. The smoother the air, the better the performance of aerobatics.

You'll find that g tolerance increases with experience. Remember, it will be decreased with decreased physical fitness on your part.

The discussion of g effects on the pilot should be kept relatively simple for the average aerobatic trainee, as they were explained here. However, you'll likely run into those who want more detail on physiological response and the effects of combinations of g force and time, so Appendix B (*G Effects on the Pilot during Aerobatics*) is included. You should be familiar with this information before instructing aerobatics.

The following maneuvers are suggested for a primary course of 5 hr dual and 4–5 hr ground school.

720° POWER TURNS

This one is a good introduction to aerobatics since it's unlikely that most of the trainees have banked over 30° since getting their private (or commercial) certificates and yet they have done this maneuver before. You can get a very good idea of what his reactions will be to aerobatics by the way the 720s are done. Use 45° banks for the first couple of maneuvers, then have him make a couple at 60° bank.

You can check back to Chap. 6 for the analysis of the 720° power turn. The expected errors listed there are valid here, too.

Don't overdo it; a couple of 720° turns each way at 45° and then 60° banks should be plenty. With some people you'll see after the first one that their feel of the airplane is good enough to move on.

STALLS

Since an aerobatic airplane spends a large percentage of its time operating in this area or close to it, a good understanding of the 1-g stall (and recovery) is necessary. Of course, you'll also give him practice in accelerated stalls, but you first should get him used to the 1-g stalls in this particular airplane.

It was mentioned earlier that many pilots are leery of stalls, and if you overdo them in this early phase of aerobatics, you could have a trainee with a queasy stomach—or an ex-trainee.

A power-on stall (you may want to review the takeoff and departure stalls) and a power-off stall (clean) should be good enough at this stage. Many people will only lose their fear of stalls after completing an aerobatic course, which might be an indication of their improper instruction in stalls when they started to learn to fly.

The power-on stall recovery techniques are most important because this is usually the situation (cruise or full power) in which the stall is encountered in aerobatics. Later introduce the cross-control (skidding) stall if the trainee is not familiar with it.

COORDINATION EXERCISES

The trainee is better off to have as much coordinated flying as possible in his early hours of aerobatics, and the name notwithstanding, these exercises are not "coordinated" as he has known the term previously. Later he'll learn that coordination in aerobatics is making the plane do what he wants, even with controls crossed, if that is required. If you suspect at all that he could have problems with his stomach, better forget this one until he's ready for slow rolls. Coordination exercises, snap rolls, and prolonged sessions of the lazy eight seem to be the toughest on the digestive system. It's possible that the exercises could even be detrimental to his aileron roll practice, since cross-controlling is not required in that maneuver. (Look back to Fig. 4-3 and explanation.)

CHANDELLE

The chandelle is a good maneuver for anybody taking an aerobatic course. For those individuals with only a private certificate, it can help justify the expense of the course since it may be required on the commercial flight test; and even if he isn't working on the commercial now he "might someday." If he's already working on the commercial, he'll get more practice on the maneuver.

Basically, however, the chandelle as described here is a good introduction to aerobatics, because to many trainees it is the first time that a large preplanned change in airspeed *and* pitch and bank attitudes is required. They will get a feel of the airplane by flying it from cruise (or above) to just above a stall — and flying by outside references. Fig. 20-5 shows the maneuver.

■ **Preparation.** Use Fig. 20-5 plus a model for your introduction of the chandelle. A chalkboard is handy in listing the airspeeds of entry and completion, as well as for noting reference airspeeds throughout the maneuver. If you are good at drawing you might lay out the chandelle on the chalkboard, but it's a tough maneuver to show that way.

■ Explanation

What — The chandelle is often defined as a *maximum performance climbing turn* with a 180° change in direction with the airspeed varying from the maneuvering speed to just above a stall. Climb performance is lost at the last part of the maneuver when the airplane is mushing, close to the stall. The maximum climb performance can be achieved if, after the pull-up, the indicated airspeed is not allowed to get below that for the max rate of climb. It has also been suggested that the airspeed not be allowed to get below max angle of climb airspeed. Since both of these airspeeds vary with altitude it is assumed here that the sea level values are used.

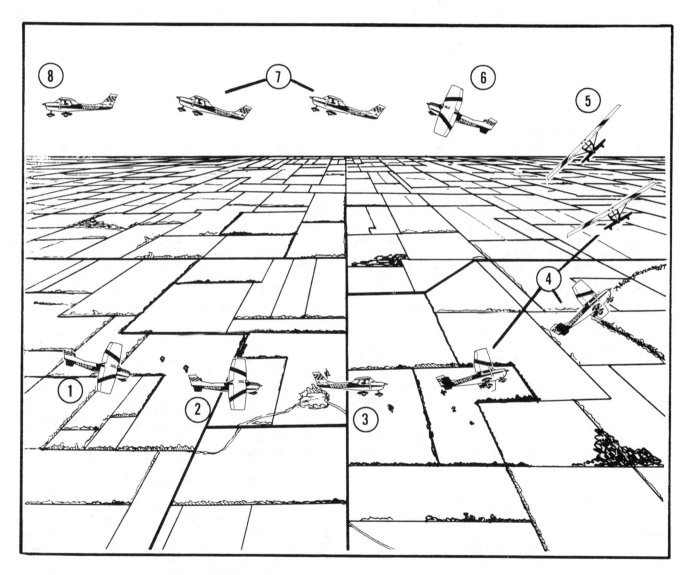

Fig. 20-5. The chandelle as seen from slightly above the initial pull-up altitude. (1) The recommended entry speed is attained (a dive, in this case). (2) The airplane is leveled temporarily. (3) The bank is established and the ailerons and rudder neutralized. (4) The pull-up and turn are made and the throttle is opened fully. (5) At the 90° position, the airplane has reached the recommended maximum pitch position and the roll-out is started. (6) The roll-out is continued with the nose kept at the pitch attitude established at the 90° turn position. (7) The roll-out is completed at exactly 180° of turn with the airspeed just above a stall. (8) Straight and level cruise is resumed.

In practicality, absolute maximum performance is not the issue; the trainee should encounter a comparably wide range of airspeed values in the maneuver, so that he will have a chance to cope with a variety of control forces while flying predetermined requirements and getting *good* performance from the airplane.

Why—The chandelle requires that the airplane be flown with varying airspeeds while following a specific path by the use of outside references. It will also require continuous attention to *orientation* and *coordination,* as well as careful *planning.* The trainee should be told that it is not only a good maneuver for an introduction to aerobatics (and for getting his head out of the cockpit), but it also leads directly to the wingover, and later, to the barrel roll.

How—A straight road is normally used as a reference, and at a safe altitude the airplane is lined up parallel to it. If possible, arrange it so the initial bank is made into the wind so that during a series of chandelles the airplane won't drift out of the local area.

Note that in Fig. 20-5 a dive is used to get the recommended entry speed. The FAA suggests that the entry speed be no more than the maneuvering speed (at max certificated weight, as listed in the *Pilot's Operating Handbook*). If the normal cruising indicated airspeed of the airplane is below the maneuvering speed the chandelle *may* be started from a shallow dive to reach that value. The bank is set up with the nose in the level-flight position, the ailerons neutralized, and the pull-up initiated. If the normal cruise IAS is above the maneuvering speed (V_A), the airplane should be slowed to V_A or below before the bank and level-flight pull-up is started.

For an introductory maneuver to aerobatics, it's suggested that a shallow dive be used, and the bank established in that dive be used for practice in transitioning from dives to pull-ups for following maneuvers.

In an airplane with a fixed-pitch prop, the rpm should be set at a cruise value that wouldn't cause overspeeding if the shallow dive is required. As the pull-up is made the power should be increased to recommended climb setting (full throttle for most airplanes) at a rate slow enough to prevent overspeeding. Full power should be applied before 90° of turn is completed. Airplanes with constant-speed props should use cruise power throughout.

The pitch-change rate should be adjusted so that the required maximum nose-up attitude is attained just as the airplane reaches the 90°-of-turn point (and the roll-out is started).

The nose is kept at this pitch attitude as the roll-out continues and if the pitch attitude is correct, the airspeed will be just above a stall as the roll-out is completed at 180° of turn.

The roll-out should be steady, with no quickening or slowing of the roll rate, throughout the last 90° of turn. Explain the effects of "torque" on the left and right roll-outs.

Fig. 20-6 shows an analysis of the requirements as the chandelle progresses. A sketch of this could be made on the blackboard or, if you're doing a lot of commercial or CFI instructing later, a series of transparencies of the various maneuvers, including the chandelle, could be made.

■ **Demonstration.** You should demonstrate a normal chandelle in each direction citing the following points:

1. The turns are made into the wind, if possible, *to keep from drifting out of the practice area* during a series of chandelles. Some pilots still believe that an airplane has a greater climb *rate* upwind than downwind, so be sure to tell them the

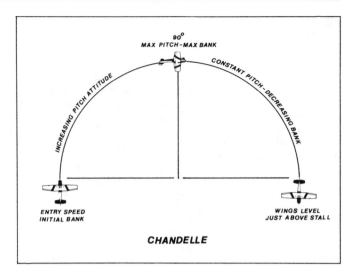

Fig. 20-6. Requirements of the chandelle. (*The Basic Aerobatic Manual*)

real reason wind is a consideration in the maneuver.

2. The dive, if required, should be shallow with a smooth entry.

3. After the bank is established (30° for most airplanes), *don't* hold opposite rudder to keep the nose from turning; some turn is expected before the pull-up starts.

4. The pull-up and increase of power is smooth; the rate of pull-up must be constant so that the proper pitch attitude is reached at the 90° point of turn. After you have instructed a number of chandelles in a particular type of airplane, you'll find that the use of reference airspeeds can help you predict the end results of the maneuver while it is still progressing. With experience you'll find, for example, that an airplane with recommended entry speed of 105 K in a good chandelle will have 90 K at 45° of turn and 70 K at the 90° position. As the airplane passes these points, take a quick check of the airspeed; if it's low, expect stall problems at the end of the maneuver. If the airspeed is high, the problem will be lack of climb performance and not an approach to the stall. He is supposed to be using outside references, but your checking of the airspeed can lead to correcting his pull-up problems. It will also increase his confidence in your instruction when he sees that you are able to foresee problems. You may want to do some solo practice and check the airspeeds at the various turn points for *your* airplane before instructing chandelles.

5. Keep the ball in the center of the turn indicator throughout the maneuver and tell him of the possible problems in the left and right chandelle because of "torque" effects. The tendency in a left chandelle is to overshoot the turn because of the left-turning tendency of the airplane in a climb; in a right chandelle the "torque" tends to stop the turn short of 180°.

6. Emphasize that in the last 90° of turn the pitch attitude is not to vary—unless he has the pitch so far off the 90° point that something has to be done to save matters.

7. In the last few seconds of the chandelle when the airplane is flying with wings level just above the stall, the common tendency is for a relaxation of "torque" correction and an otherwise perfect maneuver can get sloppy.

8. To aid in developing precision the airplane should be returned to cruise, maintaining a *constant altitude* (±20 ft).

■ **Practice.** The trainee should take his time in setting up that first chandelle. While the usual practice is to start the chandelle parallel to a long stretch of straight road, it can easily be done starting and ending perpendicular to it. You might have him try both methods. (A personal note: I prefer starting perpendicular to the road; it seems easier to keep it in sight.)

In starting and completing the maneuver parallel to a road, it's best for the trainee to have the road to his right at the beginning of a right chandelle (side-by-side seating). This will put the road to his left as the turn is completed, and since he is sitting in the left seat he'll be able to see the road better in the pitched-up—and banked—attitude in the last, say, 45° of turn when he is looking for the 180° reference. Because you are sitting in the right seat you should have the road on the left side at the start of a left chandelle. It's fairly easy to line up the initial dive with the road on the opposite side from you in the cockpit, but in the fairly radical attitudes at the *last part* of the chandelle, the road is pretty much out of the picture.

You may also want to use a reference on the horizon off one wing tip so that the maneuver is complete when the bank is rolled out, the airspeed is just above a stall, and the *opposite* wing tip is pointed at the reference.

In instrument training you may introduce the chandelle under the hood and, even now, you can have him use the heading indicator as a reference if no good outside references are available. Or more realistically, you'll probably have to cover the heading indicator to keep him *from* watching it with a glazed-eye stare, ignoring good outside references. Have him practice chandelles in alternating directions.

Common errors in chandelles are these:

1. Poor speed control in the initial dive (if required); a too-fast or too-slow entry.

2. Coordination problems on the initial bank, the usual tendency being not to use enough rudder.

3. Not neutralizing the ailerons so that the bank will be too steep at the 90° position. Also a too-shallow or too-steep initial bank.

4. Initial pull-up rate too slow. This is usually the problem when the maneuver is first practiced; expect some overcompensation later after you've told him about it.

5. Power problems. Expect that he will tend to forget to add power on some of the chandelles and then may overdo it by adding power too soon or too abruptly.

6. Trying to maintain a constant bank. The bank should increase with relation to the horizon as the first 90° of turn proceeds, even if the ailerons are properly neutralized and no roll is introduced. You can see and demonstrate this back on the ground with a model. Bank the model about 30° in the level-flight pitch attitude and move it upward *parallel to the vertical axis*. As the pitch increases, so does the bank. However, the trainee in the airplane sees that the bank is apparently increasing and uses opposite aileron and rudder to maintain the original 30°. In this case he will have too much pitch for the amount of turn and the airplane will run out of steam before the turn is completed.

You may, however, require a constant bank of, say, 30° throughout the first 90° of turn for training purposes since the FAA requires this procedure on practical tests. Explain that *these* chandelles are used as a transition to aerobatics and will have a steeper bank of 45° at the 90°-of-turn point.

7. Passing the 90° point without starting the roll-out so that the last 90° is rushed. This is quite common in the first few chandelles and is a result of lack of orientation; the airplane is ahead of the pilot. You can expect this problem to gradually work itself out with practice.

8. Allowing the nose to drop during the last 90° of turn when it should be kept at a constant-pitch attitude.

9. Stalling the airplane in the last 90° of turn. (A few may even stall it in the first 90°!)

10. Keeping the nose too low all the way around so that airspeed is well above the stall when the 180° of turn is completed.

11. Directional problems in the last part of the maneuver. The three main problems are (1) rolling out too soon in a right chandelle, (2) having to speed up the roll-out in a left chandelle, and (3) letting the nose drift to the left of the final heading in the few seconds of straightaway at the end of the maneuver. All three of these problems can be attributed to poor "torque" correction. Judicious use of right rudder is indicated for both directions of turn.

12. *And the most common error of all*—staring at the airspeed or over the nose and ignoring the ground reference(s). You'll find this to be a problem particularly with pilots who have just gotten the instrument rating and are working on the commercial certificate.

■ **Evaluation.** If the trainee is using the chandelle as an introduction to aerobatics, you should be satisfied with good coordination and orientation and should not try to perfect it as a maneuver at this time. After a half dozen chandelles he should have a good idea of how it goes and be able to perform it reasonably well without help from you.

If the practice is preparation for the commercial certificate flight test, your requirements will be stiffer and he should work until the maneuver is smooth.

■ **Review.** After you get back on the ground, use the model to review the chandelle, pointing out errors and suggesting corrections. As you do more instructing, you'll be able to note specific items during the flight and store them away for the postflight review session. You may want to make notes during the flight, but this can sometimes be disconcerting to a trainee who may be so interested in what you are writing that his flying goes downhill. If possible, make mental notes.

WINGOVER

This is a fine maneuver for smoothing out coordination during constantly changing pitch and bank attitudes. The wingover resembles part of a lazy eight, but the bank is usually steeper (the steepest bank can be anything from 1° to past 90°, but for these introductory wingovers a maximum of 60° should be used for the first few). The lazy eight at present suggests a maximum bank of 30°.

The wingover is an even better maneuver for aerobatics than the chandelle because of the more radical pitch and bank attitudes and the required use of a point on the horizon. The chandelle is a good introduction to the wingover, particularly if you have demonstrated how a point on the horizon could be used to check the 180° of turn.

As will be shown later, the wingover is a particularly good lead-in to discussion of the barrel roll.

■ **Preparation.** A model is the best aid in explaining this maneuver. You might draw a horizon and reference point on the chalkboard and do a wingover toward it with the model. If possible, use the same speed for the chandelle and wingover as would be used for the aileron roll. Naturally you'll have to take into account the manufacturer's recommended entry speeds, but if you can use, for instance, 115-K (the aileron roll entry speed used as an example in this book) for a chandelle and wingover entry, the 115-K starting technique will be pretty well smoothed out by the time the trainee gets to the aileron rolls. The Cessna Aerobat uses 105 K for the wingover entry speed and 115 K for aileron rolls, which works well, too. An entry speed of 115 K for the wingover entry in the Aerobat would use too much altitude. Work it out for *your* trainer. Make sure the maneuver is well established in your own mind, and have the trainee read about it beforehand in one of the books in the Bibliography.

A lesson plan is not usually necessary, but you might review the format just in case you decide to make one up.

■ **Explanation.** To explain the differences between the chandelle and this maneuver:

1. In the chandelle, the bank is established *before* the nose is brought up through the horizon; the wingover requires that the bank is started from a straight pull-up just as the nose passes through the horizon. (One instructor calls the chandelle "bank and yank" and the wingover "yank and bank.")

2. Both have 180° of turn, but the wingover should end at the same altitude and airspeed as at the initial point of pull-up. In the chandelle, once the bank angle is established, the ailerons are not used until the 90°-of-turn point, at which time the bank shallowing begins. The pitch in the chandelle remains the same after the first 90° of turn. *The bank and pitch in the wingover are continually changing.*

What — The wingover is a 90° climbing turn followed by a 90° descending turn in the same direction, resulting in a 180° change in direction. It is coordinated and the speed varies from 105 K, for example, in the entry, to just above a stall at the 90° point, to the entry speed of 105 K as the 180° position is reached. (Demonstrate with model.) Note that you set the power before entry and do not vary it.

Why — The wingover is used to aid coordination and orientation. The trainee is flying the airplane, using an outside reference, maintaining orientation with that reference, and flying a specific pattern. He must be coordinated through all airspeeds and attitudes during the maneuver.

It is a good introduction to aerobatic flight since he is developing a subconscious feel for the airplane in what are (to him) radical attitudes.

How — A reference is picked on the horizon and the airplane is turned to put the wing tip on it. As you describe the wingover, use the model. Go through the maneuver with your description without stopping the model. The nose is eased down to pick up the recommended entry speed; the wing tip is to be kept on the reference. As the proper speed is reached, the nose is eased up, and as it passes through the level-flight position, a *smooth* coordinated turn is started toward the reference. The back pressure is continued as the bank is increased. At the 45°-of-turn point (the airplane's heading is 45° from the original), the pitch would be highest and the bank should be approximately *one-half* that of the maximum desired. Again, it is suggested that initially 60° be used as the maximum bank, giving the trainee a definite figure to try for. Later, you may want to steepen this to a maximum of a vertical bank or slightly past. At the 90°-of-turn position *your line of sight* (not the airplane's nose) should pass through the reference, the bank should be the steepest, and the airspeed at its minimum. Bottom rudder and opposite aileron may be required to keep the ball centered in this area of the maneuver, particularly in a turn to the right.

As the 90°-of-turn position is passed, the back pressure is relaxed simultaneously as the roll-out is started.

At the 135° point, the nose is at its lowest pitch and the bank should be about one-half the maximum, as the return to wings-level flight is continued. The recovery is complete at 180° of turn, in level flight at the entry speed.

It is suggested that you repeat the above, using the model. Then demonstrate the maneuver using the model

but stopping at each 45° of turn to "freeze" the plane's bank and pitch attitude to make a point. The trainee should have a good understanding of the maneuver, and you might hand him the model and let him explain it to you after you've gone through it a couple of times. This "trainee demonstration" may be used with any or all of the aerobatic maneuvers and helps trainees to understand them better. (One of the best evidences of knowledge of a subject is the ability to explain it.) Usually the pilots you'll be working with at this point, who are more experienced, won't mind "performing."

As the trainee becomes more confident, increase the pitch attitudes to near vertical and the banks past vertical — this is assuming that you both are wearing parachutes and the airplane is in the acrobatic category. Most will enjoy the process with an increase in confidence. ("I banked past 60° and *lived!*") Don't overdo it to the point of NSMFA or take up time that could be used for aileron rolls or "real" aerobatics.

Fig. 20-7 shows a view of the wingover as seen from outside.

LAZY EIGHT. While you are discussing the wingover, you should briefly mention the lazy eight, noting that it is a required commercial maneuver and consists of a series of alternating shallow wingovers with no interruption between. The maximum bank suggestion for training is 30°, and that type of lazy eight is excellent for smoothing out coordination without introducing the trainee to the more radical aerobatic attitudes.

When working with a pilot who is preparing for the commercial flight test you should be aware that a long siege of lazy eights can induce airsickness so keep an eye out for NSMFA symptoms.

The explanation for the lazy eight is the same as for the wingover except for the less-radical pitch and bank attitudes. Instead of leveling off as shown at point *5* in Fig. 20-7, the pull-up is repeated with a bank to the right (toward the reference) with no hesitation. Smoothness in transition is important, or rather, there should appear to be no transition between the "wingovers."

Do the lazy eights with all turns into the wind so that you don't drift out of the practice area.

The lazy eight is a series of maneuvers, and the altitudes at the tops (and bottoms) of the segments should be the same throughout. You may have to adjust power if the airplane is generally losing or gaining altitude. For instance, if the airplane is gradually losing altitude, you will carry additional power and vice versa.

A top view of a lazy eight is shown in Fig. 20-8 with a discussion of points *A* through *E*. The view from the cockpit is shown in Fig. 20-9.

A good lazy eight is a joy to behold and ride through. You'll find that you or the trainee will have trouble keeping that maximum bank down to 30°. Also, both of you will have a tendency to tighten up the "segments" and do them faster and faster as the maneuver progresses.

But back to the wingover . . .

■ **Demonstration.** Getting back to the *wingover,* practically speaking, a good horizon and far references are not always available, but this and the loop-type maneuvers need such visual conditions more than do the rolls (aileron, slow, and snap). Pick the most prominent object available on the horizon and demonstrate the wingover. Explain the procedure as you do the maneuver (noting the pitch and bank attitude requirements at each 45° of heading change). Review the maneuver while flying straight and level after the demonstration.

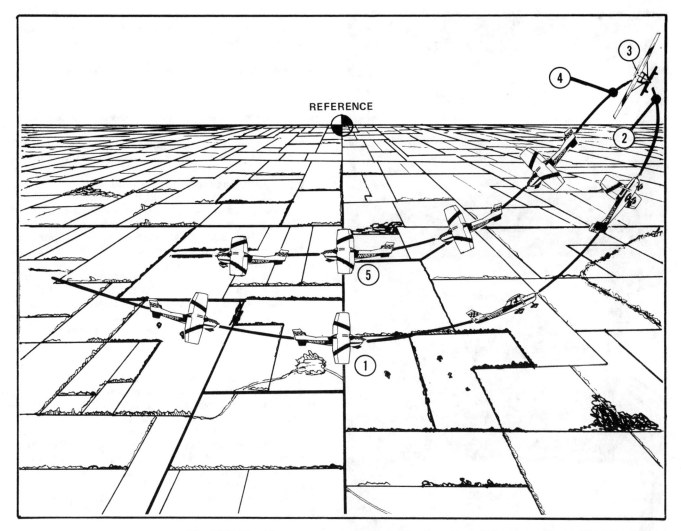

REFERENCE

Fig. 20-7. The wingover as seen from slightly above the initial pull-up altitude. (1) The pull-up is started at the recommended entry speed. (2) At 45° of turn the nose is at the highest pitch position. (3) The airplane has turned 90°, the steepest bank is reached, and the airspeed is just above a stall. (4) The airplane has turned 135° and the nose is at the lowest pitch position. (5) Level-flight attitude is reached at (it is hoped) the entry airspeed and altitude.

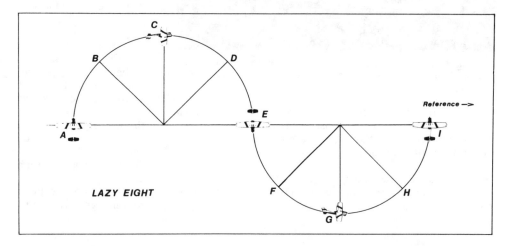

LAZY EIGHT

Reference →

Fig. 20-8. Top view of a lazy eight. (A) Beginning of the maneuver, reference point picked (point *A*). (B) At 45° of turn the airplane is at its highest pitch and one-half of the maximum bank (15° bank here). (C) At 90° of turn, longitudinal axis level, maximum 30° bank, lowest airspeed, the nose is passing through the reference. (D) At 135° of turn, lowest pitch, one-half of maximum bank. (E) Wings level, 180° of turn completed with opposite wing pointed at the reference, at initial pull-up altitude and airspeed (it says here). Points *F, G, H,* and *I* are comparable to *B, C, D,* and *E* in pitch and bank attitudes. (*The Basic Aerobatic Manual*)

Fig. 20-9. Part of a lazy eight as seen from the cockpit. Views *A, C,* and *E* are seen at those same points in Fig. 20-8. (*The Basic Aerobatic Manual*)

If you think the trainee needs another wingover demonstrated, do so, but keep demonstrations to a minimum. No follow-me-through on this one; it's a matter of feel, not mechanical use of the controls.

■ **Practice.** The new pilot may be a little timid about the maneuver, and his first wingover will probably lack "authority."

Let him complete one (don't start talking during the process) and then review his actions. If necessary, repeat the wingover, emphasizing the parts of the maneuver with which he is having problems. Don't demonstrate two in a row; let him practice a couple before (if necessary) you demonstrate another one.

Common errors made in the wingover are these:

1. Too much bank and too little pull-up initially. He is not used to such a too-high (to him) nose attitude. You can expect this with most trainees on the first few wingovers.

2. Rolling out of the turn too soon. When the roll-out is started at the 90° position, it is continued at too fast a rate, and 180° of turn is not completed. The reference point is ahead of the wing tip after the roll-out.

3. Not relaxing back pressure at the 90° point as the roll-out is begun, with the result that the airplane gains altitude.

4. Coordination problems. At the 90° point, the ball in the turn indicator may hit the down side of the tube. Usually the situation is a well-developed slip at this point, at first. As indicated earlier, inside rudder and opposite aileron may be needed to keep the ball centered, particularly in a right bank.

You will have to remind him to keep an eye on the reference point. That's the idea of the maneuver—to get him to fly the airplane while watching the reference point. The initial tendency is to stare over the nose and hope things work out right. You will probably have to remind him several times to watch the reference since it may be hidden briefly during the maneuver, depending on his pull-up rate versus roll rate, or on whether it's a high-wing or low-wing airplane. The high-wing may tend to hide it as the initial bank is entered; the low-wing may tend to give such problems just after the 90° point is passed.

■ **Evaluation and Review.** Don't expect him to be perfect with the wingover before moving on to the aileron roll and other maneuvers, but he should be able to do 180° of turn (±5°) using the reference and to recover within 100 ft and 5 K of the entry numbers.

If the trainee is obviously ill at ease and having trouble staying oriented, a chandelle or two interspersed within the wingover practice may help him.

One reason for not requiring the wingover to be "perfect" is that for many people a good wingover is harder to achieve than a good aileron roll or loop. It's a fine introductory maneuver, but don't overwork it.

After getting on the ground, discuss his errors and review the wingover. You wouldn't, of course, use a full period for the wingover, but it should be reviewed with the other maneuvers. If anything special occurred that could not be covered in detail in flight, cover it at length as necessary in this postflight briefing before he forgets.

Remember to praise where possible and discuss any errors in a constructive way.

SPINS

21

INTRODUCTION

Spins are of interest to all pilots and particularly to those involved in aerobatics. This chapter covers that area of flight training as well as points that crop up in nonaerobatic instruction for the *upright spin*.

The average private or commercial pilot has had little or no experience with spins, and even most flight instructors certificated in recent years have not had a chance to become really familiar with them.

As a flight instructor, you may be called on to demonstrate and teach spins (in a properly certificated airplane), but before you begin such training you should study and get instruction and practice beyond that normally experienced in earning the flight instructor certificate.

Maybe the average flight instructor has no plans to get involved with spins and recoveries at any time during his career, but tell that to the ham-handed (and heavy-footed) student who just put the airplane and you into a new space/time continuum during a dual takeoff and departure stall. As well as the obvious advantages of being able to extricate both of you from this situation, it's important for you not to be unsure of yourself about stalls or spins. A trainee can sense your unease in higher-angle-of-attack flight—and it's contagious.

As a case in point, when—or if—you start teaching aerobatics you'll find that most trainees are not afraid of loops, rolls, and other aerobatic maneuvers, but many are very tense when the plain 1-g stall is encountered. This may well be the result of current stall training procedures; many trainees have a feeling that once at, or past, a certain point in the stall, the airplane will spin and there is no recovery from *that*. Perhaps the instructors who taught these people were tense in the high-angle-of-attack regime because *their* instructors were tense and didn't want to get a full-stall condition, and so the cycle continued. After all, the basic purpose of stall practice is recognition of the onset and then avoidance of the stall, and this works very well until the new pilot inadvertently goes further than anticipated one day and "falls off the precipice" into a regime not encountered before.

Fig. 21-1 shows the range of instruction too many pilots are given as far as flying the boundaries of the airplane's performance is concerned. While Fig. 21-1 shows both ends of a typical general aviation airplane's spectrum, the high-speed/low-angle-of-attack area (spiral) will be discussed only briefly here (going to the airspeed red line doesn't add that much to the pilot's knowledge). In other words, the airplane's actions at the red line aren't all that much different from, say, a speed of 5 K or maybe 20 K below it, except for the control pressures and danger of bending something if that limit is exceeded.

The stall and spin conditions form a definite break from normal flying, and the suddenness of the transition from the airplane acting "right" to having a mind of its own could mean bad news for the pilot who hasn't gone that far before.

Your job will be to take the student to that precipice, and even if you don't want to actually spin the airplane, you

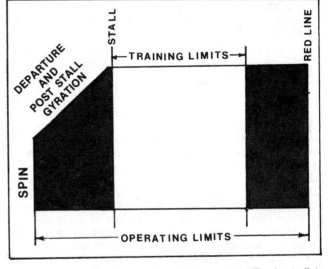

Fig. 21-1. Training limits for some instructors. The best pilots are those who know the total limits of the airplane and are able to fly to those limits and return—consistently. The flight instructor's job is to take the inexperienced pilot safely close to the limits so that he will recognize possible problems when he is flying as pilot in command.

should be able to guide him to the edge and discuss his moves for recovery. Of course, if you do decide to teach spins to your private and commercial aspirants, you must be able to handle any possible problems that could arise. You should have practice in power-on and cross-control entries, and if you plan to teach aerobatics, you should be aware of snap (accelerated) and other variations of spin entries. And you should know what to expect if the trainee wraps the airplane up into 4 (or more) turns in a botched-up snap roll or a snap at the top of a loop.

The FAA is now requiring more emphasis on stall awareness, spin entry, spins, and spin recovery techniques for airplanes for both *private* and *commercial* pilots. In neither case, however, is actual spin training *or* demonstration of spins on the flight test required.

The applicant for a *flight instructor's certificate* (airplane or glider) may present a logbook endorsement from an appropriately certificated and rated flight instructor who has provided the applicant with spin entry, spin, and spin recovery training in an aircraft of the appropriate category that is certified for spins and has found that applicant competent and proficient in those training areas. Except in the case of a retest after a failure for the deficiencies stated in paragraph 61.49(b), the person conducting the practical test may either accept the spin training logbook endorsement or require demonstration of the spin entry, spin, and spin recovery maneuver on the flight portion of the practical test (FAR 61.183).

If during the oral part of the practical test the applicant's knowledge of spin basics and how to teach spins is unsatisfactory, he'll have to take a retest and actually demonstrate both knowledge and skill in these areas in an aircraft of the appropriate category that is certificated for spins. (Paraphrased FAR 61.49—Retesting After Failure).

In other words, unlike the situation before April 15, 1991, no one can get away with what used to be termed a "spin sign-off," when two equally frightened people (the instructor and the instructor-to-be) climbed up and did a half-turn "spin," returning quickly to the airport to note in the logbook of the instructor-to-be that he "had satisfactorily demonstrated spin competency."

More about this in the *Practical Test Standards* at the end of this chapter.

THE ACCIDENTAL SPIN

The scenario: One of your students is up solo at a reasonable altitude, getting ready for the private flight test. He is practicing takeoff and departure stalls, say, in a climbing right turn. There is a good stall break, the higher left wing drops, and so a left roll is induced. The "normal" reaction is to use full opposite aileron (and maybe not enough rudder) to stop the roll. The down-moving wing is increasing its own angle of attack and is now well past the stall angle. The down-aileron on that left side further complicates the issue by increasing drag and changing the stall characteristics of the wing.

The nose is dropping in the roll, and the student's unconscious (or conscious) desire is for the stick or wheel to be held full back to bring the nose up. The student may find he is in a fix he has not experienced, or even *discussed*, before. The airplane is not reacting normally; the ailerons aren't stopping the roll as they were designed to do, and the up-elevator is not bringing the nose up as it should. A strange spiral, he's discovered! The nose is far down, and the airplane is rotating rapidly. The pilot is in a situation in which he must act against his instincts and earlier experiences in the airplane. To get out of this, the throttle should be closed, the ailerons neutralized, full rudder used opposite to the rotation, and toughest of all, he is supposed then to push the nose even farther down by brisk forward movement of the stick or wheel. The airplane is already pointing (to him) nearly straight down, and to push the nose farther down is something he'd rather not do, thank you.

The reason the airplane didn't act right during that stall "recovery" is shown in Fig. 21-2 (which looks much like Fig. 11-5). If you plotted the coefficient of lift (C_L), or relative lift of the airfoil, versus its angle of attack for the unflapped wing, it would appear as shown at *A*. As the angle of attack increases, the C_L increases until it reaches a peak and the stall occurs, with a *decrease in relative lift with any further increase in angle of attack*.

The addition of flaps at *B* would result in a higher maximum coefficient of lift, but note that as the angle of attack is increased (*solid arrow*), the flapped wing would reach its critical angle of attack and the stall break would occur "earlier" than the unflapped wing.

A down-aileron can also act as a flap, as noted in Chap. 11, even if it's farther out on the wing, and produce a change in the characteristics, as shown by *B*. As you can see in Fig. 21-2, the up-aileron, *C*, moves the critical angle of attack out to a higher value.

Here's what's happening in that takeoff and departure stall from a right bank. The two wings will be close to the

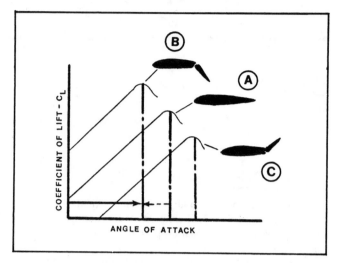

Fig. 21-2. Flap (aileron) effects on the coefficient of lift (C_L) versus angle of attack (α) curve.

peak at *A* or maybe just past the peak when the stall and roll occurs (no ailerons deflected at this point). The down-moving left wing has an even *greater angle of attack* (*solid arrow*) and *less lift,* while the up-moving right wing has decreased its angle of attack and is back in a normal flying regime.

Fine, as the airplane rolled the student did the "correct" thing and used full right aileron to stop the process (and, again, maybe not so much right rudder as would be called for) and compounded the problem. The angle of attack of the left wing is increasing because of the roll, and the down-aileron has moved (*dashed arrow*) that wing suddenly back to the curve at *B*. The two factors meet more suddenly than was expected, and because of the added drag on that left side, roll and yaw work to produce autorotation. Now that strange spiral has started and autorotation is occurring—but more about autorotation a little later.

To end the scenario, the student gropes around and with the help of the airplane gets out of the spin. After recovering (not knowing exactly how he did it), he flies shakily back to the airport, and other pursuits are substituted for flying.

One of the arguments against compulsory spin training for *all* pilots is that most spin accidents start at too-low altitudes for recovery but practicing up high is unrealistic. The biggest cause of stall/spin accidents is inattention or distraction. For instance, an engine fails after takeoff; the airplane is stalled (and starts to spin) while the pilot concentrates on getting power back and lets the airspeed decay as he tries to turn to a new landing area or back to the runway. Or the pilot has an actual high-altitude emergency and tries to stretch it to a particular field; in his concentration, he forgets the airspeed (more properly, the angle of attack) and a stall/spin accident results. Too, there's the cross-control stall when he overshoots the runway center line turning final, as discussed in Chap. 11.

Another argument against spin training is that people who want to learn to fly to use the airplane for personal transportation would not start if they knew that spins were required as a part of their training.

A third argument against this training is that gyro instruments would be damaged by continual spinning, and the result would be very expensive to the flight school operator. (This is sometimes only a good excuse to avoid teaching spins.)

The most important of all is, Are all the flight instructors qualified to teach spins?

Three arguments *in favor of* encouraging spin training in a properly certificated airplane are these: (1) A student who is taught (at the proper point in the syllabus) the various spin entries—and recoveries—at altitude might have this subconscious thought as he cross-controlled turning final: "Hey, the last time I felt *this* was that time at 5000 feet and it took us about 1000 feet to get out of it. It was a bad feeling and I'd better stop what I'm doing and take it around." (2) A student or trainee would know what was happening and not spin all the way in from 5000 ft when practicing stalls. (It has happened.) (3) It would build up the student's confidence in the airplane and himself and help all other phases of his flying.

■ **Spirals and Spins.** The student in the accidental spin described earlier thought he was in a spiral and wasted time and altitude. Many pilots are not aware of the difference between a spin and a spiral and often think in terms of one when the other is the problem.

Fig. 21-3 shows a layout that you can draw on the chalkboard to sum up the basic differences of instrument indications in spirals and spins for your low-time trainees.

SPIRAL	SPIN
A. INDICATIONS	C. INDICATIONS
HIGH AIRSPEED	LOW AIRSPEED
LOW α	HIGH α
HIGH RATE OF DESCENT	VERY HIGH RATE OF DESCENT
B. RECOVERY	D. RECOVERY
NORMAL	MECHANICAL
CONTROL	CONTROL
PRESSURES	MOVEMENTS

Fig. 21-3. Indications and recovery requirements for a spiral and a spin. You may want to put this on the chalkboard as you introduce spins.

SPIRAL. This is a steep, diving turn with a low angle of attack and high (usually increasing) airspeed. The average 2-hr student should easily be able to recover from a spiral because the controls are used normally. He levels the wings with coordinated aileron and rudder *pressures* and brings the nose up by back *pressure* as he had practiced earlier. The situation is an exaggerated version of what he had practiced before—a descending turn.

The characteristics of a *spiral* are that the airspeed is high and the angle of attack (α) is small. *The dangers existing in a spiral are these:*

1. The airspeed may well exceed the red line (V_{NE}) or even V_D, causing problems of controllability or flutter.
2. The pull-out at the high speed may exceed the limit load factor.
3. Because of confusion or other factors, hitting the ground is always a serious consideration.

The recovery is to use *normal control pressures.* The student would be using two of the levels of learning—*correlation* and *application;* in other words, either he sees the problem or the instructor nudges him ("Uh, your descending turn is, uh, getting a little steep, ha, ha.").

SPIN. To the pilot with little or no experience with spins, it may look like a spiral, but the controls aren't acting like it, even though he keeps trying a spiral recovery. The pilot has to think of mechanical control *movements* rather than pressure. And, as mentioned earlier, he has to move the elevators in a direction that would seem to make things worse.

The *spin* is at the opposite end of the airspeed envelope from the spiral as shown in Fig. 21-3. The airspeed is very low

(it may be zero or oscillating from zero to some low airspeed and back). The angle of attack is high; one or both wings will be stalled. Or as the engineers put it, "One or both wings are on the nonlinear side of the C_L versus α curve."

The dangers in the spin are these:

1. Disorientation—the rate of rotation (and descent rate) in a developed spin is very high for some airplanes, which could result in confusing which way the airplane is rotating, even if the student has done a few spins before. The fast rotation can have a hypnotic effect and, since there is little indication of the rate of descent, there may not be an urgency to "do something."
2. The requirement that the wheel or stick must be pushed, or moved "the wrong way," could be a factor that keeps the pilot who has little or no spin experience trying to pull the nose up.

The spin *recovery* depends on the *rote* level of learning. For the pilot in his first spin there is nothing in his flight experience, no matter how many hours he's logged, that can help him unless he has (1) heard or read of the recovery procedures, (2) remembers that information, and (3) is able to use the steps while under great pressure. For instance, the recovery steps for one airplane lists the following:

1. *Close the throttle.*
2. *Neutralize the ailerons.*
3. *Get the flaps up* (if they are down) for two reasons:
   a. The flaps extended speed (V_{FE}) may be exceeded in the dive after recovery.
   b. In that airplane, down-flaps tend to make the nose pitch up, which could make the spin flatter.
4. *Apply opposite rudder* to the rotation. Under stress there may be a problem in ascertaining the direction. More about this in detail later.
5. After the (correct) rudder pedal reaches the stop, *move the wheel briskly forward.* (Don't wait to see if the rudder is going to stop the rotation.)
6. When the *rotation stops, neutralize the rudder* and
7. *Pull out of the dive.*

That last admonition would seem to be unnecessary but as you instruct in spins, you'll find that some people are so glad to be out of the rotational aspects of the spin that they sit there with the nose pointed down, the airspeed increasing and altitude decreasing at an alarming rate. While, as will be emphasized, the *Pilot Operating Handbook* is the final reference, the seven steps just listed are a good general look at the various types of spin recoveries you may read about or use.

In some spin recoveries the trainee may still be unconsciously holding full opposite rudder, imposing strong twisting moments on the vertical tail as the airspeed approaches the red line.

Fig. 21-4 compares instrument indications of the two flight conditions, the spiral and the spin. In the spiral (A), you can see that the spiral has a high airspeed and the needle in the turn and slip and the small airplane in the turn coordinator are well deflected, but not as much as in a spin. The ball will probably be offset to the left for most U.S. single-engine airplanes because of "torque" correction effects. (The ball tends to be to the right in a climb, to the left in a dive.)

The spin (B) shows a much higher rate of rotation (the needle and ball are both pegged to the left). In some cases the ball may be oscillating between the center and full left deflection. Wait. You may have heard that the ball goes outside the spin so you should "kick the ball" to start the recovery. Don't—the ball is not a reliable indicator of spin direction.

The ball in the turn indicator should *not* be used to check for spin direction, because its reaction depends on the location of this instrument on the panel. If the instrument is on the left (pilot's) side, as it is in most current side-by-side air-

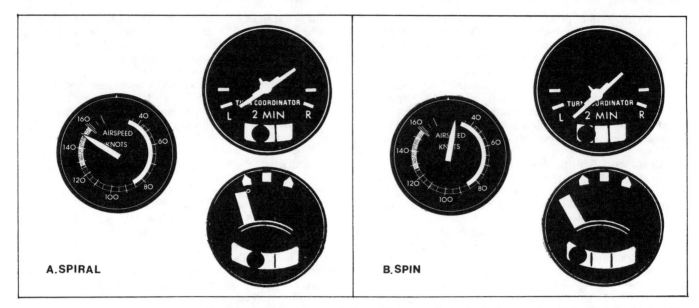

Fig. 21-4. Airspeed, turn and slip, and turn coordinator indications on the pilot's side of a side-by-side airplane. (A) In a spiral. (B) In a spin.

planes, the ball will be on the left side of the race in both right and left spins for the Cessna 150 Aerobat, Cherokee 140B, Musketeer Aerobatic Sport III, and Tomahawk. Slip indicators installed on the right side of the panels in the Aerobat and Sport had the ball going to the right side of the instrument in spins of both directions (Fig. 21-5). So don't "kick the ball" if you are in a spin. You have only a 50 percent chance of being correct in the installation mentioned. A ball in the middle of the instrument panel *may* react to move slightly "outside" the spin.

FUEL UNPORTING. The same forces that are moving the slip indictor balls to the outside of the instrument panel are also moving the fuel outboard in the wing tanks, away from the fuel tank–fuel line ports. The carburetor has a reserve in the float chamber, and the fuel lines from the tanks to the carburetor will contain some fuel so the engine will continue to run for a short time after the unporting has occurred, even if full power is used in the first few turns of the spin. If the developed spin is extended, the engine will quit and the propeller will stop. (The airplane is at such a high angle of attack and low relative velocity that windmilling will *not* occur in most cases.)

The number of turns resulting in engine stoppage depends on the amount of fuel on board (full tanks are harder to unport) and the power being carried throughout the spin. The C-152 may take 15 turns or more before the engine—and prop—stops, whereas the C-150 may quit at 7 turns. A different porting system in the C-152 gives it more usable fuel (24.5 gal) in comparison with the C-150 (22.5 gal).

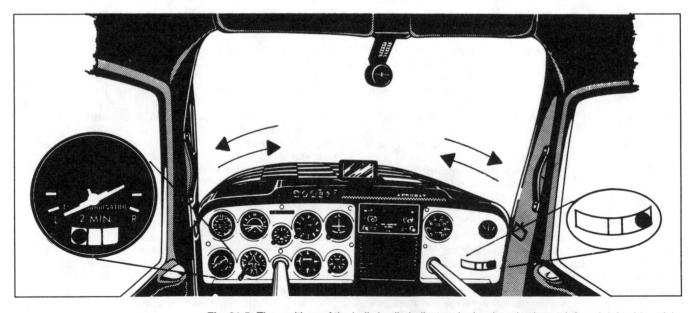

Fig. 21-5. The positions of the balls in slip indicators in developed spins on left and right sides of the instrument panel (Cessna 150 Aerobat). The same indications were found with slip indicators placed in the same relative positions on the instrument panel in a Beech Aerobatic Sport.

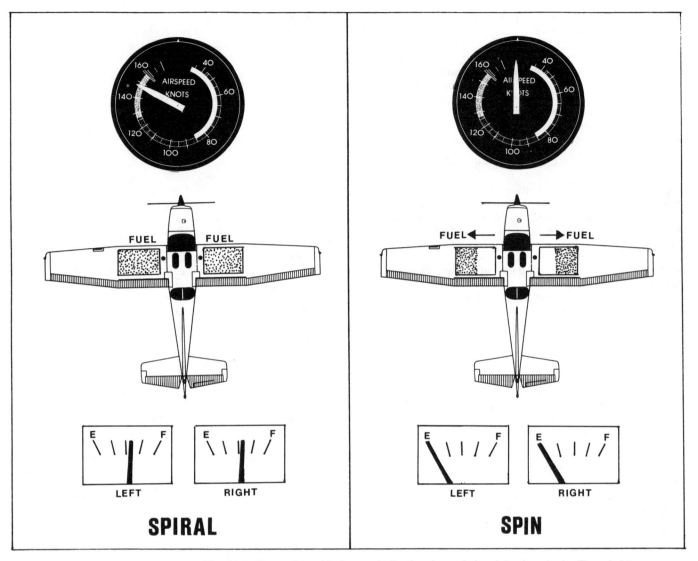

Fig. 21-6. Airspeeds and fuel gauge indications in a spiral and developed spin. The spiral is a more or less balanced turn so that the unporting encountered in a spin does not occur.

After the pull-out of the postspin dive, the starter may be used (don't forget to holler, "Clear!"), and the engine will start normally since the fuel is back into the normal distribution.

Some general information: Suppose the starter is inoperative at this point (a one in a million possibility). Perhaps you turned the key to the starting position (the mags are still hot) and something broke in there. It would take about 1200- to 1500-ft altitude and 130 K to get the prop windmilling for start, based on a 100- and a 150-HP trainer, both with midtime engines. Your airplane may vary either way from these numbers but to be on the safe side, consider them as minimums. Of course, if you are within gliding distance of an airport when this occurs, you're better off to set up the max distance glide speed and go there rather than "waste" altitude.

Fig. 21-6 shows the differences in airspeed and fuel gauge indications in a spin and spiral if both are started with the same amount of fuel (slightly over one-half in each tank in this example).

Spiral—As noted earlier the spiral is a high-speed descending turn. The airspeed, as shown here, is very high and because the turn is coordinated (or close to it) in a spiral, the fuel in the tanks will remain in the normal condition and the fuel gauges will read accurately.

Spin—In addition to the very low airspeed in the developed spin, the fuel has been unported and since the fuel quantity transmitter for light trainers is in an inboard portion of the tank, the float is not supported by fuel anymore and "hits bottom." You can tell whether you are in a spin or a spiral in this case *by looking at the fuel gauge.* Obviously that would be a good hangar flying question: "What instrument(s), besides the flight instruments, can indicate whether some airplanes are in a spiral or spin?"

Spins and spin recoveries under the hood will be covered later in the book.

A GENERAL SPIN REVIEW

In analyzing what to do about the spin, both in research and in teaching pilots spin recoveries, it is a good idea to take a closer look at the upright spin.

The spin is the result of one or both wings being stalled, that is, one or both wings being past the critical angle of attack, with one deeper into the stall and with a lower coefficient of lift. When you move the wheel or stick rearward and produce a stall, some interesting things may happen, depending on how the airplane is rigged and what control input you may have at the time of stall break. At the stall break the airplane *departs;* it leaves the realm of normal, nonstalled flight. It starts sliding down the slope (Fig. 21-1). The nose may drop, and at this point the recovery may be accomplished by a relaxation of back pressure or a forward wheel movement, with normal flight resulting almost immediately. Or the airplane could enter a *poststall gyration,* in which nonperiodic rotation and re-pitch-up occur, again depending on control input.

The *spin* is a continued condition of stall, a step beyond the poststall gyration, with autorotation occurring.

■ **Entry.** This is the preparation for the spin and includes clearing the area, use of carburetor heat, attitudes at entry, and initial application of prospin controls. Various entries you may use in demonstrations will be discussed later in the chapter.

■ **Incipient Spin.** This is the part of the spin in which the airplane is accelerating in pitch, roll, and yaw into the developed spin. The path is changing from horizontal to vertical. The aerodynamic forces (induced by the controls and asymmetric forces on the wings and other components) are overcoming the inertia of the airplane and building up the rotation rate. For most lighter airplanes, the first *two turns* are considered the incipient phase.

Steeper modes (the nose pointed farther down) require fewer turns to stabilize, because the angle of attack doesn't have to climb so high. The modified general aviation airplane at NASA took about 4 turns from the entry to get into the flat mode. For instance, if you decide to spin a *normal* category airplane illegally and don't get a developed spin at 2 or 3 turns, you might think there is no problem—until you keep on with your "experiments" and find an unrecoverable flat mode at 4 turns or so.

One demonstration you may have the trainee do later is to set up a normal power-off spin, say to the left, and at 2 turns take his hands and feet off the controls. Up to 2 turns, some airplanes will normally recover and go into a dive. (Whoever is in charge should ease it out before the airspeed gets too high.) Trimming the airplane nose up before starting the hands-off demonstration makes an even better point, as the airplane will recover and then ease out of its dive. You would have to know your trainee and the airplane thoroughly before doing this.

■ **Developed Spin.** The airplane has reached a constant rotation rate and pitch attitude or a *repeatable* pattern of rotation and pitch attitudes. Some airplanes, once the spin has developed, have a steep nose-down spin with a constant rotation rate, while others may be cyclic or oscillatory, with a part of each turn steep and the rest of it in a flatter attitude. The nose "bobs" up and down as it rotates, and the rate of rotation changes, repeating itself within each turn or so, rotation usually being faster when the nose is down farthest and slowing up when the nose rises.

Other airplanes, too, may change attitudes and rotation rates (or have cyclic modes) in the developed spin. One popular airplane approved for spins has the usual incipient spin at 0–2 turns, then has a steep mode with the nose well down and fast rotation rate at the 3- to 5-turn regime, followed by a flatter mode with the nose not so far down and a slower rotation rate in the 6- to 9-turn area. It then repeats these modes every 2 or 3 turns, each readily recoverable, but the steeper mode has an almost instantaneous recovery while the less steep mode may take 1–1½ turns to recover after aerodynamic recovery controls are applied.

For some airplanes there are *two indications* of the spin

going from the steep to a flatter mode: (1) the nose can be seen slowly rising as the spin progresses, and (2) the sound decreases as the airplane slows in airspeed at the flatter pitch (the airspeed indicator may go from a stalling speed value of, for example, 50 K down to zero).

While on the subject of altitude loss, the question often comes up as to how much the average trainer loses per turn in the spin. *It depends*—for instance, for a 1-turn spin that would include entry, incipient spin, and recovery, you'd better plan on 1000 ft from start to finish. Hold on, you say, that's a lot of altitude for a 1-turn spin. The point is that the entry and pull-out take a goodly amount of altitude, and this must be included into the "average" of that one turn. On the other hand, a 21-turn spin in one type of current airplane took 4100 ft of altitude from start to level flight again. The altitude required for entry and pull-out had much less effect in boosting total altitude as the number of turns increased. (Dividing 4100 by 21 gives an "average" of about 195 ft of altitude loss per turn.) An altitude loss of 4700 ft resulted from a 25-turn spin in a different model of the same type trainer.

Getting back to potential problems: If you are spinning with a student and unexpectedly start getting one or more of the indications mentioned, start the recovery procedure *immediately.* It's possible that a flatter mode (particularly if you had not previously encountered it or didn't know it existed for *your* airplane) could delay or even preclude recovery. (You forgot about those anvils back in the baggage compartment.)

AUTOROTATION. To get a look at what's happening in autorotation as applied to the developed spin, let's backtrack and assume that you are going to do a deliberate straight-ahead entry with no crossed controls and with the power at idle, so as not to complicate matters.

You have plenty of altitude, the airplane is certificated for spins, and you are over wide-open country away from restrictive airspace (airways, etc.). You clear the area before using carburetor heat as recommended and then smoothly close the throttle, easing the nose up to a power-off, wings-level stall attitude. Assume no flaps and no ailerons are to be used.

As the stall-warner goes off (and the wings are at the peak of the C_L versus angle-of-attack curve), lead with left rudder and move the wheel full aft at a slightly faster rate to get a definite break. (This procedure is needed for some airplanes to assure a good spin entry.)

The airplane yaws left; that left wing has less lift because of the relative decrease in velocity, and a roll is induced, further increasing the angle of attack. The left wing goes over the C_L peak and is stalled. Meanwhile the right wing has moved upward, decreasing its angle of attack, and it has more lift than the left one. Fig. 21-7 gives a schematic picture of what is happening as the airplane rolls to the left and the spin develops. Figs. 21-7 and 21-8 show the incipient phase with one wing stalled and the other basically still "flying." The flatter developed spin modes would have both wings well past the stall angle of attack.

Fig. 21-8 shows the coefficients of lift (C_L) and drag (C_D) plotted against angle of attack for the plain wing. Point *1* is the point of the approaching stall just before the left rudder is pushed and the wheel is moved rearward at a slightly faster rate. The dashed lines show the C_L and C_D values for *both* wings (since the airplane is in balanced flight at this point before the stall break).

As the left wing drops, it has less lift and more drag than the right one. In this left spin, the left wing is aerodynamically pulled "down and back," while the right wing is chasing it like a dog after its tail.

The left wing has a *higher* coefficient of *drag* and *lower* coefficient of *lift* than the right, so the forces are leaning to-

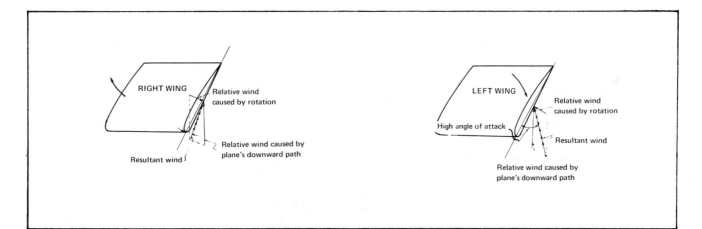

Fig. 21-7. Differences in angles of attack in a spin. (*The Student Pilot's Flight Manual*)

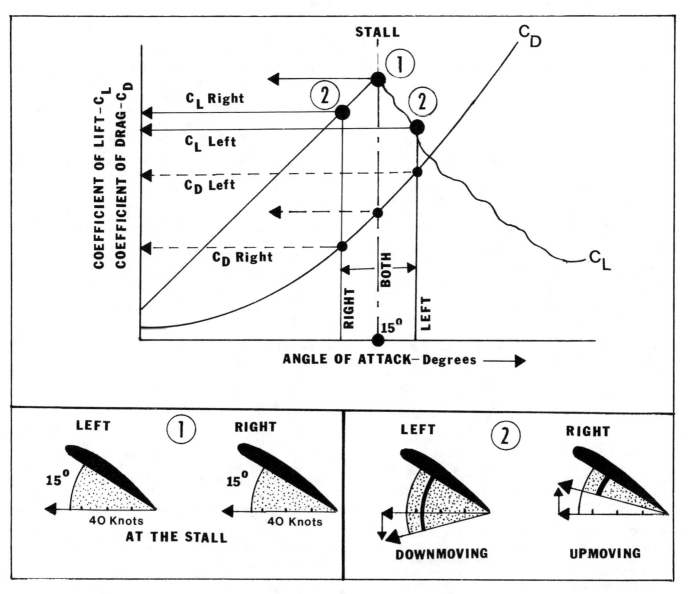

Fig. 21-8. (1) As the stall warner goes off, both wings are at the stall angle of attack (exaggerated 15° in this example). The airspeed is 40 K for each wing. (2) As full left rudder is applied the airplane yaws and the left wing has a relatively lower velocity and loses lift. The right wing is speeded up and gains lift, the combination setting up a roll to the left. The roll causes the left wing to increase its angle of attack, moving it past the stall. The right wing has a decreased angle of attack and moves back away from the stall at this point. Compare the coefficients of lift and drag for each wing, as shown in (2), as the airplane rolls and yaws into the spin.

ward autorotation. You can illustrate this difference in lift and drag by putting a cone on a string about 6 ft long on each wing tip and observing the vortices as the airplane is stalled and spun. Fig. 21-9 shows the vortex action in normal flight. The wing cones are rotating "inward" toward the low-pressure (or lifting) side. Vortices as shown by cones and strings on the tail tips will rotate in opposite directions from the corresponding wing streamers if the tail-down force is present; the streamers would be rotating toward the low-pressure (or lifting) side of the tail.

Fig. 21-9. Normal vortex action at wing and tail if a tail down-force exists.

If you have a chance, watch the actions of the tail streamer as the airplane is stalled and the recovery made: (1) the rotation is normal as the stall is approached, (2) at the break, the streamer will not be rotating, and (3) as the nose drops, it will start rotating in an opposite direction (an upload exists on the tail as the airplane "falls"). The process described is then reversed as the airplane moves back into the normal flight regime. Your trainer may have different reactions, and you might want to check it. Don't have the trainees looking back at the tail too long, because it may cause queasiness.

In developed spins in one low-wing trainer, the string and cone (kitchen funnel) on the inside wing showed that vorticity had broken down, while the outside wing still had weak vortex action. At the recovery, both cone-string indicators showed strong vorticity.

If you really want to go into the spin-demonstration business and have a *low*-wing airplane approved for spins, putting strips of yarn or tufting on the wings gives a good view of what's happening to both wings in the spin. In steeper spin modes the tufting on the inside wing may be reversed (pointing forward) while the tufting on the outside wing is acting more "normal" (it may be erratic but is blowing more or less toward the trailing edge). In flatter modes the outside wing's tufting will also be tending to reverse. Movies of tufting show that as the spin is recovered, the yarn strips quickly lie flat and stop the erratic behavior induced by spoiled flow.

Don't have your trainees spending too much time watching the wing streamers.

SPIN MODES. Okay, some airplanes have cyclic modes, changing angle of attack, rates of rotation, and rates of descent. Fig. 21-10 shows the steep and flatter modes as might be seen in a Cessna 150 Aerobat. Panel 1 shows the steep mode, usually found in the first 3 or 4 turns of the spin in that airplane. Note that like Figs. 21-7 and 21-8, at this point the left wing is stalled and the right one is not. As the spin progresses the nose moves up, increasing the angle of attack and *both* wings are on the nonlinear side of the C_L versus alpha curve, in panel 2. In extended spins, the airplane moves in an area between panels 1 and 2, repeating the process as the spin continues.

ROTATION RATES. Fig. 21-11 is the plot of rotation rates versus turns for a 21-turn right spin in a Cessna 150 Aerobat.

Notice that the steeper the pitch attitudes the faster the rate of rotation, these changes being cyclic in action.

Fig. 21-12 is a family of *right* spins for one Cessna 150 Aerobat. This particular airplane spun better to the right than to the left; that is, the spin developed earlier and was established in fewer turns in that direction.

Fig. 21-13 shows the right spins of Fig. 21-12 plus two left spins of 8 and 9 turns each. You'll notice that the rotation rates for the left spins do not increase so rapidly or attain as high a value as for the right spins. You may find that other airplanes of the same model may have different left- and right-spin characteristics from those shown here; for instance, a particular airplane may have control surfaces that no longer meet the full deflection limits as required in the Service Manual (or surpass them) or may have suffered "hangar rash" and/or slackening of control cable tension. (More about this later.)

RATES OF DESCENT. Fig. 21-14 shows altitudes and rates of descent for three left spins as tracked by radar (Cessna 152 Aerobat). Notice that while the altitude is tracking downward at a fairly constant rate, the rate of descent is changing with highest or lowest rates occurring about every 5 sec.

1. *First Spin*—The entry was normal at 10,000 ft MSL. Note that it took about 42 sec to go from 10,000 down to 5000 ft MSL with an average rate of descent of about 7150 ft per minute. There are a couple of anomalous points showing rates of descent of nearly 12,000 fpm. The power was left full on throughout the three spins and the engine stopped at about 17 turns in this one. Prospin rudder and elevator were held, but tended toward prospin positions after 4 turns (ailerons *held* neutral).

2. *Second Spin*—The attempt was made to duplicate as closely as possible the entry and control positions of the first spin, but there are differences in times and descent rates. The "average" rate of descent in this spin was approximately 6300 fpm. The engine again stopped at approximately 17 turns.

3. *Third Spin*—Full left (prospin) ailerons were applied at entry and held until recovery controls were applied. The rate of rotation and side forces on the pilot were noticeably higher than for the neutral-aileron spins. The engine stopped at approximately 15 turns, earlier than the first two spins, probably because of a greater initial unporting effect caused by the increased rotation rate.

In the C-150 and C-152, the ailerons are effective enough in the developed spin so prospin ailerons (left aileron applied in a left spin) will speed up the rate of rotation and delay recovery.

At the time of the spins in Fig. 21-14 this writer had 2756 spins logged in N7557L and had hoped to establish constants in entries and recoveries. It was interesting to note that there were noticeable differences in the profiles of spins 1 and 2. This enhanced the belief that while a particular airplane's spin characteristics may *generally* be predicted, there will be variations resulting from different accelerations and entry controls application and, as noted earlier, differences in characteristics between airplanes of the same make and model. When an airplane is operating on the nonlinear side of the C_L versus α curve, it may approach almost unpredictable conditions or, as one instructor put it, it may be in a "chaotic mode." Because of this, when you are instructing spins, always have plenty (at least 4000 ft) of altitude AGL when you enter even a 1-turn spin and plan on the recovery from *any* spins to be completed at least 3000 ft AGL.

■ **Recovery.** This is the portion of the spin process from the first input of aerodynamic controls for recovery until the

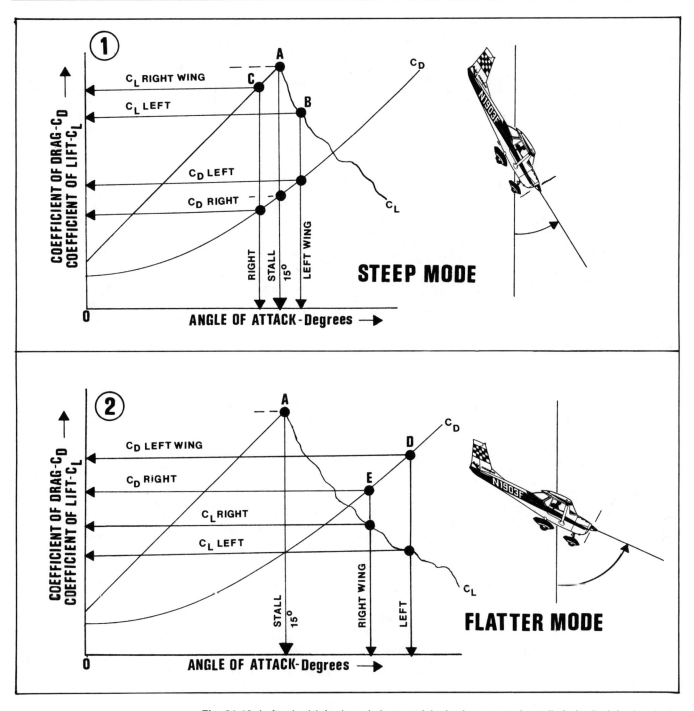

Fig. 21-10. Left spin. (1) As the spin is started, both wings are at the stall, *A*. As the left wing drops (also shown in Fig. 21-7) its angle of attack increases, *B*. The right wing, *C*, under this condition is unstalled with the unequal lift and drag forces of the two wings producing autorotation in the steep mode. For most airplanes, this is a transitory condition; a part of the incipient phase as the airplane moves into the developed spin. In others it may be a part of the oscillatory sequences or mode changes. (2) The airplane's nose has pitched up as the spin progresses and *both* wings are past the stall point. Because of the reverse slope of the curve c_L *versus* angle-of-attack curve, and the increased coefficient of drag with increased angle of attack, the left wing, *D*, *still* has *lower lift and higher drag* than the right wing, *E*, and autorotation continues.

autorotation and stall are broken. One pilot, experienced in spins, was asked if he liked an airplane that spun well, and he replied, no, he preferred one that *recovered* well.

The recovery in the older trainers (Champion and Cub, for example) was a three-step process: (1) stop the autorotation by using opposite rudder so that both wings are equally stalled (at that instant the airplane isn't tending to rotate but is moving vertically downward in a straight stall, buffeting), (2) neutralizing the rudder, (3) break the stall by forward motion of the stick or wheel, and (4) then pull out of the dive.

While the *Pilot's Operating Handbook* for a particular airplane will *always* be the final criterion for spin recovery

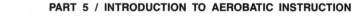

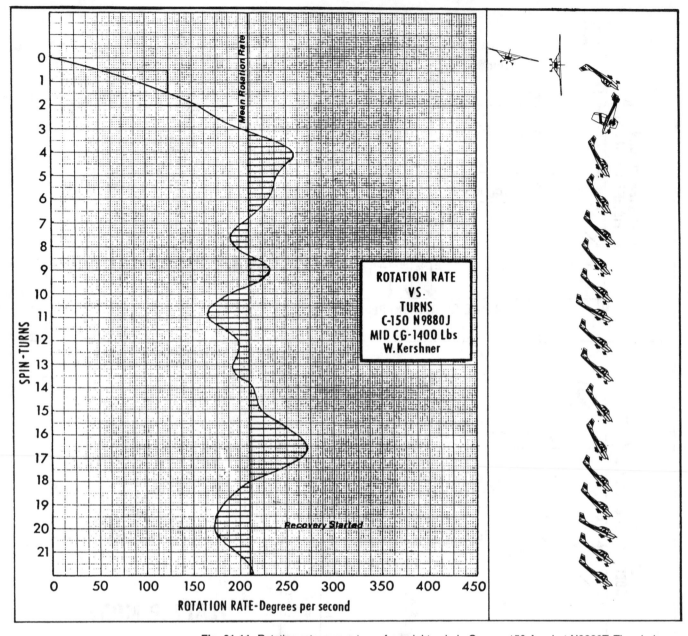

ROTATION RATE
VS.
TURNS
C-150 N9880J
MID CG-1400 Lbs
W. Kershner

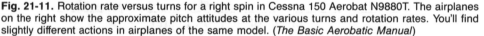

Fig. 21-11. Rotation rate versus turns for a right spin in Cessna 150 Aerobat N9880T. The airplanes on the right show the approximate pitch attitudes at the various turns and rotation rates. You'll find slightly different actions in airplanes of the same model. (*The Basic Aerobatic Manual*)

procedures as indicated earlier, but here is a general look at the problem.

If you're out doing stalls with a student or aerobatic trainee and the airplane has apparently worked its way into a spin (high rate of rotation, nose well down, and airspeed very low), the following steps are indicated:

1. *Power off.* For many airplanes, power hurts the recovery process. The nose tends to pitch up when power is applied and stay up higher while it is on. Airplanes with some of the standard tail configurations with power on seem to resist the pilot's pushing the nose down during the recovery.

Of the 29 airplane types and models spun by this writer, only the Navy turboprop trainer T-34C Flight Manual recommends the addition of power as an aid to spin recovery. If recovery does not occur within 2 turns after applying antispin controls in an erect spin, the pilot is to confirm that full opposite rudder is being applied, that the stick is forward of neu-

tral, and that there is no further indication of recovery. He is then to apply maximum (allowable) power while maintaining the antispin controls.

In extended spins in many airplanes, power may not be available because of fuel unporting, as noted earlier.

2. *Neutralize the ailerons.* If the trainee is like most pilots, he's trying to stop the spin with the ailerons, as mentioned earlier in the scenario, which makes matters worse for some airplanes. Since he may not have thought of what ailerons do to this airplane in a spin (whether they help or hinder recovery), it's better not to take any chances; they should be neutralized. More about aileron effects later.

3. *Full rudder opposite to the rotation and hold it.* Simple enough, except there's an outside possibility that he (or you) could get confused in an accidental spin and forget which way the airplane is rotating. Sure, if you did a deliberate spin and are holding, say, full left rudder throughout the

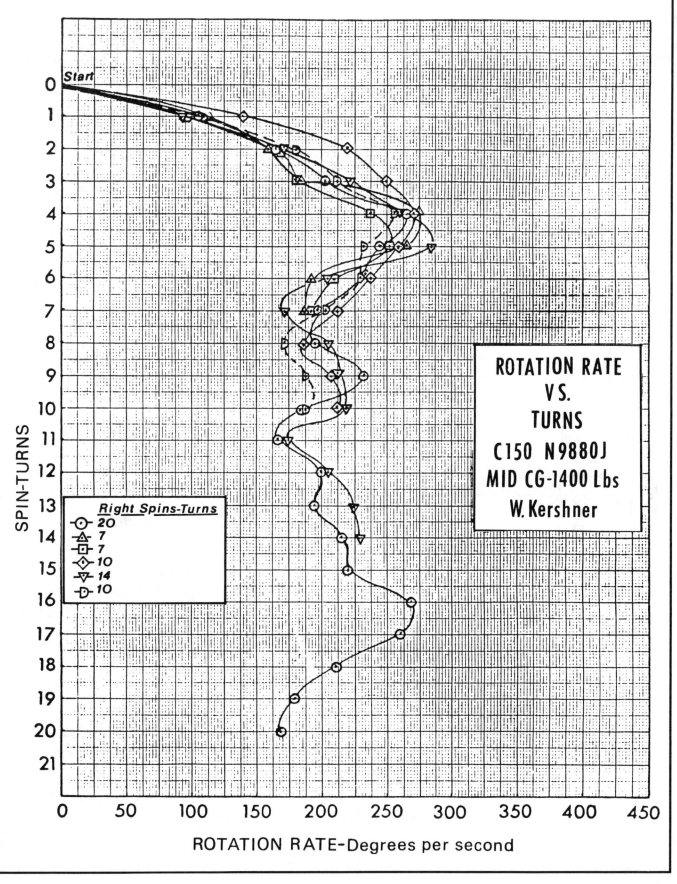

Fig. 21-12. Rotation rate versus turns for a family of right spins for one Cessna 150 Aerobat. The numbers by the symbols indicate the number of turns for that spin. (*The Basic Aerobatic Manual*)

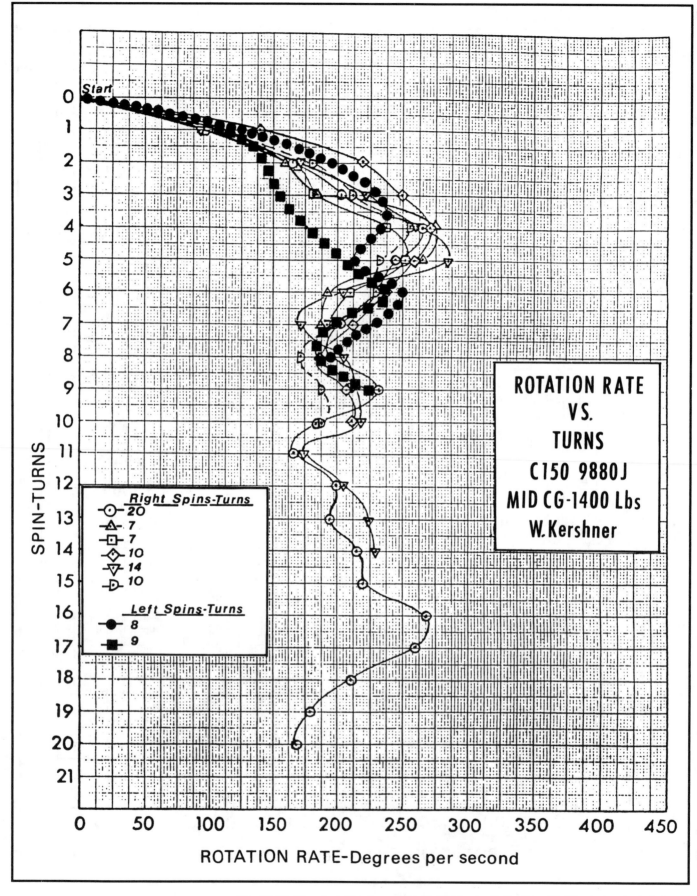

Fig. 21-13. The six right spins of Fig. 21-12 plus two left spins. (*The Basic Aerobatic Manual*)

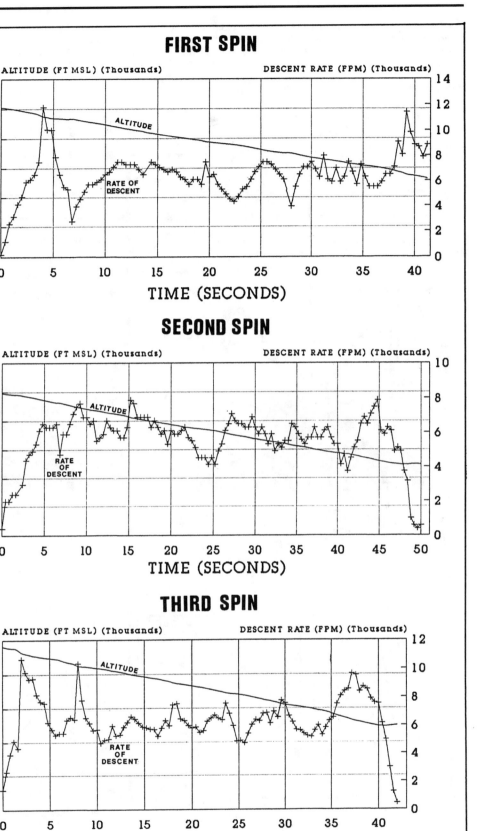

Fig. 21-14. Three left spins from 10,000 down to 5000 ft MSL. Full throttle was used throughout the spins. The engine stopped at 17, 17, and 15 turns respectively. Prospin ailerons were applied at the entry of the third spin and held throughout. Unlike the airplane in Figs. 21-12 and 21-13, this airplane spun better to the left; that is, a developed classic spin could be attained and held. There was a strong tendency to spiral out of a right spin.

Fig. 21-15. A general comparison of some (A) older (lighter) and (B) newer trainers' spin recovery techniques and airplane actions. In these examples the airplanes have glass floors so that a point on the ground directly below could be seen. The turn coordinator (B) is shown as it would be indicating in the spin itself. As will be noted, the small airplane and ball (turn coordinator) shown in the newer airplane will overshoot and may be oscillating during the recovery but will settle down as the pull-out is completed.

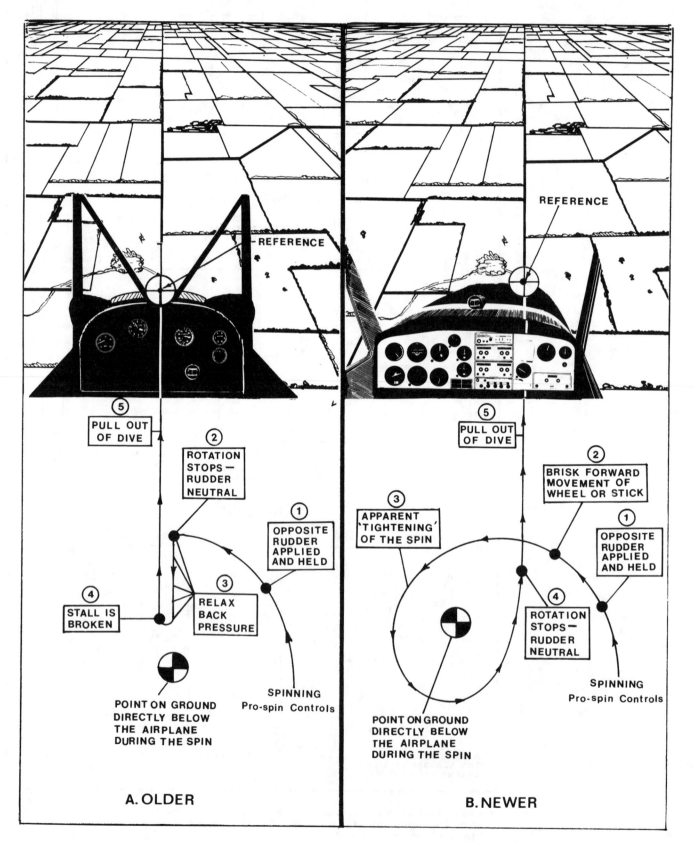

process, you should know enough to push the opposite or right rudder, but people are funny. The developed spin in that airplane may have a very fast rotation rate and the ground may look like a blur. The direction may not be so readily discernible as you might think, and you may not remember which rudder is being held—if any.

Again, the reference to use for turn direction, if you need it, is the needle in the turn and slip or the small airplane in the turn coordinator. The needle (or the small airplane) is leaning in the direction of the spin. *Rudder should be applied opposite to that indication and held.*

4. *Apply brisk forward motion on the wheel or stick immediately after the application of rudder.* For some newer airplanes, the rate of application of forward wheel is as important as how far it's moved. Again, this is different from the older, lighter trainers. Some pilots, then and now, are so glad to be out of the spin that they leave the nose down too long and the airspeed goes too high with an excessive altitude loss. (Diving into the ground while congratulating yourself on a

fine spin recovery can be as fatal as spinning all the way in.) Incidentally, as a ballpark figure, you might expect rates of descent in the steep mode of some trainers of 7000–8000 fpm; in flatter modes the rate might fall off to a mere 5000 fpm or so. Either rate would dent your airplane if you hit the ground. Some pilots hold the opposite rudder after the rotation stops and start a spin in the other direction (progressive spin), which requires a new recovery, including using rudder opposite to the opposite rudder just applied, which can be confusing.

Fig. 21-15 shows a general look at spin recoveries for some older and newer trainers. The line traces the projection of a point on the cowling to the ground, showing two types of recovery for the different vintages or, more accurately, some differences in density ratio and mass distributions.

RECOVERY FACTORS. Fig. 21-16 shows factors that have affected spin recovery procedures in earlier two-place general aviation trainers and some later "heavier" trainers and four-

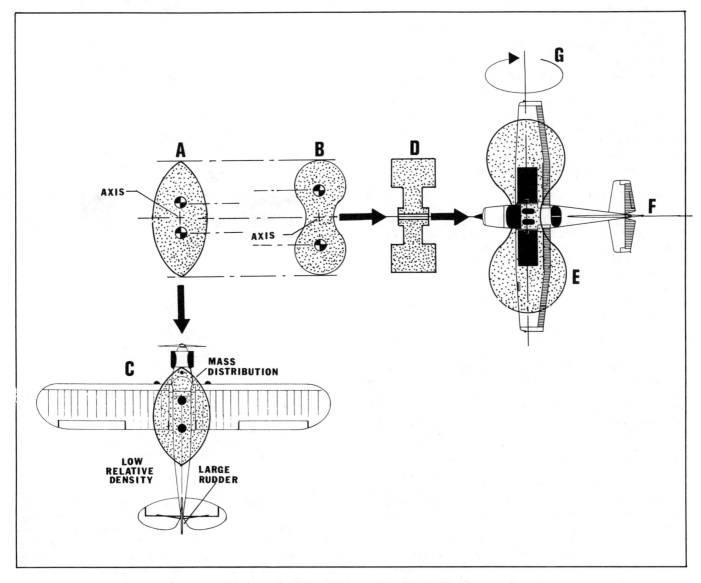

Fig. 21-16. Two objects, *A* and *B*, showing the mass distribution differences in old and new trainers. The older airplanes, *C*, tended to have the mass located as shown and were "lighter" for their size (volume). *B* shows a distribution more like a gyro wheel, *D*, and later trainers, *E*. If the rudder cannot stop the rotation, down elevators, *F*, are a more important factor in the recovery, so that the airplane is rotated around the lateral axis, *G*, with the lower angle of attack unstalling the airplane. At the same time the rudder becomes more effective because of that lower angle of attack, stopping the rotation.

place general aviation airplanes. Suppose you took the two objects, *A* and *B*, and inserted a shaft in each, as shown. For this example, both have the same diameter and mass and are set to spinning at the same rpm around its axis.

Object *A* would be easier to stop because the center of gravity of each half is located closer to the axis. (This is the same principle that is used in the rotors of gyros; once it's up to speed it tends to maintain that rpm and it takes a while for the gyro wheel to stop after you've shut the airplane down.)

Notice at *C* that the "mass distribution" of the older trainer (a Cub here) has the "closer" outlines of object *A*. In these older airplanes, opposite rudder would stop the spin (and this made a spin recovery basically a two-step process):

1. *Make sure the throttle is closed.*
2. *Stop the rotation with the rudder, then neutralize.*
3. *Unstall the airplane by relaxing back pressure or moving the stick or wheel forward.* (Look back at Fig. 21-15A.)

Since a J-3 Cub normally landed at 35 mph, it needed a comparatively large rudder for lifting a wing as it settled at that low value of dynamic pressure (opposite rudder was the best way to raise a wing in some of those earlier airplanes) and to also help the tailwheel in maintaining directional control during the landing roll-out.

The newer-model trainers land at 50 mph (as a comparison) and higher (40+ K) and are tricycle-gear types, so a smaller rudder does the job very well. There's little point in having a rudder that is effective well below the airspeed that the airplane can fly, and with the nosewheel, good directional control can be maintained throughout the landing roll. A further analogy is that a jet fighter probably would have a rudder not much bigger than a C-172 because the fighter has tricycle gear and touches down at 100+ K.

In addition to the mass distribution, the later airplanes are heavier per cubic foot of volume. That is, the interior volume (the inside of the fuselage, tail, and wings) of a J-3 Cub, weighing 1200 lb, is probably close to the interior volume of a C-152 weighing 1670 lb. This means that the C-152 has a higher *relative density* and would take perceptibly longer to have its spin momentum stopped by the aerodynamic controls, even if the two airplanes were spinning at the same rotation rate and had the same control surface areas. More details about this later.

So, the spin recovery procedures in older airplanes were different because of these factors:

1. *Larger control surfaces.*
2. *More-centered mass distribution.*
3. *Lower relative density.*

For instance, in the recovery from a developed spin (particularly to the left) in a C-152 you would *not* apply rudder and "hold it until the rotation stops or slows" before applying forward wheel or stick (as is suggested in some texts). The rotation will continue at a constant rate *until forward wheel is used,* particularly in a left spin. The student is often reluctant to use enough forward pressure and sits staring at the ground, mesmerized, believing he is pushing very hard indeed.

So rudder alone is not enough to stop the rotation in some later-model trainers. The final criteria of a spin recovery are these: (1) the rotation is stopped and (2) the airplane is unstalled. The steps shown back in Fig. 21-15B are the logical sequence for those later airplanes.

In other words, if the now-smaller rudder cannot aerodynamically stop the more widely distributed and heavier wing mass, the elevator can be effective in rotating around the *lateral* (pitch) axis, resulting in both (1) decreasing the angle of attack to get out of the stall and (2) increasing the ability of the deflected rudder to stop the rotation as it regains smoother flow (the blanketing effect of the horizontal tail is decreased).

Fig. 21-17 shows the mass distribution changes from an

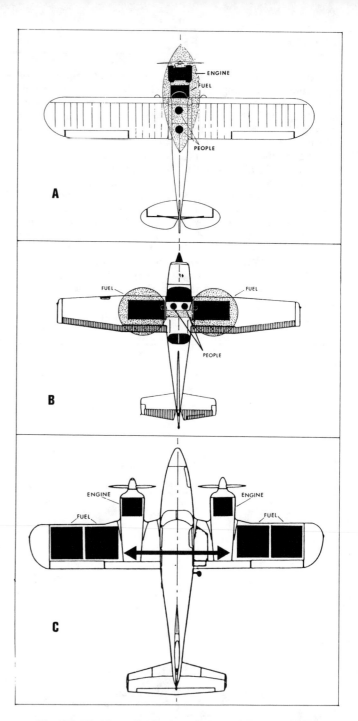

Fig. 21-17. Mass distribution changes from an older light trainer in (A), to a current single engine airplane in (B), to a light twin in (C). In many cases opposite rudder alone will not stop the rotation as mass is added outboard along the wings, but the combination of rudder *and* elevator is needed to stop the rotation and complete the recovery.

older trainer through a current single-engine airplane to a light twin. With each step, the elevators become more important in the recovery process. Mass distribution and other physical and aerodynamic factors will be covered again later in the chapter. It should also be noted that the forces causing the fuel unporting discussed earlier will result in a movement of mass toward the wing tips (Fig. 21-6).

■ **A Spin—Entry to Recovery.** Figs. 21-18 through 21-21 show a spin generally and do not refer to a particular type of airplane but are intended to show and discuss the factors you might encounter in your spin instructing.

SPIN MODES. The aerodynamicists speak of the spin attitude in terms of angle of attack. A completely flat spin would be at an angle of attack of 90°, with the airplane rotation totally around the vertical axis, moving straight downward with the nose transcribing a line around the horizon. A pilot would likely think of this as a "zero-degree pitch-down attitude" or "level-flight attitude."

NASA has set guidelines for defining the various spin modes:

Spin Mode	Angle of Attack—Degrees	Nose-down from Horizontal—Degrees
Flat	65–90	0–25
Moderately Flat	45–65	25–45
Moderately Steep	30–45	45–60
Steep	20–30	60–70

The point is not that the airplane's nose is pitching up and down within the values given, but it's assumed to be in a steady state at some approximately constant value within that range. For instance, an airplane spinning smoothly at 55° angle of attack would be in a moderately flat spin.

As a general approach, the steeper the spin attitude, the more easily the recovery is made. Some pilots don't like the steeper modes because they are looking at the ground in a (to them) vertical nose-down attitude. They prefer the flatter modes because yawing motions are apparently not as balance-disturbing as rolling motions.

During the recovery from a flatter mode, pilots may be disturbed by the nose dropping and the rotation rate apparently increasing. (Look at Fig. 21-15B to get a general idea of this situation.) Actually, it is normally a good sign; the airplane is going back through the steep mode as it moves toward recovery. Not realizing this, they could release the control application, thinking the spin is getting worse, and thereby delay the actual recovery.

However, an airplane certificated for spins will move through this stage so rapidly in a recovery that the pilot doesn't have time to react anyway.

One problem with pilots inexperienced in spins is that they may apply the proper control, but after a half-turn (or 1 turn) with no apparent recovery, they become anxious and may ease off the deflection to try something else. This naturally will delay any recovery as they try different combinations, hoping for the best, since the recommended recovery "didn't work." Even a pilot with much experience in the near-instant recoveries in the current trainers in the incipient phase (0–2 turns) may be taken aback in a developed spin by the fact that instead of the eighth-turn, quarter-turn, or half-turn required to stop the rotation after control application, it may take 1 or 2 turns before good things start happening. So, the first time this happens, the surprise may cause the pilot to back off to try something else—what he's not sure.

An improper recovery procedure could mean a delay in stopping the spin. Perhaps the opposite rudder was not put to the stop and/or the wheel or stick was moved ahead slowly, or halfheartedly. A good move in this case would be to go pro-spin controls again (rudder with the spin, and wheel full back) to get a "running start" in using the controls for recovery. Rapid and full opposite rudder followed quickly by a brisk forward movement of the wheel, *and held,* will be a move effective for recovery.

This brings up the point that if an airplane is hard to get into a spin, it could be even more difficult to get out.

Assume that the angle of attack when the stall occurs for a particular airplane is 15°. This is not such a large angle, and only a small area of the fin and rudder may be blanketed by the horizontal tail as the spin is entered. (More about this later.) Also assume that the rudder power was weak even at this low angle and the pilot barely had enough yaw force available to get the spin started.

The spin develops and gets into a flatter mode where the angle of attack is 45° or more—even more of the rudder,

Fig. 21-18. Entry to a right spin. The area is cleared, carburetor heat used as recommended, and the throttle retarded to idle. Some power may be left on to help get things started, or a blast of power may be used on some airplanes, particularly in left spins. The throttle should be closed after the entry.

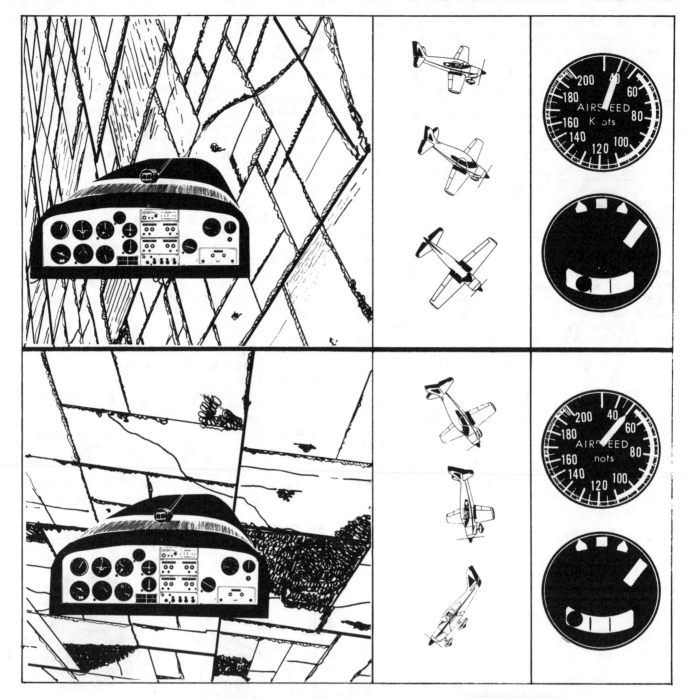

Fig. 21-19. Incipient spin. *Top:* The airplane has rolled and rotated 90°. The airspeed will temporarily decrease from the stall value (50 K) as the airplane drops momentarily, increasing the angle of attack above that of the stall. The ball is moving to the left side of the instrument, and the needle is fully deflected in the direction of rotation. *Bottom:* The airplane has rotated approximately 180° and the airspeed has now increased back to the stall value. The needle remains fully deflected, and the ball is deflected to the left side of the instrument. (The instrument is on the left side of the panel in this example.) Some airplanes go inverted within the first turn, but this doesn't mean that an inverted spin is starting, since the stick or wheel has been held back throughout the entry. You'll know if you are starting an inverted or negative-g spin by the way you are shoved into the belt and harness.

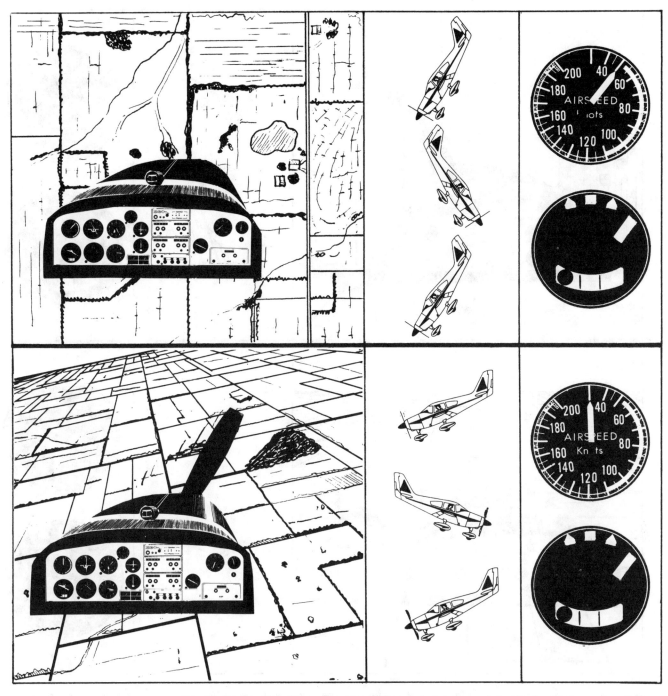

Fig. 21-20. Developed spin. *Top:* The airplane is in the steep pitch attitude with the nose well down and a high rate of rotation, in some cases 280° per second or more. The descent rate is high and may be in the vicinity of 7500 ft per minute. Usually recovery is prompt in this condition, taking an eighth turn to a half turn. *Bottom:* The nose has risen as the airplane and the rate of rotation have slowed, but the turn needle is still fully deflected in the direction of spin. The rotation rate may, for some airplanes, be in the vicinity of 160° per second in this spin phase. The airspeed is at zero. You may be getting a rate of sink in the vicinity of 5000–5500 fpm in this phase for some airplanes. The vertical speed indicator will be pegged at its maximum of 2000 fpm, so you won't know how much faster you're coming down. (The VSIs for higher-performance airplanes are marked for up to 6000 fpm up or down, but they still wouldn't show the full story in the steeper phases of a spin.) When you first get the three indications of a spin getting flatter (sight, feel, and sound), start the recovery procedure immediately, unless you have had instruction in that particular airplane and know that it's recoverable from a flatter condition.

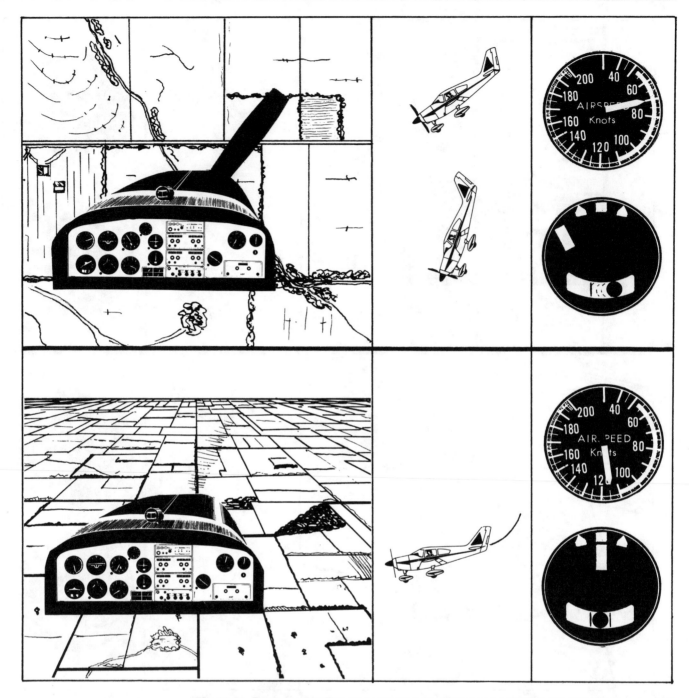

Fig. 21-21. Recovery. *Top:* The prop has stopped, which has no effect on the spin recovery (except psychologically on you). As recovery controls are applied (full left rudder followed immediately by a brisk forward motion of the wheel or stick), a glance at the turn and slip would probably show an "overshoot" as the needle and ball go past center. The airspeed is increasing, showing that the recovery is proceeding. *Bottom:* When the rotation stops, get off the rudder and start easing out of the dive. In this theoretical look at the turn and slip, the needle and ball are nicely centered, but in actuality you'll likely see oscillations to each side of center by both as the recovery continues. Immediately after leveling off, use the starter; the engine will normally start easily, as the airflow helps it windmill. Get the carburetor heat off after the engine is developing good power. Don't push carb heat off immediately after starting because you could have picked up ice in the spin under some conditions and you should give the engine (and carb heat) time to clean out residual ice.

Some experiments using 100- and 150-HP trainers with medium time on the engine have shown that to dive the airplane to get windmilling effects for starting requires approximately 1200 ft and 120–130 K (you discover after the recovery that the starter wasn't working). Consider those numbers an absolute minimum. Never deliberately spin an airplane to the point where the engine stops, because you could have trouble getting it started again. You can see the propeller slowing up from idle and should initiate immediate spin recovery. Usually the increase in airspeed during the recovery doesn't allow it to stop fully, so you can pull out normally and then open the throttle to cruise without going through a starting procedure.

which is a primary recovery control for the majority of general aviation airplanes, is blanketed by the horizontal tail wake, and it may not be effective for breaking the built-up forces of inertia.

In addition to the warning just made, consider the following: Never spin an airplane that you know is in the normal category, even if a placard against spins isn't visible. (The placard could have fallen off the panel and disappeared.) As will be shown, the manufacturer of a normal category airplane is only required to demonstrate a 1-turn spin, or a 3-sec spin, whichever is longer, with a recovery being made within 1 turn after normal antispin controls are applied. To repeat, the *incipient* spin normally consists of the first 2 turns after prospin controls are used. Using that as a criterion, you can see that the FAA doesn't require the manufacturer to demonstrate a *developed* spin for certification in the normal category. Manufacturers of multiengine airplanes aren't required to do anything concerning spins!

In some *Pilot's Operating Handbook*s for twins you may see a suggested method of spin recovery with a warning that the airplane has not been spun and the procedure is based on the best judgment of the manufacturer.

SPIN CERTIFICATION. While on the subject of what's required for spins by the manufacturer, it would be well to look at more details of certification, since your students will have questions concerning whether a particular airplane can be deliberately spun or not.

FAR 23.221 covers spin certification:

§ 23.221 Spinning.

(a) *Normal category.* Except as provided in paragraph (d) of this section, a single-engine, normal category airplane must demonstrate compliance with either the one-turn spin or the spin-resistant requirements of this paragraph.

(1) One-turn spin. The airplane must recover from a one-turn spin or a three-second spin, whichever takes longer, in not more than one additional turn after the controls have been applied for recovery. In addition—

(i) For both the flaps-retracted and the flaps-extended conditions, the applicable airspeed limit and positive limit maneuvering load factor must not be exceeded;

(ii) There must be no excessive back pressure during the spin or recovery;

(iii) It must be impossible to obtain unrecoverable spins with any use of the flight or engine power controls either at the entry into or during the spin; and

(iv) For the flaps-extended condition, the flaps may be retracted during the recovery, but not before rotation has ceased.

(2) Spin resistant. The airplane must be demonstrated to be spin resistant by the following:

(i) During the stall maneuvers contained in §23.201, the pitch control must be pulled back and held against the stop. Then, using ailerons and rudders in the proper direction, it must be possible to maintain wings-level flight within 15 degrees of bank and to the direction of turn in the most adverse manner. Power or thrust and airplane configuration must be set in accordance with §23.201(f) without change during the maneuver. At the end of 7 seconds or a 360 degree heading change, the airplane must respond immediately and normally to primary flight controls applied to regain coordinated, unstalled flight without reversal of control effect and without exceeding the temporary control forces specified by §23.143(c); and

(iii) Compliance with §§23.201 and 23.203 must be demonstrated with the airplane in uncoordinated flight, corresponding to one ball width displacement on a slip-skid indicator, unless one ball width displacement cannot be obtained with full rudder, in which case the demonstration must be with full rudder applied.

(b) *Utility category.* A utility category airplane must meet the requirements of paragraph (a) of this section or the requirements of paragraph (c) of this section if approval for spinning is requested.

(c) *Acrobatic category.* An acrobatic category airplane must meet the following requirements:

(1) The airplane must recover from any point in a spin, in not more than one and one-half additional turns after normal recovery application of the controls. Prior to normal recovery application of the controls, the spin test must proceed for six turns or 3 seconds, whichever takes longer, with flaps retracted, and one turn or 3 seconds, whichever takes longer, with flaps extended. However, beyond 3 seconds, the spin may be discontinued when spiral characteristics appear with flaps retracted.

(2) For both the flaps-retracted and flaps-extended conditions, the applicable airspeed limit and positive limit maneuvering load factor may not be exceeded. For the flaps-extended condition, the flaps may be retracted during recovery, if a placard is installed prohibiting intentional spin with flaps extended.

(3) It must be impossible to obtain unrecoverable spins with any use of the flight or engine power controls either at the entry into or during the spin.

Okay, that's the official word. One thing to note is that, contrary to popular opinion, a utility category airplane is *not* automatically certificated for spins. As stated in 23.221, to be certificated the utility must meet either normal or acrobatic spin requirements.

Suppose a manufacturer has designed a utility category airplane that meets the 1-turn spin and the within-1-turn recovery required for the normal category, but wants it to have spin-training capabilities. However, after 3 turns, deployment of the spin chute (under the tail) is required every time to recover. Although the airplane is in the utility category, it cannot meet the acrobatic (aerobatic) spin requirement and must have a placard in clear view of the pilot stating: "Spins Prohibited" (FAR 23.1567[b][2]).

During the general aviation airplane spin program at NASA Langley it was indicated that "all configurations tested, even those which *had unrecoverable flat spins when fully developed,* recovered from a 1-turn spin within 1 additional turn."

The italics are added, and the unrecoverable spins referred to were those that required the deployment of the antispin parachute for recovery. The normal aerodynamic controls were not sufficient for spin recovery under certain loadings and configurations of the test airplane. The point is that *the sooner you move to recover, the better off you are.*

Don't ever spin any airplane with people or baggage in the back. A rearward CG could mean that the spin would be unrecoverable. And even if you are lucky with having spun and recovered with people back there, they may be sick, dizzy, mad, or all of the above.

The developed spin with a rearward CG is normally flatter, and the controls may be ineffective for recovery.

For Pete's sake, don't deliberately spin an airplane that has a big fat placard against spins. If you do it and get out of a well-developed spin, write or call the manufacturer and tell how you did it.

SPIN THEORY— A REVIEW

While the following items give some ideas of what might be expected in spins for various design inputs, the problem still remains of later rigging (or hangar rash), *resulting in two airplanes of the same model having different spin and spin recovery characteristics*. Research has shown that comparatively minor alterations to various components have made significant changes in the spin performance.

But you have to start somewhere. The effects cited here are general and intended to give you information of value in discussing spins with your students.

■ **Tail Design.** This is the first factor usually looked at and in earlier years was considered *the major* (and sometimes only) one to look at as far as spin characteristics are concerned. Certainly it is a major factor in spins, and some general effects should be noted.

VERTICAL TAIL. The rudder, for most lighter airplanes, is still a primary recovery control for spins. As mentioned earlier, the airplane that is hard to spin may possibly be even worse when recovery time arrives because the effectiveness of rudder depends on how much dynamic pressure (uninterrupted airflow) it's getting.

The horizontal tail at a low position is considered to blanket a larger portion of the rudder, particularly in flatter spin attitudes. Therefore, that airplane would be *expected* to have lower rudder power and not as effective a recovery as, say, an airplane with a T-tail. The T-tail allows the full length of rudder to be in relatively undisturbed airflow, and the horizontal tail can act as an end plate, giving a greater aspect ratio effect and adding to the effectiveness of the rudder. Again, this is assuming that poor wing design or other factors haven't canceled out this advantage.

The T-tail appears to be a big step for spin recovery, but in earlier years some T-tailed jet fighters and jet transports with the engines at the rear of the fuselage encountered deep stall problems. The vertical location of the horizontal tail in those airplanes made it subject to interference from wing, fuselage, and engines. The elevator became ineffective at high angles of attack, and the pilot was unable to get the nose down for recovery. In effect the nose stayed high (in relation to the flight path) with a very high rate of descent occurring. Fig. 21-22 shows the idea.

Swept wings can produce pitch-up characteristics at higher angles of attack because the tip areas tend to stall first, moving the center of lift forward, which can also contribute to the problem (Fig. 21-23).

The swept fin and rudder may have more area blanketed by the horizontal tail and therefore be less effective in recovery than straight surfaces.

The rudder effectiveness in recovery also depends on how much area is *below* the horizontal tail and hence is not blanketed (Fig. 21-24).

The vertical tail is in a much better position for recovery in an inverted spin for most airplanes because the fin and rudder aren't blanketed by the horizontal tail (Fig. 21-24).

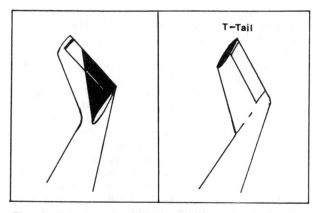

Fig. 21-22. A comparison of a conventional tail and a T-tail in a steeper spin attitude. The black area shows spoiled flow. Interference from the fuselage could also cause problems for the T-tail in some flatter pitch attitudes.

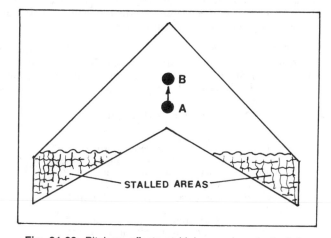

Fig. 21-23. Pitch-up effects at higher angles of attack for a swept wing. The center of lift of the unstalled wing is at *A*. At high angles of attack the tip areas are stalled, and this moves the center of the (smaller) lift area forward to *B*.

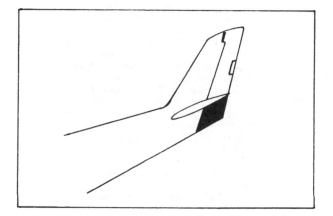

Fig. 21-24. The rudder area available below the horizontal stabilizer is important for spin recovery.

In an upright flat spin the horizontal tail and elevators may be stalled and ineffective (Fig. 21-25). A T-tail may be suffering from turbulence created by the fuselage at that pitch angle.

The *simultaneous* use of opposite rudder and down-elevator (or stabilator) for recovery may be ineffective because the down-elevator may further blanket the rudder. Using the

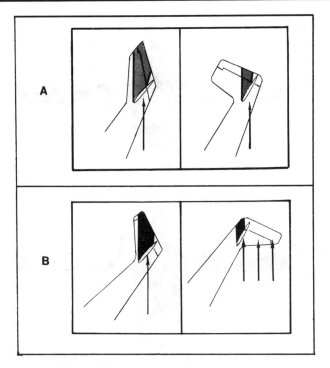

Fig. 21-25. (A) Vertical tail configurations and blanketing effects. (B) Vertical tail effectiveness in upright (positive-g) and inverted spins.

elevators *before* the rudder could delay or preclude recovery. Again, a general recovery procedure for single-engine light trainers is applying opposite rudder first and then immediately applying forward wheel briskly (depending on the airplane). In extreme conditions, "popping" the wheel forward too far too soon could result in the spin going inverted in some airplanes, but a *brisk* movement of the wheel forward is the best use of elevators for recovery for most airplanes. So the rudder is the initial recovery control in most light airplanes, with brisk forward stick or wheel movement to follow immediately.

The addition of a ventral fin will improve the flat modes (making them steeper) *for some airplanes,* but addition of dorsal fin area has shown little effect on helping recovery for several airplanes tested (Fig. 21-26).

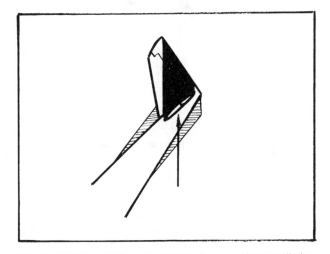

Fig. 21-26. The addition of ventral tail area will generally improve spin recovery characteristics, but the addition of dorsal fin area has little effect.

HORIZONTAL TAIL. The position of the horizontal tail surfaces has an effect on the reaction of the rudder (and fin), but the area of the fuselage under the horizontal tail is also a factor in spin and recovery. Spin fillets may be added to an airplane to increase this area. They are not effective for many airplanes but may be used for a particular model having comparatively mild problems of recovery (Fig. 21-27).

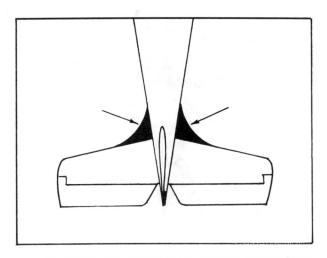

Fig. 21-27. Spin fillets (or more properly, antispin fillets).

The horizontal tail follows the wing as far as airfoils, aspects, ratios, and other measurements of efficiency and effectiveness are concerned.

One possible aid in getting good airflow past the vertical fin and rudder would be a chordwise slot near the stabilizer root. Some of the older airplanes of the 1930s, such as the Fleet trainer (a biplane), had a gap in that area, but whether that was deliberate for spin recoveries or a factor of manufacturing is not known (Fig. 21-28). The drag of such a slot would be high, and an arrangement like the one shown in Fig. 21-28B might help in this regard.

This is all theoretical, of course, since such slots have not been tested at this writing, and besides, as you learn more about spins you'll find that "predicted" response or design factors don't always work as predicted. You'll no doubt think up more ideas as your spin experience increases.

■ Fuselage Factors

AFT FUSELAGE. The cross section of the fuselage behind the CG could have an adverse effect on the airplane's reaction in the spin. The aerodynamic forces produced plus the comparatively long arm of the center of this cross section could have a significant prospin effect. Small boom (skinny) aft fuselage sections are usually "better" for spin recoveries. Fig. 21-29 shows some sample cross sections.

FORWARD FUSELAGE. Probably the first really good look at the effects of the cross section of the fuselage ahead of the CG came with the advent of jets, particularly the newer "long-nose" models. In a spin, the airplane is rotating around the center of gravity so that the long nose may be producing prospin aerodynamic forces at *certain rotation rates.* Fig. 21-30 shows the general idea for a jet. The idea could also apply to a lesser extent to a general aviation airplane. For example, Fig. 21-31 shows the cross section of the nose of a fictitious four-place long-nosed airplane that is having some minor spin recovery problems. Many factors have been examined (tail design, airfoil section, adding a ventral fin, and others), but the problem is still there. The situation in Fig. 21-31 is exaggerated to show a point, but lesser variations of this problem could occur in new designs.

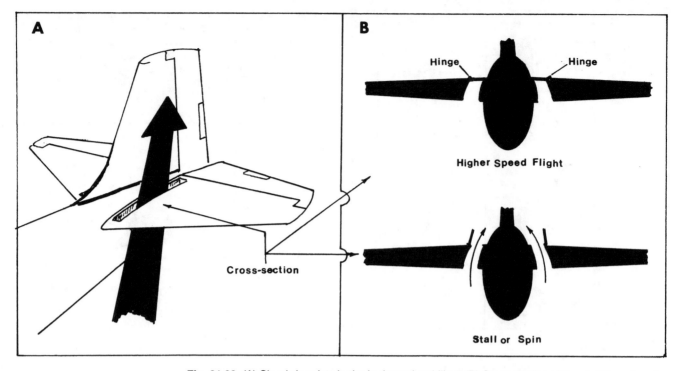

Fig. 21-28. (A) Chordwise slots in the horizontal stabilizer. (B) Cross-section of the stabilizer. Aerodynamically operated covers could be designed for the slots to decrease their drag.

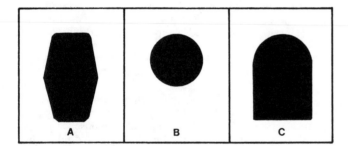

Fig. 21-29. Sample aft fuselage cross-sections.

Strakes may be of value on some airplanes and have no effect on others—depending on the other design factors.

■ Wing Design Factors

WING POSITION. NASA has indicated that all other factors being equal, a high-wing airplane generally has better spin recovery characteristics. It's possible, of course, that a low wing of a particular design could be more easily recovered than a specific high-wing design.

AIRFOIL. The wing is the driving force of the spin, and the C_L versus α curve of a particular airfoil/planform can be a significant factor in the airplane's spin characteristics. For instance, NASA spin research found in some instances that certain configurations (planforms) of the drooped leading-edge airfoil caused an airplane to enter a flat spin, even with tail configurations that did not previously have flat spin modes. The test airplane readily entered the flat spin without special control movements.

The drooped leading edge was designed to improve stall resistance and/or stall characteristics, but in certain types of installations it could adversely affect the airplane's resistance

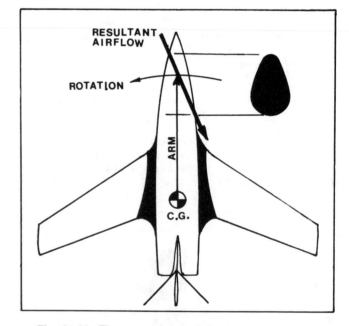

Fig. 21-30. The cross-section of the forward fuselage may result in large prospin moments at certain rotation rates. The resultant airflow is a combination of forward and rotational speeds for the point selected. The black area is a cross-section of the fuselage along the resultant airflow.

to spin and result in more critical spin and spin recovery characteristics. In this, it is like the airplane that is hard to spin and may have problems recovering, once it does spin. As an instructor, you may have little control over the design, but you could be called on by a home builder of little flight experience

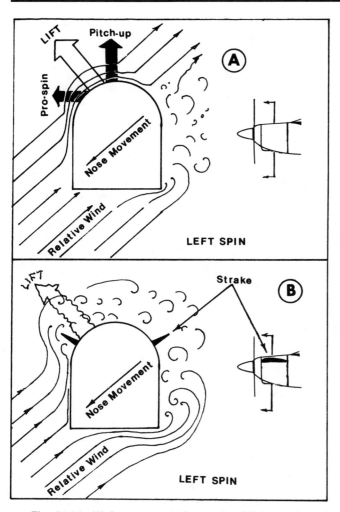

Fig. 21-31. (A) An exaggerated example of the prospin and pitch-up aerodynamic force components created by the forward fuselage of a fictitious light airplane. In this case only the airflow acting *perpendicular* to the longitudinal (X) axis is being considered. (B) Strakes are added to decrease or direct this force.

to do some unofficial stall/spin testing, and you should be aware of factors that could affect recoveries. A suggestion to you is to avoid spin testing of homebuilt airplanes, because the builder may have allowed inaccurate tolerances for incident angles or control deflections to creep in.

LATERAL CONTROLS. As indicated earlier, if you don't know exactly what effects aileron deflection may have on recovery, it's best to keep them neutral.

For instance, it would seem that ailerons *against* the developed spin would always make it worse because the down-aileron would move that inside wing more deeply into the stalled area (Fig. 21-2).

For two particular low-wing airplanes this is true. For two high-wing airplanes, applying ailerons *with* the spin caused them to speed up in rotation and then move up into a flatter mode. So, would you assume all low-wing airplanes spin worse with aileron against, and high-wing airplanes are just the opposite? *Not at all.* Another high-wing airplane might react exactly like one of the low-wing models just mentioned as far as aileron use is concerned.

Ailerons have been with aviation almost since the beginning of flight (the Wright brothers warped the wings for lateral control in their earlier airplanes), and the problem of

adverse yaw at low speeds has been the cause of much engineering work.

Differential aileron movement (more deflection up than down) has helped and other aileron designs make control better at higher speeds. The rudder is needed more in using ailerons at *low* speeds for two reasons: (1) adverse yaw *increases* with a decrease in airspeed (q) and (2) the rudder effect *decreases* with a decrease in q (dynamic pressure). So you're back to the scenario at the beginning of the chapter; the student has to consciously use much more rudder with that aileron he's using unconsciously to raise that wing in the stall.

There is the argument (and a valid one for some older airplanes) that the rudder alone should be used to raise a wing in or near the stall. In more recently certificated airplanes the lateral control requirement during the stall entry and recovery is that normal control usage be effective and this is the way it should be. Suppose, for example, you teach your student to use rudder alone to maintain lateral control throughout the stall entry and recovery. You've drilled him well and as he gets his private certificate he can raise a wing with rudder (no ailerons) very well. He flies for several years (studiously avoiding stalls since he never did "like" them) but one day gets into a cross-controlled (skidding) stall situation turning final. Does he go back to the earlier rudder training? No, he'll try to raise that wing with aileron (and rudder) as he's been doing in all the years of flying since he left you. It's better to design the system so that unconscious reactions will work in a bind. FAR 23 states it under *design stall requirements:*

> **§ 23.201   Wings level stall.**
>   (a) For an airplane with independently controlled roll and directional controls, it must be possible to produce and correct roll by unreversed use of the rolling control and to correct yaw by unreversed use of the directional control, up to the time the airplane pitches.
>
>   *   *   *
>
>   (e) During the recovery part of the maneuver, it must be possible to prevent more than 15 degrees of roll or yaw by the *normal use of controls.* [Italics added]

WING PLANFORM. Rectangular wings usually have a more roll-resistant stalling pattern (root stall first) than do tapered or elliptical wings. This can make rectangular-wing airplanes less prone to wing drop and *starting* autorotation.

■ **More About Mass Distribution.** Figs. 21-15, 21-16, and 21-17 plus the accompanying discussion of mass distribution show why there may be differences in spin recovery techniques. Not only mass loading along the wings, but added distribution along the longitudinal axis, have a significant effect in some "stretched" airplanes. For instance, the addition of a heavier engine in a particular airplane may require that the battery be placed in the rear portion of the fuselage to assure that the CG is moved back to the proper range for landing. Under the dynamics of the spin, the longitudinal distribution of mass (the dumbbell shape of *B* in Fig. 21-16) could affect the spin and recovery characteristics of the airplane.

Looking further at effects of mass distribution:

An airplane rotating in a spin is similar to a large gyroscope, and moments of inertia may exist around all three axes. When the weight is distributed along the wings, as in the case of light twins with full wing tanks and tip tanks, the recovery characteristics are not as good as when the weight is concentrated at the center of gravity. A six-place single-engine airplane with full passengers and baggage would have not-so-good spin characteristics. The actions from this type of weight distribution are sometimes called a "dumbbell effect" (see *B* in Fig. 21-16), which also describes the mental process of a person who would practice deliberate spins with a full load of passengers.

The mass distribution along the longitudinal and lateral axes may still result in a CG exactly where it's "supposed to be," but once a good rotation rate is set up there could be trouble.

Asymmetric wing fuel loading can have a decided effect on spin properties and recoveries.

The Cessna T-37 jet trainer carries a total of 309 gal of usable fuel internally in a fuselage tank (87 gal) and two wing tanks of 111 gal each. Sometimes fuel imbalance between the two wing tanks occurs and the Flight Manual indicates that under this condition the pilot can expect the spin to be oscillatory with varying yaw rates. The Flight Manual also notes that the heavier the airplane is, the slower the entry and spin rate. With very low fuel remaining, the spin tends to be initially flat and the nose remains above the horizon for as many as 3 turns. The left spin in the T-37 is more oscillatory and takes slightly longer to stabilize than the right one. The variation in spin actions from left or right spins can be noted in nearly every airplane, by the way. NASA checked (as did this writer) to see if a rotating propeller had an effect by stopping it and spinning the airplane, but there was no evidence to support propeller influence.

Along with the effects of unbalanced fuel in the T-37 spin, NASA found in spin-testing a low-wing general aviation trainer that a slight asymmetry in lateral mass distribution made a marked difference in the spin and recovery characteristics. Putting extra weight on one wing tip (you might read it as a fuel imbalance) made the spin toward the lighter wing flatter, with a higher rate of rotation, and it took over twice as many turns to recover ($3\frac{1}{8}$ turns versus $1\frac{1}{8}$ turns). As a general rule, even if the airplane you're using is approved for spins and has selectable wing tanks, you're less apt to be surprised in the spin if the fuel is balanced.

■ **Relative Density.** This subject was noted back in the section on Recovery Factors, comparing a J-3 and a C-152 volumes and weights, but here's a little more detail: The spin recovery characteristics of an airplane depend on how effectively the aerodynamic controls are able to overcome the moments of inertia existing in a developed spin. A heavy airplane in a spin at a high altitude would have problems in recovering. In fact the recovery might be delayed until the air is dense enough to make the aerodynamics controls effective, which could mean no recovery from the well-developed spin even when it reaches low altitude. The airplane could move into the ground before recovery could be completed. For the altitude variations and ratio of mass to wing area and span of most general aviation airplanes, it is not a critical factor; however, if the gross weight of a particular model were to be significantly increased over the years, or the tail area and other factors were to be changed, possible spin problems may appear.

For your more technically minded trainees, the equation for relative density is

$$RD = \frac{M}{\varrho Sb} = \frac{\text{Mass}}{\text{Air Density} \times \text{Wing Area} \times \text{Span}}$$

The wing area and span are in square feet and feet respectively. (Rho, ϱ, is atmospheric density in slugs per cubic foot and mass is the weight divided by 32.2.)

The factor that can change is, of course, the air density, ϱ. An airplane that weighs 1700 lb and has a span, *b*, of 35 ft and a wing area of 175 ft² would have, at *sea level,* $\varrho = 0.002378$, a relative density of 3.6. At a density-altitude of

10,000 ft ($\varrho = 0.001756$), the relative density would be 4.9. Looking at it simply, *added weight delays the spin recovery* because the aerodynamic controls have to overcome a greater inertia.

For general aviation trainers, altitude, ϱ, is not a factor as it is with jets, which may be operating from 40,000 down to 8000 ft, a good difference in air density. Variations in mass and the wing factors *S* and *b* are more important here.

In discussing this subject with a trainee who isn't interested in math of any kind, you might use the analogy that the relative density is a comparison of how dense (per cubic foot) the airplane is compared with the air surrounding it. Looking at it in nonmath terms, the trainee might be able to see that it's a trade-off between airplane weight, span, wing area, and density-altitude. For instance, an airplane with small wing area and short span would have a relative density at sea level as great as an airplane weighing less and/or with larger wing factors at some higher altitude.

The point, again, is that if you did plenty of spinning some years ago in lighter trainers, don't think you can take up where you left off in newer airplanes without thinking about the different spin characteristics. (Better yet, always get some dual with an instructor well experienced in spinning that *particular* airplane.)

■ **Some Added Points.** Another factor that affects spins and recoveries is the center of gravity position, mentioned earlier in the chapter. An aft CG could make a normally docile airplane difficult to recover. After you get experience in spins in one airplane, you can tell the difference in time required for recovery even by moving your seat back one notch.

And even another factor touched on earlier was power use for recovery. Generally, for most single-engine airplanes with low-position horizontal tail configurations, power tends to pitch the nose up and delay recovery. But if nothing else seems to work, the dynamic pressure added by the slipstream may help the rudder in the recovery. NASA analyzed four jet configurations and found that added thrust had small effects on spin recovery in general but did have a slight effect *toward* recovery on relatively nonoscillatory spins. For oscillatory spins the results were inconclusive.

In the case of twins, the use of asymmetric power *against* the spin is considered an aid to recovery, but the general recommendation is to bring both throttles back to idle to avoid the possibility of opening the wrong throttle.

In some airplanes in extended spins you won't have power available. (See Figs. 21-20 and 21-21.)

INSTRUCTING SPINS

Earlier in the chapter spins and spin theory were covered generally. As a flight instructor, you also need to be aware of the step-by-step points in teaching them.

Your approach will naturally depend on whether the individual wants a brief demonstration and practice for his own peace of mind for general flying, or whether the training is for aerobatics and/or a more in-depth look at the subject. You may want to offer a short spin course as a part of your instructing program, where you'll talk an hour on the background and do a few and then come back to review, followed by further discussion and another spin session.

■ **Preparation and Explanation**
PRESPIN GROUND CHECK AND BRIEFING. There will probably be occasions when somebody will bring his own airplane to you for spin instruction. Assuming that the airplane is legally spinnable and even a type or model you've spun many times before, as mentioned earlier, control cable tension may

have slackened or other factors may cause the airplane to have different spin and recovery characteristics from the one(s) you've had experience with. One quick check is to have the pilot sit in the cockpit on the ramp or tie-down spot (with the engine dead) and strongly hold the stick or control wheel as far forward as he can, while you go back and manually try to raise the elevator. If the elevator can be moved from the full-down position under such conditions, aerodynamic forces may prevent getting the required down-elevator for spin recovery, because of cable slackness.

The same test may be made with the rudder for some airplanes, but with others this couldn't be done because of direct connections of rudder and nosewheel.

Parachutes. In spin training (or any aerobatics) it's important to brief the student on the procedures for bailing out of the airplane (assuming, of course, that you are both wearing parachutes—if he or she is a CFI candidate, parachutes are not required, as noted by FAR 91.307[d][2]).

If parachutes are required for that spin flight, a good briefing at the airplane on how to jettison doors or open the airplane canopy (or canopies) and the recommended methods of getting out, plus parachuting background, should be covered thoroughly. If it has been several days or more since the last spin or aerobatic flight, the procedures should be reviewed.

It's very important that you state the minimum altitude for bailout and that there will be no jokes about the subject in the airplane. If you say, "Bail out," it's no drill. The minimum AGL altitude for bailout depends on the particular airplane type, but you should have it firmly in mind.

The trainee is to leave first, but in your preflight briefing you might add that after you give the word he has about 3 sec to comply or he'll be pilot in command while you go for help.

Be sure the trainee has a good understanding of the maneuver before you do it, and if possible let him do the spin and recovery from the beginning. If you've covered the subject well, he should be able to do a reasonable spin entry, spin, and recovery the first time without too much help from you.

Don't require that the trainee's first few spins be precise; use them as an introduction to the general idea. You'll oversee his entry and tell him when to recover. Warn him that he isn't to continue holding opposite rudder after the rotation is stopped, a common error for most new people.

What—*The spin is an aggravated stall with autorotation occurring.* You might draw Fig. 21-8 on the board and discuss it. The trainee can see that the "lesser" stalled wing has less lift and more drag, which tends to maintain the rotation. The type of airplane you'll be using will probably have to be held into the spin (if it's properly loaded). An airplane model will be a good way to demonstrate the actual spin entry. You should explain that from an aerodynamic standpoint it usually takes about 2 turns for the spin to develop.

Why—The spin is a good confidence building maneuver because it is considered to be a dangerous and often deadly action. The trainee may subconsciously, or consciously, feel that the spin is the hairiest maneuver an airplane can get into and that his chances of recovering from one are low indeed. Conquering this fear can be used by the flight instructor to increase the trainee's feeling of well-being in all phases of his flying as mentioned earlier. Spin training also teaches the trainee to recognize an approaching stall/spin and to see the airplane's resulting actions if the situation is not corrected quickly. When instructing spins, don't tell hangar stories and increase the trainee's anxiety level. After he has done a few, his confidence will increase; hasn't he conquered the deadly

spin? Afterward he'll probably mention it in the airport office—not directly, but he'll probably say something like, "I didn't think we'd get it to spin on that fourth one!"

When—*When* to introduce a maneuver can be as important as *why* it's introduced or *how* it's done. It's suggested that in an aerobatic course spins be introduced as soon after the loop as possible so that the trainee will have had experience in a fairly radical nose-down attitude and yet will have enough time remaining to practice a number of spins during the course.

In addition to deciding when to introduce the spin in an aerobatic syllabus, you should consider when to do them *during the flights,* once they've been brought into the picture. With trainees who are doing very well or who might have had spins before, you might demonstrate one at the end of the first flight, but spins will more likely be saved for the end of the second flight, since spinning is a good way to lose altitude. You'll find that doing spins in the middle of a flight in a low-powered airplane can cause more climbing back up to altitude than is practicable.

However, you'll run into people taking aerobatics as a safety course, so spin recoveries will be a major factor in their thinking, but who unfortunately tend to get queasy as a flight progresses. In these cases you may want to do spins early in the flight to ensure that spins are included. Also, as you teach aerobatics you'll find that some pilots have a sickness problem with aileron rolls (or loops) but not with spins (and vice versa) so a change in the timing for spins in a flight is worth a try.

You might plan on *1* spin each direction at the end of each flight (more, if possible—depending on the trainee—after the maneuver has been introduced). These will give him a fair amount of experience with spins by the time he's finished the aerobatic course.

How—The entry procedure, of course, depends on the airplane model and the manufacturer's recommendations. Some require the use of opposite aileron during some part of the entry; others use a normal entry of just full rudder in the direction of spin, as the stall occurs.

RECOVERY

To review the explanation of the spin entry and recovery, step by step:

1. Clear the area carefully. Look in all directions, not just below; an airplane at a lower altitude could be flying into your spin zone. Be prepared to recover *at least* 3000 ft above the surface—higher is better.

2. Use carburetor heat as recommended by the manufacturer and close the throttle.

3. Set up a straight-ahead, power-off stall with the nose at approximately the landing attitude or slightly higher.

4. As the stall is approached (the horn or light comes on), move the wheel or stick smoothly back at a slightly faster rate and simultaneously apply full rudder (smoothly) in the desired direction of spin. Hold the wheel full back and hold full rudder deflection throughout the spin. Sometimes you might give the airplane a burst of power to help salvage an entry. Explain beforehand that this might be the case. You'll find that most trainees are naturally on the shy side in their spin entries, and a spiral often results—until they are reminded to recover and try it again. Some airplanes may go slightly inverted during the first ¾ turn of the entry but will establish the proper attitude (Fig. 21-19). Tell the trainee about this during the preflight briefing so he'll know it's normal.

5. Keep a count of the turns by half turns (1 turn, 1½, 2, etc.). The first few spins should be of approximately 1½–2

Fig. 21-32. The first 2 turns of a spin and recovery (incipient spin) as seen from above. (1) The area is cleared, the airplane is lined up with a road, and the power is reduced. The nose is raised to stall attitude, and as the stall is approached, left rudder and full up-elevator are used. Some airplanes may require opposite aileron at the entry; others may need a burst of power (then reduced to idle) at that point. (2) The airplane is in the transient condition; some airplanes may move over to a slightly inverted position during this phase. (3) The spin is developing. (4) The rudder and elevators are held in the full deflected position. (5) Note that the path of the airplane is quite different from the spiral in that the airplane is flying a helix angle and is *also* rolling and yawing about its center of gravity. (6) The recovery may be started here or at (7), depending on the airplane being used. (8) The opposite rudder and forward motion of the control wheel starting to "bite." (9) Rotation has stopped and the angle of attack is back in the flying range. (10) The rudder is neutralized and the pull-out is started. (11) The airplane is brought back to straight and level cruising flight. Be sure that the recovery is complete at least 3000 ft above the surface.

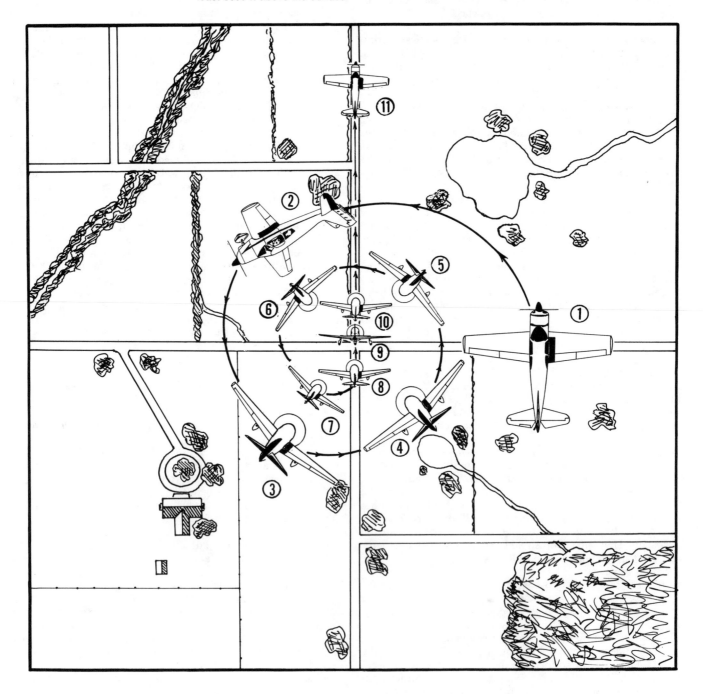

turns (incipient spins). This allows the spin to get roughly started but does not carry the matter so far as to cause the trainee to lose count of the turns. Later you can have him do 2½-turn and finally 3-turn spins, if he's progressing. You should hold the line at a maximum of 3 turns for the nonaerobatic trainee, because even though he becomes more relaxed with spins, his anxiety level may start rising rapidly after 2½ turns.

Emphasize in the preflight briefing and in the air that the trainee should not look along the top or side of the cowling during the spin but farther "up" or out toward the horizon. Checking a road or a similar distant feature (he'll see a section of it twice per turn) and counting the turns will help allay queasiness. In spins trainees tend to count as a turn each time they see the road so they'll be giving you a count of 2 turns for every actual spin turn.

During the spin itself, some people at first will tend to relax back pressure and/or rudder pressure, and so a spiral results in some airplanes. It takes quite a bit of concentration to remember to hold these pressures when the airplane is pointed at the ground and "whirling," although most trainees will be pulling hard back on the wheel. Fig. 21-32 shows a different view of a spin.

6. Start the recovery by using opposite rudder. *In any recovery, as soon as rotation has stopped, neutralize the rudder.* (The trainee will tend to continue using opposite rudder even after the airplane has stopped spinning and is in a dive.) If your airplane needs only neutralization of the rudder to recover, say so, but mention that this is only for a particular airplane type; others may need full opposite rudder *held*.

7. At this instant, some older and lighter airplanes have stopped rotating, but they are still in the stalled condition even though the nose *appears* to be pointed straight down. The back pressure must be relaxed or a brisk forward motion exerted on the wheel so that the airplane is pointed the way it's going. Using a model, you may show that this is also the situation in the straight-ahead stall (Fig. 21-33).

Again, it's generally recommended that the ailerons be neutral in the spin itself and in the recovery, to avoid complications in yaw and roll.

To review, the best *all-around* recovery procedure is to apply full opposite rudder followed immediately (or ¼–1 turn later, depending on the manufacturer's recommendations) by brisk forward movement (maybe to full forward, as required) of the stick or control wheel (see Fig. 21-15B). This form of recovery works on later types as well as older airplanes. (It may be a trifle "brisk" for a J-3 or Champ but will stop the spin.) The earlier recovery of using rudder and waiting until the rotation stops may not work on most newer types.

The back pressure is sometimes relaxed too much (or the wheel is moved forward too abruptly past the neutral position), and you both get light in your seats. Warn the trainee beforehand about this, because if it happens unexpectedly he could think that the airplane suddenly has undesirable characteristics. He probably won't realize that he did it.

8. Ease the airplane out of the dive. Don't get a secondary stall but don't dally either, since airspeed will be building up at a fast rate. Expect to pull 2½–3 g's during the pull-out.

9. Return to cruise airspeed and power.

■ **Demonstration.** It was stated earlier that the best method is to have the trainee do the spin himself from the beginning, but for a particular individual, you might decide that it's better for you to demonstrate the first one. Review the procedure as you clear the area and talk about it as you do the spin. Since the spin is a mechanical maneuver, it won't hurt to have him follow you through on the controls. Talk calmly as the procedure is followed. There's something about a spin that tends to get you talking faster and faster as you go around. You will have to deliberately slow down your rate of chatter — and it isn't easy.

One demonstration of 1½ turns of a spin (incipient) should be sufficient. Let the trainee do the next ones.

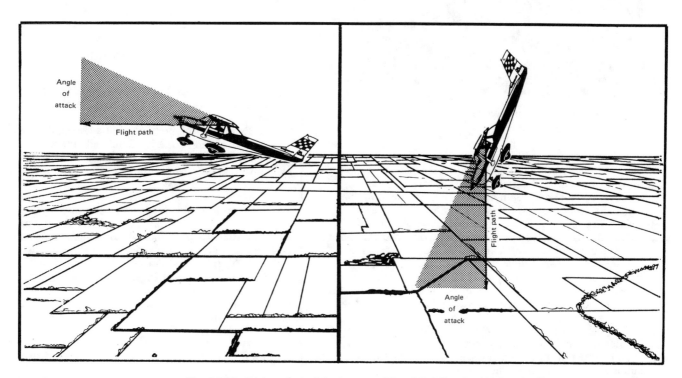

Fig. 21-33. Angles of attack in the normal "level-flight" stall and in a spin. To recover from either, you have to point the airplane where it's going.

■ **Practice.** The trainee will do some palm wiping before he does that first spin himself. You'll also note that clearing the area thoroughly has become a newly acquired virtue (it takes more time that way). There will likely be some throat clearing on his part, and he may decide the area should be cleared *again,* just to make sure. You can expect to see these delaying tactics—he may not realize he's doing them—and a little humor will ease the situation. ("I used to clear the area eight or ten times before doing a spin.")

His first entry may be halfhearted, with the result that he enters a spiral. You can help by reminding him that the controls must be held fully deflected. (You may give physical help, as well.)

Your own indications of the spin turning into a spiral will be best shown by an increase in airspeed, if you happen to be looking at the indicator at that time. However, you'll be able to sense immediately that the spin is no longer with you after you've had experience in a particular airplane, and you can stop the spiral before it becomes a problem.

The trainee's tendency will be to stare over the nose and not see anything but a blur of earth as the spin gets going, so remind him to keep his eyes moving (Fig. 21-34). He will get better oriented as more spins are practiced. As soon as he has the idea of the spin in his mind and can enter and recover with authority, start requiring recovery on a specific heading, ±10°. You might ask for 1½, 1¾, 2, or 2¾ turns, etc.

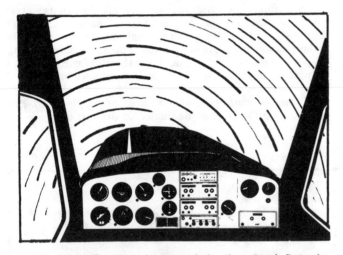

Fig. 21-34. The ground as seen during the trainee's first spin.

As you teach more spins you'll note an interesting phenomenon: In nearly every pull-up from the recovery dive the trainee will bank in the direction of the spin rotation; that is, after recovering from a *left* spin he will unconsciously pull up in a *left* bank (and vice versa), even with a clear horizon for reference. Without going into a lot of physiological detail, the reason for this is what you learned during your instrument training about inner ear functioning. The fluid and hairs at the beginning of the spin indicate, for instance, that turning (rotation) is occurring. After a period of turning or spinning at a constant or near-constant rate, the fluid accelerates to the same speed as the semicircular canal wall and the sensation of turning ceases.

When the rotation is *stopped* in the spin recovery, the fluid deflects the hairs in the opposite direction (indicating a *right* roll or turn) so the pilot unconsciously corrects for that by banking left, even though his eyes should be telling him that the airplane is already banked in that direction. The high rotation rate in a spin makes this more of a problem than in a

gentle turn. Spins under the hood, which tend to aggravate the situation, will be covered in Chap. 24.

Stay alert during the spin, even though the trainee has been doing well. He's only done a few and might have been lucky during those. Be ready to move fast to straighten things out.

Okay, suppose you do need to straighten things out. You tell him to recover and he still sits there with full rudder into the spin and the wheel all the way back, staring hypnotized at the ground whirling and coming up. You wouldn't start a discussion here on the rate of descent, spin theory, and history of spin accidents. You would have set the stage for this possibility during the preflight briefing. Emphasize there that you may verbally nudge him during his recovery by saying "rudder," meaning use that opposite rudder, and "push," telling him to get that wheel or stick forward. (How *far* forward for your airplane should be covered.) Of course, a calm classroom discussion is quite different from being in a spin.

Don't shout or act excited when he's not doing a proper recovery. An "Okay, Joe, *push!*" with emphasis but with a chuckle (if your mouth isn't too dry) will generally keep things in control. Brief him that if he's not moving fast enough you may move on the rudder and/or wheel to show him the right way. Prepare him for the fact that you may have to take over the recovery to save altitude; he should expect an "I got it!" if necessary. *Your* excitement may tend to tighten him up and hold prospin controls. There is not enough room in the airplane to get his attention as shown in Fig. 2-2.

You may have to raise your voice if the first method doesn't seem to be working, and a "*Push!*" as you move the wheel forward will make him respond with you. You probably won't have any problem getting him to apply opposite rudder; it's pushing that nose *farther* down that goes against the grain, and that's usually the point where your assistance may be needed.

In instructing spins, altitude is the greatest safety factor. At higher altitudes both you and the trainee will feel a lot better and not quite so rushed.

It can't be overemphasized that the trainee's attitude, his belief in your competence, is the main factor in his reactions in spins. And this attitude is initially and primarily established on the ground before that first spin session; a preflight briefing is vital so that you and he can get a sense of how the other will act in the air.

Common errors in doing spins include these:

1. Not getting a clean spin entry; spiraling from the beginning.

2. Relaxing back pressure after a good entry; letting a spiral get started with a resulting high airspeed.

3. Failure to close the throttle (to aid recovery) after an inadvertent power-on entry from an aerobatic or other maneuver.

4. Holding opposite rudder even after the rotation is stopped, imposing side loads on the fin and rudder.

5. Not relaxing back pressure or not moving the wheel or stick forward enough after input of opposite rudder; the stall is not broken and buffeting results, with the airplane still losing altitude.

6. Too much forward pressure, or holding it so long that negative g's result.

7. Pulling back too soon after the stall is broken, getting a secondary stall and possibly a progressive spin. This is close to error 5.

8. Hesitation in pulling out; building up the airspeed too high and putting excess stress on the airplane.

MORE ON INTRODUCTION OF SPINS. You may wonder why it was suggested that the trainee do a 1½-turn spin on his first try. Why not start out with, say, a spin of ½ turn, then do 1 turn, then 1½, etc.? It seems that the first part (entry) of the spin is the hardest on the individual; more than the spin itself, that initial rolling and nose dropping causes many people to clutch. If you did a ½ turn and then recovered, during that climb back up and clearing the trainee could have a change of heart about the whole idea. ("If that little amount was bad, it's obviously going to get much worse, so let's forget it.") You'll find that many trainees, after going through a turn or so, will be relieved; it wasn't as bad as they thought. In fact, you'll run into people who like the spin itself better than some stalls.

Of course, you aren't encouraging them to hold the airplane into a 1½-turn spin in an accidental situation when they should avoid the spin in the first place. You'll teach them the best method of avoiding the spin and the most expeditious way to stop it in the very early stages. With some people the segmented approach might be best, but for those who have built up their courage and girded their loins for the spin, the step-by-step introduction would be akin to cutting off a dog's tail by sections.

After they've recovered from the 1½-turn or even a 2-turn spin (with your moral help and maybe a little of your physical help), take a break. Don't overdo it. You might even decide that one spin that first time is plenty. As indicated earlier, after the trainee has been introduced to it, on following flights in an aerobatic course try to get in at least one spin in each direction at the end of each session. Remember, too, that Fig. 20-1 also applies to spins, whether they are being done as a confidence building maneuver or as part of an aerobatic course.

This talk about introducing spins doesn't mean that the student is suddenly thrown into spins without some thorough stall training as a lead-in. Power-on and power-off stalls in different configurations, such as takeoff and departure stalls, approach to landing stalls, cross-control stalls (skidding), and accelerated stalls should have been part of his training and should be reviewed as necessary before starting spin training.

After the trainee is comfortable in spins you should have him recover (1) just as the airplane rolls into the spin and (2) at ¼ and ½ and 1 turns.

■ **Evaluation and Review.** After a half-dozen spins the trainee should be able to enter (and recover from) a normal spin situation safely without any help from you. If you plan on letting him do solo aerobatics, make sure he has had experience in power-on as well as power-off entries—from various attitudes.

He may have questions concerning inverted spins, and you should make it clear that the airplane's attitude at entry has nothing to do with whether the spin will settle down to be normal or inverted. (The usual fear is that if the airplane is stalled in the inverted position, a resulting spin will always be inverted.) Explain that if the stall is the result of moving the wheel back toward the pilot and holding it there, the spin will settle down to be a normal one, even if the whole process was started while inverted at the top of a loop. (This requires that he continue holding rudder and up-elevator, or the spin could turn into a spiral.) If the prespin stall is a result of the control wheel being moved *forward,* the spin will be inverted, even though the airplane was in an upright flight attitude at the time (an unlikely possibility—getting into an inverted spin from an upright attitude—even though it could happen). An inverted spin is most likely to begin with the airplane in a nose-high, inverted attitude, as at the top of a sloppy Immelmann with excessive abuse of forward elevator and rudder input.

■ **Various Spin Entries.** The straight-ahead, power-off entry is not the usual way that an accidental spin happens. You'll want to show over-the-top and under-the-bottom entries (explaining under what circumstances each could occur), and for your aerobatic students you should demonstrate and let them practice upright (positive-g) spins from the top of loops or spins resulting from an "extended" snap roll. Another good demonstration for some airplanes (and some trainees) is a "feet-on-the-floor" spin entry.

The point of these various types of entries is to show that the *recovery* is basically the same for the developed upright spin. Pilots don't realize that after a certain number of turns, most lighter airplanes have forgotten just how they got into it. Even an entry from a snap roll will usually, after about 3 turns, settle down to a normal spin.

Following are some brief notes to pull the various types of entries together. You won't teach all of your trainees all the entries; certainly a student pilot shouldn't be spinning out of a loop or snap roll, but look them over to see where they might fit into various phases of your flight instruction.

It's assumed in every case that the airplane is properly certificated and loaded, that you have plenty of altitude, and that you clear the area before each spin entry.

OVER-THE-TOP SPIN ENTRY. This is the stall in which the higher wing stalls first; the result was discussed earlier in this chapter and shown in Figs. 9-6 and 11-3.

The takeoff and departure stall discussion in Chap. 9 indicated that while you could usually expect the higher wing to stall first, some airplanes and/or some situations could fool you.

In this case you'll let the trainee know that you'll get the spin started by injudiciously holding top rudder (and maybe a little aileron into the bank, to help the slip) as back pressure is used. You'll normally get a better demonstration by turning to the right, with power on.

As the airplane rolls and the high wing drops, hold the wheel or stick back. Move in with *full* rudder into the spin if you didn't already have it in all the way. Neutralize the ailerons and close the throttle smoothly. Hold it in the spin for about a turn and a half and then use normal recovery procedures.

One problem is that you are carrying a high power setting because it's often difficult to get a good spin entry over the top from idle, and when you close the throttle you may get rapid engine cooling. You don't want to be carrying carburetor heat at climb power at the beginning—this also is bad for an engine. Think about this for your airplane and work out a system that helps take care of the engine and still gives a good spin entry.

UNDER-THE-BOTTOM SPIN ENTRY. This is basically the continuation of the cross-controlled stall (rudder into the turn, aileron against). After the stall break and roll you'll neutralize the ailerons, close the throttle, and allow about a 1½-turn spin. The recovery is standard. Doing the spin to the left and carrying some power usually helps the entry. To make it totally realistic, flaps would be used as on an actual approach, *but* deliberate spins with flaps extended are usually prohibited by the manufacturer because of adverse recovery characteristics. Also, as indicated in Chap. 11, the flap-extended speed could be exceeded with the flaps down if somebody isn't quick to get them up.

NO-RUDDER SPIN ENTRY. The common belief is that the rudder is the only control that can develop the spin, and the student may get complacent about its use. That is, he may figure that as long as he doesn't use the rudder too much in the stall regime, he's okay. This was mentioned briefly in the scenario in which the student doing takeoff and departure stalls used full aileron and a *little* rudder in trying to pick up that down-moving wing.

With some airplanes you may demonstrate that using co-ordinated controls with *proper rudder* is the best route to take in lateral control near the stall. One demonstration might be to set up a takeoff and departure stall to the right. In this case, however, keep both feet off the rudder pedals (but be ready to use them as necessary—in other words, *don't* pull your feet back and set them flat on the floor). The right bank will allow "torque" to help roll the airplane to the left at the break.

As the break and roll occurs, use full right aileron in an "attempt" to raise that wing. With certain airplanes, a good left spin results with no rudder input to get it started. You may also find that here, as in other demonstrations, things don't work as discussed on the ground; the airplane may roll to the left and set up a shuddering low-speed, shallow-banked spiral—but no spin. Well, practice makes perfect (it says here).

HANDS-OFF RECOVERY. It was indicated earlier that a good demonstration of recovery would be to have the trainee set up a power-off, wings-level spin entry and at 1–2 turns have him quickly remove hands and feet from the controls. (You'll have *your* appendages ready to move, though.) You can count the turns and prebrief him to let go at "TWO" or "NOW" or whatever signal you both agree on.

The response of one airplane, hands off, at 2 turns in a left spin is a movement of the control wheel to the left (the airplane gives itself left aileron) and then a recentering of the wheel as it moves forward and the spin recovery is completed. As noted, if the airplane is trimmed nose up before the spin, after the recovery it will pull out of the dive by itself. You'll then adjust power and trim for cruise, climb, etc. It's a good confidence builder if done properly and shows that in the incipient spin the trainee is better off letting go than holding the wheel full back.

SPIN FROM A LOOP. If you are teaching aerobatics, have the trainee spin the airplane from the top of a loop at some time during the training, so he'll know what will happen if he pulls back too hard to "tighten it up." The spin will be an upright one, even if the airplane goes through some wild gyrations before settling down.

Fly a normal loop and as the airplane reaches a 30°-nose-up (inverted) position have the trainee pull the stick or wheel all the way back while moving in with full (left or right) rudder. Agree beforehand that you'll say "NOW" or something like that when he is to start to spin. Things will be happening so fast that he'll have to apply prospin controls without delay. He is to hold this control input so that a spin occurs. For recovery, the throttle is closed and normal procedures used. Fig. 21-35 shows the idea.

SNAP-TYPE SPIN ENTRY. This one usually occurs when the aerobatic trainee is late in applying recovery controls in a snap roll (with cruise or climb power). The airplane continues the roll with the nose gradually dropping as the spin progresses.

You'll note in most cases that the power is tending to keep the nose up, and the first turn or so can make you think that the airplane is going to stay flat.

Close the throttle, make sure the ailerons are neutral, and make a normal recovery.

You can work out other combinations of spin entries, but naturally you would not do the loop or snap roll spin entries except in an aerobatic airplane. You should get instruction in these procedures before going out on your own and "discovering new knowledge."

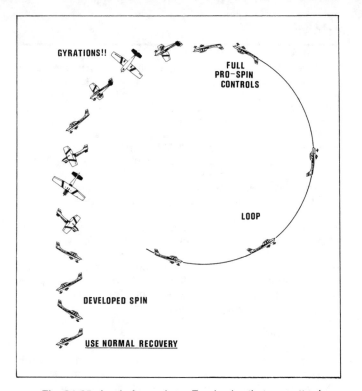

Fig. 21-35. A spin from a loop. Emphasize that no matter *how* the spin was entered, the recovery for the Aerobat will be the same after 2 turns. One suggestion: If the spin entry seems laggardly with the normal 120-K loop entry speed, you might start the loop at 130 K to get a crisper break.

SOME PERSONAL NOTES ON TEACHING SPINS. These notes are based on spinning 29 different types and models over a 47-year period but are primarily based on logged spins in six C-150/152s:

N9880J—624 spins
N4896A—17 spins
N61205—160 spins
N6534L—27 spins
N7349B—168 spins
N7557L—3693 spins

Some of these spins were up to 25 turns. At no time during the 4689 spins of various turns was there any problem in recovery using the *Pilot's Operating Handbook* procedures.

Since earlier it was noted that even airplanes of the same make and model may have different spin and/or recovery characteristics, the references to different recovery or entry procedures are given by a specific N number here.

On dual flights, 20-lb parachutes were worn by both occupants and a *maximum* limit of 210-lb trainee weight was used in Aerobat N7557L (my weight was 160 lb) for a total of 410 lb without fuel. The useful load was 479 lb, allowing 69 lb, or just under 12 gal of fuel. The practice area was 7 mi from the airport and for most flights 11–12 gal of usable fuel were carried even at lower trainee weights. For aerobatic flights fuel consumption (Aerobat) was planned at 7.0 gal per hour. The fuel level was checked by dipstick and refueling was done every flight. The flights were directly over an uncontrolled airport with the home airport in sight and FARs 91.303 and 91.307 were always followed.

The weight and balance data showed that the weight approached, but did not exceed, the maximum and the CG was well forward in the envelope.

For *any* aerobatics, minimum weather requirements were a 5000-ft ceiling and 5 mi visibility; for spins a 6000-ft ceiling was required.

No spins were started below 4000 ft AGL (1–2 turns);

extended spins were started at 7000–8000 ft AGL, allowing for a 200-ft per turn loss. All spins were recovered above 3000 ft AGL.

In the spin recoveries almost all trainees used opposite rudder, but a significant number had to be reminded to push the wheel forward more. In these cases, several additional turns resulted because of this hesitation and up to an additional 2000 ft were lost. (That's why altitude is so important when *you* do spin training.) In some instances, you may have to keep reminding the trainee or "assist" him to move the wheel or stick forward.

Some trainees were hesitant to pull out of the postspin dive and had to be reminded, as was mentioned earlier. (One not-so-serious theory is that they are so glad to be out of the "whirling" condition that they sit there with the nose down, not realizing that the recovery is not completed.)

Be prepared also to stop a too-enthusiastic forward movement of the control wheel. Usually, after the trainee has been reminded several times that more forward wheel is necessary to stop the spin, he is determined to get it right and both occupants are strained against the harness. You can remind trainees that if they get light in the seat during a recovery, the airplane is out of the spin and the wheel should be moved rearward for the pull-out.

In N7557L (3693 spins) a 2-turn hands-off recovery was demonstrated starting the *left* spin with full power, closing the throttle at 1 turn and at 2 turns releasing the wheel and getting off the rudder. The airplane was trimmed nose-up before the entry for "show business" purposes (this was always explained to the trainee) so that the airplane would pull out of the dive after the spin was stopped (still hands off) without an excessive loss of altitude. If there was a delay of more than ¼ turn of closing the throttle *or* releasing the prospin controls, the airplane would not recover on its own but the wheel would turn left into the spin and ride back slightly, requiring the recovery steps for a developed spin unless the back pressure had been relaxed during the entry process and a spiral had started to develop.

The airplane (N7557L) had a tendency to pitch down at 4 turns with a temporary increase in rotation rate at 5 turns. The hands-off recovery at 2 turns was demonstrated once, followed by trainee practice. The airplane would *not* recover hands off from 2¼ to 3½ turns, but if the trainee delayed releasing the controls until 4 turns, in many cases the airplane would drop its nose and recover, leaving the writer to reexplain that in the incipient spin (0–2 turns) the airplane would recover hands-off and at the 4-turn point *for this airplane* it would *sometimes* recover but at no other point in the developed spin.

In developed left spins in N7557L, opposite (right) rudder *alone* would not alter the rate of rotation or start a recovery. Making sure that the throttle was closed and *releasing the wheel* while holding full opposite rudder resulted in a faster rotation because the ailerons, when freed, would move to a prospin condition. The C-152 ailerons were still effective in the spin and the "automatic" prospin deflection increased the roll rate in extended left spins. The control wheel also tended to move slightly rearward, but there was never a problem in recovery.

Some of the C-150s spun better to the right than to the left (see Fig. 21-13 again) whereas the C-152s spun by the writer entered and spun well to the left, but tended to spiral out in a right entry. This was noted by an increasing airspeed. An immediate recovery from the spiral was accomplished.

It was found that closing the throttle alone just as an inadvertent spin starts was as effective as applying opposite rudder alone (though you should do both in that case).

Fig. 21-36 sums up the spin recovery steps for the C-150/152.

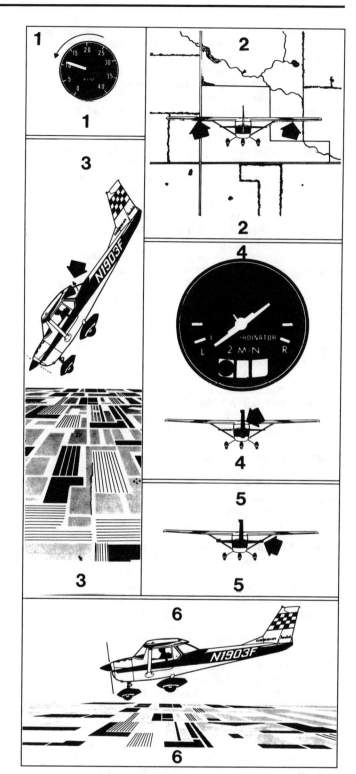

Fig. 21-36. Steps recommended for spin recovery in the C-150/152.
   1. Close the throttle.
   2. Neutralize the ailerons.
   3. Get the flaps up (if they were down in that inadvertent spin).
   4. Apply full rudder opposite to the rotation.
   5. Then immediately apply brisk forward motion of the control wheel. (Full forward and hold may be necessary at aft CG.) When the rotation stops, neutralize the rudder.
   6. Pull out of the dive.
(*The Basic Aerobatic Manual* and the Cessna 152 *Information Manual*)

Extended spins were given to CFI candidates to build their confidence that the Aerobat can be recovered as well at 15 turns as at, say, 4 turns, and these were given only at the applicant's request or agreement. It was suggested that extended spins not be given to, for instance, a private or commercial pilot wanting a safety spin course; 1, 2, or 3 turns are sufficient to recognize and recover from incipient and developing spins.

For CFI candidates, demonstrations and practice of recoveries using the elevator alone (throttle closed, ailerons neutral, feet off the rudders) were given to build confidence and to show the increased importance of the elevator as the mass distribution is moved out along the wing. This was a good confidence builder, but recoveries were not as quick as the *POH* recovery, as the applicants were reminded, and it was emphasized that the manufacturer's procedure was the safest and most effective method.

Recoveries using aileron alone against the spin were also demonstrated and practiced (N7557L). In each case, the throttle was closed, feet off the rudders, and wheel *full back,* to illustrate that if the trainee doesn't want to push the wheel forward for recovery and is a King Kong type, the spin *in this airplane* can be stopped by *opposite* aileron alone. (This would be a very bad move for some airplanes; opposite ailerons could delay recovery or set the scene for a flat spin and this was shown *to be only a last resort, since the recovery took 4 turns or more.*) This was of particular importance to smaller female CFI candidates, whose stronger male trainees might be "clutched" and holding the wheel back. Again it should be emphasized that for *other* type airplanes, opposite aileron during the spin could cause serious problems in recovery.

The C-150s and C-152s listed had no memory after *two turns* of *how* the airplane got into the spin. That is, the recovery steps cited in the *POH* were always effective, whether the spin was initiated with (1) a slow deceleration, (2) from a snap roll, or (3) from a loop (see Fig. 21-35). The relative density of the airplane is comparatively low so the inertial forces and moments decay to the "average" spin by the end of 2 turns.

■ **Practical Test Standards.** You should review the Spin Task in the Practical Test Standards before *you* go for the CFI-A practical test to be sure that you've covered all knowledge requirements for spins, including the ability to instruct spins. Later, when you are sending up one of your CFI-A candidates you should be sure that he has used that outline as a checklist.

Remember, you won't just present a logbook sign-off from a fellow instructor certifying that you have shown the ability to recover from spins, but will have to prove to the examiner that you can teach spin entries, spins, and recoveries. Following are the required spin knowledge areas in the Practical Test Standards. *Most of this was covered earlier in detail but is pulled together as a summary here.*

1. You'll have to *exhibit instructional knowledge of the element of spins by describing:*
   (a) *Aerodynamics of spins.* Check this chapter with particular attention to Figs. 21-7, 21-10, 21-16, and 21-17 and the accompanying text.
   (b) *Airplanes approved for the spin maneuver based on airworthiness category and type certificate.* FAR 23.221 covers the requirements for spin certification for normal-, utility-, and acrobatic-category airplanes and was discussed earlier in the chapter.
   (c) *Relationships of various factors such as configuration, weight, center of gravity, and control coordination to spins.*
      Configuration—Flaps down may delay or preclude recovery from a developed spin for some air-

planes. Power tends to flatten out spins for some airplanes.

Weight—The heavier the airplane, the more inertia exists in a developed spin if other factors are equal. This would mean that the recovery aerodynamic forces and/or moments would have to work harder and take longer to stop the spin.

Center of gravity—an aft CG causes delays in recovery and may require full-forward wheel or stick, and held. A CG at the rear limits or out of the envelope can cause some airplanes to go flat and be unrecoverable.

Control coordination—Since a spin results from a yawing, rolling input as the airplane approaches or reaches the stall, keeping the ball centered in coordinated flight decreases the tendency to spin.

   (d) *Flight situations where unintentional spins may occur. Distractions* may act to set up an inadvertent spin such as the cross-control (skidding) turn from base to final (see Chap. 11 and the section on under-the-bottom spin entries in this chapter) or in an emergency landing situation where the pilot is so intent on getting to a field that attention to the airplane is neglected. Also, students who are out practicing solo takeoff and departure stalls in a turn, particularly to the right, may let the "high" wing stall and the airplane roll toward it (see Chap. 9 and the section on over-the-top spin entries in this chapter).

   (e) *How to recognize and recover from imminent, unintentional spins.* The airplane is set up for the start of a spin anytime it is close to a stall and a roll or yaw is occurring. Close the throttle, use rudder against the roll or yaw, and move the control wheel or stick forward. For the vast majority of general aviation airplanes, this will break the stall and impending spin.

   (f) *Entry technique and minimum entry altitude for intentional spins.* For most airplanes the best way is to use the technique of clearing the area, using carburetor heat as recommended by the *POH,* power at idle, easing the nose up to the stall (not too high a pitch attitude or the spin entry is delayed) and, when the stall-warner sounds, applying full-back wheel or stick, as full rudder is being applied. However, a blast of power at the entry may help get a clean start; also, the *POH* may have specific suggestions for best procedures.

   While the Regulations call for a minimum recovery altitude for aerobatics (including spins) of 1500 ft, this is much too low for spin training. Again, plan on having a full recovery at or above 3000 ft AGL for any spin.

   (g) *Control techniques to maintain a stabilized spin.* As noted in the common errors for spins, one of the problems is that back pressure is relaxed *before* the spin is developed or, in those airplanes that have to be held into the spin, relaxing back pressure *during* the developed spin.

   (h) *Orientation during the spin.* Again, the trainee should not be staring straight down the nose of the airplane. All he'll see is a blur. The technique of watching the reference (road, railroad, or section line) and counting turns out loud is a real help in maintaining orientation and decreasing the chances of nausea.

   (i) *Recovery technique and minimum recovery altitude for intentional spins.* If the examiner asks on the oral, "How do you recover from a spin?" your answer should be more questions: (1) What airplane? and (2) What does the *Pilot's Operating Handbook* require? While the spin recovery technique described earlier in this chapter (Fig. 21-36) is about the most universal spin recovery this writer has encountered, there may be special procedures required for a particular airplane. The *POH* will always have precedence over any general suggestions. *Minimum recovery altitudes* were covered in (f).

(j) *Anxiety factors* associated with spin instruction. There is no question that some of your trainees will have high anxiety levels before and during the first few spins. Most will have increased confidence in their general flying after they've recovered from several spins. As an instructor you will see their relief that spins were not as bad as they had anticipated.

2. The Spin Task in the Practical Test Standards also requires that *you exhibit instructional knowledge of common errors related to spins by describing:*

(a) *Failure to establish proper configuration prior to spin entry.* For Pete's sake, don't deliberately spin with the flaps down.

(b) *Failure to achieve and maintain a full stall during spin entry.* As covered earlier, you can expect that the trainee will, in the first few spins, tend to relax back pressure as the spin starts. The airplane will slip into a spiral as shown by an increasing airspeed; the "partial" back pressure being held should be relaxed as the wings are leveled and the pull-out completed.

(c) *Failure to close the throttle when a spin entry is achieved.* Leaving power on after the spin is developed will usually result in a *flatter spin and a delay in recovery for most general aviation airplanes approved for spinning.* (It would probably have the same effect on airplanes *not approved* for spinning.)

(d) *Failure to recognize the indications of an imminent, unintentional spin.* When the airplane is approaching the stall and rolling or yawing is occurring, it is set up for a spin. *Recognize it.*

(e) *Improper use of flight controls during the spin entry, rotation or recovery.*

Entry—You'll find that trainees will unconsciously apply aileron into the spin to "help the entry." You'll have to remind them that the ailerons should be neutral throughout the process.

Rotation—When you are instructing spins in the Cessna 150 and 152, you'll see that the trainee is gradually allowing the control wheel to turn into the spin in a 3-turn, or more, spin, increasing the rotation rate and delaying the recovery. The aerodynamic forces acting on the ailerons are causing this. This will be less of a factor if your training environment restricts spin practice to 2 turns.

Recovery—The vast majority of spin trainees are more than willing to use opposite rudder (you may have to remind them *which* is the opposite rudder), and you'll find, as noted earlier, that sometimes they don't push the wheel or stick forward far enough for recovery and a great deal of altitude is lost while you remind them—several times—to push harder. The use of opposite ailerons to stop rotation is a "natural" reaction but is *not* the thing to do for some airplanes. You'll also expect that the trainee will unconsciously tend to hold opposite rudder during the postspin dive and pull-out.

(f) *Disorientation during a spin.* This was covered earlier in section 1 (Elements of Spins). You'll see that the trainee is not counting and doesn't start the recovery at the agreed number of turns, since he has no idea of the number of turns completed. Expect disorientation after the recovery from the first few spins, even if the trainee has been keeping up with the turns. This is particularly a problem on hazy days without well-defined references; he may pull up from the postspin dive

and let the airplane assume a very nose-high and/or steeply banked attitude because of inner ear problems.

(g) *Failure to distinguish between a high-speed spiral and a spin.* If strong back pressure is required to "hold the airplane into the spin" and/or the airspeed is starting to increase, a spiral is occurring; *relax* the back pressure, then recover from the spiral. Check Fig. 21-6 again for another indication of spirals/spins.

(h) *Excessive speed or accelerated stall during recovery.* Excessive speed is the result of a delay in pulling up after the airplane is out of the spin (some trainees can become fascinated by the ground rushing up at them). The C150/152 airplanes can be pulled out of the dive just about as soon as the airspeed starts picking up; other current trainers require that the nose be kept down a second or so longer to avoid an accelerated stall or possibly a progressive spin.

(i) *Failure to recover with a minimum loss of altitude.* You can check problems of this by reviewing (h).

(j) *Hazards of attempting to spin an airplane not approved for spins.* Normal category airplanes are *never* approved for spins, since as noted in FAR 23.221 they are required to do only 1 turn and recover within another turn. If you spin a normal category airplane more than 1 turn (and recover), call or write the manufacturer and tell those folks what happened—they'd be interested to know.

3. *You could be required to demonstrate and simultaneously explain a spin (1 turn) from an instructional standpoint.*

4. *You'll be expected to be able to analyze and correct simulated common errors related to spins.*

And to repeat a point about the spin portion of the oral exam, as noted earlier in the chapter, FAR 61.49(b) says it best:

> An applicant for a flight instructor certificate with an airplane rating, or a flight instructor certificate with an airplane rating, or a flight instructor certificate with a glider category rating, who has failed the practical test due to deficiencies of knowledge or skill relating to stall awareness, spin entry, spins, or spin recovery techniques must *during the retest, satisfactorily demonstrate both knowledge and skill in those areas in an aircraft of the appropriate category that is certificated for spins.* [Italics added.]

Do it right the first time and you won't have the hassle of getting a spinnable airplane and taking another flight test on spins.

SUMMARY

An airplane's spin characteristics still cannot be predicted as well as other areas of performance or stability and control, and if you teach spins you should stick strictly with the manufacturers' recommendations for entry and recovery. There would be nothing worse than deciding on the spur of the moment to take a trainee into a spin situation where *you* haven't been and discover that the airplane isn't responding to the recovery controls as it did before. That extra turn or two (or more) for recovery can take a very long time. Again, do spins with an instructor who has experience in the airplane you will be using before embarking to give dual in it. One factor often overlooked is that control cables may have slackened since the airplane came from the factory. Check all cable tensions to ensure full recovery control.

You should strongly emphasize to a trainee that (1) deliberate spins with flaps down are not to be done and (2) if he gets into an accidental spin with flaps extended he should get them up as soon as practicable to avoid exceeding the max flaps-down airspeed. Also, extended flaps could cause the airplane to have bad recovery characteristics.

Inverted spins are only briefly mentioned in this chapter. They are, by their required entry technique and physical effects on the pilot, usually avoidable in the average light (normal/utility category) airplane. The upright or positive-g version is the one that results in the stall/spin fatality statistics that continue to be published.

If you decide to have a spin program as an extra to your normal flight instruction, set up a ground and flight syllabus, giving background and spin theory as well as how-to information. Your problem will be to walk that narrow line between making your trainees complacent about spins and scaring them so that even practice stalls are avoided.

As a suggestion, before sending your primary students out to practice stalls solo, demonstrate and let them execute a recovery from a ½-turn or 1-turn incipient spin so that at least they'll have seen and/or done one before getting into a stall/spin situation. As you instruct more, you'll find that students do scare themselves while flying solo, and the scenario given in the early part of this chapter is not as rare as you first figured.

In Chap. 24 spin recoveries on instruments will be covered in detail.

REFERENCES

The following sources are suggested for further reading.

Bihrle, William, Jr., and Bowman, James S., Jr. *The Influence of Wing, Fuselage and Tail Design on Rotational Flow Aerodynamics Data Obtained beyond Maximum Lift with General Aviation Configurations.* AIAA Paper No. 80-0455. March 1980.

Bowman, James S., Jr. *Summary of Spin Technology as Related to Light General-Aviation Airplanes.* NASA TN D-6575. 1971.

Burk, Sanger M., Jr.; Bowman, James S., Jr.; and White, William L. *Spin-Tunnel Investigation of the Spinning Characteristics of Typical Single-Engine General Aviation Designs. I—Low Wing Model A: Effects of Tail Configurations.* NASA Technical Paper 1009. 1977.

Chambers, Joseph R., and Grafton, Sue B. *Aerodynamic Characteristics of Airplanes at High Angles of Attack.* NASA TM 74097.

Chambers, Joseph R.; Anglin, Ernie L.; and Bowman, James S., Jr. *Effects of a Pointed Nose on Spin Characteristics of a Fighter Airplane Model Including Correlation with Theoretical Calculations.* NASA TN D-5921. 1970.

DeLacerda, Fred. *Surviving Spins.* Iowa State University Press, Ames, Iowa. 50010. 1989.

Grafton, Sue B. *A Study to Determine Effects of Applying Thrust on Recovery from Incipient and Developed Spins for Four Airplane Configurations.* NASA TN D-3416. June 1966.

Hazards Associated with Spins in Airplanes Prohibited from Intentional Spinning. FAA AC-61-67. Washington, D.C.: USGPO. Feb. 1, 1974.

Kershner, William K. *The Basic Aerobatic Manual.* Ames: Iowa State University Press. 1987.

Mason, Sammy. *Stalls, Spins and Safety.* New York: Macmillan. 1982.

Stall Spin Awareness Training FAA AC61-67B. Washington, D.C.: USGPO. May 1991.

Stough, H. P., III, and Patton, J. M., Jr. *The Effects of Configuration Changes on Spin and Recovery Characteristics of a Low-Wing General Aviation Research Airplane.* AIAA Paper No. 79-1786. August 1979.

22

THE THREE FUNDAMENTALS OF AEROBATICS

BACKGROUND

The three basic maneuvers—the roll, loop, and snap roll—are to aerobatic flying what the Four Fundamentals (turns, climbs, glides, straight and level) are to everyday flying. Nearly every aerobatic maneuver is a variation or combination of these. Get these Three Fundamentals well established, both in theory and technique, and there won't be problems with maneuvers such as the Immelmann or Cuban eight.

THE AILERON ROLL

This maneuver is ideal as an introduction to the "real" part of aerobatics training. Since the trainee has seen rolls at airshows, on TV, and at the movies, he is impressed if *he* can do one. Some instructors may advocate using the loop as an introduction to aerobatics because it is assumed to be an extremely simple maneuver. It isn't. The loop is a maneuver that exerts g's to an extent the trainee may not be prepared for. It's a little too "active" for a first-timer.

The aileron roll has the advantage of being a simple, seemingly spectacular, easy-on-the-body maneuver with no negative g's or excess positive forces. You should, for instance, pull a maximum of 2.5–3.0 g's on the pull-up to entry, and then maintain 1 g throughout the roll. The trainee's probable reactions to his first successful aileron roll will be relief and confidence that it *is* possible for him to roll an airplane around to a normal attitude after it's been over on its back even for a short while. Your first demonstration aileron roll can set the pace for the whole course for him, so make it smooth.

Why not slow rolls? To the beginner, a slow roll is an

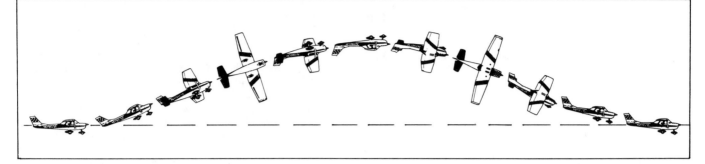

Fig. 22-1. Aileron roll. (*The Basic Aerobatic Manual*)

uncomfortable maneuver; the need for crossed controls (you'll be holding top rudder as the airplane approaches 90° of roll) and forward stick in the inverted position all add up to a bad feeling for the person who has never done aerobatics. Later, fine.

■ **Preparation.** Have the trainee read the references in the Bibliography before flying. You may want to work up a written lesson plan, but usually this is not done for a particular aerobatic maneuver. Have the maneuver clear in your mind; review your introduction a couple of times to yourself, because in order to successfully impart your knowledge of the aileron roll (or any aerobatic maneuver) you should be able to "stop the airplane" (model) and cite the control positions at any point during the discussion.

Most of the information on aerobatics in Part 5 is based on the Cessna Aerobat but can be used for other aerobatic airplanes with the proper entry speeds.

■ **Explanation**

PREFLIGHT BRIEFING. Make a point of assuring the trainee that the aileron roll is a maneuver no more violent to him or the airplane than a 720° power turn.

A model is definitely required for explaining aerobatic maneuvers and you should use it, going through the aileron roll completely as you explain it and then giving a step-by-step approach until he can get an idea of the control movements. *The explanation might be as follows:*

1. In level flight, set the power (at cruise airspeed) to a value that will not result in a prop overspeed at the recommended entry speed.

2. Pick a reference point on the horizon. After making sure that the area is clear, ease the nose over to obtain the recommended entry speed. Assume the roll will be made to the left.

3. Pull the nose up smoothly (wings level) to 30° above the horizon. Relax the back pressure just enough to maintain this pitch attitude as you start applying left aileron and rudder as if entering a steep turn. Continue to rapidly (but smoothly) increase the aileron deflection until the maximum is obtained. Use rudder as necessary to coordinate.

4. When the airplane rolls past a 45° bank, less rudder is used, to avoid pulling the nose down below the horizon (and pulling it off heading). At 90° of roll, start relaxing the back pressure more.

5. As the airplane reaches the *inverted* position, nearly all back pressure has been released to keep the nose from being pulled down well below the horizon. The nose will gradually move down as the maneuver progresses, even with no back pressure being applied. Had the original back pressure needed to pull the nose up been continued, the nose would have been "pulled" down well below the horizon as the airplane approached the inverted position.

6. After the airplane has reached the vertical bank (270° of roll), back pressure is *started* slowly and smoothly again. At this point, sufficient left rudder must be reapplied, along with increasing back pressure, to keep the nose from dropping. The airspeed value will be lower than earlier in the maneuver; hence more rudder deflection will be needed to get the desired result.

The full aileron deflection is held until the time to recover. A summation of probable control usage during the aileron roll to the left is as follows:

1. Initial pull-up—back pressure only. (Relax as necessary to maintain the proper pitch attitude.)

2. Start of the roll—slight back pressure, left aileron, and left rudder. (Continue to increase aileron deflection.)

3. At 90° of roll—full aileron, little rudder, less back pressure.

4. At 180° of roll—full aileron, very little back pressure, light rudder pressure.

5. At 270° of roll—full left aileron, left rudder being reapplied more strongly, transition to back pressure beginning.

6. Last 45° of roll—full left aileron, more left rudder, begin the increasing of back pressure. As the airplane approaches wings-level flight, the aileron, rudder, and elevator deflections are decreased as necessary to return to normal cruising flight.

Check the accelerometer to see that no more than 2.5–3 g's were imposed on the initial pull-up.

One of the problems a trainee has with the first couple of aileron rolls is that when he gets inverted he is so awed with his new position that he relaxes the aileron deflection, which, of course, slows down or stops any roll tendencies and you both are left there (for a short while) to contemplate his shortcomings. Remind him to continue the roll. Impress on him that if you are inverted, *always roll* as a matter of recovery and never pull back on the control wheel to do a half loop; it would take several hundred feet to recover, a decided disadvantage if he is on final approach when such trouble occurs.

Another problem he might have is neglecting to release back pressure when the airplane is inverted. The nose is allowed to drop and the airplane starts making knots earthward. Roll it out.

Fig. 22-2 gives an analysis of control positions for the aileron roll. Of course, the trainee will be more interested in control *pressures,* but drawing this on the chalkboard and discussing it will help clear up any hazy areas about how the controls are used in an aileron roll (and the slow roll) to the left. The solid lines are the ones indicating the aileron roll control positions.

After you've drawn the diagram on the board, use a pointer to combine it and the model to illustrate your briefing. (After a while you'll get to where you won't drop either the pointer or the model doing this.)

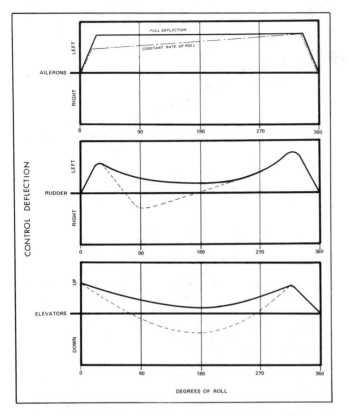

Fig. 22-2. Control deflection analysis of the aileron roll.

"Time zero" is when the airplane has reached the 30° pitch position but no roll has started.

USE OF THE AILERONS. Considering the aileron deflection throughout the roll (Fig. 22-2), it's usually best to have the beginner move on in with full aileron deflection from the start, maintaining full deflection until the airplane approaches the normal wings-level attitude at the end of 360° of roll. (Check the solid black line.) The reason for this is to have a good rate of roll going for him from the beginning, to set up the impetus to keep the airplane rolling so that he gets the 360° completed more quickly. For an airplane which uses, say, 115 K as a roll entry speed, the speed at the completion may be down to 85 K. If the deflection is as given by the solid line, the roll rate will be faster at first and will slow down as the speed decreases. The roll rate for a particular airplane is a function of aileron deflection and indicated (calibrated) airspeed, and since the ailerons are already deflected fully, the roll rate must decrease with airspeed decrease.

A diagram of the roll rate of a particular airplane shows that it increases in a straight line with calibrated airspeed up to some point at which the dynamic pressure is too great for the pilot (or system) to maintain full aileron deflection, and then the roll rate decreases again. (See Fig. 23-30.)

Later, when the trainee has the aileron roll pretty well in hand, the aileron deflection can be as shown by the dotted-and-dashed line, for a more constant roll rate and smoother maneuver. This would apply to both aileron rolls and slow rolls.

RUDDER USE. Looking at the use of the rudder in the aileron roll (solid line, Fig. 22-2), you can see that as aileron is applied, rudder is also needed; since the maneuver is coordinated (hopefully) the rudder is *always* applied in direction of roll, though to a varying degree.

Note that the curve has two peaks, *the second one higher*

than the first. This is because the airspeed is decaying and the control effectiveness is decreasing. In addition, the slower speed results in greater adverse yaw of the ailerons. Between the increased adverse yaw and less rudder effectiveness, nearly full rudder may be needed for some airplanes at the second "peak" shown. At the end of 360° of roll when the ailerons are neutralized, the rudder is no longer needed.

The dashed line shows the probable rudder deflection for a *slow roll* and indicates that opposite rudder is used at certain points around the roll. The adverse yaw and its required rudder deflection at the latter part of the maneuver are about the same for both rolls.

Improper rudder use in the aileron roll may cause problems that seem to originate from improper elevator use. For instance, if the initial amount of rudder is held as the airplane approaches the 90° roll position, the nose is pulled down and will be low at that point. The roll is completed in too low a nose position, with the elevator (or poor use of the elevator) as the apparent culprit.

If you note that the nose is high enough at the initiation of the roll and he's relaxing the elevator pressure when inverted, have him do another roll, and you check the nose position at 90° of roll. (Also subtly "fly the rudders" with him.) His rudder use may not have the saddle shown in Fig. 22-2. You may be emphasizing the use of elevator when rudder is the problem.

ELEVATORS. Note in Fig. 22-2 that the elevators are deflected up at the start of the roll, when the nose is at the 30° nose-up position. As with the other controls, the solid line shows the relative positions of the elevators in the aileron roll.

The elevators are never on the "down" side of the neutral position in the aileron roll, and the g forces should always be positive. The back pressure is eased as the 90° position is reached, is at its minimum at the inverted (180° of roll) position, then is increased as necessary to bring the nose to the level-flight position as the roll is completed. Check that 1 g is being pulled during the roll itself. The dashed line shows the relative elevator deflections during a *slow roll*.

Figure 22-2 is particularly useful for your trainees with a technical bent, and as mentioned before, is best used with a model. You may want to draw a large version on poster board, to avoid drawing it on the blackboard every time.

You might have the trainee use the model and explain the maneuver to you.

If the trainee is flying an Aerobat from the *left seat,* the elements in Fig. 22-3 might be shown on the chalkboard.

Looking at Fig. 22-3 you can note the following factors noticed by the *pilot:*

1. In the *left roll.*
   a. Torque is helping the roll and is a plus factor. "Torque" is used here as a short term for all left-turning forces and moments.
   b. With the pilot in the left seat, in a left roll he is "inside" the roll and feels more comfortable.
   c. The control wheel (Aerobat) must be turned well past 90° to get full aileron deflection and this is awkward to do, usually resulting in a less-than-full rate of roll.
2. In the *right roll.*
   a. Torque is working against the roll, resulting in a slower roll (for instance, the roll portion takes 8–9 sec instead of the 6 sec normally required for the roll portion of the left roll).
   b. The pilot in the left seat tends to feel more uncomfortable in the right roll because of the feeling that he is on the "outside" of the roll. In fact, you should avoid doing a large number of right rolls on the first flight because they may induce nausea. Start out with left

```
                    LEFT SEAT - TRAINEE

     LEFT ROLL              RIGHT ROLL

  + TORQUE                - TORQUE

  + LEFT SEAT             - LEFT SEAT

  - CONTROL WHEEL         + CONTROL WHEEL
```

```
                  RIGHT SEAT - INSTRUCTOR

     LEFT ROLL              RIGHT ROLL

  + TORQUE                - TORQUE

  - RIGHT SEAT            + RIGHT SEAT

  + CONTROL WHEEL         - CONTROL WHEEL
```

Fig. 22-3. Advantages and disadvantages for trainee and instructor in left and right aileron rolls in a side-by-side aerobatic trainer.

rolls and, after the trainee has the idea well in hand, move on to right rolls, but watch for nausea problems. The left rolls give better success chances and so are good from a psychological standpoint.

    c. For the pilot in the *left seat,* the primary advantage of a *right* aileron roll is that the control wheel is not in such an awkward position when the ailerons are deflected fully in that direction.

As the *instructor in the right seat* you will note the following:

1. In the *left roll.*
   a. Torque is working for the roll.
   b. Sitting in the right seat puts you on the "outside" of the roll, but you'll soon get used to it.
   c. The control wheel is easy to turn to the left with your right hand and you can get full play.
2. In the *right roll.*
   a. Torque is working against the roll.
   b. You are on the "inside" of the roll, but this will not be a particular advantage as you gain more experience.
   c. The control wheel is harder to fully turn to the right because of the "arm twisting" required. This is the same problem the person in the left seat has with a left roll.

DEMONSTRATION. Depending on the trainee (and you will learn with experience what route to take in instructing aerobatics to each individual), one approach is to demonstrate an aileron roll on that first flight as soon as altitude, airspace, and traffic permit. This is a good move if you think he'll spend the whole period sweating it out. Remember, some trainees will think that the aileron roll is going to be an alarming maneuver, even after you've carefully explained how simple and easy it is. Tell him what you are going to do, clear the area, and do it. Not only will it put his mind at ease (he didn't fall out of his seat, his hat didn't fly off, and he didn't get sick), but it will also give him something to look forward to. You might remind him that after he's comfortable doing chandelles and wingovers, he'll be introduced to aileron rolls by the end of *this first flight.*

Keep your demonstrations to a minimum.

■ **Practice.** When you feel that the trainee is ready to start doing aileron rolls in earnest, demonstrate one, explaining the

maneuver. (You'll soon learn to talk fast.) It may be that lack of time and his ability make it impossible to practice more than a few aileron rolls, but if at all possible he should do at least one aileron roll himself the first flight. After doing it, during the rest period of straight and level flying following, review your actions in the maneuver, making sure that he understands that he must not neutralize the ailerons as the back pressure is relaxed. Don't worry, at some time or another he will—and probably more than once.

The follow-me-through method is normally a poor way to instruct flying, but in the aileron roll it is a good aid in getting the picture of control *movement.* (He will work out the control pressures later himself.) The first approach to some aerobatics *is* mechanical; natural feel will develop with practice. The main reason for the follow-me-through approach is to impress upon the trainee the necessity of relaxing that back pressure as the airplane becomes inverted, *without* neutralizing the ailerons. The varying rudder action around the roll can also be felt.

The main idea at first is for him just to complete 360° of roll, without your covering him up with details. Sure, he's slipping in one part and skidding in another, but if he's making it all the way, that's 90 percent of the problem whipped.

After each roll (with a proper interval of rest between) you can discuss errors, gradually becoming more detailed as he gets better at it.

Some of your flight instructor trainees will enjoy analyzing their own errors and after a while may beat you to an analysis. This is fine, if they are right, and it makes the flight more valuable since they are improving their own ability at maneuver analysis. Also, you might set the small fuel sampler cup or other light object on top of the instrument panel and let the trainee see that it will stay in place when he does good aileron rolls. He'll get a kick out of seeing it still there after a roll. (Don't use any heavy objects for this experiment—his rolls may not be *that* good.)

A trainee is beginning to *think* as he flies when he accidentally relaxes aileron pressure at some point in the roll and immediately corrects it without any comment from you.

Fig. 22-4, A–H shows the aileron roll as seen from the cockpit and from behind the airplane.

Common errors made in aileron rolls are these:

1. Not raising the nose high enough initially. This can carry on through the maneuver so that the nose is too low at the completion, even though the back pressure was relaxed the usual (or proper) amount. Following right along with this type of problem is a common tendency to *not* bring the nose up to the 30° attitude *fast enough.* He shouldn't yank the nose up abruptly, but when the proper entry speed is reached, bring it up smoothly and fairly rapidly; otherwise, the speed will decay before the roll is started. If you note that the speed is too low as he starts the roll, break off the maneuver. At the beginning of the course he will be concerned about the maneuver, and a too-slow, sloppy, falling-through type of roll won't help at all.

2. Lack of coordination of aileron and rudder at the beginning of the roll (the most common error is not enough rudder).

3. Relaxing the aileron deflection as the airplane becomes inverted.

4. Neglecting to release enough back pressure when inverted. It will seem to the trainee that he has to "shove forward," an unnatural action to the nonaerobatic pilot, and he will be uneasy about it at first. Emphasize that the *relaxation* of back pressure is not a sudden move as the airplane becomes inverted, but a subtle action that begins well before the 180° of roll position and continues well

after that point. In a good aileron roll, the "passenger" should not be able to pinpoint the beginning or end of the relaxation of the back pressure. Back pressure and forward pressure can have a definite effect on heading control, particularly when the plane is vertically banked. You can expect that after a few times of your telling him to relax the back pressure, the deed will be overdone. A relaxation that's too brisk or too soon can be felt as well as seen by the nose "oscillation" and heading error as the roll continues.

5. *Not using enough top rudder* on the final part of the roll so that the airplane turns or dishes to one side. If in a left aileron roll, for example, the nose consistently ends up pointing to the right of the reference, check for this as well as too soon or too much relaxation of back pressure. While this particular problem of heading is being discussed, it should be noted that slacking off the aileron pressure during the last 45° of roll can also result in an error in heading to the right (in a left roll) since the airplane will be tending to turn as the roll is delayed. Reverse the above directions for right roll problems.

6. Not enough back pressure in the last 45° of roll — the nose allowed to be too low. This, following an initially nose-low beginning, can result in a finale that is nose-low indeed.

7. Holding rudder after the roll is complete. ("Get off that blankety-blank rudder!") The trainee can see the nose swing back to the reference after he releases the rudder.

THE RIGHT ROLL. Don't have the trainee try any aileron rolls to the right until he is able to do a reasonable (no hesitation, proper techniques, etc.) left roll. The right aileron roll is more difficult for the average pilot. Since traffic patterns are predominantly left-hand and other training procedures seem to favor turns (and rolls) to the left, most people have a stronger tendency to neutralize the ailerons and let the nose drop in a *right roll*. You might also warn them that more rudder is needed all the way around (in U.S.-engine-powered planes). Because of this, more heading problems may be encountered in the right roll. You might consider adding 5 K to the entry speed when he first practices the right roll.

■ **Evaluation and Review.** The average trainee should be able to do a reasonable aileron roll before the end of the first lesson. If he needs extra work, plan to repeat this lesson the next flight.

You should review each roll after it's been done and review the maneuver generally after the flight is completed. Emphasize any continuing problem — for instance, he still tends to have the nose too low at the beginning, etc.

Have the trainee go through the maneuver again on the ground, using the model, if he doesn't seem to have the procedure clear.

■ **Summary.** If possible, the trainee should be introduced to and allowed to practice at least two left aileron rolls the first flight. This is his first "real" aerobatic maneuver and how he reacts to it will set the pace for his future flights. Don't overdo

Fig. 22-4, A–H. Aileron roll attitudes as seen from the cockpit and from behind the airplane. (Imagine that you are in a following airplane in the illustrations on the right, matching the roll.)

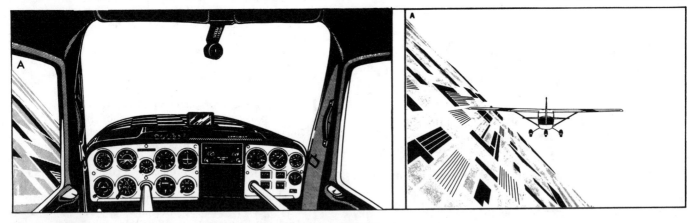

(A) 45° of roll.

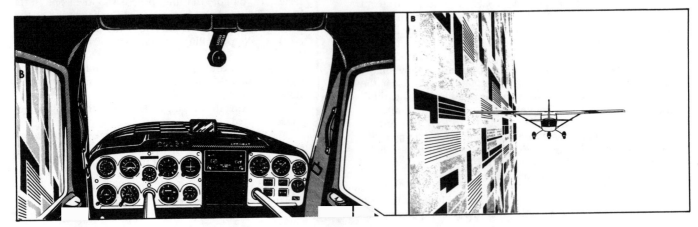

(B) 90° of roll.

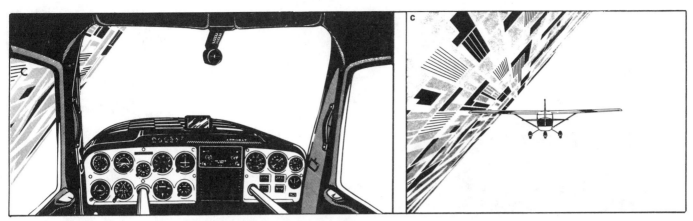

(C) 135° of roll.

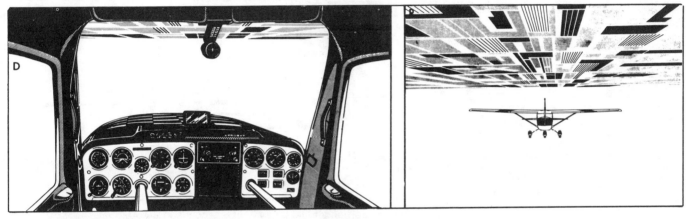

(D) 180° of roll—fully inverted.

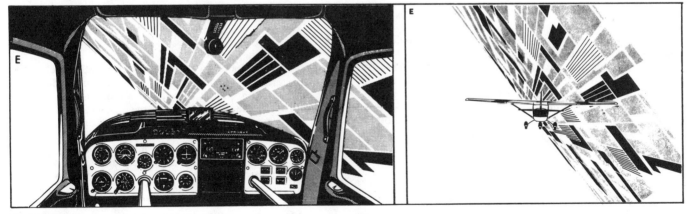

(E) 225° of roll.

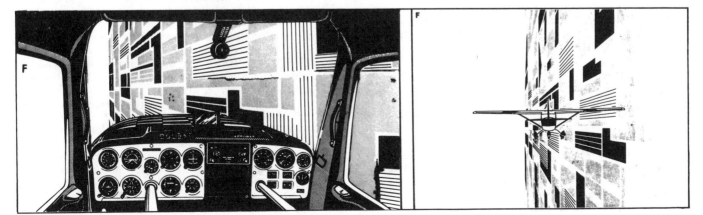

(F) 270° of roll.

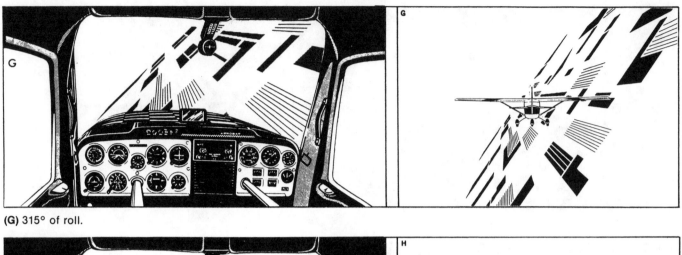

(G) 315° of roll.

(H) 360° of roll—wings-level, normal cruising flight. The roll portion should take about 6 seconds in the Aerobat.

aileron rolls, but move on to loops in the next flight and immediately follow that maneuver with an aileron roll—you'll see an improvement as he starts doing the roll unconsciously. The aileron roll builds confidence for the trainee because he will get the feeling that if the airplane becomes inverted, with sufficient altitude even he, who had not banked over 60° before this flight, could get it around and upright. In other words, he knows what to do and that it can be done (if somewhat roughly at this stage of his flying).

THE LOOP

■ **Background.** The loop is considered one of the simplest maneuvers to do, but to do it well requires more planning and coordination than is apparent at first thought. The loop is basically a 360° "turn" in a vertical plane, and the trainee may have heard from his experimenting, nonaerobatic pilot friends that "all I do is dive to a high speed and keep pulling back on the wheel." It's your job to give him the *correct* picture.

This maneuver is the one in which he will encounter the effects of 3 g's or more for the first time and it would be better for you to handle the throttle for his first few loops, since the average trainee initially has trouble coordinating throttle and elevators.

■ **Preparation.** The model and chalkboard are a must in introducing this maneuver. (You should review the loop yourself if you haven't taught it recently.)

The trainee should be assigned reading before the flight in one or more of the books listed in the Bibliography.

■ **Explanation**

What—"So, the loop is a 360° turn in a vertical plane" is one explanation. The maneuver is so well known that a description of *what* it is usually is unnecessary, but you might use the model to get an overall look at it. Usually, the loop is not demonstrated until the second flight; however, if you have an extra-sharp person who's been doing very well at aileron rolls, you may start earlier.

Why—This is one of the fundamental maneuvers of aerobatic flight and should be practiced not only to learn to do it properly but also to smooth it out as a part of more-advanced maneuvers. After he has practiced a number of loops, you'll notice that his orientation has improved radically. It's also very good for getting the feel of the elevator pressures with changes in airspeed.

The loop is also good for the trainee's general flying because the pitch attitudes are greater than he's encountered and it shows that *under the proper conditions* the airplane can be temporarily pointed straight up or straight down in perfect safety. It also teaches that outside references (nose, wing tip, through the skylights and nose) must be used and those required references change rapidly.

How—Everybody, student pilot or ATPC pilot, knows "how to do a loop," as was mentioned earlier. It is simply a maneuver (they think) requiring a lot of "pull," that's all; this will probably be the trainee's attitude at first. It would be better not to mention this to him at an early point in his training, but the loop, if improperly done, can result in some rather hairy reactions. In extreme conditions of inattention and general all-around control fumbling, a whip stall *could* result. However, don't tell him this — instead, use a positive approach in telling (and showing) him how to do a loop properly from the beginning.

You might use the model and give him an explanation of the loop as follows:

1. Clear the area and line up over a long, straight section of road or railroad. Ease the nose over to pick up the recommended entry speed. Throttle back as necessary to keep from exceeding the rpm red line. As the recommended speed is reached, initiate a smooth pull-up. Ease the throttle to full open as the nose approaches vertical, making sure that no engine overspeeding occurs. The trainee may ask, "How far do I pull the wheel (or stick) back?" (A response, though not always a good one, might be, "How far did you turn the wheel of your car when you turned into the airport parking lot?") In other words, it will be a matter of experience to get the required response from the airplane.

2. The wing tip is the reference to check when the horizon disappears under the nose, since looking over the nose would only show blue sky (or clouds) through nearly all the top 180° of the loop, and orientation would be difficult.

3. As the airplane nose reaches the vertical, the airspeed continues to decrease rapidly and more up-elevator deflection is needed (the control *pressures* are getting lighter). Add full power.

4. As the airplane approaches the inverted attitude (when the nose is an estimated 30° above the horizon), the back pressure is relaxed slightly to avoid buffeting and to maintain a more symmetrical pattern. Some loops look like an *ℓ* because pilots neglect this. Keep the back pressure relaxed until the nose has passed below the horizon on the "back side" of the loop (about 30° down). The back pressure is gradually increased during the final part of the dive recovery. A proper pull-up at the beginning of the loop will probably result in about a 3.0- to 3.5-g indication on the accelerometer. The final part of the loop will probably result in an acceleration of 2.5–3.0 g's and the airspeed will be at or slightly below the entry speed. (Make sure that the trainee doesn't stare at the accelerometer as an indicator during the initial pull-up; use it as a check afterward.)

5. Retard the throttle as the airplane approaches the vertical (nose-down) attitude on the "back side" of the loop so that the engine limits are not exceeded, but don't close the throttle because at high speeds engine-prop roughness could occur.

6. Recover to level cruising flight.

You should have positive-g forces acting on you all the way around the loop; the forces will be less at the top of the loop than during the pull-up and recovery, but the occupants should be comfortable (no hanging in the belt and harness). You both may feel a little light in your seat on one or two of the first loops but should be able to work it out without much trouble.

Fig. 22-5 shows an analysis of the loop in an Aerobat at various points (*A–J*) around the maneuver.

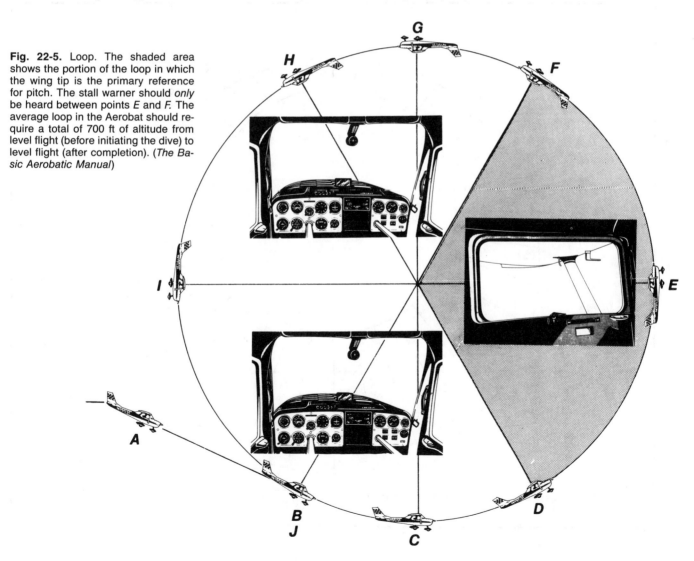

Fig. 22-5. Loop. The shaded area shows the portion of the loop in which the wing tip is the primary reference for pitch. The stall warner should *only* be heard between points *E* and *F.* The average loop in the Aerobat should require a total of 700 ft of altitude from level flight (before initiating the dive) to level flight (after completion). (*The Basic Aerobatic Manual*)

Point *A*. Have the trainee clear the area and line up over a section line or a long, straight stretch of road or railroad. Ease the nose over to pick up the proper entry airspeed.

Point *B*. Continue to watch for other airplanes as the entry speed is approached. He should throttle back to avoid exceeding the rpm red line.

Point *C*. Start the pull-up at the recommended entry speed.

Point *D*. The airplane should be pulling 3.5 g's and the trainee should start checking the wing tip.

Point *E*. At the vertical pitch attitude, the throttle is opened fully and the Aerobat stall-warner starts sounding. Continue with the back pressure. Check the wing tip.

Point *F*. Relax the back pressure just enough to stop the stall-warner and start looking through the skylights for reference.

Point *G*. Continue the relaxation of back pressure; the load factor should be 0.7+ g, or just below the normal 1-g value.

Point *H*. Start reapplying back pressure (easy!) and start checking over the nose for the ground reference. The tendency here is for most trainees to pull too hard, sounding the stall-warner or actually getting a stall buffet.

Point *I*. Retard the throttle to avoid exceeding the tach red line and make corrections to ensure that the wings are perpendicular to the ground reference line.

Point *J*. Continue pull-out (about 3.0+ g's) and climb to regain altitude or set up an aileron roll to immediately follow the loop (Chap. 23).

The arrows in Fig. 22-6 show the relative amount of g's and back pressure being applied around the loop. For this example, the recommended entry speed is 120 K, and the minimum speed at the top of the loop is 60 K, one-half of the maximum. It is assumed that 1 g is being pulled at the top, and the loop is perfectly symmetrical; that is, the radius is the same all the way around. (This is easily done on paper, anyway.) Assume these are "net" g's.

The lower part of this "ideally symmetrical loop" should have the same radius as the 1-g top, so an interesting point arises: The centrifugal force in a constant-radius turn (or loop) is a function of the square of the airspeed. If you double the airspeed, the g forces required to maintain that same radius will be *four* times that found at the lower speed. So, in the theoretical approach here, a maximum of 4 g's is needed at the bottom of the loop to maintain 1 g at the top. It's complicated slightly in the practical application by (for one item) the change in power used during the loop, so that a pull-up of 3–3.5 g's may be the maximum required.

The rate of "pull" should vary (shown by the arrows in Fig. 22-6), but things aren't as cut-and-dried as that. There are other factors involved—it is not humanly possible to be able to vary the back pressure so precisely, so a more mechanical procedure is used, as shown by points *A* and *B* in Fig. 22-6.

At point *A* the nose is at 30° above the horizon and the

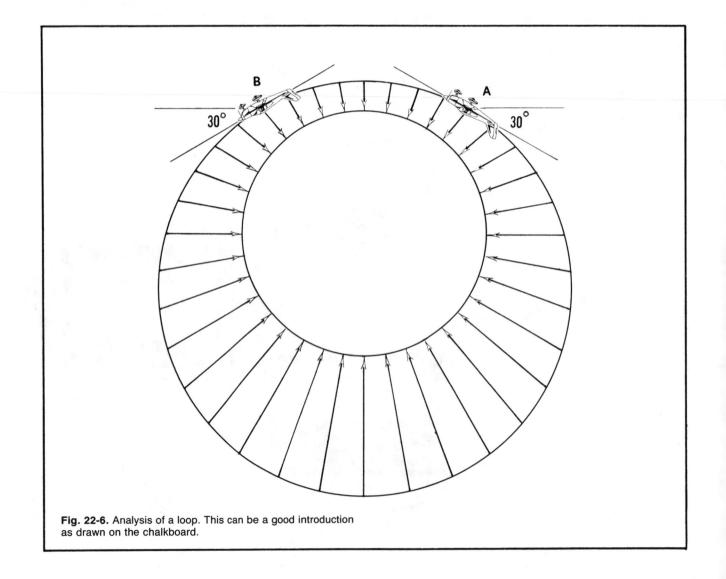

Fig. 22-6. Analysis of a loop. This can be a good introduction as drawn on the chalkboard.

Fig. 22-7. Five points for checking the wing attitude in the loop. (*The Basic Aerobatic Manual*)

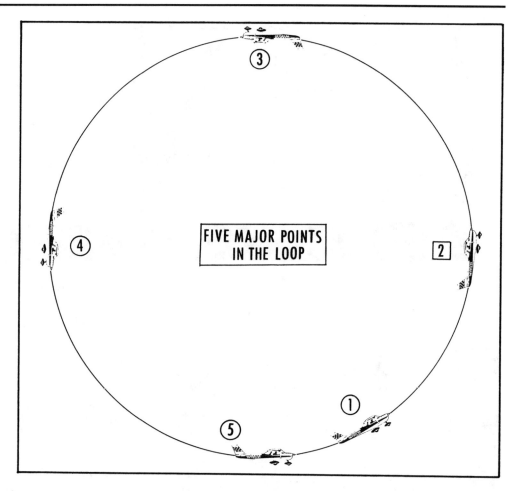

FIVE MAJOR POINTS IN THE LOOP

back pressure is relaxed. At point *B* (30° below) the back pressure is started again and gradually increased. While you described the loop in total earlier, it is well to repeat the information about the points of relaxation of back pressure when covering the theoretical side of the maneuver. One indication of a symmetrical loop is that the altitude at the bottom (completion) is the same as the altitude at the initial pull-up.

A good all-around check for a good loop in the Acrobat is a 3.5-g pull-up and about 0.7 g at the top. This gives a steady, not-too-light-in-the-seat maneuver. Check the active hand of the accelerometer at the pull-up and at the top of the loop during the trainee's efforts.

WING ATTITUDE. Fig. 22-7 shows the five major points for checking the wing attitude in doing the loop. You should cover this in explaining the loop, but don't overemphasize it at this early point in training. Like the aileron roll, 90 percent of the maneuver at first is just getting it around without too many casualties. But this should definitely be discussed in detail later. At point *1* in Fig. 22-7, the wings should be level, although you may find that some rugged individualists are already well banked by this time.

Usually the problem of wing attitude starts between points *1* and *2*. The pilot is "pulling," and probably a banking force is being exerted on the wheel or stick as well.

The wing attitude should be noted at point *2*, but it will take experience to tell whether a wing is "down" or not (Fig. 22-8).

Fig. 22-9 shows the right and wrong wing attitudes as seen from a *high*-wing airplane.

It's pretty easy to see where you stand at point *3* in Fig. 22-7 because the horizon is a good reference. A correction should be made here, if necessary. One problem is that the controls are not very effective at the slow airspeed. Also that point is often passed too quickly to get a good correction set up. However, you should discuss it for future references. The trainee will consider it very difficult to understand *which way* to roll to correct for a wing-down problem here, but you can assure him that he'll do what comes naturally when it's required. Here would be a good place to re-emphasize that the airplane flies relative to *him* when the controls are deflected, regardless of its flight attitude.

Point *4* is easy because the airplane will be pointing straight down at the prechosen road. If possible, always line up over a road or other long reference for instructing loops, Cuban eights, or cloverleafs. The throttle should be reduced at this point.

You should be able to see wing alignment problems at points *1, 2,* and *3* in Fig. 22-7, but assuming that you missed wing attitude errors at the first three points, you can still analyze any asymmetric pull on the stick or wheel at point *4*. *If the airplane has to be rolled to the left at point 4* (headed straight down) *in order to make the wings perpendicular to the road, the initial error on pull-up was a roll to the left.* In other words a *required* correction to the left meant that the initial error was to the left (and vice versa). Take a model and deliberately roll it to the *left* as you pull it up into a loop (use a desk edge or other straight line as your road reference). You'll see that the loop is apparently "tilted" and has to be rolled *left* to be aligned (wings perpendicular) to the reference.

When the airplane is at point *5* it's a little late to save the maneuver, but you still should have him check it because it's good for air discipline and if the loop is to be followed by another maneuver (particularly another loop) the proper wing attitude should be looked at here.

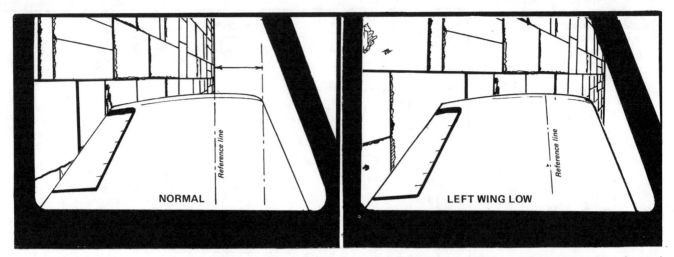

Fig. 22-8. Checking the wing position at the vertical (nose-up) pitch position—a comparison of normal and "wing low" attitudes at point *2* of Fig. 22-7. Use the left wing for *your* reference as well even though you are in the right seat; by doing this, you can check the trainee's reactions as well as see if *he* is watching the left wing instead of staring over the nose.

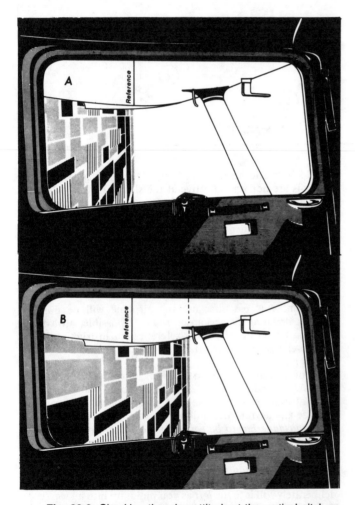

Fig. 22-9. Checking the wing attitude at the vertical pitch-up attitude. (A) The proper wing position. (B) The wing is low, either because of an input of left aileron during the initial pull-up or lack of "torque" correction as the airspeed decays (at full power). Some few people have a "high" wing at this point, caused by input of *right* aileron during the pull-up. (*The Basic Aerobatic Manual*)

RUDDER USE IN THE LOOP. This, like the wing attitude discussion, shouldn't be covered too much in detail at first, but later when he has a good grasp of elevator handling you should introduce the proper use of the rudder in the loop. Rudder use has a lot to do with wing attitude since yaw can induce roll problems. Basically, for most airplanes left rudder is needed in a dive, and right rudder is needed in a climb and at any low speeds with high power settings.

Figure 22-10 gives a look at a hypothetical situation in an airplane that uses 120 K as a loop entry speed.

■ **Demonstration.** Fly that first demonstration loop as smoothly as possible. Brief the trainee to tense his stomach muscles as you do it. As noted in Chap. 20, when *he* is doing the pull-ups he'll automatically take care of this. Talk it up as you go through. You'll find you'll have to talk pretty fast, and if you've properly briefed the trainee, you'll soon be able to be cryptic and still have him understand. (Each of *his* loops should be reviewed after completion, too.) This is a pretty spectacular maneuver to him at first and you should use a combination of enthusiasm and calm as you go through it. Make sure you are coordinated on the rudder, too.

■ **Practice.** In first trying the maneuver, the trainee will usually be somewhat shy about using enough back pressure at the beginning of the loop. You'll have to remind him to do this. You'll handle the throttle his first few loops.

Fig. 22-11, *A–J,* shows a loop in an Aerobat.

Common errors in a loop include these:
1. Neglecting to keep an eye on the tachometer during the initial dive so that the engine limits are exceeded. This, of course, is later when the *trainee* is handling the throttle—you should know better.
2. Too easy an initial pull-up so that the airplane is too slow as the top of the loop is approached. (It is possible to decrease speed too rapidly by too much back pressure as well.)
3. Another problem that may arise as the trainee does more loops is a too abrupt or jerky application of back pressure initially. This is usually an overreaction to his earlier timidity at this point in the maneuver. The back pressure should be that necessary to do the job, but applied smoothly.

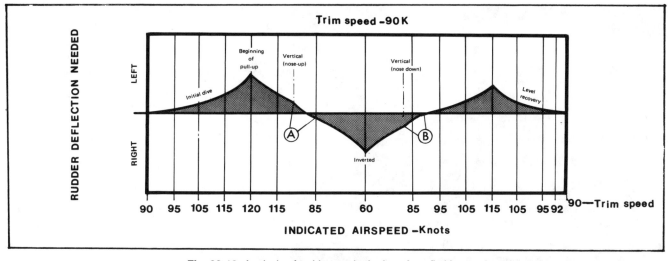

Fig. 22-10. Analysis of rudder use in the loop for a fictitious trainer. At point *A* the power was applied, resulting in less demand for left rudder. At point *B* the power was reduced.

4. Inadvertent aileron application as back pressure is being exerted on the pull-up; the lateral axis does not stay parallel to the horizon (the airplane is banked). Watch for this when you are in the inverted position; the wings should be parallel to the horizon. Later he will be able to correct for any deviations handily. One indication of good control of wing attitude is the feeling of a "bump"

(encountering your own wake turbulence) as the loop is completed. This usually startles the trainee, who may ask, "What did I do wrong?" It's good to mention this during the preflight explanation of the maneuver. Good downward paths in the cloverleaf and Cuban eight have this bump, also.

5. Neglecting to relax the back pressure as the in-

Fig. 22-11, A–J. Steps in the loop, as seen from the cockpit and alongside the airplane. (*The Basic Aerobatic Manual*)

(A) Clear the area before starting the loop, *then* ease the nose over (wings level).

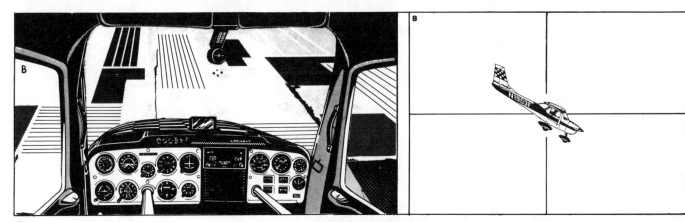

(B) Here the airplane is at the recommended entry speed (120 K for the Aerobat).

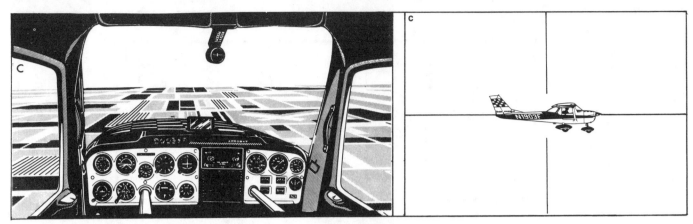

(C) The pull-up has started and the airplane is passing through the level-pitch attitude.

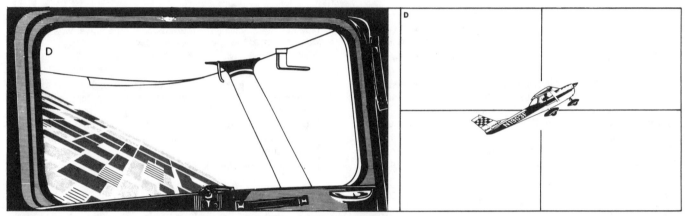

(D) At this point 3.5 g's are being pulled. Start looking at the wing tip.

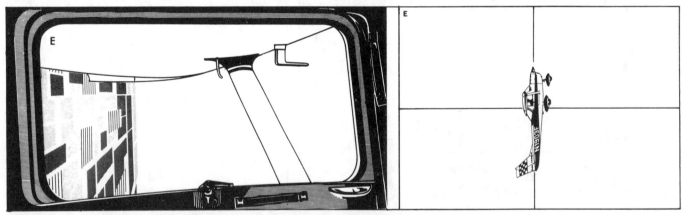

(E) The pitch attitude is vertical. Check the wing tip reference on the horizon and open the throttle fully. The stall warner will start sounding at this point.

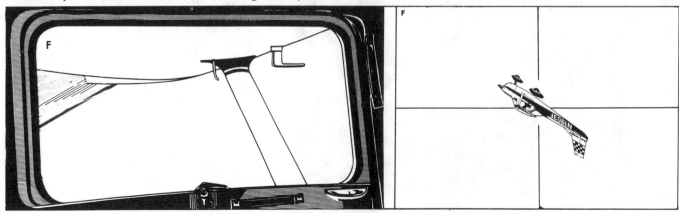

(F) At this point (nose 30° above the horizon), relax the back pressure just enough to silence the stall warner and move your view back from the wing to the nose (and through the skylights in the Aerobat).

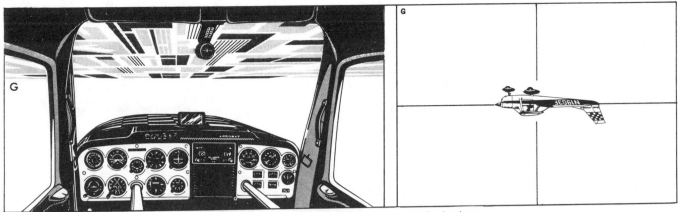

(G) Still relax back pressure slightly. Check that the wings are level in reference to the horizon.

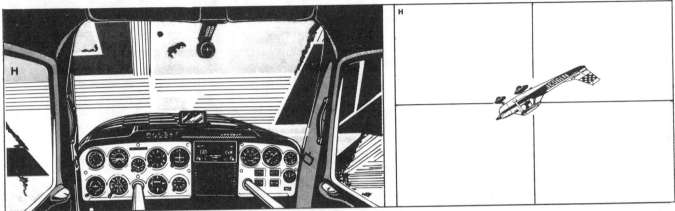

(H) At this point (nose 30° below the horizon), start smoothly reapplying back pressure.

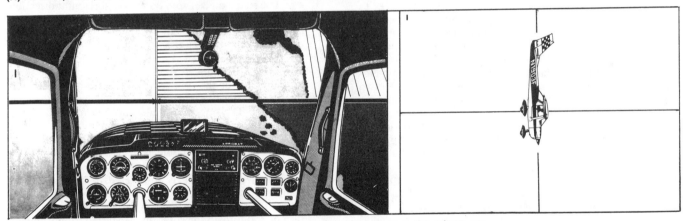

(I) As the pitch attitude reaches vertical, reduce power and make sure that the wings are perpendicular to the reference line (the light-colored road, here). Continue the back pressure. (Some pilots subconsciously believe that the airplane can't be stalled with the nose straight down and are able to do interesting maneuvers when the back pressure is applied too enthusiastically.)

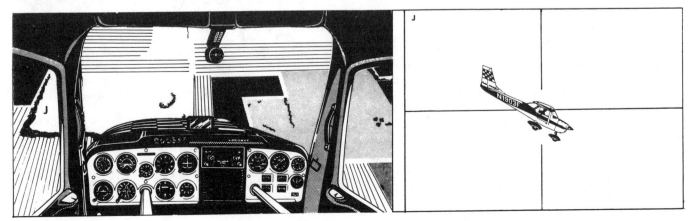

(J) Continue the pull-out. The airplane will be pulling about 3.0 g's, and you may feel the bump at this point.

verted position is being approached so the airplane makes a tight rotation at the top—with buffeting and a possible stall. One reason for this is that right after the pull-up of 3+ g's, 1 g may seem too light so he unconsciously tries to keep up the g forces.

6. Too positive relaxing of back pressures so that the occupants may tend to hang in the belt and harness. This is usually an overcompensation for error 5.

7. Forgetting to retard the throttle in the last part of the loop and exceeding the engine rpm limits.

8. Too abrupt back pressure on the back side of the loop so that buffeting or a stall occurs. This usually occurs between points *H* and *I* in Figs. 22-5 and 22-11.

9. If there is a strong crosswind at your altitude the loop may drift slightly away from the road during the 20 sec or so required to complete the maneuver. The trainee may try to "point" the airplane over to the road during the last 90° of the circle, changing or bending the plane of the loop instead of paralleling the road. If it appears that this is a problem, you'd better pick a road parallel with the wind, or do the loops perpendicular to the original road.

After he has had practice with the loop and can do the maneuver reasonably well, you should then let him handle the throttle. Expect a slight deterioration in performance when he starts this. Usually the problem is that he is so carefully watching the tachometer that he prematurely relaxes the back pressure as the nose is pointing straight up; this makes for a certain looseness at the top of the loop, and you will have to remind him about it.

This brings up the question: Should he use both hands on the wheel in the loop? There are a couple of advantages to this: (1) The back pressure on the wheel will be symmetrical, that is, there will be less of a tendency to introduce aileron deflections during the maneuver; and (2) for a small person, two hands are virtually required during the first few loops practiced in certain airplanes. If possible, use the one-hand-on-wheel, one-hand-on-throttle procedure, because when two hands are used on the pull-up a change in back pressure results as one hand is taken from the wheel to add power and then returned to the wheel. (Or there may be a slight change, as well, as power is reduced on the back side of the loop.) The airplane equipped with a stick usually doesn't present a problem.

However, if the particular airplane and physical strength of the trainee are such that both hands are required to make the initial pull-up and the loop is otherwise okay (and the trainee is enjoying it), you could discourage a potentially good aerobatic pilot by insisting on his using one hand.

Fig. 22-12 is a simplified version of Fig. 22-5 with explanations that can be used for a quick sketch on the chalkboard for your pre- and postflight briefs on the loop.

■ **Evaluation and Review.** After a dozen or so loops (not in succession!) the average trainee should be able to complete the maneuver, recognizing and correcting for elevator-and-aileron-use errors. You might expect an occasional feeling of being "slightly light" at the inverted position, but after a dozen loops, with your advice each time, he should be able to see his mistakes and start analyzing each one himself. You will want to make sure that he is able to do the basic loop well before moving on to the cloverleaf, Cuban eight, or Immelmann, since these maneuvers are more complicated variations of it.

Immediately following each loop you'll *review* his actions and note errors. You'll also make an *evaluation* of whether to let him start using the throttle, follow the loop with an aileron roll (see Chap. 23), etc.

After you are on the ground, do a general review of the loop and aileron roll. Introduce and explain the cloverleaf, Cuban eight, and Immelmann as maneuvers combining *two* of the Three Fundamentals practiced thus far.

Again, if anything of note requiring a lengthy discussion occurred in flight, here's the place to talk about it (after the flight).

THE SNAP ROLL

In discussing the Three Fundamentals, this one should be mentioned briefly but for most people should not be demonstrated until the third flight (or even later for "problem" trainees) after they have done a number of cloverleafs and Cuban eights.

■ **Preparation.** If the trainee has not had spins before, an explanation of the snap roll should be preceded by an explanation of the normal spin. If everything is proceeding very well during the second flight, as was indicated in the last chapter, you might explain briefly and demonstrate one spin and do one or two at the end of each following flight to lose altitude.

■ **Explanation**

What—The snap roll is basically a timing maneuver that is a good aid in developing orientation ability. Although coordination of rudder and elevator is required, it is the type of coordination not encountered in normal flight.

The snap roll is a spin done in a horizontal direction (Fig. 22-13). The airplane is stalled at a higher-than-normal speed and an imbalance of lift is created when the rudder is fully deflected as the stall is introduced. The result is an accelerated stall with a rapid rotation in the direction of the applied rudder. To recover, use the same technique as for the spin recovery—opposite rudder and smooth, yet positive, firm forward movement of the control wheel. The opposite rudder and the forward movement of the wheel stop the rotation and break the stall. If timed properly, starting at about three-fourths of the way around, the recovery should be to wings-level flight at a speed 15–20 K below the entry speed.

The true snap roll is considered to be a maneuver using only the rudder and elevator, but you'll find later that for some airplanes aileron can be used in the direction of roll to help the rotation get started (opposite aileron also can be an aid in making precise recoveries). The trainee will soon be subconsciously using the ailerons as necessary to help start and stop the maneuver.

The use of *prospin* ailerons in a developed spin in the Aerobat noticeably increases the rate of rotation and can be used to get a sharper, higher rate of roll maneuver. The use of pro-ailerons in that airplane must be delayed until the roll induced by the rudder and elevator has started. Premature use of ailerons will "unstall" the inside wing (that up-aileron decreases the effective angle of attack) and stop the roll, with the airplane sitting in a slightly banked attitude, buffeting, but not getting on with the snap roll.

Why—It's an excellent maneuver for the improvement of timing and orientation, and since it's a Fundamental it is a basis for more complicated maneuvers as well as the combination of maneuvers to follow.

How—Ease the airplane's nose up in a shallow climb at the recommended speed (use full power). Smoothly and quickly pull the control wheel straight back to the full aft

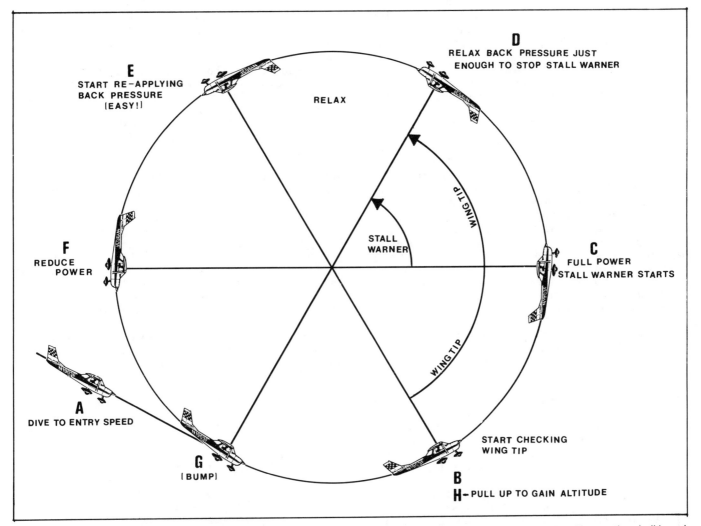

D
RELAX BACK PRESSURE JUST
ENOUGH TO STOP STALL WARNER

E
START RE-APPLYING
BACK PRESSURE
(EASY!)

RELAX

WING TIP

STALL
WARNER

F
REDUCE
POWER

WING TIP

C
FULL POWER
STALL WARNER STARTS

A
DIVE TO ENTRY SPEED

G
(BUMP)

START CHECKING
WING TIP

B

H—PULL UP TO GAIN ALTITUDE

Fig. 22-12. A simplified look at Fig. 22-5 for sketching on the chalkboard.

Fig. 22-13. Snap roll. Make the point that the airplane appears to be in a horizontal spin in this illustration.

position while simultaneously applying full rudder in the desired direction of roll.

At about three-fourths of the roll completion (270° of rotation), apply opposite rudder and move the control wheel forward to break the stall. The point to start recovery depends on roll rate and so will vary among aerobatic airplanes. The chances are good that the recovery from the trainee's first few snap rolls will not be in exact wings-level flight, and he may need to do several before gaining the ability to stay oriented. The first time a snap roll is demonstrated to him, the maneuver will likely be completed before he realizes it has started. Later he'll be surprised at how the action seems to slow down; orientation will be no problem and the snap roll will no longer be a blur of ground, horizon, and sky gyrating in (apparently) all directions.

■ **Demonstration.** This is definitely a maneuver for which demonstrations should be held to a minimum. Here's the point where time as a factor in g-force effects makes itself known. Even though you'll pull 3 g's, the same as in parts of a loop, the time factor makes them seem much less. Ask the average trainee after that first loop to estimate how many g's were pulled and he'll say, "About 12, I reckon." Ask him after a snap roll and he'll probably estimate 1½. This is a slight exaggeration, of course, and it's not the g force that tends to cause queasiness in the snap but the "uncoordinated" rolling.

It's suggested that you do one snap, discuss it, and, after a short rest period, let the trainee do one. (One instructor used to tell his trainees, "Okay, now you try it." He was reminded that it is better psychology to tell them to *do* it, not *try* it.) Review his snap roll, and if possible (after a rest) have him do another. Usually the trainee's problem is one of timing (he's late in starting his recovery at first and it's a matter of your reminding him to start an earlier recovery next time).

■ **Trial and Practice.** Fig. 22-14, A–E gives a look at a right snap roll in an Aerobat.

> *Common errors* made doing snap rolls are these:
> 1. Not moving the control wheel aft fast enough; the "snap" is more of a "mush."
> 2. Trying to use the ailerons instead of the rudder to get rotation started. This is particularly evident after he's been doing a number of aileron rolls.
> 3. Not enough rudder or elevator deflection at the initiation so that the airplane is in a slow, yawing, non-stalled roll.
> 4. Poor timing on recovery. Expect that he will overshoot the stopping point on the first few recoveries. Later, for a change of pace, he'll probably start the recovery too soon but will get the corrected timing after a few more snaps. If you find yourself inverted from an accidental half snap or snap and a half, remember to remind him to *roll* out.
> 5. Shoving the wheel forward on the recovery; that is, making the occupants feel light on the seat, pushing the nose well down below the horizon, and emptying the ashtrays. Although positive forward pressure is necessary, it can be overdone. Usually the problem is that the forward pressure is held after the recovery has been completed.
> 6. Not moving the wheel or stick forward after applying opposite rudder. The stall is not broken and the airplane may rotate or snap back to a banked attitude with the nose down (and a lot of buffeting occurring). The trainee doesn't realize that he is holding the back pressure; the more alert ones will ask, "What's going on?"—and you'll tell them.

■ **Evaluation and Review.** After about 10 snaps, the average trainee should be able to stop the maneuver with the wings being within 10° of level. You'll find occasional lapses of skill and misuse of the controls, but the maneuver will get clearer and he will make some of his own analyses. It's always best to have a few snap rolls and/or spins at the end of the flight rather than a long series of them *during* the flight. As was mentioned earlier, the trainee can work this out, and for some of them *any* maneuver can be done anytime they want to. (If *you* can hack it.) So, there's one good thing about being an instructor—if a certain maneuver starts getting close to

you, you can say, "Well, that's fine. Your snap rolls are okay, but let's see how your straight and level is for a while. I'll just open my ventilator some more to cool things off a bit." (See Fig. 3-3.)

The snap roll is important as the Third Fundamental, but other than as a fun demonstration of an accelerated stall, it's not of particular value to the average pilot taking aerobatics to make his other flying safer. (Most, though, would like to see one demonstrated.) After you are on the ground, cover in detail anything that should be corrected. If he is getting the idea, say so!

A NOTE ON THE ACCELEROMETER. The accelerometer is a very important aid in analyzing and evaluating the trainee's performance of a particular maneuver. The g requirements of each type or model for a good loop, for instance, may be slightly different, but you should know the numbers for the airplane in which you're instructing aerobatics. Looking at the g meter after a loop is completed will help you check the trainee's problems, such as a too-light pull-up (which makes the airplane too light on top of the loop) or aileron or snap roll problems. Fig. 22-15 shows good postmaneuver g-meter indications for an aileron roll, loop, and snap roll in an Aerobat.

Emphasize that after every maneuver any excess airspeed (above best rate of climb) should be used to gain any lost altitude—and for most of the civilian aerobatic trainers used today, there will be an altitude loss. Convert the "energy altitude" to real altitude. Too many trainees level off at a high speed (such as following a loop or a spin recovery) and let the airspeed dissipate without any altitude gain. *Then* the climb to get back up is started.

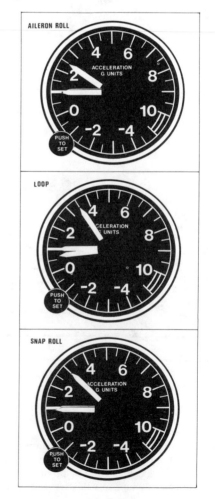

Fig. 22-15. Expected accelerometer readings after a good aileron roll, loop, and snap roll in a Cessna Aerobat.

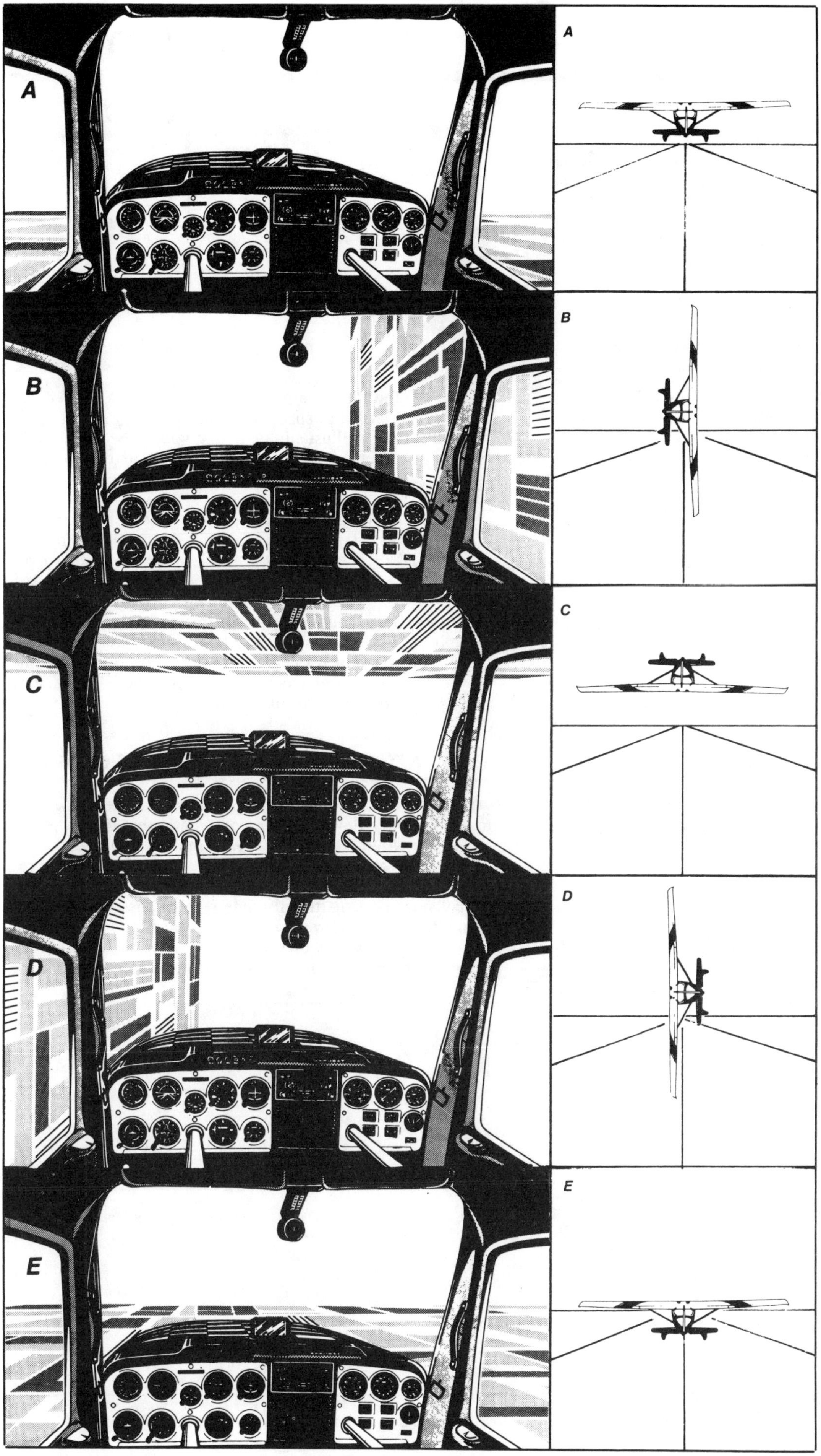

Fig. 22-14. Right snap roll as seen from the cockpit and from behind the airplane. (*The Basic Aerobatic Manual*)

(A) The airplane is at the recommended snap roll airspeed (80 K for the Aerobat) with full power, the pitch attitude slightly nose-up.

(B) Full right rudder and full aft wheel have been applied. After rotation has started, full right aileron is added.

(C) Right rudder, aft wheel, and right aileron are still held.

(D) At the 270° position, opposite (left) rudder and aileron are applied and the wheel moved smoothly forward.

(E) When the rotation has stopped, the controls are neutralized and normal cruise resumes.

COMBINATIONS AND VARIATIONS OF THE THREE FUNDAMENTALS

LOOPS FOLLOWED BY AILERON ROLLS

After the trainee has practiced several loops and is comfortable using the throttle in the maneuvers, you should have him do an aileron roll immediately out of a loop. That is, he completes the loop, pulls the nose up to 30° above the horizon, and completes a left or right aileron roll, alternating the direction after each loop. The Aerobat will have the proper aileron roll airspeed at the completion of the loop; other airplanes may have different required numbers.

As the trainee becomes more proficient in this combination you should have him reopen the throttle in the roll when the airplane is inverted. By this time the airspeed should have decayed so that no overspeeding occurs (fixed-pitch prop). The sequence of steps in throttle use in the loop and roll should be as follows:

1. Initial dive for loop—reduce power.
2. Vertical pitch-up attitude (loop)—full power.
3. Backside of loop (vertical pitch-down)—reduce power.
4. Inverted in the roll—add power.

Rather than require a continued practice of only aileron rolls after that first flight, you'd do better on the second flight to have the trainee warm up with several aileron rolls and then move on to loops and loops followed by rolls. The trainee's rolls will actually improve during the loop-roll practice.

> *Common Errors:*
> 1. The usual loop errors. (Not enough initial "pull," and/or inadvertently applying asymmetric back pressure.)

2. Too much back pressure on the back side of the loop so that the airspeed is too low for a good roll. (The stall-warner will usually be letting you know about this.)

3. Not raising the nose high enough following the loop, so that there is a noticeable loss of altitude as it is completed.

4. Hesitating to start the roll. (The nose is pointed up and the airspeed decays so that the roll is sloppy.)

5. Not completing the roll headed parallel or over the ground reference (road or railroad). (This is caused by the usual aileron roll common errors; see Chap. 22.)

6. Still thinking about problems in the loop while doing the roll, assuring that the roll will have problems, too. For all your trainees, tell them to complete the combination and *then* analyze mistakes as the airplane is climbing for the next maneuver.

LOOP WITH A QUARTER-ROLL RECOVERY

■ **An Introduction to the Cloverleaf.** The maneuvers most commonly flown using a combination of the loop and roll are the *cloverleaf, Cuban eight,* and *Immelmann.*

The cloverleaf is four loops, each with a quarter-roll recovery (the four quarter rolls are made in the same direction). The rolls are done as the airplane reaches the vertical nose-down attitude. The cloverleaf, disregarding the physical effort involved, is the easiest of the three maneuvers, because the airplane is pointed straight down and the airspeed is picking

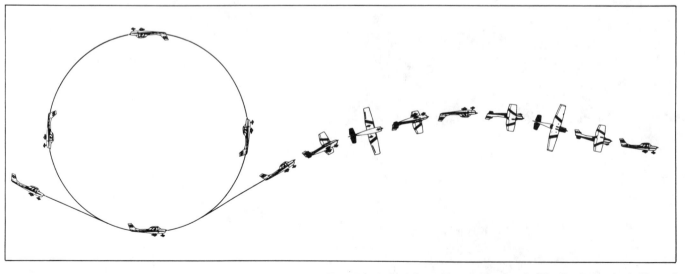

Fig. 23-1. Loop followed by an aileron roll. (*The Basic Aerobatic Manual*)

up, making the controls more effective; also the references for the 90° roll are readily seen. Because of these factors the Cuban eight and Immelmann are more difficult, in the order mentioned, for the new trainee—particularly in a low-powered aerobatic airplane.

Introduce the loop and the quarter-roll recovery, doing only one at first; then, after he's able to make the 90° roll and pull-up to level flight, bring in the idea of combining two, then three, and finally four to complete a cloverleaf. The cloverleaf is so named because, when seen obliquely, the path of the plane traces the outline of a four-leaf clover.

The full cloverleaf can be a physically tiring maneuver, so you should probably stick to a maximum of two loops with quarter rolls until later in the training.

■ **Preparation.** The complete cloverleaf is very hard to draw on the chalkboard, so a model airplane is the most important aid in explaining this maneuver. Have the trainee read about the cloverleaf in the references. If you haven't done one lately you'd better review it yourself to have it well in mind before this flight.

■ **Explanation.** Explain that, although it looks complicated, the loop with a quarter-roll recovery is no more than a combination of two of the Three Fundamentals, which the trainee can do well individually (at least he *should* be able to do them well before you start him on this maneuver). See Fig. 23-2.

You should review some of the problems he might have, such as the tendency to let the nose come up during the roll,

which results in the speed being too low for another loop segment. When having him practice a single loop and quarter roll, make sure that he completes it with enough speed to enter another loop if desired or he'll start a habit that will follow over to the full cloverleaf.

■ **Demonstration.** A demonstration of *more than two* loops and quarter rolls is not necessary.

■ **Practice.** It's best to line up directly over a road so that the loop portion can be checked. As for the roll portion, the trainee doesn't have to be directly over a road to judge a 90° roll; he can estimate the roll very well from a parallel position.

You can expect the first few times that the trainee will bring the nose up as soon as the roll is completed, and it will take a definite effort on your part to convince him to keep the nose down.

After you've admonished him a few times about the too-early (low-speed) pull-up, you'll find that in the cloverleaf he may start letting the speed build up so high that the altitude loss at the completion of even one loop is unacceptable. Watch for this.

For the loop with a quarter roll, have him lead the next loop entry speed by about 10 K for the average aerobatic airplane currently in use (since the nose is—or should be—pointed straight down as the quarter roll is completed). For the Cuban eight, where the nose is pointed down at a 45° angle when the half roll is completed, a lead of 5 K on pull-up is recommended. Of course, you will work out the best lead

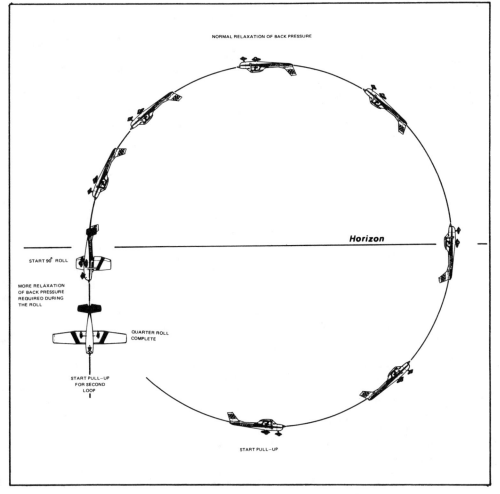

Fig. 23-2. Loop with a quarter-roll recovery (one-fourth of a cloverleaf). The cloverleaf is the first maneuver in this book to look at two of the Three Fundamentals *within* a maneuver. (The loops and rolls were two fundamentals stuck together.) (*The Basic Aerobatic Manual*)

NORMAL RELAXATION OF BACK PRESSURE

Horizon

START 90° ROLL

MORE RELAXATION OF BACK PRESSURE REQUIRED DURING THE ROLL

QUARTER ROLL COMPLETE

START PULL-UP FOR SECOND LOOP

START PULL-UP

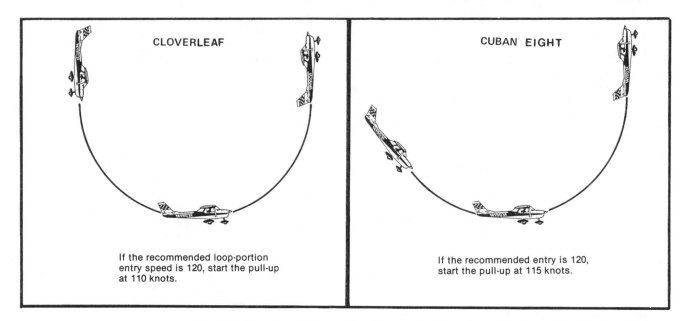

Fig. 23-3. Examples of airspeed lead for starting the pull-up (after the roll) for the cloverleaf and Cuban eight. (*Cessna Aerobat*)

for your particular airplane and maneuver. Fig. 23-3 gives an example.

The quarter roll should be brisk, not dragged out. The fairly rapid roll will make the maneuver easier to do because he completes the roll and then is free to concentrate on the airspeed and pitch attitude.

Fig. 23-4 shows a view from the cockpit at the start and completion of a quarter roll to the right.

> *Common errors* in the cloverleaf are these:
> 1. Problems with loop or roll because the trainee may not be able to concentrate on each part. (He's thinking about the roll while doing the loop, or vice versa, as was mentioned earlier about the loop followed by a roll.)
> 2. Pulling the nose up too soon after the quarter roll is completed; the airspeed is too low for the next loop.
> 3. Keeping the nose down too long, losing too much altitude.
> 4. A too-slow quarter roll.
> 5. Poor throttle handling; power applied too soon and retarded too late.

You should indicate that this is a good maneuver for air discipline. The airplane is headed straight down and the trainee's subconscious is telling him to pull out *now,* but he has to complete a specific task (a quarter roll) before doing it. This might be tied to a situation in which the airplane is in an accelerated stall condition close to the ground and he wants to continue pulling (a usually fatal mistake) instead of *thinking* and temporarily relaxing the back pressure even at the short-term cost of more altitude.

■ **Evaluation and Review.** The average trainee will tend to get faster with each loop portion of a cloverleaf. You can figure that he is proceeding very well if, with a typical airplane, he only loses 100 ft for each additional loop portion of the cloverleaf, or 300 ft in addition to the initial dive (which might take about 700 ft). You'll know the average for the particular airplane you are using and can use this as an evaluation of his progress in the loop-type maneuvers.

Keep checking the loop portion of this maneuver; the average person tends to get a little sloppy in this portion as he

practices more and more cloverleafs and gets involved in the whole maneuver. You may have to remind him concentrate on each part.

A brief postflight review with a discussion of his errors in the loop with a quarter roll and the cloverleaf as a whole should be done.

THE CUBAN EIGHT

■ **Background.** The Cuban eight is also a maneuver that combines parts of the loop and the aileron roll. The airplane's path describes a figure eight lying on its side (Fig. 23-5). Basically, it is three-fourths of a loop and a descending half aileron roll (the nose will be down at an angle of approximately 45° when the half roll is completed) to "right-side-up" flight. The nose is kept down at the 45° angle until the recommended entry speed is again attained (assume a proper airspeed lead); then the second loop portion of the maneuver is started. When the airplane is again inverted with the nose down at about a 45° angle, the half roll is completed again to an upright attitude. Cruising flight is regained after the second loop and half roll.

The Cuban eight is easily drawn on the board, but a model is the greatest aid.

The Cuban eight is a little more difficult than the cloverleaf because the pitch attitude at the roll is not straight down. This means (1) the reference is a little harder to see, plus (2) the airspeed (and control effectiveness) is not picking up so fast. Also (3) 90° more roll is required.

This maneuver is one of the best for getting the trainee involved in aerobatics. He's seen the Cuban eight at air shows and competitions, and this one grabs his imagination. The cloverleaf is a fairly complex maneuver but is basically four loops in succession, whereas to him, the Cuban eight is a *real* exercise in precision (which, of course, it is).

If an airplane gets into wake turbulence and is rolled inverted, the nose will tend to drop and the pilot (assuming that there is enough altitude available) must roll upright and get the nose up to avoid hitting the ground. In some cases, trying to *push* the nose up above the horizon to stop any descent could overstress the airplane—and most pilots would

Fig. 23-4, A–C. Vertical quarter roll to the left. (*The Basic Aerobatic Manual*)

(A) A brisk roll is started as the nose reaches the vertical (down) position.

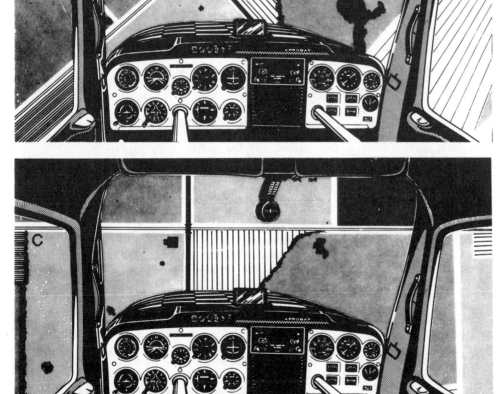

(B) Use sufficient rudder with the aileron so that the roll is "brisk" but coordinated. The 45° of roll position is shown here.

(C) The quarter roll is complete and the pull-up just starting.

neither think of it or be able to accomplish such a procedure. (They're more apt to *pull* than push the wheel or stick.)

Here you can explain that although the nose is down in the wake turbulence situation a push-and-roll effort would still be the best thing to do. The pilot would roll upright and pull up immediately, but in the Cuban eight he would keep the nose down to get the proper lead airspeed for the second half.

■ **Explanation.** Review the loop briefly, citing such points as throttle handling, where to look, and points of relaxation of back pressure. Use the model to go through the Cuban eight completely at least once, but concentrate on explaining just half of the maneuver. You should emphasize, however, that even though he's doing only half of the maneuver at first, it should be ended so that he could go immediately into the

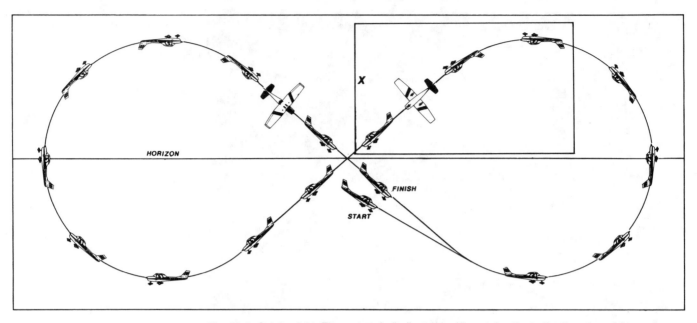

Fig. 23-5. Cuban eight. The rectangle (indicated by *X*) contains the half-roll portion of the maneuver. (See also Figs. 23-6 and 23-7.) (*The Basic Aerobatic Manual*)

second half. He should complete the first part headed down at a 45° angle and maintain this until the proper lead airspeed is reached.

You may want to handle the throttle for him at first, and as before, you can expect a temporary slight decline in performance when he starts using the throttle. It probably won't be so noticeable as was the case earlier, since he's had more practice using power in the loop.

THE NEED FOR FORWARD PRESSURE. Mention that he can expect a requirement for an unusual (to him) amount of "forward" pressure at the beginning and throughout the roll. Discuss the three main reasons for this, given here in italics (the other information is for your use):

(1) *As the nose reaches the position of 30° below the horizon, the back pressure should be relaxed even more than it was for the loop portion—in order to stop the loop.* (The roll is begun at this point.) Late or too little relaxation will allow the nose to be too low when the roll is completed and then a Cuban eight suddenly turns into a cloverleaf. Demonstrate with the model.

(2) *As the roll continues and the wing attitude approaches vertical, the airplane will want to turn, and the relaxation of back pressure or use of forward pressure, as required for a particular airplane, stops this.* If, in a half roll to the left, the nose is pointed to the right of the road at the completion, one of the major causes is the lack of forward pressure (or decreased back pressure) in the "knife edge," (90° bank) allowing a turn. The other major cause in this case is a lack of left rudder in the final 45° of roll. Since the problem of lack of forward pressure occurs halfway in the roll, it is sometimes overlooked by the instructor as a major factor in heading problems at the end. As will be discussed later in the Immelmann, a good aid in maintaining heading is the application of opposite rudder just as the half roll is started, followed by rudder *with* the aileron. This is more of a half slow roll than an aileron roll and is the best way to keep the nose lined up precisely, but the *half aileron roll is usually the best introduction.* As an instructor, you'll find a problem at the end of the maneuver is often the result of a slipup at the beginning, particularly in aerobatics. In other words, if the trainee is

ending the maneuver off-heading or having other trouble, don't ignore earlier factors that could have started the problem.

(3) *When the roll is completed and the nose is pointed down at a 45° angle, it will tend to pitch up as the airspeed increases.* You should mention that he can expect at one time or another his initial forward pressure to be too great and the engine to quit temporarily, but the roll should be continued. The engine will start again, assure him.

Fig. 23-6 shows the half roll of the maneuver as indicated by the rectangle in Fig. 23-5. Fig. 23-7 shows what the left-seat occupant will be seeing during the half roll.

STEP-BY-STEP EXPLANATION. You might explain the maneuver as follows:

1. Pick a straight stretch of road or section line and line up over it. Look around to make sure that somebody else hasn't decided to practice Cuban eights on this same stretch.

2. Lower the nose (wings level) to a 45° angle to attain the recommended speed. Ease the throttle back to prevent engine overspeed. Initiate a smooth pull-up and smoothly open the throttle to full power as the nose reaches the vertical pitch position. A check will show 3.0–3.5 g's resulting from a proper pull-up. (Again, don't let the trainee read the g meter while doing the action.)

3. Continue the loop (pulling positive g's) until the nose is pointed down at a 30° angle (the plane is inverted). Release more back pressure and do a half aileron roll to the upright attitude. A half roll to the left will be easier.

There will be time to complete the roll and *then* reduce power before the rpm gets too high. The nose should be kept down at the 45° angle until the airspeed is again at the entry speed (check that rpm). Repeat the loop and half roll.

He'll probably be so involved with finding himself inverted, headed downward at a 45° angle, and with the half roll, that the throttle may be the last thing to cross his mind— unless he's reminded—so *you* watch the rpm.

■ **Demonstration.** Demonstrate only one full Cuban eight before you allow him to try one. After you've completed the demonstration, review it as you climb back to the starting

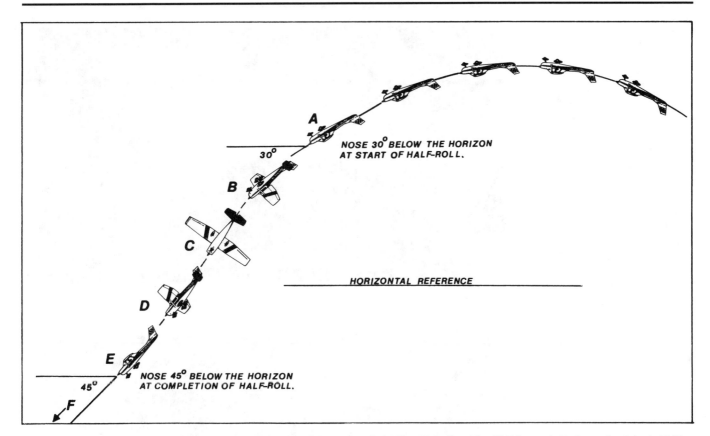

NOSE 30° BELOW THE HORIZON
AT START OF HALF-ROLL.

30°

HORIZONTAL REFERENCE

NOSE 45° BELOW THE HORIZON
AT COMPLETION OF HALF-ROLL.

45°

Fig. 23-6. Details of the rectangle in Fig. 23-5. (See Fig. 23-7 for cockpit views of points *A–F*.) The nose traces a helix from 30° down (inverted) to 45° down (upright). At the point where the back pressure would be started again in the loop (*A*), start a half roll and release *more* back pressure as the airplane rolls. As the roll continues (*B*), slightly increased forward pressure may be needed. Most aerobatic trainees tend to slack off on the forward pressure at *C*, and the nose is allowed to move to the right (in a left roll) of the reference line. This slacking off is one of the major causes of being off-heading at the end of the maneuver. In addition to the forward pressure being an important factor for directional precision, the increasing of top (left) rudder is needed to maintain heading as the roll is completed (*D*). Forward pressure is needed at *E* to keep the nose from pitching up, which would result in a too-low airspeed for the next loop. As the airspeed reaches the proper value (*F*), bring the nose up through the horizon for the second loop portion of the maneuver. (*The Basic Aerobatic Manual*)

altitude. As an aerobatic devotee, you can generally think of the rest period between maneuvers also as *a time to climb*.

■ **Practice.** At the end of his first half Cuban eight you should be pleasantly surprised if he is within 20° of the desired heading. The use of a long, straight stretch of road, railroad, section line, etc., is *essential*. The trainee can *see* if he is getting off-heading, and can start making his own corrections early. Work on the problems mentioned earlier, allowing him to rest between maneuvers.

If he is doing reasonably well with one-half of the maneuver, try to have him complete *at least one* full Cuban eight before the period is over.

Common errors in the Cuban eight are these:
1. Problems associated with the loop, such as not enough back pressure at the beginning and/or too much back pressure at the top of the loop so that buffeting occurs.
2. Applying back pressure or not enough forward pressure throughout the half-roll portion of the maneuver, causing deviation from the reference heading. The other side of this problem is pushing *too hard* during the half-roll portion, resulting in the trainee (and you) being pushed against the belt and shoulder harness. This makes

him inadvertently push harder on the wheel or stick, making him lighter in the seat, resulting in a harder push on the wheel, and so the cycle goes. (The trainee will often ask, "What happened?" or "What's doing this?"— not realizing that *he* has created the problem.) If the airplane does not have an inverted system, the engine's hesitation with the zero or negative g's adds to the excitement.
3. Not completing the roll at the 45° nose-down attitude; having the nose too high or low in the dive. Usually, after he's done a session of cloverleafs it will tend to be too low.
4. Too much early aileron deflection, resulting in a buffet or aileron stall.
5. Not enough top rudder as the roll is completed, allowing the airplane to get off heading at that point.
6. Neglecting to retard the throttle as the airspeed picks up in the dive(s).

Indicate that forward pressure is needed starting at the 30° nose-down (inverted) position for three reasons: (1) to stop the loop, (2) to stop any turn away from the reference (the airplane will want to turn the way it's banked), and (3) to stop the pitch-up as the roll is completed and the airspeed is high.

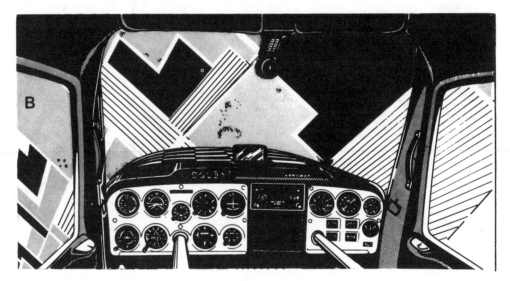

Fig. 23-7. Half roll of a Cuban eight as seen from the cockpit. Views A–F would be seen at points *A–F* in Fig. 23-6. (*The Basic Aerobatic Manual*)

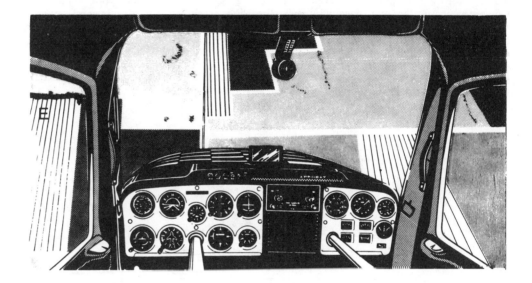

■ **Evaluation and Review.** Don't expect great things from the trainee with the Cuban eight even after he has done, say, two dozen of them. His later tendencies will be to cheat a little on the 45° pitch-down attitude; he'll try to have the dive too shallow, particularly if he is doing a series of maneuvers and/or has a minimum altitude restriction.

If he is getting "bumps" as he passes through his own wake turbulence in the second dive of the Cuban eight, he generally has good control of the vertical plane of the maneuver. Of course, you will run into an individual who does a *three*-dimensional Cuban eight all over the sky and somehow manages to cross his wake turbulence. This should be obvious, as is the example of the 720° power turn in which the altitude goes wild but wake turbulence is somehow encountered as the airplane recrosses an altitude.

You will have a good idea that progress in the Cuban eight—and aerobatics in general—is going great when the trainee is heading earthward, inverted, at a 45° angle, and you can see that the only thing he's worried about is keeping lined up with that road as he starts, and continues, the roll.

After the flight, review general problems with the maneuver. Go through the entire maneuver using the model; you may be surprised to find a trainee who's been doing the maneuver mechanically, and seemingly well, throughout the flight but isn't able to "see" it. Have him use the model and explain the Cuban eight to you, so that when he leaves the airport after this flight he will have the idea firmed up. Other-

wise, any wrong concepts he has may jell for a few days.

After reviewing his problem areas, find something good to say about his flying that period.

THE IMMELMANN

As you now know, and the trainee will learn shortly, the Immelmann is a maneuver named for a World War I German pilot named Max Immelmann. (Whether Immelmann did this maneuver or even heard of it is a point for conjecture. At any rate it's the popular name and will be used here.)

As he will soon learn, an Immelmann is more difficult to do correctly than a Cuban eight. For most underpowered aerobatic airplanes, a good Immelmann is hard to find. The maneuver consists of a loop with a half roll at the top so that the flight path is changed 180° with (hopefully) a gain in altitude. It is excellent for developing good planning and coordination. Fig. 23-8 shows a side view of the maneuver.

When you start teaching Immelmanns you will be introduced to maneuvers you haven't seen before. Of course, when you were learning the maneuver *you* introduced a few variations of your own.

For instance, there is the *Sky-Suspended-Immelmann*. This is the maneuver in which the trainee stops the nose (inverted) at a too-high attitude, speed rapidly disappears, and the ailerons are centered (in awe). The earth and sky appear

Fig. 23-8. Immelmann. See Fig. 23-9, A–E, for the views from the cockpit. The insert shows the "classic" maneuver.

suspended (inverted in position, of course). The engine quits (if you don't have an inverted system), and it becomes your duty to put things back to rights; the feeling is that the airplane is fluttering down like a clean-shot duck. It takes a little altitude, plus a couple of minutes, for airplane and occupants to get back to normal.

Another variation is a *Combination-Cuban-Eight-and Immelmann*. The nose position at the end of the maneuver is too low for an Immelmann and too high for a Cuban eight. *You'll run into a lot of these.*

■ Explanation

What— The Immelmann is a combination of a loop and a half roll, and while more powerful airplanes may gain altitude during its execution (many trainees *will* literally *execute* it, as in putting it to death), in most trainer aircraft you will be satisfied if the initial altitude is again reached. But it is more important to develop the proper techniques than to worry too much about altitude gain. Like the Cuban eight, a plane drawn through the flight path of the airplane should be perpendicular to the earth's surface.

Why— This maneuver is one of the best for flying the airplane through a wide range of airspeeds and doing a precise half roll at speeds approaching the stall. The Immelmann is one of the more difficult maneuvers for the low-powered airplane because of the low speed at the point of roll.

The Immelmann is a good maneuver to back up your discussion of wake turbulence recoveries. The roll at the top of the loop is at a very low airspeed and the airplane's control effectiveness would be close to that on final when it's slow and dirty. The "push and roll and pull" required for the Immelmann is like recovery from a wake turbulence upset. The roll performance here may be a sobering experience for the trainee who thinks that after an aerobatic course "wake turbulence will no longer be a factor."

How— Line up over a road or section line so that a reference for both directions of flight can be used. If a long, straight reference is not available, pick a prominent object on the horizon and turn the airplane so the tail is lined up with it. As the 180° of "turn" is completed, the reference point is used for the half roll. Check the area for other airplanes and lower the nose to establish the recommended airspeed. Retard the throttle as necessary to stay below the red line on the tach. Initiate a smooth pull-up and gradually open the throttle so that full power is developed by the time the vertical position is reached. Check for the horizon as the airplane pitches past the vertical attitude. Note that about 3.5 g's result from a proper pull-up.

There are two (or more) ways to make a half roll in the Immelmann. Anyway, here are two:

1. Start the roll when the nose is approximately 30° above the horizon and allow the nose to move smoothly down to the horizon as the half roll is completed. Use very slight forward pressure and rudder opposite to the aileron application. Then as the airplane's bank becomes vertical, apply full rudder *with* the ailerons.

2. As the airplane approaches 45° above the horizon in the inverted position, relax all the back pressure and use *full* aileron deflection (and rudder with it as necessary) to complete the half aileron roll. Again, a roll to the left will be more easily accomplished for most airplanes.

■ Demonstration.
One thing you'll discover as you instruct more aerobatics is that a good horizon is required for the trainee who's starting any of the loop-type maneuvers. Things are happening pretty fast for him at first, and if it's a hazy day with only a faint horizon, he will most likely miss it. This isn't quite so critical in the loop, Cuban eight, or clover-leaf because he will also be using ground references to complete the maneuver. With the Immelmann, however, the horizon is very important since he will be using it to stop the loop and start the roll.

Fig. 23-9 shows the roll portion of the Immelmann as seen from the cockpit.

■ Practice.
Don't push too hard at first on his rolling out on an exact heading. Let him get used to stopping the loop and rolling out at the lower speed. You'll see some funny maneuvering; one instructor always managed to name one of the gyrations after the trainee in question. "Well, that was a fine Johnsonmann. Now try an Immelmann." An aside like that, with good humor, can ease tension.

> *Common errors* in doing the Immelmann are these:
> 1. The usual problems associated with the loop and Cuban eight (throttle handling, back pressure, etc.).
> 2. Too abrupt forward pressure at the top of the loop, creating a "sky suspended" situation. (This is the point at which you find where the mechanic left the pliers.)
> 3. Too early or too late initiation of the half roll. Even if the roll is smooth, the timing may be off; this is particularly a problem when the horizon is not well defined. If a reference point is "put on the tail," as mentioned earlier, the trainee may "lose" it when time comes for the roll-out.
> 4. Lineup problems after the roll-out. After he's practiced a number of Immelmanns using coordinated aileron and rudder all the way, show him that the application of opposite rudder at the beginning of the roll (with coordinated rudder and aileron later) can be the best aid in heading control.

■ Evaluation and Review.
Your evaluation of the Immelmann, as of other maneuvers, should be on the side of safety rather than precision at this early stage. Before you finish with *any* maneuver, be sure that he is safe, because he may come out on your day off and go up for some solo aerobatics.

In addition to safety, you should evaluate his performance by judging whether he is improving his flying of the maneuver. There will always be temporary setbacks (maybe he's done five good ones and the next one can only charitably be called "lousy," but you can see that progress is being made).

Always have specific items to discuss about a maneuver or the flight. If nothing was greatly outstanding either way, say something like, "It went well and I particularly liked your approach (or landing). We may have to work a little on your climb," etc. (How about, "Your preflight check was especially good today.")

Cover the Immelmann and review his problems with a repeat of your suggestions given in the air. It's your job to remember items from the flight that should be covered. If you take a knee board along during your aerobatic instructing you can make notes of items for review on the ground, but this could be distracting for some people, as noted in Chap. 20.

Don't make excuses for his problems; cite errors that were the result of his mishandling *as well as* those resulting from idiosyncrasies of the particular airplane. Make sure he knows where he stands in *his* flying of the maneuver.

Note again that you are introducing the combination maneuvers in order of difficulty as far as visual sighting of a

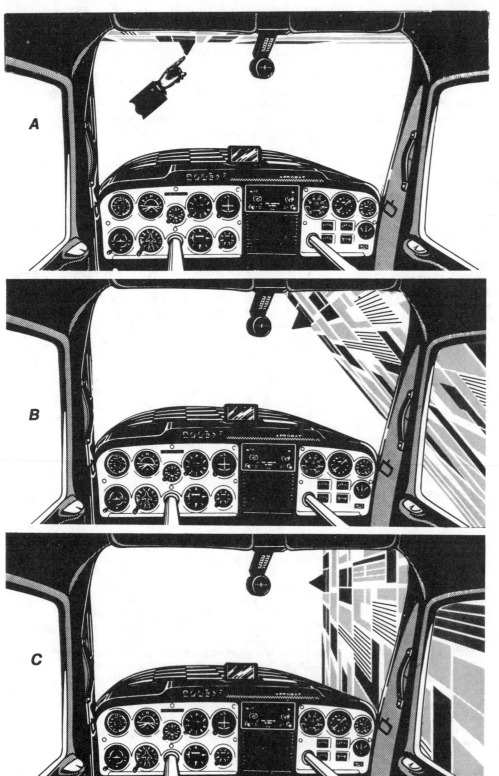

Fig. 23-9, A–E. The Immelmann, as seen from the cockpit during the 180° aileron roll. Compare the airplane positions *A–E* in Fig. 23-8 with *A–E* here. (*The Basic Aerobatic Manual*)

(A) As the airplane reaches the 30°-nose-"up" position (inverted) and the "prominent object" comes into sight, an aileron roll is started (push and roll). You may not be so lucky as to have a giant hand pointing out the reference. It's possible, also, that *your* practice area does not have a large, black triangle on the horizon for a reference as shown in this series. As you gain more experience, you can look "up" through the skylight at the 45° pitch (inverted) position and so pick up the reference early.

(B) The roll continues.

(C) You may have to concentrate on forward pressure here.

reference and airspeed pickup or control effectiveness are concerned. You might sketch Fig. 23-10 on the board to show the trainee the *why* of your procedures.

SPINS AND COMBINATIONS

Aerobatic trainees enjoy combining various maneuvers; it indicates that progress is being made and is one of the best ways to establish orientation skills. Fig. 23-11 shows a good combination of maneuvers to use after the spin has been introduced. One can do a 1½-turn spin using a road as reference, followed by a loop, aileron roll, and snap roll. You can, for instance, start off by having him do a spin followed by only a loop or an aileron roll and then gradually work him up to the full system.

You can combine the spin with any of the maneuvers he'll have in the course. It doesn't have to be first in the combina-

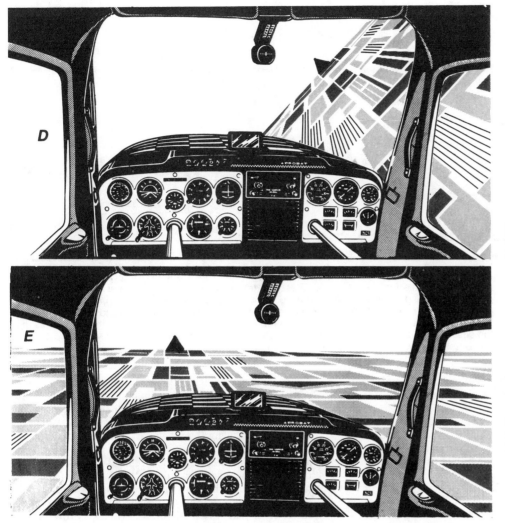

(D) Start applying more top (left) rudder and *back* pressure at this point.

(E) The airplane is just above a stall as the wings are leveled. Here, normal cruise flight is being resumed.

tion but it is a good one to use in picking up extra airspeed after the recovery because of the nose-down attitude. For instance, you would not likely follow a loop or aileron roll with a spin because there would be too long an interval required for the airspeed to decay down to a safe spin-entry value—and the sequential concept would be lost. A spin *following* an Immelmann (and it sometimes does, whether wanted or not) or a snap roll would make better sense.

THE BARREL ROLL

■ **Preparation.** It's almost impossible to draw a barrel roll on the chalkboard, but a model will give the desired results. Have the trainee study the references.

■ **Explanation.** The barrel roll is a precise maneuver in which the airplane is rolled around an imaginary point 45° to

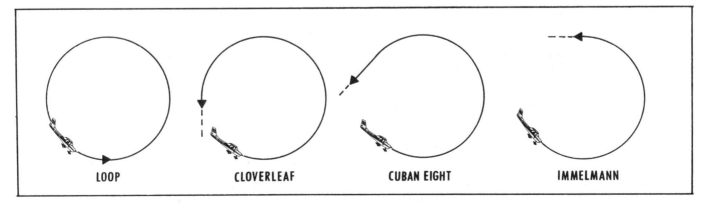

LOOP   CLOVERLEAF   CUBAN EIGHT   IMMELMANN

Fig. 23-10. Loop-roll combinations in the order of difficulty (*left to right*). (*The Basic Aerobatic Manual*)

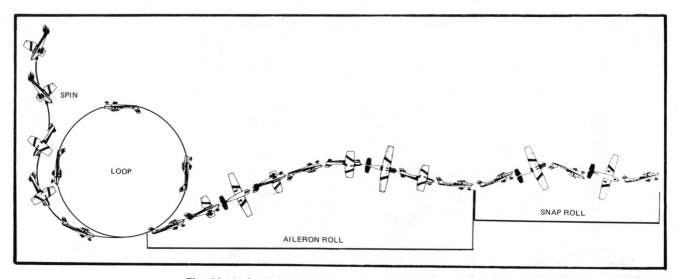

Fig. 23-11. A combination of maneuvers using the spin as a starter. It's recommended that you handle the throttle for the first few of these the trainee does. Set up a normal one-and-a-half-turn power-off spin and recover, holding the nose down to a loop entry attitude. Reapply power smoothly as the dive proceeds but don't overspeed the engine. Complete a normal loop and bring the nose up for an aileron roll to the left. Complete the aileron roll with the nose slightly higher than normal so that the airspeed will be that required for a snap roll. Complete a snap roll to the right. The directions of roll for the last two maneuvers may be reversed for later series.

the original flight path. A positive-g level is maintained throughout the maneuver, and the ball in the turn indicator should stay in the middle.

You may wonder why the barrel roll is taught this late, since it appears to be so simple. Well, it is a precise maneuver requiring particular airplane attitudes at particular reference points, which is difficult for the average trainee to do properly at first.

This maneuver might be considered an exaggeration of the wingover, but instead of starting to shallow the bank at the 90° position, the pilot must steepen it continually until the airplane has *rolled* 360° and is back on the original heading. The rate of roll must be much greater than that used for the wingover because the airplane must be in a vertical bank at 45° of turn, and it must be inverted at 90° of turn. The roll and turn is continued until the airplane is headed in the original direction with the wings level. Compare the barrel roll in Fig. 23-12 with the same view of the wingover in Fig. 20-7.

From behind the maneuver looks as though the airplane is being flown around the outside of a barrel. This is a very good maneuver for gaining confidence and keeping oriented while flying inverted in balanced flight. Good coordination is required to do the barrel roll properly and the trainee will show an improvement in that area after a session of barrel rolls. The barrel roll is generally more difficult and precise than the aileron roll, and he may have to work on this one awhile.

Why—The barrel roll is one of the best maneuvers for improving orientation. Unlike the other aerobatic maneuvers covered thus far, the barrel roll requires a *constantly* changing bank *and* pitch (with attendant changing airspeed) and a radical change in heading (90°) while the airplane is rolling. The average trainee probably will be looking at the wing tip at a time when he should be checking the nose, or vice versa. When he is able to stay well oriented in the barrel roll, he is ready to move on to the reverse Cuban eight or reverse cloverleaf.

How—You might use the following explanation, or develop your own: (1) Make sure the area is clear, then pick a reference on the horizon off the wing tip as in the

wingover and lazy eight. (2) Set the throttle to low cruise rpm and ease the nose over to pick up about 10 K more than used for the wingover or set up the airspeed used for a loop, whichever is higher. Power adjustment should not be necessary during the maneuver. You might have some of your sharper trainees apply full power as the airplane approaches inverted and then remind them to throttle back as the airspeed picks up in the last part of the maneuver. (3) Smoothly pull the nose up and start a coordinated climbing turn (note that it will have to be at a much faster rate than was used for the wingover) toward the reference point. (Assume that at first the roll will be to the left.) (4) When the nose is 45° from the original heading, it should be at its highest pitch attitude and the left bank should be vertical. (5) When the nose is at 90° from the original heading, you should be looking directly at the reference point that was originally off the wing tip—from a completely inverted position (momentarily). (6) When the airplane heading is again 45° from the original, the bank is vertical but you will be in a *right* bank as far as the ground is concerned; that is, the right wing is pointing straight down at this instant of roll. The nose will be at its lowest pitch attitude at this point. (7) The roll is continued to wings-level flight as the nose is raised back to the cruise attitude.

The maneuver must be symmetrical; the nose must go as far above the horizon as below. The barrel roll requires definite checkpoints to ensure that the airplane is at the correct attitude throughout. It is interesting to note that if the barrel roll is to the left, all of the airplane's path is to the left of the original line of flight and the airplane's nose is always pointed to the left of the original flight line (until it merges again at the completion of the maneuver). The opposite occurs, naturally, for the barrel roll to the right.

Another method of doing a barrel roll is to pick a reference on the horizon, turn the airplane 45° to the reference point, and proceed to make a wide roll around this *real* point. One disadvantage of this method for the newcomer is that it depends on the pilot's own judgment of how large the orbit around the point should be. For an

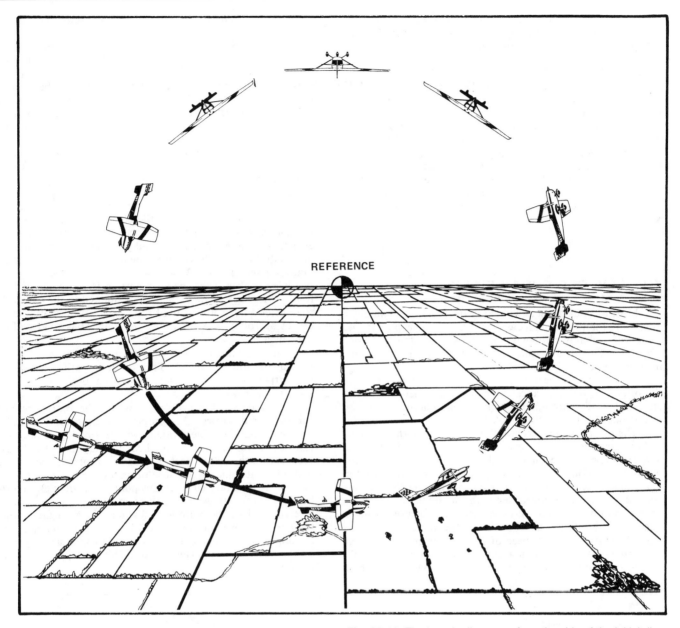

REFERENCE

Fig. 23-12. The barrel roll as seen from the side of the initial dive.

introduction to the maneuver, the first method is usually better, but you may prefer the second and work out your own techniques of instructing it.

■ **Demonstration.** Try not to lose the reference point yourself while demonstrating this one. You may find your explanation is not keeping up with the airplane, which usually results in sputtering and stuttering while the maneuver proceeds to its foregone conclusion—and then you have to do a new demonstration. Don't worry, this will happen plenty of times during your career of instructing aerobatics—when your mouth can't keep up with your brain or the maneuver—and it can ease tension if you react to it with humor.

Usually the trainee is surprised to see the same wing tip back on the reference point and may confess that, like the first snap roll, the earth and sky were blurred and he had no idea where the reference was during the maneuver.

■ **Practice.** You may rest assured the trainee will "lose" the reference point during the first couple of barrel rolls. He'll usually stare over the nose, seeing nothing but blue sky or ground and not really seeing the point at all.

Common errors during barrel rolls include these:
1. Not pulling the nose high enough in the first 45° of the maneuver, which means that the highest and lowest nose positions are not symmetrical to the horizon.
2. Not maintaining a *constant* rate of roll. Usually things are fine at the 45° position; the nose is at its highest pitch and the bank is vertical. As you approach the position of 90° of turn you will probably find that he is not going to be completely inverted at that point and will have to rush things a bit to make it. The usual reason is that he did not maintain a constant rate of roll. Remember that the nose is up and the airspeed is slower in this segment of the maneuver, so the controls must be deflected more to get the same rate. This is where coordination comes in. Watch for it in particular.
3. Letting the nose drop after passing the 90° point; losing too much altitude and gaining excess airspeed.

4. Failure to roll out on the original heading; having the wing tip well ahead, or well behind, the reference when the maneuver is completed.

■ **Evaluation and Review.** Review each barrel roll briefly in the air, and have the trainee use the model on the ground. This one can be hard to "see," so go over it again as necessary after getting on the ground.

By the time a half-dozen barrel rolls have been practiced, the average trainee should be oriented throughout the maneuver even though he may still have minor problems of heading and symmetry. After a dozen rolls he should be starting to work on a constant roll rate and starting to ease his heading problems. After several hundred, he may begin to be satisfied with his barrel rolls but will realize that constant practice is required.

THE REVERSE CUBAN EIGHT

Most of your trainees will enjoy the reverse Cuban eight even more than the Cuban eight, and as you are explaining it, some will probably comment that it looks like a much classier maneuver. The idea of rolling while climbing adds a new dimension to aerobatics.

While it may appear at first to be a better maneuver than the original, you can expect problems with lineup after the airplane is flown through the loop portion each time. It is best at this point to take a close look at it.

The reverse Cuban eight differs from the normal Cuban eight in that the half roll is executed as the airplane is *climbing* at a 45° angle. Use the model and demonstrate the difference; it should seem obvious but is not always so. As is the case for many aerobatic maneuvers, after the trainee sees the maneuver illustrated (or demonstrated with the model) and flies it, he may still need another ground discussion before getting a good understanding. Fig. 23-13 shows the maneuver.

The entry (pull-up) speed of this maneuver should be at least 10 K *lower* than that recommended for the "normal" Cuban eight so that the airspeed will be very low after the half

roll is completed and the "fly through" (loop portion) is started.

You should mention that one of the more common errors in this maneuver is the failure to bring the nose up to the 45° position before the roll is started. The wing tips will be useful aids for checking pitch attitude, but he should get back to the nose until the 180° roll is complete (use the horizon). He should then check through the skylight (Aerobat) or look down out of the open-cockpit airplane at the road/railroad reference.

If he uses the aileron-type half roll, expect a problem in heading at first. The trainee may not be lining up with the reference road as the nose is "pulled down" through the loop portion, a common contributing factor to heading problems.

In the aileron roll the nose projects an orbit, using a reference point, and while at the completion of the 360° of roll the nose is again pointed exactly at the beginning reference (it says here), at the 180° roll position it will be off laterally to the "left" of the original line if a left roll is being made (and vice versa). Fig. 23-14 shows the idea. You may suspect this is a problem if, as the loop portion is continued and the road comes back in view, the nose is pointed to the *left* of the road after a *left* roll was used (Fig. 23-15). You may want to use a slow roll type of entry after the trainee has an idea of the maneuver.

Another cause of directional error is not having the wings level as the roll is stopped. Since the loop path is (or should be) perpendicular to the lateral axis, a problem like that shown in Fig. 23-16 results.

A third possible problem, one of the more probable causes of error at first, is improper correction for "torque" in the initial pull-up. The airplane will be slowing rapidly and full power will be used, so the heading may have a left error even before the roll is started. This is a problem because the high pitch attitude of the nose obscures the horizon. Clouds may be used as references at this point (if available), and the heading can be flown using them in relation to the ground reference. Since there are very few days in most areas when the horizon is well defined, you'll have to use references that are not on the horizon but are projected to an imaginary point on the unseen horizon (Fig. 23-16). This can apply to all maneuvers.

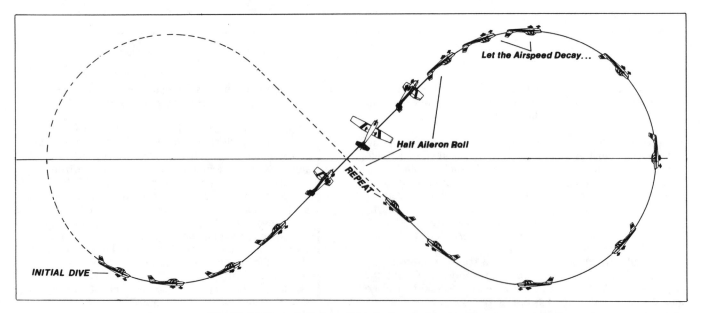

Let the Airspeed Decay...

Half Aileron Roll

REPEAT

INITIAL DIVE

Fig. 23-13. Reverse Cuban eight. As the airplane becomes inverted, look over the nose to level the wings with the horizon, then check the road/railroad reference by looking "up" through the skylights for last-second lineup corrections. (*The Basic Aerobatic Manual*)

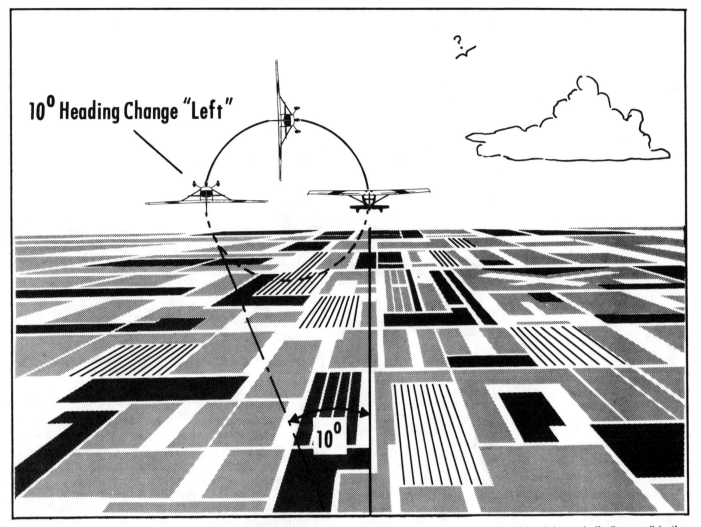

Fig. 23-14. Possible errors in lineup in the loop portion caused by doing a *half aileron* roll in the reverse Cuban eight. The dashed portion of the circle shows that the airplane would have returned to the original heading had the roll been continued. (*The Basic Aerobatic Manual*)

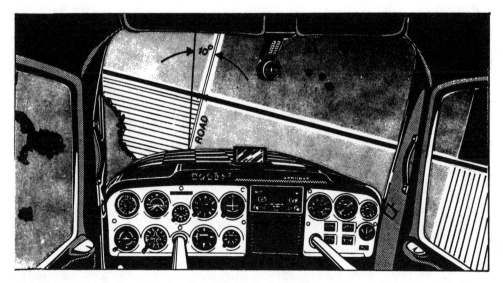

Fig. 23-15. If the airplane is off-heading at the beginning of the pull-through, the same original error will be shown by the reference (road). The error will be in the direction of roll. (*The Basic Aerobatic Manual*)

Fig. 23-16. If the wings are not level at the completion of the half roll, lineup problems will occur in the loop portion. (*The Basic Aerobatic Manual*)

■ Procedure

1. Choose a straight stretch of road, and after clearing the area, ease the nose over to pick up the recommended airspeed. Retard the throttle to keep the rpm from going over the red line.

2. Bring the nose up smartly to a pitch of 45° above the horizon; add full power as the airspeed decreases (the power at this point minimizes a loss in altitude).

3. As the airspeed decreases, complete a half roll in either direction. Looking back at Fig. 23-14 and the discussion, you can see that the *half aileron* roll can result in a heading error. To avoid or minimize this problem, use forward pressure (smoothly please) as the airplane approaches and passes the 90° roll point. This will help to keep the nose lined up with the road/railroad reference as the roll is made. Usually a brief hesitation before "pulling through" makes for a better rounded loop portion, keeps the airspeed from being excessive at the bottom, and stops excessive altitude loss. The hesitation can allow time for any correction of wing attitude before the loop starts.

4. Fly the loop, using the procedures such as throttle handling, control pressures, etc., as in the loop itself.

5. Bring the nose back up to the 45° pitch position and repeat the process.

Common errors in the reverse Cuban eight are these:

1. Not raising the nose high enough (45° above the horizon), resulting in a poorly executed maneuver.

2. Having too great a roll rate—"overrolling" and going past the inverted position.

3. Forgetting to look at the road/railroad before the pull-through is started.

4. Lineup problems after the half roll is completed. The reverse Cuban eight may also be done using a half snap roll at the 45° pitch position. Such variations will be good for your more adept trainees.

5. Too high an airspeed when the loop portion is started. (A high airspeed at the bottom of the loop results.)

6. Failure to retard the throttle as the airspeed picks up in the dive.

7. Too much altitude loss is a result of not having the nose high enough initially and too much airspeed when executing the loop portion.

VARIATION—1½ SNAP ROLL. Another variation (for more advanced trainees) is a 1½ snap roll at that position (Fig. 23-17). This maneuver is covered here because it is a good follow-up to the reverse Cuban eight but for most people should not be

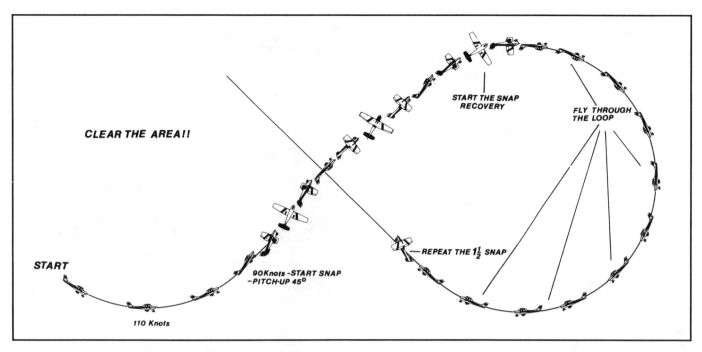

Fig. 23-17. One-half of a reverse Cuban eight, using a 1½ snap roll entry. (*The Basic Aerobatic Manual*)

done until the snap at the top of a loop is well in hand. Needless to say, the 1½ snap entries can be a problem of orientation at first.

In this maneuver, as in the snap at the top of a loop, you remind him to look for the reference in the direction of rotation in order to pick it up as soon as possible. Too many times trainees look over the nose, and the reference (usually a road or runway) is seen too late for precise recovery—and a new maneuver is born.

SNAP AT THE TOP OF A LOOP

Use the recommended airspeed, which will normally be 10–15 K higher than that used for the loop alone, and fly the loop. You may find the best airspeed for this after you've instructed a few of them in a particular airplane. The trainee should pull approximately ½ g more than for the normal loop, but one of the major errors is to "haul back" with continuing excessive g forces, getting a buffet as the inverted position is approached.

The snap portion should be symmetrical, that is, with the snap beginning and ending at the proper points on the loop circle (Fig. 23-18).

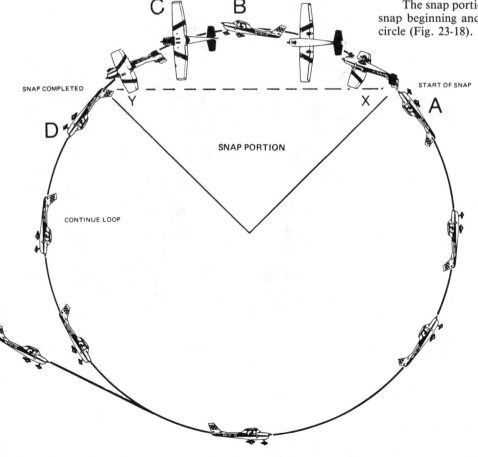

Fig. 23-18. Snap at the top of a loop. See Fig. 23-19 for cockpit views of points *A–D.* (*The Basic Aerobatic Manual*)

As was mentioned in the section on the reverse Cuban eight, as the snap progresses the trainee should be reminded to look for the reference (lean over, as necessary) in the direction of roll. Fig. 23-19 shows where you should be looking at points *A–D* in Fig. 23-18. Start the snap recovery when you see the road reference out the left side. The recovery should be complete at point *D* and the loop continued.

Usually the snap at the top will not be as "brisk" as the normal snap roll because the rate of speed decay sometimes catches the trainee by surprise. For lower-powered airplanes with a slower snap rate you may find that starting the snap with the nose higher (45°) may give a more symmetrical pattern (Fig. 23-20).

One problem the trainee will have is to *not* relax back pressure enough as the snap roll is completed, which will contribute to a nose-low position at the end of the snap. Another error is to wait too long to initiate the snap action with the result (again) that the nose is too low when the snap roll is complete.

A difficulty you may expect some trainees to have is that, for some reason, a snap roll started from an inverted position takes "longer" than the usual one and they'll start the recovery much too soon. And, of course, you will encounter the rare individual who will get so engrossed in the snap roll that he continues well past the recovery point—and you both get further practice in spin recoveries.

To review some *common errors:*

1. Not attaining enough airspeed for entry. This will probably be the situation for nearly every trainee, since most tend to be somewhat reluctant to get the higher speed required. (Some may comment on this.) You went through this in the normal loop with some people, and you may have to work on this for several of the maneuvers.

2. Poor loop pattern. The added airspeed can throw off the back-pressure pattern and the loop is not "efficient."

3. Overspeeding of the engine. He's tense in the new maneuver and this—plus the fact that he will be 10 K or more faster at this pitch attitude than the same attitude in which he had been opening the throttle for the loop—can cause this problem. Warn him about it, but *you* be prepared to keep the rpm below the red line.

4. Too late in starting the snap. The nose is too low (inverted) and the recovery nose position is too low. (The airspeed will have decayed further because the airplane is nearer the top of the loop, and a sloppy nose-low recovery is encouraged by this, too.)

5. Loss of orientation during the snap. Most trainees have this problem to some extent, and for some it will cause a great deal of trouble. As you will see after instructing aerobatics for a while, a split-second delay in starting the recovery from a snap maneuver can mean that precision is lost.

6. Poor transition from the snap back to the loop. Not flying the airplane in the loop. Some people will be so pleased at a good snap that they'll spend some time mentally congratulating themselves while the rest of the maneuver falls apart.

7. Failure to reduce power on the back side of the loop. Things are occurring in rapid order and this may be forgotten during the melee.

8. Starting the recovery too soon, feeling that a snap roll started from an inverted position takes "longer" than the usual one.

■ **Chalkboard and Model Use.** As mentioned briefly in the cloverleaf section, loop-type maneuvers should be drawn

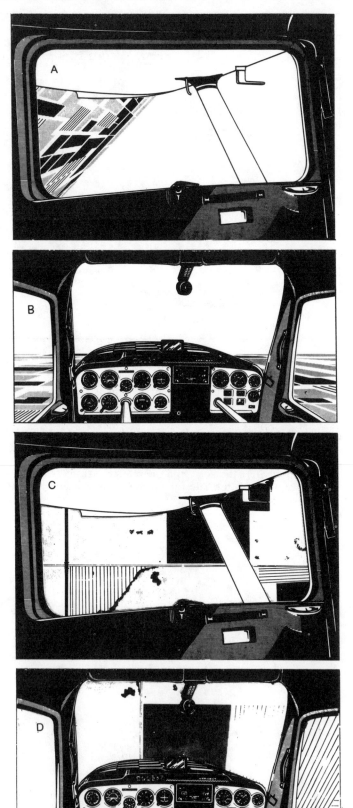

Fig. 23-19. Snap at the top of a loop as seen from the cockpit of an Aerobat. Views *A–D* are seen at those same points in Fig. 23-18. (*The Basic Aerobatic Manual*)

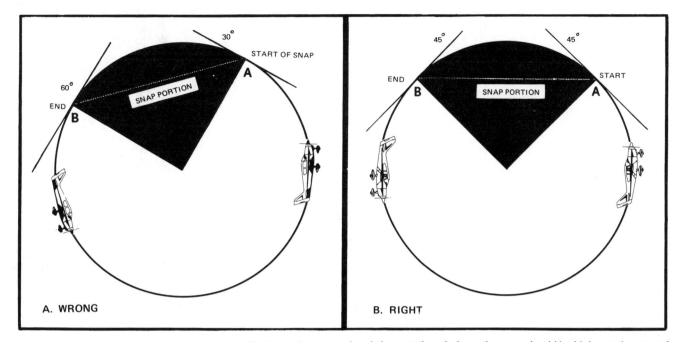

Fig. 23-20. For the underpowered and slow-rotating airplane, the nose should be higher at the start of the snap, in order to get a more symmetrical pattern.

on the chalkboard and the model used, following the lines, rolling the model at the proper places while you explain what is happening. This makes it easier for the trainee to see what is to be done, compared with using the model alone (which would leave no "trace" of the maneuver as a whole).

THE FOUR-POINT ROLL

Hesitation rolls appear to be the ultimate in aerobatic flying to the beginner, and the eight- and sixteen-point rolls *are* difficult. But the four-point roll can be done reasonably well by a person who has had 4 or 5 hr of aerobatic instruction.

If you haven't a recommended airspeed for the four-point roll, start off at about 10 K higher than given for the straight aileron roll or slow roll. *You* may be able to bring it back down to the regular speed, but usually the trainee can use the extra airspeed.

For most of the airplanes without inverted systems, the "hesitations" should be short. It's possible to have four good points without the engine hesitating but you'll find that the average person will have one problem — being too slow.

You'll use the horizon for reference and think of the maneuver as, "Left aileron, left rudder. Stop!" for each of the four points.

You may prefer to have him bring the nose up higher in the beginning of the roll so that it can be "falling" throughout the maneuver. Work it so that there is no excess amount of forward pressure used to keep the nose up (Fig. 23-21).

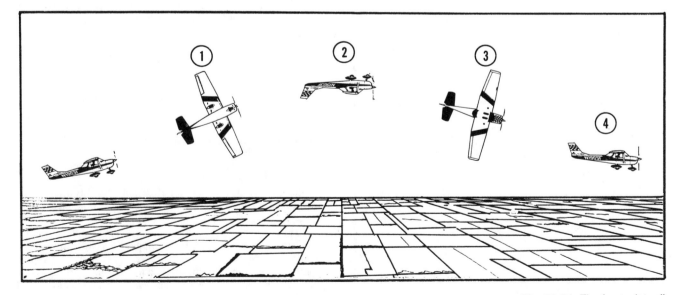

Fig. 23-21. The four-point roll.

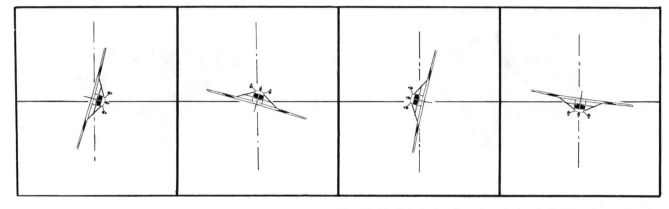

Fig. 23-22. A typical pattern for a trainee's first few four-point rolls.

Common errors in the four-point roll are these:

1. Not stopping precisely at the points. Usually the first point is less than 90° of bank, and the other points are also behind (Fig. 23-22).

2. Coordination problems; having trouble at the first point because of the required neutralizing of the controls.

3. Hesitating too long at the first two points so that the rapidly decreasing airspeed (or rapidly lowering nose, with attending loss of altitude—and a possible too-high airspeed) makes it necessary to rush the last point.

4. Poor heading control. This is usually the result of letting the airspeed decay excessively.

You may introduce the eight-point roll later. Fig. 23-23 shows how the points might be remembered. Of course, the airplane is not flying the pattern shown (it's rolling on a point, but some trainees might more easily visualize it as flying on the sides of an octagon).

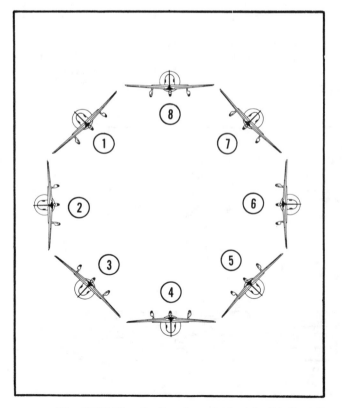

Fig. 23-23. Visualization of an eight-point roll to the left.

The eight-point roll will naturally require better control and a higher airspeed than the four-point. Usually the problems begin at point *5* in Fig. 23-23, with points *6, 7,* and *8* not so well defined, as he hurries to complete it before the maneuver turns into a humorous situation.

The common errors of the eight-point roll are those of the four-point roll—only magnified.

THE REVERSE CLOVERLEAF

It's doubtful that any of your trainees in the introductory course would be ready for this maneuver, but you may want to demonstrate it and perhaps teach it to someone who is advancing at a rate far beyond the norm.

It's a more graceful maneuver than the cloverleaf. (Some say the *other* maneuver is the reverse cloverleaf, but since this one is more difficult it would seem that it is the variation of the original.)

What—This maneuver requires that the airplane complete the quarter roll while climbing rather than diving. It is still a combination of four loops and quarter rolls and the pattern has essentially the same cloverleaf shape when seen obliquely (Fig. 23-24).

Why—This maneuver reveals immediately the ability of a trainee to maintain orientation. In the pull-up and roll there is very little but sky to be seen, so timing and a constant rate of roll are important. Have the trainee use all references available. (Check through the skylights, canopy, etc.)

How—As a start, you can use the recommended loop entry if no other information is available, but you may find with practice that this speed may be changed slightly. The throttle is handled as in the loop, adding and reducing power as necessary to stay under the rpm red line.

This maneuver is in one sense a follow-up to the wingover and barrel roll in that a wing is placed in line with a point on the horizon and the entry dive begun. Review the wingover *and* barrel roll and note that the beginning of the reverse cloverleaf is very similar to the barrel roll—which he should already be able to do before starting this one. *Remember to always tie in new material with old—it helps the transition.*

If possible, in addition to the reference point on the horizon, start the maneuver lined up over a road. Your attitude relative to the road (point *A,* Fig. 23-24) as the plane is pointed down (and pulling through the loop portion) gives an idea of how the last loop and roll went and you can check for problems. Fig. 23-25 shows how a road should appear as each

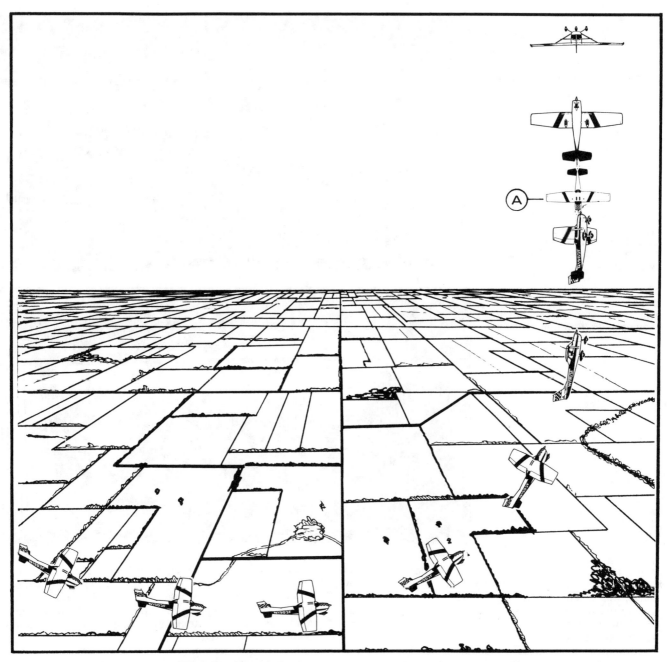

Fig. 23-24. One-fourth of a reverse cloverleaf as seen from above and the side of the initial dive. Four rolls and loops complete the maneuver. The four quarter rolls are made in the same direction.

portion of the loop as point *A* of Fig. 23-24 is reached.

One way to check the trainee's understanding of the maneuver is to ask him which way the airplane will be headed after the end of the first loop. (If you are rolling to the left, it will be 90° to the *right* of the original heading, and this would be the case for each quarter of the maneuver.)

The points on the horizon off the wing tip have to be quickly chosen as each dive is entered, and since the chances of having four really outstanding references exactly at positions perpendicular to or in line with the road are not good, he may have disorientation problems.

You will naturally have him initially practice only one part, adding the rest gradually as his technique improves.

Fig. 23-26 shows the reference at the initial dive and as the roll portion is completed. Unlike the barrel roll, the roll is

stopped when the wings are level (inverted) and the nose is pointed at the reference. The nose is brought smoothly (wings level) through the point and the road reference is picked up.

The most precise form of the maneuver would be a "true" reverse cloverleaf with the airplane completing 90° of *roll* while its nose is straight up, then back pressure applied to fly through the loop portion. However, more practically, in lower-powered training airplanes the roll and loop are combined to make sure that it's not a situation of running out of airspeed, altitude, and inspiration before the reverse cloverleaf gets a good start. It's a smooth continual motion of loop and roll.

Back pressure is being used as the roll is made and helps "turn" the nose toward the point; the usual tendency is to turn more than 90° from the original heading. Fig. 23-27 shows the

Fig. 23-25, A–D. Reference road as seen from the cockpit at each part of the loop with the pitch attitude at point *A* in Fig. 23-24. The arrows indicate the direction of the initial dive and pull-up. (*The Basic Aerobatic Manual*)

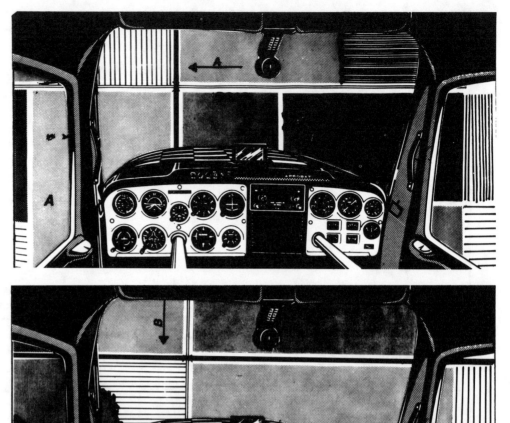

(A) The backside of the first loop, the airplane's pull-up perpendicular to the road (it was lined up with it in the initial dive, as indicated by the arrow).

(B) The backside of the second loop, the airplane again lined up with the reference road but the heading 180° from that of the initial dive.

results in a pull-up and left roll. In the situation shown, *left* rudder may be used to yaw the nose back to the point. Of course, this is supposed to be a reasonably coordinated maneuver (and with practice can be a *fully* coordinated maneuver); however, the trainee might salvage a bad situation by "subtle" use of rudder.

Continual back pressure should be used, and there should be no hesitation or stopping of the nose at the inverted position, but a *slight* relaxation of back pressure at about 30° above the horizon will give the loop portions more symmetry.

One common error you'll encounter is that of too much back pressure throughout the top half of the loop portion and not enough at the bottom, resulting in high airspeed and excessive loss of altitude.

You'll have to analyze the trainee's errors as the road reference comes back into view and also be able to "see" the

Fig. 23-26. Reference for the reverse cloverleaf as seen from the Aerobat. (*The Basic Aerobatic Manual*)

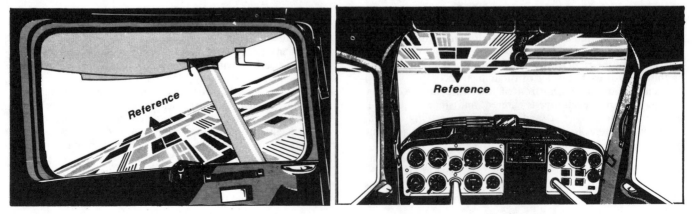

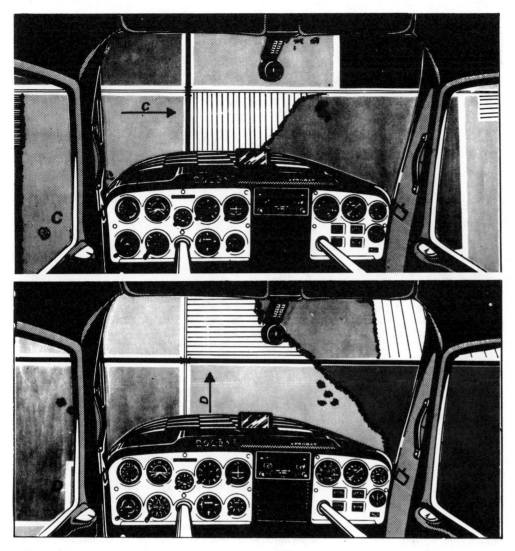

(C) The backside of the third loop, the pull-up path again perpendicular to the reference road (but in a reverse direction from view **A**).

(D) The backside of the fourth loop, the airplane being pulled up to level flight and headed in the same direction as at the beginning of the maneuver.

maneuver during any part of it. By doing only one roll and loop combination at first, you'll be able to stop then and analyze his problems.

Some people "gain" a heading of nearly 45° per unit of the maneuver and wonder why, when it's completed, they are

Fig. 23-27. View from the cockpit when the airplane has turned more than 90° from the original heading, overshooting the reference. (If this picture looks backward, get a model and exaggerate the turn in the pull-up and left roll.) (*The Basic Aerobatic Manual*)

headed almost 180° from the original heading—this causes added disorientation for a few seconds after the completion of the full maneuver.

By analyzing the maneuver using a model, you will be able to answer the trainee's questions. You will find that in this maneuver, the airplane's attitude at a particular point of the action can later affect its path in a manner that is not always easily predictable—without prior analysis.

RECOVERY FROM UNUSUAL ATTITUDES

In setting up an introductory aerobatic course, you should definitely include discussion and practice of recoveries from unusual attitudes, which result from wake turbulence as well as from deliberate aerobatic attempts. By including this training you will not only help the trainee to have a better chance at survival should he encounter wake turbulence at low altitudes, but also list another item with which to justify his taking the course.

First of all, avoid conveying the idea that after taking an aerobatic course the trainee should lose his fear of wake turbulence. If conditions are right, even a championship aerobatic pilot could have an accident caused by wake turbulence (Fig. 23-28).

Be familiar with Advisory Circular AC-90-23D (or E), *Wake Turbulence*. A few minutes' talk about it at a chalkboard

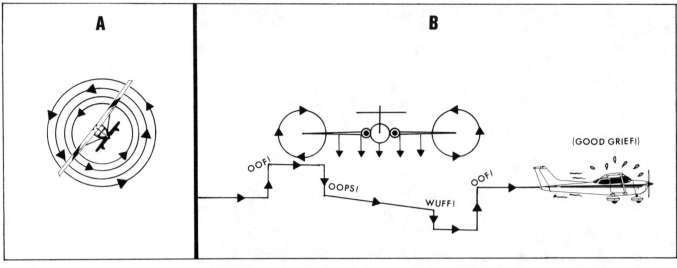

Fig. 23-28. Two types of wake turbulence encounters. (A) Following the same path of the airplane that made the vortices. Usually this is on takeoff or approach and the problem is that of loss of control and hitting the ground. (B) Crossing behind a large airplane at cruise. The path shown is the result of the up and down effects of the vortices and downwash. The problem here may be of overstressing the airframe if the encounter is at a high (your) cruising speed.

can clear up a lot of gray areas for the trainee. (AC-90-23D is sent free from the FAA, but if you have a copy and need more right away, you have permission to reproduce it on a copy machine.) The *Aeronautical Information Manual—Basic Flight Information and ATC Procedures* has a good section on the subject, and it is continually updated as more is learned.

Review with him the various conditions in which he might encounter this menace and how, by knowing the actions of the vortex, he might avoid an accident. Emphasize that *if in doubt, he should delay the takeoff and landing rather than encounter wake turbulence.*

■ **Explanation.** Using the information you've obtained from AC-90-23D and other sources, review with him at the chalkboard the actions of wake turbulence. Discuss the effects of flying into—from directly behind—a vortex. You should also cover the effects of flying through a pair of vortices at a 90° angle (Fig. 23-28).

Needless to say, if the plane is inverted by wake turbulence at a low altitude, a half loop is *not* the best method of recovery (Fig. 23-29).

Impress on the trainee that if the airplane is inverted before he can stop it, it would be best to continue the roll, rather than try to roll against the vortex, since the roll rate of the vortex may equal or exceed the roll rate of the airplane at a particular airspeed.

Assuming full deflection of the ailerons, the rate of roll increases in a straight line with airspeed until a speed is reached where the pilot's strength is no longer sufficient to maintain full deflection. Naturally, strength varies among pilots, but for airplanes with a wheel control the maximum force is 50 times the diameter (inch-pounds) "exerted on opposite sides of the wheel and in opposite directions."

Fig. 23-30 shows the rate of roll versus velocity for a fictitious airplane. The peak rate of roll is 90° per second at 130 K; above that speed the ailerons cannot be fully deflected, because the control forces required exceed the maximum allowed for certification.

Notice that while the airplane has a fairly good *maximum*

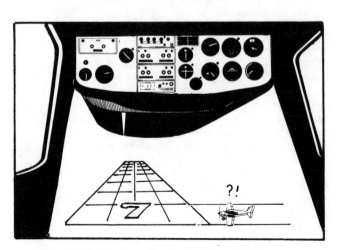

Fig. 23-29. This is no time to be doing a half-loop recovery.

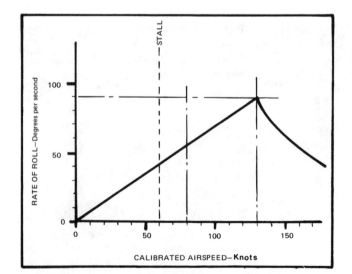

Fig. 23-30. Rate of roll versus velocity for a fictitious airplane.

rate of roll, at the approach speed of 80 K the rate is not so good, leading one to believe that the vortex may be the deciding factor as to which way the airplane will roll, whether the pilot chooses that direction or not. If the airplane turns during the recovery process, it may move out of the vortex core areas, since they are relatively small and close together.

■ **Demonstration.** Don't set up a simulated wake turbulence problem until the trainee has at least 3 hr of aerobatic instruction, and even then don't do it less than 3000 ft above the surface. Wait until a later part of the flight before demonstrating or practicing these recoveries. A half snap is a good, unexpected entry for you to do for his recovery attempt, but his stomach may not be acclimated for a half snap at the beginning—he may need a "warm-up."

Your steps of instruction might be like this:

1. Discuss wake turbulence and methods of recovery before the flight.

2. Brief him (in the airplane) that you are going to do one-half of an aileron roll and you want him to take over and (1) complete the roll or (2) reverse the roll as you may choose.

3. Note that, starting immediately, during any of your demonstrations of any maneuver, he is to get the airplane back to level, right-side-up flight as soon as possible, if he's told to take over. (*Roll,* not split-S.) You can test his understanding of this by beginning to "demonstrate" or "review" a loop and saying "all yours" when the plane is inverted at the top. Granted, with the very low airspeed at the top of a loop, his roll will be sloppy; but at least he's rolling it and not continuing the loop or doing a split-S.

4. Set up conditions of approach, but don't use flaps, since the airplane could be overstressed if he lets it get away from him and funny stability problems might arise if he gets into a spin. Let him know what you are about to do, set some altitude as ground level, and do the half roll at about 400 ft above that reference. You would probably be better off to use power and extra airspeed to get the half roll, and then after the airplane is inverted retard the throttle and slow it to approach speed before you say, "You got it!" He'll get some idea how difficult it is to recover under these conditions and will likely go below "ground level" the first time or two (or more).

You'll be (1) giving him a better chance of survival if he encounters wake turbulence (at least he has a better chance if he hits the ground right side up), and (2) showing him that deliberate aileron or slow rolls at the recommended speed are quite different from a low-speed, low-altitude situation and *if at all possible he should avoid wake turbulence.*

Review with him the fact that the nose must be kept up, which means he must apply more forward pressure than he's used in aileron rolls as the throttle is opened and the roll started, then back pressure as it is completed.

In an actual situation, unlike now when you the instructor are quietly sitting there competent and able to take over should trouble arise, the trainee's nonpilot passengers may be screaming and trying to hang on to him, and it would be the wrong time for that. Passenger noise and actions would contribute to the confusion, particularly if he holds much forward pressure and runs into zero- or negative-g flight. Also, that right control wheel would seem to be a good handle—the passenger thinks. The passenger will always *pull* on the wheel as he hangs on and this is not good when you are inverted. You'll have to fight this, too. All in all, it's infinitely better to avoid the problem in the first place.

5. At a safe altitude, your next step will be to divert his attention and do a half snap (retarding the throttle), having him recover. Note the altitudes of entry and recovery.

In this training you will be trying to instill a controlled response rather than a natural one. The natural response to an inverted attitude will be to pull back on the wheel to "remain

comfortable," which is the wrong response; the pilot must learn to be uncomfortable for a short while in order to keep altitude loss at a minimum.

In summation, make the point that when in the future actual wake turbulence is encountered, the airplane he'll be flying will likely be near gross weight and have a lesser rate of roll than the airplane you are using, and he probably will not have thought about wake turbulence for weeks—adding a surprise element. The airplane will be slow on takeoff and landing, and, in addition, on approach the power will be low and the airplane dirty. You should emphasize that on *every* takeoff or landing at a large airport the possibility of wake turbulence should be considered. (It should be considered at small airports, too.)

SOLO AEROBATICS

The trainee will naturally want to practice solo aerobatics and you will have to be sure that he can recover from any attitude (and power) situation. At the end of 5 hr of flight instruction, he will most likely be safe in the Three Fundamentals and the Cuban eight. If there is any doubt, you should set up a check ride including these, plus other combinations and variations, and should also include recoveries from unusual attitudes caused by inattention during these maneuvers. You should restrict his solo practice to maneuvers you've checked him on; if he decides that a Lomcevak is probably a good maneuver to try since he's up there anyway, things could get a bit uncertain.

If you have been giving aerobatic instruction strictly to introduce the concept you'd also better get him checked in the particular airplane and its systems before sending him up alone. If you don't, you might find that his aerobatics are fine but he keeps landing in outlying spots because he runs out of fuel and doesn't know about switching to another tank. Make sure that your insurance covers solo aerobatics.

■ **The Hammerhead Turn.** Assuming that it's legal for your airplane, the trainee should have dual practice in doing this one deliberately so he'll know what to do if he finds the airplane pointing straight up with the airspeed rapidly decaying to zero.

You should make the point before—and during—the flight that if he finds himself in the above-mentioned situation his recovery (rudder application) should be to the *left* so that he'll be helping "torque" move the nose. If right rudder is used, it may just be strong enough at low speeds to counteract "torque" so the airplane remains pointed straight up until it runs out of airspeed. In this situation, what follows is called a whip stall or a "tail slide," a condition that could possibly result in structural damage, and definitely would result in getting the attention of whoever is in the airplane. The airplane actually moves down and backward, the elevators may be forced up or down to the full stop, and it (finally) whips over violently to assume a full nose-down attitude. A whip stall, according to those who have encountered it unexpectedly, takes approximately 10 years of one's life to complete.

The hammerhead is a good way to reverse course in a very small horizontal area and is often used in air show demonstrations. If the trainee encounters it accidentally while doing aerobatics, his correction of using left rudder will also bring the airplane over on its back because the "outside" (right) wing is getting more lift because of the yaw effect. Demonstrate this.

In the well-flown, deliberate hammerhead, the airplane will pivot precisely about its vertical axis with no roll occurring. This requires cross-controlling of ailerons and rudder and judicious use of forward pressure at the peak of the maneuver.

Emphasize that no matter in what attitude the airplane is placed, it will always end in a nose-down attitude, assuming that a reasonable amount of altitude is available. The pilot's job, if he gets into an unusual attitude like the approach to a whip stall, is to make the transition to the nose-low attitude as easily and smoothly as possible. The trainee may later do hammerheads and whip stalls in airplanes that have them listed as approved maneuvers, but he shouldn't practice them solo in the average aerobatic trainer.

■ **Vertical Snap Roll.** You'll find that this one is not mentioned for most aerobatic trainers, but if it's legal for yours, this maneuver demonstrates that despite some rather wild gyrations, the airplane will (finally) recover in a nose-down attitude of its own accord. This is not to be tried in airplanes of unusual design or those with people in the back seat, or with baggage, etc. (It's illegal to do aerobatics with people in the back seat of a four-place aerobatic airplane, or to have baggage back there.)

Impress upon the trainee that if he gets into a situation like the vertical snap and he's not sure what's happening, he should neutralize the controls until the airplane assumes some reasonable attitude—and then recover using the controls as

necessary. If it's okay for your airplane to do vertical snaps, you'll find that it can be a good confidence builder. But it causes nausea if too many are done.

■ **Spins.** You should be well satisfied with the pilot's ability to recover from any possible accidental spins he might get into. By this time, he should be able to handle stalls and spins, perhaps not with enthusiasm, but with competency, at least. *You* might review Chap. 21 for ideas on different spin entries he could encounter.

■ **Summary of Solo Aerobatics.** If possible, have him practice in an area close to the airport where you can see his performance and critique it afterward. It's a good idea to give him a list of the maneuvers in the order that they should be done. This doesn't mean that he does one aileron roll, then one loop, etc.; he should, for instance, work on the aileron roll before the loop, and so on. You'll need an orderly list because some people's loops look like aileron rolls and vice versa.

Sure, your schedule most likely wouldn't allow you to stand on the ramp watching aerobatics and making notes but, *if possible,* you can give him a chance of building his confidence by doing solo aerobatics—while still getting the benefit of your instruction.

INTRODUCTORY SYLLABUS

Following is a syllabus for an advanced instruction course containing 5 hr of dual and 5 hr of ground school. The items *suggested* to be covered are just that, suggestions, and you may refer to the Bibliography for books covering, for example, stability and control. You may also find that your local situation does not require coverage of anything other than aerobatic maneuvers, but the listing of additional instruction topics is included, anyway.

■ **First Flight**

GROUND INSTRUCTION (1:00)

___1. *Discussion of the course,* aims, and general procedures. (Always clear the area before starting any aerobatic maneuver.)

___2. *Discussion of the effects of g forces* on the airplane and pilot.

___3. *Discussion of limit load factors* (g's) of normal (+3.8, −1.52), utility (+4.4, −1.76), and aerobatic (+6.0, −3.0) airplanes. (Why aerobatics should not be done in normal or utility airplanes.)

___4. *Discussion of probable g forces* to be encountered in the course (+4, 0). (How the accelerometer works.)

___5. *Discussion of pilot physiology.* (How to avoid or cope with nausea problems by stopping at the first "loss of enthusiasm"—the NSMFA syndrome—staying up in smooth air versus immediately returning to the airport, and avoiding certain types of food or drink before flying aerobatics.)

___6. *Discussion of introductory or review maneuvers and common errors.*

___ a. The 1-g stall (power-on).

___ b. The 720° constant-altitude steep turns (45° or 60° banks).

___ c. The chandelle, as a maneuver introductory to aerobatics and as required for the commercial certificate.

___ d. The wingover (60°, then vertical pitch and past-vertical banks at the steepest point).

___ e. The lazy eight, as a series of shallow wingovers and as required for the commercial certificate.

___7. *Introduction of the aileron roll* (one of the Three Fundamentals of aerobatics).

___ a. Procedure (the maneuver as a whole, power setting and airspeeds).

___ b. Breaking the maneuver into the usage of each control.

___ c. The aileron roll again as a whole maneuver.

___ d. The left and right aileron roll with expected differences in effects of "torque."

___ e. Common errors.

___8. *Preflight check of the airplane* with special emphasis on checking for structural damage. (Emphasize that the same type of preflight check should be done for *all* airplanes.)

___9. *Briefing at the airplane* concerning the parachute and procedure in exiting the airplane.

FLIGHT (0:45)

___1. *Climb to a safe altitude in a clear area.* Instructor shows trainee practice area limits.

___2. *Takeoff and departure stalls,* left and right.

___3. *Demonstration and practice of 720° turns* (45° and 60° banks).

___4. *Demonstration and practice of chandelles* (30° initial bank) *and wingovers* (60° and 90° banks).

___5. *Demonstration and practice of aileron rolls.* Trainee does at least two left rolls (and one right roll if conditions and time permit).

___6. *Postflight briefing* and critique of maneuvers.

Comments_____

_____

_____

Instructor_____

Date_____ Ground School Time_____ Flight Time_____

■ Second Flight

GROUND INSTRUCTION (1:00)

___1. *Review of chandelles and wingovers.*

___2. *Review of the aileron roll* (techniques and common errors).

___3. *Introduction of the loop* as the second Fundamental—the maneuver as a whole, step-by-step procedure, and common errors.

___4. *Discussion of the loop* followed by an aileron roll.

___5. *Introduction of the cloverleaf* as a simple combination of the two Fundamentals, and common errors.

___6. *Introduction of the Cuban eight* as the next combination of the two Fundamentals.

___7. *Discussion of the Immelmann* as the next step in the loop-roll combination and as an introduction to recoveries from inverted flight.

___8. *Brief discussion of spin entries and recoveries,* including the 2-turn, hands-off recovery.

FLIGHT (0:45)

___ 1. *Aileron rolls* (left and right).

___ 2. *Demonstration and practice of loops* with instructor handling throttle.

___ 3. *Practice of loops* with trainee handling throttle.

___ 4. *Loops followed by aileron rolls* with instructor, then trainee, handling throttle.

___ 5. *Instructor demonstrates* one or two parts of cloverleaf.

___ 6. *Trainee practices cloverleaf* with instructor handling throttle.

___ 7. *Instructor demonstrates Cuban eight.*

___ 8. *Trainee practices Cuban eights* with instructor handling throttle.

___ 9. *Instructor demonstrates hands-off spin recovery* at 2 turns.

___10. *Trainee does a hands-off recovery at 2 turns* followed by a 3-turn left spin and standard recovery.

___11. *Postflight review.*

Comments_____

_____

_____

Instructor_____

Date_____ Ground School Time_____ Flight Time_____

■ Third Flight

GROUND INSTRUCTION (0:45)

___1. *Review of aileron rolls,* loops, Cuban eights, and spins.

___2. *Discussion of spin theory* (airplane spin certification, factors that affect spin characteristics).

___3. *Discussion of spins* to be practiced at end of flight (at least one 3-turn spin in each direction).

FLIGHT (0:45)

___1. *Practice of aileron rolls* left and right.

___2. *Loops followed by aileron rolls* with the trainee handling throttle.

___3. *At least one two-part cloverleaf.*

___4. *Practice of Cuban eights* with trainee handling throttle.

___5. *Demonstration and practice of the Immelmann.*

___6. *Practice of a 3-turn spin* in each direction.

___7. *Postflight review.*

Comments_____

_____

_____

Instructor_____

Date_____ Ground School Time_____ Flight Time_____

▪ Fourth Flight

GROUND INSTRUCTION (0:45)

___ 1. *Discussion of airplane certification and categories.*

___ 2. *Coverage of maneuvering* and gust envelopes and airspeed indicator markings.

___ 3. *Discussion of the effects of airplane weight* on maneuvering speed.

___ 4. *Discussion of the effects of weight* on the reactions of the airplane to vertical gusts.

___ 5. *Review of normal stalls* and introduction of accelerated stalls.

___ 6. *Review of spins* with a discussion of various entries.

___ 7. *Review of Cuban eight.*

___ 8. *Introduction of the reverse Cuban eight.*

___ 9. *Introduction of the snap roll* as a "horizontal spin."

___10. *Discussion of snaps* at the top of a loop.

FLIGHT (0:45)

___1. *Practice of cloverleafs and Cuban eights.*

___2. *Demonstration and practice of snap rolls.*

___3. *Practice of spins* with over-the-top, under-the-bottom, and normal entries (at least one entry of each type).

___4. *Demonstration* (and practice) of a snap at the top of a loop.

___5. *Postflight review.*

Comments_____

_____

_____

Instructor_____

Date_____ Ground School Time_____ Flight Time_____

▪ Fifth Flight

GROUND INSTRUCTION (0:45)

___1. *Discussion of longitudinal stability,* static and dynamic.

___2. *Discussion of the FAR requirement for stability.*

___3. *Discussion of forces* and moments on an airplane in flight.

___4. *Discussion of weight and balance effects* on longitudinal stability.

___5. *Discussion of the ground effect* (takeoff and landing performance and stability).

___6. *Review of aileron roll, loop, Cuban eight, and Immelmann.*

___7. *Discussion of wake turbulence* and methods of recovery. (Avoidance is the best policy because no aerobatic course can ensure a successful recovery. Emphasis on *roll* rather than split-S, or half-loop, recovery.)

___8. *Discussion of aileron roll, loop, and spin recoveries under the hood.*

FLIGHT (1:00)

___1. *Warm-up with loop, aileron roll, and snap combination.*

___2. *Practice of one or more Cuban eights.*

___3. *Demonstration of the airplane's longitudinal stability.*

___4. *Demonstration and practice of half-roll recoveries* from wake turbulence upsets, rolls with and against initial upset to compare effectiveness of recovery and altitude loss.

___5. *Practice of aileron rolls* and loops under the hood (optional).

___6. *Recovery from inverted flight* under the hood (optional).

___7. *Practice of spin recovery* under the hood (optional).

___8. *Postflight review.*

Comments_____

_____

_____

Instructor_____

Date_____ Ground School Time_____ Flight Time_____

▪ Sixth Flight

GROUND INSTRUCTION (0:45)

___1. *Introduction of the barrel roll,* reviewing the wingover.

___2. *Discussion of the four-point roll* (optional).

___3. *Description of eight-point roll,* and reverse cloverleaf (optional).

___4. Effects on spin of aileron use and of power. (Why the ailerons should be neutral and the throttle closed for the most effective recovery.)

___5. *Review of maneuvers* covered in course (as requested by trainee).

___   a. Aileron roll.

___   b. Loop.

___   c. Cloverleaf.

___   d. Cuban eight.

___   e. Immelmann.

___   f. Snap roll.

___   g. Spin.

FLIGHT (1:00). Review and practice of as many of the briefed maneuvers as time allows, with special emphasis on spins, including demonstration of power and pro- and antispin ailerons on rotation rate and recovery, if *Pilot Operating Handbook* lists approval.

- -

Comments_____

_____

_____

Instructor_____

Date_____ Ground School Time_____ Flight Time_____

- -

You will vary the syllabus to meet the abilities of the trainees. With some you might have an instrument take-off and climbout and simulate a PAR approach (hooded) back to the airport every flight. You may teach some trainees aileron rolls, loops, and spins under the hood; others may still be struggling with simple rolls and loops at the end of the course, but with experience you'll find that many can be moved along faster than you had at first thought.

SOME ADDED NOTES

In every phase of flight instruction the most important thing for the student to learn is judgment or decision making, and this is the hardest to teach. The pilot who on one day makes the right decisions in aerobatics (or cross-country flying or instrument approaches) may the next day, because of different factors, push things too far for safety. Your attitude here will affect the trainee long after he's left your direct influence. For instance, some of your aerobatic trainees may go into competition and/or air show work. The pressure of wanting to complete a maneuver when it has started out wrong (maybe at a low altitude) at an air show may push him into trying to complete it, with fatal results. There are hundreds, maybe thousands, of people watching and he would "look bad" if he broke it off. If he did abort the maneuver the vast majority of the crowd would not know the difference and other pilots watching would congratulate him on his good judgment, but that's not the way it seems when he's up there in front of everybody.

The point you should make is that aerobatic flying requires the most self-discipline of any phase of flying and that *you*, as his instructor, would not fly at all or would stop part of a flight rather than push past a *predetermined limit*. It's been hit hard in this book, but it's worth saying again: The first few hours of his introduction to aerobatics is the time when the trainee's habits and judgment patterns will be set— just as they are in all the new phases of his flying.

Set your own limits and stick to them so that the trainee will have a guideline for judgment.

6 INSTRUMENT INSTRUCTION

24

BASIC INSTRUMENT INSTRUCTION

INTRODUCTION

In Chap. 12, instrument instruction was introduced to the student pilot for emergency purposes. Because of the limitations in hood time, a rather shallow, but direct, approach was made. It is hoped that the recipients of that training never have to use it; if they do, it most likely means that poor judgment caused them to get into such a situation. Chap. 12 was presented as a guide for teaching that type of instrument flying by the instructor who may never teach for the instrument rating. That chapter will give you some background on basic instrument flying and should be reviewed, particularly for the unusual-attitudes work. These next chapters on instrument instruction are based on the requirements for training people who will have to cope with the ATC system and actual instrument conditions for perhaps hours at a time and who must have a wider knowledge of their airplanes and systems.

It's suggested that you read straight through Chaps. 24 and 25 first before reading for details. The material is a general guide with suggestions and information that might help you set up a syllabus, if you so desire.

When you start instructing instruments, you'll find that this phase of teaching, like primary, advanced, or aerobatic instructing, has its own peculiar demands. When instructing under actual IFR conditions you will have to watch the trainee's errors and instruct him while you are monitoring the instruments to keep the airplane under control on the assigned course and altitude, as well as being finally responsible for clearances and instructions from ATC.

When in VFR conditions (the trainee under the hood) you are responsible for keeping clear of other airplanes while still monitoring (and instructing) the trainee and checking the instruments. You'll find that it's all too easy to spend a lot of time with *your* head down in the cockpit while explaining or demonstrating a new exercise.

Encourage early use of communications with ATC. Do as much basic work (as well as ATC work) under actual conditions as is legally possible. Too many newly trained instrument pilots have their first encounter with actual conditions when they are out on their own with their family in a heavily loaded (and perhaps marginally stable) airplane and find that turbulence, as well as the strange gray world they're in, can be extremely disconcerting. Vertigo can be another problem not encountered before.

Your philosophy in instrument instruction should be, as in other instruction, to explain simply and start easily and to keep the trainee moving on as his ability allows. Doing basic exercises too long can hurt his morale and set back learning, but be sure he is ready before moving on. An instrument training flight also requires three parts: (1) the preflight briefing, (2) the flight, and (3) a postflight briefing and review—and PEDPER works here, too.

You'll have to be familiar with the requirements for an instrument rating and so should know the parts of the FARs that are applicable. The AIM—*Basic Flight Information and ATC Procedures* will be one of the most valuable publications for work on the instrument rating. See that a current copy is available for your trainees.

It's likely that your instrument trainees have worked with and visited Flight Service Stations, towers, and approach control positions but haven't had any experience with the Air Route Traffic Control Center. On one of the required cross-country trips, plan it so that you can visit a Center. Talk to the people who worked your flight, and if they have time, you might go over with them the route you'll take returning home. Telephone well ahead of time and set up an appointment for any tours or films.

The same instructing procedure should be used for instrument work as for the visual maneuvers; that is, *the ground instruction should parallel the flight instruction.* During the basic instrument phase you should discuss instruments and their operation, and as you introduce navigation equipment, cover the operation, advantages, and disadvantages of the equipment. You would *not* at the basic instrument stage introduce weather theory and services or give a lengthy ground session on ATC procedures. As the need arises in ground or flight sessions, fill in specific points on weather or ATC, but remember the idea of learning given in Chap. 2—starting off with a concentrated dose of nonessential subjects for a particular level of training can bring on boredom.

The material included here should be used to get ideas for your own airplane and airway traffic situation; the steps given are listed in what is believed to be the best general order of introduction for instrument training. As you'll see, the coverage of instruments parallels that of introducing visual flight in the first part of the book; the first and second flights will be covered in detail (as in Chaps. 4 and 5), with the following material given generally and in the recommended order of introduction.

FIRST FLIGHT—GROUND INSTRUCTION

■ Preparation

WORKING SPEEDS—DESCENTS, CLIMBS, AND HOLDING. In setting up an instrument course you should establish working airspeeds for the airplane to be used. You'll have to consider several factors.

In considering a *descent airspeed,* look at the green arc of the airspeed indicator, which is the area of operations that will take the strongest vertical gusts without the airplane being overstressed. This would be the safest area to operate in as far as turbulence is concerned. However, at the higher airspeeds of the green arc, higher gusts *could* cause structural damage before the airplane stalls; at the lower airspeeds the airplane could be stalled and loss of control could occur (for instance on an approach). So, after looking at the green arc, pick an airspeed that's right in the middle (Fig. 24-1).

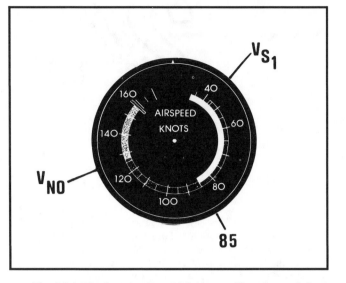

Fig. 24-1. The first step in establishing working airspeeds is to look at the green arc of the airspeed indicator of your airplane. The mid-arc value of 85 K would be used here as a starting point.

This midpoint value would be the safest point to use as far as the airplane and elements are concerned, *but* it may not be the most efficient for climbs, approaches, or holding (these will be the conditions in which the airplane won't—or shouldn't—be using cruise values of airspeed or power). For cruise you'll be using 65 or 75 percent as required for faster en route movement.

Probably the most important consideration in deciding on a good all-around airspeed is finding one best suited for the approach. It should not be so fast that you would break out and be unable to land, nor so slow that time is wasted on final (or turbulence could cause a loss of control). Other fac-

tors are (1) an airspeed at which flaps could be used, either after breaking out or during the approach and (2) an airspeed that would be found on approach charts (the 90-K value for time-to-station is the best here). So, as Fig. 24-2 shows, you now would consider both the top of the *white* arc and the approach chart numbers.

The midpoint airspeed of 85 K is also the top of the white arc (V_{FE}) but, again, looking at the idea of the 90-K airspeed on the approach charts, you might want to move the approach speed up to that value. This means that after breakout at minimums you would have to slow the airplane 5 K before putting the flaps down. Other airplanes may have the midpoint of the green arc at a higher value than 90 (or 120) K, and you could decrease the approach speed to compromise for a flap speed and a selected speed on the approach chart.

If the max rate *climb speed* is close to the descent figure decided on (say the climb speed is 85 K), you could also use 90 K for climb with very little loss of efficiency with easier remembering (90 K for climb *and* approach). However, if the approach speed you decided on is 10 K or more from the recommended climb speed, it's better to use the manufacturer's number for the climb, even if it makes an extra item to remember. (The best rate of climb airspeed for this particular airplane is 75 K and it should be used.) Having working climb and descent airspeeds at the same value isn't as important for the full-time instrument trainee as it would be for the nonrated private pilot who may have to remember some numbers if he inadvertently flew into instrument conditions. If you do decide that you want the climb and descent airspeeds to be the same, *don't pick a number below the best rate of climb airspeed*—performance will suffer.

While holding is not as much a required skill as in earlier times (it says here), you'll want to give the trainee dual practice in the art and should find a *holding speed* for the airplane.

From an aerodynamic and propulsive efficiency standpoint, a holding airspeed of 1.2–1.3 times the bottom of the green arc (if it's marked as CAS) would be the area to use at *max certificated weight*. As a rule of thumb, this speed would

Fig. 24-2. The original figure has now been increased to 90 K. (Assume IAS = CAS here.)

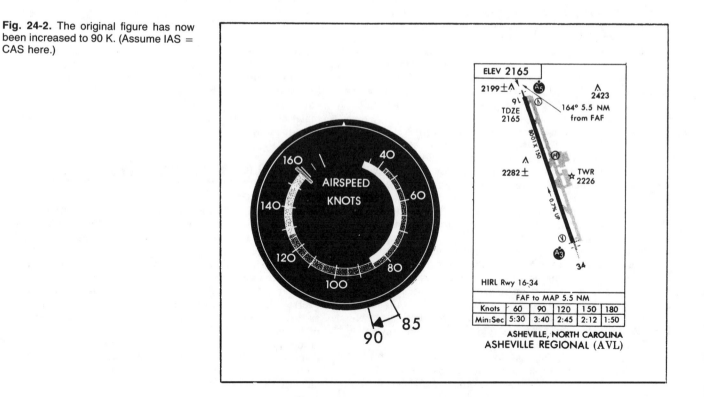

be decreased by one-half of the percentage of weight decrease to get the maximum endurance time possible. This is obviously not practical; the IFR pilot is too busy to work out the best airspeeds as he continues to hold for several minutes or even an hour or more. Also, this could be too low an airspeed for an aft-loaded airplane in turbulence. A look at another airplane with a value of 60 KCAS at the bottom of the green arc would show that a multiple of 1.3 times that figure would be 78 K, and a holding speed of 90 K (for an approach chart speed of 90 K) would not make an appreciable difference in holding endurance in a practical situation. If the airplane is low on fuel or is to be held for an unexpectedly long time, the manufacturer's figures (if available) or the thumb rule for weight decrease could be used. Again, because of large differences in IAS and CAS at the bottom of the green arc in some airplanes, if the ASI is marked as IAS, convert this to CAS. Use the multiplier (1.2 or 1.3) on that and then convert back to indicated airspeed for your reference.

The airspeed indicator in Figs. 24-1 and 24-2 is marked as IAS; at an indicated stall speed of 44 K (bottom of the green arc) the CAS is 51 K. At 1.2 times the stall CAS, the number for holding at max weight is 61 K CAS or 59 K IAS. This could be too low a figure for turbulent air and ATC would not appreciate your using 60 K as an approach speed. In this case you might want to hold and approach at 90 K, but remember that a lower speed would be used to obtain max endurance if you're in a real bind.

You'll work compromises for *your* airplane for the three important airspeeds, but in working them out, it's suggested that you follow the general steps given here, which can be applied to a large percentage of general aviation prop (reciprocating engine) airplanes.

POWER SETTINGS. You should establish general power settings for holding (clean) and descents (clean and with the gear down, if applicable) and make the point to the trainee that they may vary slightly with altitude and loading and that he will adjust power as necessary to get the required result. In holding, the airspeed is established and power is set up to maintain altitude. In the descent, the chosen airspeed is maintained and the power is decreased from the level-flight setting as necessary to get the desired rate of descent (usually 500 fpm, but more about that in Chap. 25).

The climb power will usually be that recommended by the manufacturer for climb; it may be full power for some airplanes or a less-than-maximum manifold pressure and rpm for others. You won't normally use anything other than that recommended value, unless you want to demonstrate and have him practice 500 fpm climbs at the climb airspeeds for training exercises at reduced power. (This is assuming that the normal climb rate is *more* than 500 fpm.) You might set up a checklist of power settings and airspeeds for instrument work.

Have these numbers set in your mind, written on a piece of scrap paper, or in a more formal setting before starting your program with a trainee. You might be interested in setting up a graph or graphs for your airplane as shown in Fig. 24-3. This is not to say that you couldn't work with some people the first part of the first flight in their own airplane (that you haven't flown before) to establish power settings for *the working airspeeds you chose at your leisure on the ground*. The technically minded owner would enjoy doing a little test flying. Don't take him up and grope around for *airspeeds;* you'd waste his time and look bad as a flight instructor, too.

■ **Explanation.** Lay out briefly and in general terms the route you'll take in his instrument training; it won't take long and will give a good idea of the interim goals you'll be working on.

This first (and the second) flight will, for most trainees, be their first real look at the flight instruments. Sure, they had emergency flying by reference to instruments, but that may have been years ago and they probably didn't *look* at and understand the workings of each instrument the way an instrument pilot does. A statement of, "We'll now look at the instruments from a rated instrument pilot's standpoint," will usually make the trainee feel that he *is* moving on up with the pros.

Generally, it's best not to spend a lot of time at this point in discussing theory and actions of the flight instruments and systems. However, you should definitely cover such information at some point early in his basic instrument instruction. He'll look at the instruments with new eyes after he understands their systems better.

THE PITCH INSTRUMENTS. This first flight will usually consist

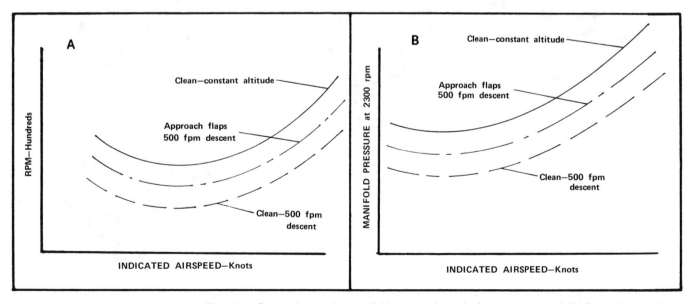

Fig. 24-3. Graphs for rpm (or manifold pressure) required, versus airspeed. (A) For a fictitious fixed-pitch-prop airplane. (B) For an airplane with a constant-speed prop.

of using the pitch instruments (attitude indicator, altimeter, airspeed indicator, and vertical speed indicator), which are gyro and pitot-static instruments, and this wouldn't follow the discussion as set up in the ground school for the two systems. After you've used all of the instruments in flight they should be discussed with their respective systems, as given in Chap. 2, Ground Instruction Techniques.

You should discuss the four pitch instruments and their general characteristics so that the trainee will know what to expect this flight. A sketch on the chalkboard like Fig. 24-4 would be a good aid in the discussion.

1. *Attitude indicator.* The attitude indicator uses the principles of "rigidity in space" and is the only instrument that indicates pitch and bank attitudes. You don't need to go into detail such as pitch and bank limits at this stage (save that for the ground coverage of the instruments later) but you should give some idea of how it should be set, on the ground or in flight, and some indications the trainee will see for climbs, straight and level, and descents. If your attitude indicator has any special face markings or has a caging knob, these should be mentioned. Discuss the pitch changes to be used this flight, but don't take much time on theory (and gimbals and such). You probably prefer to use the pitch reference lines on the attitude indicator, if available, rather than talk about bar-width changes.

2. *Altimeter.* Usually you won't have to go into much detail about this instrument before this first flight but will go into its use as a pitch indicator. You might note that it not only tells the pilot how high he is at the time, but also how high he'll be shortly. It will show him the need for pitch change; the rate of its deviation from the chosen altitude in straight and level flight will be an indirect indicator of the pitch change required for correction. He will use the altimeter as an aid to "seeing" the airplane's pitch attitude and will get practice in correcting pitch attitude without excessive altitude changes.

3. *Airspeed indicator.* The trainee has used the airspeed indicator to maintain pitch attitudes in climbs and glides but usually has not considered it an aid to controlling pitch in straight and level flight and in turns. He will be shown that it will indicate a change of pitch as well as be useful in maintaining a constant pitch attitude.

4. *Vertical speed indicator.* This instrument has been neglected by most VFR pilots and many don't know whether the airplanes they rent have a vertical speed indicator or not. It's not required for IFR flight (FAR 91).

Give the trainee the full story about how the instrument is not only good for setting up a rate of climb or descent but also is a good *trend* instrument.

TRIM. Maybe the trainee has been flying up to now without using the trim except for takeoff and level flight, but for instrument flying, proper trim use is vital at all times. Some pilots say, "I like to feel a little pressure against my hand"; this technique is an awkward way to fly at best and could cause real problems if they are distracted by ATC or other outside factors they'll encounter later.

You should have the trainee use the trim at *every* pitch change so that at any time, once a new attitude or airspeed is established, the airplane is trimmed for it as soon as possible. As one experienced instrument pilot put it: "Trim the airplane so that at any time, if necessary, you can (1) pick up the chart that fell behind the seat, (2) whomp a skyjacker upside the head, or (3) fend off (?) the advances of that great looking member of the opposite sex who is riding shotgun in the right seat." Maybe the trainee won't have such problems to overcome but it's always best to be prepared. *The airplane should be trimmed so that a temporary lapse of attention does not result in loss of control.* This doesn't mean that he can figure on trimming the airplane and taking a nap while proceeding cross-country; most airplanes aren't that stable even in the

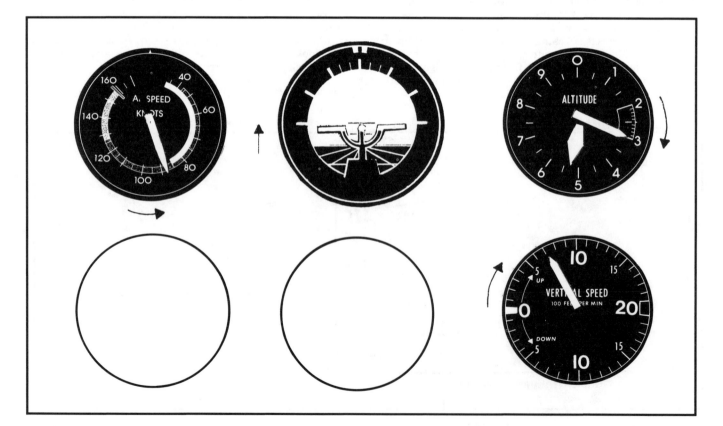

Fig. 24-4. The pitch instruments.

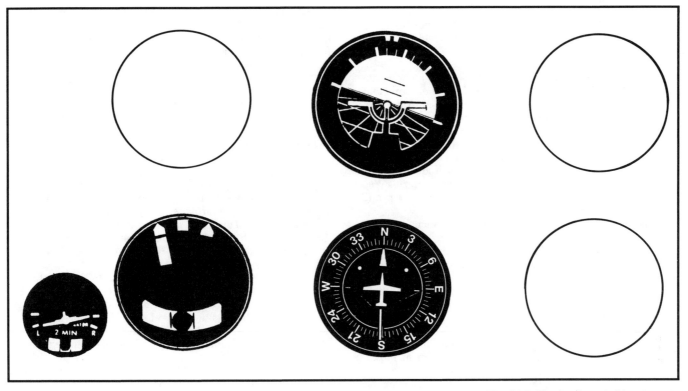

Fig. 24-5. The bank control instruments.

smoothest air. He may say to you, "I'm going to have an autopilot in the airplane I'll be using for IFR work." Don't count on it; instead teach him to use the trim properly (a skill many VFR-type pilots never manage to attain).

BANK CONTROL. Depending on how much time you used in the explanation of the four pitch instruments, you'll probably want to cover the bank instruments as well (Fig. 24-5). This would be a good idea, because, as always, you should keep your explanation ahead of the flying. Moreover, it allows you leeway if the trainee is doing unusually well and is ready to move on to the next area of flight. ("We probably won't get into this, but I'd like to cover it as a tie-in with pitch control.")

While discussing the *attitude indicator* you might note that it has several other names (attitude gyro, gyro horizon, and artificial horizon). The terms "gyro horizon" and "artificial horizon" are usually reserved for the older-type instruments but old-time instrument pilots often call all attitude instruments "artificial horizons."

It's likely that the trainee will have fewer problems using this instrument for bank control than he did in pitch control, because bank indices are easier to read and he has probably been using it as a bank reference in (too much of) his VFR practicing for the private and/or commercial certificate check ride.

Some pilots seem to have trouble when first using this instrument under the hood; they tend to read it wrong. If the instructor banks the airplane and asks for a correction to wings-level flight, the trainee often misreads the instrument and *increases* the bank. Make sure you work out such problems early.

Some older attitude indicators have caging knobs so that if the instrument is tumbled by exceeding the pitch and/or bank limits, it may be caged (with the airplane being flown straight and level by use of the other instruments) and uncaged to be used again. Emphasize that if there is no caging knob, as is the case for the majority of newer instruments, the

airplane will have to be flown straight and level for several minutes before it re-erects itself.

Note that the *attitude indicator* is the *only* instrument used for both pitch *and* bank control, so it will be the center of the normal scan.

The *heading indicator* has also been called the "directional gyro" (DG) and the "direction indicator." Affirm that it will make a good bank reference as well as indicating the direction, since if the airplane is banked it will normally turn. Most instrument pilots, if they had the use of only one of the gyro instruments, would prefer to have the heading indicator.

The older types of heading indicators are considered to have limits of pitch and bank of 55°. Above these values the instrument may "slip," but you can save this for later discussion.

Note that one advantage of the heading indicator is that it can be reset accurately immediately after it has tumbled (and the magnetic compass has settled down).

Like the vertical speed indicator, the *turn and slip* often is not given the proper attention in an instrument training program. It does not directly indicate the bank, which is probably one reason it is often shunted aside.

The turn coordinator is being used in most later airplanes and there are arguments pro and con concerning that instrument. Some say that because the rear view of an airplane (rather than a needle) is represented the pilot might get it confused with the attitude indicator, but others argue equally well that it helps the pilot to "see" the plane's general bank attitude. Whichever type you use, have the trainee familiar with it so that in partial-panel work he will still have the airplane under control.

Discuss the standard-rate turn and its indications on two types of turn and slip instruments (2- and 4-min turns) and the turn coordinator. Also cover the rule of thumb for setting up the required bank on the attitude indicator (balanced turn) for a standard-rate turn (TAS/10 + 5, for mph and TAS/10 + ½ of the initial result, for knots). Note that this rule is most accurate in the 90- to 170-K range.

The term "turn indicator" is used in this book to cover either the turn and slip or the turn coordinator, whichever you are using.

■ **The Instrument Scan.** Helping to develop a good scan for the trainee will be one of your most important contributions to his instrument flying. For people who are rated but well out of currency, the redevelopment of the scan is one of their biggest problems.

Review with the trainee the "Basic T" arrangement of the flight instruments, pointing out the position of the attitude indicator at its "central" position. The attitude indicator (when using the full panel) should be checked at every sweep because it is the only instrument that gives both pitch and bank *attitude* information.

The basic requirements for a good instrument scan are these:

1. *Cross-check.* The flight instruments must be covered continuously (again with the attitude indicator being included in every sweep of the other instruments).

2. *Interpretation.* Looking at the instruments without seeing what they indicate is not the answer. Watch for *trends*. The airspeed is starting to decrease. What does this mean? (Confirm with the other instruments.)

3. *Control.* After the cross-check and interpretation comes the control of the airplane through instrument indications and trends. There will be one of two major problems occurring with developing the trainee's scan (and some people manage to have both during the process of learning): (a) *a too-fast scan* (The eyes are moving rapidly from one instrument to another without registering the information; the problem here is a lack of *interpretation*.) and (b) *fixation on one instrument*. (Suddenly the altimeter or vertical speed has become fatally attractive. The heading indicator and/or other instruments are ignored and problems result. The *cross-check* is being neglected.)

PRIMARY AND SUPPORTING INSTRUMENTS. For many instrument trainees the method is complicated and confusing. Sometimes an instrument is "primary" as a maneuver is initiated and then becomes "supporting" as it continues.

In a climb, for instance, the attitude indicator is a good initial pitch reference, but for more precise information, the airspeed indicator is used. In smooth air, the airspeed may be flown to within 1 K of that desired, but it's not as easy to do this with the attitude indicator.

If you got your instrument rating using the primary-supporting method, prefer it, and would be able to impart the information to the trainee, then by all means use it.

CONTROL AND PERFORMANCE INSTRUMENTS. This method of analysis of setting up a scan and controlling of the airplane is considered by many to be the most logical and easiest to learn.

The *control instruments* are the *attitude indicator* and *manifold pressure gauge* (at a given rpm) and/or *tachometer*. With these indicators you control the airplane's pitch and bank and altitude performance (straight and level, climbs, and descents). Your control of the airplane through these attitude and power indicators will establish the performance shown by the remaining flight instruments.

The *performance instruments* show what is happening from your control of the airplane through the attitude indicator and power instrument(s). The *altimeter, turn indicator, airspeed, heading indicator,* and *vertical speed indicator* show what the airplane is doing in the various performance regimes.

The advantage of knowing attitude and power settings for various performances is that they can be repeated and are

pretty well constant for a particular airplane and configuration, with some *minor* changes occurring because of altitude and weight changes. In other words, you can count on *attitude plus power equals performance* (Fig. 24-6).

FIRST FLIGHT—FLIGHT INSTRUCTION

■ **Preflight Check and Start.** Emphasize that as an instrument pilot the trainee will have more to check (antennas secure and other items that would assume more importance in instrument flight). While it might seem that checking all of the antennas during the basic part of the instrument instruction is not necessary, you should establish the proper attitude from the beginning. Don't ignore some antennas or the deicing equipment as the airplane is inspected. Use a brief explanation such as: "This is the marker beacon antenna. Check it for security (with the other antennas) each time. We'll be using it on the ILS approaches later in the course, and I'll go into more detail about it then."

Like in the preflight check for his first flight lesson, he may not be able to identify every antenna or other special IFR item on the airplane, but after a couple of times he'll know and will treat them as an integral part of his operation. You may not choose to do it the first flight, but you should discuss the effects of communications or navigation equipment antennas on the top or bottom of the airplane; one radio with the antenna on top may be efficient for communication when parked under the tower while the one on the bottom is blanketed out. (That antenna on top might give problems when flying directly over a facility.) Make sure that sometime in the program he's made aware of which COMM/NAV (number one or two) works with which antenna(s) and emphasize that he should determine this information for any other airplane he flies after getting the rating.

Starting is, of course, a function of the airplane used rather than whether it is going on a VFR or IFR flight, but because the airplane being used for instrument training has much more electronics equipment than the trainers he's been using, a thorough check should be made to ensure that all such equipment is OFF before the start is attempted. Depending on your avionics setup, you might establish a left-to-right check across the panel (he only reads on the checklist "all avionics equipment off") or you may actually list each item to be checked on the checklist. This latter approach could mean an extra-long checklist, and it's been found that the more items listed, the greater the possibility of one being overlooked as the pilot tends to start skimming the page. *The best idea is to have an avionics master switch in the airplane you're using for instrument instruction; this could save you some radio problems later.* Your first project when you start instructing instruments should be to set up a checklist that not only covers specifics of your training airplane but could also be converted for use in IFR work for any airplane. (You might be generally specific or specifically general.)

After starting, point out items that the trainee might not have paid much attention to in his VFR training, such as suction and ammeter indications.

Set the heading indicator (with the magnetic compass) and set the attitude indicator.

■ **Taxi.** As he taxies out, point out how the turn indicator might be checked by S-turning while taxiing. The calibration cannot be checked until in the air, but a check of operation should be made; and in the case of an electrically driven instrument with an OFF-ON switch you might discover that "somebody" forgot to put the instrument in operation. Sure, on the local training flights in VFR conditions, it wouldn't be a tragedy if you took off without it on, or with it out of action

Fig. 24-6. Control and performance instruments.

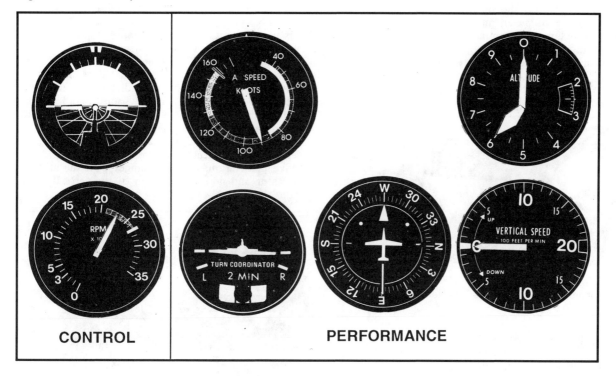

CONTROL   PERFORMANCE

for some other reason, but check it to establish good habits. In actual instrument conditions, if there is a loss of the other two gyro instruments after takeoff, a fatal time lapse could occur before he realizes that the turn indicator is centered not because he's flying straight, but because it's not getting power.

■ **Pretakeoff Check.** If you decide not to go through the check of every piece of electronics equipment during the first pretakeoff check, you should at least point out each one as you check across the panel. If you go into too much detail at first, the trainee won't see the point since he won't need this information for several hours yet.

It's a good training procedure to start issuing practice ATC clearances in the basic course, but it's best not to do it the first flight (more about that later).

Check the heading indicator against the magnetic compass for evidence of creep during the taxi and run-up. Reset the attitude indicator as necessary.

After the pretakeoff check is complete, get clearance from the tower or check the area carefully at an uncontrolled field and taxi into position for takeoff. Check the heading indicator against the runway heading *and* magnetic compass. You might expect some small difference in heading indicator and compass indications when you line up with the runway because of deviation differences on different headings. (You initially set the heading indicator with the compass on one heading right after the start and the runway heading is 90° to that.) But you are really looking for significant "slip" of the heading indicator.

■ **Takeoff and Climb (Visual).** The start, taxi, and preflight check this first flight may take 10–15 min, and the takeoff and climb to practice altitude will probably use approximately another 10 min.

Call the trainee's attention to the indications of the pitch instruments during the climb. Refer to the heading and bank instruments as necessary to maintain the proper course. Compare the attitude of the airplane with the indications of the pitch instruments. It's best to get up to smooth air for the first couple of basic flights.

■ **Demonstration and Practice.** You should spend about 15 min maximum for each instrument; about 10 min of that time the trainee will be under the hood using each one. (At first he should set up pitch attitudes using outside references and compare the instrument indications with the pitch attitude of the airplane.) The following exercises will use cruise power unless otherwise noted:

1. *Attitude Indicator.* Keep your eyes open for other traffic. Don't require close tolerances of heading, because you are concentrating on pitch. However, you'll have to occasionally remind the trainee to mind his bank.

   The trainee should set up normal cruise speed and power at a specific medium altitude (5000 ft, for example) and adjust the miniature airplane for zero pitch. Have him do the following:
   a. Maintain level flight (wings level).
   b. Make minor changes up or down (using the pitch lines or bar widths, as applicable) and note effects on the indications of other instruments.
   c. Return to normal cruise maintaining a constant-level-pitch attitude. (Have him make turns as necessary to stay in the local area.)

2. *Altimeter.* Break for a couple of minutes before introducing the altimeter work; then again have the trainee set up normal cruise speed and power setting. Emphasize that in addition to the instrument indications he should note the pitch attitude of the airplane during the following exercises (without the hood at first):
   a. Maintain level flight by reference to the altimeter as a pitch control.
   b. Ease the nose up for minor altitude gain. (Make sure that the trainee understands that pulling the nose up to gain altitude can only work on the front side of the power curve.)

c. Ease the nose down for minor altitude loss.

d. Verify the pitch attitude shown on the altimeter by cross-checking the altimeter and attitude indicator (Fig. 24-7). Have him ease the miniature aircraft on the attitude indicator well above the horizon bar and note the rapid change in altitude. Correlate the reactions of the altimeter to the attitude indicator pitch settings and to the actual attitude of the airplane.

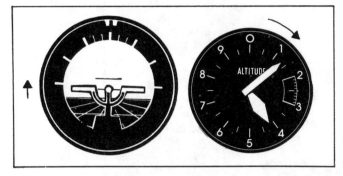

Fig. 24-7. Compare the reactions of the altimeter to changes in pitch (cross-check).

e. Check the altimeter reactions by the rate of pitch change. A rapid pitch change gives an apparent lag. A small pitch change allows the altimeter to give an immediate indication.

f. Have him correct the pitch and altitude by the altimeter and change the pitch to stop the altimeter, then make a smooth return to the desired altitude. *The altimeter is naturally the main instrument for constant-altitude flight.*

g. He should practice the following:
  (1) Holding a constant altitude by use of the altimeter.
  (2) Maintaining level flight by use of the attitude indicator *and* the altimeter.
  (3) Losing or gaining 200 ft of altitude by no more than 1-bar-width or a selected marker pitch change. Have him return to the original altitude using the same technique. He should practice cross-checking the altimeter and attitude indicator during this exercise. Repeat until the cross-check is smooth. You'd better hold to a ± 20-ft tolerance for straight and level flight; the trainee should initiate the pitch changes for corrections.

3. *Airspeed indicator.* This may be the first time some trainees will have noticed the actions of the airspeed during very *minor* pitch changes in straight and level flight.

Introduce and have him practice the following exercises:

a. From normal-cruise level flight (65 percent power) make small changes in pitch and note the slow changes in the airspeed. Make large changes in pitch and note the reaction of the airspeed indicator. Note that there is an apparent lag, but that this is due to inertia of the airplane rather than a function of the instrument.

b. Fly level normal cruise and maintain a constant airspeed, referring to the airspeed indicator primarily (also check wings level).

c. Fly level normal cruise and maintain a constant airspeed, cross-checking all four of the pitch instruments, and if his progress warrants it, go to exercise d.

d. Introduce the use of trim as an aid to pitch control. Maintain level flight at various airspeeds and power settings, using all pitch instruments (Fig. 24-4). Use trim for every speed and power change.

4. *Vertical speed indicator (VSI).* The average person usually will need more work with the vertical speed indicator than the first three instruments because of lack of familiarity with its use. You may have to put more emphasis on the instrument at this early stage in order to bring the trainee on a par with his use of the other instruments. At first, have him make 1-bar-width or reference lines changes on the attitude indicator and note the indications of the vertical speed indicator. You might use some, or all, of the following exercises. (*Note the attitude of the airplane in each exercise.*)

a. Maintain a constant altitude at normal cruise using the vertical speed indicator as a *trend* instrument.

b. Make predetermined pitch changes with the attitude indicator and note the reaction of the vertical speed indicator (Fig. 24-8).

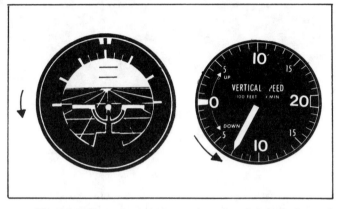

Fig. 24-8. Have the trainee compare the pitch attitudes and the indications of the VSI.

c. Practice setting up 500-fpm climbs (or what performance allows) and descents by reference to the VSI. Note that the instrument is not to be "chased," but small pitch corrections should be made and the indications checked.

d. Set up a descent of 1000 fpm and then pull up immediately into a climb. Note that the needle reverses itself with no delay (but there will be a lag before the climb rate is indicated if it is not the "instantaneous-type" instrument).

e. After normal cruise at a constant altitude is resumed, practice climbs and descents 100 ft from the base altitude. Do not use over 200-fpm indications on the VSI for these altitude changes.

f. Cross-check attitude indicator, altimeter, airspeed indicator, and vertical speed indicator in maintaining level flight.

g. Set up climbs and descents of 100 fpm and note that the vertical speed indicator shows the pitch change more quickly than the altimeter.

h. Practice getting back to the chosen altitude from deviations of ±50 ft (200 fpm or less is to be used for altitude corrections). Use all four instruments (attitude indicator, altimeter, airspeed indicator, and vertical speed indicator).

NOTES TO YOU. If the *maximum* of 15 min per pitch instrument is used, you'll spend more than an hour (counting breaks) on the airwork at altitude. Probably you will use much less time on each instrument since the trainee may start losing interest after 6 or 7 min of using a particular instrument. If in doubt, break off the use of each pitch instrument after you see reasonable progress, and plan to repeat the practice on the second flight.

This first lesson on instruments can be fatiguing, so as an overall recommendation you'd probably do better to stop after all of the pitch instruments have been used and he's getting the idea unless you feel that he's really ready for bank work.

On the way back to the airport have him practice various rates of descent at a constant airspeed and varying power settings. Keep *your* eyes open as the airport is approached. Later you'll probably want to simulate an ASR or precision approach at the end of each flight during the basic instrument phase.

■ **Postflight Discussion.** After the flight you should analyze and grade the trainee (either verbally or on paper) on the following items:

Preflight check ____
Start ____
Taxi ____
Pretakeoff check ____
Takeoff ____
Climb ____
Clearing procedures ____
Pitch control ____
Altitude control ____
Cross-check ____
Interpretation ____
Control ____
Smoothness ____
Orientation ____

Don't stretch this discussion out; but cover his problems, and if progress merits it, note that the bank instruments will be flown the next flight.

SECOND FLIGHT

■ **Preflight Discussion.** If you didn't get to discuss the bank instruments before the first flight you'd do it now.

Give the trainee a chance to ask questions about the pitch *and* the bank instruments and review any problems he had with the first flight. (A grade sheet or record of each flight is even more helpful in an instrument course than in other flight courses. There are too many details in teaching instruments for you to remember easily.) Discuss a simple departure clearance.

THE SCAN. Review the instrument scan, noting the cross-check to ensure that the flight instruments are covered properly without fixation.

You might want to sit in the airplane with the trainee and review the scan for the Four Fundamentals (straight and level, turns, climbs, and descents). If a simulator is available at your school, it would be good to alternate flights and simulator sessions; the simulator work can be most important for you in ensuring that the trainee has developed the best scan for him. If you sense problems of fixation, the simulator exercise may be "run back" so that you can watch him and analyze the problem. Maybe the altitude was right on but his heading control was well out of limits during a particular exercise. You'll have to ensure that the heading indicator (or turn indicator, as applicable in partial-panel work) is put back into the scan.

One point: When using alternating simulator and airplane sessions, it is not intended that the lesson sequence be *exactly* simulator, then airplane, then simulator, etc. There may be times in your syllabus when maybe even three airplane or three simulator "flights" in a row may be required.

Discuss with the trainee a simple clearance to the practice area, giving headings and altitudes. Note that this clearance will be given to him at the warm-up area.

■ **Preflight Check.** Review any shady areas from the first lesson concerning special considerations for instrument flight.

■ **Starting, Taxiing, and Pretakeoff Check.** Again, emphasize that all electronics equipment must be off before starting. Set the heading indicator with the magnetic compass. (Check the turn indicator while taxiing.) Check all instruments and equipment, using the checklist. Complete the run-up and have him copy and read back the simple clearance discussed before the flight.

Emphasize at this point that the gyro instruments require a minimum of 5 min to "spool up" to efficiency. Accidents have happened when a pilot started up, taxied out with little or no pretakeoff check, and took off into low IMC.

■ **Takeoff and Climb.** The trainee should make a visual takeoff and traffic departure; then make hooded *straight* climbs as practicable to the prechosen altitude and practice area. You naturally will be responsible for clearance from other traffic and may have the trainee make occasional small shallow turns to help you do this. Have him go under the hood well before reaching the selected altitude so he can get settled down.

■ **Airwork**

BANK CONTROL. Plan to spend no more than 15 min (10 min hooded) on each bank instrument:

1. *Attitude indicator.* Have the trainee establish straight and level flight at normal cruise speed at a particular altitude. (You might have him practice exercises a–d visually, noting the bank altitude of the airplane compared with the natural horizon.) Require that he stay within ±50 ft of the prechosen altitude during these maneuvers. Have him do the following:
   a. Set up a 5° constant-banked turn for approximately 90° of turn. Roll out level, then turn in the opposite direction.
   b. Practice constant banks of 10, 15, 20, 25, and 30° in opposite directions. Maintain a constant altitude. (Check both pitch and bank.) The banks should be maintained within ±5° of that desired.
   c. You should put the attitude indicator in various banks (shallow and medium, left and right) so that a proper reading of the instrument and proper response will become second nature.
   d. Practice both pitch and bank control using the attitude indicator. Have him repeat exercises a through d under the hood.
2. *Heading indicator.* Note that the heading indicator gives an indirect indication of bank, since it's assumed that the airplane will be in a balanced turn. Slow movement of the heading indicator means a shallow bank; fast movement means a steep bank.

   Emphasize that except for practice exercises, when correcting to headings, the pilot should use the degrees of bank equal to the degrees to be turned, *up to the bank required for a standard-rate turn.* In other words, if the heading is off 5°, the maximum bank to correct should be 5°; for 10°, bank 10°. He should understand that a standard-rate turn is not to be exceeded in normal instrument flight.

   The heading indicator is always a big reference for bank control in straight and level flight. You might have him practice the following exercises visually, noting the relationship between the bank attitude of the airplane and the reaction of the heading indicator:

    a. Straight flight using the attitude indicator and heading indicator.

    b. A 180° turn in each direction (10° bank) using the attitude indicator and heading indicator. Repeat the turns, using 20° bank. Note the comparative rates of movement of the heading indicator.

    c. Straight and level flight by the use of all instruments except the turn indicator.

    d. Practice 90°, 180°, and 360° turns (15° banks) using all instruments except the turn indicator.

    e. Repeat exercises a–d under the hood.

3. *Turn indicator.* Point out that the turn needle or reference airplane also indirectly indicates the bank attitude of the airplane. When the airplane is banked (ball centered), it is turning, the rate of turn being proportional to the angle of bank. When the heading indicator is not available, the turn indicator is the reference for straight flight. Have him do the following exercises at normal cruise under the hood:

    a. Keep the ball centered. Check that the wings are level and the heading is constant.

    b. Note the heading, then cover the attitude and heading indicators and have the trainee fly for 2 min. Uncover the heading indicator and check any heading deviation.

    c. Reverse course and cover the attitude and heading indicators again.

    d. Repeat exercise b until the heading is within ±5° (smooth air).

■ **Combinations of Flight Instruments.** You might not get into the following hooded exercises this flight, but they should be used after the trainee has had a chance to fly each of the instruments and is smoothing out his hood flying. This will be the first time you'll see any possible problems with scan. Most instrument trainees will have some trouble when they first have to cope with the full array of instruments. If he's been doing average or above-average work with the individual instruments or combinations of two, but his performance deteriorates when the full panel is used, you'll have to put particular emphasis on his scan (Fig. 24-9).

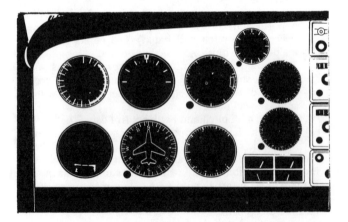

Fig. 24-9. The average trainee will often have a tendency to see only one instrument.

Have the trainee fly *straight and level* at normal cruise and at a prechosen altitude and heading, using the following combinations (require him to make turns as necessary to stay within a particular area), as shown in Fig. 24-10.

If the situation permits, you should combine these exercises with a simple vector and cruise letdown (500-fpm) problem to the vicinity of the airport as an introduction to radar

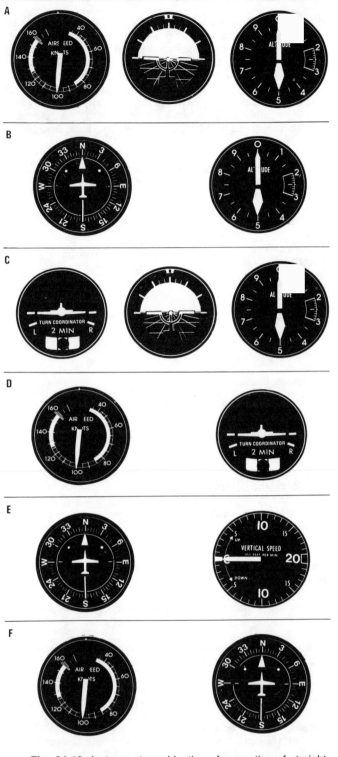

Fig. 24-10. Instrument combinations for practice of straight and level flying. Uncover the other instruments after several minutes of using the various combinations.

    (A) Airspeed, attitude indicator, and altimeter to show relationships among three of the four pitch instruments.

    (B) The two flight instruments absolutely necessary for flying IFR. All other flight instruments are aids to these.

    (C) The turn indicator is used for turn indications (what else?) and the attitude indicator and altimeter are used for pitch control here.

    (D) Flying staight and level with airspeed for pitch (after equilibrium has been established) and the turn coordinator for heading control. Uncover the altimeter and heading indicator to check progress after 3–4 min of flying by these two instruments.

    (E) Heading indicator and vertical speed indicator.

    (F) Airspeed for pitch and heading indicator.

(The Instrument Flight Manual)

vectoring and preparation for an Airport Surveillance Radar (ASR) approach.

After he's had practice in using all instruments and has completed the exercises listed, you should take a look at his performance and grade on the following segments:

Heading control:
    All instruments ____
    Attitude indicator only ____
    Heading indicator only ____
    Turn indicator only ____
Altitude control:
    All instruments ____
    Attitude indicator only ____
    Altimeter only ____
    Airspeed indicator only ____
Cross-check ____
Interpretation ____
Control ____

When having the trainee fly using the individual instruments, check to see which one(s) he does an extra good job with—he may be concentrating on that one when a full scan is required.

■ **Instrument Ground School.** As was noted earlier, hold off on going into detail about instrument operation until the trainee has had a chance to use all of them. Sometime after the second flight he should get a session on information he can use as an instrument pilot, and the sample lesson on the instruments in Chap. 2 might be checked for some ideas.

You should discuss effects of precession in the turn on pitch indications of the attitude indicator as well as acceleration effects, as might be encountered right after takeoff.

Talk about the other flight instruments, including the *magnetic compass* and its foibles, as applicable to flight. A comment of, "You noticed that when we were doing those turns, the attitude indicator (or heading indicator or other instruments) did thus and so, etc.," will bring out the idea that this is information that can be used.

THE FOLLOWING FLIGHTS (NORMAL OPERATIONS)

The information in the rest of this chapter is in the suggested *order* of introduction and use of the material for basic instrument flying. To repeat, the main thing you will have to guard against is spending too much time on one item or one instrument. A common failing of new instrument instructors, in addition, is to move into radio and cross-country work too soon and they find that the basics are weak when holding or approach practice is started. A setback can occur if the trainee has to go back to basics, and all too often the instructor keeps him working on the approach and missed approach procedures when practice in turns and climbs and descents is needed. The biggest temptation you will have after several dual hood sessions during the day is to let that last trainee get away with heading or altitude problems, figuring that "he has plenty of time left to work it out." Again, as with VFR training and aerobatics, the first 5 or 6 hr in this type of instruction pretty well establish a person's attitude and habits in instrument flying.

■ The Four Fundamentals
PITCH AND AIRSPEED CONTROL (CONSTANT ALTITUDE— FULL PANEL). The exercises discussed earlier were initiated at cruise airspeed, and you'll add another "distraction" when he is able to cope with it.

You can use the following exercise to check on scan and cross-check efficiency:

Have the trainee (hooded) trim the airplane and fly straight and level (check the heading indicator) for 1 min. Slow to 10 K below cruise, trim and adjust power, and fly for 1 min maintaining a constant altitude and heading. Repeat the procedure, slowing down 10 K each time, retrimming and adjusting power as necessary, down to 10 K above the stall. Have him make turns as necessary to stay in the area. It's good practice also to have him move in increments back up the airspeed scale, maintaining a constant altitude and heading. You may want to use 5-K increments.

Check on: Altitude control, heading, power use, trim, smoothness, and transition. This could be the point where you'd want to introduce the idea of changing power greater than required to maintain altitude at the new airspeed. For example, if 2300 rpm is being used for 100 K and you want the trainee to fly at 90 K (requiring, say, 2200 rpm), have him make an initial reduction to *2100* rpm to expedite the transition and then readjust the power as the airspeed reaches the desired value. Some instructors recommend an "overchange" of power of 100 rpm (over or under as required) for the fixed-pitch-prop airplane and 3 in. of manifold pressure for the constant-speed-prop when airspeed changes are made. At the higher airspeed range for a particular airplane, these margins will be cut down as the engine(s) has little or no excess power.

Common errors you can expect when the trainee does the transition for the first time:

1. Poor altitude control. Expect him to gain altitude when slowing, and lose altitude when increasing the airspeed. This is the result of too-rapid pitch changes.

2. Heading control problems. Generally, you'll have to keep on the trainee about the heading at first; altitude seems to be the thing he cares most about.

3. Neglecting to use trim after each transition. He may feel that he "can hold it" for a minute and trimming is a lot of trouble, but you'll have to keep working so that he automatically trims each time the airspeed is changed. To repeat, good trim can keep the airplane under control if his attention is distracted. Have him use the rudder trim at low speeds if it is available.

4. Poor power control. This is where the trainee, who has to have the rpm or manifold pressure *exactly so* during the power change part of the transition, can drive you up the side of the cabin. (His concentration on the exact power setting lets the altitude and heading run wild.)

STANDARD-RATE TURNS (CONSTANT ALTITUDE—FULL PANEL). Note that at this point you are still working with a constant-altitude situation. You'll be tempted to start combining climbs and descents and turns, but you'll find with more experience that most people need to get settled down and use the instruments in fairly simple exercises to establish their scan before working on combinations. You'll have to set up a delicate balance between boring the trainee with overly repetitive, relatively simple exercises and pushing him along too fast, with resulting problems later.

The trainee may consider the standard-rate turns as the first "real" instrument work he's done. Even though skill in transitioning from one airspeed to another is essential, the standard-rate turn is a term he's associated with instrument flying.

Discuss these turns on the ground, and mention the methods of establishing them with the turn indicator or attitude indicator. The preflight session before this turn work should include talking about calibrating the needle or turn coordinator. In the air, you should demonstrate this calibration before starting the turn exercises.

To get the trainee used to the idea of standard-rate turns and the bank required at various airspeeds, have him do the following exercises under the hood (full panel):

1. Set up power and trim to maintain a constant altitude at normal cruise. Choose a cardinal heading. Make a 90° standard-rate turn to the left, fly straight and level for *1 min,* and then make a 90° standard-rate turn to the right.

2. Slow 10 K, adjust trim and power to maintain a constant altitude; make a 90° standard-rate turn in each direction with a *1-min* straight and level flight between each turn.

3. Slow the airplane in 10-K increments down to 10 K above the stall, repeating the exercises 1 and 2 above. Again, you may prefer 5-K increments.

As the exercise continues, point out the decrease in bank required as the airspeed is decreased.

Expect some overcontrolling at first in setting up the standard-rate turn at each airspeed. He has already practiced the airspeed changes, but you should make corrections as necessary, since he is apt to be concentrating on the turn part of the exercises at the cost of the transition work.

He'll likely have problems with altitude and heading during the exercises; the usual problems are loss of altitude in the turns and overshooting the 90° roll-out point. If the airplane is off-heading at the end of a turn, have him correct it back to the cardinal value.

Note that the clock has been added to the scan. You might find it better the first couple of transitions or turns for you to keep up with the 1-min straightaways, but by the end of this period you should leave the timing to the trainee.

Fig. 24-11 shows some combinations you could use in these turn exercises *after* the trainee is comfortable with full-panel turns.

DESCENTS. Fig. 24-12 has some suggested exercises for working out problems in descents. (To repeat, these are to be done after the trainee has had practice in full-panel descents.)

CLIMBS. Fig. 24-13 notes several climb exercises you might use as you feel they would fit into your syllabus.

When his control is smoothing out you should act as approach control on the way back to the airport and descend him under the hood at a cruise airspeed to the altitude to start an ASR, as progress permits.

■ **More about the Working Speeds of the Airplane.** The previous exercises basically maintained a constant altitude throughout, and you should spend part of a preflight period reviewing with him the working speeds of the airplane. Note that the power settings will vary with airplane loading and the density-altitude. While you will require some exercises such as the vertical S-1 and S-2 (to be discussed shortly), try to keep the basic instrument airwork realistic, and after the working speeds are introduced you can set up simulated climb or descent requirements as might be issued by ATC.

Have the trainee fly the working speed exercises and confirm the power settings as they apply. Note that the engine-out speeds and power settings are not included in the exercises but will be covered later when they apply. You may give *all* of the numbers at this time to make the list complete if you are training him in a twin.

Fig. 24-14 is a table you might set up to confirm some numbers you gave him on the ground. If time permits it's best to have the trainee fly the numbers himself because it will usually help his scan and he'll try to fly more precisely.

■ **Instrument Takeoffs.** About this stage in training (five to six flights) you should introduce the instrument takeoff, warning that it is more of a training maneuver than of practi-

TURNS

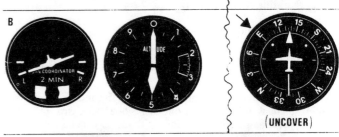

Fig. 24-11. Instrument combinations for practice of turns.

(A) Attitude indicator and altimeter. Set up the proper bank to maintain a standard-rate turn at the chosen airspeed. Start at a particular heading, then cover the heading indicator. Roll out at the time required to turn to a predetermined heading and uncover the heading indicator.

(B) Do the same exercise using the turn indicator and altimeter.

(C) Airspeed and heading indicator combination. Note that the airspeed has been decreased by 3 percent for the pitch required to maintain altitude in a standard-rate turn, as mentioned earlier in the chapter. Here you are using the rate of direction change to maintain a standard rate of turn (or as close to it as possible).

(The Instrument Flight Manual)

cal use. If the weather is bad enough to require an ITO, it would be below landing minimum for the type of airplane he'll be flying, and if an emergency situation occurs shortly after takeoff he would be in trouble as far as landing back on the airport is concerned. There may be times later in his flying career when an instrument takeoff is justified, but don't encourage it here.

You might bring up the point that the ITO is good if he will go into instrument conditions immediately after takeoff (assuming he's decided that he has to go) because problems can arise if he transfers from purely visual references to the instruments at a low altitude.

Go over the procedure on the ground step by step before the flight, mentioning such items as taxiing forward a few feet on the runway to straighten the nosewheel (or tailwheel). Be sure he understands that you'll be giving directions and when you say "Right!" during the process it means he must give it *right* rudder, not that he is going to the *right* of the center line.

DESCENTS

Fig. 24-12. Descents.

(A) Straight descent without the attitude indicator or turn indicator. The power is set to get 500 fpm at the chosen descent speed.

(B) A timed straight descent without the attitude indicator or heading indicator. Check (uncover) the heading indicator after leveling off. Use an altitude "lead" of 10 percent of the rate of descent; for a descent of 500 fpm, start to level off 50 ft above the chosen altitude. (For 1000 fpm, use 100 ft, etc.)

(C) A timed straight descent using the attitude indicator, vertical speed indicator, and clock (plus power for 500-fpm descent). You'll find that by using the proper power setting and maintaining a 500-fpm rate of descent on the VSI, if the airspeed indicator is uncovered during the descent, the airspeed will be very close to 90 K for this example airplane. After the proper lapsed time and leveling, uncover the altimeter to check the descent accuracy. Also check the heading indicator to see how the trainee fared on heading. Theoretically, if you kept the wings level, the heading should be close on, but check it anyway.

(D) A timed descending turn to a predetermined altitude and heading. After rolling out and leveling off, uncover the altimeter and heading indicator.

(The Instrument Flight Manual)

Give commands for correction, not information on what he's doing.

Mention, too, the acceleration effects on the pitch indications of the attitude indicator—the nose will appear higher than it really is, and he may lower it, to his chagrin, in an actual situation.

It's a good idea for *you* to demonstrate and describe the ITO with the trainee observing under the hood. The trainee can take over at a safe altitude and continue the climb (hooded) to the practice area. He should be able to do the next ones with advice from you, and you should have him come out from under the hood at a safe altitude (say, a couple of hundred feet) to look back and check his lineup and position with respect to the runway. After that, depending on the individual, you may put him under the hood at the beginning of the ITO and keep him there until it's time for a break after some exercises at altitude. It's possible, too, that on rare occasions you may have a trainee fly under the hood from ITO to "breakout" at the end of a simulated ASR approach at the end of a period. But, if it won't interrupt your training situation, a short visual break every 15–20 min during a flight is a good idea for most people.

Use the ITO every flight following its introduction, or at your own discretion, but grade him on the later ones.

Fig. 24-15 shows the steps in an instrument takeoff for one light two-place trainer.

■ **Introduction to Holding Patterns.** After the various working speeds are clear in the trainee's mind, introduce the idea of a simple constant-altitude holding pattern using two 1-min legs over an imaginary VOR. The clock was introduced earlier, but this will be a good way for him to pin it down as part of his scan without undue pressure. You won't be using a VOR as a fixed reference, so keep in mind that a strong wind can move the airplane out of the area after several patterns. Fig. 24-16 shows the general idea.

Before the flight discuss the reason for the holding pattern and where it might be used in actual IFR work.

As an additional exercise, you may have the trainee descend or climb in the pattern at 500 fpm, but note that *climbing* in a holding pattern is not normally done. If you are using a twin, as he progresses in this exercise you might throttle

CLIMBS

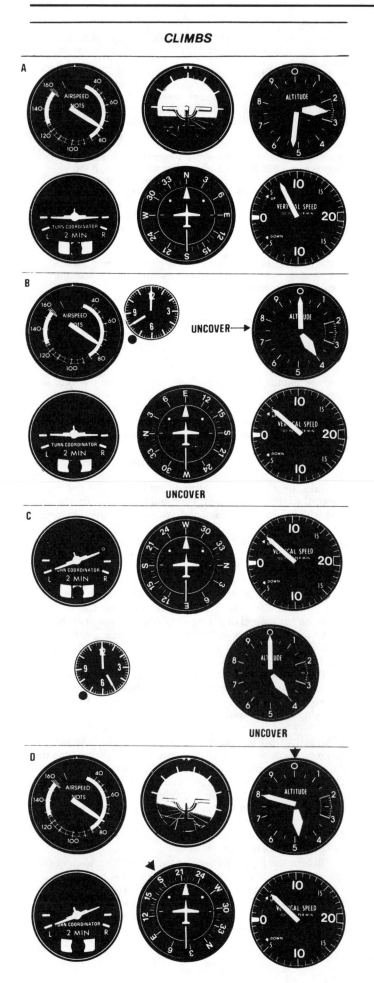

Fig. 24-13. Climbs. This airplane used 75 K as the best climb airspeed.

(A) Straight climb, using all flight instruments (plus power).

(B) Straight timed climb without the attitude indicator, heading indicator, or altimeter. Uncover the heading indicator and altimeter at the end of the prechosen time period to check your accuracy.

(C) Timed climbing turn to a predetermined heading and altitude (airspeed, attitude indicator, and altimeter covered). At the end of the time period uncover those instruments and check your altitude. The heading should be good, since you have use of the heading indicator during the climbing turn.

(D) A climb (500 fpm) and turn to a prechosen altitude and heading using all flight instruments. Stop the climb and turn as the prechosen indications are reached.

(The Instrument Flight Manual)

back one engine to zero thrust and have him maintain altitude (if possible at that altitude and loading), or descend in the pattern at 500 fpm. In this case, don't pull a surprise engine-out situation; tell him about it well beforehand. You might indicate that with an engine out an emergency exists and that his airplane will have priority, but it is possible that the traffic situation could require a short period of holding. Besides, this exercise is *good* for him.

One procedure to show the principle of holding in a crosswind would be to set up a visual holding pattern over a road at, say, 600 ft, using the road as the inbound leg and having a house or crossroad as the "fix." He can check what kind of correction is needed and see the pattern. (Use a 1-min inbound leg for this.)

■ **Progress Check.** After the working speeds have been used and the simple holding pattern work has been practiced, you should take a look at his progress in those areas. You may wonder why in the introduction of working speeds no mention is made of "slow" or "high" cruise speeds and power settings. These items were omitted because the average instrument pilot chooses one power setting (usually for either 65 or 75 percent) and uses this for cross-country flying; the other working speeds apply to actual phases of IFR work. You'll have to occasionally use exercises or conditions that are strictly for training purposes, but try to be able to apply the work to actual requirements as much as possible.

> The *common errors* you'll find in the full-panel practice of airspeed and holding exercises will fall under the following areas, and you should *check each when looking at the trainee's progress:*
>
> 1. *Normal cruise.* Check altitude control, heading, and his cross-check of instruments.
>
> 2. *Cruise climb* (500 fpm). Airspeed (pitch) control, heading, rate of climb, smoothness, and cross-check.
>
> 3. *Cruise descent* (500 fpm). Airspeed (pitch) control, heading, rate of descent, and cross-check.
>
> 4. *Best rate of climb* (normal climb). Airspeed (pitch) control, heading, "torque" correction (ball centered), and cross-check.
>
> 5. *Holding* (constant altitude). Altitude control, airspeed (pitch) control, heading, rate of turn, timing, and cross-check.
>
> Note that cross-check is a factor to examine in each of the exercises and a poor one can be the instigator of problems in any phase of instrument training and later actual operations.

■ **Vertical S-1 (Hooded, Full Panel).** This may seem just an exercise, but it is useful for working on the develop-

WORKING SPEEDS AND POWER SETTINGS FOR THE AIRPLANE

	AIRSPEED	POWER SETTING
A. Normal Cruise	_____	_____
B. Cruise climb (500 fpm)	_____	_____
C. Cruise descent (clean--500 fpm)	_____	_____
D. Best rate of climb (2 engine, if applicable)	_____	_____
E. Best rate of climb (1 engine out if applicable)	_____	_____
F. Holding (clean and maintaining altitude)	_____	_____
G. Holding (engine-out, clean)	_____	_____
H. Approach speed (clean and maintaining altitude	_____	_____
I. Approach (gear down) descending 500 fpm	_____	_____
J. Approach (engine-out, gear down) at 500 fpm	_____	_____

Fig. 24-14. A table to be used for setting up working speeds and power settings of the airplane. Twin-engine numbers may be included as applicable. If you have an engine out (G), and notify ATC, you "probably" won't be required to hold, but it's a good confidence building exercise.

Fig. 24-15. Instrument takeoff for a particular two-place trainer (nosewheel).

(A) Set the heading indicator when the airplane is taxied forward a few feet and lined up with the runway and apply full power smoothly, watching for heading deviations.

(B) At 50 K, the nose is raised to the first reference line (10°) and held there. Maintain the exact runway heading with the rudder.

(C) At 65 K the airplane will lift off. Maintain heading and watch for a tendency for the nose to pitch up as the airplane leaves ground effect.

(D) Continue the climbout at 65 K (best rate of climb airspeed for this example airplane).

(*The Instrument Flight Manual*)

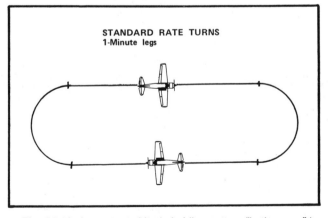

STANDARD RATE TURNS
1-Minute legs

Fig. 24-16. A constant-altitude holding pattern "in the open" is a good way to introduce the idea.

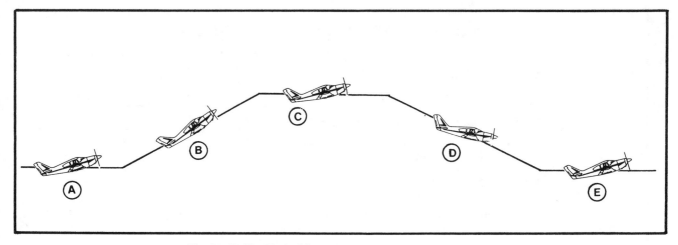

Fig. 24-17. The Vertical S-1.
(A) Two minutes of straight and level at recommended holding speed.
(B) Climb 1000 ft at best rate of climb speed but adjust power for 500-fpm climb (or 400-fpm, etc., as performance allows).
(C) Straight and level for 2 min at holding speed, and then
(D) A 500-fpm descent for 1000 ft at holding speed.
(E) Repeat as necessary.

ment of a good cross-check and it can be a first-rate introduction to precision, timed climbs, and ascents (and the first real introduction to later approach practice). Basically, it is a straight and level, straight climb, and straight descent combination. Start by

(A) having the trainee fly straight and level for 2 min (clean) at the recommended holding speed, followed by

(B) a 2-min straight climb at 500 fpm for 1000 ft at the best rate of climb speed, then

(C) 2 min of straight and level at holding speed, and finally

(D) a 2-minute 500 fpm straight descent at holding speed (clean).

Repeat steps A–D until the maneuver is smooth or you feel that lack of progress or fatigue makes it wise to quit. You can always come back to this one. Fig. 24-17 shows the idea; you should sketch the maneuver on the board and discuss it before flying. You may later want to cut the times to 1-min intervals to get more action.

Common errors in this exercise are these:

1. *Straight and level portion.* Watch for poor altitude, heading, and airspeed control (he can be flying at the proper altitude but has let the airspeed and power vary so that the holding speed isn't maintained).

2. *Climb.* The transition to the first climb or two may be delayed slightly so that the climb is behind the time. If, for instance, the climb gets started 15 sec late but still requires 2 min to complete, at first you should cut the following straight and level portion 15 sec short so that the trainee can start the descent on a cardinal time point (such as the 12 o'clock position of the sweep second hand). Watch for airspeed chasing in the climb, particularly if the transition was delayed. Check the rate also, and expect heading problems, as "torque" may not be corrected because of his concentration on time and altitudes.

3. *Descents.* Check his airspeed and rate of descent control. Cross-check problems show up more readily, it seems, in the descent than the climb because the average person tends to feel more rushed when he's losing altitude. Look for a hooded version of the problem as shown back in Fig. 5-7.

TRANSITIONS. Probably the most common error in the transition from the *climb to level flight* is to lead the altitude by the proper amount but to ease the nose over too quickly and reduce power. You'll see cases of trainees using up the 2-min straightaway staggering along just under the chosen altitude, trying to sneak back up on it. You'll also see people who'll want to get right back up to the altitude and add full power and radical pitch attitudes, which also can result in not getting the airplane stabilized on altitude during the 2 min of straightaway.

The transition to the descent is usually underdone at first, and the airplane is not nosed over and power reduction is delayed so that the rate of descent starts late or is not enough for timing purposes.

Expect a delayed reaction the first few times in the transition from the descent to the level-flight portion. The average person's reactions will be behind the airplane (most tend to be timid in control usage at first, as is the case for your low-time preprivate students). While it's important for all transitions to be firm, the transition from the descent to level flight (and subsequent climb) must be good; in a 200-½ approach situation the airplane could be allowed to go well below the *decision height* (DH) before stopping. You should mention this to emphasize the importance of a good transition and also to stress the idea that the practice will be put to use in actual approaches and other instrument work.

■ **Vertical S-2 (Hooded, Full Panel).** This exercise is good as a next step aad is particularly useful when you are instructing in a complex airplane. Emphasize that the practice is applicable to an approach and missed approach procedure, which he will be doing on practice approaches later. You might explain it before the flight as follows:

The airplane will be flown straight and level at normal cruise for 2 min. Then, the transition will be made to approach airspeed and configuration (gear down, set prop(s), use flaps as recommended for an approach as applicable to the airplane). A 2-min, 500-fpm descent will be made and the airplane cleaned up and a climb made at the best rate of climb speed and power as the minimum altitude (1000 ft below the altitude of the start of the descent) is reached. Climb back to the original altitude and repeat as time allows. (Have the trainee make turns as necessary to stay in the area.)

During the *straight and level* part check for a wandering heading; usually altitude won't (or shouldn't) be a problem at this stage of experience.

S-2 TRANSITION. Check for heading, altitude (again, he'll usually gain altitude in slowing up), proper approach airspeed, and procedures. Expect some fumbling around the first few times that gear, flaps, and prop are manipulated. At first the trainee will be concentrating on these items and heading, altitude, and airspeed will go awry. Here again, you can make the points that (1) if the airplane is trimmed properly, a *temporary* lapse of attention doesn't result in a quick loss of control and (2) the pilot should not spend a lot of time concentrating on items other than the flight instruments, but might, for example, put the gear handle down, check the flight instruments, and then come back to the gear to confirm that it's down and locked. Or if he has to check some approach chart numbers he is not to get so engrossed in reading the whole chart that the flight instruments are ignored. Even if the airplane is well trimmed, he should read a little, then check the instruments. You'll have to be working on this with some people throughout their training.

S-2 DESCENT. Airspeed control, heading, and rate of descent may give him trouble at first. Usually, if the transition is good a reasonable descent follows, but you'll find that some people are still "transitioning" when they get to the altitude to execute the missed approach.

S-2 MISSED APPROACH. Expect some delay and slight confusion in cleaning up the airplane and establishing the climb the first few times. Heading is usually a problem here as "torque" is a factor and the trainee may be distracted.

Fig. 24-18 shows the Vertical S-2.

■ **ASR Approach.** You should go into more detail on the simulated ASR approach using the skills the trainee learned in the vertical S maneuvers, and do one at the end of each flight during this part of the basic instrument work. Set up minimums of 500 ft and 1 mi, and work him in from the practice area to minimums (assuming that your airport doesn't have an approach control and an ASR approach). If you are introducing an actual ASR, you handle the transmissions the first time unless he is extra sharp, but he should do the full routine on following ASRs. Let him concentrate on the flying without distractions at first.

If you intend to instruct instruments at a small airport without tower or approach control, you may be able to set up a good simulated ASR by checking landmarks on a detailed map of the area around the airport and checking the trainee on each mile of an approach. The actual ASR approach uses an approach angle of about 3° and the airplane should descend 300 ft per mile on the approach until reaching minimums. Fig. 24-19 shows a setup for a fictitious airport. Naturally the figures given are for altitudes above the airport and you would work in your own MSL figures and check points.

You will enjoy setting the trainee up on the (wide) downwind leg and "flying" him to minimums and having him come "contact" and see how well you both did. Use the procedures and terminology throughout as is used by the controllers on *your* ASR approaches.

■ **Timed Turns (Standard Rate).** The use of timed turns is good practice in pinning down the instrument scan.

Following are some exercises you can have the trainee do:

1. Perform 90° and 180° level timed turns using all instruments. Calibrate the turn needle.

2. Make 90° level timed turns (attitude indicator covered).

3. Make 90° level timed turns with the attitude and heading indicators covered and then check for heading accuracy (use prime headings for first practice, then start at "off" headings, such as 057°, etc.). Correct the headings as necessary.

4. Make 180° and 360° timed turns as in exercise 3.

You may decide to have your more progressive trainees practice the hooded timed-turn exercises with the altimeter and attitude, heading, and vertical speed indicators covered, leaving only the turn indicator for bank information and the airspeed for pitch information. This may sound like asking for trouble and it *will* require smooth air and careful supervision at first, but it's a good confidence building effort and can teach that the airspeed is a good reference for pitch in a turn as well as straight and level. His elevator handling will smooth out.

Following are some numbers obtained for a Cessna 172 at light weight and at a medium altitude (5000–6000 ft MSL). These numbers will be close for other airplanes of that type (fixed-gear) but you should check for the airplane you're using.

The figures on the left represent the wings-level, level-flight speeds, and the minus numbers on the right are the decreases in airspeed (pitch-up) required to maintain altitude in a particular standard-rate turn, after rolling in at each airspeed (60, 70, 80, etc.).

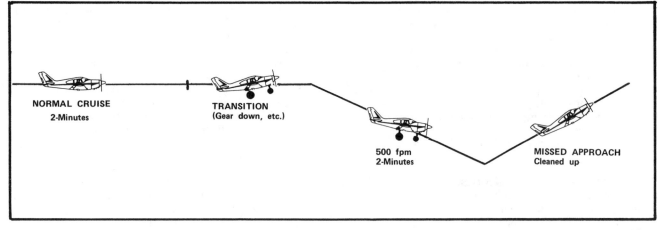

NORMAL CRUISE
2-Minutes

TRANSITION
(Gear down, etc.)

500 fpm
2-Minutes

MISSED APPROACH
Cleaned up

Fig. 24-18. The Vertical S-2.

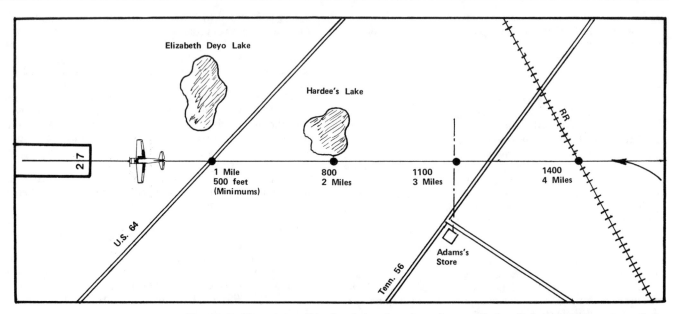

Fig. 24-19. The trainee will be hooded and won't see how you (unhooded) can set him up so well. As a controller you would say as you are abeam of Adams's store (or your reference), "You are three miles from touchdown, altitude should be _____ *MSL* (1100 ft above the airport), turn left, heading 265°."

Straight and Level	Level Standard-Rate Turn
60 K	−2 K = 58 K
70 K	−2 K = 68 K
90 K	−3 K = 87 K
105 K	−3 K = 102 K

In a standard-rate turn for the type of airplane mentioned, the back pressure or trim required to decrease the airspeed about 3 percent (2–3 K, depending on the airspeed) will be sufficient to maintain altitude.

Have the trainee set up a particular airspeed at or near normal cruise and establish straight and level flight using all instruments (let the airspeed stabilize and have him note it). Note the heading and altitude and cover the altitude and heading indicators, and altimeter. Have him set up a 180° or 360° standard-rate turn and use back pressure, or trim, to decrease the airspeed by about 3 percent (or your number) in the turn. Timing should be done and the roll-out completed on the prechosen heading. As the airplane is rolled out of the bank, the back pressure should be released to increase the airspeed to its straight and level value. Uncover the heading indicator and altimeter to check heading and altitude.

The biggest problem most people will have is the tendency to stare at the airspeed indicator and neglect the turn indicator. You should also have him do the exercise using only the airspeed and attitude indicators, using the rule of thumb bank for a standard rate of turn and checking the accuracy of the attitude indicator pitch with the airspeed change.

The decrease in airspeed required to maintain altitude in the turn is approximately proportional for one-half standard rate (subtract 1½ percent) and for double standard rate (subtract about 5 percent) but, again, check it for your airplane. You might want to review Fig. 24-11 and build your own combinations.

■ **Timed Climbs and Descents.** This exercise should follow the timed-turns practice. The trainee may set up the following exercises under your supervision:

1. Using all flight instruments and from a particular altitude (5000 or 6000 ft AGL, etc.), set up a 500-fpm straight descent at holding speed (clean, if applicable) for 2 min, starting on a cardinal position of the clock sweep second hand. (You should point out that at 15-sec intervals the altitude should be down another 125 ft.) Use trim as necessary.

2. Level off at holding speed and fly straight for 1 min or more, then establish a 500-fpm cruising climb (if performance allows), checking the altimeter and clock to climb 1000 ft. (Repeat exercises 1 and 2 as necessary to smooth out technique.)

3. Repeat 1 and 2, covering the altimeter and attitude and heading indicators immediately after commencing descent or climb. Check all instruments after leveling and make corrections as necessary.

Review Figs. 24-12 and 24-13 for some ideas on other instrument combinations for descents and climbs.

■ **Combined Timed Turns and Altitude Changes.** This starts putting the timed turns and climbs and descents together, but note that you will be emphasizing the final result that must be attained, as in actual instrument operations. While the pilot tries to set up a standard-rate turn or rate of descent, his goal (clearance or nav requirements) will be to reach a specific altitude and heading. The timing doesn't always work out evenly, so corrections will have to be made at the end of the time period allotted. This is good for scan development.

Have the trainee do the following exercises:

1. Using all flight instruments and starting at an even altitude (5000 or 6000 ft, etc.), make a 360° standard-rate turn and 500-fpm descent at the chosen descent speed (2 min). *Complete exactly 360° of turn and 1000 ft descent, even if the timing is off.* Fly straight and level at holding speed for at least 1 min.

2. Repeat the maneuver, making compensation for previous errors in turn rate and/or descent rate.

3. Climb 1000 ft at 500 fpm and turn 360° at standard rate (or climb 800 ft at 400 ft per minute, etc., as the airplane's performance allows).

4. Repeat the climb and turn, compensating for errors in turn rate and/or climb rate.

5. Repeat exercises 1 through 4, alternating directions of turn.

Again, in the descent, watch for too fast a turn and too high a rate of descent. In the climb the problem is usually poor "torque" correction; the tendency is to also pull the nose up too high and get too slow.

Depending on his progress, you may want the trainee to do a couple of these exercises partial panel as an introduction to unusual situations.

As noted in exercise 1 and in Chap. 12, the maneuvers are complete when the airplane is on the prechosen altitude and heading. Fig. 24-20 shows a climb and turn to an altitude and heading, but the same principle would apply for a descending turn.

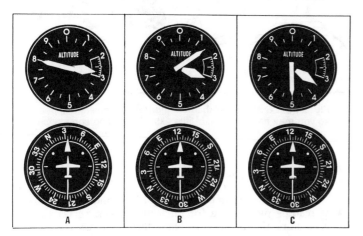

Fig. 24-20.
(A) Control of the airplane is recovered on the heading (035°) and altitude (2800 MSL) shown. The desired heading is 125° and the required altitude is 3500 ft. A standard-rate turn and a climb are started.
(B) The heading is reached and the turn is stopped. A straight climb is continued to the required altitude.
(C) The requirements are complete when the airplane is on both the required heading and altitude.
(The Student Pilot's Flight Manual)

■ **Introduction to the VOR.** The moves so far have been to introduce the flight instruments one by one, adding the power instruments and the clock to the trainee's scan (Fig. 24-21). As he gains skill in the basics, the next step should be to start using one of the VOR receivers in doing simple exercises.

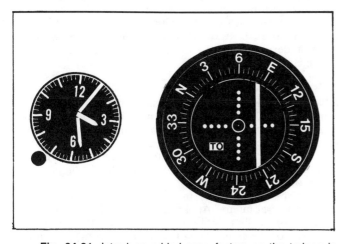

Fig. 24-21. Introduce added scan factors as the trainee is ready for them.

Most of the pilots you'll be working with will have had some experience with tracking to (or away from) a VOR as an aid to finding a strange airport, but like the flight instruments, have probably never really "looked at it." On the ground, brief the trainee on the VOR and how to track *precisely* to or from the VOR and how to intercept inbound and outbound courses. Discuss holding over a VOR on a particular radial at this time if you feel that his basic flying skill warrants it.

The first VOR work (and VOR work should be under the hood from the beginning) should entail constant-altitude simple intercepts and holding patterns. Don't bring in the second VOR or the use of cross-bearing (or intersection holding) until he is proficient at dividing his scan. Remember, this is still a part of the basic airwork and is building his scan, as well as introducing the idea of dividing his attention between maintaining control of the airplane and following a predetermined geographic route or pattern.

You can introduce the procedure turn during this phase.

After the trainee can handle the constant-altitude work with the VOR, you should require, for instance, shuttling down in the holding pattern and the 1-min inbound leg.

■ **More Clearances.** As the training progresses, continue to give pretakeoff clearances, gradually adding to their complexity. The holding or intercepting of radials may involve assignment of altitudes and other clearances "issued" as you feel necessary.

UNUSUAL ATTITUDES AND SITUATIONS

■ **Introduction and Preflight Discussion.** The following items should be covered in the preflight discussion:

1. *Steep turns* (full panel). Turns of at least 45° of bank should be discussed, with special attention given to altitude control, the usual big problem for the trainee as steeper banks are introduced.

2. *Stalls* (full panel). Any expected stall characteristics of the airplane should be covered ("In a power-on stall, the left wing tends to drop first," etc.).

3. *Unusual attitudes* (full panel and attitude and heading indicators covered). The recovery technique for spirals and approaches to stalls should be covered fully.

4. *Engine-out procedures* (full panel). V_{YSE}, V_{SSE}, V_{MC}, and engine-out procedures should be reviewed, if the trainer is multiengine.

5. *Magnetic compass.* Acceleration (and deceleration) errors and northerly turning errors should be reviewed.

6. *General review.* Items concerning basic instrument flying may be cleared up at this point if the trainee has questions.

Since this flight concerns loss of operations of some of the flight instruments, during the preflight discussion and others following you should present possible in-flight problem areas (icing of pitot tube, etc.) and alternate courses of action.

Do thorough preflight and pretakeoff checks, as always.

■ **Takeoff and Climb.** If on this flight you have the trainee make an ITO, after a safe altitude is reached cover the attitude and heading indicators and have him continue the hooded climb to a nearby VOR to a prechosen altitude. Watch for "torque" problems.

■ **Straight and Level Flight (Attitude and Heading Indicators Covered).** After he has leveled off (uncover the attitude and heading indicators to help the process, then recover), have the trainee fly straight and level for 2 min without those two instruments (partial panel). Acceptable performance limits later as he is getting ready for the flight test are ±10° in heading and ±100 ft in altitude. Work on smoothness, coordination, and accuracy because this will be judged on the flight test.

■ **Turns, Climbs, and Descents (Partial Panel).**
Work on the following limits as he progresses:

1. Climb or descent to assigned altitude—within 10 sec of estimate.

2. Heading on recovery from timed turn—within 20° for each 360° of turn.

3. Airspeed—within 5 K of assigned speed.
Keep on with the idea of good trim use.

■ **Steep Turns (Full Panel).** Have the trainee practice 360° turns with 45° of bank at an airspeed not more than the maneuvering speed of the airplane being used in the training program.

Later acceptable performance will be (1) altitude ±100 ft of the entering altitude, and (2) heading within 10° of the assigned recovery heading. (He may be required to do more than 360° of turn on the flight test.)

Watch for altitude control problems.

■ **Stalls (Full Panel).** Be sure that the trainee has a good idea of the types of stalls to be done and the best method of recovery before practicing the following stalls. The nose should be lowered *below* the level-flight attitude in order to avoid a secondary stall but this should not be overdone. You'll know the best recovery technique for your airplane but remember that you will have to present the information so that the technique can be applied to other airplanes as well.

You might review the material on approach to stalls as given in Chap. 12 and look again at Figs. 12-16 through 12-20 before setting up your ground and flight sessions on this subject.

Stalls should be practiced from climbs and descents in climbout and approach configurations. Require coordinated recoveries with the least loss of altitude consistent with safety. Have the trainee practice the following exercises as time permits:

1. Power-off stall, clean configuration, straight ahead. Recover with full power.

2. Power-off stall, approach configuration, straight ahead (use gear and flaps as applicable). Recover using full power. Clean up the airplane.

3. Repeat exercises 1 and 2 using 30° banks during the stall entry.

4. Power-on stall (climb power), clean configuration. Recover without additional power.

5. Power-on stall (climb power), approach configuration (gear *and* flaps as applicable).

6. Repeat exercises 4 and 5 using 30° banks during the stall entry.

Require that the heading after the stall recoveries should be within 20° of the heading at the stall "break."

> *Common errors* you'll encounter include these:
> 1. Timidity in setting up the stall; slow flight occurs, but no break.
> 2. Not relaxing back pressure enough after the break so that a secondary stall occurs. Expect this to happen at first with nearly every instrument trainee. He'll be tense, and one manifestation of tenseness, whether under visual or instrument conditions, is a tendency to increase or hold back pressure (the person is "holding on to," and consequently pulling, the control wheel). Doing stalls, particularly under the hood, at first turns out to be *not* the favorite occupation of a lot of people, but most trainees feel better about it after doing a few (Fig. 24-22).

Fig. 24-22. Stalls under the hood don't have universal appeal, so you may have to work a little extra in building confidence.

> 3. Neglecting to add full power on recovery if partial power is used in setting up the stall.
> 4. Disorientation in using the attitude indicator, such as reading pitch down for pitch up (or vice versa) and misreading the direction of bank during the recovery.
> 5. Forgetting the base altitude and heading to be returned to.

RECOVERIES FROM UNUSUAL ATTITUDES (FULL PANEL AT FIRST)

(Review Chap. 12.)

At a safe altitude do the following exercises:

1. Put the airplane into a diving spiral in various configurations and power settings. Have the trainee practice recovering with the minimum loss in altitude and then return to a predetermined altitude and heading. (Make sure he cleans up the airplane before trying to climb.) The trainee should not look at the instruments during the entries to the unusual attitudes.

2. Put the airplane into a nose-high turn (stall approach) and have the trainee recover. Note that these exercises are done first with full panel.

Repeat exercises 1 and 2 with the attitude and heading indicators covered. Recoveries from unusual attitudes without the attitude and heading indicators may be a flight test requirement.

■ **Spin Recoveries**

A REVIEW OF THE SPIRAL. The most common result of loss of control in IMC is the *spiral*. The pilot is distracted and/or panics and the airplane is allowed to set up its own spiral with the possibility of several things occurring:

1. The aircraft could break up in flight as the dynamic pressure at the high airspeed causes structural failure while still in the spiral.

2. Excessive g forces in an attempted pull-out could cause structural failure. An airplane that stalls at 1 g at 60 K would theoretically have 16 g's imposed on it if the airplane broke out of the clouds at 240 K and the (strong) pilot used enough back pressure to stall it at that airspeed. (The g's imposed in a stall at a higher-than-normal stall speed are a func-

tion of the *square* of the ratios; hence, stalling an airplane at 4 times the normal-g stall speed of 60 K, or at 240 K, would exert 4², or 16 g's.) Of course, the average airplane would not approach the 16 g's of the example; structural failure would occur at less than one-half that load.

A *rolling* pull-out is worse than a straight one. The twisting moments on the wings will cause failure before the expected number of maximum g's occur. A rule of thumb for military airplanes is that a rolling, or asymmetric pull-out must be limited to two-thirds of the maximum listed acceleration forces. A normal category airplane with a maximum (structural failure) g force of 5.7 g's could possibly have structural failure in a rolling pull-out of more than 3.8 g's. Since few general aviation airplanes have g meters, the advice is to avoid rolling pull-outs.

3. The aircraft in the IMC spiral might not recover and hit the ground before in-flight structural failure occurs.

These three spiral scenarios are usually fatal for all occupants.

You should instruct your trainees that flying the airplane is a number one priority. You'll be surprised during your ATC dual sessions that if a trainee drops an Approach Chart book he'll tend to give *full attention* to the book until it has been picked up and the proper page has been reselected. (Meanwhile the airplane, neglected, has slipped off into a spiral and altitude and airspeed are not where they should be.)

A suggestion, if the Approach Chart book or a chart has fallen to the floor:

1. The trainee should check the instruments to be sure the airplane is flying properly.

2. He picks up the chart or book.

3. He gets back to flying the airplane.

4. He opens the chart or book to the approximate area of the approach or route required.

5. He gets back to flying the airplane.

6. He finds the required approach plate or route segment.

7. He gets back to flying the airplane.

A REVIEW OF THE SPIN. The *spin,* while not as prevalent in fatal IMC accidents, has a rate of descent varying from 6000–10,000 ft per minute, depending on the general aviation airplane, and you should at least discuss, if not demonstrate, the instrument indications in the spin and give a procedure for

recovery using the available flight instruments. The factors involved should be made known to the trainee; the following is based on several popular general aviation trainers. (You might want to review Chap. 21 before introducing spins.)

Looking at the instruments:

1. The airspeed will be very low or at zero. There may be some oscillation between a low airspeed and zero.

2. The turn indicator will be pegged in the direction of the spin rotation.

3. In most side-by-side general aviation airplanes, the ball on the pilot's side will be fully deflected to the left for spins, both left and right. (The ball should not be used for indications of spin direction.)

4. The altimeter will be unwinding rapidly.

5. The vertical speed indicator will be pegged for instruments with limits of 6000 fpm or less.

6. The attitude and heading indicators should be ignored, even if they "seem" to be acting properly. (This writer found in one aerobatic airplane that the attitude indicator was useful in seven out of ten spin recoveries—that's not good enough if you happen to be in one of the other three spins.) Fig. 24-23 shows instrument indications of a developed left spin in a general aviation airplane.

SPIN RECOVERY STEPS. Here is a general look at the steps in spin recoveries; the *Pilot's Operating Handbook* may have the best information for your particular airplane.

1. *Close the throttle.* For the majority of general aviation airplanes, closing the throttle tends to pitch the nose down, making recovery easier. In the incipient phase of the spin (0–2 turns), just getting the power off may be enough to stop the spin. (But you should do the other steps cited here to assure the quickest recovery.)

2. *Apply full opposite rudder* to the turn indications of the needle, or the small airplane in the turn coordinator.

3. *Move the stick or wheel forward* as soon as the rudder hits the stop.

4. *Check the airspeed.* As soon as the airspeed starts to increase, neutralize the rudder pedals and start a smooth, symmetrical pull-out.

5. *Continue the pull-out.* The airspeed will continue to increase until the airplane is approximately in the level-flight attitude.

6. *When the airspeed hesitates or starts to reverse,* stop

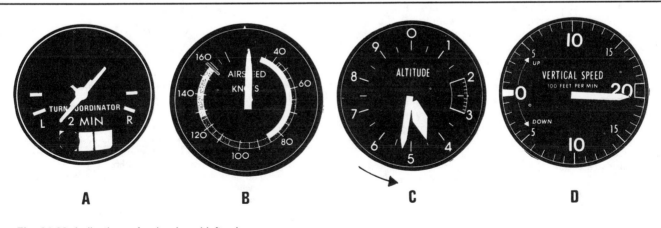

A   **B**   **C**   **D**

Fig. 24-23. Indications of a developed left spin.
(A) The turn and slip needle or turn coordinator (small airplane) will probably be deflected to the stop. In this side-by-side seating example the ball is pegged to the left (and will be so for left *or* right spins).
(B) The airspeed is very low, and in some spin modes may indicate *zero*.

(C) The altimeter will show a rapid loss of altitude.
(D) The vertical speed indicator will probably be pegged, since rates of descent (depending on the airplane type and spin mode) may be in the 6000- to 10,000-fpm range.

(The Instrument Flight Manual)

the back pressure and pick an altitude to concentrate on (use the 100-ft hand) and maintain that altitude with the elevators. Keep the wings level with the turn indicator. Forward pressure may be needed at first to keep the nose from rising above the level-flight attitude(s).

7. As the airspeed moves back to the cruise value, add cruise power, continuing to keep the wings level with the turn indicator as you get the attitude and heading indicators back in action.

8. Climb and turn to regain altitude and avoid obstructions, as shown back in Fig. 24-20.

In step 4, it may be argued that the needle or the small airplane will center when the rotation stops, showing that the airplane is out of the spin. However, any variation of input of the rudder as the airplane is being pulled out of the dive may cause the needle to oscillate or peg to one side or the other. The small airplane in the turn coordinator responds to inputs of rudder *or* ailerons, and either type of turn indicator may cause the pilot to have a rolling pull-out or to set up conditions for an entry into a spiral as the spin recovery is being completed. Fig. 24-24 shows steps as the spin recovery is started. Fig. 24-25 shows the steps and instrument indications after the pull-out is started and the recovery is completed.

It's suggested that the recovery be well covered on the ground beforehand so that the trainee will know the steps required.

You do the first spin demonstration (with the trainee *not* under the hood), explaining the recovery as it is being done. Before demonstrating the spin and recovery, review the procedure. (Things *will* be happening rapidly during the process and you may find yourself stammering trying to explain the rapid sequence of events.)

It's best not to have the trainee under the hood at first because of possible nausea problems. Whether you want to put him under the hood at all and how many instrument-reference spins (if any) you will have him do will depend on mutual agreement.

Also remember that to be legal, parachutes must be worn unless the trainee wants to use this training as a part of his CFI-A certification (FARs 91.303 and 91.307).

■ **Flying the Magnetic Compass (Heading Indicator Covered).** When reviewing the magnetic compass on the ground (including variation and deviation, if you think it's needed), particularly discuss the reaction of the instrument to turns in the northern hemisphere (northerly turning error) and acceleration (on an easterly or westerly heading, *a*cceleration of the airplane results in a more *n*ortherly indication, *d*eceleration a more *s*outherly heading—ANDS). Brief the trainee on the practice to be done, such as turning to cardinal headings and the use of timing for turning to headings in between (using a cardinal heading as a timing point).

In flight demonstrate and then have the trainee practice straight and level flying and turns to cardinal and noncardinal headings. Your sharper people would enjoy the challenge of a short session of holding or tracking to and from a VOR on a preselected radial.

Demonstrate that the magnetic compass may act as a low-budget attitude indicator when used on a heading of south. (The compass turns in the proper direction and at a slightly greater rate than on other headings.)

If the situation allows, have the trainee do work with the magnetic compass at higher altitudes in smooth air and then down low in bumpy conditions—to see how much tougher it is to fly a heading when the air is rough. The use of a hood may or may not be of value in this practice, depending on the location of the compass; for one that is located well above the panel the hood would be of little value, and you may decide to

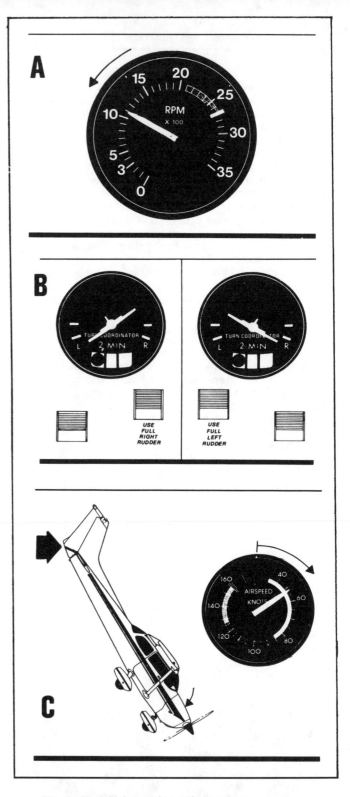

Fig. 24-24. Initial steps in a spin recovery.
(A) Close the throttle. Make sure the ailerons are neutral.
(B) Check the turn indicator. Apply full opposite rudder to the needle or small airplane indication. (Note: This is a side-by-side airplane with the instruments shown on the pilot's side. The ball will be to the left in this arrangement for both left and right spins.)
(C) After applying full rudder and moving the wheel briskly forward, the airspeed moving from zero (or a very low airspeed) shows that the recovery has started. Neutralize the rudder and start applying (centered) back pressure on the wheel. During the recovery the increasing airspeed tells you that the airplane is out of the spin. Don't fixate on the turn and slip or the turn coordinator after checking the direction of rotation or during the forward motion of the wheel *or* during the pull-out; the turn indicator(s) could, in some cases, be jammed to one side or the other, even as a wings-level pull-up is being accomplished.

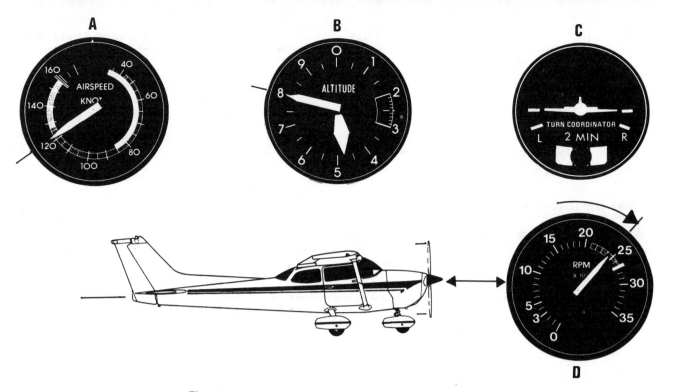

Fig. 24-25. After the pull-out is started, the following steps are recommended to complete the spin recovery.

(A) Continue the back pressure (or allow the nose to rise), watching the airspeed. When the airspeed hesitates or stops increasing, the pitch attitude is approximately level.

(B) Look at the altimeter immediately and pick the closest altitude to "fly." (Don't let the altitude increase or decrease—"fly" the altimeter with the elevators as the airspeed bleeds back to the cruise regime.)

(C) The turn coordinator or turn and slip is again dependable once the airplane is in the straight and level condition. Keep the wings level.

(D) As the cruise airspeed is approached, add power to cruise value.

(The Instrument Flight Manual)

use the compass demonstration and practice as a break from the hood work. Don't hit the use of the compass very hard for one flight and then forget it, but have short sessions as a change of pace in later flights, too.

Fig. 24-26 shows the application of turning to a predetermined heading using the magnetic compass.

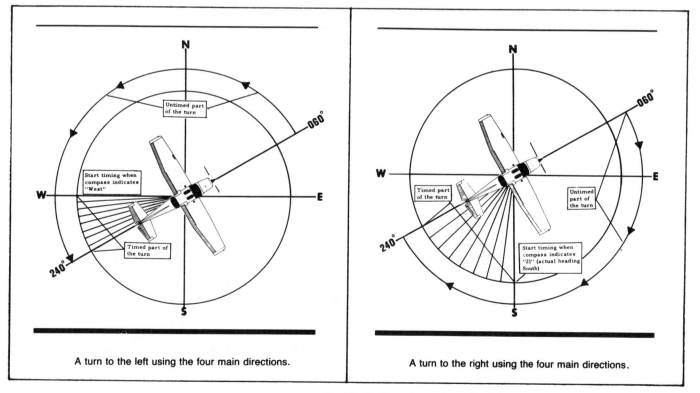

Fig. 24-26. Turns to predetermined directions using the magnetic compass.

■ **No-Gyro ASRs.** As the trainee's basic instrument flying skill develops, you should act as a controller for no-gyro ASR approaches, or request one or more practice ones if the home airport or nearby big airport has the approach control facilities. Brief him before the flight about the "start turn, stop turn" directions he'll get and that turns will be expected to be standard rate in the pattern except on *final,* where they will be one-half standard rate.

REVIEW OF BASIC INSTRUMENT FLYING

While you will be continually reviewing the trainee's progress, you should plan to have his basic instrument flying checked by another instructor, if possible, before moving on to concentrated navigation and communications (ATC) work. You should review the basic instrument instruction covered thus far, including recoveries from unusual attitudes. You and the trainee should review all grades and comments for the flights up to this point. Of course you've been working on his weak spots all along, but you should set a definite review period to put it all together. You might see a definite trend by going over them with the trainee. Fig. 24-27 is part of a sample grade and comment sheet.

A "Charlie pattern" or "pattern C" is a good single, if somewhat complicated, exercise for covering and checking a trainee's progress in basic instrument flying (except for recoveries from unusual attitudes). You might get some practice, as

---

TURNS AND ALTITUDE-CHANGE COMBINATIONS

Turns and descents (Full Panel)

Bank control	1	2	③	4	5
Heading	1	2	③	4	5
Rate of descent (500 fpm)	1	2	3	④	5
Airspeed control	1	2	3	④	5
Use of power	1	2	③	4	5
Orientation	1	2	③	4	5
Cross-check	1	2	3	④	5
Use of trim	1	2	3	④	5

Descents may be made clean or gear-down as applicable to the training airplane and the instructor's analysis of previous problems:

Turns and climbs (Full Panel)

Bank control	1	②	3	4	5
Heading	1	2	③	4	5
Rate of climb control (use best rate)	1	2	3	④	5
Airspeed	1	2	3	④	5
Orientation	1	2	③	4	5
Cross-check	1	2	3	④	5
Use of trim	1	2	3	④	5

The exercises above should be practiced with the H/I and A/I covered if the trainee needs more partial-panel work.

COMMENTS *Trainee needs work on airspeed control (chases it). Tends to concentrate on one instrument too long. Does not use trim after transitions until reminded. G. V. Yellowhammer CFI-442723*

Fig. 24-27. A reference and grading sheet for a particular phase of basic instrument instruction. In the grading system used here, 1 is best, 2 above average, 3 average, 4 below average, and 5 unsatisfactory. You probably will want to set up your own syllabus and grading system.

well, when some of your trainees in the pattern C let things get sloppy. If he gets into problems with spirals (and such) in a basic instrument flying exercise, you'll at least be aware that more work is required before moving on. Fig. 24-28 illustrates a pattern C.

Again, don't continually use practice patterns as part of your training process, because after working at them for a while the trainee will begin to lose interest. Some people can't see the application to "real" instrument flying. If, however, you decide that for your purposes you need some extra exercises, Fig. 24-29 shows A and B patterns. Note that you would have to set up specific low, normal, and high cruise airspeeds and power settings for these exercises.

Following are some suggestions for the *basic instruments review flight* you should fly with each trainee as he is getting ready to move on to ATC work.

■ **Before the Flight.** Lay out flight requirements and give him a short oral examination of instrument operations and the working speeds of the airplane being used. Ask questions about the area covered thus far in the course.

The review flights by you (like the one described next) and interim check flights by other instructors will help alleviate checkitis on the FAA check. He'll have had recent experience in flight check procedures when he gets to the final.

Be sure that the trainee knows exactly what is expected of him on this flight.

■ **The Flight.** If the trainee has been doing instrument takeoffs you should require one on this flight.

TAKEOFF AND CLIMB. During the climbout, the headings should be within ±5° of that desired and banks within ±5°

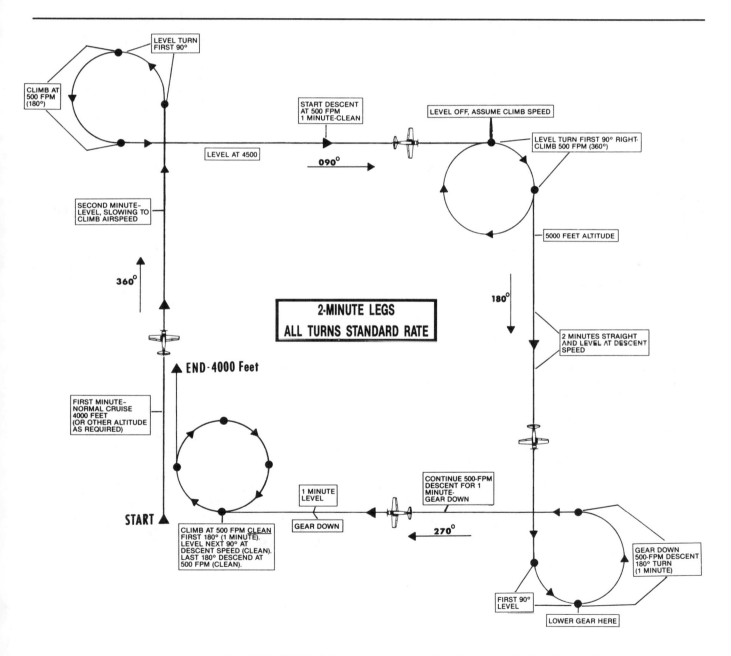

Fig. 24-28. Pattern C. If you use it, discuss it well beforehand but don't expect the trainee to remember the details. Give it to him step by step as he flies it. (*The Instrument Flight Manual*)

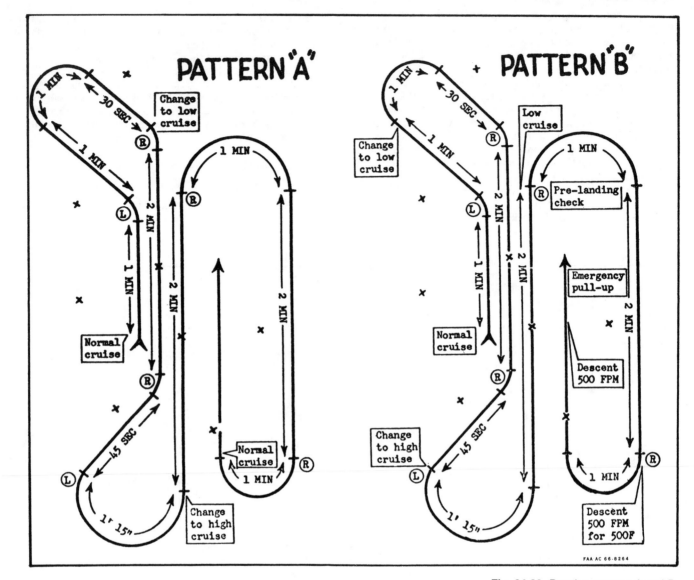

Fig. 24-29. Practice patterns A and B.

of that required for a standard-rate turn. Steeply banked turns (5° more than required for a standard-rate turn) during the climb are unsatisfactory.

Have him practice climbing 90° standard-rate turns or standard-rate turns to headings as specified by you until the prechosen altitude is reached.

Check him on heading control, bank, pitch, cross-check smoothness, orientation (turns initiated in the correct direction), and use of trim.

THE FOUR FUNDAMENTALS. *Straight and level flight.* This should be practiced clean at various airspeeds suggested by you. Particular attention should be paid to pitch and heading during transitions. One minute of straight and level flight should be done at each speed. (Again, watch for problems with heading control during transitions.)

Straight and level (full panel):
    Heading control ____
    Altitude control ____
    Cross-check ____
    Use of trim ____
Transitions (full panel):
    Heading ____

    Altitude control ____
    Use of power ____
    Cross-check ____
    Use of trim ____

If time permits, a shortened version of the requirements should be done with the attitude and heading indicators covered.

Turns and altitude change combinations (full panel). Ask for combinations of turns and altitude changes to predetermined headings and altitudes. Proper airspeeds and power settings should be used. Any loss of control or obvious overshoots of altitude or heading work require extra time before taking the progress check.

Turns and descents (full panel):
    Bank control ____
    Heading ____
    Rate of descent (500 fpm) ____
    Airspeed control ____
    Use of power ____
    Orientation ____
    Cross-check ____
    Use of trim ____

Descents may be made clean or gear-down as applicable to the training airplane and your analysis of previous problems:

Turns and climbs (full panel):
    Bank control ____
    Heading ____
    Rate of climb control (use of best rate) ____
    Airspeed ____
    Orientation ____
    Cross-check ____
    Use of trim ____

You might have him do some partial-panel climbs and descents if he needs more work in this area.

RECOVERIES FROM UNUSUAL ATTITUDES. Have the trainee recover from spirals and approaches to stalls using the full panel, followed by recoveries with the heading and attitude indicators covered. A base altitude and heading to return to should be chosen before each maneuver is started. Each maneuver is ended when the airplane is at the predetermined altitude and heading. The recovery and return to the base altitude and heading should be made without undue hesitation or loss of control.

Emphasis should be placed on turning in the direction nearest to the preselected heading. The climbs should be at best rate, descents at 500 fpm (clean).

Recognition of the problem (spiral, approach to a stall) ____
Recovery techniques:
    Power control ____
    Bank recovery ____
    Pitch recovery ____
    Altitude control after recovery is initiated ____
    Heading control after recovery is initiated ____
    Return to base altitude and heading ____
    Cross-check ____
    Use of trim ____

Watch for any tendency to "clutch" or react the wrong way, particularly in the partial-panel work. He should be able to recognize the problem (spiral, approach to a stall, etc.) and move promptly and in the proper direction.

Spin recoveries may be done if the airplane is certificated and by mutual agreement, plus ensuring that requirements of FARs 91.303 and 91.307 are met.

CHARLIE PATTERN. This pattern takes a minimum of 16 min to complete, and if the required extended practice of the other basic instrument flying exercises does not take up the last part of the period, it could be flown as a check of his ability to think and fly the airplane. You can read aloud the required headings, configurations, and turns, or have him refer to a diagram on his knee board.

POSTFLIGHT REVIEW. Discuss any problems the trainee may have during the review flight and if necessary set up more dual sessions to iron out any deficiencies. He may do slightly less-smooth work during the review flight because of tenseness, but he should have good control of the airplane at all times (if the issue was ever in doubt).

Following is a sample basic instrument flying progress check. You might get some ideas from it for setting up such a check for your instrument program.

BASIC INSTRUMENT FLYING PROGRESS CHECK

■ **Review and Discussion of Check.** Earlier lessons should be reviewed briefly and this check discussed by the check pilot. This will cover as many different maneuvers as possible in the time allotted. However, if the review flight and previous records indicate problems in a particular area, the check pilot should emphasize these in this review and during the flight check.

■ **Preflight Check, Start, and Taxi.** The check pilot will grade the trainee on the thoroughness and understanding of the items on the preflight check.

The trainee will be graded on starting and taxiing procedures and use of the checklist:

Preflight check	1	2	3	4	5
Starting procedure	1	2	3	4	5
Use of checklist	1	2	3	4	5
Taxi technique (check of instruments during taxi, etc.)	1	2	3	4	5

■ **Pretakeoff Check, Clearance, ITO, and Climb.** The trainee will be graded on use of the checklist during the preflight check, clearance copying, the instrument takeoff, and the hooded climb.

All maneuvers during the check will be hooded unless otherwise noted.

Pretakeoff check:					
Use of checklist	1	2	3	4	5
Thoroughness of check	1	2	3	4	5
Clearance readback	1	2	3	4	5
Instrument takeoff:					
Heading control	1	2	3	4	5
Pitch control (rotation)	1	2	3	4	5
Lift-off	1	2	3	4	5
Post-lift-off procedures	1	2	3	4	5
Climb:					
Heading control	1	2	3	4	5
Pitch control	1	2	3	4	5
Bank	1	2	3	4	5
Use of trim	1	2	3	4	5

■ **Basic Instrument Flying (Full Panel).** The trainee will be checked on his ability to fly the airplane using all of the flight and power instruments. It is suggested that the vertical S maneuvers be used for a check on pitch and power control, and the Charlie pattern used as a check for combination of pitch, bank, and power control. At least 1 steep turn (constant-altitude) of at least 360° duration and 45° bank should be done in each direction. The trainee will demonstrate his stall recovery techniques during this phase.

Vertical S:					
Transitions	1	2	3	4	5
Pitch control	1	2	3	4	5
Use of power	1	2	3	4	5
Heading	1	2	3	4	5
Use of trim	1	2	3	4	5
Charlie pattern:					
Heading control	1	2	3	4	5
Bank control (standard rate)	1	2	3	4	5
Pitch	1	2	3	4	5
Altitude control	1	2	3	4	5
Use of trim	1	2	3	4	5
Timing	1	2	3	4	5
Climb and descent rates	1	2	3	4	5
Cross-check	1	2	3	4	5

The check pilot may use other patterns to check the trainee's skill in using all instruments.

Steep turns:

Bank control	1	2	3	4	5
Altitude control	1	2	3	4	5
Roll-out on predetermined heading	1	2	3	4	5
Use of trim	1	2	3	4	5

■ **Stalls (Full Panel).** The airplane should be stalled in climbs and descents in climbout and approach configurations. Special attention should be paid to coordinated recovery with a minimum loss of altitude (no secondary stall). The check pilot will be more interested in safe and positive stall recovery than specific altitude and/or heading limits on this check. A predetermined altitude and heading should be returned to after each stall.

Entry	1	2	3	4	5
Recovery:					
Heading control	1	2	3	4	5
Altitude control	1	2	3	4	5
Use of power	1	2	3	4	5
Postrecovery technique:					
Clean up airplane	1	2	3	4	5
Return to prechosen					
altitude and heading	1	2	3	4	5

■ **Partial-Panel Flying (Attitude and Heading Indicators Covered).** Straight and level flying, turns, climbs, and descents will be completed within the following limits:

Straight and level: ±100 ft and ±10° heading (smooth air).
Climbs or descents: ±10 sec of estimate to a prechosen altitude.
Timed turns: ±20° for each 360° of turn.
Airspeed: ±5 K of assigned speed.

The check pilot may use various other combinations of flight instruments as time allows.

■ **Unusual Attitudes (Attitude and Heading Indicators Covered).** The trainee will recover from high-angle climbs that are approaching a stall, power-on spirals, and other unusual attitudes set up by the check pilot. A base altitude and heading should be chosen to return to after recovery is effected. The heading indicator may be reset after the recovery.

The check pilot may ask for a brief demonstration of the use of the magnetic compass in turning to headings. If the airplane being used is a twin engine, the trainee may demonstrate the simulated loss of an engine (retard throttle) at cruise (full panel).

Spiral recovery	1	2	3	4	5
Approach-to-stall recovery	1	2	3	4	5
Return to base altitude and heading	1	2	3	4	5

You may want to use this part of the Progress Check to do any spin recovery procedures.

■ **Return to Airport.** A simulated (or real, if applicable) radar vector and ASR approach will be made to the home airport (500 and 1 minimums). The trainee will be judged on orientation, response to instructions from actual or simulated approach control, and basic flight techniques.

Heading control	1	2	3	4	5
Altitude control	1	2	3	4	5
Rate of descent	1	2	3	4	5
Response to instructions	1	2	3	4	5
Use of checklist	1	2	3	4	5

Comments:_____

_____

_____

Check Pilot_____

Date_____

SUMMARY

As you do more instrument instructing, you'll establish criteria for each flight. At first, as is (or was) the case for visual instructing, you won't know exactly what standards to set up. Basically, in checking a trainee's progress in any phase of flying you should consider three main ideas:

1. Does the trainee understand what he is doing? Use questions to check, or have him explain.

2. Can he do the maneuver? If he understands the why of it, can he fly the airplane safely and accomplish the job? (You'll find people who understand *why* and *how* of an exercise or flight segment, but don't have—as the psychology/physiology people might put it—the motor skills to accomplish it.)

3. Can he correlate this with other instruction? You shouldn't have to review earlier work *in detail*. For instance, later in an instrument course you certainly should be able to describe a complicated approach or exercise without going into details on how to do the standard-rate turn or set up a 500-fpm descent.

A suggestion: Don't have the trainee fly at exactly 5000 ft, but instead assign him altitudes like 5140 or 4680, which will require a closer scan to hold them right on. The same goes for headings; as he progresses, ask for turns to 133° or 277°, rather than 090°, 180°, etc. You'll see a better scan after a session of this type of work.

The pressure will be on to move into ATC work as soon as possible, but as stated before, the pilot must be able to fly the airplane well on instruments before he is able to handle the distractions of communications and navigation.

COMMUNICATIONS, NAVIGATION, AND ATC

INTRODUCTION

This is the part of the instrument course that the trainee has been waiting for (after all that repetition of the basic instrument practice). He may not learn until later that (1) the basic instrument training can be put to good use in navigation work and (2) sometimes cross-country work can get a little old, also.

Review possible COMM/NAV and electrical problems after he's had a chance to work with all of that equipment in the airplane. A short session of unusual attitudes could be inserted in various lessons not only to give a "break" but to also make sure that he is still able to cope with such a problem. Too often, emphasis is kept on cross-country flying and approach work, and unusual attitudes may be given only a lick and a promise in the final review flight.

It is assumed here that when you start instrument instructing you are fully familiar with all phases of instrument work, so step-by-step discussions of how to do various approaches are not included. Instead, this chapter, like the last, is to be used as a guide to how to instruct instruments with recommended orders of introduction to exercises, approaches, and cross-country work, plus the expected common errors you'll encounter. For review work on details of charts, approach plates, and weather services available, check the books in the Bibliography.

For many if not most of your trainees, the exercises and requirements using all of the instruments also may be done partial panel, which can be a challenge for your sharper people and raise the confidence of the more timid (if you don't let things get out of control).

VOR AIRWORK

■ The First Full VOR Flight

PREPARATION. Have the latest en route charts available so that you can point out the airways structure and VOR and other symbols as you get more involved in this work. The first VOR flight should be a simple short cross-country (hooded) to a nearby VOR with a holding pattern or specific instructions for departing a radial; but bringing the trainee right into the system with a chart is a good idea when you start navigation in earnest.

EXPLANATION. In your ground instruction before the flight, review the introductory VOR work covered during the basic part of the instruction and lay out some simple problems on the chalkboard that you'll do on this and following flights. You might review intercepting a particular radial and how it is necessary to "see" his position in relation to the station.

Indicate that the first work is to be done at a constant altitude, but go to work on climbs and descents in the VOR practice as soon as practicable. As was noted in the last chapter, some of your trainees may have already done varying-altitude work with the VOR and descents in the holding pattern, plus the types of procedure turns and descents in procedure turns. (Discuss the "old" type of procedure turn as well as the 90°–270° or 80°–260°.)

When discussing holding and procedure turns, break out an en route chart or approach chart and show the application of what he will be doing; however, save actual detailed chart work for later as he begins to use more of the information.

You could give a (very) brief idea of the VOR's advantages and disadvantages (line-of-sight reception, etc.), but the main goal of this first full navigation flight is to be sure that he understands the equipment and how it's used in the airplane. Most instructors tend to overestimate the instrument trainee's knowledge of VOR—after all, the trainee must have a couple of hundred hours—but you'd better cover it well during the training (not all at once!) so that no gaps exist.

At some time during the periods of work with the VOR you should discuss the 30-day-check IFR requirement and limits of ground and airborne checks, but in most cases you should avoid it this first flight.

You should also talk about and later demonstrate, if possible, the use of the VOR receiver part of the radio magnetic indicator (RMI) and horizontal situation indicator (HSI).

■ **Preflight Check and Start.** Point out the VOR antenna(s) again as the usual thorough preflight check is done, and mention the possibility of "blanking" as the airplane is banked in certain relative positions to the VOR station, and that you will demonstrate the idea shortly.

■ **Taxi and Pretakeoff Check.** You could give the trainee a clearance to the nearest VOR at an assigned VFR altitude, and then act as tower, departure control, and the Air Route Traffic Control Center during the takeoff and climb. Continue to require a thorough check of instruments (and radios as applicable).

■ **ITO and Hooded Climb.** The ITO should naturally always be full panel, but if the trainee is progressing well, you should have him climb out in the general direction of the VOR, with the attitude indicator and heading indicator covered.

■ **VOR Tracking and Holding (Full Panel).** If the location of the VOR is such that several minutes of flight are necessary to reach it, you should require tracking toward it on several different radials to avoid too long exposure to one condition. For example, track toward the VOR on the 090 radial, and then establish a track in on the 085 and 080 radials, etc. (Fig. 25-1).

This part of the flight should include as much tracking to and from the VOR on different selected radials as possible (Fig. 25-2). Stick to easily remembered numbers at first, but

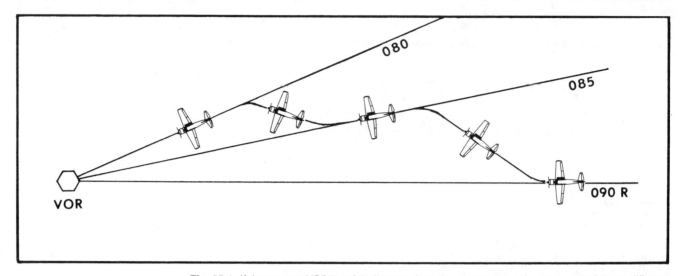

Fig. 25-1. If the nearest VOR is a fair distance from the airport, have the trainee track in on different radials to break the monotony and to get practice in intercepting a prechosen inbound course. (The angles have been "stretched apart" here.)

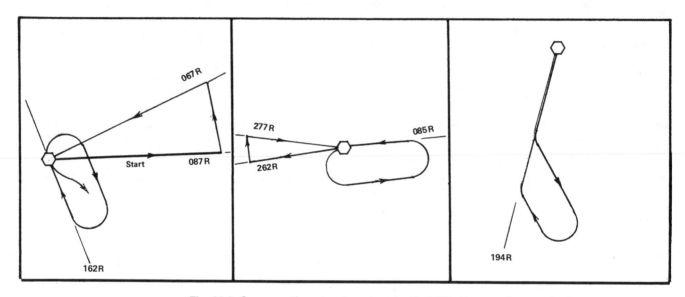

Fig. 25-2. Some practice procedures in using the VOR. You can diagram simple exercises such as these and let the trainee fly them as an introduction to the use of charts and approach plates. Use the preselected holding speed.

as the trainee improves go to more "realistic" numbers. Some examples:

1. Fly outbound on the 087 radial until tracking is established, track in on the 067 radial, and hold on the 162 radial (standard pattern) making at least one full pattern.

2. Track out R-262, inbound on R-277, hold on R-085 (nonstandard). Maintain a constant altitude.

3. Out on R-194, make the old-type procedure turn, and track in on R-194 (014° inbound track).

Remember that *you* are responsible for staying clear of other traffic.

You can simulate VOR airways or VOR approaches using your "local" omni station; it's a good idea to use different setups to give experience and to give a change. Fig. 25-3 shows how the isolated VOR such as the Arnold Air Force Station at Tullahoma, Tenn., may be used to simulate the airways setup and VOR approach at Owensboro, Ky.

The Arnold VOR is a simple system (no DME), and various VOR/DME approaches around the country could be used

by your acting as the "DME" as you look out and spot geographic references at various distances from your station. You could also later use ADF bearings from area commercial broadcast stations as intersections and make up your own fictitious VOR approaches (for training purposes *only*). Make sure that the information you present on those approach charts is complete.

If the approach you are simulating has, for instance, a 1120-ft Minimum Descent Altitude (MSL), you can add 3000 ft (or whatever is necessary in your part of the country for safe clearance) and have the trainee execute a missed approach when reaching, say, 4120 ft indicated. It's good for the

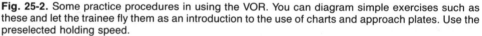

Fig. 25-3. You can use a nearby VOR to simulate specific airways networks or VOR approaches at similarly equipped facilities anywhere in the country. You might want to select a new "ground level" of 3000 or 4000 ft AGL to avoid traffic problems.

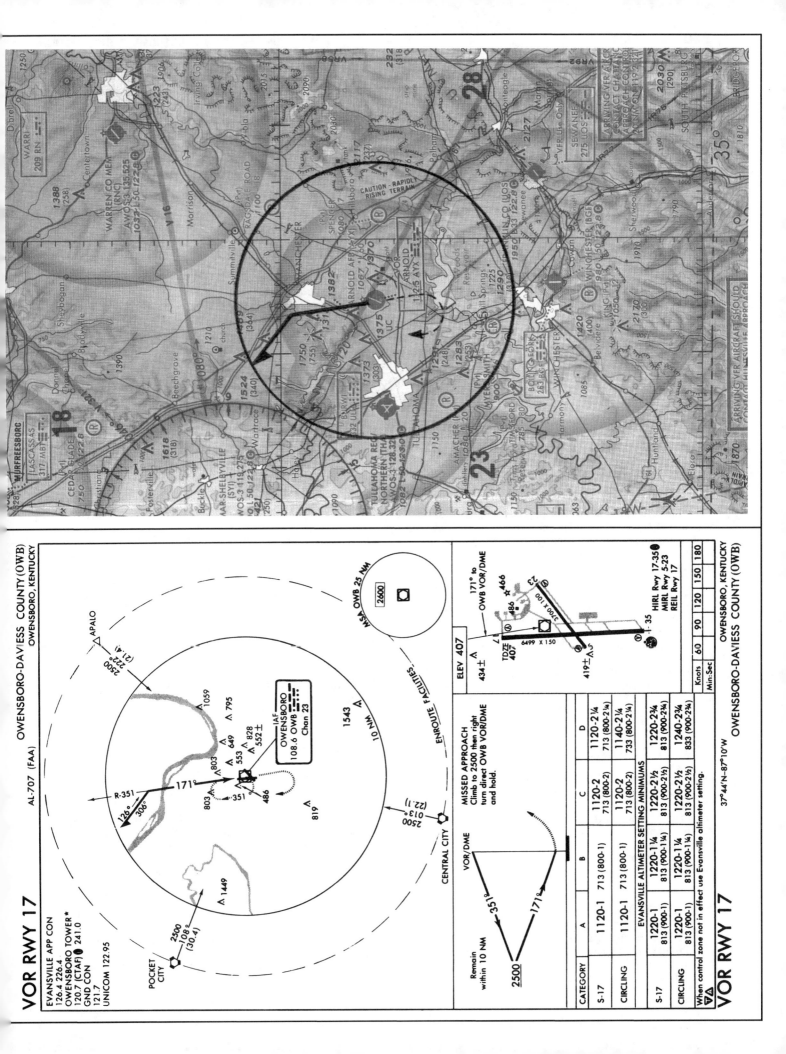

first simulated approaches to be up at altitude and away from a lot of traffic to give the trainee a chance to get the idea without too many distractions for both of you. If the nearest VOR is associated with an airport you still might want to do the introductory work up high out of everybody's way. Approach control should know of your presence, but you won't have so much VFR traffic to contend with.

Naturally, you'll do plenty of approaches using real facilities and altitudes later in the training, but start easy.

■ **VOR Tracking and Holding (Partial Panel).** This exercise is good to check the trainee's scan and control of the airplane while navigating without the use of all of the flight instruments. Have him track to and from a VOR on selected radials at a constant altitude. In addition, standard holding patterns should be practiced on selected radials using various combinations of flight instruments, such as these:

1. Airspeed and attitude indicator covered.
2. Attitude and heading indicators covered.
3. Turn and heading indicators covered.

At the end of each period of VOR work, have him use the VOR in getting back to the airport, if feasible; and later you can end each flight with a hooded VOR (or other) approach, if your airport has one.

During one flight demonstrate the effects of "antenna blanketing" in various banks with respect to the VOR station. You should also show (in a fairly tight 360° turn) that heading has nothing to do with the needle (Course Deviation Indication, or CDI) indication, but note that there may be a needle deviation in some parts of the turn because of blanketing of the airplane antennas. You may be able to demonstrate that certain propeller rpm settings will cause CDI fluctuation.

As the trainee progresses and starts coping well with the VOR, clock, and flying the airplane, you should introduce the idea of the 1-min inbound leg and expected further clearance time or expected approach time with the technique of shortening the holding pattern as necessary to meet these timing requirements.

During a chalkboard session, discuss holding at an intersection (with the use of one or both VOR receivers). At this point bring in the use of the Distance Measuring Equipment in holding at an intersection. Start using the DME on exercises also now, if available.

You'll want to know when to start moving on from strict VOR exercise-type flying and should look at the following areas for grading after a couple of full sessions of VOR (and DME) tracking and holding. Analyze his progress using a list as a reference.

VOR tracking ____
Orientation ____
Use of cross bearings ____
Timing ____
Holding pattern ____
Wind correction (as applicable) ____
Use of DME ____
Use of two VORs (without confusion) ____
Altitude control ____
Heading control ____
Airspeed control ____
Use of trim ____

ADF AIRWORK

■ **Relative Bearings and Magnetic Headings.** Explaining the use of the automatic direction finder (ADF), using only a chalkboard, is one of the toughest projects you'll have in teaching instrument ground school. A number of

training aids of all sizes are available, and you should use one to help clarify matters.

One problem is the "face" of the ADF indicator currently used. The dial, calibrated from 0° to 359°, can be confusing at first, especially when relative bearings of more than 180° are involved. For instance, suppose the airplane is headed 060° and the relative bearing is 315°; most people would have an easier time subtracting the 45° that the needle is pointing to the left of the nose than adding 315° to 60° and then subtracting 360° to get the answer of 015° to the station (Fig. 25-4).

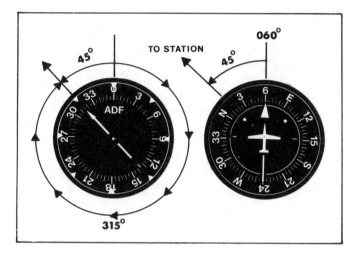

Fig. 25-4. For some people it's easier to subtract the degrees of left bearing than to add a large number to the heading.

Of course, the rotatable ADF card and the Radio Magnetic Indicator (RMI) give a good picture of the airplane in relation to the station, but since you can't guarantee that the trainee will always have such help in every airplane he'll be flying in IFR conditions, you'd better see that he has an understanding of relative bearings.

Like the VOR work, progress in using the ADF starts when the person can "see" his relative position to the station, but the ADF is usually more difficult and it takes more time before this happens.

Whether you want to go into the theory of the operations of ADF (and MDF) equipment before the first flight could depend on the student, but you should certainly note the frequency range of operation (190–1750 kHz). Most radio beacons (including compass locators associated with the ILS) are in the 190–535 kHz frequency range.

Even if your airport is well away from a radio beacon or compass locator, you can make good use of nearby commercial broadcast stations and as for the VOR can set up a simulated ADF approach for any approach chart you have, using "base" altitudes and headings as listed on the chart.

But getting back to the problem of trainee orientation with the ADF, Fig. 25-5 compares the standard presentation with two other types of relative bearing indications.

Indication *C* in Fig. 25-4 might be a tough one for some people to read but *B* makes it clearer for the majority. In fact, many pilots don't think of the actual numbers but only the degrees that the needle varies from nose or wing or tail.

Discuss tracking and homing to a station. Spend some time locating the airplane through the use of cross bearings from two stations. You might go through a couple of prechecked sample problems showing the trainee how to convert from compass to magnetic to true bearings and draw the lines on a sectional or IFR enroute chart. Of course he likely will

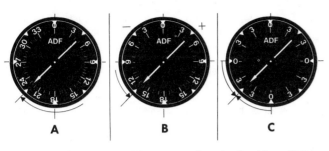

Fig. 25-5. Three possible presentations for fixed-face ADF indicators with the needles pointing in the same relative position (*assume that the airplane's heading is 270°*).
    (A) The relative bearing is 225°; 270° + 225° = 495° − 360° = 135° to the station.
    (B) The relative bearing is 135° left; 270° − 135° = 135° to the station.
    (C) The relative bearing is 45° behind the left wing; 180° − 45° = 135° to the station. Or the needle is 45° to the left of the tail; 090° (tail heading) + 45° = 135° to the station.
                                                                    (*The Instrument Flight Manual*)

not have to locate himself this way but it's a different approach that may help him in the use of ADF.

Mention that because of different deviation values for different headings he may turn 90°, based on the ADF needle, to head to a station but the heading indicator (as set to the compass) may show this angle of turn as more or less. He also must take into account that the airplane takes up space in turning that 90°; this can cause an apparent error between the initial and final ADF needle indications. (The station had a relative bearing of exactly 090 but he had to turn 95° or 100° to the right in order to bring the needle right on the nose.) Mention that the closer to the station he is the more of a problem this is, for a given radius of turn.

You may go into the theory of operations as deeply as the trainee would like, but remember the first priority is the practical aspects of what he will be doing in everyday IFR work. If he is going to fly in some remote areas where a radio beacon is the only available navaid, he should have a thorough background in all its possible uses.

■ Some First-Flight Suggestions for ADF Work. After you've given a ground session and the trainee is beginning to be able to work out headings to the station from problems you've given him, set up a simple constant-altitude flight for ADF familiarization or orientation. Don't plan on as much progress as with the VOR. Emphasize that the VOR radials are geographically fixed and the airplane's heading has nothing to do with them. The radio beacon may be considered also to have 360 "radials," but the pilot has to consider his heading when finding those radials.

START, TAXI, PRETAKEOFF CHECK, AND CLEARANCE. After the pretakeoff check is complete you should issue a clearance to the nearest nondirectional beacon, citing route, altitude, and post-lift-off instructions. If a commercial broadcast station is being used, the "clearance" should be to a preselected approach plate NDB. (Set up the proper local broadcast frequency.)

ITO AND CLIMB (HOODED). Have the trainee make an instrument takeoff and full-panel climb toward the nearest NDB or commercial broadcast station, preferably following the clearance issued earlier.

If the facility is far enough away, the clearance may then be discarded and the airplane tracked to the station on any various geographic bearings you choose.

Have the trainee find the course to the station from various geographic positions and relative bearings. (Note: Here's where you will reemphasize that with the heading indicator set to the magnetic compass, the relative bearing plus heading will give the *compass course* to the station at that instant but a crosswind may require a different *compass heading* to *track* directly to it.) Again, note that the distance required to make the turn to the station and a different deviation value for that new heading might mean that the final bearing to the station could be different after the turn is completed.

Check your winds aloft (by estimate) and demonstrate a short procedure of homing and tracking to the station flying as directly crosswind as possible to make your point more clearly. You might have the trainee watch this without the hood so that you can point out the transmitting tower(s) and he can see the crab and/or relative position of the airplane to the station.

ADF TRACKING (FULL PANEL). Following are a few exercises you might use (other numbers work just as well).

Upon reaching the facility, the airplane should track outbound on a bearing of 045° for 5 min, then turn to 155° to intercept and track inbound on a bearing of 245° *to* the station. Other tracking exercises might include (5 min outbound, or later use times such as 4 min 37 sec, etc., to introduce the idea of approach requirements):

1. Outbound track 330°, turn to 230° to intercept, and track inbound 130° (to the station).

2. Outbound 225°, turn to 360°, then intercept the 270° bearing from the station and track 090° inbound.

3. Make 45° interceptions of prechosen outbound and inbound bearings.

Fig. 25-6 shows tracking exercises 1 and 2.

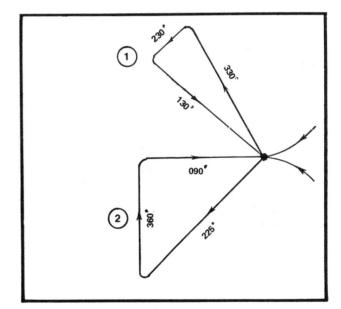

Fig. 25-6. Some tracking exercises (make up your own, also). Don't be too tough at first on requirements when tracking outbound.

If you feel that the trainee is able (and isn't tired), have him hold on a couple of different inbound bearings using 2-min inbound legs if the current wind situation isn't too tough. Use the recommended holding speed and configuration. The 2-min inbound legs allow more time to set up tracking.

You'll judge progress on this flight by considering the following items:

1. *Orientation.* Is he beginning to "see" his relative position to the station? This doesn't mean that he can instantly come up with an exact figure as far as the course to the station is concerned, but if he knows after looking at the relative bearing and heading that the airplane is northeast of the station and that the inbound course would be somewhere in the vicinity of 225°, the details can be worked out shortly. You'll find that some people have a great deal of trouble at first with orientation and will be far off in their analysis of where they are. ADF work can cause mental fatigue more quickly than almost any other phase of instrument training because the trainee is working with numbers to orient himself and also is flying the airplane (Fig. 25-7). Don't push too hard on the first couple of ADF training flights; let him have more visual breaks than usual.

Fig. 25-7. Concentrated ADF work is very fatiguing to the instrument trainee. His mind will start fogging after a long session.

2. *Tracking.* Tracking *to* a station is much easier than tracking outbound for most people, and you shouldn't be too tight on requirements on tracking outbound at first. After a trainee is able to do a reasonable job of holding an inbound track, you should start concentrating on outbound practice.

3. *Timing.* The clock will be a part of the scan, but the tendency will be to overconcentrate on the ADF at first and ignore it. Remind him as necessary.

You'll also be working on good altitude control in tracking and holding and should watch for excessive airspeed variations in the holding pattern.

One thing that will come to mind when instructing people who are having extra problems coping with the ADF is that you are putting a lot of time on procedures that are more difficult to do than VOR work and there is just not that much requirement for skill with the ADF. After all, the pilot can always *home* into the outer marker locator. Besides, if there are thunderstorms in the area, the ADF loses much of its effectiveness. In fact, ADF training *does* seem to take too large a portion of the training program, but sometime later the pilot may have to make an instrument approach with his family in a deteriorating weather situation when an NDB (ADF) approach is the only one available—you'd like him to be able to do it safely. A big point, too, is that the ADF practice is one of the best ways to teach him to think *and* fly the airplane and so affects all other phases of training and later flying. You'll find that people who can orient themselves and make a good NDB holding pattern and approach also can generally do an above-average job using the VOR, but the reverse is not necessarily true.

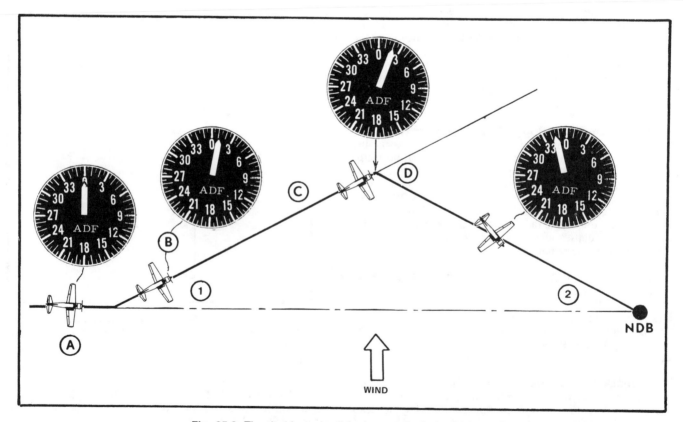

Fig. 25-8. The double-angle-off-the-bow method of getting an ETA to the station. (A) Tune in and identify the station. Turn the airplane to get a course to the station (relative bearing 000°). (B) Turn 10° (left in this example), which will put the ADF needle in the 010° position. Start timing. (C) Fly this heading until the relative bearing has doubled (020°). Get the time and turn to *track* into the station. The time for the second leg should be close to that used to fly the first but will be longer in the wind situation indicated. You might set up a couple of problems on your computer using specific true airspeeds and winds to see this effect. (Angles *1* and *2* are equal but exaggerated.)

ADF TRACKING AND TIME TO THE STATION. Of course, the requirement for finding the time to the station by either the "off-the-wing" method or the double-the-angle-off-the-bow method are seldom used these days, but both exercises are good for orientation and timing. (As stated in Chap. 24, don't overdo exercises that don't pertain directly to actual IFR requirements.)

As a review, take a look at the double-the-angle-off-the bow method of checking the time to the station. The airplane will fly two legs of an isosceles triangle, shown in Fig. 25-8.

When you're demonstrating or letting the trainee practice this exercise, you'll get better results in showing the idea if that initial turn of 10° is made away from the wind, as shown in Fig. 25-8. There are two reasons for this: (1) If a strong crosswind component exists, the 10° cut into the wind might be exactly the right correction for *tracking* directly to the station from point *A* and you and the trainee could still be timing that first "leg" when station passage occurs; and (2) by turning away from the wind, the time on leg *1* will be less than that required to track into the station on leg *2*, and it's better to be slightly late at a fix than to reach it before the ETA.

You might also have the trainee do some time-to-station work and tracking around the station as a method of developing his orientation (after discussing the procedure on the ground).

Fig. 25-9 shows an idea of getting the time to the station and tracking around it to a particular inbound bearing. While this is good for teaching orientation, expect some confusion

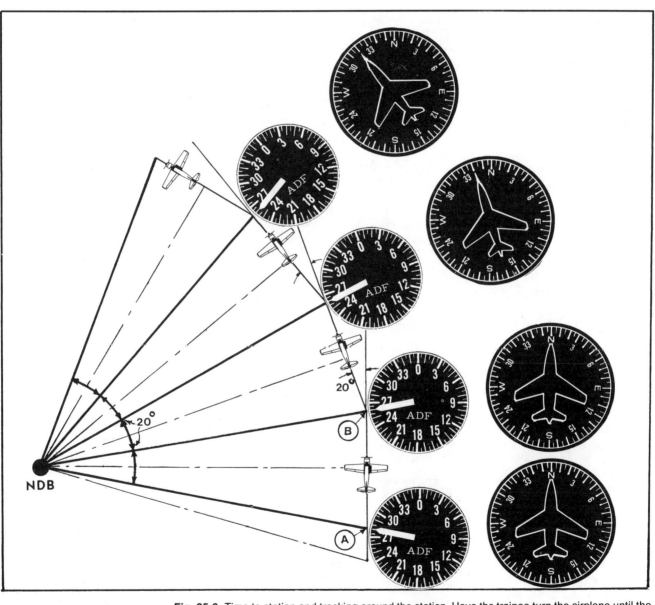

Fig. 25-9. Time-to-station and tracking around the station. Have the trainee turn the airplane until the station is at about 285° (or 075°) relative bearing (285° here) and hold the resulting heading (360° here). The timing will start when the needle has moved back to the 280° position (A); the extra 5° is needed by the trainee to be ready to start the timing. (B) When the needle has moved back to the 280° (or 100°) position, the time is noted. Under no-wind conditions the time required from *B* to the station would be *3* times that used to fly the *A–B* leg. If a 10° arc is used, the multiplier is *6*. A rotatable ADF card makes it easier to see.

on his part on the first few legs; you'll have to repeat your explanation and probably will have to demonstrate again.

The thing you must emphasize in these exercises is that the relative bearing is controlled by the heading and that a constant heading is necessary on the legs if timing is to be accurate. Also, the degrees of heading change will change the relative bearing by the same amount when each turn is made to a new leg.

The trainee may be asked to do IFR navigation by ADF systems on the flight test, so he should be ready.

APPROACHES

If possible, issue each of your trainees copies of the approach charts to be used during this phase of instruction early so that they may have time to look them over and can ask both general and specific questions. They should also have access to legends for Instrument Approach Procedures charts. You might make up booklets composed of the legends and approach charts you'll be using in training. More about this later in the chapter.

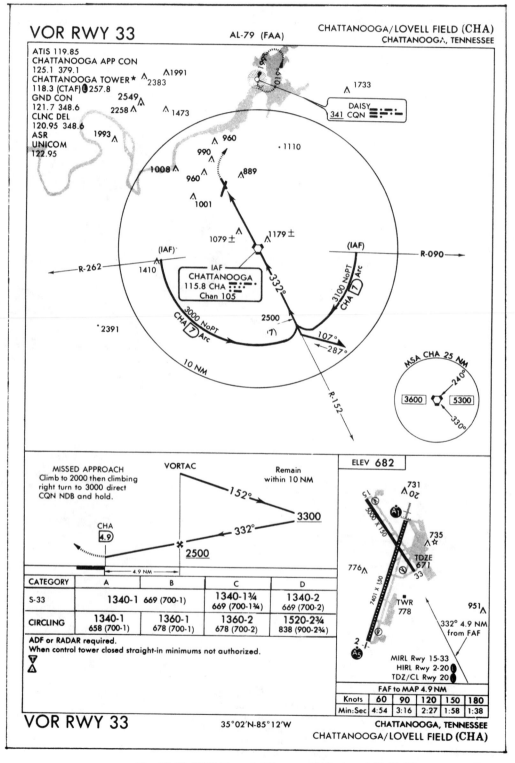

Fig. 25-10. VOR Runway 32 approach for Lovell Field, Chattanooga, Tenn.

■ **"Real" VOR Approaches.** Common sense dictates that the straight or normal VOR approach be the first "real" approach practiced by the trainee. It's not as fast moving as the ILS or as difficult to see as the ADF action and, hence, makes a good introduction to the closer tolerances of approach work without putting on too much pressure at first. (Fig. 25-10 shows a typical VOR approach.) Before each flight you should discuss the approaches to be done that period. After the first session or two of VOR approach work you might also introduce "local" ADF approaches so that if progress permits during a VOR flight you can move on to ADF approach practice too. As always, you should stay slightly ahead with your ground briefing so that if the flight is going well you can move right on to the next exercise.

In your preflight briefing(s) on the VOR approach, go over the selected approach chart step by step and give the trainee a chance to see the specific requirements for that procedure, as well as items to consider on others. Cover the profile view, plan view, minimums section, and other data.

Here the trainee will start putting all of the basic instrument flying and the VOR airwork together, and you'll have to watch for overconcentration on the procedures at the cost of handling the airplane. Stay with him on requirements of heading, airspeed, and altitude as laid out on the approach plate.

An average VOR approach from the point the airplane is set up to cross the VOR the first time to completion of the missed approach procedure for another go at it (or to go home) will take something in the vicinity of 10–15 min, so *your* demonstrating the procedure would waste good training time. He's able to do the basics by this time—or should be—and you've discussed the approach fully on the ground, so let him do this one from the beginning under the hood, with your close supervision. You should handle the communications that first time.

If you are making simulated approaches to a busy airport, actually using approach control and the tower, things will appear to be getting out of hand the first few times as you monitor the trainee's instrument flying, nav work, and timing and make transmissions as necessary to avoid entanglements with VFR traffic, while also looking for other traffic, controlled or otherwise. It will seem, too, that the only day you can work with a trainee on much-needed VOR approaches is when the wind requires traffic to take off into your final approach (that is, the VOR approach is for Runway 32 and the wind is from 140° at 20 K, requiring *everybody* to take off that way). In that situation, you, the other traffic, and the tower will soon be ready to knock it off. You'll run into this problem particularly when practicing a back course ILS approach. Often you'll have to break off an approach shortly after getting set up on final, and both you and the trainee will be uptight during the session (Fig. 25-11).

One common error you will encounter on the trainee's first, or second, "real" approach is that he will tend to be late in starting the first descent (and will not have slowed up to approach speed or looked at the checklist). Some people will persist in descending below the altitude required in various segments, although the biggest problem will most likely be "getting down" on the final approach.

Don't plan on teaching the various types of approaches in blocks. Don't bore the trainee with VOR approaches and then move on to ADF approaches only, neglecting VOR work until time for final review for the flight test. Get him going well on VOR work, then do some ADF, later some ILS approaches, and come back to VOR work.

These VOR approaches, particularly when the trainee starts doing the communicating and nav work, will give you an indication of his reactions later when things might be coming hot and heavy. As you evaluated the student's reactions to new or unusual situations before solo to set the *approximate*

Fig. 25-11. The only day you can get a chance to shoot some VOR (or other approaches) at the big airport the wind conditions require traffic to take off the "wrong way." After a couple of simulated approaches this way, both you and the tower will be more than glad to call it quits.

solo time and to get an idea of the type of flying he'd do later, you'll be continually watching and extrapolating the instrument trainee's reactions to judge his ability to cope with the ATC system after he's on his own.

As work progresses on VOR approaches you should judge the following major areas and make verbal or written comments on details as necessary:

Preapproach procedures (checklist, slowdown, etc.) ＿＿
Orientation ＿＿
Holding ＿＿
Procedure turn (as required) ＿＿
Final approach:
    Airspeed control ＿＿
    Heading ＿＿
    Altitude control (check each segment of the approach) ＿＿
Missed approach procedures ＿＿
Headwork (ability to cope under pressure) ＿＿
Basic instrument flying (general) ＿＿

Some *common errors* you'll see, in addition to the ones just mentioned, will be these:

1. Having trouble keeping up with the times in the holding pattern.

2. Forgetting to reset the OBS in the procedure turn so that the needle is reading backward (and confusion follows) on final.

3. Neglecting voice reports, particularly when asked to report out of the procedure turn.

4. Overcorrecting when tracking in on final.

5. Failure to note the time over the final approach fix. (This is a *most common* common error.)

6. Going below MDA.

These are the most common errors. As you gain instructing experience on VOR approaches, you may encounter uncommon errors as well.

Discuss circling approaches using the VOR, and practice at least two if possible.

■ **VOR/DME Approach.** If practicable for your area and training airplane, you should follow the VOR approaches with a discussion and practice of VOR/DME approaches. If an airport with such an approach is not readily available for practice, an en route VORTAC facility fairly close may be used in tracking at a constant radius or you may use it to practice a specific VOR/DME approach using higher altitudes for all segments, as was done for the early VOR approaches.

Note back in Fig. 25-10 that the DME could be used as an *aid*. It's a requirement for a VOR/DME approach.

If practicing any kind of a VOR/DME approach is out of

the question, you should at least break out an approach chart and go over it with your trainee(s) during one of the ground school sessions.

The practice of tracking around the station using the ADF can be an aid for the trainee to visualize what he's doing here. You can do it with a VOR, too.

■ **ADF Approach.** When the trainee's high work with the ADF is smoothing out and he's had several VOR and (hopefully) VOR/DME practice approaches, start the full ADF approach work. Fig. 25-12 shows a typical NDB approach. Get

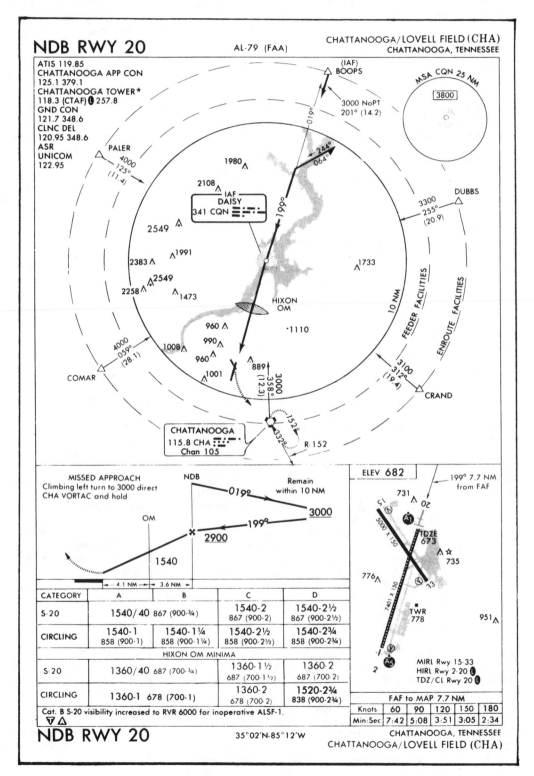

Fig. 25-12. An NDB approach to Lovell Field, Chattanooga, Tenn.

such a chart for a nearby airport and discuss the general and unique features of these approaches.

While you are working on NDB approaches, base your grading on the following major areas; you may find particular errors within these divisions but such a listing can act as a memory jogger after the flight.

Orientation ____
Tracking ____
ETA accuracy (if applicable) ____
Approach:
    Procedures (checklist, use, etc.) ____
    Tracking outbound ____
    Procedure turn____
    Inbound track ____
    Timing ____
    Altitude control ____
    Airspeed control ____
    Voice procedures ____
Missed approach:
    Procedures (cleaning up airplane) ____
    Aircraft control ____
    Voice procedures ____

Again, watch for possible problems tracking outbound. Timing (outbound, or inbound from a final approach fix) can be neglected on any approach (but it seems to be more prevalent in ADF approaches), and you may have to remind the trainee on several approaches that the clock is a part of the scan. At first he will tend to concentrate on the ADF indicator to the neglect of airspeed and altimeter.

If there is any *one* common error that occurs in approaches, it's neglecting to slow down before crossing the fix initially. You'll work with people who will roar across the fix outbound at cruise speed (or faster, it seems sometimes) and fly, say, 3 min outbound—usually in a stiff tailwind. They realize as they get near the procedure turn that the airplane should be slowed (and do so); the trip inbound at approach speed into that headwind is a lengthy one indeed. The ship's log of one such approach reads: "Thursday, March 6. Cruising as before. Inbound from final approach fix Briarwood Airport. Drinking water gone. Crew growing mutinous. Scurvy aboard. Airport appears to be gaining on us."

Discuss and have him practice circling approaches as applicable.

■ ILS Approach. This is the approach the average trainee has always associated with instrument work. The others are okay, but when he gets to this one the real business of instrument flying has begun.

In the preflight discussion, cover the instrument landing system in general and brief him on the specific approach to be used in practice. Fig. 25-13 shows the components of the ILS.

The ILS is the most accurate approach system, and normally on an instrument flight to an airport with ILS and other types of approaches, the pilot may expect to be cleared to the ILS unless winds or other factors make another type more feasible.

Okay, if the ILS is considered *the* approach, why so late in introducing it? There are two main reasons: (1) its components include the use of the omnihead(s) *and* ADF and the trainee should be used to including these in his scan, and (2) the ILS localizer is 4 times as sensitive as the VOR and he should have pinned down tracking on that before working on this one.

In your ground school work, discuss each component,

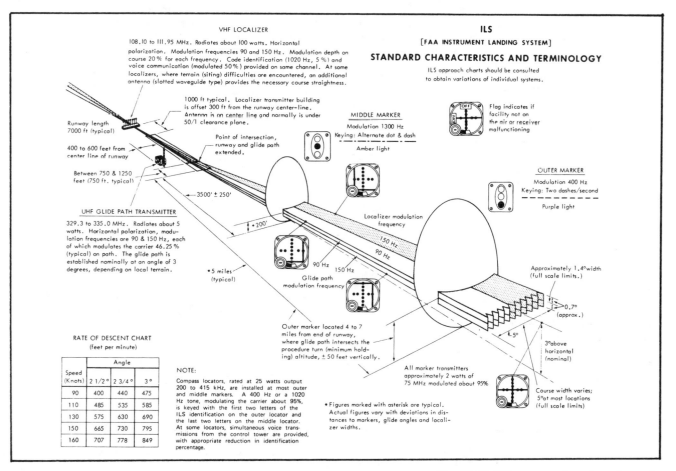

Fig. 25-13. ILS components and requirements.

giving information such as glide slope angle(s) and thickness, localizer width, marker beacon(s), compass locators and their expected effective ranges, and lighting. You should give a good coverage of the system he's about to use, but save the theory of how it operates for later. Be sure to explain the needle action (works properly on front course, but in reverse on back course unless special equipment is in the airplane). Emphasize that the reactions of both needles are more sensitive than the VOR, and easy corrections should be made in pitch and heading.

The first few times you introduce the ILS (letting the trainee do the flying and you working the communications), you can expect certain things to happen. The trainee may be off to one side of the center line and turns the airplane until the localizer needle centers, which, of course, means a bad overshoot, so he turns back until the needle centers, repeating this problem throughout the approach. The path may look like Fig. 25-14. Stress that the heading indicator is the big aid and set up limits of heading change for correction, based on your airplane and the trainee's experience.

The first ILS approaches should be done without the glide slope so that he can get used to the tracking procedure and not be distracted. After he is able to start pinning down the center line, add the use of the glide slope. Try to schedule so that he has smooth air on the first couple of approaches.

Don't let the trainee depend on radar vectors by approach control to get set up on final; make him use the ADF, reminding him that the outer marker locator is a good backup for approach if the localizer or heading instruments fail on an approach.

When the glide slope is introduced, expect overshooting, although the average trainee will not go up and down through the slope as noticeably as in problems with the localizer. Usually, he will get either below or above the slope and stay there, trying to get back to his proper place as time (and distance) runs out. You will have some who overcorrect enough to be above *and* below the glide slope several times during an approach, but the most common error is getting stuck off the slope.

You will choose the best approach speed for a particular airplane, based on the factors discussed in Chap. 24, and then find the general power settings required to get the proper rate of descent to stay on the glide slope. Review "The Power-required Curve" in Chap. 6.

You should set definite limits for airspeed variation to stay on the glide slope; you might figure on a ±5-K correction to get up or down to the glide slope before changing power.

During these approaches and in all practice or actual approaches, make it clear that *a missed approach should be initiated at any point during the approach if conditions be-* *come unsafe, or even if he thinks there is a good possibility that conditions could become unsafe.*

Work in some holding practice at the outer marker with an expected approach time if practicable (there'll usually be other traffic about). You can set up power settings for holding and the approach. For some airplanes, lowering the gear at the holding speed and power setting is enough to set up an approximate 500-fpm descent; for others a power adjustment is necessary even after the gear is extended. A fixed-gear airplane power setting for holding will require a reduction in power for the descent at that speed, since you aren't able to suddenly increase parasite power required by putting the gear down.

For the airplane with a constant-speed prop you should note that the throttle(s) must be retarded during the descent to keep the manifold pressure from building up.

If you are teaching instruments in a twin, fly several one-engine ILS approaches (set up a no-thrust condition, *don't* feather) and get a rough figure for the manifold pressure required for level flight and clean and gear-down descents in that condition. Your job here, as in other instructing, is to take the trainee into areas that would be dangerous for him to encounter for the first time alone. Imagine the problem the new instrument pilot would have in his twin, making an actual ILS approach after losing an engine if he had never practiced this. The strain of an actual emergency, plus the fact that he probably would have no idea of the power required to make a single-engine approach, could cause a go-around (on one engine) and perhaps result in a fatal accident. In a real situation he will be less apt to panic if he has done it before and has numbers to use.

A rule of thumb would be to use an approach speed for the twin at or above the all-engine best rate of climb speed so that if there's a failure on final there will be a buffer while the pilot gets organized. However, you'll need to ge the best numbers for each twin you instruct instruments in.

In any ILS approach you should check the trainee on the following items:

Preapproach procedures (including use of checklist) _____
Orientation _____
Holding:
    Airspeed control _____
    Altitude control _____
    Heading _____
    Orientation _____
    Timing (make EAC time) _____
Approach:
    Voice procedure _____
    Airspeed control _____

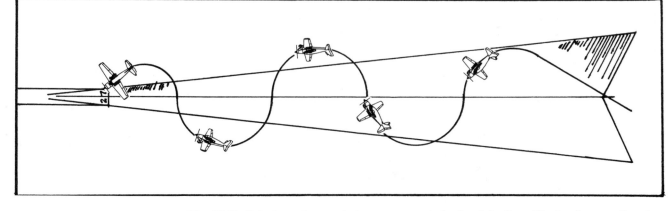

Fig. 25-14. At first, nearly every trainee overcorrects for the deflection of the localizer needle unless you insist on limits of heading correction.

Localizer precision ____
Glide slope control ____
Missed approach:
Cleanup ____
Voice procedures ____
Control of airplane ____

BACK COURSE ILS. Back course ILS approaches are being replaces by front course approaches around the United States. If there is a back course ILS in your area a few approaches would be a good training for staying oriented when the needle works "backward." If you are using older VOR/ILS indicator heads with blue and yellow markings, you should point out that the "blue" (150 Hz) is always on the same side of the approach (front or back) with 90 Hz on the other.

Fig. 25-15 shows an older-type presentation. Probably the airplane you'll be using for training will have a straight ILS receiver and the needle will read in "reverse" when inbound on the back course.

You'll have to keep the reminding the trainee of this needle situation as he practices the approaches. A few instrument instructors neglect to tell their trainees that *the needle always indicates which sector (blue or yellow) the airplane is in, whether on a front or back course approach, and if the pilot is looking at the approach chart he can "see" where he is at all times* (Fig. 25-16). (Left indication equals blue; right indication equals yellow.)

You'll sometimes see that the airplane is well off to one side of the center line in a back course approach and the trainee is diligently correcting *into* the needle as he did on the front course approach and for VOR work. Under certain conditions, it's a good move to let him continue and goof up the approach, particularly if you have already done a great deal of reminding about it. Would he try to salvage a hopeless and dangerous situation (on actual instruments) or would he break it off and call for a new approach? His judgment will be a big factor in your decision to recommend him, and it doesn't hurt

Fig. 25-15. An older-type VOR presentation. Approaching on the *front course* of the localizer, the indication is that the airplane is in the blue (150-Hz) "right-hand" sector and a correction to the left is needed to get back on the center line. On a back course, on final approach, the needle is indicating that the airplane is in the 150-Hz sector, but the pilot would have to correct *against* the needle to get to the center line. In both cases the airplane is on the same geographic side of the center line (in the 150-Hz or blue sector). (*The Instrument Flight Manual*)

every once in a while to let the trainee get in a situation that requires some cool thinking.

Coming out of the procedure turn on the back course is a spot that causes many a trainee to bite the bullet. He's turned off into the yellow sector and is now headed back to the course line. When he looks up he suddenly realizes that the needle is to the *right* of center (it's in the yellow, though) and figures that the center line has been overshot; a rapid turn back is started. Fig. 25-17 shows the problem.

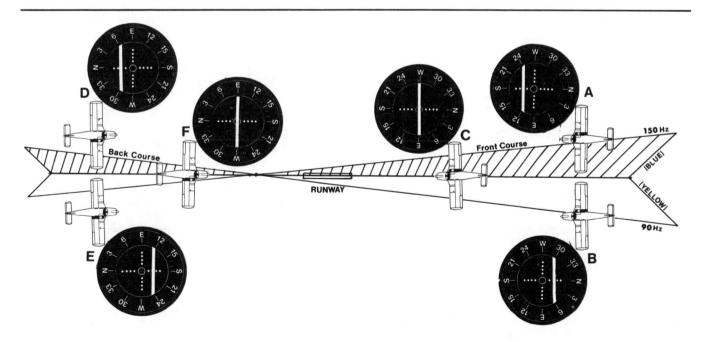

Fig. 25-16. The ILS localizer. The setting of the omnibearing selector has no effect on the needle indications when the set is tuned to a localizer frequency. However, many pilots set up the published inbound course on the OBS as a quick reminder of the base course when on approach. (*The Instrument Flight Manual*)

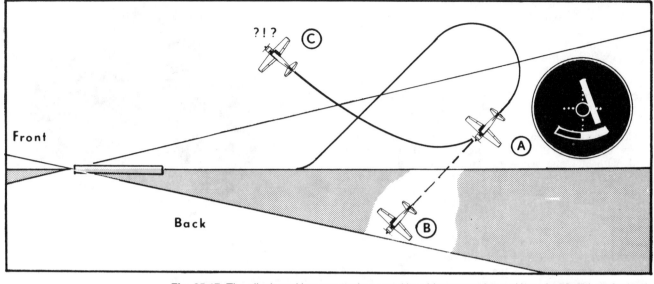

Fig. 25-17. The pilot is making a procedure turn (the old-type turn is used here for clarity) on the back course. At point *A* he suddenly realizes that the needle is to the *right* (it's still pointing out that the airplane is in the 90-cps yellow sector). He thinks in terms of the front course needle reactions and figures that he's past the center line at position *B*. Approach control radar wonders what is going on as he takes off on a cross-country *C*, looking for that elusive center line.

Many back course approaches use an intersection with the VOR as the final approach fix, and trainees seem to have more trouble getting the time at the intersection than with a VOR or ADF station passage, and certainly more trouble than when passing marker beacons.

The problem arises because the pilot has *two* needles to keep up with and one is "working backwards." He may get so confused he's looking at the wrong one for cross-bearing information; if he is off course the problem is compounded as he tries to figure out which one is telling him what. You'll save many a practice approach by reminding the trainee about an intersection passage and that he needs to note the time. Don't let him get dependent on you for this, though.

Of course, if you want to really test his mettle, have him do a partial-panel back course ILS, using only one VOR receiver for course and cross-bearing information. In turbulent air this has been known to reduce the strongest soul to a shattered hulk.

At this stage you might want to discuss the simplified directional facility (SDF) and cover the similarities between it and the ILS localizer. A review of AIM—*Basic Flight Information and ATC Procedures* would be good for the trainee at this stage of training.

■ **Approaches in General.** When you first start instructing instrument flying, sit down and analyze what the trainee needs to know not only to pass the written and flight tests but also to cope with the ATC and weather systems later. *Don't* mechanically follow a preset syllabus that someone else wrote. However, do use one as an aid to set up instruction based on *your* situation and airplane, one that can be expanded to general instrument flying.

Take a good look at the approaches you'll be using in your instrument course. Fly through them in your mind and make notes for reference in your ground and flight instruction. For instance, you might decide it would be a good habit to extend the landing gear on all approaches at the point the airplane has completed the inbound turn and is headed for the airport on final. Although it would be better to wait on an ILS approach until starting the descent on the glide slope, even then you might decide that the extra "dragging up" from the procedure turn to the glide slope is worth it in order to

have *one* point to always put the gear down.

Always try to tie in one procedure with another if possible. The gear procedure above might keep one of your trainees from making a later gear-up approach in actual instrument conditions when he's under pressure. You may have good arguments against a procedure of gear extension and reject it for your course, but you should plan generally to have a minimum of variables. Remember that in the first few hours of actual IFR on his own the new pilot will be tense and probably forgetful. (You should help him set up a good checklist for the airplane he'll be using later.) Be sure he understands the various parts of the different approaches.

The procedure in instructing instrument approaches is, like other phases of instruction, to discuss them before the flight so that the trainee understands what he'll be doing, start off with simple requirements with few distractions (and smooth air), and then move on into more complex situations and rougher air.

USING THE SYSTEM— GROUND SCHOOL INSTRUCTION

Before starting this phase of ground instruction on using the ATC system you should sit down by yourself and lay out the requirements for an orderly presentation of needed information. Don't just start mechanically passing out facts that are not applicable to the IFR training course. They only take up needed room in the trainee's memory cells.

Of course, you'll have been paralleling the flight work with appropriate ground instruction and the suggested route is an advanced version of that set up in Chap. 13 for the pre-private student. As the trainee gets into the cross-country portion of the program he'll need to review and expand his knowledge of airplane performance, weather, weather hazards (icing, turbulence, fog), and weather services available. Following are a few suggestions.

■ **En Route Charts and Approach Charts.** Make sure that the en route charts and approach charts are covered thoroughly in a classroom situation and have the trainee study them at home, so he can learn their details without thinking about a flight he's about to make. Sure, you'll go over them before the flight(s), but the primary purpose then will be to cover specific routes and approaches, with some general information included as well. The main thing is not for the trainee to memorize every symbol used on the charts, but to be well informed about the common symbols and know where to look for the less common ones. Don't forget that details on circling approaches should be covered, too.

■ **The Computer.** Usually a good review on using all information available from the computer (mechanical or electronic) is in order, because the average instrument trainee probably hasn't used some of the functions (such as making an altitude correction for a nonstandard atmosphere) since he got his private or commercial certificate.

For most people you won't have to set up a course for the computer, but encourage them to review the manual and practice some or all of the problems therein and require its liberal use for the cross-countries in the course. Or you could set up some computer problems to be worked by those who are particularly rusty.

■ **Communications and Navigation.** Before and during the cross-country IFR instruction it would be best to have one or more ground discussions of the ATC system so that you can give the trainee(s) an overall look at what the IFR pilot can expect. If you are running a more formal ground school with a number of students, try to have controllers from towers, FSSs, and the nearby Air Route Traffic Control Centers come and talk to the group. This will alleviate much anxiety in the trainees who've been sweating being under (to them) "complete control of a benevolent—or not so benevolent—dictatorship." Too many instructors never stop to show their IFR trainees the whole picture of the system; they conscientiously instruct in bits and pieces and fail to show *how* and *why* things happen on an IFR flight. People learn the parts better if they understand the whole.

Before getting deeply into general cross-country IFR, you should discuss at leisure the steps involved, including planning a particular flight, filing the flight plan, what happens to the flight plan, the clearance and the actions of ground control, tower, departure control, the Center, then approach control, tower and ground control. Follow a flight from start to finish with possible communications procedures and problems during the trip. Use the charts and have the trainee(s) follow on his. Needless to say, this could take a couple of hours or one night ground school session, and it should be done before the long cross-country. Here would be a good time to bring in FARs *applied to instrument flying.* Don't set up a block of FARs for a ground school session; but you can bring in items during the cross-country discussion, such as VOR receiver testing, required airplane papers, static system checks, and alternate airport requirements.

■ **Other Points.** Here is a good place to expand on the physiological factors of instrument flight with particular emphasis on vertigo and its effects on the pilot. Try to induce it on one of his (hooded) flights.

■ **Weather.** To start teaching weather as a block while the trainee is working on basic instruments means a loss of learning. The principle of a need-to-know aid to learning applies here as it did for the student pilot.

Weather sequences and other reports, charts, and forecasts can be dull for people who just sit and listen. Bring them into the discussion. Make sure that everybody has copies of the information you are discussing and use local or nearby reporting station reports for samples. Tie this in with the cross-country discussion mentioned earlier.

Brief him (them) on weather hazards such as icing and turbulence and how to cope with them—or stay on the ground.

If you are operating from a larger airport where a Weather Service Office is located, have the trainee do his weather planning there for each cross-country flight. Sometimes it seems easier to just use the phone, but this is a training course and he should be able to look at the weather charts. Actual visits to Flight Service Stations should be done as much as possible too, and if you're flying from a small airport you'll have to set aside time for WSO and FSS visits during your cross-country work. Stay up to date on the modernization (computer) plans for the Flight Service Stations. A visit to an ARTC Center should be worked into one of the cross-country flights. More about this later.

■ **Knowing the Airplane.** Discuss the *Pilot's Operating Handbook* and review power setting charts.

Weight and balance information and a couple of weight and balance problems should be covered in the ground school periods. Use numbers for the airplane being flown. Emphasize that proper weight and balance is even more important for IFR work because turbulence and icing can add to control problems and that even without these menaces, flying a marginally stable airplane on the gauges for a long period of time can induce fatigue and cause an accident on approach.

The electrical system will certainly take on more importance than it had when the trainee was strictly a VFR pilot. You'll find that many people don't understand the comparative power required by various components in use. Fig. 25-18 shows a sample composite list of power required for various electrical components. If you can get the actual loads for your airplane, you should do so.

Such things as the lights (beacon and navigation) use over twice as much power than the pitot heat for this airplane, and if you're IFR in icing conditions at night the pitot heat is a lot more important than lights.

Cover the other systems (hydraulic, fuel, etc.) with emphasis on IFR operations.

■ **Advanced IFR Systems and Instruments.** During the ground school sessions (preferably toward the last of the course) you should discuss the principles and properties of area navigation, and if the equipment is available, demonstrate the procedure in flight.

Later in the course show the trainees more sophisticated equipment, such as flight directors, radar, the horizontal situation indicator (HSI), and autopilots (Fig. 25-19). But even if these items are available in your airplane, first teach your people the hard-core basics and make sure they could fly the airplane using "standard" instrumentation. They can get acquainted with the more complex equipment later in the course or after they get the rating.

Fig. 25-20 is an RNAV approach for Tullahoma, Tenn. You'll want to stay on top of the latest types of approaches for your area, including the possible introduction of LORAN C.

Fig. 25-21 shows a LORAN RNAV Runway 18R for the New Orleans Lakefront Airport. You might discuss generally the possibility of LORAN approaches in your area in the future.

The use of GPS (Global Positioning System) is becoming more widespread and you should be familiar with it. While the principle of operation is the same for all GPS equipment, there are different modes of operation that apply to the various makes. The chances are good that in the future you'll be doing GPS navigation and approaches as replacements to some, or most, VOR and ADF approaches.

CROSS-COUNTRY FLYING—AIRWORK

The first part of the navigation and communications instruction consisted of simple flights directly to nearby facili-

ELECTRICAL LOAD ANALYSIS CHART

	AMPS REQUIRED
STANDARD EQUIPMENT (Running Load)	
Battery Contactor	0.5
Fuel Indicators	0.1
Flashing Beacon Light	6.0
Instrument Lights	0.7
Position Lights	2.5
Turn Coordinator	0.3
OPTIONAL EQUIPMENT (Running Load)	
Altitude Blind Encoder	0.1
Strobe Lights	3.0
ADF	1.0
Nav/Com	1.0*
	2.25**
Transponder	2.0
Glide Slope	0.5
Marker Beacon	0.1
Autopilot	2.5
Encoding Altimeter	0.1
Nav/Com (720 Channel)	2.9
DME	1.2
Pitot Heat	2.9
Post Lights	0.6
RNAV	0.65
Interphone System	†
Avionics Fan	1.0

ITEMS NOT CONSIDERED AS PART OF
RUNNING LOAD

Cigarette Lighter	7.0
Clock	†
Control Wheel Map Light	0.1
Courtesy & Dome Lights	1.2
Flap Motor	8.5
Landing and Taxi Lights	9.0 ea
Map Light (Door Post)	0.2
Air Conditioner (High Blower)	6.7
Ventilation System Blower (High Speed)	5.0

† Negligible
* 1.0 Receiving
** 2.25 Transmitting

Fig. 25-18. Electrical load analysis chart. The values given were picked for typical equipment in a four-place airplane. The equivalent items for *your* airplane may have slightly different electrical power requirements, but the main idea is to compare the various demands of different components. (For instance, compare the landing/taxi lights with the turn coordinator or ADF.)

(The Instrument Flight Manual)

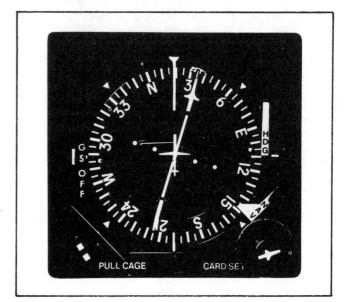

Fig. 25-19. The HSI. *(Narco Avionics)*

lowed by a 1-hr straight flight home, wouldn't be nearly as valuable. Of course, the ideal arrangement of airways and airports with different approaches near your home base probably won't be available, so you'll have to be flexible.

Fig. 25-22 shows how a short (but active) cross-country may be made from an airport with no instrument facilities. Look at the En Route Low Altitude chart for your area to set up your own IFR cross-countries for various flights. They should be simple at first with increasingly complex routes and approaches. For instance, a first X-C flight might simply be a flight from Franklin County Aiport (*1*) to Arnold VOR (*2*) and return. You might want to do the full route shown in Fig. 25-22 on a VMC day, with the trainee hooded and with a preflight phone call to Huntsville tower to see how the schedule looks there for the next couple of hours. You might even have the trainee make a flight log and file a "flight plan" for training purposes. After he has gotten at ease with the system you should file an actual flight plan for this trip or another one and have him fly it VMC (hooded) or actual IFR. It's assumed that the airplane is equipped for VOR, NDB, and ILS approaches.

Refer to the numbers in Fig. 25-22:

Point *1*: The departure and return in this example will be at Sewanee–Franklin County Airport (Tenn.). The trainee would make a flight log, getting winds and weather as necessary and filing a "flight plan" with you. You would be ATIS, clearance delivery, ground control, local control (the "tower"), and departure control. You would act as ATC until Huntsville approach control, local control, and departure control are contacted. In this example you might want to make a stop at Huntsville to visit the tower and restart the flight back to home base using a real ATIS, clearance delivery, ground control, local control, and departure control.

Point *2*: After takeoff the trainee would climb to the assigned altitude and track directly to the Arnold VOR. You may want to impose an altitude limitation for the first 5 min, then with a clearance to climb to the assigned en route altitude.

Point *3*: Fly V67 to the Shelbyville (Tenn.) VOR at the "assigned" altitude and make a VOR approach, checking with Nashville approach control and/or Shelbyville Unicom with an announcement of your intentions.

Point *4*: Depart the Shelbyville VOR and fly V321 toward

ties and setting up requirements such as holding and simple letdowns and approaches. After the trainee has had some tracking and practice approaches, put it all together with reasonably short triangular flights, having him plan and file, get all clearances, and generally handle all action under your supervision. While good straight and level flying is extremely important in IFR work (it's like VFR flying; he'll spend much more time in straight and level than climbing, turning, or making approaches), you should plan these flights so that the periods of straight and level are relatively short, if possible. A triangular cross-country of about 30-min legs and a VOR approach at one airport and an ILS approach at another using the full ATC system would be a good setup for learning. A 1-hr flight to an airport where a VOR approach is made, fol-

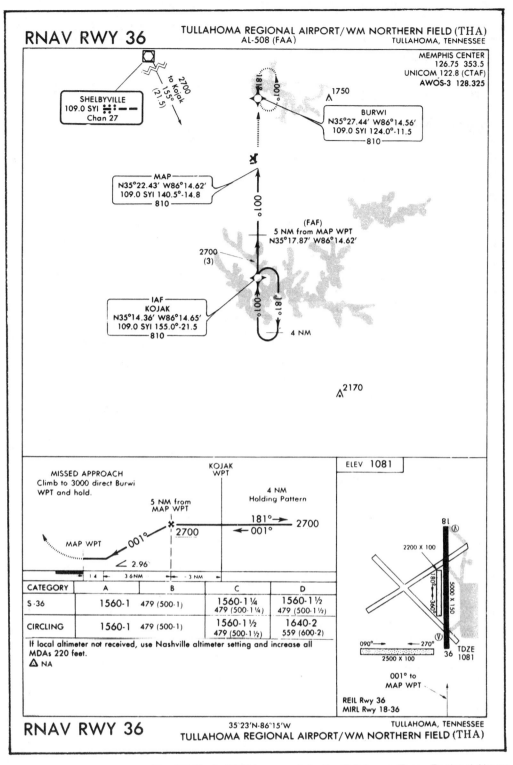

Fig. 25-20. An RNAV approach for the Tullahoma, Tenn., Regional Airport.

Huntsville (Ala.), contacting HSV approach for vectors for an ILS.

Point 5: Shoot an ILS approach at HSV.

Point 6: Depart HSV, tracking by VOR and then ADF to Winchester (Boiling Fork NDB—BGF *263*).

Point 7: Make an NDB approach to Winchester Municipal Airport, executing a missed approach. You could fly direct (VFR) to Franklin County Airport (*1*) or vector the trainee for a "PAR approach" under your guidance to the airport (if he isn't too fatigued by this time).

Needless to say, the route and approaches cited here would take a considerable amount of time and effort and wouldn't be used as a first cross-country. Note that the trainee would get cross-country tracking using VOR and NDB and would make VOR, ILS, and NDB approaches. As noted earlier, take a look at *your* chart to see some combinations for the best training steps during the cross-country phase of instruction.

Fig. 25-23 shows the three approaches done on the sample trip. You might want to set up a packet for each of your planned training trips with the applicable en route chart(s), approach plates, flight log, and flight plan forms, at least for a few of the more simple routes and approaches.

So at this later stage of training the whole routine of planning, flying, and approaching should be done. As indicated earlier, on at least one of these flights a Center visit and/

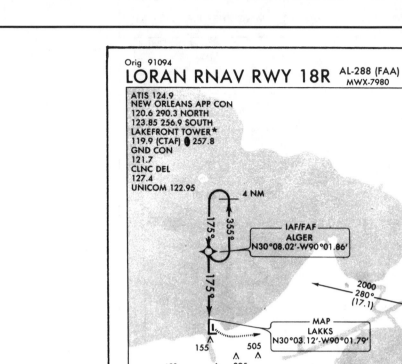

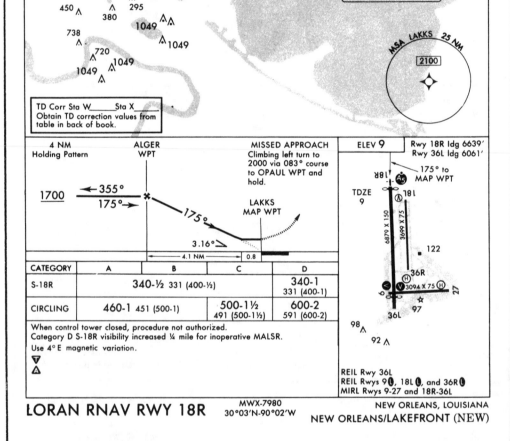

Orig 91094

LORAN RNAV RWY 18R

AL-288 (FAA)
MWX-7980

NEW ORLEANS/LAKEFRONT (NEW)
NEW ORLEANS, LOUISIANA

ATIS 124.9
NEW ORLEANS APP CON
120.6 290.3 NORTH
123.85 256.9 SOUTH
LAKEFRONT TOWER ★
119.9 (CTAF) 257.8
GND CON
121.7
CLNC DEL
127.4
UNICOM 122.95

IAF/FAF
ALGER
N30°08.02'-W90°01.86'

MAP
LAKKS
N30°03.12'-W90°01.79'

OPAUL
N30°03.83'-W89°42.77'

2000
280°
(17.1)

083°
263°

MSA LAKKS 25 NM
2100

TD Corr Sta W_____Sta X_____
Obtain TD correction values from
table in back of book.

CATEGORY	A	B	C	D
S-18R	340-½ 331 (400-½)			340-1 331 (400-1)
CIRCLING	460-1 451 (500-1)		500-1½ 491 (500-1½)	600-2 591 (600-2)

When control tower closed, procedure not authorized.
Category D S-18R visibility increased ¼ mile for inoperative MALSR.
Use 4° E magnetic variation.

4 NM
Holding Pattern

1700

355°
175°

ALGER
WPT

175°

LAKKS
MAP WPT

3.16°

4.1 NM   0.8

MISSED APPROACH
Climbing left turn to
2000 via 083° course
to OPAUL WPT and
hold.

ELEV 9

Rwy 18R ldg 6639'
Rwy 36L ldg 6061'

175° to
MAP WPT

TDZE 9

6870 X 150
3699 X 75
3094 X 75

REIL Rwy 36L
REIL Rwys 9, 18L, and 36R
MIRL Rwys 9-27 and 18R-36L

LORAN RNAV RWY 18R

MWX-7980
30°03'N-90°02'W

NEW ORLEANS, LOUISIANA
NEW ORLEANS/LAKEFRONT (NEW)

Fig. 25-21. A LORAN RNAV approach for the Lakefront Airport at New Orleans. There are only a few of these types of approaches available at this printing, but the numbers are expected to increase as more flight testing is done.

Fig. 25-22. A sample cross-country from an airport with no instrument approach facilities. The text explains the route and numbers.

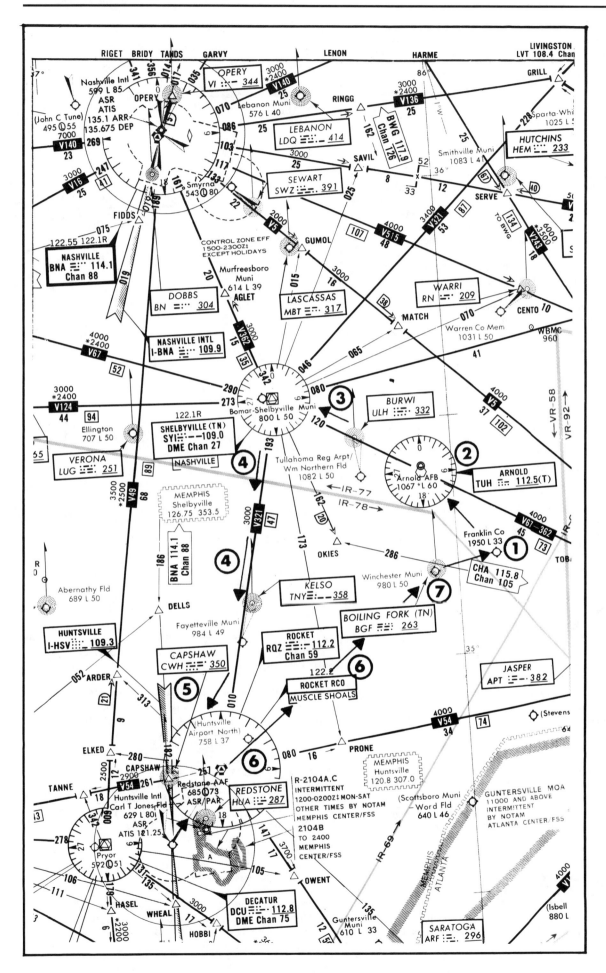

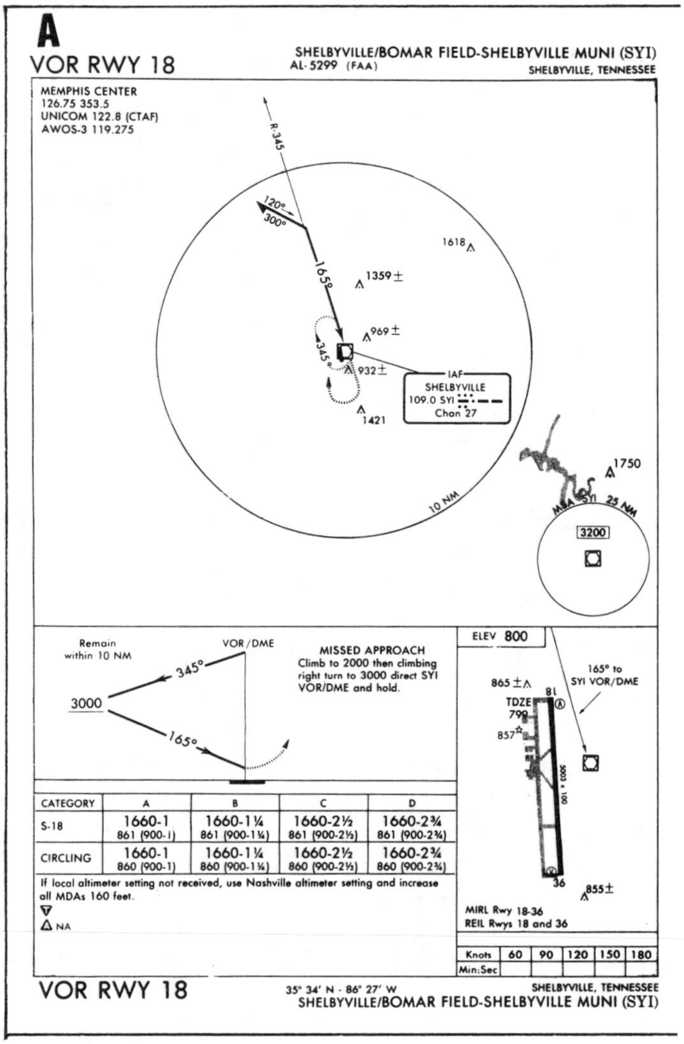

Fig. 25-23. The three approaches mentioned for the example cross-country in Fig. 25-22. (A) A VOR 18 approach at Shelbyville. (B) An ILS RWY 18L at Huntsville. This might be a little complicated for an early part of the training, but is presented as a possibility. (C) An NDB approach to Winchester.

or a Weather Service Office or Flight Service Station stop should be made assuming that these facilities aren't available at "home."

One of the cross-country flights should be at least 250 NM in simulated or actual IFR conditions on Federal Airways or as routed by ATC, including VOR, ADF, and ILS approaches at different airports.

In these cross-countries, as in those in the preprivate certificate instruction, you should get flights to as many different airports as possible.

Following are some major items you should consider on the cross-country flight, as applicable:

Preflight planning ____
Preflight inspection ____
Clearance readback ____
Instrument takeoff ____
Posttakeoff procedure ____
En route:
    Altitude control ____
    Tracking ____
    Communications procedures ____
    ETA accuracy ____
    Holding ____
ILS approach:
    Preapproach procedures ____
    Procedure turn ____
    Final-course tracking ____
    Glide path control ____
    Use of ADF ____
    Missed approach procedure ____
VOR approach:
    Preapproach procedures ____
    Procedure turn ____
    Final-course tracking ____
    Descent on final ____
    Use of clock ____
    Minimum descent altitude (recognition) ____
ADF approach:
    Preapproach procedures ____
    Orientation ____
    Procedure turn ____
    Tracking ____
    Descent on final ____
    Use of clock ____
    Missed approach procedure ____
Approach communications ____

The trainee should not take the Knowledge Test until he/she has had at least one cross-country and has used the ATC system from start to finish.

PROGRESS CHECK AND RECOMMENDATION FLIGHT

As the trainee approaches the level of skill and has the hours to take the flight test (and the written test has been passed), you should use one or perhaps two check rides to review all of the material covered and to see if he is ready for recommendation. The information is listed as an aid to checking the requirements of the flight test. Under the current testing procedures the material may not be required in the order listed (at the check pilot's discretion), but the trainee should be knowledgeable in the areas mentioned below in preparation for the flight test.

■ **Oral Operational Test.** On the flight test the trainee may be quizzed on instrument flight planning; preparing and filing of an instrument flight plan; aircraft performance, range, and fuel requirements; and required instruments and equipment and their proper use.

INSTRUMENT FLIGHT PLANNING

Check and analysis of weather information ____
Check of NOTAMs ____
Flight log preparation ____
Knowledge of en route charts ____
Knowledge of approach charts ____
Equipment available (charts, computers, hood, *Airport/Facility Directories,* etc.) ____

REQUIRED INSTRUMENTS AND EQUIPMENT. The trainee should be asked to explain the use of required instruments, avionics equipment, and any special systems installed in the

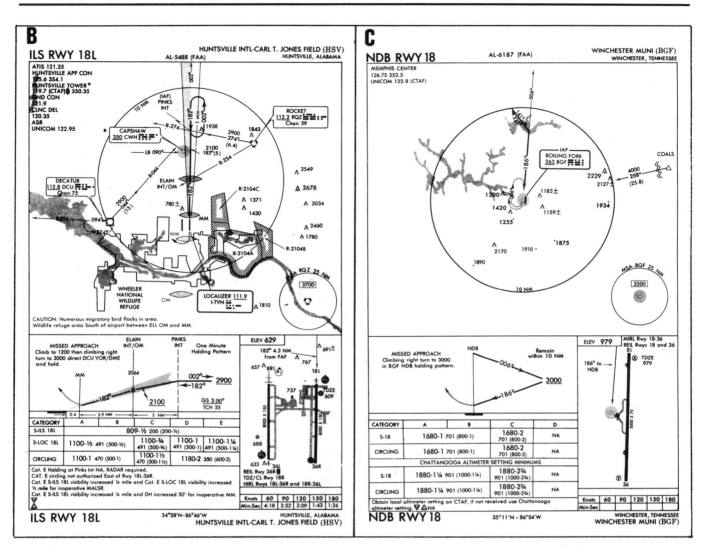

airplane being used. He should be able to explain and use all equipment.

The trainee should know the date of the last check of the altimeter/static system for the airplane and should be able to run a VOR accuracy check on the ground or in the air.

The limitations of the various flight instruments and systems should be known. (Failure to check the operation of flight instruments before taking off is disqualifying.)

He may be checked on knowledge of the following flight instrument limitations and/or malfunction corrections:

Gyro system:
    Altitude indicator operations and limitations ____
    Heading indicator operations and limitations ____
    Turn and slip or turn coordinator ____
    Static system ____
    Altimeter ____
    Airspeed ____
    Vertical speed indicator ____
    Magnetic compass ____
Avionic system malfunctions:
    Reports to be made to ATC (avionics loss) ____

AIRCRAFT NUMBERS—PERFORMANCE, RANGE, AND FUEL REQUIREMENTS. The trainee should have the information on the airplane limitations as well as performance, range, and endurance data (or know exactly where to obtain it) and apply it effectively in planning an instrument flight. The *Pilot's Op-*

erating Handbook and other sources may be used as references to confirm:

Stall speed at max weight (approach and landing configurations) ____
Max gear extension speed ____
Max flap extension speed(s) ____
Max certificated weight ____
Useful load ____
Turbulence penetration speeds ____
Range at 65 and 75 percent, various altitudes ____
Best rate of climb speed (IAS) at various altitudes ____
Recommended speeds for max endurance ____
Total fuel capacity ____
Gallons of usable fuel ____
Type of fuel required (80-, 100LL, etc.) ____
Oil type required ____
Location of avionics circuit breakers or fuses ____
Physical limitations of the equipment (line of sight, etc.) ____
Location of various antennas and equipment each serves ____
Deicer and anti-icing equipment (if applicable):
    Location and volume of fluid reservoir ____
    Deicer boot operations ____

PREPARING AND FILING AN INSTRUMENT FLIGHT PLAN. The flight plan may be filed by telephone or in person at the FSS. The trainee should accept and read back clearances and request and accept amended clearances in flight.

Knowledge of flight plan requirements (including alternate airports, fuel, etc.) ____
Preparation of the flight plan ____
Clearance readback ____

■ **Instrument Flying Test.** Have the trainee perform all maneuvers hooded.

STRAIGHT AND LEVEL FLIGHT (ATTITUDE AND HEADING INDICATORS COVERED). Expected tolerances will be ±100 ft in altitude and ±10° in heading. The stability of the aircraft used and the existing turbulence should be taken into account.

Heading control ____
Altitude control ____
Use of trim ____

TURNS, CLIMBS, AND DESCENTS (ATTITUDE AND HEADING INDICATORS COVERED). The trainee should make turns of 180° and 360° and constant-rate climbs and descents. These may be performed separately or combined with other maneuvers demonstrated during the flight check.

Climbs:
    Rate control ____
    Heading ____
    Airspeed control ____
Descents:
    Rate control ____
    Heading ____
    Airspeed ____
Turns (timed):
    Bank control ____
    Altitude control ____
    Roll-out on heading ____

Tolerances of Performance:
    Climb or descent to assigned altitude—within 10 sec of estimate
    Heading on recovery from each timed turn 20° for each 360° of turn
    Airspeed—within 10 K of assigned speed
    Any disorientation or loss of flight control is disqualifying

STALLS AND MANEUVERING AT APPROACH SPEEDS (FULL PANEL). Stalls should be done from climbs and descents in climbout and approach configurations, wings level, and turning. Recoveries should be effected to straight flight with coordinated control usage and the least loss of altitude consistent with safety.

Stalls:
    Entry ____
    Recovery ____
    Postrecovery procedures ____

Heading on stall recovery should be within 20° of entry heading.
    Straight flight and turns should be performed in climbing, descending, and level flight, in approach configuration of gear and flaps at the airspeed used for instrument approaches in the airplane used.

Maneuvering at approach speeds:
    Transition to approach speed and configuration ____
    Airspeed control ____
    Altitude control ____
    Heading ____

Use of trim ____
Procedures ____

Tolerances of performance:
    Altitude— ± 100 ft
    Approach speed— ±10 K
    Altitude during transition— ±100 ft
    Heading during transition— ±10°

STEEP TURNS (FULL PANEL). Turns of at least 45° bank at a speed of not more than the maneuvering speed should be made for at least 360°. A stable rate of turn and degree of bank should be held:

Altitude control ____
Constant bank ____
Entry and roll-out precision ____

Tolerances of performance:
    Altitude— ± 100 ft
    Assigned recovery headings— ±10°

RECOVERIES FROM UNUSUAL ATTITUDES (ATTITUDE AND HEADING INDICATORS COVERED). The trainee is expected to recover expeditiously and safely from unusual attitudes set up by the check pilot. The trainee will resume flight in the configuration assigned using proper power and trim settings.
    If the check pilot has to take over to prevent exceeding the airplane's operating limitations, it is considered disqualifying:

Recognition of unusual attitude ____
Spiral recovery ____
Approach-to-stall recovery ____
Postrecovery procedures ____

ENGINE-OUT PROCEDURES (MULTIENGINE AIRPLANE—FULL PANEL). The trainee will demonstrate hooded flight with a simulated engine loss (or actual feathering, if conditions permit). During the retrimming operation heading should be maintained within ±20° and altitude within 100 ft of prechosen values:

Promptness of proper action ____
Heading control ____
Altitude control ____
Procedures ____

■ **Radio Navigation and Approach Procedures**
RADIO NAVIGATION AND ORIENTATION. The trainee will be expected to demonstrate the use of radio navigation aids for a simulated IFR flight. The trainee will be expected to be able to pinpoint his position by use of cross bearings or by a time and distance computation based on change in relative bearings. During cruising flight the altitude should be held within 100 ft of that selected.

VOICE COMMUNICATIONS. The trainee will be graded on use of communications in normal IFR operations. On the flight check by the FAA examiner actual communications with aeronautical ground stations rather than simulation will be used, if possible:

Proper frequencies ____
Accuracy and clarity ____
Proper phraseology ____

STANDARD INSTRUMENT APPROACHES, INCLUDING HOLDING. The trainee will be expected to do ILS (front and back, and localizer) approaches, VOR and ADF approaches (including circling approaches), and stay within prescribed limits.

Altitude errors below prescribed altitudes during initial approach or after passing the final fix should be considered disqualifying for recommendation.

Holding patterns should conform to standard procedures or clearances:

Preapproach procedures ____
Holding ____
Procedure turn (if required) ____
Altitude control ____
Heading control ____
Final approach:
    Tracking ____
    Altitude control ____
    Use of clock ____

MISSED APPROACH PROCEDURES. At least once during the recommendation flight the trainee should be requested to execute a missed approach using the procedure prescribed for the airport involved. All radio reports, contacts, and clearances will be demonstrated in addition to the appropriate maneuver. Any descent below the authorized approach minimum altitude will be disqualifying unless the check pilot advises him that the field is in sight:

Recognition of the situation ____
Altitude control ____
Heading control ____
Initial execution (power, cleanup, etc.) ____
Procedure after initial execution ____
Voice reports ____
Decision (alternate, new approach) ____

EMERGENCIES (RADIO OR INSTRUMENT MALFUNCTIONS). The check pilot may induce simulated malfunctions of radio, instruments, and other systems. The trainee should identify the problem promptly and take appropriate action. The use of alternate air sources and the resetting of circuit breakers are examples of the types of action required:

Recognition of problem ____
Appropriate action ____
Reaction to emergency ____
Control of airplane during emergency ____
Heading control ____
Altitude ____

COMPLIANCE WITH ATC INSTRUCTIONS AND PROCEDURES. The FAA flight test itself will be conducted in accordance with an IFR flight clearance. The check pilot may simulate ATC if conditions warrant. The trainee should react to all clearances as if on an IFR flight (some cannot be accepted due to equipment limitations or other factors):

Accuracy of clearance compliance ____
Acknowledgment ____
Familiarity with ATC procedures ____

Upon completion of this flight the check pilot will recommend the trainee for (1) the FAA instrument rating check ride or (2) extra time and recheck.

1. Recommended for FAA flight test
    Instructor _____
    Checklist:
        Current medical and pilot certificate ____
        Written test grade slip ____

        Application form for flight test complete (recommendation signed) ____
    Logbooks complete ____
2. Needs extra time and recheck
    Instructor _____

This checklist was based on an overall check of the trainee's readiness to take the FAA flight test; the FAA check pilot may use only a few of the maneuvers listed or may choose some variations to check knowledge of the fundamentals. See the latest Practical Test Standards for specifics of the flight test and for a Practical Test Checklist.

MORE ABOUT USING A FLIGHT SIMULATOR

The minimum of 40 hr of training required for the instrument rating can include up to 20 hr of ground trainer time given by an authorized instructor. The flight simulator or a flight training device is a valuable aid in instrument training, and if you have a chance to use one in setting up your syllabus you should do so. While the whole idea here is to teach a person to fly an *airplane,* the simulator has an advantage over the airplane under certain conditions. If your school is really going in for instrument training, an FAA-approved simulator or flight training device can be a good investment because of the variety of approaches and cross-country navigation work available. To equip an airplane with equal avionics to do all of the approaches would be a sizable investment.

Some of the older refresher pilots you'll be instructing will have a look of real dismay when you tell them that part of the training will consist of time in a ground trainer. They're remembering the World War II–type ground trainers they flew that wheezed and jerked around, with little realism. The newer models can be integrated with the airplane training to make a good syllabus for the new instrument pilot. You'll have to set up the syllabus carefully; it's not just a matter of simulator, airplane, simulator, airplane, etc., until the 40 hr are filled up. In some cases, you'll introduce new material in the airplane and use the simulator to work out techniques and procedures. On the other hand, in certain parts of the syllabus you would use the simulator first to make sure the trainee knows what is expected before going out and coping with, say, ILS approaches in traffic. The biggest advantages of the simulator are that it can be used in an environment without noise and *too* much pressure and it can be stopped at various points so you can discuss any problems and then "fly" on.

However, it doesn't matter how well the trainee can fly the simulator or flight training device; if he can't fly the airplane it's all wasted. If the trainee can fly the simulator but not the airplane, his instrument rating could always be stamped "Instrument Rating—Ground Operations Only."

SUMMARY

No doubt it appears that basic instrument flying (Chap. 24) was given more coverage than cross-country and approach work. You'll be grading the trainee's basic instrument flying all during the instruction given in this chapter, however; including it again would make *this* chapter the long one. Also, it's easier to go into detail on basic instrument flying because the exercises can be set up exactly, whereas the proximity and types of navigation and approach aids vary widely from airport to airport.

The main thing is to parallel both ground and flight instruction so that the maximum learning conditions exist. Again, don't let the trainee's bad habits get a chance to "set." And keep *your* cool!

PUTTING IT ALL TOGETHER

GETTING READY FOR THE FLIGHT INSTRUCTOR'S CERTIFICATE

As with other certificates, you'll have to pass the Knowledge and Practical Tests; and there will be much emphasis on teaching techniques and how people learn.

Earlier in the book it was suggested that you start a library when working on the flight instructor's certificate. You should order Advisory Circular AC-00-2.5, or its latest equivalent, from U.S. Department of Transportation, Utilization and Storage Section M433.2, Washington, D.C. 20590. This free publication is a checklist of current Advisory Circulars as of a certain date. If you subscribe, you'll be sent the latest AC-00-2.5 one to two times a year.

The Bibliography lists some books that could be put in your library, and there will be others coming along throughout your career.

■ **Knowledge Test.** The current Knowledge Test airplane rating is based on AC 60-14, *Aviation Instructor's Handbook,* which at this printing is undergoing revision. The questions on the Knowledge Test are based on the *Flight and Ground Instructor Written Test Book,* FAA-T-8080-18A (expired as a book September 1, 1995), which is supplemented by periodic computerized test question updates. All Knowledge Tests are done by computer at selected computer testing center locations. Lists may be obtained through Fed World (703) 321-8020 in the FAA library file names TST_SITE. The latest question supplements to the original Written Test Book may be obtained through Fed World sources.

For the flight instructor certificate with an instrument rating (airplane) you'll need to have already passed the Fundamentals of Instructing Knowledge Test. The Knowledge Test for this rating is based on the *Instrument Rating Written Test Book,* FAA-T-8080-20A (expired as a book on September 1, 1995, but with latest questions updated on the computer system by Fed World).

Your flight instructor will have the latest information available.

■ **Practical Test.** For the Practical Test, you should have the latest issue of FAA-S-8081-6 (A, B, or C, etc.), which lays out the requirements for the flight instructor (airplane), and FAA-S-8081-9 (A, B, or C), which gives the Practical Test requirements for the instrument instructor (airplanes and helicopters).

By this time you've had enough experience in taking flight tests so that you'll know *what papers are needed,* know that checkitis is seldom fatal, and also are well aware that the oral part of the practical test can be very important in setting the scene for the rest of the examination.

You can practice instructing with people who are at a particular stage of training (under *your* instructor's supervision), such as a nonpilot in his very first flight and a private pilot with no instrument experience in that first instrument instruction flight, etc., as noted in Chap. 2.

AFTER THE FLIGHT TEST

No matter how much experience you had before getting the certificate, you'll really start learning about flying when you are instructing. Working with students will lead to some situations (good and bad) you may not have encountered before.

Attend a flight instructor's seminar once a year to keep up with the latest FARs and instructing techniques.

There may be a period in your experience (usually after 300 or 400 hr of instructing) when you may feel very superior to those "private pilot clowns" who make lousy takeoffs and landings in an airplane you're flying very well every day. That's the difference—the amount of time flown; as you gain more instructing experience you'll start wondering how some of them fly as well as they do for the hours put in.

■ **Flight Reviews.** You'll do flight reviews and should take an easy approach to each one. Remember that they are *reviews,* not flight checks, and also that most of your local private pilots will not be looking forward to the procedure with anticipation. How you handle the situation can mean the difference as to whether a low-time private pilot who doesn't fly very often will take the review or give it up. A certain amount of dual required within a given time period might be better than the review type of approach, so that the instructor will have a lever to help a pilot when it's needed. ("Charlie, you have to get two hours of dual in the next twelve months anyway, so why don't we use some of that time working on those crosswind landings of yours.") Some of the people you'll review will be nervous and defensive; do your pre- and postflight discussions out of earshot of other people.

The review should consist of three parts, for any lesson.

PREFLIGHT DISCUSSION. Review FAR Parts 61 and 91 and recent changes, or Notices of Proposed Rule Making. Again, give the pilot a chance to ask any questions in private. Many people are afraid to ask questions where they might be overheard by others, particularly if they're certificated pilots and the question is one that should have been asked when they were students. Discuss the items to be covered in the flight portion and let him bring up any points or make suggestions as to what *he* feels he needs to work on.

THE FLIGHT. Even though the pilot has a commercial certificate, he may not have done a chandelle or lazy eight for 15 years, *so generally, with some exceptions noted later, you'll stick to basics on the review.* You could waste a lot of time getting him smooth in those maneuvers when actually he's shaky on stalls or slow flight. Sure, you can let him try a chandelle just as a change of pace, but move on to the fundamentals. Following are some basic items you might take a look at:

1. *Preflight check.* Stay with him on the check and discuss the items as he goes around the airplane. The check should be thorough and he should be *looking* at the airplane. A lot of pilots walk around the airplane but really don't see it.

Others, you'll find, literally "kick a tire and light the fire" and need reinstruction on how to preflight an airplane. If this is the case, stop everything and make sure he knows what to do before going on with the flight. Some pilots tie their airplanes down for a week and then jump right in and fly without knowing whether someone has stolen the fuel or whether rain might have contributed a great deal of water to the fuel tanks. (In an actual case, a pilot ran out of fuel shortly after leaving an airport, landed in a field, with resulting injuries to members of his family, as well as his airplane. He had either flown more since the last fueling than he figured, or someone had stolen fuel while the plane was tied down. Either way, he didn't look in the tanks before flying, so he didn't know.)

2. *Starting.* Does he use a checklist and does he check to make sure the prop is clear before starting?

3. *Taxiing.* Watch for too fast taxiing or brake riding.

4. *Pretakeoff check.* He should use a checklist and *be aware;* some people automatically go down a checklist and might as well not use one. *Safety* is the primary thing you'll be looking for.

5. *Takeoffs.* If he nearly kills you both running off the runway or taxiing out in front of landing traffic, he needs work.

6. *Pattern work.* Check for good *clearing procedures* during the pattern, and work on any sloppy entries, legs, or departures.

7. *Airwork.* Slow flight, 720° power turns, and stalls are a good exercise for pilots who haven't had instruction since getting the private (or commercial) certificate many years ago. A good move in reviewing a private pilot might be to introduce and let him (briefly) practice an advanced maneuver such as a chandelle or on-pylon eight to encourage him to work on his flying. This reinforces the fact that it is an instruction and review flight rather than a flight check and he's gotten something for his money. Mentioning in the preflight discussion that such a new maneuver will be introduced can defuse any resentment about the review. With some pilots, in a properly certificated airplane, you might even discuss and demonstrate a spin if they haven't done one. However, the major goal is to take a look at the pilot's fundamental flying and safety practices. (Introducing an advanced maneuver can let you see his reaction to a new situation and give a check on his judgment in a strange situation, too.)

8. *Hood work.* A short session of hood work with emphasis on the Four Fundamentals (the 180° turn, or whatever turn and descent and/or climb is needed to get out of the weather) and recoveries from unusual situations would be good. You might act as approach control and have him track to or from a VOR and/or vector him back to the airport.

9. *Landings.* Try to get some crosswind takeoff and landing practice and a no-flap approach and landing. You'll probably notice now as was indicated in an earlier chapter that many private and commercial pilots tend to "putter around" opening windows, turning off rotating beacons, etc., during the landing roll, a point of the flight where a large percentage of accidents occur. Discuss this *after* the airplane has been parked and the engine shut down.

POSTFLIGHT DISCUSSION. Review the flight and make suggestions concerning his flying and safety habits, and if you feel it necessary, suggest dual and/or solo practice in certain areas of flying. He may get enthusiastic about going on to the next certificate after being introduced to an advanced maneuver or two and you should encourage this, since he'll be working to become a safer and more precise pilot. But, again, the primary purpose of this flight is to review the basics. Since this is a review flight he can't be "failed," and he should be aware of this from the beginning. When the review was introduced it was felt that a few instructors might take advantage of the review flights and "require" more dual to make extra money, but such people would soon be found out and their business would start falling off. Use your own judgment about added dual; some instructors offer to give an extra flight without charge, if needed, but if you honestly feel that extra dual time is necessary for his safety, recommend it, and charge for your time with a clear conscience.

Even the few who start off resenting the review flight will be glad they did it afterward—if *you* approach it in the right way.

. . . AND LONG AFTERWARD

You'll realize that being a flight instructor was really worth all the work when some years from now you get on a 767 (or its future rocket equivalent) and see a familiar name above the flight deck door, or hear the flight attendant announce that "Your Captain is Hosea Z. Flyte." You know that there can only be one person with a name like that and you go back some years, remembering the airport kid who washed and fueled airplanes and who was your student from the first flight and how you worked with him. He did a good job on the private flight check and went on with his learning. (You must have answered a million questions before he left to fly at Pensacola.) It's human nature—you'll have to tell your seatmate or a flight attendant that *you* were his first instructor.

It's a good feeling.

FAA APPLICATION FOR AN AIRMAN CERTIFICATE AND/OR RATING

TYPE OR PRINT ALL ENTRIES IN INK

Form Approved OMB No: 2120-0021

Airman Certificate and/or Rating Application

U.S. Department of Transportation
Federal Aviation Administration

I Application Information

- ☐ Student
- ☐ Recreational
- ☐ Private
- ☐ Commercial
- ☐ Airline Transport

- ☐ Instrument
- ☐ Additional Aircraft Rating
- ☐ Airplane Single-Engine
- ☐ Airplane Multiengine
- ☐ Rotorcraft

- ☐ Glider
- ☐ Lighter-Than-Air
- ☐ Flight Instructor _____ Initial _____ Renewal
- ☐ Flight Instructor Reinstatement
- ☐ Additional Instructor Rating

- ☐ Ground Instructor
- ☐ Medical Flight Test
- ☐ Reexamination
- ☐ Reissuance of _____ Certificate
- ☐ Other _____

A. Name *(First, Middle, Last)*	B. SSN	C. Date of Birth Mo. Day Year	D. Place of Birth

E. Address	F. Nationality ☐ USA ☐ Other_____ Specify	G. Do you read, speak and understand English? ☐ Yes ☐ No

City, State, Zip Code	H. Height In.	I. Weight Lbs.	J. Hair	K. Eyes	L. Sex ☐ Male ☐ Female

M. Do you now hold, or have your ever held An FAA Pilot Certificate? ☐ Yes ☐ No	If yes, has certificate ever been Suspended or Revoked ☐ Yes _____ Date ☐ No	N. Grade Pilot Certificate	O. Certificate Number	P. Date Issued

Q. Do you hold a Medical Certificate? ☐ Yes ☐ No	R. Class of Certificate	S. Date Issued	T. Name of Examiner

U. Have you ever been convicted for violation of any Federal or State statutes pertaining to narcotic drugs, marijuana, and depressant or stimulant drugs or substances, or motor vehicle operation involving alcohol related offenses? ☐ No ☐ Yes	V. Date of Final Conviction

Glider or Free Balloon Pilots only:	*Medical Statement: I have no known physical defect which makes me unable to pilot a glider or free balloon*	W. Signature	X. Date

II Certificate or Rating Applied For on Basis of:

☐ A. Completion of Required Test	1. Aircraft to be used *(if flight test required)*		2a Total time in this aircraft hours	2b Pilot in command hours
☐ B. Military Competence Obtained in	1 Service	2 Date Rated	3 Rank or Grade and Service Number	
	4 Has flown at least 10 hours as pilot in command during the past 12 months in the following military aircraft.			
☐ C. Graduate of Approved Course	1 Name and Location of Training Agency			2 Agency School Number
	3 Curriculum From Which Graduated			4 Date
☐ D. Holder of Foreign License Issued By	1 Country	2 Grade of License		3 Number
	4 Ratings			

III Record of Pilot time *(Do not write in the shaded areas.)*

	Total	Instruction Rec'd	Solo	Pilot in Command	Second in Command	Cross Country Instruction Received	Cross Country Solo	Cross Country Pilot in Command	Instrument	Night Instr Rec'd	Night Take-off/ Landing	Night Pilot in Command	Night Takeoff/ Landing Pilot in Command	Number of Flights	Number of Aero-Tows	Number of Ground Launches	Number of Powered Launches	Number of Free Flights
Airplane																		
Rotor-Craft																		
Gliders																		
Lighter than Air																		
Training Device Simulator																		

IV Have you failed a test for this certificate or rating within the past 30 days? ☐ Yes ☐ No

V Applicant's Certification I certify that the statements made by me on this application are true.	A. Signature	B. Date

FAA Use Only

EMP	REG	D.O.	SEAL	CON	ISS	ACT	LEV	TR	S.H.	SRCH	#RTE		RATING (1)	

FAA Form 8710-1 (6-89) Supersedes Previous Edition

Instructor's Recommendation

I have personally instructed the applicant and consider this person ready to take the test.

Date	Instructor's Signature	Certificate No:	Certificate Expires

Air Agency's Recommendation

The applicant has successfully completed our _____ course, and is
recommended for certification or rating without further _____ test.

Date	Agency Name and Number	Official's Signature
		Title

Designated Examiner's Report

☐ Student Pilot Certificate Issued *(Copy attached)*

☐ I have personally reviewed this applicant's pilot logbook, and certify that the individual meets the pertinent requirements of FAR 61 for the pilot certificate or rating sought.

☐ I have personally reviewed this applicant's graduation certificate, and found it to be appropriate and in order, and have returned the certificate.

☐ I have personally tested this applicant in accordance with pertinent procedures and standards, with the result indicated below.

　　　☐ Approved—Temporary Certificate Issued *(Copy Attached)*

　　　☐ Disapproved—Disapproval Notice Issued *(Copy Attached)*

Location of Test *(Facility, City, State)*	Duration of Test	
	Ground	Flight

Certificate or Rating for Which Tested	Type(s) of Aircraft Used	Registration No.(s)

Date	Examiner's Signature	Certificate No.	Designation No.	Designation Expires

Evaluator's Record For Airline Transport Certificate/Rating Only

	Inspector	Examiner	Signature	Date
Oral	☐	☐		
Simulator Check	☐	☐		
Aircraft Flight Check	☐	☐		

Inspector's Report

I have personally tested this applicant in accordance with pertinent procedures and standards, with the result indicated below.

　　　☐ **Approved**—Temporary Certificate Issued　　☐ **Disapproved**—Disapproval Notice Issued

Location of Test *(Facility, City, State)*	Duration of Test	
	Ground	Flight

Certificate or Rating for Which Tested	Type(s) of Aircraft Used	Registration No.(s)

☐ Student Pilot Certificate issued　　☐ Certificate or Rating Based on　　☐ Instructor　☐ Flight　☐ Ground
☐ Examiner's Recommendation　　☐ Military Competence　　☐ Renewal　☐ Approved
　☐ ACCEPTED　☐ REJECTED　　☐ Foreign License　　☐ Reinstatement　☐ Disapproved
☐ Examiner Recommends Retesting　　☐ Approved Course Graduate　　**Instructor Renewal Based on**
☐ Reissue or Exchange of Pilot Certificate　　☐ Issued　　☐ Activity　☐ Training Course
☐ Special medical test conducted—report forwarded to Aeromedical Certification Branch, AAM-130　　☐ Denied　　☐ Acquaintance　☐ Test

Training Course (FIRC) Name	Graduation Certificate No.	Date

Date	Inspector's Signature	FAA District Office

Attachments:

☐ Student Pilot Certificate (copy)　☐ Airmans Identification (ID)　　☐ Notice of Disapproval
☐ Report of Written Examination　　　　　　　　　　　　　☐ Superseded Pilot Certificate
☐ Temporary Pilot Certificate (copy)　Form of ID _____　☐ Answer Sheet Graded
　　　　　　　　　　　　　　　Number _____　☐ Answer Sheet Graded
　　　　　　　　　　　　　　　Expiration Date _____　(Foreign Instrument)

G EFFECTS ON THE PILOT DURING AEROBATICS (FAA)

FAA–AM–72–28

G EFFECTS ON THE PILOT DURING AEROBATICS

Stanley R. Mohler, M.D.
Office of Aviation Medicine
Federal Aviation Administration
Washington, D.C. 20591

July 1972

Prepared for
DEPARTMENT OF TRANSPORTATION
FEDERAL AVIATION ADMINISTRATION
Office of Aviation Medicine
Washington, D.C. 20591

TECHNICAL REPORT STANDARD TITLE PAGE

1. Report No.	2. Government Accession No.	3. Recipient's Catalog No.
FAA-AM-72-28		

4. Title and Subtitle	5. Report Date
G Effects on the Pilot During Aerobatics	July 1972
	6. Performing Organization Code

7. Author(s)	8. Performing Organization Report No.
Stanley R. Mohler, M.D.	

9. Performing Organization Name and Address	10. Work Unit No.
Aeromedical Applications Division Federal Aviation Administration Washington, D.C. 20591	
	11. Contract or Grant No.
	13. Type of Report and Period Covered

12. Sponsoring Agency Name and Address	
Office of Aviation Medicine Federal Aviation Administration 800 Independence Avenue, S.W. Washington, D.C. 20591	OAM Report
	14. Sponsoring Agency Code

15. Supplementary Notes

16. Abstract

Sport, precision, and competitive aerobatics, and especially air show and demonstration flying are enjoying a rebirth of interest exceeding that of the 1930's. Improved aerobatic airplanes and power plants are in the hands of more civilian pilots than ever before. These aircraft enable the pilot to easily initiate maneuvers which exceed human tolerances, yet not over-stress the aircraft. Military aircraft reached this point in World War II and the G-suit was perfected to protect the pilot. The military groups still use the G-suit but this equipment is impractical for most civil aero-batic activities. This paper provides information on (1) the nature of aerobatic G forces, (2) human physiology in relation to G forces, (3) human tolerances to various levels and times of exposure to G forces, and (4) means by which tolerance to G forces may be increased in terms of (a) the general physical condition and (b) the time during the maneuver when the G forces are imposed.

17. Key Words	18. Distribution Statement
Aerobatics G Effects	Availability is unlimited. Document may be released to the National Technical Information Service, Springfield, Virginia 22151, for sale to the public.

19. Security Classif. (of this report)	20. Security Classif. (of this page)	21. No. of Pages	22. Price
Unclassified	Unclassified	19	$3.00

Form DOT F 1700.7 (8-69)

G EFFECTS ON THE PILOT DURING AEROBATICS

Many prospective aerobatic trainees enthusiastically enter aerobatic instruction, but find their first experiences with G forces to be unanticipated and very uncomfortable. The uninformed student may actually lose consciousness at three (+) G's and incorrectly assume that he is unfit for aerobatics. If the aerobatic instructor does not have a basic understanding of the physiology of G force adaptation, he will be unable to clearly explain the phenomenon, and the student will likely be lost to further aerobatic activity.

No airplane pilot is "complete" without training in stalls, which is why proficiency in stall recovery must be demonstrated prior to solo. Similarly, unless the pilot has instrument training and an instrument rating, he is limited to fair weather operations and is a potential hazard should he get into loss-of-outside-reference conditions in haze, fog, rain, or darkness. This is why all airline transport pilot certificate applicants must demonstrate instrument proficiency. Likewise, the complete pilot is proficient in spin recovery, and demonstration of this capability is required for the instructor certificate. Many feel that the complete pilot also must have some aerobatic training, especially today when wake turbulence upsets are potentially more severe than in past years because of the introduction of large jet aircraft.

Push-over Forces

Let us assume that an aircraft is flying along a straight course with wings level and that the pilot pushes over into a 70° dive (Figure 1). The force (f) necessary to deflect the aircraft from its prior straight path through the curved path is directly proportional to the product of the mass (m) of the aircraft and the rate of change of velocity (a for acceleration) experienced. In physics this is stated by the formula $f = ma$.*

---

*Acceleration (a) is the time-rate of change of velocity. It is a vector quantity, i.e., it has direction and magnitude.

We define one (+) G as the strength of the gravitational force (which tends to accelerate a mass toward the center of the earth) all of us experience when stationary at or near the surface of the earth (note that a body is not accelerating when it is standing still or is moving at constant velocity in a straight line—if it makes a curve, even at constant velocity, it accelerates because it moves away from the straight path). This force may be expressed in terms of an individual's "weight".

If an elevator begins to move up while one is standing in it, one experiences a "heaviness" feeling during the acceleration phase. If the upward acceleration is great enough to double one's weight should he be standing on scales at this time, we define the accelerative force as two (+) G's. When the elevator reaches constant velocity as it moves up, the acceleration returns to zero and the individual is back at one (+) G. As the elevator begins to decelerate as it moves toward the top of the building (if not, it might shoot through the roof), the individual, not strapped to the floor, tends to continue going up, and the scales show less than the individual's weight (if the weight shown is half that at one (+) G; convention refers to this as an accelerative force of 0.5 (+) G). If the elevator rapidly reversed course and began accelerating downward so that the individual and the elevator fell solely by the accelerative pull of gravity, the individual would experience during this period zero G (since his body would register zero on the scales). If the

---

With a constant velocity, one can have acceleration due to a change in the direction of motion of a moving mass. The case of a uniform change in the magnitude only of velocity (as in the case of a freely falling body) is called uniformly accelerated linear motion. When the direction only is changed this is called uniform curvilinear motion or translatory motion in a circle.

Acceleration is equal to V^2 over R and is directed toward the center of the circle. R is the radius of the circle. If the path of the airplane or object does not follow a perfectly circular route, the motion is not uniform curvilinear motion.

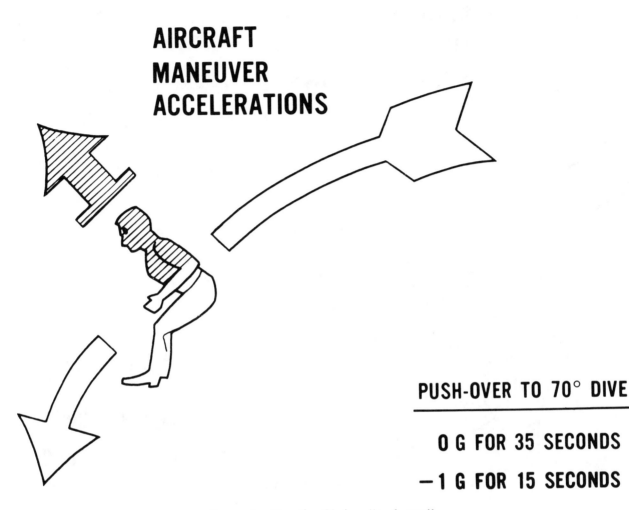

AIRCRAFT MANEUVER ACCELERATIONS

PUSH-OVER TO 70° DIVE

0 G FOR 35 SECONDS

−1 G FOR 15 SECONDS

FIGURE 1.—Negative G's in a "push-over".

elevator were forced to accelerate downward faster than the acceleration caused by the pull of gravity alone, and if the individual were tied to the elevator, so that he would be pulled with it, he could experience "negative" G's. If a spring scale were placed between the individual and the floor, the footward tug in pounds on his body could be measured, and if this tug were to be exactly equal to his weight, we would call this an acceleration of one (−) G.

Figure 1 illustrates that a push-over from straight, wings level flight to a 70° dive can produce forces on the aircraft (and the strapped-in pilot) of zero G for 35 seconds to one (−) G for 15 seconds (depending upon how hard the pilot pushes over).[13] Other G/time combinations are also possible, of course. If the pilot were not snugly strapped in, the one (−) G acceleration

would result in his departing the aircraft at a tangent to its curved flight path, and through a "parabolic arch" free-fall, reaching a terminal velocity at one (+) G (not counting the effect of wind blast, which, depending upon the pilot's speed at exit, exerts a certain drag force). At the zero G acceleration, the aircraft and the pilot are both falling solely by the pull of gravity and the pilot has the sensation of floating. Note that at one (−) G, the blood and body organs, especially the heart, liver, and intestines, tend to move toward the head of the pilot (more about this later).

Pull-up Forces

Figure 2 reveals possible forces in a 70° dive and pull-up. These are four (+) G for three seconds to six (+) G for one second. Note how

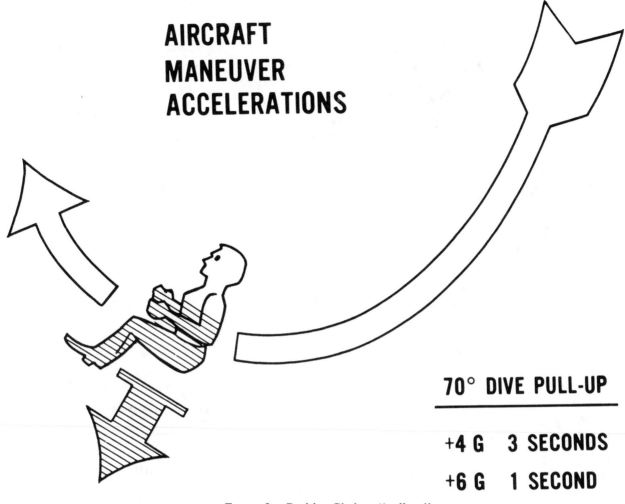

AIRCRAFT MANEUVER ACCELERATIONS

70° DIVE PULL-UP

+4 G   3 SECONDS

+6 G   1 SECOND

FIGURE 2.—Positive G's in a "pull-up".

the blood and body organs tend to pool toward the lower part of the body. Obviously, since the human brain requires essentially continuous blood circulation from the beats of the muscular pump, the heart, for maintenance of an adequate oxygen supply, and since the circulatory system is a complex network of flexible vessels of which the major vessels run lengthwise in the body, there is a physiological limit to the time the pilot can withstand these higher G forces before losing consciousness. A brief loss of consciousness in a maneuver can lead to improper control movement, causing structural failure of the aircraft or collision with another object or the terrain.

Steep Turn Forces

In steep (+) G turns, the centrifugal force tends to push the pilot through the floorboards,

and as shown in Figure 3, a steep turn of 180° change of direction will yield two (+) G for 35 seconds and if made in 15 seconds can yield five (+) G. Every private pilot has been taught that (by reason of geometry and vector forces) all 60° banked turns held at constant altitude pull two (+) G during the turn.[7,8] The difference in this maneuver between fast and slow aircraft is that the faster aircraft covers more area during the maneuver.* The same interrelationships exist in all aerobatic maneuvers, with aircraft capabilities

*Assuming constant acceleration, if one doubles the velocity of an airplane in a 60° banked turn, the radius of curvature becomes four times greater. If one were to triple the velocity, nine times greater. For a constant acceleration the radius of curvature varies as the square of the velocity. The same applies, of course, to the diameter of a loop. High speed loops take a great deal of vertical airspace.

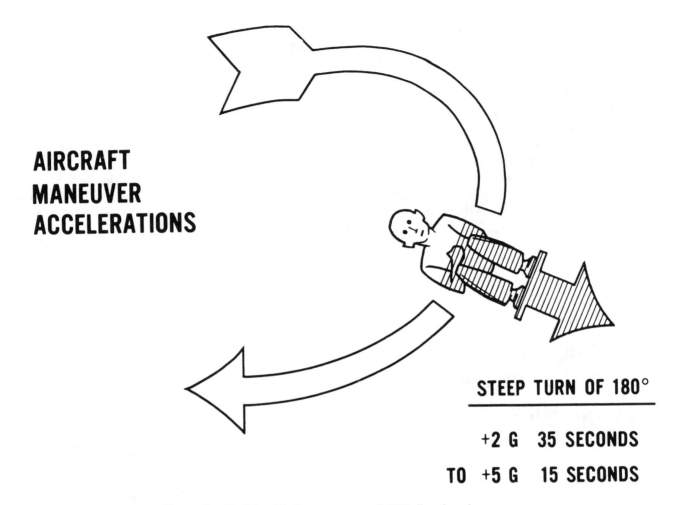

AIRCRAFT
MANEUVER
ACCELERATIONS

STEEP TURN OF 180°

+2 G 35 SECONDS

TO +5 G 15 SECONDS

FIGURE 3.—Positive G's in steep turns of 180° direction change.

varying according to design and power plant characteristics.*

Aircraft G Limits

Most aerobatic maneuvers for demonstration are variations of the loop, slow roll, and

---

*When an airplane is banked 60° with coordinated controls (left stick, left rudder) and kept at a constant flight altitude, the airplane describes a circular path to the left. The resultant vector from the center of lift of the aircraft is perpendicular to the lateral and longitudinal axes of the aircraft and is twice as long as the lift vector from the same point. The result is that a 30°/60°/90° triangle exists, having a hypotenuse of two and an opposite side (with respect to the 30° angle) of one. The sine of 30° is ½. Therefore, the force acting on the pilot "through the floorboards" in a 60° banked constant altitude turn is twice that in straight and level flight. For further details see Kershner's and Hurt's publications.

snap.[6] If the maneuver is accomplished to impose a (+) G load, it is referred to as an "inside" maneuver, while (−) G load maneuvers are termed "outside".

The FAA has established G load design limit factors for civil certified aircraft that weigh under 12,500 pounds, and these are as follows (each limit is supplemented by a times "1.5" safety factor in case of occasional accidental excess loading):

Category	+ G	− G
Normal	3.8	0.4 times 3.8
Utility	4.4	0.4 times 4.4
Acrobatic	6.0	0.5 times 6.0

The above limits are minimum design requirements under Federal Aviation Regulations

23.337 (see Code of Federal Regulations, Title 14, Aeronautics and Space, FAR Part 23, 1972).

Terminology for Plus and Negative Accelerations on the Pilot

In the same fashion as with aircraft design limits for G, the human body has upper limits, as modified by the necessity to keep the physiologic functions operational. In order that a clear delineation of the human limits to G can be presented, certain arbitrarily established parameters must be utilized. Figure 4 provides a standardized terminology for describing G forces acting on the body of a pilot during aerobatic maneuvers.[1] It will be noted that since the top of the heart elastically hangs in the chest from the aorta and certain nearby tissue sheets, the direction in which the heart moves relative to the skeleton when the whole body is accelerated determines the name given to the force. For example, an inside loop pulls (+) G's and tends to move the heart toward the pelvis, and since the long axis of the spine is referred to as the G_Z axis, this maneuver is said to pull (+) G's in the Z axis, and, if 2 G's were pulled at a point, this would be written: 2 (+) G_Z. An outside loop pulling one (−) G's at a point would be indicated by 1 (−) G at that point.

The G_Y axis runs through the shoulders and has some applications in snaps and spins (but requires a specially mounted G meter for measurement). The G_X axis runs through the chest and

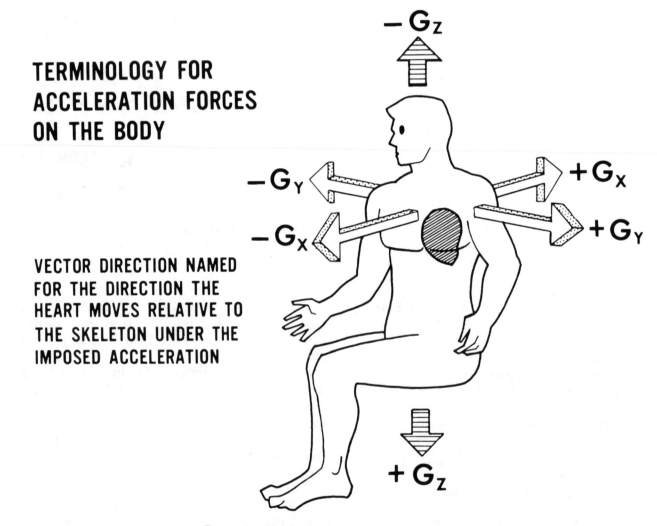

TERMINOLOGY FOR ACCELERATION FORCES ON THE BODY

VECTOR DIRECTION NAMED FOR THE DIRECTION THE HEART MOVES RELATIVE TO THE SKELETON UNDER THE IMPOSED ACCELERATION

$-G_Z$

$-G_Y$ $+G_X$

$-G_X$ $+G_Y$

$+G_Z$

FIGURE 4.—Standardized acceleration terminology.

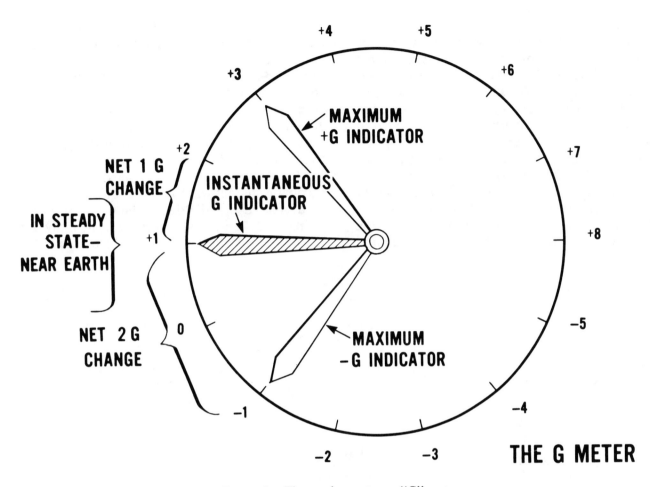

FIGURE 5.—The accelerometer or "G" meter.

has a certain bearing on tail slide maneuvers and recoveries and crash impacts, where seat belt and shoulder harness prevent excess movement toward the $(-)$ G_X direction with sudden decelerations. Physiologically, because of its design features the body can adapt much more readily during aerobatic flight to G_X and G_Y axis accelerations than with the G_Z axis; hence, we will direct our major attention here to the G_Z axis accelerations $(+)$ and $(-)$.

The G Meter

Figure 5 portrays the instrument panel G meter, which indicates the accelerative forces experienced by the aircraft, that, by virtue of the pilot's cockpit seating arrangement, are exerted through the G_Z axis of the pilot.

Note that one needle stays at the maximum $(+)$ G's experienced during a maneuver, one needle stays at the maximum $(-)$ G's, and one fluctuates continuously with the imposed ac-

celeration force (in straight and level horizontal flight this will read one $(+)$ G). An instant reset button is present allowing centering to one $(+)$ G of all three needles prior to the next maneuver.

Note also that the pilot experiences a net change of one G during a maneuver from a one $(+)$ G baseline that pulls two $(+)$ G's, while he experiences a net change of two G's during an outside maneuver starting from a one $(+)$ G baseline and going to a one $(-)$ G level.

If the pilot performs a maneuver from a one $(+)$ G baseline to 4 $(+)$ G's, his net change is 3 G's, while if he performs a 4 $(-)$ G maneuver, from the same baseline, his net change is 5 G's. Negative maneuvers of the same numerical value as positive maneuvers but with opposite sign therefore impose greater physiological net changes on the pilot. In addition, as discussed later, the body's physiologic adaptive mechanisms to combat G load effects in the G_Z axis are designed primarily to adapt to $(+)$ G_Z changes (since we evolved in a (+) G_Z environment) and function poorly against (−) G_Z accelerations.

Maneuvers

Typical aerobatic maneuvers were filmed by the author using an on-board movie camera and were also timed by a stop watch. Light aerobatic aircraft of the Beech Musketeer Aerobatic Sport III and the Beech T-34 type were used. In addition, the G forces experienced were recorded at critical points in the maneuvers. The data would be modified somewhat by other types of aircraft and by varying the force of control application.[4] Appreciation is extended to Mr. William Kershner who assisted with most of the maneuvers.

Figure 6 reveals that the inside loop in the above type aircraft takes about 15 seconds and pulls 3.5 (+) G's for 1 second at the 4–5 o'clock position and at the 7–8 o'clock position. The rest of the loop averages about 1 (+) G. The build-up to (and drop-off from) 3.5 (+) G is necessary to change the direction of the relatively fast aircraft as it enters the 6 o'clock position (f=ma) to an upward direction, and the pilot's pull-back on the stick is, accordingly, of correspondingly greater force at the 3.5 points. The pull-out entails the 3.5 (+) G level because the accelerating aircraft must be changed in direction from straight down to the horizontal against the pull of gravity.

The average "naive" person or novice aerobatic student will "gray out" (lose vision due to decreased blood flow through the brain and retina) at 3.5 (+) G's, especially if the individual doesn't expect the imposition of the accelerative

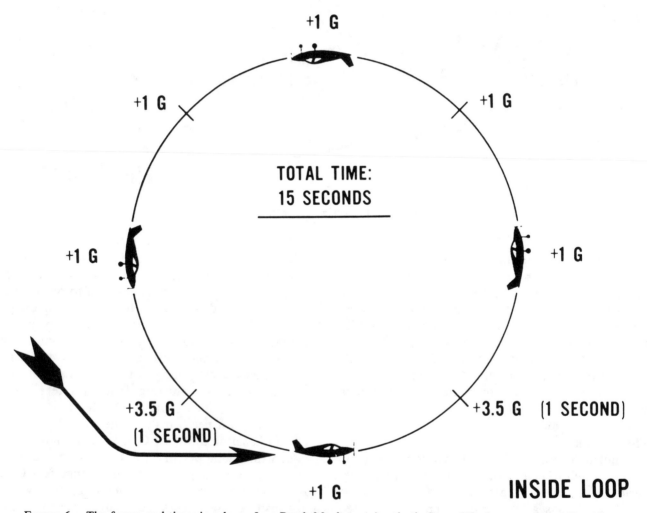

FIGURE 6.—The forces and times in a loop. In a Beech Musketeer Aerobatic Sport III, the entry is at 140 mph to get one G at the top at 70 mph.

force and is unprepared or does not know how to adapt. Accordingly, this simple maneuver can frighten the initiate if some explanation and guidance is not given by the instructor or demonstrator ahead of time. The experienced pilot in good health knows how to adapt and what to expect, and at the levels of G shown for the times given in Figure 6, derives pleasure from the physiological and optical sensations. The chances are he wants more. Years ago, for example, pilots vied with one another for the maximum number of consecutive inside loops. Charles "Speed" Holman made 1,453 consecutive loops in 1928 at St. Paul, Minnesota, taking five hours (this gives an average time of a little over 12 seconds per loop).[2]

It is noted also that, initially, persons may become airsick during aerobatics because of the unusual stimulations of the inner ear plus the anxiety of not knowing exactly what is happening. This tendency is usually rapidly lost (apparently through "habituation" of the semicircular canals) and for most persons never becomes a serious problem again.

The inside aileron roll is a maneuver between the slow-roll (which pulls zero to one ($-$) G) and the barrel roll (which is sort of a spiral loop and pulls plus G's all the way around). The aileron roll (see Figure 7) takes six seconds for completion, either to the left or the right, and reaches a maximum of 2.5 ($+$) G's.

The inside snap (Figure 8) takes three seconds to the left and pulls 2.5 ($+$) G's. Since the snap to the right in aircraft having propellers that rotate clockwise (when seen from the rear of the aircraft) is entered at a slightly higher airspeed, the snap takes 2.9 seconds and recovery pulls three ($+$) G's. The propeller spiral slip stream tends to

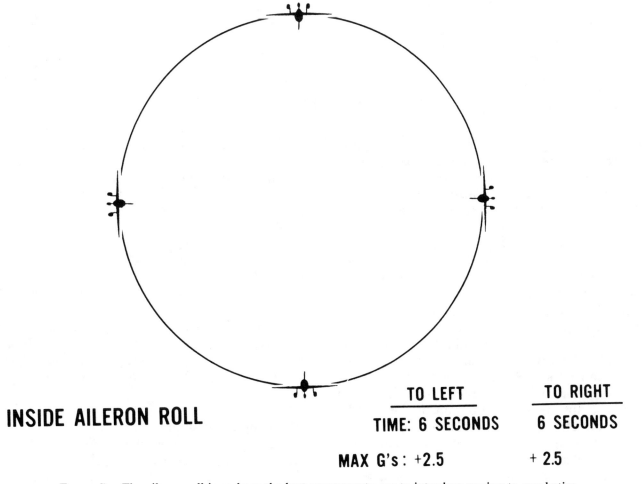

INSIDE AILERON ROLL

	TO LEFT	TO RIGHT
TIME:	6 SECONDS	6 SECONDS
MAX G's:	+2.5	+2.5

FIGURE 7.—The aileron roll is perhaps the best maneuver to use to introduce novices to aerobatics.

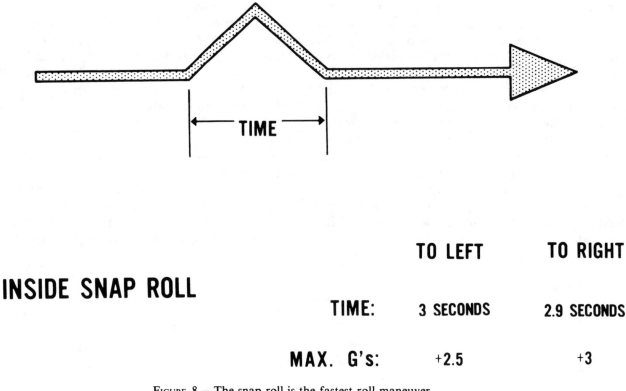

INSIDE SNAP ROLL

	TO LEFT	TO RIGHT
TIME:	3 SECONDS	2.9 SECONDS
MAX. G's:	+2.5	+3

FIGURE 8.—The snap roll is the fastest roll maneuver.

yaw the aircraft to the left, hence, the slightly higher entry speed in right snaps to compensate for this effect. This maneuver is so rapid that it is over before the novice realizes what happened. He may be confused and claim that a left snap was actually to the right.

Three turn spins (Figure 9) require a loss of altitude of about 1,100 feet in the aircraft referenced, taking 12 seconds, and leading to 3.5 (+) G's on pull-out for three seconds. Some aircraft pitch partially upside down, nose low, on entry, a frightening experience to some novices, and the increasingly rapid roll rate in certain aircraft plus the nose low attitude is also frightening if not understood. Subtracting three seconds for recovery, one can see that each spin turn is three seconds, approximately that of one snap (the snap is, in effect, a horizontal spin, entered from an accelerated stall at a speed somewhat higher than that of spin entry).

Figure 10 shows the inside square loop, which takes 24 seconds and contains four 4-second legs and four 2-second vertical ninety degree turns. Note that the abrupt change of attitude from level to nose-up and nose-down to level pull 4.2

(+) G's for as high a speed as is safely attainable by the aircraft (f = ma). This level of G's will definitely black out the unprepared person. For this short period of time, the body of a 170-pound pilot weights 714 pounds. The arms and head are four times as heavy and the novice feels very uncomfortable. A portion of the top of the square loop records zero G's in aircraft without inverted fuel flow capability and an oil system suitable for inverted flight, since the aircraft actually follows a somewhat parabolic falling maneuver because the engine cannot sustain negative G's for more than a few seconds. In aircraft with systems suitable for inverted flight, one (−) G would be recorded across most of the top of the square loop.

Physiologic Adaptation to G Forces

Figure 11 illustrates the arterial blood flow pattern as the blood leaves the heart. The heart pumps the oxygenated blood upward through the body's largest artery, the aorta. The aorta arches 180° and sends a column of blood downward to the trunk and lower limbs. The carotid arteries exit from the arch of the aorta and serve the

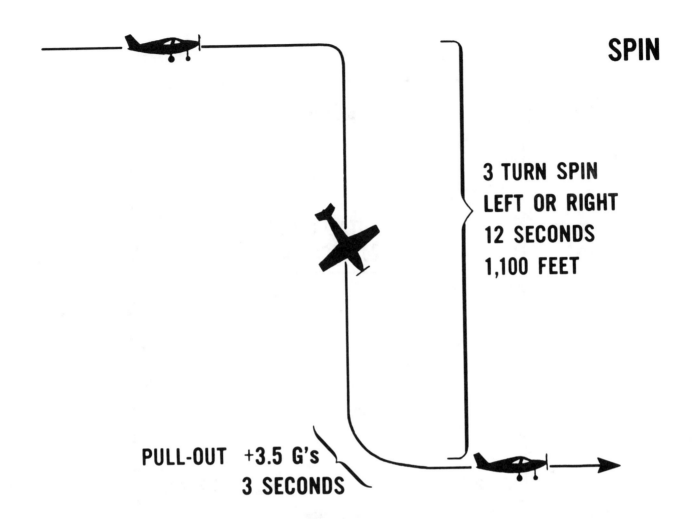

SPIN

3 TURN SPIN
LEFT OR RIGHT
12 SECONDS
1,100 FEET

PULL-OUT +3.5 G's
3 SECONDS

FIGURE 9.—Spins consume a great deal of altitude in a relatively short period. Therefore, start these at a safe altitude.

head, including the brain and eyes. The subclavian arteries also receive blood from the aorta and serve the arms.

When the mean blood pressure rises above a certain "normal" level in the aorta and carotid arteries, the diameter of these elastic vessels increases, resulting in the stretching of their walls. "Stretch receptors" in the aortic arch and carotid artery detect the extent of this stretch and send nerve signals through the "visceral" nervous system into the central nervous system where the signals are processed in the lower brain and result in outflow signals that are carried by the parasympathetic nervous system (in this case, the vagus nerve portion) to the heart. The impulse is inhibitory and the heart rate slows. The blood pressure then tends to lower toward "normal".[1]

The above reflex arc represents a "feedback" mechanism and is a "cybernetic" governor in the complete sense of the term. In addition to the adjustment, which keeps the arterial pressure from rising excessively, there is a parallel balancing mechanism that comes into play if the mean blood pressure is too slow. This causes a reflex nerve arc to send impulses through the sympathetic nervous system to the heart, resulting in a faster rate and an increase in blood pressure. The reciprocal relation between blood pressure and heart rate as controlled by the above two reflexes is known as "Marey's Law". These reflex adjustments require about five seconds to come into play. Therefore, if one experiences a rate of onset of one G per second, it is obvious that unless some other interim method is used to compensate, a force generated by five G's can be reached before physiologic reflex compensation results. In the positive G_Z direction, blackout occurs unless some other compensation mechanism

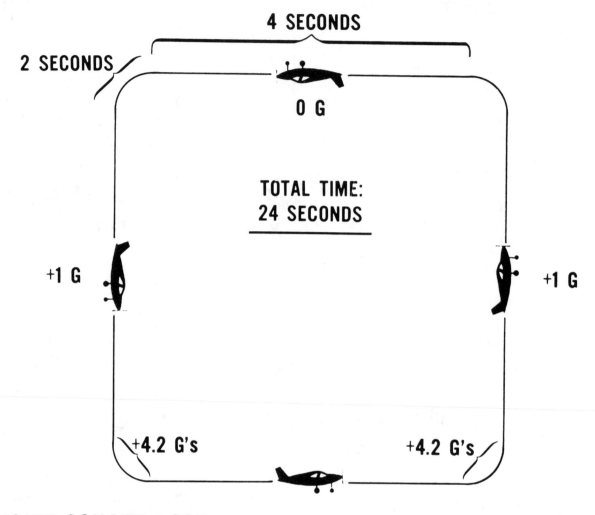

INSIDE SQUARE LOOP

FIGURE 10.—The square loop requires a high level of G's on entry and recovery.

is used. (Other compensation methods include holding the breath and forcefully exhaling and/or leaning forward to decrease the distance of the column of blood between the heart tip and the eye level (base of the brain)[1,5].

In addition to the stretch "pressure sensors" (also called baroreceptors or pressoreceptors) described above, there are certain additional regulator mechanisms including the closing or opening of the tiny arteries just before the capillary network. The vast numbers of these tiny arteries (arterioles), their sensitivity to nerve stimulation, and their critical point of location in the arterial system make them powerful additional controllers of blood pressure. Also, adrenalin released into the blood in highly stressful $(+)$ G_Z circumstances can raise blood pressure through its cardiac stimulation effect. In very stressful $(-)$ G_Z maneuvers, this should have an adverse effect by forcing the heart to send blood to an already overengorged brain.

Figure 12 shows, in the center top, a simplified schematic diagram of the one $(+)$ G_Z "normal" situation, illustrating the tip-of-the-heart to eye level distance, the position of the liver between the diaphragm and the heart, and the relative size of the aorta and major outflow arteries. Large veins (jugulars) return the blood from the head to the heart and a large vein (the vena cava) runs parallel with the aorta and returns blood from the lower limbs and trunk to the heart.

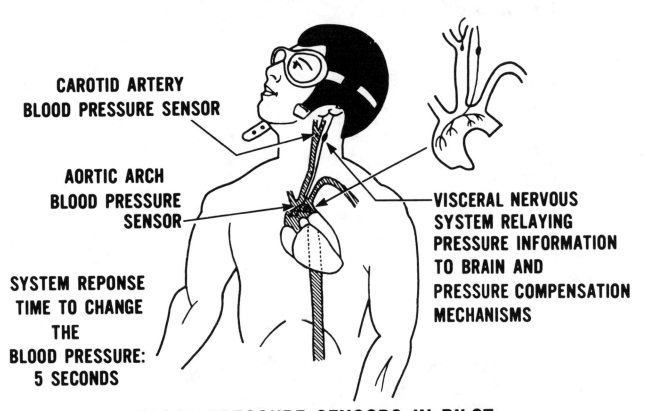

CAROTID ARTERY BLOOD PRESSURE SENSOR

AORTIC ARCH BLOOD PRESSURE SENSOR

SYSTEM REPONSE TIME TO CHANGE THE BLOOD PRESSURE: 5 SECONDS

VISCERAL NERVOUS SYSTEM RELAYING PRESSURE INFORMATION TO BRAIN AND PRESSURE COMPENSATION MECHANISMS

BLOOD PRESSURE SENSORS IN PILOT

FIGURE 11.—Aerobatic pilots should be familiar with the physiologic mechanisms which compensate for G force imposition during aerobatics.

At the right is shown a four $(+)$ G_Z force that results in a pulling away of blood from the vessels supplying the brain. The heart is smaller because it cannot fill well since the lower trunk and leg vessels now have pooled blood. The veins especially dilate because of their thin walls relative to arterial walls, and considerable blood pools in the abdominal "splanchnic" veins. The liver has slid toward the pelvis. Gray-out to unconsciousness may occur in the experienced aerobatic pilot because of decreased blood flow through the brain and eye retina if a four $(+)$ G_Z acceleration is experienced steadily for one minute.

At the left, a three $(-)$ G_Z force is shown. The blood in the major arteries and veins to the head is forced toward the head, and the brain and eyes are engorged. The experienced pilot feels head discomfort and fullness sensations, and some have reported that the lower lid creeping over the cornea gives a "reddish" visual aensation. The whites of the eyes become bloodshot and at higher $(-)$ G_Z forces, tiny capillaries may rupture (due to high arterial pressure and high venous pressure on both ends of the thin walled capillaries), causing small red hemorrhages. Some persons have experienced tiny retinal capillary hemorrhages, also, resulting in several days of "spots before the eyes". Note that the liver has pressed up through the diaphragm against the heart and lungs, making breathing difficult. Note also that the aorta to the lower trunk and legs is low on blood volume. There is no effective way to compensate for the above $(-)$ G_Z changes as there is for the $(+)$ G_Z changes. Unconsciousness may occur if 4.5 $(-)$ G_Z is imposed for five seconds, but as with $(+)$ G_Z unconsciousness, as soon as the accelerative force is removed, a rapid return to consciousness occurs with restoration of circulation (due to the highly oxygenated status of the arterial blood—assuming the pilot is not exposed to a hypoxic environment).

The lower center diagram of Figure 12 il-

CIRCULATION DYNAMICS DURING G$_Z$ ACCELERATION

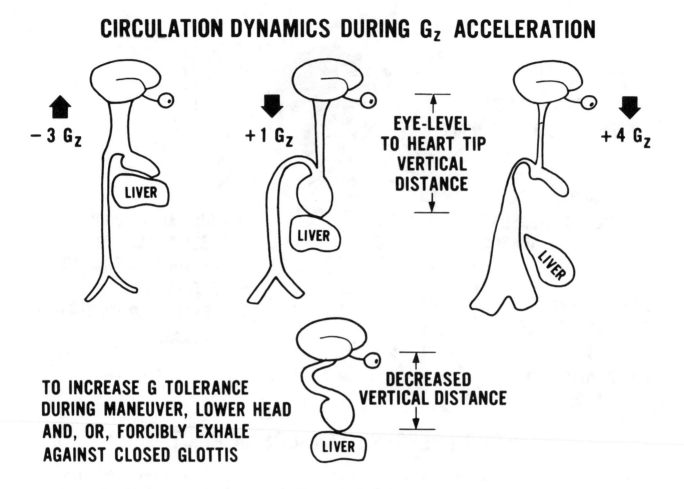

TO INCREASE G TOLERANCE
DURING MANEUVER, LOWER HEAD
AND, OR, FORCIBLY EXHALE
AGAINST CLOSED GLOTTIS

FIGURE 12.—The human body is a relatively soft and flexible structure, hence the specific effects of vertical axis accelerations upon it.

lustrates that if the vertical distance between the heart and the brain is decreased, and/or the thoracic and abdominal pressures are increased prior to or during a (+) G$_Z$ acceleration by attempting to exhale against a closed glottis (Valsalva maneuver), an increase of (+) G$_Z$ tolerance can be obtained of an additional 1 to 2 (+) G's.[1,5]

Figure 13 illustrates the vertical "8", a maneuver that when entered from the inverted position and performed well in competitive aerobatics is good for 40 points (Aresti system).[9] If the upper loop is "inside" and the lower "outside", 36 points are possible. The maneuver in Figure 13 combines high aerodynamic performance requirements with pilot physiologic limits. The figure is shown in the "Aresti" system of "aerocryptographic" portrayal, with the broken lines indicating (−) G$_Z$ forces. The competitive aircraft with inverted fuel systems,

favorable power/weight ratios, and fairly light wing loadings are particularly suited for this maneuver. Examples are the Moravan National Corporation Zlin 526A or 526AS (160 hp), the DeHavilland DHC–1B–2–S3 Chipmunk (200 hp) and the Yakovlev YAK–18PM (300 hp). The Pitts Aviation Enterprises Pitts S–1 Special also performs the maneuver well.

In communication with various pilots who have practiced the maneuver, it has been found that periods of unconsciousness occur at the 7–9 o'clock position on the inside loop that follows the outside loop.[10] If the inside loop is performed first, followed by an outside lower loop, the unconsciousness does not happen, but, as previously stated, the maneuver is then worth fewer points.

The mechanism of unconsciousness experienced in the maneuver as portrayed in Figure 13 is as follows. The maneuver is entered from

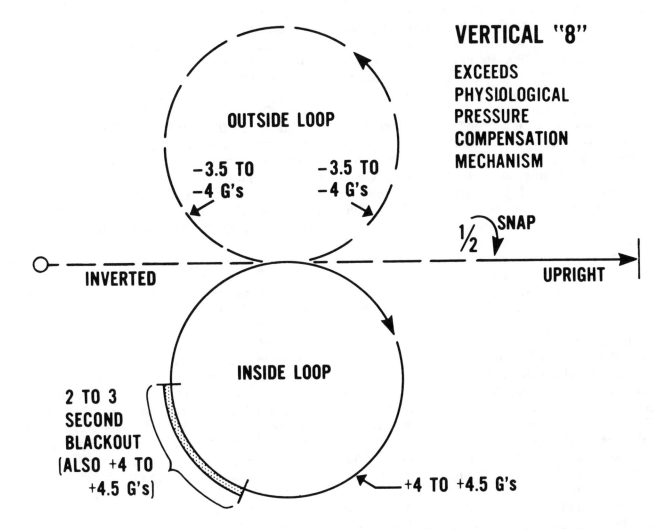

FIGURE 13.—This maneuver is very stressful from the physiologic as well as the aerodynamic standpoints.

the inverted position, which has already resulted in a certain congestion of the brain and eyes with blood since there will be a one (−) G_Z force. The pressure receptors in the aortic arch and carotid arteries (Figure 11) will have sensed the blood engorgement and will have sent signals to slow the heart. In addition, the large veins of the neck will be stretched by the slowing of the venous blood due to the (−) G_Z force tending to retard the return of the blood from the head to the heart. Normally the venous blood returns to the heart in a gravity-feed system. In the (−) G_Z position, this venous return is, therefore, somewhat impaired. There is a reflex stretch mechanism in these great neck veins, which is designed to speed up the heart and cause it to pump more blood in the presence of venous overload (this is known as the "Bainbridge"

reflex). Stretch receptors in the great neck veins detect excessive venous blood pressure and send signals through the visceral nervous system to the central nervous system, with return signals through the sympathetic nervous system to stimulate the heart. This reflex is designed for the (+) G_Z circumstance, and has an adverse effect in a (−) G_Z maneuver since it results in the heart attempting to pump more blood into an already overloaded head circulation having decreased venous outflow. Because of the engorged slow-flowing blood circulation, the brain becomes somewhat hypoxic since it extracts much of the oxygen from the slow moving blood during the outside maneuver at the 4–5 o'clock and 7–8 o'clock positions.

Following the completion of the outside loop, which required at the 7–8 o'clock position 3.5 to

4 (−) G's along the G_Z axis to round-out horizontally, the inside loop is abruptly begun. Here the (+) G's begin to be imposed on the G_Z axis, an imposition for which the blood pressure physiology is not attuned. On the contrary, all reflexes have been working against the (−) G_Z axis. There is a five second delay in reflex response from one (+) G_Z to increased (+) G loadings, and when the loadings begin from the (−) G_Z side, this response is increased by several seconds. When the first strong (+) G_Z forces hit at the 4–5 o'clock position, the nervous mechanisms are still trying to catch up with the imposed (+) loading and all of the compensations have to be made in the completely opposite direction. These are possibly aided by the Valsalva maneuver and by head lowering. The blood flow slows to the brain and at the 6 o'clock position the brain has extracted much of the oxygen. Some slight increase in circulation may be gained at about the 6 o'clock point as the G forces reduce somewhat. However, the acceleration against the pull of gravity at the 7–8 o'clock position imposes 4–4.5 (+) G's upon an already overstressed, barely compensating, cardiovascular system. This new load decreases the circulation once again and, this time, loss of consciousness occurs at the 7 o'clock–9 o'clock position. Since the large G forces begin easing off after the 8 o'clock position, and the physiological pressure maintenance reflexes begin to have an effect, circulation to the brain is reestablished and consciousness returns.

Physiologically, no permanent harm occurs to the healthy individual in the above maneuver. World War II dive bomber pilots found that if they did a (−) G push-over they were more likely to black-out when the (+) G recovery was made than if they rolled over and made a (+) G entry to the dive.[1] In view of their loss of consciousness during the pull-up when the aircraft itself was under great aerodynamic stress, they feared inadvertent control movements that might structurally damage the aircraft under these conditions during the visual loss of reference "black-out", prior to and just after the unconsciousness. Also, the relaxed hold on the stick while unconscious could lead to a dive into the ground. Accordingly, they trimmed for a predetermined amount of nose-up flight prior to the point of possible loss of consciousness. The aircraft, thus, tended to do the proper thing while the pilot was an "inert passenger".

Physiologic Tolerance

It appears desirable that, depending upon the

pilot and aircraft, no maneuver be routinely performed that leads to unconsciousness at any point. Physiologically, humans progressively adapt within limits to imposed strains and stresses, and with practice, any maneuver will have less and less of an effect (again, within limits, depending upon human physiology and the individual pilot). A layoff of some days or weeks can result in a lowered G tolerance, but, normally, this returns rapidly with practice. One old-time (1930's) aerobatic pilot (known as the "batman") fixed a harness in his garage and hung upside down for a few minutes each day. This would give a one (−) G_Z acceleration and, possibly, he did maintain a slight increased physiologic tolerance in the (−) G_Z axis. However, (−) G_Z adaptation does not increase as effectively as (+) G_Z adaptation, since, as already mentioned, the physiologic mechanisms are designed to counteract (+) G_Z accelerations.

Other factors that affect G tolerance are (1) the skeletal anatomy, (2) the cardiovascular architecture, (3) the nervous system, (4) the quality of the blood, (5) the general physical state, and (6) experience and recency.

Short, squat individuals would inherently have an edge toward G tolerance maximum levels over tall, long-bodied, long-necked individuals, although superb aerobatic pilots are, of course, found in the latter category. A highly efficient heart, free of coronary disease, and capable of raising the blood pressure rapidly upon demand, is a prerequisite to safe, prolonged aerobatics. Some very young individuals have such elastic arteries that attempts to raise blood pressure are partially thwarted by lateral distension of the vessel walls. Normal aging results in a decrease in elasticity in arterial walls and, in this respect, acts to increase the tolerance to (+) G_Z accelerations. The nervous system involves temperament (some persons will never emotionally adapt to aerobatics because of fears instilled in childhood or because of an inability to see the point of such activity). The quality of the blood relates primarily to its hemoglobin content (this is the oxygen carrying component and should be between about 13–17 grams %—females should be especially wary of iron deficiency anemia), its salt content (be sure to have adequate Na^+Cl^- in hot weather) and water content (excessive dehydration lowers blood volume and the ability to change blood pressure).

The general physical state includes adequate sleep (at least seven hours prior to aerobatics), absence of infections (never conduct aerobatics with an illness—viral or bacterial), absence of hangover or drug effects (never undertake aerobatics with a hangover or while taking alcohol or any drug) and good physical fitness. Aerobatic pilots should not become obese or "out-of-shape", or undertake "crash diets". Experience leads to knowledge and understanding and the development of additional tolerance to G forces (1–2 G's). Recency leads to physiologic adaptation and fitness and an increase of perhaps one to 1.5 additional G's. This is lost in a week or so of layoff, but comes back rapidly after warm-up.

With respect to the physical and mental fitness of aerobatic pilots, one female pilot during the mid-1960's had severe anemia and lost consciousness in a maneuver while practicing for the international competition. The plane nosed over and dove into the ground with fatal consequences. Also, severe emotional upsets preceding competitive aerobatics can lead to excessively low maneuvers and crashes.[11] Documented loss of consciousness during $(+) G_Z$ G's in closed course air racing due to a weakened condition caused by diarrhea exists (pylon racers can pull 4–5 $(+) G_Z$ G's for several seconds).[12] In the same study, a pilot who suppressed information concerning an earlier heart attack died during the stresses of the race. If at any time the pilot does not feel mentally or physically up to par or if his aircraft, the environment, or the general circumstances seem wrong, postponement or cancellation should be accomplished. A word of caution on alcohol: no alcohol for 24 hours prior to aerobatics, to avoid the hangover effect, and, once again, never perform with a hangover. Do the celebrating after the performance!

One additional point of caution. Use only aircraft designed for aerobatics in conducting these maneuvers (unless the pilot is a highly skilled test pilot who knows the limits of the aircraft and performs for demonstration purposes). Always wear a parachute during aerobatic maneuvers and plan ahead concerning escape from maneuver points of possible structural failure. In inverted flight with $(-) G_Z$ loadings, a double seat belt is a good idea, as is a shoulder harness. Never leave loose cushions or other objects in the aircraft, which can jar loose and jam the controls. The above and related points were also stressed by Duane Cole on 4 August 1972 at the International Aerobatic Club Meeting, Experimental Aircraft Association 20th Annual Convention, Oshkosh,

Wisconsin. With respect to the duration of aerobatic routines, the late Bevo Howard used 12 maneuvers during performances in his Buecker Jungmeister. These included a series of slow rolls in a 360° circle, a hammerhead stall and turn, an 8-point sectional roll, a 1½ snap roll, an inverted snap, a double snap, a double snap on top of a loop, a square loop, a vertical 8 (inside loop to outside loop), a vertical snap, a spin, inverted flight with hands off the controls and an inverted ribbon pickup. The above maneuvers required approximately 15 minutes and obviously would be quite fatiguing at the end. Howard kept in good physical condition by swimming.

At the 26 May–4 June 1972 "TRANSPO" at Washington, D.C.'s Dulles International Airport, the following times were clocked for aerobatics by the indicated pilots or groups:

Scotty McCray (Glider)	7 minutes
Bob Hoover (Shrike)	10 minutes
Mary Gaffaney (Pitts)	10 minutes
T. Poberezny, C. Hillard, G. Soucy (Pitts—formation)	15 minutes
Walt and Sandi Pierce (Dual aerobatics)	10 minutes
Hughes and Kaizian (Wing riding)	10 minutes
Dawson Ransome (Pitts)	10 minutes
Bob Hoover (F-51)	15 minutes
Art Scholl (Chipmunk)	10 minutes
Average:	10.8 minutes

It is apparent from the above that by historical precedent and present practice, the average series of aerobatics covers about 11 minutes. Eleven minutes of consecutive aerobatics is quite fatiguing and illustrates the necessity for maintaining good physical conditioning and health.

Figure 14 shows the maximum limits of human tolerance to G's as determined in large centrifuges.[1,14] Note that for $(+) G_Z$, the point where gray-out begins for the uninitiated is three G for five seconds. Eight $(+) G_Z$ for 15 seconds causes blackout and temporary unconsciousness, even in the most experienced individual. A military G-suit (sometimes referred to as the "Anti-G" suit) can provide a tolerance of 4.5 $(+)$ G's in the Z axis for five minutes.*

---

*During World War II, German Stuka dive bomber pilots were able to adapt to 8 $(+)$ G's for a few seconds on pull-out by a combination of the Valsalva maneuver and bending over to a rather extreme angle (personal communication, Dr. Harald von Beckh, former Stuka pilot).

The $(-)$ G_Z tolerance for students is two to five seconds and causes subjective discomfort (as noted earlier, two $(-)$ G's is a net change of three G's from one $(+)$ G_Z). The experienced pilot can tolerate 4.5 $(-)$ G_Z G's for five seconds before head pain (headache—referred to as cephalalgia), breathing discomfort (the lungs are pressed by the liver) and other subjective unpleasant sensations lead the pilot to terminate the maneuver. Note that the $(-)$ G's cause enough discomfort to lead the pilot to terminate the negative maneuver prior to loss of consciousness. In this sense, outside maneuvers are physiologically safer than inside maneuvers. In fact, the first pilot to ever perform a loop, Adolphe Pegoud in France in the summer of 1913, performed both outside and inside loops.[3,15]* He repeatedly and routinely demonstrated outside loops (flying "over the top", down-under, and back up) in a Bleriot monoplane. The airplane, which used wing warping for bank, was modified by adding stronger "landing wires" for $(-)$ G to the wings and a heavier horizontal tail. The outside loop, as noted, was entered from a dive with the aircraft pushed forward over on its back and allowed to curve up and around to the starting point.

*Actually, Pegoud was the second pilot to perform this feat. The first was Lt. Peter Nesterov at Kiev (Imperial Russian Air Service) in a Nieuport IV, 20 August 1913. W.K.

Pegoud also performed two other types of $(-)$ G_Z maneuvers and tail slides.

In vertical 8s and other maneuvers that lead to unconsciousness, the pilot should discontinue these at the point where visual sensation begins to change. Consideration should be given to logging the periods and points of unconsciousness in various extreme maneuvers in order that long range corrective action can be taken in terms of modifying the repertoire.

Figure 15 lists the G_Z tolerances in the $(+)$ and $(-)$ directions of the average healthy experienced human. Note that the one $(+)$ G_Z axis tolerance is about 20 hours in the immobile sitting position, and that after this period, the desire to slouch, to drop the head to one side or the other, or to lean on a table, not to mention to lie down, becomes overwhelming. All of these changes lead to diminishing the height of the blood column from the heart to the brain. After two hours of standing straight with no movement (as soldiers at attention), Marey's Law and its backup, the Bainbridge reflex, begin to fail, and the subject will faint unless he can move around (reestablishing blood circulation—partly through the leg and arm muscles squeezing the veins, the latter provided with one-way valves directing the pooled blood toward the heart), sit, or lie down.

Figure 15 contains data on the upper tolerances of fit, experienced, aerobatic pilots,

HUMAN ACCELERATION TOLERANCE LIMITS

TYPE OF G	DIRECTION OF BODY MOVEMENT	AIRCRAFT MANEUVER	ONSET OF INITIAL SYMPTOMS	EXPERIMENTAL HUMAN MAXIMUM EXPOSURES	PHYSIOLOGICAL LIMITS
+	HEAD TO FOOT	PULL OUT OF DIVE OR TIGHT LEVEL TURN	STUDENT: 3G 5 SECONDS GRAYOUT	8G FOR 15 SECONDS (4.5G FOR 5 MINUTES WITH G SUIT)	BLACKOUT TO UNCONSCIOUSNESS
−	FOOT TO HEAD	PUSHOVER	STUDENT: 2G 5 SECOND SUBJECTIVE DISCOMFORT	4.5 G FOR 5 SECONDS	PAIN (HEADACHE)

FIGURE 14.—This chart is for application to aerobatic students.

ACCELERATION TOLERANCES
(EXPERIENCED INDIVIDUALS)

G	TIME	SYMPTOM
+8	15 SECONDS	BLACK-OUT TO UNCONSCIOUSNESS
+7	15 SECONDS	BLACK-OUT TO UNCONSCIOUSNESS
+6	20 SECONDS	BLACK-OUT TO UNCONSCIOUSNESS
+5	30 SECONDS	BLACK-OUT TO UNCONSCIOUSNESS
+4	1 MINUTE	GRAY-OUT TO UNCONSCIOUSNESS
+3	3 MINUTES	GRAY-OUT TO UNCONSCIOUSNESS
+2	13 MINUTES	GRAY-OUT TO UNCONSCIOUSNESS
+1	20 HOURS (SITTING)	2 HOURS-STANDING, NO MOVEMENT
0	21 DAYS (MAXIMUM EXPOSURE TO DATE)	
−1	10 MINUTES	SUBJECTIVE DISCOMFORT TO UNCONSCIOUSNESS
−2	1 MINUTE	SUBJECTIVE DISCOMFORT TO UNCONSCIOUSNESS
−3	30 SECONDS	RESPIRATORY DISTRESS ADDED TO UNCONSCIOUSNESS
−4	6 SECONDS	PAIN (HEADACHE) ADDED TO UNCONSCIOUSNESS
−5	4 SECONDS	PAIN (HEADACHE) ADDED TO UNCONSCIOUSNESS

FIGURE 15.—This table represents a spectrum of G limitations for the average experienced aerobatic pilot.

with unconsciousness as the ultimate end point. Centrifuge studies by the U.S. Navy, Johnsville, Pennsylvania, facility, reveal the somewhat lower tolerance in adult male research subjects when "gray-out", the loss of peripheral vision, is the end point for $(+) G_Z$ and throbbing headache for $(-) G_Z$ (personal communication, Dr. Harald von Beckh). These are as follows:

Time	$(+) G_Z$	$(-) G_Z$
3 seconds	12.0	4.0
10 seconds	4.2	3.3
15 seconds	3.6	3.2
30 seconds	3.4	2.5

The $(+) G_Z$ tolerance is so high for three seconds because the brain can operate on the oxygen diffused in its tissues for this brief period. The 4.0 $(-) G_Z$ tolerance represents a net change of five G's from the one $(+) G_Z$ point (as compared with three G's to the four $(+) G_Z$ point).

Incidentally, one reason we lie down to get the most restful sleep is that we place the G_Z axis parallel with the pull of gravity, thus minimizing the work requirements to raise the blood against the field of gravity. By periodically turning during sleep around the G_Z axis, the gravitational pull in the less significant G_X and G_Y axes is also averaged out during the sleep.

Some four-legged animals can sleep in the standing position, but note that their G_Z axis is parallel with the horizontal and they keep the head approximately at heart level. Also, some cranes sleep on one locked leg, and here too the head is tucked near the horizontally held body at heart level. Animals that hang upside down have special adaptative structures to control the blood flow in the $(-) G_Z$ position.

Concluding Comments

Aerobatics for the experienced pilot is the true elixir of flight. Too many pilots, however, have been lost through inadequate knowledge of the physiology of aerobatic flight. All desiring to practice this advanced form of flight should assimilate basic knowledge of the phsyiologic aspects of aerobatic flight.

REFERENCES

1. Christy, Ralph L.: "Effects of Radial, Angular, and Transverse Acceleration", in Aerospace Medicine (Randel, Hugh W., Ed.), Williams and Wilkins Company, Baltimore, Maryland, 1971, p. 167.
2. Feature: "Famous Flyers—'Speed' Holman", *FAA Aviation News,* Federal Aviation Administration, Washington, D.C., June 1972, p. 12.
3. Feature: "Pegoud's Remarkable Performances", *Scientific American,* 18 October 1913.
4. Gerathewohl, S. J.: *Die Psychologie des Menschen in Flugzeug,* Johann Ambrosius Barth, Munchen, Germany, 1954, p. 128.
5. Henry, James P. (Chairman): "Sustained Linear Acceleration", *Report of Working Group 67, Committee on Hearing, Bioacoustics, and Biomechanics,* National Academy of Sciences—National Research Council, Washington, D.C., 1 April 1971.
6. Howard, Beverly, "The Art of Aerobatics", *Flying,* December 1961.
7. Hurt, H. H.: *Aerodynamics for Naval Aviators,* NAVAIR 00-80T-80, U.S. Navy, Washington, D.C., 1965, p. 177.
8. Kershner, William K.: *The Private Pilot's Flight Manual,* Iowa State University Press, Ames, Iowa, 1965, p. 80. Also, *The Advanced Pilot's Flight Manual* (same address), 1970, p. 80.
9. Opp, Ernest: *Aerobatic Handbook,* Bethalto, Illinois, 1966, p. 38.
10. Poage, Jack B.: Letter to author, 30 March 1971.
11. Siegel, P. V., and Mohler, S. R., "Medical Factors in U.S. General Aviation Accidents", *Aerospace Medicine,* Vol. 40, No. 2, February 1969, p. 180.
12. Snyder, R. C., and Davis, A. W.: "Medical Factors in Unlimited Class Air Racing Accidents", *Aerospace Medicine,* Vol. 43, No. 5, May 1972, p. 512.
13. Van Sickle, N. D.: *Modern Airmanship,* D. Van Nostrand, Princeton, New Jersey, 1957, p. 260.
14. Webb, Paul (Ed.): *"Bioastronautics Data Book",* NASA Report SP-3006, National Aeronautics and Space Administration, Washington, D.C., 1964.
15. Winchester, Clarence: *Wonders of World Aviation,* The Waverly Book Company, England, Great Britain, 1939, p. 72.

——— BIBLIOGRAPHY ———

(See also Bibliographies at the ends of Chapter 21 and Appendix B.)

Aviation Instructor's Handbook. FAA. Washington, D.C.: USGPO, 1977.

Civil Pilot Training Manual. Civil Aeronautics Administration. Washington, D.C.: USGPO, 1941.

Cole, Duane. *Conquest of Lines and Symmetry.* Milwaukee, Wisc.: Ken Cook Transnational, 1970.

De Lacerda, Fred. *Surviving Spins.* Ames: Iowa State University Press, 1989.

Dogan, Peter. *Instrument Flight Training Manual.* 2d ed. Santa Clarita, Calif.: Aviation Book Company, 1991.

Flight and Ground Instructor Written Test Book. FAA-T-8080-18A. Washington, D.C.: USGPO, 1993.

Flight Instruction of Foreign Students. Naval Air Training Command. Corpus Christi, Tex., 1975.

Flight Instructor Practical Test Standards for Airplane—Single-Engine and Multiengine. FAA-S-8081-6A. Washington, D.C.: USGPO, 1994.

Flight Instructor Practical Test Standards for Instrument Airplane and Rotorcraft/Helicopter: FAA-S-8081-9A. Washington, D.C.: USGPO, 1990.

Instrument Flying Handbook. AC 61-27C. FAA. Washington, D.C.: USGPO, 1980.

Instrument Rating—Practical Test Standards—Airplane/Helicopter: FAA-S-8081-4A. Washington, D.C.: USGPO, 1990.

Instrument Rating—Written Test Book. FAA-T-8080-20A. Washington, D.C.: USGPO, 1993.

Kershner, William K. *The Advanced Pilot's Flight Manual.* 6th ed. Ames: Iowa State University Press, 1994.

———. *The Basic Aerobatic Manual.* Ames: Iowa State University Press, 1987.

———. *The Instrument Flight Manual.* 4th ed. Ames: Iowa State University Press, 1990.

———. *The Student Pilot's Flight Manual.* 7th ed. Ames: Iowa State University Press, 1993.

———. *The Student Pilot's Flight Manual Syllabus.* 2nd ed. Ames: Iowa State University Press, 1998.

Langewiesche, Wolfgang. *Stick and Rudder.* New York: McGraw-Hill, 1944.

Mason, Sammy. *Stalls, Spins, and Safety.* New York: Macmillan, 1982.

Mohler, Stanley R., M.D. *G Effects on the Pilot during Aerobatics.* Washington, D.C.: FAA, Office of Aviation Medicine, 1973.

Pilots' Handbook of Aeronautical Knowledge. FAA. Washington, D.C.: USGPO, 1980.